FISSION PRODUCT TRANSPORT PROCESSES IN REACTOR ACCIDENTS

PROCEEDINGS OF THE INTERNATIONAL CENTRE FOR HEAT AND MASS TRANSFER
NAIM AFGAN, EDITOR

ICHMT
Belgrade

1. Afgan and Schlünder **Heat Exchangers: Design and Theory Sourcebook**
2. Afgan and Beer **Heat Transfer in Flames**
3. Blackshear **Heat Transfer in Fires**
4. deVries and Afgan **Heat and Mass Transfer in the Biosphere**
5. Denton and Afgan **Future Energy Production Systems**
6. Hartnett **Alternative Energy Sources**
7. Spalding and Afgan **Heat Transfer and Turbulent Buoyant Convection**
8. Zarić **Thermal Effluent Disposal From Power Generation**
9. Hoogendoorn and Afgan **Energy Conservation in Heating, Cooling, and Ventilating Buildings**
10. Durst, Tsklauri, and Afgan **Two-Phase Momentum, Heat and Mass Transfer in Chemical, Process, and Energy Engineering Systems**
11. Spalding and Afgan **Heat and Mass Transfer in Metallurgical Systems**
12. Jones **Nuclear Reactor Safety Heat Transfer**
13. Bankoff and Afgan **Heat Transfer in Nuclear Reactor Safety**
14. Zarić **Structure of Turbulence in Heat and Mass Transfer**
15. Taborek, Hewitt, and Afgan **Heat Exchangers: Theory and Practice**
16. Metzger and Afgan **Heat and Mass Transfer in Rotating Machinery**
17. Afgan **Measurement Techniques in Power Engineering**
18. Soloukhin and Afgan **Measurement Techniques in Heat and Mass Transfer**
19. Abramson **Guidebook to Light Water Reactor Safety Analysis**
20. van Swaaij and Afgan **Heat and Mass Transfer in Fixed and Fluidized Beds**
21. Radovanovic **Fluidized Bed Combustion**
22. Mori, Sheindlin, and Afgan **High Temperature Heat Exchangers**
23. Sheindlin **High Temperature Equipment**
24. Bougard and Afgan **Heat and Mass Transfer in Cryoengineering and Refrigeration**
25. Afgan **Transient Phenomena in Multiphase Flow**
26. Spalding and Afgan **Heat and Mass Transfer in Gasoline and Diesel Engines**
27. Markatos **Computer Simulation for Fluid Flow, Heat and Mass Transfer, and Combustion in Reciprocating Engines**
28. Kline **Near-Wall Turbulence: 1988 Zoran Zarić Memorial Conference**
29. Bergles **Heat Transfer in Electronic and Microelectronic Equipment**
30. Rogers **Fission Product Transport Processes in Reactor Accidents**
31. Hanjalić **Mathematical Modeling and Computer Simulation of Processes in Energy Systems**

FISSION PRODUCT TRANSPORT PROCESSES IN REACTOR ACCIDENTS

Edited by

J. T. Rogers
Carleton University
Ottawa, Ontario, Canada

HEMISPHERE PUBLISHING CORPORATION
A member of the Taylor & Francis Group

New York Washington Philadelphia London

FISSION PRODUCT TRANSPORT PROCESSES IN REACTOR ACCIDENTS

1 2 3 4 5 6 7 8 9 0 EBEB 9 8 7 6 5 4 3 2 1 0

A CIP catalog record for this book is available from the British Library.

Library of Congress Cataloging in Publication Data

Fission product transport processes in reactor accidents / edited by J. T. Rogers.
p. cm.—(Proceedings of the International Centre for Heat and Mass Transfer ; 29)
Papers from the Seminar on Fission Product Transport Processes in Reactor Accidents held in Dubrovnik, Yugoslavia May 22–26, 1989.
Includes bibliographical references.
1. Nuclear reactors—Accidents—Congresses. 2. Fission products—Congresses. 3. Heat—Transmission—Congresses. 4. Mass transfer—Congresses. I. Rogers, J. T. II. Seminar on Fission Product Transport Processes in Reactor Accidents (1989 : Dubrovnik, Yugoslavia)
TK9153.H394 1990
621.48′31—dc20 90-4243
ISBN 0-89116-876-1 CIP

Contents

4. TRANSPORT INTO CONTAINMENT

5. CONTAINMENT TRANSPORT PHENOMENA

7. CLOSING SESSION

Preface

The negligible release of radioactive iodines and cesium from the containment in the Three Mile Island accident in 1979, in spite of gross meltdown of the reactor core, stimulated significant re-evaluations of, and new research into, fission-product behavior in nuclear reactor accidents. The results of these efforts have already permitted more realistic assessments of reactor accidents and have also resulted in improved procedures for accident management.

The Chernobyl reactor accident in 1986 again focussed attention on fission-product behavior in severe reactor accidents and emphasized the need for the world-wide nuclear power community to co-operate further and to exchange the results of research in this field in the most open and effective manner.

Recognizing the very important roles of heat and mass transfer in all aspects of fission-product behavior in severe reactor accidents, the Executive Committee of the International Centre for Heat and Mass Transfer, in the wake of the Chernobyl accident, decided at its meeting in Tashkent in April, 1987, to organize a Seminar on Fission Product Transport Processes in Reactor Accidents. The Executive Committee believed that the Centre, because of its international status bridging east and west and its established role in the fields of heat and mass transfer, was uniquely positioned to undertake such a task.

The Seminar, planned and organized by an International Organizing Committee of experts in the field, was held in Dubrovnik, Yugoslavia from May 22 to May 26, 1989. Twelve invited lectures and 47 invited papers were presented. Also, in a closing session, a panel composed of the session chairman, aided by lively discussion from the floor, summarized the issues raised, conclusions reached and problems remaining. Eleven of the lectures and all of the papers are included in these Proceedings. In addition, the Proceedings include four invited papers which were not presented orally at the Seminars as well as a summary of the closing session, prepared by the session chairmen.

The Seminar was attended by 130 persons from 21 countries and two international organizations.

The Executive Committee of the Centre and the Organizing Committee of the Seminar acknowledge, with many thanks, the assistance provided to the Seminar by the sponsoring organizations, the United Nations Educational Scientific and Cultural Organization, Paris; Carleton University, Ottawa, and Boris Kidric Institute of Nuclear Sciences, Belgrade. We are also indebted to the co-operation of the International Atomic Energy Agency in Vienna.

The hard work and co-operation of many people contributed to the organization and operation of the Seminar and to the preparation of these Proceedings. A major load was carried by, and special thanks go to, the chairmen of the sessions: D. F. Torgerson, Atomic Energy of Canada Ltd., Pinawa, Manitoba, Canada; F. C. Iglesias, Atomic Energy of Canada Ltd., Chalk River, Ontario, Canada; A. P. Malinauskas, Oak Ridge National Laboratory, Oak Ridge, Tennessee, USA; E. F. Hicken, Gesellschaft fur Reaktorsicherheit, Garching, FRG; F. Abbey, UKAEA Safety and Reliability Directorate, Culcheth, Warrington, UK; D. F. Ross, U.S. Nuclear Regulatory Commission, Washington, D.C., USA, and V. A. Sidorenko, State Committee for the Supervision of Nuclear Power Safety, Moscow, USSR. Dr. Ross and Professor Sidorenko also acted as vice-chairmen of the Seminar.

Other members of the Organizing Committee who assisted in the organization of the seminar are: N. Afgan, Yugoslavia; V. S. Asmolov, USSR; M. Hache, France; M. S. Kazimi, USA; E. G. Lowenhielm, Sweden; M. Matausek, Yugoslavia; V. S. Osmachkin, USSR; A. Suzuki, Japan; W. O. Schikarski, FRG, and C. Wood, Canada.

Particular thanks go to Dave Torgerson, who is also chairman of the Working Group on Source Term and Environmental Consequences of the Committee on the Safety of Nuclear Installations of the Nuclear Energy Agency, for providing the initial contacts that were so important in establishing the program for the seminar.

Of course, it would have been impossible to organize and run the Seminar without the able support of the ICHMT Secretariat under the direction of Naim Afgan, the Secretary-General of the Centre. Special thanks go to Jovica Riznic, the Seminar Secretary, whom we thank for his enthusiastic and timely support. Thanks are also extended to the members of the Secretariat staff who ably assisted him: Natasa Riznic, Stanislava Djuric and Radisa Rajkovic. Finally my personal thanks go to Catharine Coletelo and Deborah Turcotte of the staff of the Department of Mechanical and Aerospace Engineering, Carleton University, Ottawa, for secretarial assistance.

J. T. Rogers

1. INTRODUCTORY SESSION. OVERVIEW LECTURES

A Generic Overview of Severe Accident Phenomena

H. J. TEAGUE
Safety and Reliability Directorate
United Kingdom Atomic Energy Authority
Wigshaw Lane, Culcheth
Warrington WA3 4NE, UK

D. F. TORGERSON
Atomic Energy of Canada Research Company
Whiteshell Nuclear Research Establishment
Pinawa, Manitoba, ROE 1LO Canada

1. INTRODUCTION

In the course of the nuclear industry's development, there has been a continuous evolution of safety philosophy and technology. From the beginning, the highest importance has been attached to safety and every effort has been made to foresee any possible incident having safety implications, and to make appropriate provision for the nuclear plant to survive without causing harm to the operators or to the public. This first emerged as the Maximum Credible Accident approach. Designers were required to show that they had identified all credible accidents, and to demonstrate for each one a satisfactory combination of procedures and engineered safety features (ESFs). The principal ESFs are automatic protection to shut the reactor down and a reliable means of decay heat removal.

That approach led to the concept of Design-Basis Accidents (DBA), which provide specifications for the performance required of the safety equipment and procedures. For such an approach to succeed, all DBA fault sequences must be identified and accurately analyzed. Furthermore, the ESFs must be extremely reliable. Given those conditions, analysis of a severe accident (conveniently defined as one that lies beyond the DBA envelope) did not form part of the safety case.

The DBA approach contains weaknesses that led to questions concerning completeness, understanding of postulated sequences, and equipment reliability. If an unforeseen accident should occur, or if a postulated accident develops in an unexpected way, or if an ESF fails, confidence that safety will be preserved is seriously undermined.

Containment is an ESF of particular significance. Some DBAs result in a limited release of radioactivity from the reactor's primary circuit and containments are designed to prevent such activity from escaping. It was realized that the containment would also provide an additional means of protection against unforeseen incidents and for those that were inadequately covered, for whatever reasons, by other ESFs.

The accident at Three Mile Island in 1979[1,2] provides a striking example of the effectiveness of containment. First, the sequence-oriented safety strategy, for a combination of reasons, led to a failure to intercept successfully what should have been a minor incident. Second, although the reactor was virtually destroyed, the containment successfully prevented any significant harm to plant personnel or the public.

The intense debate that followed the TMI-2 accident led to a much clearer appreciation of safety strategy in general and severe accidents in particular. An important consequence was a more realistic analysis of ESF reliability. Of more significance to this seminar is a recognition of two important features. First, even if an incident progresses beyond the DBA, there is a useful margin available in the ability of the ESFs to cope with demands well in excess of those for which they are designed. Second, to exploit that capability, it is necessary to adopt a flexible, symptom-oriented approach to accident management. A corollary of that is the need for a high degree of generic phenomenological understanding that needs to be disseminated appropriately among all plant personnel who would play a part in controlling a severe accident. These questions are addressed in greater depth in this session.

Progression beyond the DBA necessarily entails a divergence from the well-defined conditions of the design basis. In the early stages, with virtually undisturbed geometry, it is possible to analyze the course of events with some precision and confidence. Increasingly, as conditions change, the range of possible events grows and with it the difficulty of further prediction. Moreover, the relevant phenomena interact with each other and cannot properly be treated in isolation. The physical and chemical conditions are beyond those encountered in normal plant conditions and are correspondingly difficult to reproduce and study.

The only practicable way to deal with this problem is a combined strategy of studying separate effects and developing, in parallel, sophisticated models that enable accident progression to be calculated, taking into account the interdependence of the relevant phenomena. The separate-effects studies produce data and confirm the fundamental scientific modelling, whilst the codes enable that understanding to be applied to specific plants and accidents.[3]

Because of the number and range of parameters to be embraced, it is not practicable to reproduce experimentally all the possible accident sequences that need to be addressed. Such experiments are expensive and extremely time-consuming to set up and interpret. Nevertheless, there is a growing stock of integral experiments that are sufficiently close to the conditions of interest to provide a check on the validity of the computer models. The models can then be used to evaluate risks and countermeasures in relation to severe accidents. It is important to recognize that, since exceeding the DBA conditions is a very low probability event, it may not be necessary to achieve the same standard of qualification required within the design basis.

In what follows, we summarize severe reactor accident phenomena and point out some of the complexities asscciated with them. Figure 1 is an overview of the various processes that may occur during a severe accident in a PWR.

2. IN-CORE/VESSEL PHENOMENA

2.1 Core Damage Progression

An incident becomes a severe accident when it is no longer possible to cool the fuel by normal means. The first response will be to attempt to maintain cooling by whatever means is available, but such attempts need to be guided by an understanding of how core damage may develop. If the

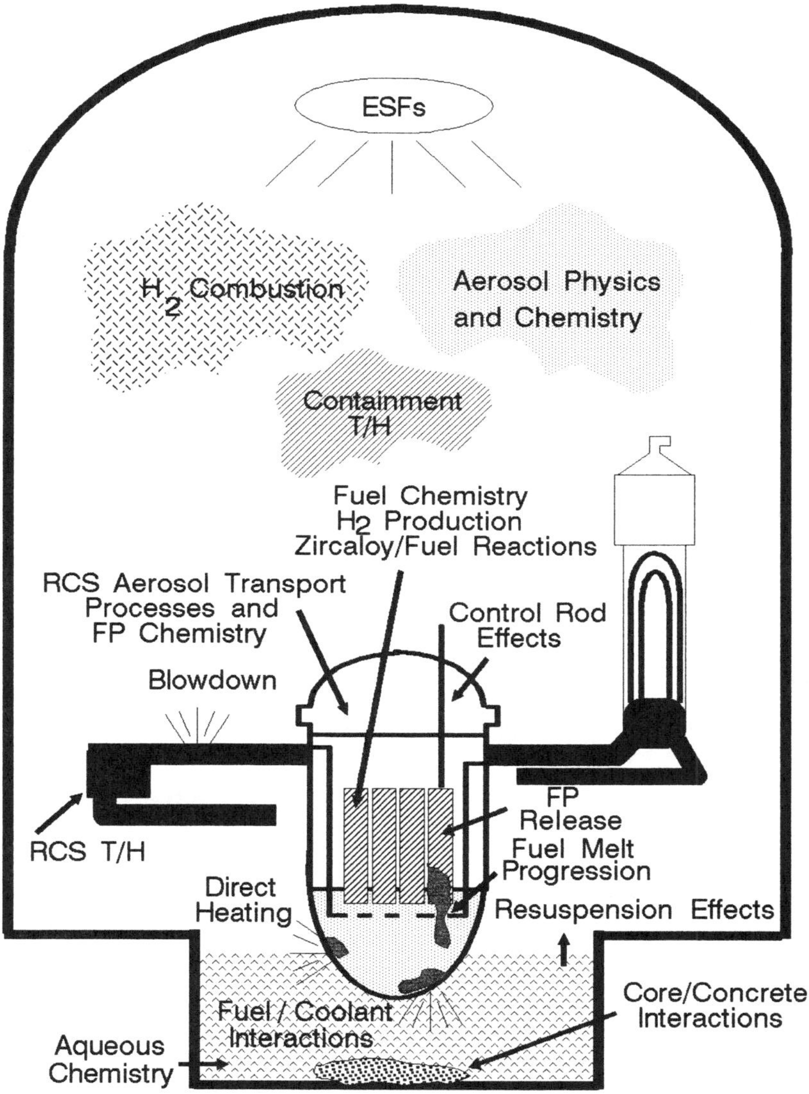

FIGURE 1. PWR Severe Reactor Accident Phenomena

attempts fail, the effects of further damage and attempts to mitigate those effects will also depend on the details of damage progression.

The initial heating of the fuel is not difficult to predict, at least for as long as the original geometry is retained. The codes used in the design basis will adequately represent the thermal-hydraulics and the temperature distribution in the core. The source of heat will be fission product decay within the fuel, with an additional term arising from the oxidation of Zircaloy cladding as the temperature rises.

The first major problem is assessing when and how the core geometry will alter significantly. The melting points of the individual materials within the core as designed are well known, but chemical changes, such as oxidation by steam and eutectic formation between neighbouring materials, radically affect the composition and properties of the different components.

The oxidation of Zircaloy depends on the availability of steam and, below the Zircaloy surface, on solid-state diffusion of oxygen, which is the determining process at that stage. Unoxidized Zircaloy will melt at about 2050 K and UO_2 at 3100 K. However, the relocation under gravity of an overheated fuel pin depends on the detailed history of the heating process. At slow heating rates (up to 0.5 K/s), the fuel cladding will be completely oxidized to ZrO_2 before reaching the melting point of Zircaloy. At heating rates above 5 K/s, the cladding will start to melt when there is only a thin shell of ZrO_2, which will be too weak to prevent relocation of the cladding. For intermediate rates, the oxidized cladding may have enough strength to hold the molten Zircaloy in place for a time while it begins to attack, by chemical dissolution, both the solid UO_2 and ZrO_2.

A further complicating factor arises from the control rods. In a PWR the silver-indium-cadmium absorber material melts at 1100 K, so it will be molten when the stainless steel cladding fails, probably by localized eutectic interaction where it contacts the Zircaloy of the guide tube. The relocated liquid alloy would then become free to attack Zircaloy cladding and accelerate the propagation of such damage even at temperatures as low as 1500 K.

Similar processes would occur in boiling water reactors (BWRs), where boron carbide absorber will rapidly attack stainless steel cladding at 1500 K. Alumina-boron carbide may be used, clad in Zircaloy, as a burnable poison. A strong interaction will occur at 1750 K, resulting rapidly in a liquid phase, which will in turn produce a liquid phase in both UO_2 and ZrO_2 at temperatures above 2200 K.

Thus, a simplistic model postulating a single core-slumping temperature is quite inappropriate. As regions melt, or become too weak to support themselves, their relocation must be considered realistically, recognizing that some of the liquid phases, such as those incorporating absorber material, will be extremely mobile and will resolidify only at extremely low (in the context of overheated cores) temperatures. Moreover, effects will be highly localized (e.g., near control-rod positions) and may result in opening up chimneys through the core, with marked implications for the supply of steam.

It can be seen that the core-melt progression is extremely complex and will occur in different locations in different ways. Material will tend to migrate downwards by a series of cycles of overheating, melting, cooling and resolidifying. Each cycle is likely to change the conditions governing the course of the next cycle, and it is clear that the uncertainties multiply rapidly as the process advances. The phenomena are in principle reasonably well understood and provide a good basis for developing a strategy for severe accident management, but it is unlikely that precise mechanistic descriptions of specific accidents will be possible for some years, if ever.

2.2 Fuel-Coolant Interaction

The phenomenon of vapor explosions has long presented intractable problems, particularly in the foundry and paper-making industries. Vapor explosions sometimes occur when hot molten material in bulk is quenched in a quantity of vaporizable liquid. However, more often than not, the transfer of heat between the materials is effected comparatively inocuously. Enormous efforts have gone into seeking a reliable explanation of the phenomenon, with only partial success.

Simple thermodynamic considerations show that the proportions in which the two fluids are mixed are crucial. A large excess of water could absorb all the available heat without necessarily producing any steam at all. Very limited quantities of water would reach very high temperatures, but the damage potential would be limited by the volume of steam that could be formed. Maximum conversion of heat to mechanical energy occurs when the volumes of interacting materials are roughly equal. Very damaging events have occurred, irrespective of the initial proportions, in which the components become rapidly mixed in approximately equal proportions by volume and on such a fine geometrical scale that vapor is formed with explosive violence. Because of the obvious relevance to severe reactor accidents, in which substantial quantities of molten fuel may fall into pools of coolant, the nuclear industry has devoted extensive efforts to understand how such explosions occur.

It is found helpful to treat the process in 4 stages:

1. Coarse mixing. In this stage the fuel is dispersed in the form of irregular drops, of perhaps a centimetre or so in size, in a metastable configuration. If violent boiling occurs on first contact, the materials are prevented from interpenetrating. It is likely that the metastable state consists of a stable vapor film surrounding each drop that limits heat transfer.

2. Triggering. The equilibrium is then disturbed by a trigger. This may be an applied pressure pulse or it may occur spontaneously. Experiments have shown triggering taking place when a drop wanders into a region confined by solid surfaces, such as a corner of the experimental vessel. The precise nature of the process is not at all understood, but it is probably the local collapse of the vapor film, resulting in a small-scale interaction that causes a pressure pulse.

3. Propagation. This stage is analogous to detonation of a chemical explosive and is set off by the trigger. It produces a self-sustaining wave that advances through the coarse mixture, causing fine fragmentation of the coarse drops. Heat transfer from the fine fragments supplies energy for maintaining the progress of the wave. Thus, extremely rapid heat transfer is brought about in a rapid coherent way. The precise mechanism for this self-propagating fragmentation wave is not completely understood, but there is impressive experimental evidence for its existence.

4. Expansion. The mixture of fuel, steam and water expands in this stage as an energetic source of pressure capable of producing great damage.

This division into four stages is an approximation to physical reality. In fact, the four stages are not entirely separate and each can be influenced to some extent by the others. In particular, the degree of confinement by external constraints can greatly influence the process. The general features of this model, however, are reasonably well established, and the model has been a powerful aid in assessing the significance of steam explosions.

Attempts have been made, by considering each process in detail, to estimate the likelihood of a damaging interaction within the reactor vessel in the course of meltdown. One major potential mode of failure is the generation by a steam explosion of energetic missiles, such as the pressure vessel head, which might penetrate the containment (alpha-mode in the terminology of WASH-1400).[4] By assigning plausible probability distributions to the events that make up a steam explosion, there is a rather clear consensus that its occurrence would have a conditional probability of at most 0.01, and probably much less. At present, this estimate is based on expert judgement and some experts would prefer to maintain a strong program of research to establish a stricter mechanistic argument. Most experts would wish to see effort continue towards confirming the scientific basis but would find it difficult to accord such work very high priority.

2.3 Natural Convection

Natural circulation plays a part at many stages of severe accident progression and often has a dominant part in determining the course of events. The process is, in principle, strongly self-stabilizing and dependable, but there are several complicating factors in severe accidents.

First, there is the thermal-hydraulics in the early and core degradation stages of the accident. Heat transfer will be strongly influenced by convection, as will the transport of steam and hydrogen, and consequently chemical behaviour. In the cases of station blackout and certain small-break LOCAs, the primary circuit will be filled with a mixture of steam and hydrogen at the relief valve pressure setting. Natural convection around the circuit may well be the prime determinant of when the circuit will fail, with important implications. If failure is by melting in the pressure vessel lower head, a violent and dispersive ejection of molten debris would result, pressurizing the containment through rapid heating of the containment atmosphere. Such an event is known as direct containment heating (DCH). There is at least a possibility that this failure mode would be forestalled by failure in the upper pipework leading to the steam generators as a result of overheating by natural convection.

Codes that have been developed to analyze the process give good predictions of scale-model experiments. It must be remembered that in reality there will be important disturbing influences, such as the effect of relief valve opening and closing, water injection, and major disturbances as debris relocates or crusts collapse. It has also been pointed out that hydrogen may separate, forming an upper layer of such a low density that thermal density changes are not sufficient for the steam of higher molecular weight to penetrate the hydrogen layer. One or both of these processes may explain why the temperatures reached in the upper structure of the TMI-2 reactor were lower than might have been expected from heat transfer by natural circulation.

The transport of fission product aerosols both within the primary circuit and in the containment is governed largely by natural convection. The analysis is difficult because the movement of the atmosphere transporting the fission product is partly determined by the distribution of those fission products since they are the source of the heat. This is another instance of the intimate interdependence of different physical processes so common in severe accident progression.

2.4 Fission Product Releases from Fuel

The progression of severe fuel damage and fission product release mechanisms with increasing temperature is shown in Figure 2. Below 1300 K, fission products in the fuel matrix are relatively immobile, and only gas formed at external surfaces of the pellet can escape, either by direct recoil or by knockout. Up to about 1600 K, the fuel sheaths strain, oxidize, and rupture, thus releasing the gases collected at the pellet surfaces. From 1600 K to 2050 K, sheath oxidation occurs more rapidly, and release from the fuel matrix due to diffusion becomes important. Above 2050 K, UO_2 grain growth becomes a dominant release mechanism, along with the dissolution of UO_2 by molten Zircaloy. Above 2700 K, the ZrO_2/UO_2 reaction may lead to further fuel dissolution and fission product release. Finally, fuel melting can occur, as indicated in Figure 2, at temperatures that depend on the stoichiometry of the fuel.

The releases associated with severe fuel damage will depend on such parameters such as the composition of the hydrogen/steam mixture in the core (i.e., the oxidation potential), the extent of chemical reactions of fission products with the sheath and control/structural materials, and the concentrations of fission products. A further consideration is that the core is unlikely to heat up uniformly, and different regions could experience different conditions at any point in the sequence. The exact conditions in the core are sequence-specific and are affected by such phenomena as the extent of Zircaloy oxidation, flows and pressure. For the more volatile fission products (Kr, Xe, I, and Cs), however, releases may be essentially 100% prior to core slumping. This simple picture may be modified following completion of the TMI core examination. For other species the releases are highly dependent on chemical phenomena. This can be illustrated by two examples: the effect of sheath oxidation and the effect of steam pressure.[5,6]

The degree of sheath oxidation, which depends on flow conditions in the core, affects the releases of Te, Ba, and Sr. In the case of Te, stable zirconium tellurides are formed, and Te is only released following oxidation of the sheath. For Sr and Ba, unoxidized Zircaloy can reduce the oxides to the relatively more volatile metals. Thus, sheath oxidation would retard the release of these species.[7] For sequences that are characterized by relatively high vessel pressures (for example, the LWR TMLB′ sequence), the high steam pressure could convert the involatile oxides of Ba and Sr to the more volatile hydroxides. Other species that have low volatility under reducing steam conditions (e.g., La, Ru, Tc) are likely to be retained in the melt until the time of vessel failure.

In addition to the more volatile fission products, volatile inactive species, such as Sn from the sheath and Cd from the control rods, are likely to be released. Core slumping would lead to the release of any remaining iodine and cesium, along with a small amount of less volatile

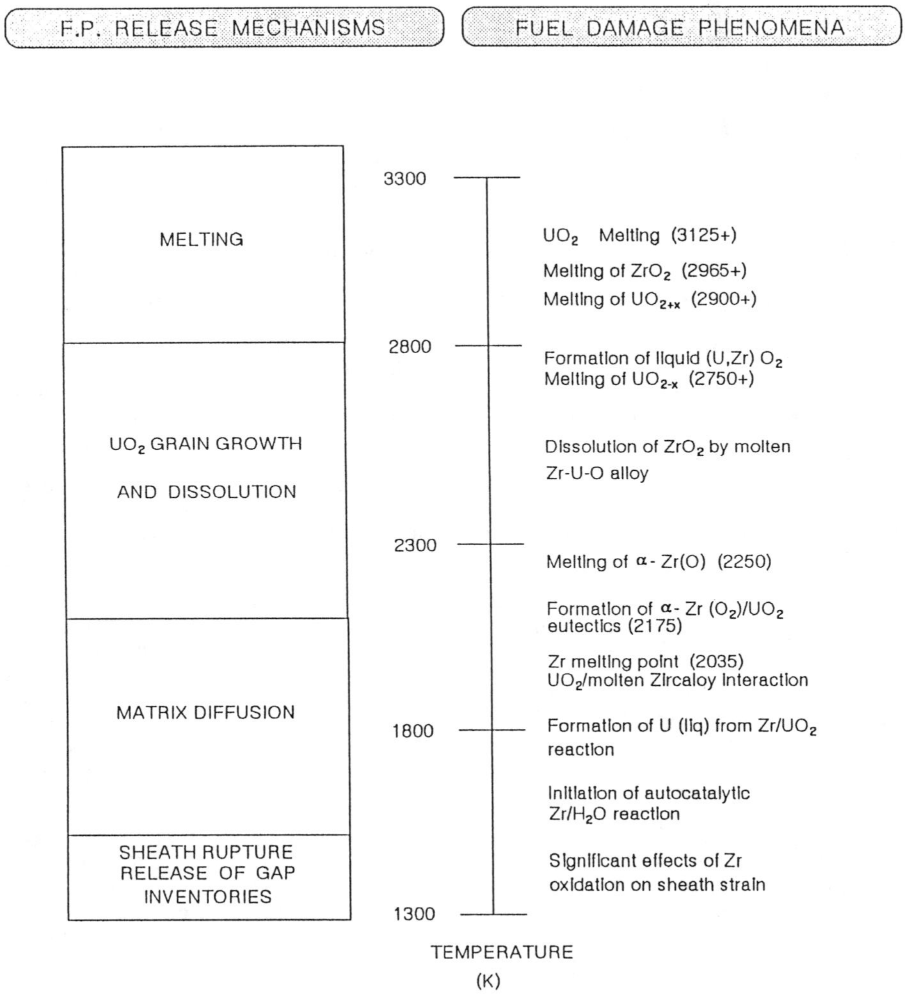

F.P. RELEASE MECHANISMS
FUEL DAMAGE PHENOMENA
MELTING
UO2 GRAIN GROWTH AND DISSOLUTION
MATRIX DIFFUSION
SHEATH RUPTURE RELEASE OF GAP INVENTORIES
3300
2800
2300
1800
1300
TEMPERATURE (K)
UO2 Melting (3125+)
Melting of ZrO2 (2965+)
Melting of UO2+x (2900+)
Formation of liquid (U,Zr) O2
Melting of UO2-x (2750+)
Dissolution of ZrO2 by molten Zr-U-O alloy
Melting of α - Zr(O) (2250)
Formation of α - Zr (O2)/UO2 eutectics (2175)
Zr melting point (2035)
UO2/molten Zircaloy interaction
Formation of U (liq) from Zr/UO2 reaction
Initiation of autocatalytic Zr/H2O reaction
Significant effects of Zr oxidation on sheath strain

FIGURE 2.

species, e.g., Sr and Ba, and the relocation of control rod materials, e.g., Ag and In, which are relatively abundant in the core. The release of these materials could have significant impact on subsequent transport in the RCS.

It should be mentioned that highly oxidizing conditions would result in significantly different releases. At low temperatures (about 800 K), in air, fuel rapidly fragments to form a powder with an average particle size of 2 μm.[8] The mechanism involves the attack of oxygen along the grain boundaries of the fuel to form U_3O_8. The resulting expansion separates the grains from the matrix, and the grains are rapidly oxidized to form the small particles. At higher temperatures (1500-1700 K), grain destruction does not occur, and the fuel oxidizes due to oxygen diffusion through the matrix. In this case, the resulting particle sizes are much greater (10 μm). Fission products that are not usually considered to be volatile would be released as fuel aerosols, along with enhanced releases of species expected to be more volatile under oxidizing conditions, such as Ru. Many of the qualitative features of the Chernobyl releases can be characterized by considering fuel and fission product behaviour as a function of oxidation potential and temperature in the core.[9]

2.5 Hydrogen Generation

The main process producing hydrogen in a severe accident is the oxidation of core materials by steam. A variant of that may occur in the late stages of the accident in reactions between molten core debris and the water contained in concrete. Analysis and prediction are comparatively straightforward in the early stages, when the core geometry is still in its design state, but becomes increasingly difficult and uncertain as the accident proceeds and possibilities multiply. It is convenient to consider hydrogen generation in three stages: early, intermediate and late.

In the early stage, when the core is more or less intact, the oxidation process is governed by solid-state diffusion in Zircaloy. As has been pointed out in considering core degradation, the heating rate is strongly influential. A slow rate gives time to form a strong outer shell of ZrO_2 that will delay fuel relocation.

The intermediate stage introduces further variables. If the fuel fails by brittle collapse, a rubble bed will be formed, with a large effective surface area. UO_2 will be exposed to the steam, giving an increased formation of hydrogen. Although this could increase the capacity for hydrogen formation by as much as 60% of that from Zircaloy alone for pure steam at high pressure, even a small concentration of hydrogen would reduce the extra potential production to 18%. When melting occurs in the rubble, the reaction may be changed by the opening or closing of channels admitting steam. Certainly, progression to a molten pool will reduce the effective surface area and, hence, the rate of hydrogen production.

The late stage may involve release of molten debris from the primary circuit with further reaction between debris and steam in the containment atmosphere. Chemically produced hydrogen from finely dispersed fuel may augment the DCH effect (see below). Debris may fall into water in the space below the core, giving a sharp increase in hydrogen production. Finally, hydrogen will be produced as a result of reactions with water in the concrete structural material.

3. RCS TRANSPORT/FAILURE PHENOMENA

3.1 Fission Product Transport

The release of material from the core region leads to the formation of appreciable aerosol concentrations near the core/RCS interface. Aerosol agglomeration processes and vapor condensation lead to growth of the aerosols during transport through the RCS. Most aerosol transport codes assume an initial particle size distribution and do not treat the initial formation of aerosols from vapor condensation. Studies assessing the formation of primary particles predict that, under some conditions, particle growth occurs by heterogeneous condensation as opposed to agglomeration.

Estimates of the amount of material released to the RCS from a melting core range from 400-5000 kg.[7,10,11] Much of this release would be associated with structural and control-rod materials, which produce aerosols having a total surface area that exceeds the surface area of the primary system.[12] For example, typical PWR control rods are composed of an 80% Ag, 15% In, 5% Cd alloy, having a melting point of 1100 K, in a 304 stainless steel sheath. Experiments have shown that significant quantities of aerosols are formed from control rods having this composition when the sheath melted (about 1700 K).[13,14,15] Thus, there is considerable scope for physical and chemical reactions between fission product vapors and aerosols.

There is a growing body of evidence to suggest that the composition of RCS aerosols may be complex. For example, the control rod vaporization experiments mentioned above showed that the emitted aerosols had a bimodal size distribution, with one mode being dominated by cadmium, while the other mode was dominated by indium oxide. Also, evidence from the Marviken and LACE tests suggests that the aerosols will not be homogeneous. In the Marviken experiments,[16] where mixtures representing core material were vaporized using a plasma torch, two general types of aerosols were found that consisted of Ag-rich centres and MnO crystalline particles.[13] During transport, some of the MnO particles deposited on the Ag, along with the simulated fission products, resulting in an aerosol in which Mn, Cs, and Te were predominant on the surface. Also, it is likely that there were chemical reactions between some of the species: Cs was found to be associated with Mn, Cs with Te, and Te with Ag. Chemical changes may also have occurred over time on the surface of the aerosols, as it was found that the tellurium became more soluble as the aerosols aged.

The LACE experiments have also provided some insights into the nature of the aerosols.[17] In these experiments, two-component aerosols, consisting of water-soluble CsOH and insoluble MnO, were generated separately and mixed before entering piping leading to a containment vessel. The resulting aerosols were found to have an MnO core with a Cs-rich outer layer.[13] This may have implications for aerosol transport in containment, since it was found that the particles that were rich in surface CsOH experienced preferential steam condensation and faster settling rates than the MnO-rich particles.

There are a number of mechanisms by which fission product vapors and aerosols would deposit on RCS surfaces. Fission product vapors would be removed by physical and chemical adsorption processes, while aerosols would deposit by sedimentation, thermophoresis, turbulent and Brownian motion, and impaction. If steam condensation were to occur, then diffusiophoresis would also be a likely removal mechanism. Sedimentation appears to be the dominant process for removal of aerosols from the gas phase.

In many reactor designs, the fission product/aerosol pathways would include water pools, such as the BWR suppression pool and the PWR quench tank. The efficiency of such pools to attenuate fission product transport would depend on particle sizes, steam mass fractions, and pool temperatures.

The revolatilization of fission products and resuspension of aerosols is highly dependent on thermal-hydraulic conditions. Processes that could lead to revolatilization/resuspension include rapid RCS depressurization, small-scale steam explosion events, other events that result in high-velocity flows, and heating due to radioactive decay. The effect of deposition/resuspension processes is to delay the ejection of material from the RCS to containment, and could result in time-dependent changes in the size distribution and composition of the aerosols. The prediction of these phenomena requires codes that can produce realistic heat and mass-transport conditions and, at the same time, accurately describe fission product and aerosol behaviour.

Iodine, cesium, and tellurium are considered to be the most important fission products released in the early stages of a severe accident, with similar volatilities and diffusion properties. Other fission products are unlikely to be released in any quantities until core melting occurs, at which point a fraction of the Ba and Sr inventory would likely be released under the expected reducing steam conditions. As discussed in the section on fuel releases, Ba and Sr might exhibit increased volatility in a high-pressure steam environment owing to the formation of the more volatile hydroxides. Although there are little data on the transport of these species, their compounds have relatively low volatility and it is unlikely that there would be any extensive reactions with RCS materials, other than co-agglomeration to form larger aerosols.

The predominant form of cesium in the RCS is likely to be CsOH. While the high steam concentrations would tend to stabilize the CsOH, there is potential for the CsOH to react with surfaces to form stable compounds. Therefore, there is a possibility that there could be considerable retention of cesium in the RCS.

Thermodynamic analyses,[18] chemical kinetics analyses,[19] and experiments[20] have demonstrated that the predominant iodine species transported in the RCS would be CsI. This could be modified, however, if there is an opportunity for the CsI to react with compounds that form stable cesium salts, such as the formation of chromates, silicates, and borates. The formation of these compounds would depend on such parameters as the oxidation potential and temperature. If CsI were to decompose, then there would be considerable scope for other reactions owing to the reactivity of iodine species. For example, it is likely that HI or elemental iodine would react with cadmium (from the control rods) to form CdI_2, or with other aerosol materials, such as tin and silver, to form less volatile salts.

The potential effects of boron compounds on iodine and cesium transport should also be mentioned. A relatively large quantity of boron is present in PWRs (typically 12 000 moles, compared to 95 moles of I and 1200 moles of Cs). The volatility of boron depends on the amount of steam present, which tends to stabilize H_3BO_3 and HBO_2 in the gas phase. Experiments have shown that CsI reacts with boric acid to produce HI and cesium borates.[21] A more limited reaction occurs when CsI vapor is passed over boric-acid-coated steam surfaces.[22] Since most of the Cs released into the RCS is expected to rapidly form CsOH, the hygroscopic properties of aerosols containing CsOH could be modified if there are reactions with H_3BO_3 or B_2O_3 to form cesium borates. Also, as cesium borates are less volatile than CsOH, the transport of cesium could be retarded.

For a rapid depressurization sequence, much of the boric acid would be lost from the RCS in the ejected coolant. In high-pressure sequences, most of the boric acid would be expected to remain in the core and RCS. As the core becomes uncovered, boric acid/oxide would volatilize and plate out on surfaces, depending on the degree of steam starvation. Thus, there may be considerable potential for reactions between cesium and boron species in high-pressure accidents.

The third fission product likely to dominate during the early stages of a severe accident is tellurium. (As discussed previously, the release of this fission product is dependent on the extent of sheath oxidation.) Thermodynamic calculations show that Te and Te_2 are the most stable species in the neutral or reducing steam environments expected,[18] and at RCS temperatures, zirconium or tin tellurides would likely be converted to elemental tellurium. At lower temperatures, as shown by the Marviken results, tellurium may combine with other species to form more complex compounds. For example, cesium tellurides may be formed at lower temperatures in reducing steam conditions if excess cesium is present. The stability of cesium tellurides, however, is less than that of the other predominant cesium species (CsOH and CsI), and therefore it is not clear that large quantities of tellurium salts would be formed. The only highly volatile form of Te that could occur is H_2Te, but the formation of this species requires high temperatures and very reducing conditions. At lower temperature, H_2Te would likely react to form other, less volatile species.

3.2 Vessel Melt-Through

A possibility that has received a great deal of attention in recent years is that the pressure vessel may fail at a point in the lower regions when still pressurized. Molten debris would then be violently ejected into the containment atmosphere in such a way as to become finely dispersed, providing a large area for rapid heat transfer and possibly hydrogen-producing reactions. As a result, there is a danger of a rapid increase of pressure, which could lead to serious damage of the containment. This effect has been termed direct containment heating (DCH).

The pressure rise due to rapid heat transfer to the atmosphere will be less if some of the heat is transferred to other materials. These include liquid water and solid structural materials. Thus, the significance of DCH depends on a complex interplay of conflicting processes. Impaction of released molten debris on concrete or steel surfaces will be more or less effective in providing a heat sink not associated with a pressure increase, but there is considerable uncertainty depending on the droplet size, time of contact with solid surfaces, wetting and so on. The process is clearly highly specified to the detailed arrangement of the plant.

There is a possibility that primary circuit failure would occur in the upper regions as a result of overheating by natural convection (see above). In that case failure could occur in one of the hot legs of the primary circuit, depending on the local stress distribution and mechanical properties in the circuit boundary, both determined by the temperature field.

It will be seen that there is a rather wide range of possible outcomes of primary circuit failure. The prediction of which one will actually occur requires a great deal of knowledge and understanding of some complex and highly interactive phenomena. Moreover, the application of that knowledge and understanding will need to take account of many specific features of the plant design. Therefore, the reliance that can be placed on predictions in this area is limited, which has led to a growing interest in exploring possibilities for controlled depressurization of the primary circuit (see below).

4. CONTAINMENT PHENOMENA

4.1 Aerosol Transport

Since an accident usually begins with a discharge of coolant into the containment system, the initial containment atmosphere would contain aqueous aerosols formed by rapid evaporation of water, shearing action, and jet instabilities. Typically, about 25% of the discharged water would be converted to steam, and as much as 50% of the water could form aerosols. The initial distribution of the aerosols may be bimodal, with modes at about 1- and 10-μm mean diameter.[23,24] These emissions would include, or would be followed by, gases and aerosols composed of fission products, fuel components, structural materials, and reaction products such as hydrogen. As emphasized in the previous section, the mass of material released to containment in a severe accident would eventually be dominated by structural components. For the most part, these components would not react chemically with water. Thus, much of the emphasis in analyzing this stage of an accident has been on aerosol physics and thermal-hydraulics. There are chemical phenomena, however, that may affect the behaviour of some of the more important fission product species, and we shall return to these phenomena later.

Owing to the presence of steam, solid aerosol material rapidly becomes contained in water droplets. The description of aerosol behaviour is thus "reduced" to the growth, agglomeration, and deposition of water droplets in a steam/air atmosphere. Since the particles are spherical, some of the uncertainty associated with the shape factors for solid aerosols is removed. Thus, the transport properties of containment aerosols are considered to be one of the more developed disciplines in the field of source term assessment.[25] It must be emphasized, however, that thermal-hydraulic uncertainties can have a large impact on predictive capability.

If DCH, as discussed in Section 3.2, could occur, then large quantities of containment aerosols would result. Experiments have shown that 0.3-6% of the ejected material forms aerosols.[26] If a large fraction of the core were involved in DCH, then the releases of aerosols would be comparable to those associated with the in-vessel phase of a severe accident.

Aerosols would be removed from the containment atmosphere by gravitational settling (large particles), diffusion by Brownian or turbulent motion to surfaces (small particles), thermophoresis (transport through a temperature gradient from a hot gas to a cold surface), diffusiophoresis (transport to surfaces by means of steam condensation), and by containment leakage. These transport properties will be strongly affected by the rate of particle growth due to steam condensation and particle agglomeration.

Particle growth due to steam condensation can be extremely rapid. For example, a supersaturation of only 1% causes a 1-μm radius drop to grow at a rate of 4.65 μm/s.[27] This illustrates the extreme importance of correctly coupling the thermal-hydraulic conditions in containment to aerosol behaviour. Aerosol agglomeration rates are a function of the net forces acting on the aerosol particles. The usual practice, however, is to calculate separate agglomeration rates due to gravitational, diffusive, and turbulent processes, and to sum them to obtain an overall rate. This approach appears to be particularly valid when one process dominates.

If the accident is terminated before failure of the reactor pressure vessel, then the important species in containment would be the noble gases and aqueous aerosols, including inorganic forms of the more volatile fission products, such as iodine, cesium, and tellurium. Owing to the rapid growth of the aerosols, gravitational settling would likely be the dominant removal mechanism.

There is evidence that the chemical composition of the aerosols released to containment could influence their subsequent behaviour. For example, if the ejected aerosols contain hygroscopic surfaces (as was the case for some of the LACE aerosols described previously), then there could be two effects on the containment thermal-hydraulics: a reduction of the equilibrium vapor pressure in containment, and the release of the heats of condensation and solution. In general, the two effects would tend to counteract each other as far as water condensation is concerned. For example, an analytical study by Clough et al.[28] showed that the amount of water condensing on a hygroscopic aerosol increased by a factor of about 3.5 over inert aerosols when the latent heat was ignored. Including the latent heat reduced the increase to 10%. However, the latent heat effects would be highly dependent on containment thermal-hydraulics, since heat losses would have considerable effect on the condensation. This illustrates again the importance of coupling thermal-hydraulics to aerosol transport.

An important caveat to the above considerations is that proper account must be taken of other possible events during a severe accident, and that phenomena cannot be considered in isolation. For example, the hygroscopic properties of aerosols would be affected by the extent of conversion of CsOH to cesium borates, either in the primary system or in containment. Also, a containment building contains sufficient quantities of CO_2 to convert fission product hydroxides (such as CsOH, $Sr(OH)_2$, and $Ba(OH)_2$) to carbonates, although the rates of these reactions may not be rapid enough to have large-scale impact. These examples, however, serve to illustrate the importance of treating the various processes collectively.

4.2 Fission Product Chemistry

In addition to the chemistry effects discussed above for aerosol transport in the RCS and containment, there are a number of other chemical phenomena that are of interest. The chemistry occurring in containment can be extremely complex, and it is not possible to even summarize all aspects of this topic in this overview. Therefore, we restrict this discussion to some aspects of the aqueous and gas phase chemistry of the two fission products that could form volatile species in containment: iodine and tellurium.

Iodine released to containment as an iodide would quickly partition into the aqueous phase. In aqueous solution, chemical changes are affected by such parameters as temperature, pH, radiation, redox potential, concentrations, and solution composition.[29] Since the core inventory of iodine in an LWR is about 12 kg, and since a PWR contains about 4×10^5 L of water, the concentration of iodine would not likely exceed 10^{-4} $mol \cdot dm^3$ for any length of time. Iodides are relatively inert chemically, but in radiation fields, they can combine with Ag particles to form insoluble AgI. Also, in the oxidizing containment environment, iodides are unstable and tend to oxidize to reactive aqueous species, e.g., I_2 and HOI. The rate of oxidation, however, can be very slow unless the pH is low. In a PWR, the cooling water is maintained at pH 7 at 300°C, which corresponds to pH 10 at 25°C. Thus, the release of the cooling water would tend to prevent the oxidation of iodide. Also, spray systems may contain additives, such as hydrazine, sodium hydroxide, and sodium thiosulphate, that would affect the pH and/or oxidation potential in such a way as to reduce the tendency of iodides to oxidize. At TMI, the pH of the water remained in the range 7.5–8.5,[30] and relatively little volatile iodine was observed. Nevertheless, the dominant forms of the volatile iodine were organic iodides as well as I_2, indicating that some oxidation must have occurred.

Oxidation of iodide in containment aqueous solutions is a complex process.[31] For example, the rates could be very dependant on the presence of oxidation catalysts, such as copper and iron ions, and on radiation levels. Radiation produces both oxidizing and reducing species in solution whose net effect is to favour iodide at high iodine concentrations (at neutral to basic pH), and oxidized species at the low concentrations expected in containment (at more acidic pH).

There is increasing evidence that organic iodides can be readily formed in the aqueous phase.[32] Water in the containment building would be contaminated with a variety of organic impurities, such as those originating from paints, coatings, insulation, seals, lubricants, connectors, and gaskets. It is likely that the total concentration of organic impurities would exceed the total concentration of iodine. Experiments have shown that I_2 and HOI solutions react with unsaturated hydrocarbons, ketones, alcohols, aromatic compounds, etc., to form organic iodides. In addition, organic free radicals formed by radiolysis initially enhance organic iodide formation, although in the longer term at steady-state conditions, destruction of organic iodides by radiation should result in lower concentrations. (It should be noted that the formation of organic iodides in the gas phase due to radical reactions is unlikely to be a major source owing to the rapid reaction of the radicals with oxygen.) Although organic iodides are unstable towards hydrolysis, the rates are relatively slow at low temperatures. Thus, in the shorter term, I_2 may be the dominant airborne iodine species, but in the longer term, it is expected that the concentration of organic iodides would increase.

Although iodine chemistry in containment has been actively developed over the past several years, relatively little is known about tellurium behaviour. Tellurium may be released to containment in the elemental form along with various metal tellurides. The most important isotope is ^{132}Te ($t_{1/2}$ = 3.25 d), which decays to ^{132}I ($t_{1/2}$ = 2.3 h). The amount of ^{132}I reaches a maximum about 24 h after reactor shutdown and dominates iodine activity levels up to about 3 d. At TMI, it was noted that most of the airborne ^{132}I was organic,[33] which may be attributable to chemical changes due to the decay of the parent tellurium.

Aqueous tellurides would tend to oxidize in containment to form insoluble TeO_2 and oxy-anions, although the rates of oxidation may be slow. There are no stable inorganic forms of tellurium that are expected to be volatile under containment conditions. (The only potentially volatile inorganic tellurium species is H_2Te, but this species is unstable in air.) Elemental tellurium can react with organic impurities in aqueous solutions to form volatile organic tellurides, but under containment conditions, radiolysis is probably required.

4.3 Core Debris - Concrete Interaction

By the time core debris comes into contact with structural concrete, the accident will be well advanced and a large number of factors will have played a part in determining the temperature, composition and physical form of the fuel. The manner of release or ejection is also strongly dependent on earlier events. Still more possibilities are introduced in considering subsequent events; for example, local geometry of the containment floor, walls, internal structure, and size and shape of cavities, all have a strong influence. The quantity and distribution of water present when debris emerges may well be crucial. If there is sufficient water to quench the debris so that it forms a permeable bed of rubble that can be kept cool by the water, then interaction with the concrete may be unimportant. If not, then the debris will attack the concrete and a multicomponent, multilayer pool will develop that can penetrate vertically and spread laterally.

Some of the details of core-debris/concrete interactions are shown in Figure 3. In the molten region, the fluid is self-heated and strong convection patterns build up, transferring heat to the solid boundaries, which themselves move as concrete is attacked by the molten debris. One form of attack is physical erosion as pieces of concrete become detached as a result of thermal stresses and are carried away by buoyancy forces to become incorporated in the melt. Chemical reactions may produce copious evolution of gases, such as steam, hydrogen, carbon dioxide and carbide monoxide, with significant effects on flow patterns and transport of both heat and fission products in the pool. Effects of that kind depend very much on the type of aggregate used in the concrete, notably whether it is carbonaceous or siliceous in nature.

It is likely that there will be stratification within the molten pool, given the wide range of densities of the constituents. For instance, there is a tendency for high-density materials (UO_2 and metals) to migrate to a layer at the bottom. As the lighter materials in the concrete become incorporated, however, the configuration can become unstable, causing a sudden reversal.

Conditions at the top of the pool are of particular significance, since the proportion of heat transferred upwards is clearly a major influence both on the rate of growth downwards and sideways, as well as on conditions in the containment. A solid crust will form if upward heat transfer is high enough at the surface in relation to internal heat transfer. However, changes either within the pool or as a result of external disturbance, such as the addition of water, may disrupt the crust and expose the underlying molten pool.

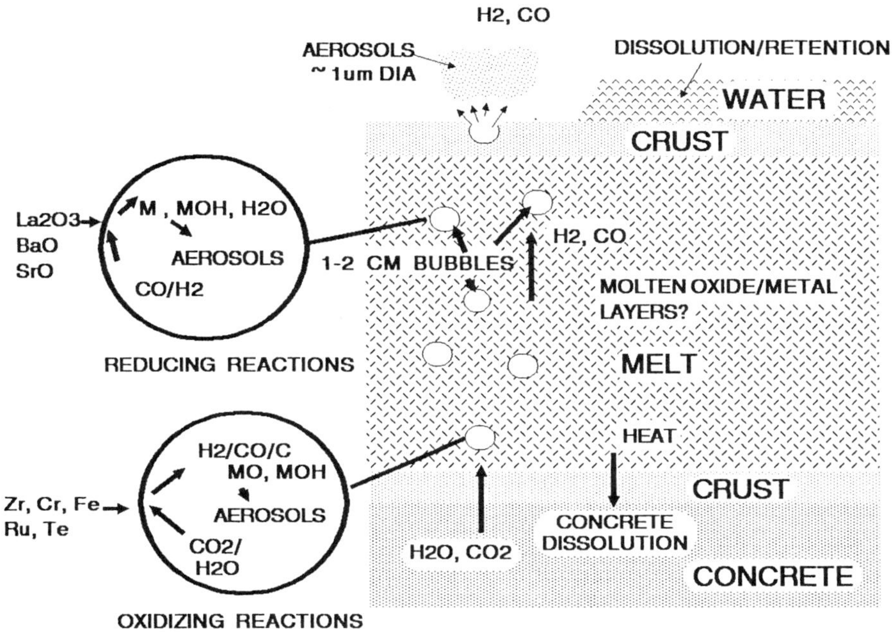

FIGURE 3. Core-Debris/Concrete Interactions

Penetration downwards into the concrete has received a great deal of attention over the last decade or two. Computer modelling, backed by experiments, has led to the incorporation of this aspect into suites of codes used to assess the significance of the risk of basemat penetration. There is however, a wide band of uncertainty reflected in considerable divergence of results from experiments, notably in the relative rates of downward and lateral growth of the molten pool. Such divergence probably reflects the sensitivity of penetration rates to a number of features. Concrete composition is probably the most significant of these, but others include the composition of the molten debris, particularly the free-metal content, and the scouring effect of material flow within the pool.

There are a number of uncertainties associated with fission product releases during core debris/concrete interactions. As indicated in Figure 3, heat from the melt results in dissolution of the concrete, with the release of H_2O and CO_2. These gases form bubbles that are 1-2 cm in diameter and are relatively oxidizing. The steam and CO_2 then react with metals in the melt (such as Zr, Cr, Fe, Ru, and Te) to form CO, carbon, and H_2, along with metal oxides/hydroxides. If the metal oxides/hydroxides are more volatile than the metals (e.g., as is the case for Ru), then the bubbles would be expected to be enriched in that element relative to the melt. Closer to the top of the melt, the bubbles would then be more reducing, and metal oxides such as La_2O_3, BaO, and SrO could be reduced to more volatile species. This would result in enhanced releases of these species compared to the releases expected from the primary system.

Since Zr from the fuel sheath would be expected to be a major reductant in the melt, the degree to which the CO_2 and water vapor react with the melt to form reducing agents will depend on the degree of sheath oxidation that occurs in the reactor vessel. If there is a sufficient quantity of unoxidized Zr, then hydrogen could be the predominant gas released during the reduction phase of the interaction.

4.4 Hydrogen Combustion

The impact of hydrogen combustion on containment performance is determined by the amount and rate of energy deposition. The total energy deposited in containment would be limited by the amount of hydrogen burnt, which, in turn, depends on the amount and distribution of the hydrogen, steam concentrations, and ignition and flammability limits. The rate of energy deposition depends on such factors as the gas composition, pressure, temperature, geometry (i.e., flame acceleration), and gas dynamics within containment.[34]

Flammability limits define the compositions for which a flame will propagate through a quiescent atmosphere. There is an extensive data-base available that includes such effects as temperature, diluents, and propagation direction.[35,36] The lower flammability limit for hydrogen/air mixtures has been measured to be in the range 4-9% hydrogen. Steam concentrations of about 60% by volume are probably sufficient to inert the containment atmosphere.[37] In the early stages of an accident, steam concentrations are likely to be high, and combustion would be suppressed. For this reason, early containment failure due to hydrogen combustion may be an unlikely event. At later stages, once steam has been condensed, hydrogen combustion is more likely to occur.

The rate of energy deposition is highly dependent on conditions in containment, and on the location of the burn. Burning velocities (which determine the rate of energy deposition) are strong functions of the gas dynamics; turbulent burning velocities can be orders of magnitude higher than laminar burning velocities. Obstacles in the flame path produce turbulence and, therefore, may accelerate the flame. (However, obstacles can also absorb heat, which would tend to reduce the peak pressures. The overall effect of obstacles will depend on their size and geometry.) Concentration gradients can cause transient burning rates that exceed steady-state rates in a zone having a particular composition. Thus, there are a number of potential mechanisms that would affect transient pressures.

Venting from one area of containment to another area that does not contain a flammable mixture will normally reduce the energy deposition rate for the same burning rate. Under certain conditions, however, turbulence and instabilities produced by the venting flow can increase the burning rate sufficiently to increase the energy deposition rate and cause large-amplitude pressure oscillations. Venting of burnt gases from one area to another area containing a flammable mixture may lead to multipoint or jet ignition, which in turn can increase energy deposition rates and even cause a transition to detonation for certain compositions.

The detonation sensitivity of hydrogen/air/steam mixtures, and factors affecting the transition from a deflagration to a detonation, are important questions with respect to containment performance. Detonations occur when flame speeds are sufficiently high to produce strong shock waves that ignite the gas ahead of the flame front by compression. This results in high local peak pressures that greatly exceed constant-volume combustion pressures. However, the direct initiation of detonations requires relatively high-energy ignition sources, which are not likely to be found in containment. A more plausible mechanism is the possible transition of a deflagration to a detonation, which is dependant on geometry and thermodynamic conditions. The thermodynamic conditions (composition, pressure, temperature) determine the detonation cell size, which must be compatible with the local propagation volume for a transition to occur, and for detonation to be sustained. Geometry conditions (e.g., obstacles and vents) determine whether transitions can occur for given thermodynamic conditions.

The major effects of hydrogen combustion relate to containment integrity and the possibility of enhanced fission product releases due to pressurization. An additional consideration is the survivability of safety components if a diffusion flame becomes stabilized in the vicinity of the components.

Overall, hydrogen combustion may not have a large net effect on aerosol transport within the containment building unless containment is impaired. The shock waves produced might de-agglomerate some aerosols, but the increased turbulence would enhance agglomeration. The latter effect would likely dominate. The combustion shock front and thermal-hydraulic effects associated with local turbulence could resuspend particles on surfaces. Only large particles (> 30 μm) would be affected, however, as the stronger adhesive forces for small particles are unlikely to be overcome. The large particles would quickly settle out following resuspension. In addition, the presence of water films would increase the adhesive forces, and such films are likely to exist in the humid containment environment. Heat from the combustion front would also have a limited effect because of the short contact time for heat transfer.

There is a possiblity that the chemistry of airborne iodine could be affected by hydrogen combustion. Dry CsI aerosols have been shown to react with intermediate combustion radicals to produce molecular iodine.[38] Also, chemical kinetics analyses indicate that hydrogen combustion would convert airborne organic iodides to I_2 and HI.[39] However, the impact of hydrogen combustion on iodine aerosol transport may not be significant owing to the likelihood that CsI would not exist as pure, dry aerosols in containment.

5. ENGINEERED SAFETY SYSTEMS

From the foregoing material, it is plain that, for severe accidents beyond the design basis, it is not possible to specify the course of events in advance with any degree of confidence. Nevertheless, the containment has a good potential for enabling the accident consequences to be controlled and mitigated, provided the operator is equipped with a good understanding of the relevant phenomena and the physical means to intervene constructively.[40]

There is a distinction between those ESFs that form part of the design basis and those that may be invoked only under severe accident conditions. ESFs for dealing with design-basis accidents are specified with precision and are required to achieve a clearly defined standard of performance in equally clearly specified conditions. Since they form part of the design basis, appropriate quality assurance procedures will be necessary to ensure compliance with the level of qualification laid down for safety equipment.

It is not practicable to proceed in the same way for ESFs that might be brought into play in severe accidents. A more flexible, pragmatic approach is called for. The general aim is to ensure that all potentially useful features can be exploited as fully as is reasonably practicable. One prerequisite to do that, is an awareness on the part of the operator as to what could be done in the particular circumstances. Therefore, he not only needs to possess an adequate understanding of the processes that interact in an accident sequence, but he also needs information about conditions inside the reactor and its containment.

It is not necessarily helpful to provide additional instrumentation since excessive proliferation of information can add to the strain imposed on the operator during an accident. The instrumentation used normally has the best chance of assisting operators, and modifications might best be aimed at extending its range to cover conditions beyond those specified in the design, and ensuring that it can withstand, to the greatest extent practicable, the more severe physical conditions of the accident. These will include, for example, high temperatures, hostile chemical environments, structural failures and the effect of water immersion or sprays.

The same considerations apply to design-basis ESFs, such as pumps, valves, and fans, that might need to be employed in unusual ways in the event of a severe accident. An additional factor enters when active intervention (as opposed to the passive character of instrumentation) is contemplated. The primary purpose of design-basis ESFs must be given overriding priority to avoid the danger that malfunction or maloperation of well-intentioned provision for what should be extreme and improbable events may actually lead to such accidents.

Beyond making maximum use of design-basis equipment, there is also scope for special measures specific to severe accidents that would not be used in any other circumstances. One that has been extensively discussed over the past few years is filtered containment venting (FCV) and a number of countries have already introduced such systems. Something of a dilemma arises because the essential purpose of a containment is to confine the contents and the provision of a system for venting generates additional dangers of defeating that purpose through maloperation or other kinds of failure. Against that, if operator intervention has failed to prevent containment pressure approaching the failure limit, controlled partial venting through filters is clearly preferable to a release that may be total and catastrophic. Proponents of FCV tend to regard it as equivalent to a safety valve on a pressure vessel. FCV is at least an option that can be considered as a component in the defences against severe accidents in association with other provisions. Arguments for and against FCV depend on specific plant design and on what other ESFs are provided.

Currently, there is keen interest in possibilities of deliberate primary circuit depressurization to forestall the potential dangers of high-pressure melt ejection (see above). Cooling the core by injecting water may also be facilitated at lower pressure.

It cannot be taken for granted that depressurization is to be encouraged without a proper survey of what other effects, possibly harmful, may result. The decision would be taken in very difficult circumstances, with a great deal of uncertainty concerning conditions within the plant. Depressurization in inappropriate conditions could even provoke more severe damage by disrupting an existing pattern of cooling. Similarly, depressurization may lead to conditions more favourable to violent interaction between water and molten core materials. Apart from uncertainties arising in the course of an accident, there is a risk of inadvertent activation of the system. This question will be debated vigorously in the next year or so.

Other ESFs for possible consideration beyond the DBA include countermeasures against hydrogen combustion, such as recombiners, igniters and containment inerting by bulk nitrogen injection. There is no clear agreement on the need or efficiency of such measures. Difficulties arise from the need to provide for very large volumes, both for ignitors and inerting, and the power requirements for igniters are not trivial. Again, there is some risk that the ESF may actually induce a new danger, particularly in the case of inadvertent operation of inerting. The balance of advantage will again be strongly dependent on specific designs. Although in appropriate circumstances igniters or recombiners may well prove to be valuable ESFs, it seems unlikely that inerting will find favour in this application.

6. SUMMARY

There has been a continuous evolution of safety philosophy and technology since the onset of nuclear power. Design-basis accidents (DBAs) have provided a specification for the performance required of safety equipment and procedures. However, unforeseen accidents that lie outside the DBA envelope (i.e., severe accidents) did not form part of the safety environment. Nevertheless, it has been recognized that the impact of severe accidents may also be mitigated by DBA safety features. The most striking example of this is TMI-2, where, although the reactor core was virtually destroyed, the containment system successfully prevented any significant harm to plant personnel or the public. To increase our confidence in the performance of safety systems and procedures during severe accidents, we must elucidate a variety of physical and chemical processes, including core damage progression, fuel-coolant interactions, natural convection, aerosol processes, hydrogen behaviour, core-debris/concrete interactions, fission product chemistry, and the performance of engineered safety features. The objective of this overview paper was to introduce these phenomena, and to point out some of the complexities associated with severe reactor accident phenomenology.

REFERENCES

1. Rogovin, M., director and Frampton, G.T., deputy director (USNRC, Special Inquiry Group), Three Mile Island - A Report to the Commissioners and to the Public, NUREG/CR-1250 (Vol. 1,2), 1980.

2. Kemeny, J.G., Report of the President's Commission on the Accident at Three Mile Island: The Need for Change: The Legacy of TMI (11 Vols.), 1979.

3. Kouts, H., Department of Nuclear Energy, Brookhaven National Laboratory, Upton, New York, Review of Research on Uncertainties in Estimates of Source Terms from Severe Accidents in Nuclear Power Plants, NUREG/CR-4883 BNL-NUREG-50261, 1987.

4. Rasmussen, N.C., Reactor Safety Study, An Assessment of Accident Risks in US Commercial Nuclear Power Plants, WASH 1400, USNRC, 1975.

5. Albrecht, H., and Wild, H., Behaviour of I, Cs, Te, Ba, Ag, In, and Cd During Release From Overheated PWR Cores, ANS International Meeting, Cambridge, 1983.

6. Osborne, M.F., Collins, J.L., Lorenz, R.A., and Strain, R.V., Fission Product Release and Fuel Behavior in Tests of LWR Fuel Under Accident Conditions, *Proc. IAEA Int. Symp. on Source Term Evaluation for Accident Conditions, Columbus,* p. 89, 1985.

7. Albrecht, H., and Wild, H., Review of the Main Results of the SASCHA Program on Fission Product Release Under Core Melting Conditions, *Proc. ANS Topical Meeting of Fission Product Behaviour and Source Term Research, Snowbird,* 1984.

8. Cox, D. S., Iglesias, F. C., Hunt, C. E. L., Keller, N. A., Barrand, R. D., Mitchell, J. R., and O'Connor, R. F., Oxidation of UO_2 in Air and Steam with Relevance to Fission Product Releases, *Proc. ACS Symposium on Chemical Phenomena Associated with Radioactivity Releases During Severe Nuclear Plant Accidents, Anaheim,* 1986.

9. Torgerson, D. F., Garisto, F., Iglesias, F. C., Hunt, C. E. L., Cox, and D. S., The Chernobyl Source Term, *Proc. ANS Topical Meeting on Anticipated and Abnormal Transients in Nuclear Power Plants, Atlanta,* vol. 2, p. XVI-3, 1987.

10. OECD/NEA/CSNI Report, Nuclear Aerosols in Reactor Safety, Paris, 1985

11. Schrol, E., and Friederichs, H. G., Calculation of Fission Product Release to the Environment for Some PWR Core Melt Sequences, *Proc. ANS Topical Meeting on Fission Product Behaviour and Source Term Research, Snowbird,* 1984.

12. Butland, A. T. D., and Kuhlman, M. R., Factors Affecting Primary System Radionuclide Retention in an LWR Severe Accident, *Proc. IAEA Int. Symp. on Source Term Evaluation for Accident Conditions, Columbus,* 1985.

13. Bowsher, R., and Nichols, A.L., The Formation of Heterogeneous Aerosols in Severe Reactor Accidents, *Proc. Water-Cooled Reactor Aerosol Code Evaluation and Uncertainty Assessment Workshop,* eds. E. della Loggia and J. Royen, p. 170. CEC/OECD Nuclear Science and Technology, EUR 11351, 1988.

14. Bowsher, B. R., Jenkins, R. A., Nichols, A. L., Row, N. A., Simpson, J. A. H., Silver-Indium-Cadmium Control Rod Behaviour During a Severe Reactor Accident, AEEW-R 1991, 1986.

15. Mitchell, J. P., Nichols A. L., and Simpson, J. A. H., The Characterization of Ag-In-Cd Control Rod Aerosols Generated at Temperatures Below 1900 K, *Proc. CSNI Meeting on Nuclear Aerosols in Reactor Safety,* ed. W.O. Schikarski and W. Schock, KfK-3800/CSNI-95, p. 37, 1985.

16. Collen, J., Unneberg H., and Mecham, D., Overview of the Marviken Experimental Procedures, *Proc. ANS Topical Meeting on Fission Product Behaviour and Source Term Research,* p. 20-1, EPRI-NP-4113-SR, 1985.

17. Bloom, G. R., Dickenson, D. R., Hilliard, R. K., McCormack, J. D., Muhlestein, L. D. and Rahn, F. J., LWR Aerosol Containment Experiments (LACE) Program, *Proc. IAEA Int. Symposium on Source Term Evaluation for Accident Conditions, Columbus*, 1985, IAEA, Vienna, p. 321., 1986.

18. Garisto, F., Thermodynamics of Iodine, Cesium and Tellurium in the Primary Heat Transport System Under Accident Conditions, Atomic Energy of Canada Limited Rept., AECL-7782, 1982.

19. Wren, D. J., Kinetics of Iodine and Cesium Reacting in the CANDU Reactor Primary Heat Transport System Under Accident Conditions, Atomics Energy of Canada Limited Rept., AECL-7781, 1983.

20. Lorenz, R. A., Collins, J.L., Malinauskas, A. P., Kirkland, O.L. and Towns, R. L., Fission Product Release From Highly Irradiated LWR Fuel, NUREG/CR-0722, 1980.

21. Bowsher, B. R., and Nichols, A. L., High Temperature Studies of Simulation Fission Products: Part IV, Interaction of Caesium Iodide With Boric Acid Over the Temperature Range 400 to 1000°C, UKAEA, AEEW-R 1973, 1985.

22. Bowsher, B. R., Bruce, D. M., and Dickinson, S., The Interaction of Caesium Iodide With Boric Acid in the Temperature Range 100 to 1000°C, *Proc. OECD/CSNI Workshop on Iodine Chemistry in Reactor Safety, AERE Harwell*, 1985.

23. Almenas, K. K., and Marchello, J. M., The Physical State of Post Loss-of-Coolant Accident Containment Atmosphere, *Nuclear Technology*, vol. 44, pp. 411-428, 1979.

24. Koestel, A., Gido, R. G., and Lamkin, D. E., Drop Size Estimates for a Loss-of-Coolant Accident, NUREG/CR-1607, LA-8449-MS, Los Alamos National Laboratory, 1980.

25. OECD/NEA/CSNI Report, Nuclear Reactor Accident Source Terms, Paris, 1986.

26. Brockmann, J. E., and Tarbell, W. W., Aerosol Source Term in High Pressure Melt Ejection, *Nuclear Science and Engineering*, vol. 88, p. 342, 1984.

27. Dunbar, I. H., Steam Condensation Modelling in LWR Aerosol Codes, CEC Report on Contract ECI-1232B 7210-84-UK, 1984.

28. Clough, P. N., Dunbar, I. H., and Knights, C. F., The Chemical Composition of Containment Aerosols and Its Effect on Steam Condensation, *Proc. Water-Cooled Reactor Aerosol Code Evaluation and Uncertainty Asessment Workshop*, eds. E. della Loggia and J. Royen, p. 312. CEC/OECD Nuclear Science and Technology, EUR 11351 EN, 1988.

29. Lemire, R. J., Paquette, J., Torgerson, D. F., Wren D.J., and Fletcher, J. W., Assessment of Iodine Behaviour in Reactor Containment Building from a Chemical Perspective, Atomic Energy of Canada Limited Rept., AECL-6812, 1981.

30. Hofstetter, K. and Baston, V. F., Water Chemistry, *ACS Symposium Series*, vol. 293, p. 108, 1986.

31. Sellers, R. M., A Data Base for the Reaction Kinetics of Iodine in Aqueous Solution, *Proc. Specialists' Workshop on Iodine Chemistry in Reactor Safety, Harwell*, AERE-R-11974, p. 91, 1986.

32. Wren, J. C., Paquette, J., Wren, D. J., and Sanipelli, G. G., The Formation and Volatility of Organic Iodides, *Proc. Specialists' Workshop on Iodine Chemistry in Reactor Safety, Harwell*, AERE-R-11974, p. 333, 1986.

33. Cline, J. E., Voilleque, P. G., Pelletier, C. A., and Thomas, E. D., Studies of Airborne Iodine at TMI-2, Sources and Filtration, *Proc. 16th DOE Air Cleaning Conference, San Diego*, 1980.

34. Tennankore, K. N., Koroll, G. W., Kumar, R. K., Lam, A. H. T., Chan, C. K., and Wren, D. J., Hydrogen Combustion Issues and Containment Integrity, *Proc. Int. Conference on Nuclear Containment, Cambridge*, 1987.

35. Coward, H. F., and Jones, G. W., Limits of Flammability of Gases and Vapors, U.S. Bureau of Mines Bulletin 503, 1952.

36. Kumar, R. K., Flammability Limits of Hydrogen-Oxygen-Diluent Mixtures, *Journal of Fire Sciences*, vol. 3, p. 245, 1985.

37. Tamm, H., MacFarlane, R., and Liu, D. D. S., Ignition Effectiveness of Thermal Heating Devices in Hydrogen-Air-Steam Mixtures, *Proc. Int. Meeting on Thermal Nuclear Reactor Safety*, NUREG/CP-0027, 1983.

38. Nelson, L. S. et al., The Behavior of Core-Simulant Aerosols During Hydrogen/Air Combustion, SNL Report SAND 85-1817C, 1985.

39. Wren, D.J., private communication, 1983.

40. Report by the Senior Group of Experts on Severe Accidents, Severe Accidents in Nuclear Power Plants, OECD Nuclear Energy Agency, Paris, Nuclear Energy Agency, p. 32, 1986.

Thermalhydraulics and Activity Transport

B. H. MCDONALD and D. J. WREN
Atomic Energy of Canada Limited
Whiteshell Nuclear Research Establishment
Pinawa, Manitoba, ROE 1LO Canada

INTRODUCTION

The potential consequences of a reactor accident, in terms of its impact on public safety, rest on the source term of radioactive fission products. The "source term", as defined by an international group of experts, is "the quantity of radioactive material which might be released in a nuclear accident: its physical and chemical form and the other quantities needed to completely specify its dispersion in the environment (e.g., energy in the plume, height of release, duration of release etc.)" [1]. Although there are a large number of physical and chemical factors that will contribute to the determination of the source term for a given accident scenario, those factors having a direct impact on the rate of transport are of obvious importance. The thermalhydraulic conditions controlling the rate of mass transport, among other things, are probably the most important factors influencing the source term.

This paper is an overview of the areas in which thermalhydraulics most strongly influences activity transport during a severe accident in a water-cooled reactor. It also includes some discussion of the areas where coupling between the physics used in separate computer models of the two phenomena must be considered in any mechanistic "best-estimate" calculations of the source term. A "severe" accident is herein defined as one in which a loss of cooling plus degraded emergency core cooling lead to a loss of core geometry and extensive fuel element failures.

The influence of thermalhydraulics on activity transport during a severe accident can be divided into six broad areas:

1. Fuel temperatures and fission product release from fuel;
2. In-core atmospheric conditions and gas flows;
3. Activity transport in the reactor coolant system;
4. Fission product chemistry in the reactor coolant system;
5. Activity transport in containment; and
6. Core-concrete interactions.

These areas (illustrated schematically in Figure 1) relate to the different environments experienced by fission products as they are transported from the fuel to a leak in containment. Within each area, there are individual phenomena that are reactor and accident specific in their thermalhydraulic behaviour and coupling to activity transport.

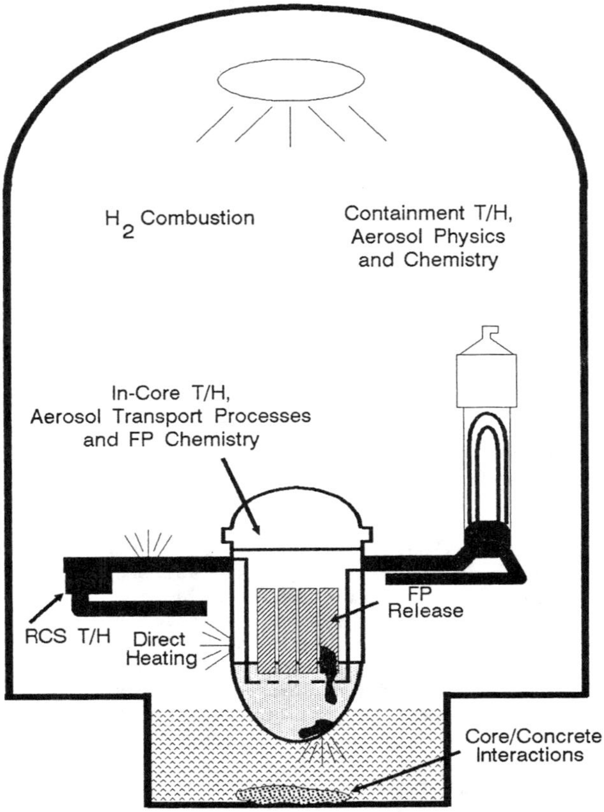

FIGURE 1. Schematic diagram of a PWR reactor showing the main areas where thermalhydraulics affects activity transport.

While the influence of thermalhydraulics on activity transport is generally obvious, there are also less obvious feedbacks of fission product behaviour on thermalhydraulics. These coupling effects are often neglected, particularly in unsophisticated source-term calculations. Several years ago, Kress [2] provided an initial assessment of some of these effects. Since then, a number of these coupling mechanisms have been included in more sophisticated computer models of fission product transport and thermalhydraulics.

This overview paper does not contain any detailed calculations of the coupling between thermalhydraulics and activity transport. That is more properly the purview of the many other, more focussed papers of this meeting. Instead, the intent here is to summarize briefly the major areas where thermalhydraulics most strongly affects the source term and to highlight particularly those areas where modelling capabilities are weak or where further consideration might be given to feedback effects.

FUEL TEMPERATURES AND FISSION PRODUCT RELEASE FROM FUEL

In an accident, the transport of activity is initiated with its release from the uranium oxide fuel matrix and the fuel element cladding (Figure 2). This is the phenomenon for which the impact of thermalhydraulics is best developed. The rate of release of fission products from fuel is directly related to the temperature of the fuel and the gas environment around the fuel (e.g., pressure, redox potential). There is an extensive database in the experimental literature on fission product release as a function of temperature. All sophisticated safety analysis codes include this dependence.

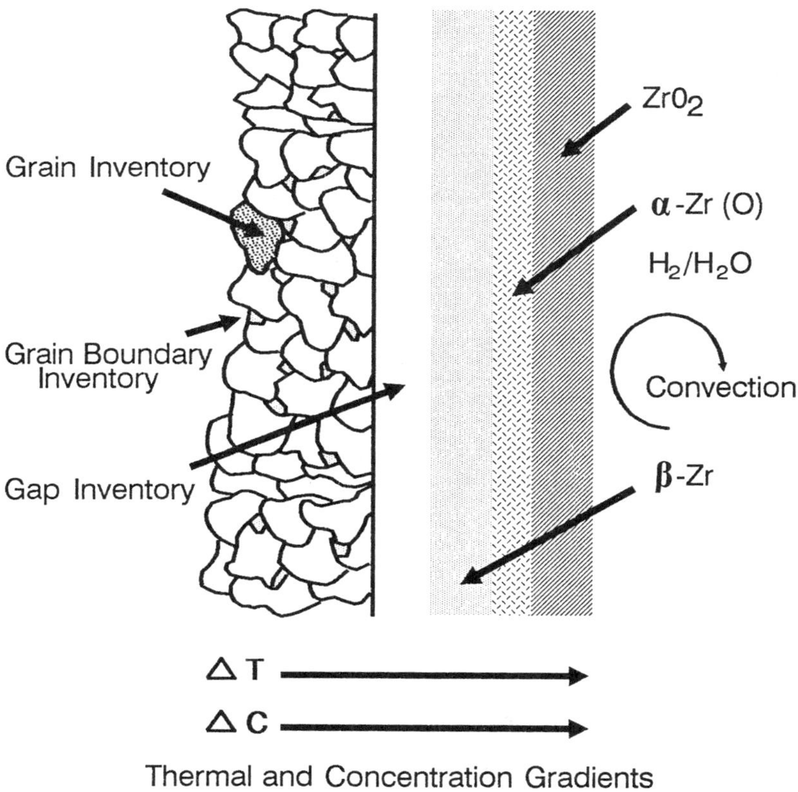

FIGURE 2. Thermalhydraulics and activity release from UO_2 fuel.

A theoretical consideration of fission product release rates yields a simple correlation of release rates with temperature [3]. Simple correlations have also been developed based on experimental data for use in accident source-term calculations [4]. More sophisticated computer models of fission product release rates have also been developed based on a mechanistic treatment of fission product diffusion, which is temperature-controlled (e.g., references 5 and 6).

The challenge facing those who would produce best-estimate predictions of source terms lies with the accuracy of the existing models and correlations for the prediction of transient events and the actual temperature history of the fuel. In a recent overview paper, Hobbins et al. [7] showed that the data obtained in experimental fission product release studies for light-water-reactor (LWR) fuel display a significant variation with temperature, and also significant scatter. In fact, the scatter, covering several orders of magnitude in the fractional release rates of volatile fission product materials at any given temperature, demonstrates the difficulty of performing such experiments, and in determining specific material release rates for specific fuel temperature conditions.

Osborne et al. [8] have obtained the most extensive consistent data on fission product release from LWR fuel at temperatures up to 2000°C. Their data cover a range of between 30 and 100% of the nominal values, demonstrating the difficulty in performing these types of experiments. They have also found that there is a correlation between the amount of fuel porosity (which is a function of the fuel history) and the fractional release of volatile fission products.

Computer codes for prediction of fission product release during severe accident conditions have made much use of the experimental data and theoretical analyses performed in support of the experiments. As with all codes however, these must be used with caution. One comparison of code predictions by Horvath et al. [9] showed that differences between results from two codes could be attributed to the different integration time steps used.

Since the temperature of the fuel is crucial in controlling the fission product release rates, the latest efforts in accident model development have coupled the core thermalhydraulics with the fission product release. One example is the integrated SCDAP/RELAP5/TRAP-MELT code package [10-13]. A code suite like this, incorporating all the state-of-the-art models, should have enormous potential. Like all "chains" however, this code package can be no stronger than its weakest link. Because the package is so complex, because the modelled phenomena are non-linear and inter-related, and because the numerical methods of three different codes are involved, great care must be exercised in interpreting results produced by the package. Nevertheless, combinations of codes such as this demonstrate acceptance of the intimate coupling of fission product release and transport to thermalhydraulics.

Direct coupling of thermalhydraulics and fission product release implicitly recognizes the problem that a reactor core cannot be considered as a single homogeneous node for the purposes of fission product release. This inhomegeneity naturally results in a gradual release, over time, for all fission products (both "volatile" and "involatile") as different regions of the core heat up at different rates. The relative timing of the peak release rates for the different classes of fission products will vary with time, but there will be no sharp separation. Models of fission product transport elsewhere in the reactor system and containment, which do not recognize this, will predict the behaviour of the bulk of the fission products correctly, but will necessarily treat the behaviour of some incorrectly.

At this point it is appropriate to pose a question that will be repeated, in essence, throughout this paper: how accurate are thermalhydraulic predictions of core temperature histories likely to become? There will always be some degree of uncertainty in core temperature predictions and there is substantial uncertainty in the release rate database, that may be resolved slowly, due to the difficulty in conducting the experiments.

IN-CORE ATMOSPHERIC CONDITIONS AND GAS FLOWS

The transport of fission products within the core volume will be largely determined by the bulk gas flows, which can be calculated using thermal-hydraulics. It is important here to distinguish between the two main reactor types: those with large, single-vessel cores (e.g., PWRs and BWRs) and those with multiple-channel cores (e.g., CANDUTH-PHWR and RBMK). The gas flows in these two core types will be very different. The large vessel cores will develop internal circulation loops and the transport of fission products from the fuel to the reactor cooling system (RCS) piping is unlikely to be direct for all the material (Figure 3). Also, the large size of the vessel will lead to nonhomogeneous effects such as fission product plateout in the upper plenum of the core. In contrast, in multiple-channel cores, the gas flow will be in a well-defined direction along the channel (Figure 4). However, there will also be a definite temperature gradient along the channel, which must be factored into consideration of fission product deposition.

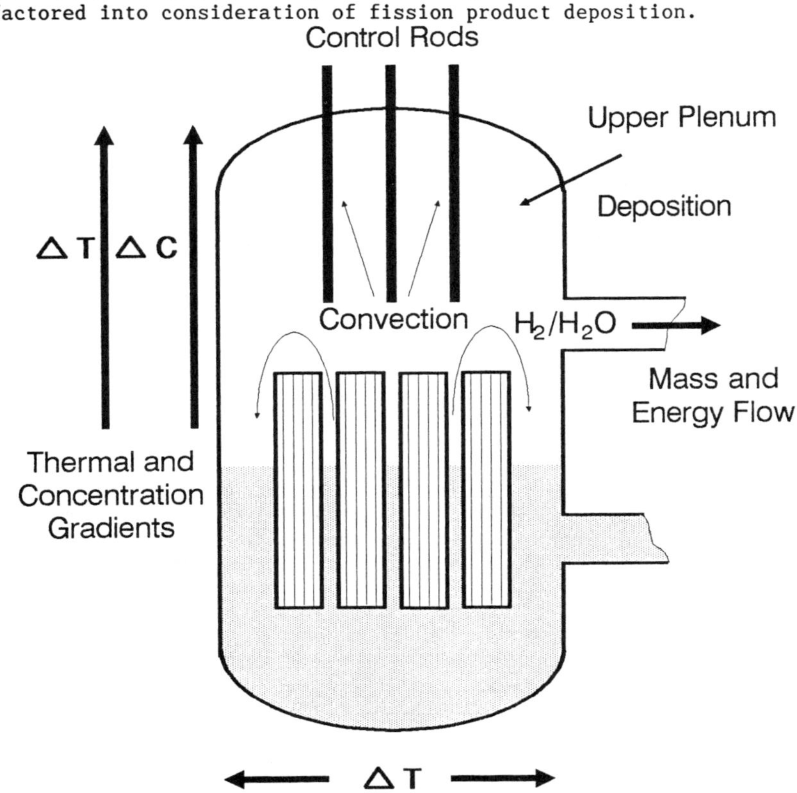

FIGURE 3. Thermalhydraulics and activity transport in an LWR reactor core.

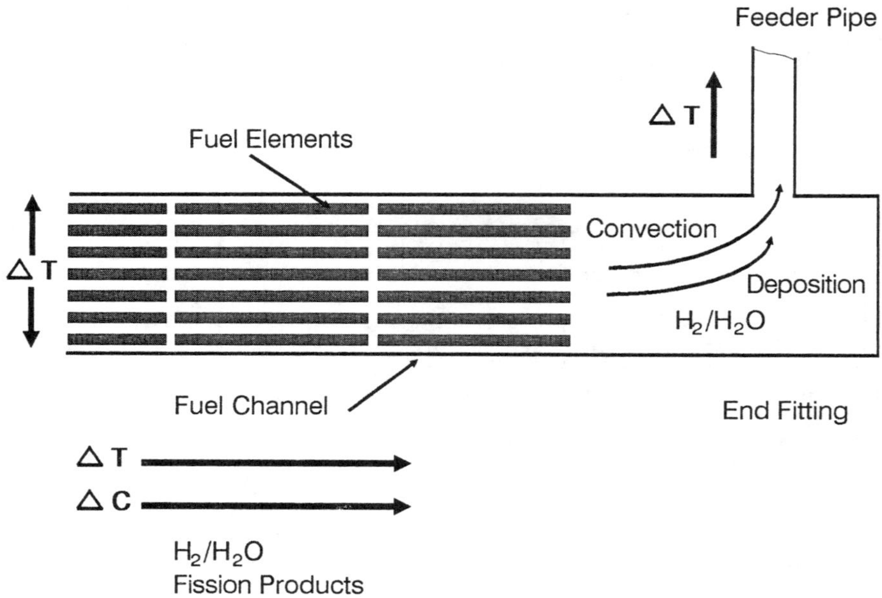

FIGURE 4. Thermalhydraulics and activity transport in a multiple-channel reactor core.

In both types of reactor cores, it is important to consider the impact of the atmospheric conditions in the core on the chemistry of the fission products. The in-core thermalhydraulics will determine the atmospheric chemistry ($H_2/O_2/H_2O$ ratios) and the concentrations of both fission products and non-fission product materials (e.g., control-rod chemicals and aerosols, boric acid). The latter may have a profound influence on the physical or chemical form of the fission products existing in the core (see, for example, Beard et al. [14]). Although this may not affect the rate of transport of activity from the core, since the mass transport via bulk gas flows will be relatively independent of fission product chemistry, it may have a significant impact on the continuing activity transport outside the core.

This latter effect was poorly handled in the early safety analysis codes, partly because of a lack of sufficient experimental data. The experimental database is now being improved and current code development is including the effects of nonhomogeneous gas phase chemistry on the chemistry of the fission products in the core. In particular, codes such as VICTORIA now specifically include equilibrium thermodynamic calculations of fission product chemistry in the core on a nodalized basis.

Several integrated experiments to study the behaviour of fission products released from fuel overheated to the point of catastrophic cladding degradation have been carried out or are planned. Examples of the former are the STEP program [15] and the ACRR tests [16], while planned programs include the BTF [17] and PHEBUS [18] test programs. In the STEP tests, heatup and steam-flow transients were used that simulated the conditions expected in operating power reactors undergoing various types of hypothetical severe accidents. Hydrogen evolution by the Zr oxidation reaction was found to be reasonably well predicted by any current reaction rate temperature correlations. This indicates that the accuracy of the oxidation correlation as a function of temperature is not critical when considering a large, relatively inhomogeneous system such as an elongated fuel-rod bundle. It may be possible to extend this result to cover an entire LWR core, especially after cladding degradation has begun. Thus, while detailed mechanistic codes describing metal/water reactions require extensive knowledge of the temperature history of the core to predict hydrogen evolution, in practice, the inhomogeneities tend to average out and such detailed thermalhydraulics information may not be needed.

The result for hydrogen production may be extended to consideration of the atmospheric chemistry in the core and its impact on fission product chemistry. Although in principle there may be large local variations in the H_2/H_2O ratios that could alter fission product chemistry, in a large system, these variations may be unimportant. Since reaction kinetics are very rapid at the temperatures in the core during a severe accident, this is quite likely. Thus, only averaged thermalhydraulic conditions for larger volumes of the core may be needed for sufficiently accurate predictions of fission product behaviour in the core.

Experiments conducted at ORNL [19] to investigate the release of materials from LWR cores during severe accident sequences indicate that during fast-melt conditions, the dominating materials released are Cd, Mn, Ag, and the volatile fission products, rather than structural or fuel materials, UO_2 or Zr. The presence of control-rod materials in the released aerosol is a feature of LWRs and BWRs; in pressure-tube reactors such as the CANDU^{TM}-PHWRs or the RBMKs, control-rod materials will not be present in the channels, but fuel cladding elements may be important [20].

The presence of large quantities of airborne fuel cladding, control rod and structural materials adds greatly to the total mass of aerosol material that may be transported from the core through the RCS. This aerosol material may serve as surfaces for fission product condensation. It must also be correctly accounted for to predict the transport of all aerosol material, including pure fission product aerosol particles, through the RCS. The thermalhydraulics should correctly predict the timing and masses of all material released from the core to ensure correct predictions of transport in the RCS.

The vaporization of large amounts of control rod and fuel cladding material, as well as fission products, can have a direct impact on the thermalhydraulics in terms of the atmospheric chemistry. As Kress pointed out for fast core-melt sequences [2], the fission products alone can form a significant fraction of the atmosphere in the core, thus affecting the thermal properties of the gas phase. This feedback needs to be incorporated in the thermalhydraulic calculations, particularly where large temperature gradients from the overheated fuel to the upper plenum lead to possible material condensation in the upper plenum.

A whole-core modelling code, called FAEREL, has been developed in the United Kingdom [21]. This code incorporates mechanistic models for the vaporization and gas phase transport of fission products and fuel, control and structural constituents to calculate limiting release rates from severely overheated LWR cores. FAEREL divides the core into a number of vertical flow channels and uses the MARCH [22] subdivision of the core into radial zones. The quantity of material released as a function of time from one flow channel in each radial zone is calculated, then multiplied by the number of channels in that radial zone to give the release history of the radial zone. An important result arising from calculations with this code [21] is that gas phase mass transfer can impose distinct upper limits on the rates of release of structural constituents in high-pressure (S_2D and TMLB') sequences that are inconsistent with the predictions of empirical codes such as CORSOR [23], particularly for the tin component of Zircaloy and for the silver component of the control-rod alloy. The authors, however, point out that their results involve numerous uncertainties, the most serious arising from the dependence on MARCH predictions of thermalhydraulic conditions in the core and that recent thermalhydraulic modelling studies suggest that in-vessel recirculatory flows may give much more uniform core temperatures than are predicted by MARCH, particularly in high-pressure sequences.

Two important points may be made: (1) the flow patterns in a reactor core during a severe accident sequence can have a significant influence on the materials released from the core; and (2) the thermalhydraulic conditions in core must be considered along with fuel, control, and structural material behaviour if realistic assessments of material release from the core are to be obtained.

Multiple-channel cores do not have the same coupling of thermalhydraulics to activity transport because of the confined geometry of the individual fuel channels. During the early phases of a severe accident, before channel failure, mass transport will occur unidirectionally along the channel. This results in very short residence times in the hot core region and large temperature gradients along the channel. These are the conditions encountered by the important volatile fission products released from the fuel early. Currently there are no models available to couple this thermalhydraulic situation with fission product release and transport, since bounding analyses are sufficient to show compliance with safety criteria. In the later stages of a severe accident, when the core geometry has been lost through multiple channel failures, the situation will become more analogous to the LWR conditions but with less heat generation and slower fission product transport from the core [24].

ACTIVITY TRANSPORT IN THE REACTOR COOLANT SYSTEM

In the RCS, fission products will be transported as gaseous species or as aerosol particles. For gaseous species, the bulk gas flow rates and temperature gradients, as they affect deposition and resuspension, are most important. The impact of thermalhydraulics on aerosol physics and transport is much more complex, because it plays a role in impaction, agglomeration, deposition, and resuspension. This section will focus on the modelling assumptions used to describe the coupling of thermal-hydraulics to aerosol physics (shown schematically in Figure 5). The next section examines the influence of thermalhydraulics on fission product chemistry.

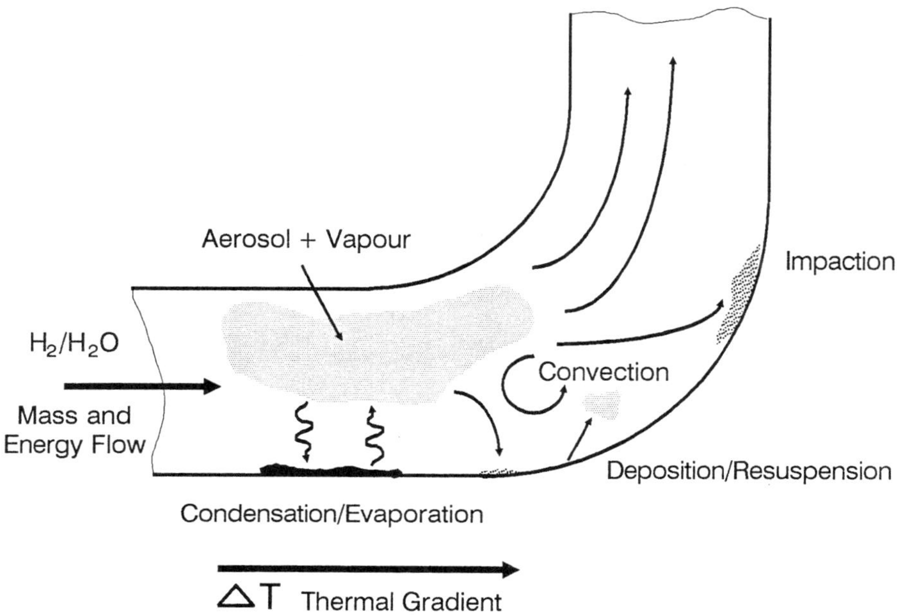

FIGURE 5. Thermalhydraulics and activity transport in the RCS.

A comprehensive review of the thermalhydraulic factors affecting fission product transport and retention in the primary circuit of an LWR during a severe accident has been given by Butland and Kuhlman [25]. Most of their comments apply to pressure-tube reactors as well. The gas flow rate and composition during the accident sequence are important in determining concentration and the residence time in the RCS of released fission products. Accident sequences may have flows ranging from virtually nothing to very high rates during periods of core slumping. Quoting from the review [25], "The gas flow has such a strong influence on the phenomena which occur during RCS transport that detailed time profiles of the flow rates are required in order to make any assessment of the retention of the transported species". Gas flow rates are strongly dependent on local pressure gradients, which in turn can be strongly influenced by local condensation/evaporation events, suggesting that there could well be a need for integrated thermalhydraulic/fission product transport analysis, particularly when concentrations of fission products and structural materials are high. The gas composition influences the chemical form of several important species, and thus the thermalhydraulic evolution of the gas properties becomes important as well.

The temperature of the gas is an important thermalhydraulic parameter in that it affects bulk gas velocities (hence residence times in the RCS) and governs the condensation of the more volatile fission products on surfaces and the co-transported aerosol particles. It is not clear what level of detail is required to provide adequate information to the RCS transport and deposition codes. Some thermalhydraulics codes such as SCDAP offer much more detailed predictions of core temperatures than other codes, such as MARCH. The impact of varying local temperatures in the core on the temperatures and flow rates in the RCS is not clear. There appears to be a need for integrated core/RCS analysis of fission product behaviour. Butland and Kuhlman [25] have also pointed out that the RCS surface temperatures, which are a direct result of the RCS coolant/gas thermalhydraulic history, are important in determining fission product retention through the mechanisms of thermophoresis, diffusiophoresis, and direct gaseous species condensation, as well as playing a vital role in the condensation/evaporation process. In all, without a detailed knowledge of the time history of the thermalhydraulic behaviour of the RCS, it is difficult, if not impossible, to determine the RCS transport and retention characteristics of any released fission products during severe accidents.

Condensation and revaporization of material in the RCS leads to coupling between thermalhydraulics and material transport. Kress has shown in unsophisticated calculations [2] that the heat of vaporization will have limited importance within the core. However, at very high concentrations of condensible materials (ca 1% by vol.) in the gas flows entering the RCS, the heat of vaporization could lead to local temperature rises of up to 50 K. This is sufficient to alter the saturation pressure for condensible species and alter the rates of aerosol formation and growth. This can affect the number and size of aerosol particles, affecting their subsequent interactions and transport physics. Clearly this type of feedback between aerosol physics and thermalhydraulics should be included in sophisticated activity transport codes.

Fission products also act as localized heat sources through radioactive decay. This can amount to megawatts for the core inventories of the volatile fission products. Again, Kress [2] has suggested that this can be neglected in terms of its impact on gas temperatures. However, this heat source can not be neglected when considering revaporization from surfaces. Jordan and Leonard have reported a study on fission product revaporization in the RCS [26]. They noted that inclusion of the decay heat from deposited fission products in the thermalhydraulic calculation perturbs the fission product condensation/evaporation physics in the RCS. In order to conduct their theoretical study, Jordan and Leonard combined the MERGE code [27], which calculates RCS thermalhydraulic behaviour, with the TRAP-MELT2 code [13], which calculates fission product behaviour, creating what they call the TRAP-MELT3 code. With this code, they have calculated that during a Surry TMLB′ sequence the structure surfaces can rise in temperature by up to 100 K, with inclusion of decay heat. This rise in temperature, however, does not produce significant additional fission product release (calculated) from the RCS before melt-through of the reactor vessel. After melt-through, structural temperatures are likely to continue to rise, resulting in revolatilization of some fission product materials in the hottest portions of the RCS. This provides the potential for significant source term release of volatile fission products to the containment after melt-through. Jordan and Leonard suggest that, in such cases, the RCS might act as more of a fission product retarder than as a fission product retainer.

Butland et al. [28] have reported on a study to validate models used to predict RCS thermalhydraulics and radionuclide transport and deposition in a PWR severe accident using a code called SPRITE (Severe Accident Primary System Thermalhydraulic Evaluation). This code was constructed from MERGE [27], but with more sophisticated models for heat conduction and radiation. They have compared the results of some calculations with data from the Marviken Program [29] and have also performed calculations for a "large modern PWR", using TRAP-MELT2 [13] as well as SPRITE. They note that the improved heat transfer models in SPRITE, as opposed to MERGE, produce much better agreement with Marviken test data, and that the improved thermal performance results in improved gas flow calculations, because of the inherent coupling of flow and temperature through the thermalhydraulics. Significant differences in flows and temperatures were also reported for SPRITE and MERGE calculations on the "large modern PWR", suggesting the need for adequate modelling of all thermalhydraulic processes for realistic code prediction.

Another study that reaches a similar conclusion was reported by Donahue et al. [30]. They used a Stone & Webster Engineering Corporation proprietary code (THREED-RCS), which also has sophisticated heat transfer models, and found that "fission product retention is very sensitive to thermalhydraulic conditions that exist when volatile fission products are revaporized from various structures". They note that for reasonable prediction of the retention/revaporization events likely to occur in the RCS of a BWR during a severe accident, a detailed treatment of heat sinks must be included, beta- and gamma-heating must be represented accurately, and free convection flow rates must be modelled accurately. Accurate thermalhydraulic representation is required to obtain accurate predictions of fission product behaviour in the RCS.

While much attention has been given to the need for modelling the thermalhydraulics together with fission product behaviour for accurate representation of the events likely to occur in the RCS during a severe accident, the discussion would not be complete without some description of experimental data that has been obtained and can be used for validating the behaviour of RCS fission product behaviour codes. A review of large-scale experiments to determine aerosol behaviour in the RCS has been given by Rahn et al. [31], in which the Marviken [29] and LACE [32] programs are presented. In the Marviken program, five tests were conducted during 1983 to 1985 to provide data on the behaviour of high-concentration aerosols and volatile species in a full-scale simulated water-cooled reactor primary circuit. Extensive experimental investigations were made of such effects as aerosol mass concentration, temperature, superheated steam, and water environment on the transport properties of aerosols and vapours. In the LACE program, inherent aerosol behaviour was investigated for postulated high-consequence accident situations, such as containment by-pass situations, impaired containment, either early in the scenario, or later, but simultaneous with a large fission product release. Both programs are multinational in scope, and the data and data analyses are proprietary. Use of the data in code comparison exercises is providing both code developers and code users with valuable information, which will ultimately result in improved code capabilities.

The need for coupled thermalhydraulic and aerosol transport analysis has been addressed by Sehgal et al. [33], who note that the two processes affect each other directly. The RAFT code [34] explicitly links the two processes by incorporating a homogeneous nucleation model. This is unique among current aerosol transport codes. Numerical simulations using RAFT have shown that within the RCS, under most accident conditions, the particle size spectrum is determined by the competition between the homogeneous and heterogeneous nucleation mechanisms, with agglomeration playing only a secondary role [34]. The importance of this result depends on its impact on transport and retention rates in the RCS for particular fission products.

Some of the remaining outstanding issues in the modelling of fission product behaviour in the RCS concern deposition of aerosols at pipe bends, the effects of surface roughness on aerosol deposition, resuspension and redistribution of deposited aerosols, decay heat and revaporization. Some of these concerns have been elaborated upon already above, others will properly await the publication of the data and the data analyses from the Marviken and LACE programs and further future work. Some of the data already available show that a considerable amount of the fission product material could deposit in the RCS piping, including some that could form a liquid film. Accurate modelling of the evaporation and transport of this material requires an accurate description of the thermalhydraulic conditions in the RCS [35], supporting the need for fully coupled thermalhydraulics/transport codes.

FISSION PRODUCT CHEMISTRY IN THE REACTOR COOLANT SYSTEM

In addition to controlling the mass transport through the RCS, thermal-hydraulics will control the chemical environment experienced by fission products. Exposure to varying chemical environments can alter the chemical speciation of the fission products. This concern has already been raised during the discussion of activity transport in the core. It is much more important in the RCS because the temperatures are lower and the equilibrium chemistry is different.

The thermalhydraulic factors influencing the chemistry are residence time, temperature and the time-dependent concentrations of fission products and other species in the RCS. For example, the bulk gas chemistry (normally an H_2/H_2O mixture) will control the oxidation state of the fission products being transported by that flow. For most of the fission products, the oxidation state is fixed over a wide range of H_2/H_2O ratios and temperatures. However, in an accident where air, which contains oxygen, becomes part of the bulk flow (such as occurred at Chernobyl), the chemistry can be very different [35].

Chemical reactions in the RCS are most important for those fission products that are transported as gas phase species, since gas phase reactions occur much faster than reactions with solids. By far the most important gas phase species are cesium and iodine. Under most chemical conditions these species will be transported through the RCS as CsI and CsOH, either in the gas phase or deposited on aerosol particles [36].

In LWRs containing boric acid and BWRs containing boron carbide control rods, there will be boric acid present in the RCS during an accident. Studies have shown that boric acid can react with CsI and elevated temperatures to release HI [37]. The latter molecule will be transported much faster through the RCS than CsI because it will not condense, thus changing the timing of iodine release to containment. Correct prediction of the relative timing of CsI and boric acid passage through the RCS and the probability of condensation in the same locations requires intimate thermalhydraulic coupling to fission product transport.

The location of fission product condensation, the condition of the condensation surface (e.g., oxide thickness) and the residence time on a surface can affect fission product chemistry. For example, studies have shown that CsOH can react with 304 stainless steel to trap the cesium [38]. Other studies have shown that CsI may also react, much more slowly, with stainless steel to trap the cesium and release the iodine as HI [39]. The reaction rates depend on the type of surface material and its history, the temperature, and the bulk gas chemistry (H_2/H_2O).

Chemical reactions of fission products in the RCS will normally make second-order perturbations to the timing and chemistry of fission product release to containment. This is because aerosol physics will dominate the mass transport rate of fission products for most accident scenarios. Detailed coupling of thermalhydraulics to fission product chemistry in the RCS does not yet exist in any code and is a strenuous demand which is not warranted for most calculations. It is worthwhile to note, however, that such couplings do exist and that transport calculations predicting situations in which chemical reactions could occur to a significant extent should be examined closely for their validity.

ACTIVITY TRANSPORT IN CONTAINMENT

Activity transport in the containment building can occur by means of a number of processes, depending on the accident scenario and the time during the accident. Major transport phenomena include: aerosol transport; gas leakage; aerosol generation and release from molten core-debris; resuspension from surfaces; pool scrubbing and revolatilization from water pools; and molten-core ejection (Figure 6).

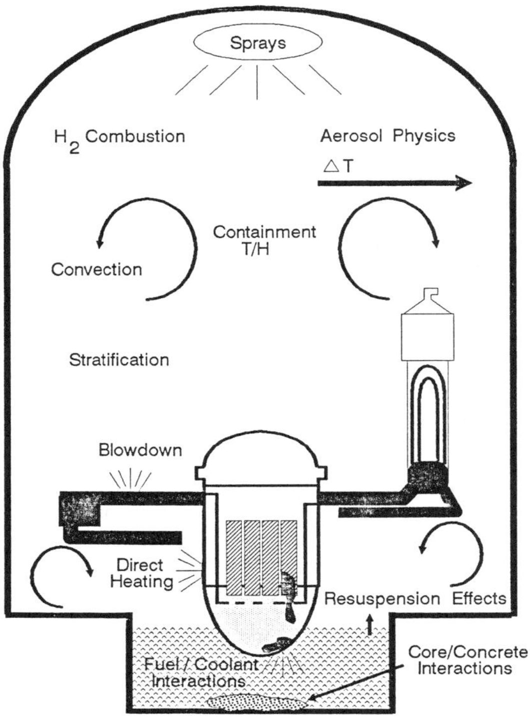

FIGURE 6. Thermalhydraulics and activity transport in a containment building.

For severe accidents, the vast bulk of fission products released from the fuel will enter the containment building atmosphere as part of an aerosol (excluding the noble gases). Computer codes to describe the aerosol physics in the containment building have been developed to predict the rate of depletion of these aerosols from the containment building atmosphere. It is the time-dependent aerosol concentration that determines an essential component of the source term to the environment through containment leakage or failure.

In this section, the modelling assumptions used in aerosol behaviour codes and the adequacy of thermalhydraulic codes in predicting the information required for fission product transport analysis are addressed. It is in the containment that the direct coupling between thermalhydraulics and fission product transport is most important, because of the high degree of accuracy required for some of the aerosol model variables.

Aerosol physics describes the time evolution of the size distribution of the aerosol particles and the time evolution of the concentration and spatial location of airborne aerosol material. This physics is controlled by the thermalhydraulic conditions present in the containment building. The physical phenomena include agglomeration/deagglomeration, condensation/evaporation, particle transport to surfaces and deposition/ resuspension on surfaces.

The agglomeration processes modelled in containment aerosol codes are Brownian agglomeration, turbulent agglomeration and gravitational agglomeration. Calculation of first two processes depends on knowledge of the local temperature and local turbulence of the gas surrounding the aerosol particles. These thermalhydraulic variables may vary considerably throughout containment depending on factors such as the presence of localized heat sinks (e.g., vault coolers) or the rate of local steam condensation.

Deagglomeration is normally associated with energetic events in which shock waves can cause loosely bound aerosol particles to break up into smaller parts. Steam explosions or rapid hydrogen deflagrations can cause such shock waves. This phenomena is not normally included in aerosol codes, partly because of limited knowledge of the criteria for breaking up particles and partly because of the difficulty in predicting the shock stresses. Wet aerosols in condensing steam environments will be resistant to deagglomeration because of strong forces binding aerosol particle components together.

In the RCS, condensation and evaporation on aerosol particles refers to the growth or depletion of the solid aerosol material itself. In the steam-filled containment, condensation and evaporation refer to the growth or depletion of liquid water on the aerosol particles. On entering containment from the RCS, the solid aerosol particles act as condensation nuclei if the steam concentration exceeds the supersaturation ratio for containment. The thermalhydraulics required to define the steam concentration and temperature (giving the supersaturation ratio) is obvious. The problem is to determine what these values are everywhere in containment as a function of time.

The condensation rate on a particle is a very strong function of the local supersaturation ratio. This is the ratio of the vapour pressure of water in the gas phase over the vapour pressure of water condensed on the aerosol particle. The latter is not a simple constant value. It increases with decreasing particle radius and diverges significantly from the bulk-water value for very small particle sizes.

A European code exercise that compared only the modelling of steam condensation on aerosol particles has been reported by Fermandjian et al. [40]. The codes examined were AEROSIM-M, AEROSOLS/B1, and NAUA Mod5. The NAUA Mod5 and AEROSOLS/B1 codes both model the effect of the droplet curvature on the saturation vapour pressure. At any one time, the steam condenses only on those particles with radii above a certain critical radius (which depends on the saturation ratio, and hence, will generally vary with time). The calculation in AEROSIM-M is based only upon the saturation ratio (although the user may input a fixed critical radius).

The chemical composition of containment aerosols will also have an effect on condensation. This is because some of the fission products and other compounds that could be released to containment have a strong affinity for water. Chemical species such as CsI or CsOH will dissociate in water to form ionic solutions, with water vapour pressures significantly less than that of pure water. This phenomenon has been addressed by Clough et al. [41]. Hygroscopic aerosols (having a strong chemical affinity for water) will grow quickly because of steam condensation, even in superheated environments [41], resulting in their early removal by gravitational settling. Modelling of such hygroscopic aerosols requires a more complete thermodynamic database than now exists [41] and signifies a need to "couple the aerosol equations to a calculation of both mass and heat transfer [aspects of the thermal hydraulics], because the release of heat during the condensation process is a major determinant of how much condensation occurs."

Complex phenomena such as water condensation on hygroscopic aerosol particles can be treated if the initial particle size distribution and chemistry of the dry aerosol particles are known and the atmosphere in containment is homogeneous. However, the problem is much more challenging if these conditions can not be met.

Aerosol condensation/evaporation influences the thermalhydraulics by way of two feedback effects. Condensation on particles will reduce the local steam concentration. Kress [2] has estimated that this effect is negligible compared to bulk condensation rates on containment walls. The heat of vaporization released during condensation will raise the aerosol particle temperature and heat the surrounding air. This will alter the local supersaturation ratio and the rate of aerosol particle growth. Clough et al. have considered this phenomenon [41].

Mass transport of aerosol particles to surfaces is governed by several physical phenomena, including gravitational settling, thermophoresis, diffusiophoresis and convective flow. Again, with the exception of gravitational settling, calculation of these phenomena requires detailed knowledge of the thermalhydraulics in containment.

A comprehensive review of fission product behaviour and retention in the containment building of an LWR during a severe accident has been given by Schock et al. [43]. Several codes have been developed for calculation of containment fission product behaviour and retention. Experimental studies, in particular DEMONA [44], have provided useful data for code evaluation and validation. Schock et al. state that the remaining areas of concern include the coupling of aerosol behaviour to the thermalhydraulics and the influence of spatial inhomogeneities in complicated containment geometries. They indicate that, although such influences may be of lesser importance than the dominating agglomeration and deposition mechanisms in determining containment fission product behaviour, they may alter source terms, and their magnitude remains to be qualified.

Jordan and Kogan [45] have used the QUICKM code [46] to explore the effects of coupled thermodynamics/aerosol calculations. They found unexpected sensitivity to the numerical treatment of the coupled models and to the degree of coupling. Their results were also sensitive to assumptions on the transition of two-phase flow from the pressurized RCS to containment. The calculations required definition of the saturation ratio, which is a sensitive function of gas temperature, and its prediction was particularly uncertain. Since the saturation ratio drives steam condensation on particles, the conclusion from their work was that "further analysis is needed before any prediction of water vapour condensation on particles is believable."

Recently, Abbey and Schikarski [47] reviewed current developments and the state-of-the-art in modelling containment aerosol behaviour. In their review, they list areas that require further work, including the neglect of possible resuspension of deposited aerosols both in the RCS and in the containment and turbulent agglomeration and turbulent deposition processes in high Reynolds number flows. They further state that the major limitation on the mechanistic modelling of aerosol behaviour in containment is the modelling of agglomeration rates whose accuracy is limited by uncertainties, which include aerosol shape factors. The multicomponent nature of the aerosols and the behaviour of residual aerosols consisting of low concentrations of small particles are also regarded by them as problems. They stated that the development of models to deal with steam condensation onto particles is still in progress and, in principle, requires the simultaneous solution of the thermalhydraulic and aerosol transport equations in a coupled mode.

The results of a recent code comparison exercise based on the LACE LA-4 containment response experiment are reported by Della Loggia et al. [48]. The general conclusions of that report include the statement: "During the course of aerosol code benchmark exercises....it became apparent that for LWR containment conditions the major remaining uncertainties in aerosol behaviour modelling were those due to the coupling to thermal hydraulic....". In short, predicted behaviour is very sensitive to the assumptions made in the containment computer codes about the condensation of steam in the bulk.

The importance of modelling the interaction between thermalhydraulics and aerosol behaviour with care was further amplified in a code comparison exercise based on certain LACE experiments and reported by Wilson et al. [32]. They describe a "numerical deagglomeration" of aerosol particles observed in the calculations of the discrete particle size (or "bin") codes NAUA and CONTAIN. The concentrations of particles in the smaller size bins were observed to increase after evaporation of water from the aerosol. The apparent generation of smaller particles turned out to be a result of the fashion in which the codes were handling evaporation. In brief, Wilson et al. explained the effect as follows: consider that all particles in a particular bin, which are all assumed to have the same diameter, are initially dry, and that during the next time step, water is predicted to condense upon them. At the end of the step, some of these wet (and larger) particles will be distributed to bins of larger size, but those wet particles that have not grown sufficiently to be redistributed will remain in the original bin, with the same, original overall size as the previously dry particles. However, the particle diameter of the solid aerosol within these wet particles must be less than the diameter of the original dry particles. If evaporation subsequently occurs, dry particles of unrealistically small diameters will appear. This demonstrates that getting the physics right is not sufficient; one must be very careful to ensure that the numerical methods are accurate as well.

Gelbard [49] has reported a new numerical method for handling aerosol particle growth (or shrinkage) due to steam condensation (or evaporation) to be incorporated in the CONTAIN code which should not only alleviate the "numerical deagglomeration" phenomenon observed by Wilson, but also lead to reductions in other potentially damaging numerical errors. Gelbard's method allows the particle bins to "move" in the direction of increasing mass (for condensation) and is a novel application of the method of characteristics.

In containment, as in the core and the RCS, the local behaviour of fission products depends on the local thermalhydraulics. In the computer models of the fission product release from the core, there is a large degree of nodalization in an attempt to deal with the impact of temperature gradients and other varying thermalhydraulic parameters. Such a high degree of nodalization may also be necessary to accurately describe activity transport in containment.

Particular issues in containment nodalization relate to natural circulation flow patterns, stratification and multi-compartment geometry effects. Some of these issues have been addressed in separate computer codes that model the behaviour of hydrogen in the containment. Many of the available aerosol transport codes, however, do not include the capacity to model the impact of these effects on aerosol behaviour.

Wolf and Kanzleiter have issued a report on DEMONA test A9, which was devoted to the study of the effects of multi-compartment geometry [50]. They examined impact of the resultant natural circulation flow patterns and additional stratification phenomena upon the thermalhydraulic containment behaviour and aerosol depletion processes. Their general conclusions are rather far reaching, as the nature of the experiment virtually disqualifies any computer code that uses the well-mixed assumption in its computation volumes, or nodes, from having much of a chance of predicting stratification and natural circulation phenomena. Nonetheless, the data from this experiment stand as a challenge for any code that purports to be able to handle multi-compartment containment buildings and stratification effects.

The implication for coupled thermalhydraulics and aerosol transport phenomena of correct nodalization cannot be avoided: subtle changes in pressure for example due to steam condensation on aerosol particles can have a notable impact on the natural circulation flow patterns. The impact on activity transport could also be significant. For example, incorrect prediction of gas flows could lead to incorrect predictions of aerosol settling in the wrong location (e.g., large containment pools vs surfaces with very little condensed water). This can have a very great impact on the subsequent predictions of fission product resuspension/ revolatilization, particularly for iodine, in the longer term following an accident [51].

Clement has calculated the impact of heat and mass transfer rates on aerosol growth rates [52]. He describes some of the general physics of condensation processes on aerosols, and also describes how practical calculations might be performed for aerosols in vapour-gas mixtures in cavities which are exchanging heat with walls through conduction, condensation and radiation. He noted that while it is necessary, in all practical calculations for large systems, to perform spatial averaging, or treat large volumes as homogeneous, the reconciliation of such procedures with the actual physics is particularly acute with aerosol formation and growth, which are often extremely localized.

The Summary, Conclusions, and Recommendations of the CEC Workshop on Water-Cooled Reactor Aerosol Code Evaluation and Uncertainty Assessment [53] provides some insight into the overall question of thermalhydraulic and aerosol coupling, as well as some pertinent remarks in related areas. Quoting from the summary: "There has been major progress in understanding and handling the problems associated with the coupling between thermal-hydraulics and aerosol behaviour since 1984. The large-scale tests have confirmed the need for this coupling, and demonstrated that a number of important modelling problems still exist, in spite of improvements to the handling of steam condensation which have been incorporated into many codes....The theory of the coupling of heat and mass transport associated with condensation processes is now well-developed, but its implementation even in the fully coupled codes is not optimized....improvements in the treatment of coupled heat and mass transfer in the boundary layers at containment surfaces, and of the influence of aerosols on bulk steam condensation are needed....whilst the presence of aerosols will have only a minor impact on the overall thermalhydraulic conditions in the containment, the reverse sensitivity is very great....many of the existing aerosol modelling codes are uncoupled, and require thermal-hydraulics data as input. Whilst these codes fulfill a very useful function, the best approach to overcoming the outstanding problems will be in terms of fully-coupled treatments of thermalhydraulic and aerosol behaviour."

Most of the focus on the coupling of thermalhydraulics and activity transport lies in the area of aerosol behaviour in containment, discussed above. Early in a severe accident, the vast majority of the airborne activity, which is readily available for release via either slow leakage from an overpressurized containment or a burst release associated with a sudden containment failure, is in the form of an aerosol. The time dependent depletion of this aerosol from the containment atmosphere dominates calculations of the source term for accidents with early containment failure. Other transport and coupling phenomena can be more important later in an accident, particularly after the aerosol production rates from the overheated fuel or core/concrete interaction have decreased. These include pool scrubbing, resuspension/revolatilization and hydrogen combustion.

Pool scrubbing is a particularly important mechanism for fission product retention in BWRs, but may also occur in other water-cooled reactor accidents (e.g., V sequence scenarios for some plants and water pools over the molten-core/concrete melts). This phenomenon has been reviewed by Morewitz et al. [54]. Water pools can act as very effective filters for the trapping and retention of fission products which are carried in gas bubbles that rise through the water. Fission products can also be released from water pools and resuspended in the containment atmosphere via mechanisms of gas sparging, boiling and flashing.

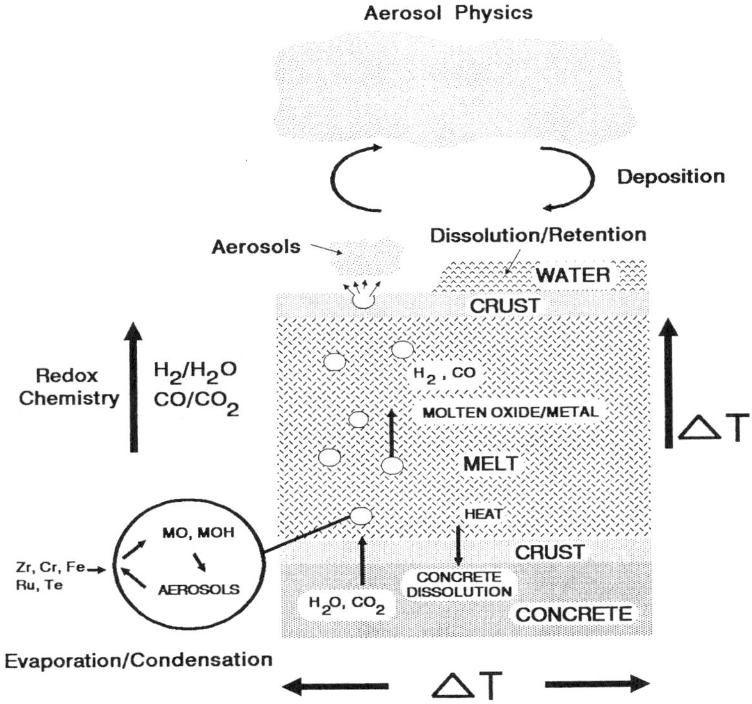

FIGURE 7. Thermalhydraulics and activity transport during molten core debris-concrete interaction

It has been explicitly recognized that there is a need to couple the thermalhydraulics to the aerosol production models in order to predict the activity release from the melt. Smith et al. [58] have reviewed the production of aerosols by core/concrete interaction during postulated severe LWR accidents and noted the need for accurate thermalhydraulic calculations. Additionally, the summary of the Specialists' Meeting on Core Debris-Concrete Interaction [57] noted the importance of directly coupling the thermalhydraulics with chemical equilibrium calculations. To be relevant, the equilibrium calculations require accurate values of the melt and gas-bubble temperatures and the gas-bubble chemistry.

Coupling of thermalhydraulics to aerosol generation is not a one-way process, since the behaviour of the activity can feed back on the core-melt temperatures. It is important to properly account for decay heat removed from the melt as a result of aerosol release. Also the generation of a very concentrated cloud of aerosols above the molten debris can affect the transfer of heat from the melt to the rest of the containment atmosphere. Some of the aerosol material must in fact resettle on the crust of the molten debris affecting its heat retention and possibly affecting continued aerosol release.

The final conclusion in the summary of the Specialists Meeting [57] stated, "For mechanistic calculations of containment loading, a coupled model of energy and mass exchange between the core debris, the cavity and the containment should be developed". A relevant question to pose is the likelihood of achieving the accuracy in the thermalhydraulic predictions required to make such a coupled code valuable.

CONCLUDING REMARKS

It is clear from the preceding review that thermalhydraulics is intimately linked to activity transport throughout the temporal and physical course of a severe accident. The conclusions of individual researchers and collective specialists meetings are that this linkage needs to be explicitly recognized through the development of fully coupled thermalhydraulics and activity transport computer codes. Such codes would be able to address the problem of predicting the source term for given accident scenarios on a best-effort basis.

The effort to produce such coupled codes is now in progress around the world. Some attempts involve coupling of existing separate codes, while others involve the creation of newer codes with the coupling "built in". These codes are not yet mature, but progress is being made. The level of the challenge faced is daunting. Sophisticated computer programming is required to deal with the complexity of the problem, the computing speed limitations imposed by the need, for example, to repeatedly calculate chemical equilibria as a function of time and space, and numerical errors which can sneak up on the unwary. There is good ground for concern that in some areas, e.g., thermalhydraulic coupling to condensing steam aerosols, the problem may prove practically intractable at the highest levels of physical sophistication. Nonetheless, the fruits of these efforts will lead to greater understanding of the source term and a more secure foundation for prediction of the potential consequences of all reactor accidents.

REFERENCES

1. "Nuclear Reactor Accident Source Terms," Report by an NEA Group of Experts, OECD, March (1986).

2. T.S. Kress, "Review of Areas that may Require Simultaneous Coupled Solution of the Thermal Hydraulics and Fission Product/Aerosol Behaviour Equations for Source Term Determination," Proceedings of CSNI Specialist Meeting on Nuclear Accidents in Reactor Safety, Karlsruhe, September (1984).

3. C.D. Andriesse, "Interpretation of Fission Product Release from Overheated Fuel," Proceedings of an IAEA Symposium on Source Term Evaluation for Accident Conditions, IAEA-SM-281, Columbus, October 1985.

4. R.P. Wichner, T.S. Kress and R.A. Lorenz, "Fission Product Release from Fuel," in "Technical Bases for Estimating Fission Product Behaviour During LWR Accidents," NUREG-0772 (June 1981).

5. J. Rest, "An Improved Model for Fission Product Behaviour in Nuclear Fuel Under Normal and Accident Conditions," J. Nucl. Mat. 120, 195 (1984).

6. L.D. McDonald, D.B. Duncan, B.J. Lewis and F.C. Iglasias, "FREEDOM: A Transient Fission-Product Release Model for Radiactive and Stable Species," Proceedings of IAEA Technical Committee Meeting on Water Reactor Fuel Element Computer Modelling in Steady-State, Transient and Accident Conditions, Preston, September 1988.

7. R.R. Hobbins, D.J. Osetek and D.L. Hagrman, "In-Vessel Release of Radionuclides and Generation of Aerosols," Proceedings of an IAEA Symposium on Source Term Evaluation for Accident Conditions, IAEA-SM-281, Columbus, October 1985.

8. M.F. Osborne, J.L. Collins, R.A. Lorenz and R.V. Strain, "Fission Product Release and Fuel Behaviour in Tests of LWR Fuel under Accident Conditions," Proceedings of an IAEA Symposium on Source Term Evaluation for Accident Conditions, IAEA-SM-281, Columbus, October 1985.

9. G.L. Horvath, K. Krinizs and P. Vertes, "Comparison of Two Fission Product Release Codes," Proceedings of an IAEA Symposium on Source Term Evaluation for Accident Conditions, IAEA-SM-281, Columbus, October 1985.

10. G.A. Allison, G.A. Berna, R.J. Wagner, T.C. Cheng, L.J. Siefken and D.L. Hagrman, "Capabilities of the Integrated SCDAP/RELAP5/TRAP-MELT Severe Accident Analysis Computer Code," Proceedings of an IAEA Symposium on Source Term Evaluation for Accident Conditions, IAEA-SM-281, Columbus, October 1985.

11. C.M. Allison, "Severe Core Damage Analysis Package (SCDAP) Code Conceptual Design Report," EGG-CDAP-5397 (1981).

12. V.H. Ransom, "RELAP5/MOD2 Code Manual, Volume 1: Code Structure, System Models and Solution Methods," NUREG/CR-4312 (1985).

13. H. Jordan and M.R. Kuhlman, "TRAP-MELT2 User's Manual," NUREG/CR-4205 (1985).

14. A.M. Beard, B.R. Bowsher and A.L. Nichols, "The Interaction of Molecular Iodine Vapours with Silver-Sodium-Cadmium Control Rod Aerosol," Proceedings of Int. Symp. on Severe Accidents in Nuclear Power Plants, Sorrento, March, IAEA-SM-296 (1988).

15. J.E. Herceg, R.L. Ritzman and R.N. Oehlberg, "Source Term Experiments Program (STEP)," Proceedings of an IAEA Symposium on Source Term Evaluation for Accident Conditions, IAEA-SM-281, Columbus, October 1985.

16. K.O. Reil, A.C. Marshall, R.O. Gauntt, R.W. Ostensen, P.J. Pickard, C. Fryer, K.T. Stalker and K. Muramatsu, "Results of the ACRR-DFR Experiments," Proceedings of ANS Meeting on Thermal Reactor Safety, San Diego, February 1986.

17. P.J. Fehrenbach and J.C. Wood, "Description of the Blowdown Test Facility COG Program on In-Reactor Fission Product Release,

Transport and Deposition Under Severe Accident Conditions," Atomic Energy of Canada Limited Report, AECL-9343 (1987).

18. "PHEBUS Fission Products Program," CEA Report IPSN-3-87, June (1987).

19. G.W. Parker, G.E. Creek and A.L. Sutton Jr, "Source Term Evaluation from Recent Core-Melt Experiments," Proceedings of an IAEA Symposium on Source Term Evaluation for Accident Conditions, IAEA-SM-281, Columbus, October 1985.

20. S.R. Mulpuru, F.B. Banks and M.D. Pellow, "Aerosol Material Release from Zircaloy-4 Clad UO_2 Pellet at Temperatures up to 2000°C in an Argon Atmosphere," Proceedings of the Ninth Annual Conference of the Canadian Nuclear Society, Winnipeg, June 1988.

21. P.N. Clough, H.C. Starkie and A.R. Taig, "Vaporization and Transport of Structural Constituents from LWR Degraded Cores," Proceedings of an IAEA Symposium on Source Term Evaluation for Accident Conditions, IAEA-SM-281, Columbus, October 1985.

22. R.O. Wooton and H.I. Avci, "MARCH (Meltdown Accident Response Characteristics) Code Description and User's Manual," NUREG/CR-1711 (1980).

23. M.R. Kuhlman, D.J. Lehmicke and R.O. Meyer, "CORSOR User's Manual," NUREG/CR-4173 (1985).

24. J.Q. Howieson, H.S. Shapiro, P.J. Allen, J.T. Rogers, T. Mostert and R.W. Van Otterloo, "A Probabilistic Risk Assessment Study of the CANDU-600," Proceedings of Int. Symp. on Severe Accidents at Nuclear Power Plants, IAEA-SM-296, Sorrento, (1988).

25. A.T.D. Butland and M.R. Kuhlman, "Factors Affecting Primary System Radionuclide Retention in an LWR Severe Accident," Proceedings of an IAEA Symposium on Source Term Evaluation for Accident Conditions, IAEA-SM-281, Columbus, October 1985.

26. H. Jordan and M. Leonard, "Fission Product Revaporization in the Reactor Cooling System," Proceedings of an IAEA Symposium on Source Term Evaluation for Accident Conditions, IAEA-SM-281, Columbus, October 1985.

27. R.G. Freeman-Kelly and R.G. Jung, "A User's Guide for MERGE," NUREG/CR-4172 (1985).

28. A.T.D. Butland, N.A. Johns, D.A. Williams, "Validation of Models Used to Predict Primary System Thermal-Hydraulics and Radionuclide Transport and Deposition in a PWR Severe Accident," Proceedings of an IAEA Symposium on Source Term Evaluation for Accident Conditions, IAEA-SM-281, Columbus, October 1985.

29. J. Collen, H. Unneberg and D. Mecham, "The Marviken Large Scale Transport Tests. Summary Report," MXE-301, Studsvik Energiteknik AB (Nov. 1985).

30. M. Donahue, M. Hazzan, J. Metcalf and E. Warman, "Analysis of Retention/Revaporization in a BWR Mark II Power Plant," Proceedings

of an IAEA Symposium on Source Term Evaluation for Accident Conditions, IAEA-SM-281, Columbus, October 1985.

31. F.J. Rahn, J. Collen and A.L. Wright, "Aerosol Behaviour Experiments on Light Water Reactor Primary Systems," Nuclear Technology, 81, 158, (1988).

32. J.H. Wilson, A.L. Wright and P.C. Arwood, "LACE Aerosol Code Comparison Activities at Oak Ridge National Laboratory," Proceedings of a CEC/OECD Workshop on Water-Cooled Reactor Aerosol Code Evaluation and Uncertainty Assessment, Brussels, September 1987, available as EUR 11351 EN.

33. B.R. Sehgal, A.T. Wassel, R.K. Ahluwalia, K.H. Im and V.E. Denny, "Coupled Thermal-Hydraulic/Aerosol Transport Analysis Capability for Severe Accidents," Proceedings of an IAEA Symposium on Source Term Evaluation for Accident Conditions, IAEA-SM-281, Columbus. October 1985.

34. K.H. Im, R.K. Ahluwalia and C.F. Chang, "RAFT: A Computer Model for Formation and Transport of Fission Product Aerosols in LWR Primary Systems," Proceedings, American Nuclear Society Meeting on Fission Product Behavior and Source Term Research, EPRI NP-4113-SR (1985).

35. D.F. Torgerson, F. Garisto, F.C. Iglesias, C.E.L. Hunt and D.S. Cox, "The Chernobyl Source Term," Proceedings of the ANS Topical Meeting on Anticipated and Abnormal Transients in Nuclear Power Plants, Atlanta, April 1987.

36. F. Garisto, "Thermodynamics of Iodine, Cesium and Tellurium in the Primary Heat-Transport System Under Accident Conditions," Atomic Energy of Canada Limited Report, AECL-7782 (1982).

37. B.R. Bowsher and A.L. Nichols, "High Temperature Studies of Simulant Fission Products: Part IV. Interaction of Cesium Iodide with Boric Acid Over the Temperature Range 400 to 1000°C," AEEW-R 1973 (1985).

38. R.M. Elrick, R.A. Sallach, A.L. Ouellette and S.C. Douglas, "Reactions Between Some Cesium-Iodine Compounds and the Reactor Materials 304 Stainless Steel, Inconel 600 and Silver, Vol.I, Cesium Hydroxide Reactions," SAND83-0395 (1984).

39. R.M. Elrick, R.M. Merrill and A.C. Ouellette, "The Thermal Instability of Cesium Iodide," NUREG/CR-5155, SAND88-1187 (1989).

40. J. Fermandjian, H. Bunz, I.H. Dunbar and G. Lhiaubet, "Comparison of European Computer Codes Relative to the Aerosol Behaviour in PWR Containment Buildings During Severe Core Damage Accidents (Modelling of Steam Condensation on the Particles," Proceedings of a CEC/OECD Workshop on Water-Cooled Reactor Aerosol Code Evaluation and Uncertainty Assessment, Brussels, September 1987, available as EUR 11351 EN.

41. P.N. Clough, I.H. Dunbar and C.F. Knights, "The Chemical Composition of Containment Aerosols and its Effect on Steam Condensation," Proceedings of a CEC/OECD Workshop on Water-Cooled Reactor Aerosol Code Evaluation and Uncertainty Assessment, Brussels, September 1987, available as EUR 11351 EN.

42. J. Jokiniemi and J. Makynen, "The Growth of Hygroscopic Aerosol Particles," Proceedings of an IAEA Symposium on Source Term Evaluation for Accident Conditions," IAEA-SM-281, Columbus, October 1985.

43. W.O. Schock, W. Schikarski, F. Abbey and I.H. Dunbar, "Behaviour and Retention of Fission Products in the Containment," Proceedings of an IAEA Symposium on Source Term Evaluation for Accident Conditions, IAEA-SM-281, Columbus, October 1985.

44. T.F. Kanzleiter, J.P. Hosemann, W.O. Schock, D.M. Haschke, M. Peehs and K. Hassmann, "DEMONA Experiments to Investigate Aerosol Behaviour Within a Containment," Proceedings of an IAEA Symposium on Source Term Evaluation for Accident Conditions, IAEA-SM-281, Columbus, October 1985.

45. H. Jordan and V. Kogan, "Aerosol Modelling in Containment," Proceedings of an IAEA Symposium on Source Term Evaluation for Accident Conditions, IAEA-SM-281, Columbus, October 1985.

46. H. Jordan and V. Kogan, "Aerosol Behavior Modeling," NUREG/CR-4388, BMI-2130 (1985).

47. F. Abbey and W.O. Schikarski, "Aerosol Behaviour and Light Water Reactor Source Terms," Nuclear Technology 81, 293 (1988).

48. E. Della Loggia, I.H. Dunbar, J. Fermandjian, R. Ricchena, and L. Wolf, "The Interface Between Thermal Hydraulics and Fission Product Transport in Severe Accident Analysis," Proceedings of an IAEA International Symposium on Severe Accidents in Nuclear Power Plants, IAEA-SM-296, Sorrento, March 1988.

49. F. Gelbard, "Aerosol Growth in a Nuclear Reactor Containment Environment," Proceedings of a CEC/OECD Workshop on Water-Cooled Reactor Aerosol Code Evaluation and Uncertainty Assessment, Brussels, September 1987, available as EUR 11351 EN.

50. L. Wolf and T. Kanzleiter, "Multi-Compartment DEMONA-Experiment A9 and Conclusion for Coupled Thermohydraulic/Aerosol Codes", Proceedings of a CEC/OECD Workshop on Water-Cooled Reactor Aerosol Code Evaluation and Uncertainty Assessment, Brussels, September 1987, available as EUR 11351 EN.

51. D.J. Wren, J. Paquette, J.C. Wren, P.N. Clough and H.C. Starkie, "A Review of Iodine Chemistry Under Severe Accident Conditions," Atomic Energy of Canada Limited Report, AECL-9089 (1987).

52. C.F. Clement, "The Calculation of Aerosol Growth Rates in Conjunction with Heat and Mass Transfer," Proceedings of a CEC/OECD Workshop on Water-Cooled Reactor Aerosol Code Evaluation and Uncertainty Assessment, Brussels, September 1987, available as EUR 11351 EN.

53. P.N. Clough, "Summary, Conclusions, and Recommendations," Proceedings of a CEC/OECD Workshop on Water-Cooled Reactor Aerosol Code Evaluation and Uncertainty Assessment, Brussels, September 1987, available as EUR 11351 EN.

54. H.A. Morewitz, F.J. Rahn and R. Borkowski, H. Bunz and W. Schock, "Resuspension/Re-entrainment of Aerosols in LWRs Following a Meltdown Accident," in Selected Source Term Topics, CSNI Report 136, (1987).

55. R. Borkowski, H. Bunz and W. Schock, "Resuspension of Fission Products During Severe Accidents in Light Water Reactors," Report to the Commission of the European Communities XII/978/84-EN (1984).

56. W.J. Smith II, F.W. Whicker and H.R. Meyer, "Review and Categorization of Saltation, Suspension and Resuspension Models," Nucl. Safety 23, 685 (1982).

57. see for example the Proceedings of the CSNI Specialists Meeting on Core Debris-Concrete Interactions, Palo Alto, September 1986, EPRI NP-5054-SR, February (1987).

58. P.N. Smith, A.T.D. Butland, P.E. Potter, M.A. Mignanelli and G.J. Roberts, "Importance of Core/Concrete Aerosol Production and Some Containment Heat Sources to the Source Term," Proceedings of an IAEA Symposium on Source Term Evaluation for Accident Conditions, IAEA-SM-281, Columbus, October 1985.

Fission Product Transport in the Primary System, Important Phenomena, and Code Status

J. A. GIESEKE, H. JORDAN, and M. R. KUHLMAN
Aerosol Science Group
Battelle Columbus Laboratories
505 King Avenue
Columbus, Ohio 43201, USA

INTRODUCTION

The purpose of this paper is to identify important issues concerning the transport and deposition of radionuclides in the reactor coolant system (RCS) under accident conditions and to examine how such issues are being treated or should be treated by the various available computer codes. In general, the RCS is a very important section of the transport pathway along which radionuclides move and by which they are attenuated as they travel after being released from the fuel. The RCS can serve as a sink for radionuclides that may deposit from the gas and react with surfaces, or can serve as a repository for materials deposited from the gas which are then available for later release into the transporting gas stream. The RCS may also have thermal hydraulic conditions that foster aerosol growth by condensation or agglomeration, and may provide an environment in which gas phase or heterogeneous chemical reactions may occur. Obviously the physical and chemical processes occurring in the RCS are extremely complex and depend on such parameters as the rates of release of radionuclides, control rod, and structural materials from the core region; gas flow rates and patterns; surface and gas temperatures throughout the system; gas composition; and accident progression. A key question which remains concerns the minimum phenomena that one must model in order to obtain a satisfactory prediction of the quality and nature of radionuclide release from the RCS throughout the course of an accident.

BACKGROUND

In a degraded core accident, fission products will start to be released from the core as fuel rod cladding fails. This will generally occur with large-scale loss of coolant in the core, such that the core is uncovered. As the core uncovers, the uncovered portion superheats the evolving steam, and this superheated steam, together with hydrogen resulting from the reduction of steam by the zircalloy cladding, constitutes the medium for fission product transport from the core to the breach in the reactor cooling system.

The flow rate of the steam/hydrogen mixture is a strong function of core melt progression, particularly in low pressure scenarios. Initially high (hundreds of pounds of steam per second) flow rates will subside as the level of the water drops. As sections of the melting core slump into residual water in the bottom vessel head, renewed surges of steam flow will occur. Eventually, the bottom head may dry out and gas flow reduces to expansion flow resulting from rising temperatures. Flow may resurge on vessel melt-through and consequent coupling of reactor cooling system with containment hydrodynamics. High pressure scenarios dampen fluctuations in gas flow because they superimpose a long term high depressurization flow.

Gas temperatures immediately above the core range from several hundred degrees Fahrenheit to more than 2000 F. The thermal inertia of the reactor cooling system structures is, however, very large by comparison to the energy content of the steam, and gas temperatures will cool to values near RCS normal operating temperatures (barring local decay heating) a short distance away from the core. This very high temperature gradient along the flow means that chemical speciation of the fission products is likely to occur in the vicinity of the core because chemical association/dissociation is highly favored by the high temperature and temperature gradient there. This rapid decrease in temperature with distance from the core also means that, except for the noble gases, fission products are likely to condense to the liquid phase within the boundary layer, or, if they are volatile enough, shortly after they leave the core. This they do by nucleation, condensation on so created nuclei, and condensation on structural surfaces. Chemisorption of vapors is a competing mechanism. Typically, PWR control rod materials yield high concentrations of nuclei in the core region, and one may expect most volatile fission product phase transitions to occur by condensation onto these nuclei.

A major role of the RCS is then in defining the source term to the containment: it acts as a chemical and aerosol reactor that conditions the fission product release from the melting fuel. Low gas flow rates (long residence time) and physical confinement lead to very high concentrations. Aerosol concentrations, for example, are orders of magnitude higher in the RCS than in the containment. Such high concentrations lead to rapid growth of these aerosol particles by agglomeration and by chemical reaction and condensation of vapors. Particle size is strongly conditioned by the RCS; the source term to the containment consists of much larger aerosol particles than would be the case in the absence of the RCS.

The other major role of the RCS lies in its potential for fission product retention. This can occur by chemical reaction with surfaces for highly reactive fission product vapors such as tellurium, or by aerosol particle deposition. The latter may be influenced by flow rate and structural characteristics of the RCS; it is probably dominated by sedimentation onto horizontal surfaces. Characteristic structures/flows that may affect deposition by mechanisms other than sedimentation are, for example, flow past tube bundles, flow in zig-zag paths (steam dryers), vortical flow (steam separators), flow through pipe bends, flow at pipe entrances/exits (tube sheets). None of these characteristic structure/flows are well enough understood to model realistically; some, however, have been investigated experimentally under a restricted set of conditions.[1]

Models of fission product behavior in the RCS are necessarily gross idealizations of a very complex system. These models, nevertheless, provide insight into the role of the RCS in conditioning the source term to containment and, with some uncertainty, quantify it. Our understanding, based largely on code calculations, suggests that the most critical or highest sensitivity (not most uncertain) aspects of fission product behavior in the RCS are:

- Chemical form of the fission products
- Chemical reactivity of fission products
- Quantity of fission product release
- Timing of fission product releases relative to each other and relative to the thermal conditions of the RCS
- The residence time of the fission products in the RCS (gas flow rate)
- Structural surface area, system gas volume, and geometry for flow
- Fission product revolatilization due to decay heating after vessel failure.

DESCRIPTION OF AVAILABLE CODES

TRAPMELT3

TRAPMELT3 models the complex reactor cooling system in an idealized way but, within this idealism, treats the major phenomena in an interactive fashion. Perhaps its greatest limitation is its inability to model possible change of the chemical forms of the radionuclide vapors during transport through the system and as a function of system changes with time. Thus iodine and cesium are assumed to be in the form CsI and CsOH from the outset and throughout their transport through the RCS. This assumption critically affects their physical states and surface reaction rates. Similarly, tellurium is assumed to transport in elemental form. All other fission products have sufficiently low vapor pressure, even under severe accident conditions, that it is reasonable to assume they nucleate near the fuel-gas interface. They are thus introduced into the system as small particles. The exact size of these particles proves unimportant because of the rapid agglomeration rates experienced by the high concentration aerosol that is present in the vicinity of the core during the major portions of severe accident sequences.

The rates of irreversible sorption, on structure surfaces, of the volatile species are based on empirical reaction rates coupled to mass transfer coefficients. The deposition rates used in TRAPMELT were based on the work of Elrick, Sallach, and others[2,3] and are 0 for CsI, 0.01 cm/sec for CsOH, and 1.0 cm/sec for Te. Thus the salt CsI is not predicted to react with metal surfaces whereas the metal Te is much more reactive than the hydroxide CsOH. The data base from which the empirical representation was derived is quite limited and uncertainties in the sorption rates are large. However, these deposition velocities are consistent with a recent review.[4] Sorption of volatile species on aerosol particles is not modeled.

Modeling of condensation of the volatile fission product vapors on RCS surfaces and existing aerosol particles is based on gas and surface temperatures and calculated mass transfer coefficients that are obtained by analogy to the heat transfer coefficients employed in the thermal hydraulic analysis that is performed by TRAPMELT3 in parallel to the mass transfer calculations. If, after a time step, supersaturation of a vapor persists, this is instantaneously nucleated to the smallest particle sizes within the distribution of sizes carried in the computations since the code does not model nucleation mechanistically.

Aerosol behavior is treated in detail. Growth can occur by condensation of volatile fission products and by agglomeration due to Brownian motion, differential gravitational settling and differential turbulent motion. Particles can deposit on surfaces by gravitational settling and by impaction from turbulent flow. They can also deposit by thermophoresis and diffusion from laminar and turbulent flows. All of these deposition mechanisms are treated in a generic way: the code does not specifically model characteristic flows/geometries, such as, for example, vortical flow in a BWR steam separator.

TRAPMELT3 also models steam and hydrogen transport through the reactor cooling system, given their evolution from the core region as a boundary condition. To do so, the equations of mass and energy transport are solved together with the equations of state of steam and hydrogen. Solution of the momentum transfer equation is circumvented by externally imposing a system pressure (which may, in general, be a function of time). Heat transfer between gas and structures occurs by convection and radiation. The fission products, both those deposited and those suspended in the gas, provide decay heating both to structures and the gas.

The main characteristics of the numerical treatment of the model are as follows:

- The physical system is nodalized into a few, characteristic control volumes that are assumed to be well mixed but may contain several, disconnected structures.
- Structures are treated as lumped capacities. Outside boundaries are perfectly insulated.
- The aerosol size distribution is discretized into 19 size classes.
- The composition of the aerosol particles is assumed to be independent of size class.
- Ten chemical groups are tracked:

  Xe + Kr
  CsOH + RbOH
  CsI + RbI
  Te + Se + Sb
  Sr
  Ba
  Ru + Rh + Pd + Mo + Tc
  La + Eu + Pr + Nd + Pm + Sm + Gd + Y + Nb + Zr(F.P.)
  Ce + Np + Pu
  Control rod + structural materials + Zr(clad).

RAFT/PSAAC

The RAFT code was the first reactor cooling system fission product transport code to attempt to model the nucleation of fission product vapors as they escape the fuel.[5] This is done in conjunction with a vapor phase, liquid phase chemical equilibrium speciation calculation on the following system:

Cs I Te Ag In Cd Sn Sr Mo Ru Ba Zr Mn H O Ar,
their oxides and hydroxides, and CsI and HI.

Classical nucleation theory, corrected for nonclassical effects and ionic nucleation is applied to the so predicted chemical species, resulting in the creation of embryonic aerosol particles that are allowed to grow by vapor condensation and by agglomeration. For the latter process, Brownian, gravitational, and turbulent mechanisms are considered.

Nucleation and condensation of vapors on particles are calculated in competition with the condensation of vapors on structures. Vapors can also react with structural surfaces based on mass transfer coefficients and chemical reaction rate data for CsOH and Te as developed by Elrick and Sallach.[2]

Aerosol particles are allowed to deposit on structural surfaces by diffusion, thermophoresis, turbulent impaction, and gravitational setting in a generic way. In addition, the code has specific models for inertial deposition in pipe bends, for inertial deposition in the vortical flow of a steam separator and for internal deposition in the zig-zag flow of a steam dryer.

The numerical treatment of RAFT is that of time dependent, axially one dimensional, once through, channel flow. The equations for the mass balance of each chemical species, as well as the aerosol population balance equations, are solved by finite difference in the time and the axial distance. The aerosol particle size distribution is discretized into a number of (flexible) size bins and these in turn are grouped into size sections for which aerosol particle composition is tracked.

RAFT requires, as input, axial distributions and time histories of pressure, gas temperature and structure surface temperatures and of the mass flow rate of steam and hydrogen in the system. These are provided by the PSAAC code that has been coupled to RAFT. This coupling also provides for fission product decay heating feedback in the thermal hydraulics of the system.

PSAAC requires, as boundary conditions, the flow rates and temperature of steam and hydrogen evolving from the core region. It then solves the energy and mass balance equations for each species throughout the reactor cooling system, modeled as a system of well mixed control volumes. Forced convection, natural convection and radiative heat transfer are considered. Condensation of steam is allowed--a water phase is considered by a homogeneous two phase flow model. Special features are:

- Buoyancy driven loop flows in hot legs and U-tube steam generators
- Partially full pressurizer with drainage to the hot leg
- Fission product scrubbing in the pressurizer.

Because the PSAAC code is nodalized as a system of well mixed control volumes, while the RAFT code uses a finite difference scheme, their coupling is difficult and requires approximation schemes. Of particular concern is the lack of the spatial resolution in PSAAC that is needed to adequately predict the temperature gradients, near the core, that drive the nucleation model in RAFT.

MCT2 Code

In general, the transport portion of the MCT2 code follows closely the TRAPMELT code from which it is derived. Some differences, however, now exist. The simulation of core melt progression during severe accidents is not an exact science, and may never be. It is necessary to examine assumptions which are made in modeling this process since some of these assumptions influence the processes occurring throughout the system. One assumption found in MCT2 is that any silver present in the inventory will remain in the core region due to its low melting point. This is not an important concern for BWR accidents, but could be very significant with respect to aerosol formation in a reactor equipped with Ag-In-Cd control rods.

This code assumes that gas flow is blocked at any molten node in the core. As a result, no flow passes through any nodes above the lowest molten node in a vertical column. This assumption is expected to significantly reduce the amount of hydrogen formation from the Zr-H_2O reaction, and will also alter the temperature profile of the core during the melting period. Reducing the amount of H_2 formation, of course, will alter the fluid properties in the RCS and therefore have some effect on fission product transport. More importantly, however, reduced H_2 production will lower containment pressures, reduce the probability of ignition, and increase the effectiveness of pool scrubbing of fission product aerosol because of the higher steam content of the gas injected into the pool.

A phenomenon which has received attention in the MCT2 code is fluid recirculation within the RCS, especially within the core and steam separators and dryers. A major reason for attending to this particular phenomenon is that great uncertainty is attached to the potential impact of flow recirculation. The large uncertainty arises from a wide range of opinions regarding the influence of recirculation on fission product deposition within the RCS, not from a wide range of possible outcomes for a given set of conditions. Large temperature differences are predicted to exist between the top of the core and the control volumes modeled above the core. Some analyses[(6)] indicate that for certain accident conditions, these temperature differences are sufficient to drive a natural convection flow which exceeds the forced flow predicted from the rate of gas generation. This code calculates the forced flow, then evaluates the extent of convective flow and performs the required redistribution of mass and energy between the volumes involved. The "natural circulation" between the volumes tends to mix the material uniformly between the two volumes and damp out differences in temperatures through the volumes.

In a sense, incorporation of intervolume convective mixing in the code serves as a renodalization of the problem based upon temperatures in the different parts of the system. A number of sensitivity studies of the TRAPMELT portion of the STCP have been performed to investigate code sensitivity to the nodalization selected to represent a reactor RCS.

Because each of the volumes considered by the code is treated as being perfectly mixed, effects attributable to the choice of a nodalization scheme are usually quite small when the RCS is taken as a whole. The choice of nodes does, however, influence the location of deposited material in the RCS. It was also found[(6)] that, for the sequences analyzed, the recirculation influenced the overall results but little.

The extent of intravolume circulation plays a role in determining the heat and mass transfer coefficients within the volume. Both the STCP and MCT2 make use of the Rayleigh number to evaluate the natural convection within volumes and to select the appropriate correlations for heat and mass transfer. The rate of deposition of fission products on RCS and particle surfaces is governed in part by the mass transfer coefficients. The mass transfer correlations used in the early versions of the TRAPMELT code calculated the mass transfer coefficients based on the assumption of fully developed turbulent flow within each volume. Even if the Reynolds number is in the turbulent regime for a control volume, the flow is not fully developed immediately upon entering the volume. Because of the finite distance required for a flow profile to become established, the use of a single heat or mass transfer coefficient will always be valid only as an approximation. Nevertheless, the coefficients should be calculated from correlations for the appropriate flow regime. This is accomplished in both MCT2 and STCP.

MCT2 has retained the STCP formulation for the mass transfer coefficient to particle surfaces, but has incorporated a set of coefficients for transfer to wall surfaces which represent an improvement over what was present in the original code. The STCP currently calculates the transfer coefficients in a somewhat different fashion, but with the effect of having the mass transfer coefficient depend upon the flow regime. The STCP calculates heat transfer coefficients which are appropriate to the flow regime and take into account the degree of natural convective mixing in the control volumes. These heat transfer coefficients are then converted to mass transfer coefficients by means of the Chilton-Colburn analogy.

To examine the effect of the differences in the formulae used to calculate the mass transfer coefficients, STCP calculations were performed for selected conditions during a TC sequence and an AE sequence in a BWR as analyzed previously.[7] The mass transfer coefficients calculated covered a range of more than two orders of magnitude, but the values calculated using the MCT2 formulation generally differed by less than a factor of three from the values calculated using the scheme in the earlier version of the TRAPMELT code. While the rate of wall deposition is proportional to the mass transfer coefficient during each time step, the overall effect of an error in the coefficient will be less than linear. One must also keep in mind that transfer of mass to the walls via condensation will occur only under appropriate conditions of temperature and vapor concentration, and that the rate of transfer will be greatest during turbulent flow through a volume. Therefore condensation rates predicted by the earlier version of STCP would have been more likely to overpredict condensation than underpredict.

With respect to mass transfer rates, there is a problem in the MCT2 code with the rate of chemisorption calculated. The chemisorption rates included in the code are based on rough laboratory scale experiments. These rates are, however, the only practically useful estimates of this process. The rate at which this process will occur in a large scale system, however, can be limited by the mass transfer rate to the surface. The mass transfer limitation on this process should be taken into account in the MCT2 code to provide valid estimates of the importance of this process. Otherwise, curious results can be obtained from the code in which nearly all of the Te and much of the CsOH are deposited via chemisorption on the RCS surfaces. This error would be especially evident in large control volumes which have long residence times associated with them.

Major Uncertainties

In the preparation of NUREG-1150 by the U.S. NRC, a panel of Technical Experts on Source Terms was asked to estimate uncertainties in the modeling of various source term phenomena and from these estimates an uncertainty distribution was developed for use in identifying the issues important to the overall uncertainty in risk.[8] Among the issues or phenomena considered in that evaluation were "In-Vessel Release", "In-Vessel Retention", and "Revolatilization from Reactor Coolant System". Some of the observations of this panel are useful in suggesting issues that are considered most important to uncertainty and those that might benefit from analysis with appropriate models.

This panel believed the extent of zirconium oxidation in the core to be important to in-vessel release of tellurium. Uncertainty in the availability of zirconium metal for reaction with tellurium leads to wide uncertainties in the predicted release of tellurium. In-vessel retention of iodine and cesium is believed to be strongly dependent on RCS pressure through its influence on residence times. Whether or not the RCS develops one or two breaches strongly affects the revolatilization of fission products, particularly iodine and cesium. This suggests that uncertainties in flow paths and patterns are believed to be important. In general, the wide range in the values of estimated release fractions suggests that better modeling would be likely to reduce uncertainties in RCS transport and deposition predictions.

An OECD (NEA) group of experts also evaluated the importance of various issues to fission product transport and deposition within the RCS.[9] They found that high impacts on source term predictions would be associated with the coupling of thermal hydraulics with fission product transport and with the modeling of fission product revolatilization from deposits due to decay self-heating. A medium to high impact of steam generator tube rupture scenario modeling was also expected. Moderate importance was ascribed to the following phenomena: chemical reactions between core materials and gases with tellurium; chemical reactions of fission products with control rod materials; hydroxide formation; late air entry into the RCS; modeling of primary particle formation; aerosol formation, transport, and growth during revolatilization; hygroscopic properties of aerosols; natural circulation in and between RCS regions; and aerosol re-entrainment or resuspension processes.

In an independent review of U.S. NRC studies, Kouts, et al[10] also cite the importance of flow paths and patterns on RCS retention of fission products. They suggest that the best possible accuracy in predicting circulating current flow rates may be plus or minus a factor of 2. That is, with current predictions, flow patterns may always be poorly predicted and as a result, fission product transport and deposition estimations are inherently limited in accuracy.

In general, the apparent large contributions to the uncertainty in risk arise from thermal hydraulic uncertainties affecting containment integrity. The importance of radionuclide retention in the RCS lies in its effect on the source term to the containment. Various reviewers have identified as major uncertainties the modeling of flow patterns in the RCS, re-evolution/re-entrainment of deposits, and the nucleation and condensation of vapors both from the core region directly and that from re-evolved deposits.

Resuspension/Re-entrainment

Late releases of radionuclides from the RCS represent a situation contributing significantly to risk. Late sources become important mainly because there are significant deposition mechanisms that create repositories of radionuclides within the RCS through accumulation of materials previously released from the fuel.

Deposits on solid surfaces can be re-entrained in the gas through physical resuspension of either liquid or solid deposits, or through vaporization of either liquid or solid deposits. It is worth reviewing these mechanisms to determine if a sufficient basis of technical information exists to adequately predict such occurrences. In addition it is worth reviewing accident conditions to assess whether or not occurrences of re-entrainment are important to uncertainty and worthy of modeling.

In the course of the Marviken and LACE experiments, it became obvious that in some cases the physical form for the transporting fission products may not be solid but may be better described as a liquid. Examination of the physical properties for some of the chemical species most likely to be present in the expected temperature and gas-composition environments further substantiates this. CsOH would not be likely to exist in other than liquid form, for example, and when combined with possibly solid or semisolid materials would probably be a slurry capable of moving along surfaces in the direction of gas flow or downward as influenced by gravity.

The Marviken experience[11,12] was that under conditions approaching those expected in nuclear reactor accidents, the CsOH, CsI, and Te transporting in a high steam content gas at about 400 C was in a liquid form. The liquid form had direct impact on the sampling system operation and on transport along the system piping.

Similar observations were made during the LACE experiments that used NaOH as a simulant for CsOH and alumina powder as a simulant for insoluble particles.[13] The NaOH and alumina transported as a slurry under the experimental steam-air flow conditions with much of the material moving along the pipe walls as in the traditional mist-annular flow regime for two-phase flow. The fraction of the total liquid mass which transports along the surface depends on the total amount of liquid relative to geometry and flow conditions.[14,15,16] It is apparent that a thin film will transport with little if any accompanying mist or droplet flow. As the amount of liquid is increased, the surface film becomes unstable and atomization from the film occurs as it becomes unstable with some increase in film thickness as the liquid rate increases further. Under steady-state conditions the deposit of droplets into the film is balanced by atomization from the film. At geometrical discontinuities such as the end of a pipe, large droplets are formed from the film but these are generally quite large.[13,14]

In general, the sizes of droplets formed under two-phase flow conditions will be much larger than the aerosol particles formed by condensation of radionuclide vapors. Nevertheless, under conditions of high aerosol concentrations or vapor supersaturation, particle sizes may increase, by agglomeration or condensation, to the extent that they are 10 μm or more and therefore become similar in size to film atomized droplets which would be expected in the same size range. Equations have been developed to estimate the droplet sizes[16] but in the absence of physical properties for molten deposits, estimates cannot easily be made. Studies of droplet formation in simulated RCS geometries for typical flow and liquid property conditions would seem to be justified in order to answer some of the many questions surrounding this possible phenomenon. Prior to such studies, simpler studies to verify the liquid nature of the fission products and to evaluate physical properties of such liquids or slurries would do much to reduce uncertainties. At the present it seems that flow rates would have to be higher than seem likely under most accident conditions and exist over time periods longer than seem possible before atomized materials can be expected to be a significant source of particles small enough to be persistent.

For the possible case of dry deposits, it is evident that resuspension is more likely. This situation has been extensively reviewed[17,18,19] with fairly consistent conclusions. Deposits of solid particles kept carefully dry will be resuspended from surfaces at rates and thresholds depending on the conditions.[20,21] Resuspension is more likely at higher velocities for thicker deposits and at lower humidities. Particle size, particle shape, particle composition, and flow geometry also affect the onset of resuspension. Estimates of resuspension under accident conditions suggest that the conditions favor retention rather than resuspension.[19] Large uncertainties exist, however, because of the choice of a conservative resuspension model. Experiments with simulated core region geometries illustrate that dust deposits are released in spurts and the aggregated masses of the released material behave in a manner very dissimilar to that of the individual particles.[22]

Models and correlations for predicting dust deposit resuspension are probably inadequate for simple geometries and are even less reliable if applied to typical geometries and postulated accident conditions. If the deposits are dry and powder-like, their resuspension is very uncertain. The result would probably be less than complete dispersion and the resuspension as agglomerates having large particle sizes such that they would quickly redeposit. Determinations of deposit characteristics as either liquid or powder should precede further study of resuspension. If it can be shown that dry powder-like deposits can occur, then further work is required to allow reasonable estimates of resuspension rates.

Recirculation

Recirculation--either within a given control volume or among two or more contiguous volumes--is a phenomenon whose effects have been the subject of much conjecture. Adequate representation of recirculation between volumes would seem to be a matter of incorporating appropriate volume definitions and flow junctions into the formulation of the problem. Subdivision of the communicating volumes into distinct sub-volumes would permit definition of the amounts and direction of flow exchange in an unambiguous fashion, and could be accomplished within the framework of most existing codes. A key problem which exists in the codes discussed above is that they were structured to treat the problem as a once-through phenomenon. The RAFT code considers the fission product transport as a one-dimensional channel flow path, and the thermal hydraulic codes which drive transportation in TRAPMELT3 and MCT2 do not permit recirculatory flow in a clean fashion. The MCT2 code does apply a fix at the end of a time step. It assesses whether conditions in adjacent control volumes are conducive to recirculatory flows, and calculates the extent of such flows based solely on the fluid temperature differences. The code then redistributes the mass and energy between the connected volumes in an attempt to approximate the natural circulation.

The effect of recirculatory flows in the RCS on fission product fate is twofold. First, recirculation will redistribute the fission products between volumes which may exhibit different retention charac-teristics. Second, the recirculatory flow itself may significantly alter fission product deposition rates by modifying the flow rates within volumes and bringing into play entrance effects between volumes. The transport and deposition codes which use a connected control volume representation of the RCS are capable of simulating those effects if the characteristics of the flows are provided by the thermal hydraulic codes.

Further study is required to better define the conditions under which recirculation within a control volume is expected to significantly affect the fate and distribution of fission products released from the fuel. It seems clear that this concern especially applies to the high-pressure accident sequences, such as TMLB, or to the portions of other sequences characterized by long RCS residence times and great temperature differences between connected volumes. The available two- and three-dimensional fluid flow codes need to be applied under a variety of conditions and validated against experimental results. Such results can then be used in the fission product transport codes and the results compared with predictions made without consideration of recirculation. Until such comparisons are performed for a variety of conditions, the issue of recirculation will remain a source of uncertainty in code predictions of fission product release from the RCS.

Aerosol Nucleation/Revaporization

The formation of aerosol particles from vaporizing fission products in the core region is expected to occur within the boundary layer at the fuel surface. The flow of steam and hydrogen past the overheated fuel provides cooling at most locations (except as the steam reacts with the zircalloy cladding) and provides gradients of both fission product vapor concentration and gas temperature away from the surface. As the vapors are released from the hot fuel they are diluted and cooled as they diffuse away from the fuel surface. Along the temperature gradient within the boundary layer, the gas is expected to become supersaturated in vapors and condense either by homogeneous or heterogeneous nucleation (on ions or on other particles). The size of the particles formed at this time is probably irrelevant since growth by agglomeration and condensation will be the major factors determining the transporting sizes.

Some models of RCS transport employ an axial temperature gradient along the flow path through the RCS piping to define a point of nucleation further downstream from the fuel or core. Such models require an unjustified assumption that nucleation does not occur in the boundary layer, but is prevented by immediate vapor dilution in the release volume. Downstream nucleation may occur in cooler sections of the RCS but again it would be expected in boundary layers rather than at a point along a one-dimensional temperature-concentration profile along the flow direction.

In later phases of an accident progression, after vessel failure, it is possible that deposits of radioactive material, in the RCS, revaporize as the result of decay heating with little convective cooling. For some scenarios,-
a large fraction of the volatile species (Cs, I, Te) is predicted to deposit in the RCS early in the sequence. This fraction then has the possibility of revaporization after vessel failure, perhaps after the containment has failed and offers little attenuation to the environmental

source term. Vapors revolatized during this period will enter an environment that may be markedly different than that which existed prior to vessel failure. In particular, there will be a far lower concentration of condensation nuclei, and revolatized vapors may transport principally as vapors rather than aerosols. To what degree they escape the RCS is, however, strongly dependent on their phase; small aerosol particles are likely to escape; vapors have the potential for reaction with surfaces. It is critical therefore to address homogeneous and ionic nucleation during this phase.

CONCLUSIONS

(1) Deposition models specific to special geometries can be used to improve predictions.

(2) Nucleation modeling is probably not important in the pressure vessel but may be needed for proper description of revaporization from deposits.

(3) Liquid flow on walls is likely to be an important redis-tribution mechanism for radionuclides within the RCS. This may have a major impact on deposit masses and locations, and therefore on local heating and revaporization.

(4) Alternate flow paths caused by RCS failure due to local heating other than meltthrough would impact heavily on transport modeling.

(5) Releases from the RCS have not been shown to be strongly sensitive to the treatment of recirculation employed in the transport calculations.

(6) Chemical forms and their prediction are expected to be important because of the effects on condensation rates and on wall-vapor and aerosol-vapor interactions.

REFERENCES

(1) Gieseke, J. A., et al, "Aerosol Deposition in BWR Steam Separators and Dryers", EPRI NP-5597 (January 1988).

(2) Elrick, R. M. and R. A. Sallach, "Fission Product Density in the Primary System", Proceedings of the International Meeting on Light Water Reactor Severe Accident Evaluation, Vol. I, p. 4.6-1-4.6-5, ANS Pub. No. 700085 (1983).

(3) Sallach, R. A., Greenholt, C. J., and Taig, A. R., "Initial Interactions of Tellurium Vapors with Reactor Materials", NUREG/CR-2921, Sandia National Labs Report No. SAND82-1145 (1983).

(4) Taig, A. R., et al., "Interpretation of Experimental Results on the Interactions of Fission Product Vapors With Reactor Materials", Proceedings of the ANS Topical Meeting on Fission Product Behavior and Source Term Research (Snowbird, Utah), EPRI NP-4113-SR, p. 38-1 (July 1985).

(5) Im, K. H., Ahluwalia, H. C., Chuang, C. F., "RAFT: A Computer Model for Formation and Transport of Fission Product Aerosols in LWR Primary Systems", <u>Aero. Science and Tech.</u>, 4:125-140 (1985).

(6) New York Power Authority, "Comments on NUREG-1150, Reactor Risk Reference Document", Attachments I, II, III, and IV to JPN-87-051 and JPN-87-045 (March 4, 1988).

(7) Gieseke, J. A., et al., "Radionuclide Release Under Specific LWR Accident Conditions, Volume II, BWR Mark I Design", BMI-2104, Vol. 2, Battelle Columbus Laboratories (July 1984).

(8) USNRC, Uncertainty Papers on Severe Accident Source Terms, NUREG-1265 (May 1987).

(9) CSNI, Report to CSNI on Source Term Assessment, Containment Atmosphere Control Systems, and Accident Consequences, CSNI Report 135 (April 1987).

(10) Kouts, H., Review of Research on Uncertainties in Estimates of Source Terms from Severe Accidents in Nuclear Power Plants, Brookhaven National Laboratory, NUREG/CR-4883 (April 1987).

(11) Collen, J., "Overview of Marviken Procedures", NAS/ENS Topical Meeting on Fission Product Behavior and Source Term Research, Snowbird, Utah (July 1984).

(12) Strom, L., Makynen, J. M., and Piispanen, W., "Aerosol Instrumentation for the Marviken Aerosol Transport Tests", J. Aerosol Sci, 15, 347 (1984).

(13) Bloom, G. R., et al., "Status of the LWR Aerosol Containment Experiments (LACE)", IAEA-SM-281, 39 (1985).

(14) Gieseke, J. A., "Pressure Loss and Dust Collection in Venturi Scrubbers", Ph.D. Thesis (1963).

(15) Wicks, M., and Dukler, A. E., "Entrainment and Pressure Drop in Concurrent Gas-Liquid Flow. I. Air-Water in Horizontal Flow", American Institute of Chemical Engineers Journal, 6, p. 463 (1960).

(16) Kataoka, I., Ishii, M., and Mishima, K., "Generation and Size Distribution of Droplet in Annular Two-Phase Flow", Jour of Fluids Engineering, 105 (1), p. 230-238 (March 1983).

(17) Morewitz, H. A., et al., "Resuspension/Re-Entrainment of Aerosols in LWRs Following a Meltdown Accident", in CSNI Report 136 (April 1987).

(18) Brockman, J. E., "Range of Possible Resuspension and Re-Entrainment of Fission Products During Containment Depressurization", SAND 84-0410, 2, Appendix G (1985).

(19) Fauske and Associates, Inc., "Resuspension of Deposited Aerosols Following Primary System or Containment Failure", Atomic Ind. Forum, Inc., IDCOR Program Technical Report 11.6 (1984).

(20) Wright, A. L., and Pattison, W. L., "Results From Simulated Upper-Plenum Aerosol Transport and Aerosol Resuspension Experiments", Proceedings of the CSNI Specialist Meeting on Nuclear Aerosols in Reactor Safety, p. 353-365 (February 1985).

(21) Braaten, D. A., Shaw, R. H., and Paw U, K. T., "Particle Detachment in Turbulent Boundary Layers", Aerosols: Formation and Reactivity, p. 370-378 (1986).

(22) SRI International, "Aerosol Retention in Nuclear Reactor Components", EPRI NP-3705 (September 1984).

An Overview of Containment Fission Product Behavior and Transport Codes

W. O. SCHIKARSKI and W. SCHÖCK
Kernforschungszentrum Karlsruhe GmbH
Laboratorium für Aerosolphysik und Filtertechnik
Postfach 3640
D-7500 Karlsruhe 1 FRG

1. INTRODUCTION

In the severe accident analysis of light water reactor power plants (LWR) fission product behavior and transport and eventually release from the containment plays a decisive role in the assessment of the radiological source terms. Great efforts have been undertaken in recent years to develop and improve codes for calculating the behavior of aerosols released into the containment space during these accidents since fission products are not released alone rather become airborne in the containment together with other materials like cladding and structural materials etc. and form nuclear aerosols. Except for the noble gases, which are not subject of this paper, and for some iodine compounds all radioactive materials are released in the form of nuclear aerosols.

Although the emphasis in modelling fission product behavior and transport in containments lies on the aerosol codes the special behavior of iodine and its compounds should be considered separately because iodine can exist in a number of different chemical states which come together with a wide variation of chemical environments to which the fission product can be exposed during a severe accident.

This article gives an overview on the recent developments in modelling both nuclear aerosol behaviour and iodine behaviour in LWR severe accidents. Finally, unresolved issues are discussed.

2. AEROSOL BEHAVIOR MODELLING

2.1 Status of Code Development

Mechanistic codes modelling the behavior of nuclear aerosols were first developed under dry conditions for studies of aerosols in fast reactor accidents. The physical models in dry codes are generally agreed state-of-the-art and are described elsewhere [1,2,3]. The codes can be grouped into discrete codes, which divide the size range up into discrete size classes, and lognormal codes, which make the more restrictive assumption that the size distribution remains lognormal at all times.

Results of dry code calculations have been compared in [1] and [4]. The discrete codes agree very well with one another. The results of log-normal codes agree equally well with one another but always diverge significantly from those of the discrete codes. Dry aerosol codes were also used to predict experimental results of large scale tests [5]. The results showed the same systematic differences between log-normal and discrete codes, and the discrete codes were in much better agreement with the experimental results.

LWR containment aerosol codes can be viewed as consisting of the dry codes together with the addition of models specific to condensing steam conditions. In an LWR the additional mechanisms of diffusiophoresis due to condensation of steam on heat sinks, mostly walls, and condensation onto or evaporation from the aerosol particles themselves have to be taken into account. Beginning in 1975 with the development of NAUA, fast reactor codes have been extended to take account of condensing steam effects on aerosols: AEROSIM-M,AEROSOLS/B1, HAA-4, the MAEROS model in CONTAIN, NAUA Mod5, QUICK-M and REMOVAL-Mod0 are the codes which are used for LWR aerosol calculations today.

From the point of view of incorporation into codes, diffusiophoresis is a simple process to model, as it involves particle size independent removal rates. Two expressions have been proposed for calculating this rate [6], one based on the steam partial pressure gradient at the wall and the other based on the total condensation rate of steam onto the wall. These two expressions are physically equivalent, but the form using condensation rate makes better use of the available thermal-hydraulic data. Most codes now use the condensation rate method.

The major difference between the codes is in their modelling of condensation onto the particles. They have different solutions to the problem of where to make the boundary between the aerosol code and the thermal-hydraulics code from which it takes its input. The thermodynamic quantity which dominates condensation is the supersaturation of the atmosphere. Unfortunately, this is not calculated by any of the existing thermal-hydraulic codes. The best one can obtain from a thermal-hydraulic code is a bulk volume condensation rate which is calculated with the assumption of instantaneous equilibrium between vapor and liquid state of airborne water. This rate is taken as an off-line input by most codes and then translated to the actual quantities with which the codes work.

But steam condensation onto and evaporation from particles is a phenomenon with both thermal-hydraulic and aerosol aspects. Not only does the aerosol behavior depend on thermal-hydraulic conditions but the thermal-hydraulic calculation strictly depends on aerosol quantities. Condensation and evaporation are fast enough to make equilibrium a good approximation. However, because of the solute and curvature effects the position of the equilibrium depends on the nature of the aerosol. If e.g., the conditions allow evaporation then equilibrium can be reached only if there is liquid airborne water still available. This in turn depends on aerosol settling rates. Stand-alone thermal-hydraulic codes ignore the curvature and solute effects and put the condensed water into droplets of uniform settling rate. The CONTAIN code was the first fully coupled thermal-hydraulic and multi-component aerosol code. Both sorts of processes interact at each time step. However, at present the code does not include the effects of curvature and solutes on the condensation.

Once one models condensation onto particles, one has to be able to distinguish between the water and non-water mass in the calculation. It is also useful to be able to follow separately radiologically important nuclides. The question is then whether to allow the aerosol composition to be a function of particle size or to leave it as particle size independent. Some codes distinguish between water and non water mass as a function of size, some have size dependent modelling of solid components. While for water a size dependent modelling is necessary in every case, to correctly take into account the curvature effects, the need for size dependent multi-component modelling depends on the scenario. In many cases the high coagulation rates result in rapid mixing of the components and a size independent treatment of particle composition is good enough. It has even been shown that in some cases, where experimental evidence seemed to indicate a size dependent composition, the observed effects could be explained by size independent phenomena and the size independent calculation reproduced the experimental data very well [7].

Subject to experimental validation and with the exception of the modelling of condensation on the particles, the stand-alone aerosol codes are virtually in their final shape. There is still scope for adding on further facilities and improving the details of physics but it is unlikely that the overall trend of the predictions will be altered. On the other hand a scientifically correct modelling of steam condensation requires a coupled thermal-hydraulic and aerosol code, because no stand-alone code will be an exact solution to the condensation problem. But existing stand-alone approximations may be adequate for the purposes of most applications in reactor safety analysis.

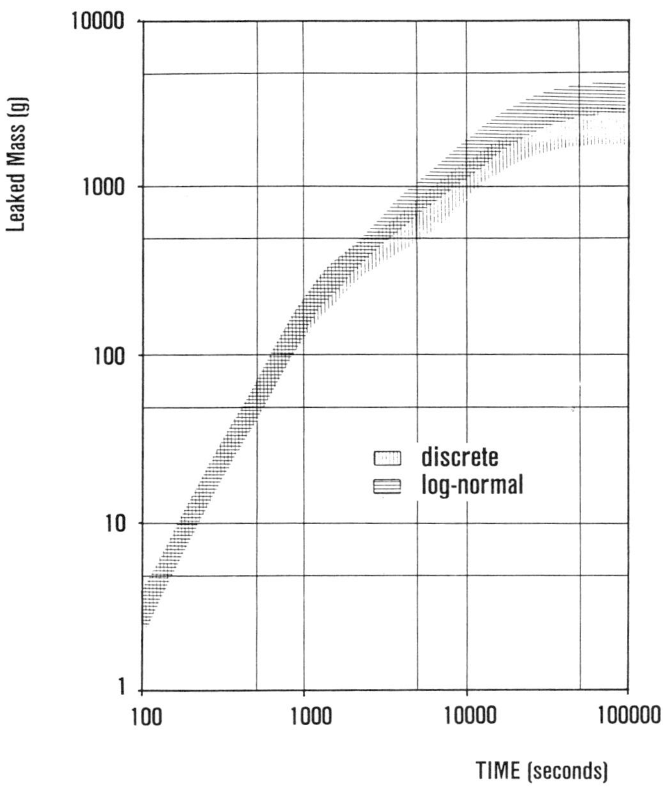

FIGURE 1: Results of a rigid benchmark code comparison

As an example of the present state of the art in aerosol modelling one result of a code comparison study is presented. The study was performed as a rigid benchmark exercise [8]. Fig.1 shows the accumulated leaked mass calculated by the codes for an AB-δ sequence in a Surry reactor. The figure shows bands within which the results of the calculations with the discrete and log-normal codes lie. Two items are worth to note. First, the lognormal codes systematically give higher predictions of the accumulated leaked mass, the source term. Second, within a code group the scatter in the predicted leaked masses is only a factor of two. This shows that all codes essentially use the same physics in their modelling and that the remaining uncertainties in the calculation are small. In fact the uncertainties in input data, which are not reflected in this rigid benchmark exercise, are believed to be larger than the uncertainties in aerosol modelling and computation.

2.2. Status of Code Validation

In order to keep the codes manageable rather than for physical or numeric reasons, most of them make, among others, the following simplifying assumptions:

- Condensation on particles, which is one of the dominant processes in LWR nuclear aerosols, requires thermodynamic input from a containment code. This input, especially condensation rates, is mostly transferred offline from the thermodynamics code to the aerosol code, which assumes that no feedback of aerosol behavior on thermodynamics takes place.

- Shape factors and particle composition are usually modelled as size independent, which implies that internal mixing processes due to coagulation are fast compared to removal time constants.

- Homogeneous spatial mixing of aerosol is assumed, at least within one control volume. The boundaries of control volumes in codes which have multi-compartment options should be chosen such that they coincide with physical walls of building compartments.

The validation of these assumptions goes beyond the capabilities of small and intermediate scale experiments and can only be done with large scale investigations.

The first large scale experimental program to investigate aerosol behavior related to LWRs was the Containment Systems Experiment (CSE) program [9]. At the time of the CSE experiments, however, no aerosol codes existed, and afterwards validation of aerosol codes using CSE data has not been attempted more than occasionally. Later, large scale experimental programs have been set up and carried out with the expressed aims to investigate details in aerosol behavior and to validate existing computer codes. To date three such programs have been completed, they are, in historical order:

NSPP, an aerosol behavior program carried out in a 38 m^3 single-volume vessel, aimed mainly at investigating the behavior of different aerosol species and mixtures thereof in condensing and non-condensing atmospheres [10]. NSPP established a broad experimental data base for code validation.

DEMONA, a large scale program simulating accident conditions in large PWR containments and investigating aerosol behavior in a 640 m^3 model containment with structured geometry [11]. DEMONA was specifically aimed at the validation of the NAUA code, other codes were included in code comparison efforts afterwards.

LACE, a large scale program investigating aerosol behavior in an 852 m^3 single-volume vessel during different scenarios, including the V-sequence and resuspension phenomena [12]. Two of the LACE experiments were aimed at containment aerosol behavior.

We cannot discuss more than a few selected items from the results of these programs in this paper, and have chosen those which relate directly to the above mentioned code assumptions.

The effect of condensation

In DEMONA the effect of condensation on aerosol removal was demonstrated by performing separate experiments under non-condensing and condensing conditions as well as performing experiments with successive non-condensing and condensing phases.

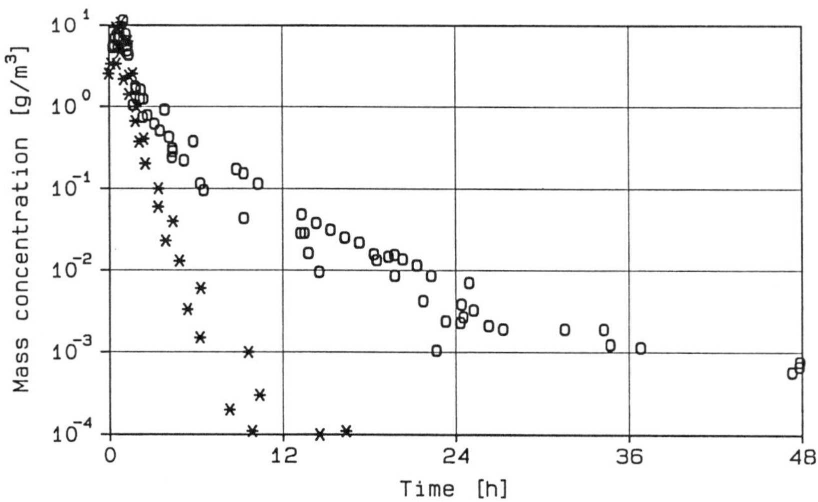

FIGURE 2: Comparison of measured time dependence of aerosol mass concentration, (*) condensing and (O) non-condensing atmosphere

In Fig.2, the measured aerosol concentration values are shown as a function of time for one experiment in non-condensing and one in condensing atmosphere. Very similar results have been obtained in all experiments with condensing atmospheric conditions. The mass concentration drops below the detection limit of the measurement (1 mg/m^3) after about 6 to 8 hours. This is in contrast to the dry experiment, without condensation, in which the detection limit is reached only after two days. These observations, which had also been made in NSPP, clearly demonstrate the influence of condensation processes. Enhanced settling and diffusiophoresis are the dominating processes of aerosol removal in condensing atmospheres. The other variations in experimental conditions are relatively ineffective.

Results of experiments in which dry and condensing phases were combined confirm this finding. In these experiments steam injection was interrupted one hour before until three hours after the aerosol injection phase. The results show the slow dry removal at the beginning as in the dry experiment and the fast removal as in the other experiments after the re-start of steam injection (cf Figure 4 below).

That these differences in the removal rate are due to condensation has been further proven by two independent measurements of liquid airborne water. In all experiments the aerosol photometers showed a large increase in their signals when condensation on particles began. Further, the droplet concentration was separately measured with droplet calorimeters, a special development for DEMONA which was also used in two of the LACE experiments. In DEMONA the calorimeters showed constant liquid airborne water concentrations of a few g/m^3 during the condensing phases of all experiments independent on the solid aerosol concentration. This shows that condensation rates are governed by the thermodynamic conditions and not by aerosol mechanics, as was expected for a system with insoluble aerosols. In the LACE experiments where soluble aerosol materials were also used, a dependence of liquid airborne water on the aerosol concentration was observed.

The effect of aerosol composition

In NSPP iron oxide, uranium oxide (U_3O_8) and concrete aerosols were used separately and in mixtures. In DEMONA tin oxide, iron oxide, silver and, accidentally in one case, magnesium oxide were used as aerosol materials separately and in different mixtures. In LACE the aerosols for the containment experiments consisted of manganese oxide and caesium hydroxide which were injected in overlapping time intervals. This variety of aerosols constitutes a good basis for examining the effect of particle composition on the overall aerosol behavior.

Looking at the time dependence of the total mass concentration data it is seen that the effect of aerosol material on the overall removal was small as long as the aerosol materials were insoluble. Only the removal rate of the aerosols containing iron oxide was slightly faster in DEMONA, and in NSPP the removal of concrete aerosols was slower.

The composition of the particles determined by analyzing impactor samples was found to be independent of the particle size in DEMONA. In agreement with the theory the explanation is that the mixing (coagulation) time constants were short compared to the other (depletion) time constants. This was confirmed in one of the experiments with different aerosols injected at different times.

In the LACE experiment LA4, which simulated aerosol behavior in the containment, the time dependent concentration of the fractions of the mixed manganese oxide and caesium hydroxide aerosol showed a diverging behavior at late times of the experiment, caesium hydroxide being depleted faster than manganese oxide. Although initial aerosol concentrations were in the range of several g/m^3, which leads to very high coagulation rates, the diverging behavior, on the first sight, gave rise to speculations on "independent behavior of different species" (the small MnO particles remaining airborne longer than the large CsOH droplets) or even "demixing of components".

This is, of course, not true because there is simply no inhibitor for coagulation. It has been shown by systematic code calculations [7] that this kind of behavior is also obtained with codes like NAUA that assume complete instantaneous and particle size independent mixing of components. In Fig.3 the result of one of the calculations is shown.

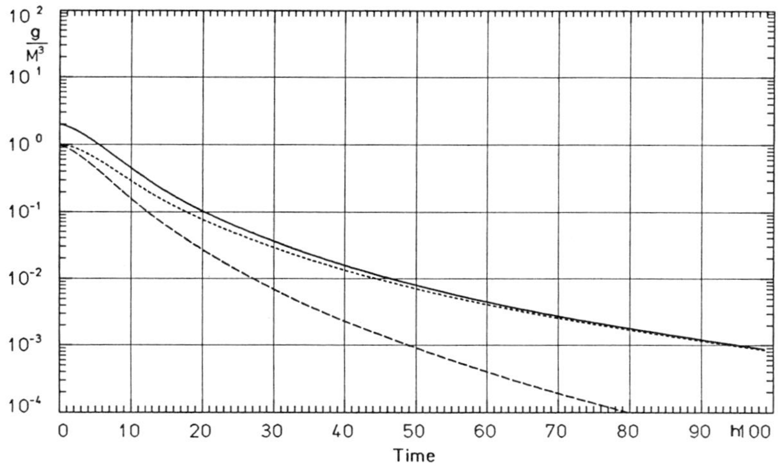

FIGURE 3: Time dependent behavior of a mixed aerosol. —— total concentration, ... and ---- fractional concentration of components

The curves represent the concentrations of the two components and the total concentration (solid line) of a mixed aerosol consisting of a fine fraction (dotted line) and a coarser fraction (broken line) of initially equal mass. Coagulation was calculated as non-selective among particles of any composition and the change in composition due to coagulation within size classes was updated every time step. In a size class all particles are equally composed. With these assumptions already, the fractions diverge, showing that the reason for the divergence is not a different behavior of individual particles but simply a matter of overall aerosol mechanics. It has to be emphasized additionally that the behavior of the initially fine fraction (dotted) is identical to that of a fine aerosol by itself. The addition of a coarser fraction does not change the behavior of the fine fraction. Instead the removal of the coarse fraction is enhanced as compared to the behavior of a coarse aerosol by itself.

<u>The effect of multicompartment structure</u>

The behavior of aerosols in structured enclosures and the related modelling is still under discussion. Having in mind that inhomogeneities do occur in certain instances of an accident or an experiment, we nevertheless believe that in many cases a single compartment treatment will result in not too large errors in the overall leaked mass, i.e.the source term, because of the limited duration and spatial extent of inhomogeneities. For a more detailed analysis multizone containment codes and multicompartment versions of aerosol codes are now available and have been used for comparison with the single-compartment codes.

In DEMONA a real multicompartment facility was used for studies of aerosol behavior. Although the major part of the experiments were conducted under nearly single-compartment conditions, spatial inhomogeneities have been observed, but they were restricted in time to the aerosol generation interval. As a final more complex test case in the DEMONA program an experiment with complex geometry was performed.

The thermodynamic behavior was very different from that of all previous experiments. Spatial differences in the gas composition developed, steam was enriched in the steam injection compartment and air enrichment developed in the annulus of the containment.

Unlike the observed thermodynamics, the measured overall aerosol behavior was very similar to that observed in the other experiments. In Fig. 4 the aerosol mass concentration is shown for the mulit-compartment experiment and compared to a single compartment experiment. Both experiments start with a non-condensing phase, condensation starts after four hours. Except for the higher initial concentration in the multi-compartment experiment, no significant differences in overall behavior were observed. Spatial inhomogeneities were seen only during aerosol injection and the subsequent non-condensing phase, which is reflected in the scatter of the data of the multi-compartment experiment. As soon as steam injection was resumed, however, spatial homogenizing occurred very quickly.

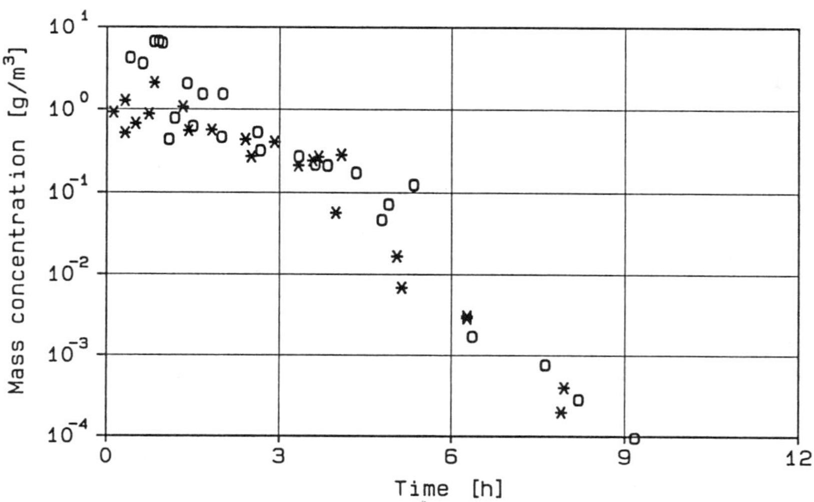

FIGURE 4: Comparison of measured time dependence of aerosol mass concentration, (*) single-compartment experiment and (O) multi-compartment experiment

This finding could be completely explained with the results of a multi-compartment thermal-hydraulic calculation, which showed that, even in the multi-compartment geometry of the experiment, two strong coupled convection loops developed and linked all compartments together. Only during aerosol injection one loop was interrupted. With the end of aerosol injection this loop started working again, homogenizing aerosol concentrations in all compartments.

The experiment showed that aerosol behavior calculations with single volume codes may be applied to certain multicompartment situations. More importantly, the experiment showed that this finding can be explained in all details by more sophisticated thermal-hydraulic investigations. Generally it appears that a refined treatment, beyond the single-volume approach, is more necessary in containment thermodynamics than in aerosol behavior.

2.3 Overall Comparison with Experiments

The first major exercise in which codes were compared to an experiment with condensing conditions was organized by CEC [13]. Eleven participants in this exercise used a variety of six different codes, the experiment selected was the reference experiment of DEMONA.

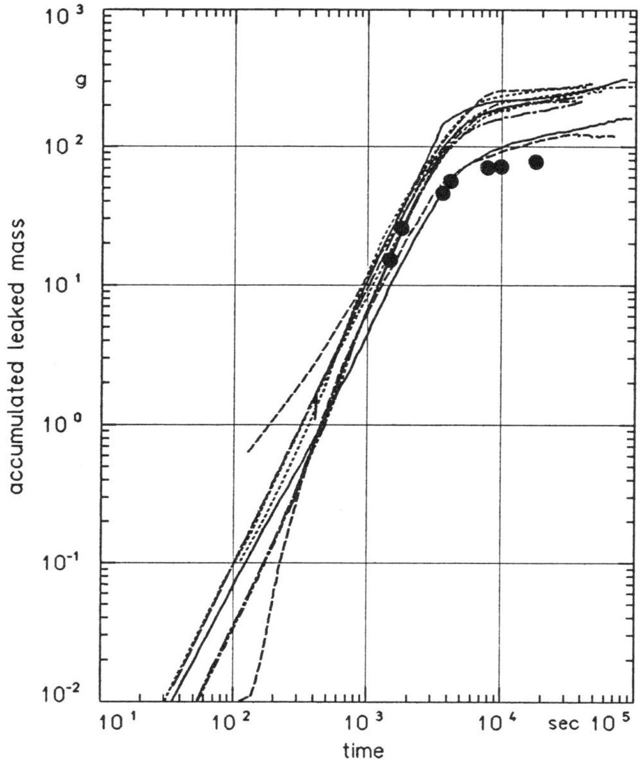

FIGURE 5: Comparison of code calculations with experimental results

In Fig.5 the calculated accumulated leked masses are compared to one another and to the experimental values. The figure shows that all calculations give very similar results, with the exception of two codes which stand out as giving a better fit to the experimental results. These two codes are not of the single-compartment discrete type, and, since convincing reasons for the better match to the experimental values have not yet been found, the results have to be regarded as more or less incidental. The agreement among the single-compartment discrete calculations is within a factor of two, the agreement with the experiment is within a factor of three.

It is interesting to note that this factor of three also appears as an upper limit in the post test comparison of calculated and experimental accumulated leaked masses for all the other DEMONA experiments [14].

The CEC study also recommended that the first priority of future development should be on the side of thermal-hydraulic modelling and the coupling to aerosol codes.

3. IODINE BEHAVIOR MODELLING

The release of iodine during a severe core melt accident is of particular significance, since for some accident sequences this isotope could be dominant for the overall radiological source term. The probable chemical states of iodine in irradiated UO_2 fuel rods are in the forms of CsI, I and I_2, but molecular iodine and iodine atoms should be present predominantly [15]. The release from fuel and the consequent behavior of the iodine species in the primary system and in the containment depends not only on the primary iodine compound present in the fuel matrix, but also on temperature conditions during and after release, the existence of other airborne materials (aerosols, gases, water vapour) and on pressure conditions. In several fission-product release experiments CsI was observed as main iodine compound also indicating that high concentrations of CsI are stable in steam up to the temperature of at least 1300 K. Also thermodynamic calculations indicate the stability of CsI. Therefore, CsI is considered to be the dominant iodine compound formed under accident conditions [16].

However, the existence of iodine and hydrogen iodide can not be ruled out under accident conditions, because these species could be formed by several possible reactions of CsI or by recombinations of iodine atoms from the hot regions of the core. Hydrogen iodide and iodine are both very reactive and interact readily with a wide variety of materials. This leads depending on the accidents sequences to a variety of iodine containing compounds formed in the primary system after release from the fuel.

Modelling of the iodine behavior in the first phase of a core melt accident - i.e. during and after release from fuel into the primary system - appears to be a complex and difficult endeavor which has not yet led to a sufficiently comprehensive and consistent mathematical description, at least for the part of iodine behavior in the primary coolant system. Since the retention function of the primary coolant system for fission-products is generally limited, it was reasonable to concentrate the efforts mainly on the iodine behavior in the containment and neglect - as a first approximation - the physics and the chemistry in the primary coolant system. This is the base for the code IMPAIR (<u>I</u>odine <u>M</u>atter,

Partition and Iodine Release) [17], which is now available as version IMPAIR 2/M [18].

The IMPAIR-2 code was written to analyse the iodine behavior in a LWR containment space with water and/or gas phases as function of time. The code serves primarily for the analysis of conditions as they may occur after a core melt accident in the atmosphere of the containment space and in the sump of the containment where water, water soluble substances and deposited or plated out aerosol material are collected. In its present version the IMPAIR-2 code contains about 20 differential equations describing the thermodynamic and kinetic interactions including phase changes and droplet entrainment. For the gas phase chemistry the iodine oxidation, the formation of organic iodides and the reversible deposition on surfaces are taken into account. The iodine chemistry in the water phase considers the hydrolysis and the oxidation. Beside of the iodine oxides the treatment and kinetics of silver iodide and methyl iodide play an important role. It should be mentioned that the code usually is operated in close connection with an aerosol code as described in the previous section (section 2), since a large part of the iodine species exist under accident conditions in the gas space of the containment in the form of aerosols (mainly CsI and AgI). The latest version of the code considers also the transport pattern between different compartments in a containment system and the source or depletion rates of various iodine species in these compartments.

The two following figures show an example of the concentrations of the iodine species in the gas space (Fig. 6) and the water space (Fig. 7), of a containment system, as taken from [18].

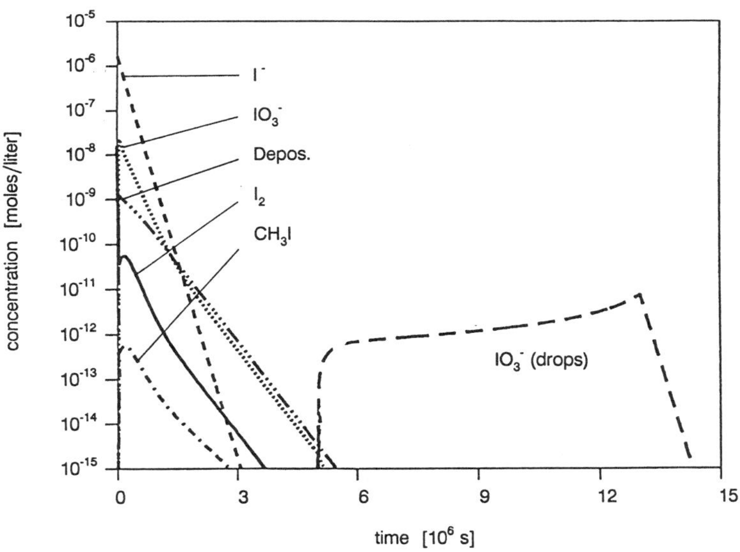

FIGURE 6: Calculated concentration of iodine species in the airborne state

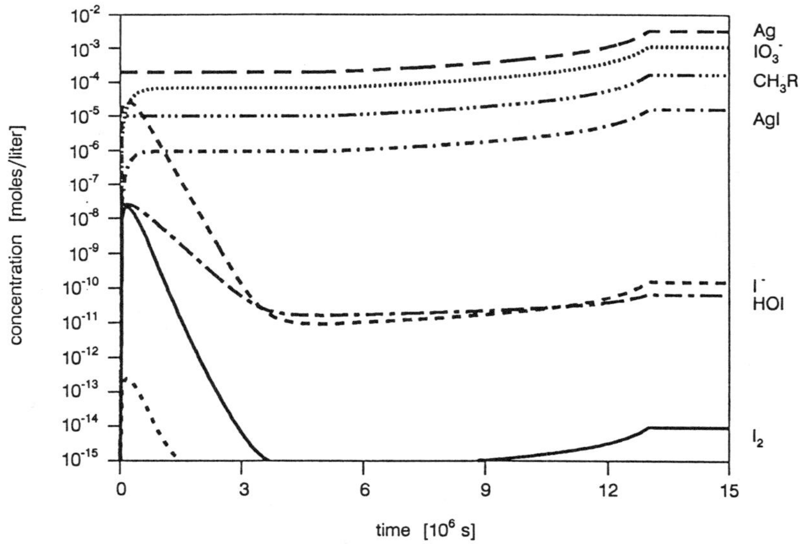

FIGURE 7: Calculated concentration of iodine species in the acqueous state

The IMPAIR code is undoubtedly an improvement for the iodine behavior modelling in LWR containment systems after severe accidents comparing earlier approaches by estimation of release fractions from core and from containment only and by neglecting the physical and chemical interactions of the different iodine species. However, a number of input data on reaction constants, influence of pH values, partition coefficients etc. are not yet verified under the special thermodynamic conditions considered. Therefore, the operation of IMPAIR-2 requires still an experienced scientist familiar with both the accident conditions and the iodine chemistry.

4. FUTURE NEEDS

Nuclear aerosol behavior modelling has always been an applied discipline of aerosol physics. The development is determined mainly by the accuracy needed in the end product, namely the radiological source term, for its use in plant analysis. There is little point in adding details to the aerosol code beyond the ability of other codes and calculations to supply accurate input, or where the eventual impact on plant safety estimates is negligible.

The long term development of aerosol codes and their extensive and ongoing comparison with experiment have led to the general consensus that levels of airborne activity in containment can be accurately calculated. If one asks for the quantitative uncertainty in calculations with discrete codes, it is believed that the error in the leaked mass from the containment during a core melt accident is of the order of or less than a factor of five.

It is this factor of five which should be used as a measure for future research needs. Phenomena which have the potential of altering source terms by this amount should be investigated further. Such phenomena may be either well known but not yet properly modelled ones or others which have not yet been taken into account in present source term assessments for different reasons.

In the first place again, coupling to thermal-hydraulics has to be mentioned. Comparison of experiments under condensing conditions with code calculations show good agreement in some cases and less good agreement in other cases. It can be shown that the reasons for the differences are insufficient thermodynamic input to the aerosol behavior codes. Even the simple question whether and when, in the real containment atmosphere, condensation on particles will occur is still subject to debate. Since the problem of thermal-hydraulic coupling is not one of lacking physical understanding but rather one of laborious and expensive work on the development of a new generation of thermal-hydraulic codes, there is still a possibility that an improvement of the situation could be achieved.

The influence of chemical reactions on aerosol behavior has been neglected in nuclear aerosol modelling so far, since the majority of the nuclear aerosols can be considered as inert under severe accident conditions. There might, however, exist some chemical reactions, e.g. in the iodine-water-system in the presence of other aerosols, giving rise to problems of the behavior of these materials under severe accident conditions. This means that all possible chemical reactions should be investigated and analyzed which can introduce deviations from the fission product behavior as modelled in the codes described.

Finally long term sources in the containment have to be investigated and modelled. In scenarios with sump water ingression resuspension of fission products from the boiling sump has been identified as a potential contributor to the source term [19]. Resuspension from the sump is a very weak but long lasting source of highly radioactive particles. Work to quantify and to model this source is underway [20]. Another long term mechanism for producing aerosols in the containment is the melt concrete interaction [21]. This source is mainly non-radioactive but accumulates to a significant amount of material [22] which may constitute an impact for vent filter systems.

5. CONCLUSIONS

Containment aerosol behavior codes have been much improved over the last decade. The standards of modelling have reached an equally high level for all the stand-alone versions of discrete codes which are in use today. The results of these codes agree very well with one another, i.e. the differences in calculated accumulated leaked mass are less than a factor of two. Numerical schemes which use incisive simplifications such as the preservation of a log-normal size distribution have been recognized as being of only limited use except for special cases as e.g. low aerosol concentration. There is little reason for the further use of such codes.

Aerosol codes have been extensively compared to experiments. Recent large scale experimental programs provided an invaluable data base for code validation. Already performed comparisons of codes against these experiments gave very promising results, the differences between calculated and experimental accumulated leaked mass was less than a factor of three. The conclusion was drawn that for source term calculations the error will be less than a factor of five for sequences which can be handled by the codes.

The existing stand-alone codes are in their final shape, further add-on improvements will not change the overall picture significantly. The real needs of containment aerosol modelling are outside the scope of the existing codes. The coupling of aerosol models to a new generation of thermal-hydraulic codes has the potential of leading to a real improvement. The demand for such work should be continuously supported.

Work on special phenomena as chemical reactions, mainly in the iodine-caesium-water-system for the airborne and aqueous states, is underway to improve and validate iodine behavior codes. Also long term aerosol sources are under investigation to improve the input data base to containment aerosol behavior codes.

REFERENCES

1. OECD ed., Nuclear Aerosols in Reactor Safety, A state of the art report by a group of experts of the NEA/CSNI, CSNI/SOAR No.1, Paris (1979)

2. OECD ed., Nuclear Aerosols in Reactor Safety, Supplementary Report to reference [1], Paris 1985

3. F.Beonio-Brocchieri, H.Bunz, W.Schöck, I.H.Dunbar, J.Gauvin, S.Miyahara, Y.Himeno, K.Soda, N.Yamano, Nuclear aerosol codes, Nuclear Technology Vol. 81 (1988), pp 193-204

4. I.H.Dunbar, J.Fermandjian, Comparison of Sodium Aerosol Codes, EUR 9172 EN (1984)

5. R.K.Hilliard, L.D.Muhlestein, The ABCOVE Program: Preliminary Results of Tests AB5 and AB6, CSNI Specialists Meeting on Nuclear Aerosols in Reactor Safety, KfK 3800, CSNI 95 (1985)

6. J.F.van de Vate, Investigations into the dynamics of aerosols in enclosures as used for air pollution studies, ECN86 (July 1980)

7. W.Schöck, Influence of aerosol process coefficients on the coagulation function and on long term aerosol behavior, Coagulation Workshop, Kernforschungszentrum Karlsruhe, 16.-18. March, 1988

8. H.A.Morewitz, F.J.Rahn, R.Sher, Results of the GREST code comparison exercise, CSNI 116 (1986)

9. A.K.Postma, B.M.Johnson, Containment Systems Experiment, Final Program Summary, BNWL-1592 (1971)

10. R.E.Adams, M.L.Tobias, J.C.Petrykowski, Aerosol behavior in a steam-air environment, CSNI Specialists Meeting on Nuclear Aerosols in Reactor Safety, KfK 3800, CSNI 95 (1985)

11. W.Schöck et al., DEMONA Jahresberichte 1984, 1985, 1986, KfK 3942 (1985), KfK 4182 (1986), KfK 4285 (1987)

12. F.J.Rahn, Summary LWR aerosol containment experiments (LACE) program, LACE TR-012 (1987)

13. W.Schöck, Post-test calculations of aerosol behavior in DEMONA experiment B3 with various computer codes, EUR 11374 EN (1988)

14. W.Schöck, H.Bunz, W.Schikarski, J.P.Hosemann. D.Haschke, A.Fromentin, R.Taubenberger, T.F.Kanzleiter, Th.Schröder, M.Fischer. H.Ruhmann, M.Peehs, G.Weber, Results of the DEMONA Program, Transactions of the 4th International ENS/ANS Conference, Geneve, 1.-6. 6. 1986, ENS Berne (1986) Vol.3, pp. 219-224

15. B.R.Bowsher, Fission-product chemistry and aerosol behavior in the primary circuit of a pressurized water reactor under severe accident conditions, Progress in Nuclear Energy Vol.20 (1987), pp 199-233

16. A.P.Malinauskas, J.T.Bell, The chemistry of fission-product iodine under nuclear reactor accident condotions, Nuclear Safety 28 (1987), pp 505-514

17. K.Hassmann, J.P.Hosemann, M.Peehs, Spaltproduktfreisetzung beim Kernschmelzen, Verlag TÜV Rheinland (1987)

18. M.Furrer, R.C.Cripps, E.Frick, Jod-Verhaltenscode IMPAIR-2, Draft PSI report, Feb. 1988

19. R.Borkowski, H.Bunz, W.Schöck, Resuspension of Fission Products during Severe Accidents in LWRs, KfK 3987, EUR 10391 EN (May 1986)

20. W.Schöck, M.Wagner-Ambs, Resuspension of fission product aerosols from the boiling sump, this conference

21. H.Alsmeyer, BETA experiments in verification of the WECHSL code - Experimental results in the melt-concrete interaction, Nuclear Engineering and Design 103 (1987), pp 115-125

22. W.Schöck, H.Bunz, Assessment of the contribution of long term sources to the overall aerosol source term in LWR core melt accidents, Severe Accidents in Nuclear Power Plants, Proceedings of an international symposium held in Sorrento, 21.-25. March, 1988 IAEA-SM-296, IAEA Vienna 1988, Vol 2, pp 37-45

2. FUEL BEHAVIOR AND FISSION PRODUCT RELEASE

LECTURE

Chemical Processes in Fission Product Release and Transport

D. A. POWERS
Sandia National Laboratories
P.O. Box 5800
Albuquerque, New Mexico 87185, USA

1. INTRODUCTION

Severe nuclear reactor accidents dominate the risk of the use of nuclear power. Nothing so distinguishes severe reactor accidents as the extensive release of radioactive materials from the reactor fuel and the potential escape of these radioactive materials into the environment. The release of radionuclides begins when the fuel is overheated and the cladding on the fuel ruptures. Release of radioactivity can continue as the fuel further degrades, melts, slumps from the core region, collapses into the lower plenum of the reactor coolant system and, perhaps, is expelled into the reactor containment. But, the mere release of radionuclides from fuel does not pose a direct public threat. Released radionuclides must migrate from the point of release to a point of escape from the plant before there can be public consequences.

The release of radionuclides, especially while fuel degrades within the reactor coolant system, has been the focus of much attention in severe reactor accident analyses. It is during this stage of a hypothesized accident that there can be extensive releases of the more volatile and more radiologically consequential radionuclides such as cesium, iodine and perhaps tellurium. The earliest attempts to estimate radionuclide release during an accident sought simply to bound the radiological consequences of severe accidents [1]. Nearly complete releases of noble gases and iodine, which was considered to be gaseous I_2, were assumed. Much more limited releases of other radionuclides as aerosols were hypothesized. These assumed releases were taken to be generically applicable and were not sensitive to the peculiar features of particular accidents at specific nuclear plants.

More mechanistic treatments of radionuclide release began with the Reactor Safety Study [2]. In this work, seven classes of radionuclides were defined: (1) noble gases such as krypton and xenon, (2) halogens such as bromine and iodine which, again, were assumed to be gaseous, (3) alkali metals such as rubidium and cesium, (4) chalcogens such as tellurium, (5) alkaline earths such as strontium and barium, (6) refractory metals such as molybdenum and

This work was supported by the U.S. Nuclear Regulatory Commission and performed at Sandia National Laboratories which is operated for the U.S. Department of Energy under Contract Number DE-AC04-76DP00789.

ruthenium, and (7) rare earths such as lanthanum and cerium. Each of these radionuclides poses its own radiological threat as shown by the relative comparisons in Table 1. Each was treated in the Reactor Safety Study as having its own release characteristics during three release regimes of in-vessel core degradation: (1) release when the cladding on fuel ruptures, (2) release as fuel melts, and (3) release when fuel interacts with water. Later, a fourth regime, diffusional release from overheated but unmelted fuel, was added [3]. Release estimates developed for the Reactor Safety Study were of a generic nature exhibiting dependencies only on the fuel temperature and whether or not fuel interacted with water. A bounding analysis of radionuclide behavior was done by neglecting any retention of radionuclides during transport through the reactor coolant system. Attenuation of the release by processes in the reactor containment was considered. To the extent containment integrity was maintained during fuel degradation, the contributions to accident risks by radionuclide releases during core degradation were reduced.

The drive to obtain more realistic estimates of radionuclide releases accelerated following the severe accident at Three Mile Island. On the one hand, this acceleration was driven by what appeared to be excessive conservatism in releases estimated with the methods developed for the Reactor Safety Study [5]. On the other hand, the acceleration was driven by the growing use of probabilistic risk assessment methods and the need for radionuclide release estimates that were sensitive to peculiarities of the plants and the accidents being analyzed.

TABLE 1. Relative radiological importance of individual elements on an equal release fraction basis [4]

Element	Typical inventory (moles per mole U)	Normalized lung dose*	Normalized total latent cancers**
Kr	$1.2x10^{-3}$	0.04	0.07
Xe	$9.6x10^{-3}$	0.005	$1x10^{-4}$
I	$4.1x10^{-4}$	1.00	0.1
Rb	$1.1x10^{-3}$	$2x10^{-4}$	$1x10^{-5}$
Cs	$4.9x10^{-3}$	0.09	1.0
Te	$8.2x10^{-4}$	0.6	0.004
Sr	$2.7x10^{-3}$	0.9	0.7
Ba	$2.5x10^{-3}$	0.6	0.2
Mo	$8.1x10^{-3}$	0.7	0.06
Ru	$5.0x10^{-3}$	3.0	1.0
La	$2.2x10^{-3}$	1.6	0.08
Ce	$4.8x10^{-3}$	8.0	2.0

* Normalized to iodine.
** Normalized to cesium.

There has been somewhat parallel development of improved methods for estimating radionuclide behavior under accident conditions by the U. S. Nuclear Regulatory Commission [6,7] and by the nuclear industry [8,9]. The most critical addition to the technology for estimating radionuclide behavior within the reactor coolant system has been the explicit consideration of radionuclide retention in the system following release from the fuel. Examples of the effectiveness of natural retention processes in the reactor coolant system calculated with the Source Term Code Package [6] are shown in Table 2. These examples show that for many accidents large fractions of the radionuclides released from the fuel may never escape the reactor coolant system or contribute to the public threat of severe reactor accidents. The retention of radionuclides in the reactor coolant system does, however, depend on the peculiarities of accidents and plants. Calculation of the retention requires detailed knowledge of both the rates and the timing of radionuclide releases in addition to the overall extents of radionuclide releases. This has mandated an improved technology for estimating radionuclide release as well as an expanded technology for estimating radionuclide behavior after release.

TABLE 2. Radionuclide release and in-vessel retention predicted for various plants and accidents with the Source Term Code Package [16,17]

	Fraction of the initial core inventory of					
Plant *	Iodine		Cesium		Tellurium	
(accident)	released	retained	released	retained	released	retained
SURRY						
(AG)	1.00	0.13	1.00	0.13	0.86	0.15
(TMLB')	1.00	0.76	1.00	0.80	--	0.69
(S_3B)	1.00	0.72	0.99	0.76	0.30	0.23
SEQUOYAH						
(S_3B)	0.99	0.46	0.99	0.46	0.86	0.77
(TB)	0.99	0.66	0.97	0.73	0.84	0.77
(S_3HF)	0.97	0.67	0.97	0.73	0.84	0.77
(TMLU -SGTR)	0.98	0.64	0.94	0.61	0.33	0.10
PEACH BOTTOM						
(TBUX)	0.93	0.65	0.94	0.65	0.32	0.31
(TC1)	0.92	0.15	0.91	0.17	0.30	0.25
(TC2)	0.96	0.04	0.96	0.17	0.69	0.49
GRAND GULF						
(TC)	0.99	0.46	1.00	0.44	0.41	0.27
(TB)	0.98	0.29	0.98	0.45	0.38	0.32
(TBS/TBR)	0.97	0.12	0.97	0.13	0.39	0.25

* Accident designators are defined in references 2, 16, 17.

Prediction of extensive retention of radionuclides in the reactor coolant system would appear to further mute the importance of in-vessel radionuclide releases to the estimation of risk. Not long after predictions of extensive retention appeared, however, it was recognized that retention of especially the volatile radionuclides might not be permanent [8,10-12]. Retained radionuclides would continue to decay and the radioactive heat could raise temperatures of the radionuclides deposited in the reactor coolant system to the point that they revaporized. Indeed, for some accidents and plants, it was predicted that cesium, iodine and perhaps tellurium would be vaporized extensively from the reactor coolant systems and might again contribute to the public threat of severe accidents. Worse, the revaporization was predicted to occur late in accidents when it was more likely that the reactor containments would have ruptured and the revaporized radionuclides could escape the nuclear plants and enter the environment. Prediction of revaporization has revitalized interest in in-vessel release and transport of radionuclides because these processes directly affect the availability of material to revaporize and the rate at which revaporization can occur.

The technology of estimating radionuclide behavior within the reactor coolant system has evolved to consider three issues:

- release of radionuclides from fuel,

- transport and deposition of radionuclides following release, and

- revaporization of deposited radionuclides.

Analyses of these issues require that the rates and not just the overall extents of processes be understood. Further, to meet the objectives of risk analyses, the rate processes must be understood in terms that are sensitive to the differentiating features of both reactor designs and particular accidents.

Though the methods for estimating in-vessel radionuclide behavior have improved since the time of the Reactor Safety Study, significant uncertainties must still be ascribed to predictions obtained with these methods. An attempt to quantitatively characterize these uncertainties is underway [13]. Uncertainty estimates for radionuclide release [14] and revaporization [15] prepared for this study are shown in Figures 1 and 2, respectively. These uncertainty estimates are cast as cumulative probabilities that values of the release or revaporization fractions will actually be achieved in an accident. The probability distributions show that it is likely very large fractions of the volatile radionuclides such as cesium and iodine will be released or revaporized in an accident. It is also probable that release or revaporization of more refractory elements such as barium will be low. These indications are consistent with predictions obtained with the currently available methods for estimating in-vessel radionuclide behavior. On the other hand, the distributions also show that there are significant probabilities that predictions obtained with available methods are wrong. Phenomena can arise which may inhibit release and revaporization of the volatile radionuclides or enhance release and revaporization of the more refractory radionuclides.

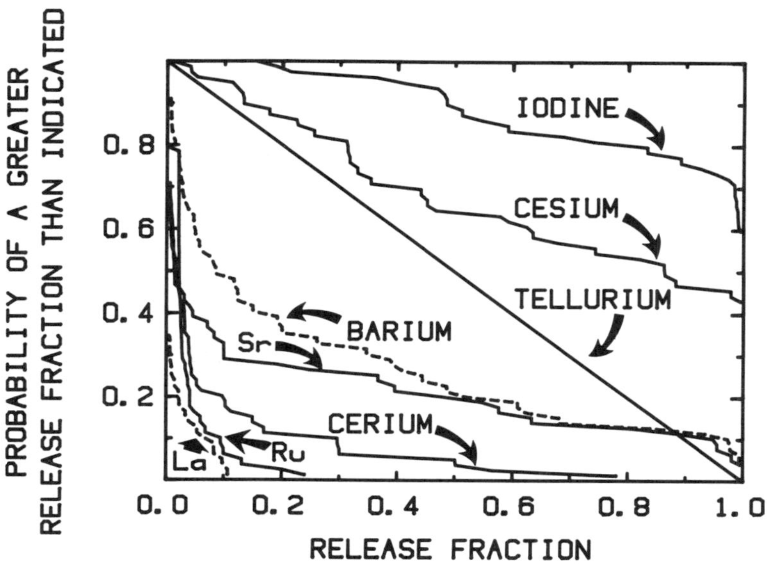

FIGURE 1. Uncertainty distributions for the releases of radionuclides during core degradation.

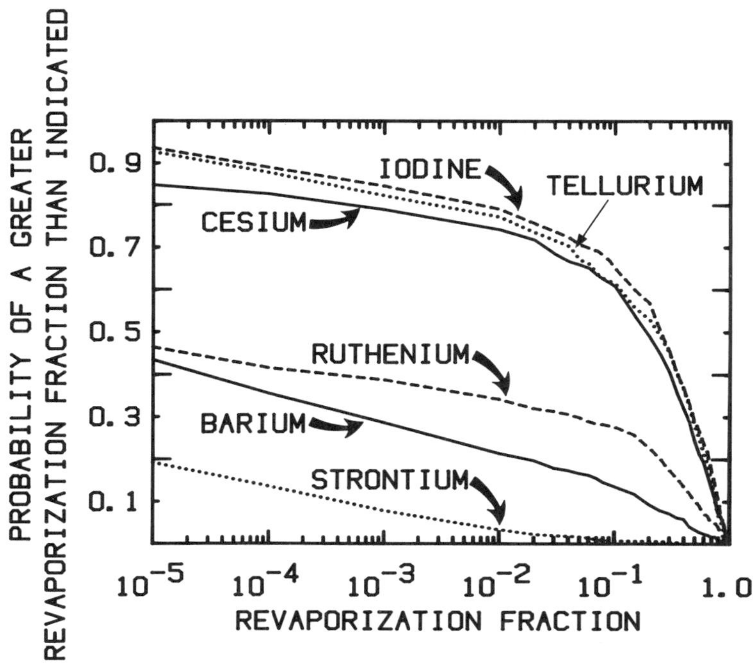

FIGURE 2. Uncertainty distributions for the revaporization fractions of deposited radionuclides. Revaporization fractions for La and Ce were always less than 10^{-7}.

In this paper, current trends in the continuing development of the technology for estimating the in-vessel release, transport and revaporization of radionuclides under accident conditions will be described. Some of the more important sources of uncertainty and possible approaches to the resolution of these uncertainties will be discussed.

2. RELEASE OF RADIONUCLIDE IN-VESSEL

Release of radionuclides during a reactor accident begins with the so-called "gap release" of vapors and some small amount of particulate material [18] when the fuel cladding ruptures. This initial release is miniscule in comparison to releases that take place as an accident progresses and fuel heats perhaps to melting. The continuing release is an example of classic vaporization of trace species from a host material. A formal expression for the rate of vaporization is (terms in this and other equations are defined in Section 6, NOMENCLATURE, of this paper.):

$$\frac{dN(i)}{dt} = (-1/RT) \sum_{j=1}^{M(i)} Ak(i,j) \left[\frac{P_{eq}^{o}(i,j)\gamma(i)N(i)}{\rho_m V} - P(i,j) \right] \quad (1)$$

Vaporization is driven by the thermochemical disequilibrium between the reactor fuel and the ambient atmosphere as reflected by the terms $P_{eq}^{o}(i,j)\gamma(i)N(i)/\rho_m V - P(i,j)$. Net vaporization ceases once equilibrium concentrations of vapor species are established in the atmosphere and these driving force terms become zero.

Calculation of the driving force terms entails two difficulties. Selection of appropriate chemical forms for the radionuclides in the fuel is the first of these difficulties. In truth, any chemical form could be selected for a particular radionuclide, provided an appropriate activity coefficient, $\gamma(i)$, could be defined. Only now are data being obtained that permit definition of the activity coefficients (see C. A. Alexander and J. S. Ogden in these Proceedings). Usually, simple, intuitive, species selections such as BaO for barium are made. More complicated schemes have been developed for some species. For instance, Tam et al. [19] determined the condensed phase speciation of cesium and iodine from the equilibria:

$$Cs(g) + I(g) \rightarrow CsI(g) \rightarrow CsI(l) \quad (2)$$

$$2Cs(g) + UO_2 + O_2 \rightarrow Cs_2UO_4(c) \quad (3)$$

$$2Cs(g) + Mo + 2O_2 \rightarrow Cs_2MoO_4(c) \quad (4)$$

The speciation of the chemical form of tellurium in fuel is an unresolved issue. Tellurium is quite volatile and might be expected to have a facility for release from fuel similar to that of cesium. Yet, tellurium is released with difficulty in experiments. Lorenz et al. [34] have hypothesized that tellurium is released from fuel, but rapidly reacts and binds with unoxidized Zircaloy cladding rather than entering the bulk gas flow. Alexander and Ogden [35] have found, however, that the vapor pressure of SnTe over tellurium-contaminated Zircaloy is high enough that release should be observed in the experiments. An alternate possibility is that tellurium reacts and binds to Mo-Pd-Ru metal inclusions in the fuel [24].

The second difficulty in determining the driving force for vaporization is the identification of vapor species that contribute to the release process. At the high temperatures of interest in reactor accident analyses, vapor phase speciation can be complex and not obvious [20]. Consider, as an example, the release of barium. Vaporization of barium can involve at least the vapor species Ba, Ba_2, BaOH, Ba_2O_2, $Ba(OH)_2$, BaH and Ba_2O in the reactive environment of the reactor coolant system under accident conditions:

$$BaO(c) \rightleftharpoons BaO(g) \quad (5)$$

$$2BaO(c) \rightleftharpoons Ba_2O_2(g) \quad (6)$$

$$2BaO(c) + H_2 \rightleftharpoons Ba_2O(g) + H_2O \quad (7)$$

$$BaO(c) + H_2 \rightleftharpoons Ba(g) + H_2O \quad (8)$$

$$2BaO(c) + 2H_2 \rightleftharpoons Ba_2(g) + 2H_2O \quad (9)$$

$$BaO(c) + 1.5H_2 \rightleftharpoons BaH(g) + H_2O \quad (10)$$

$$BaO(c) + 0.5H_2 \rightleftharpoons BaOH(g) \quad (11)$$

$$BaO(c) + H_2O \rightleftharpoons Ba(OH)_2(g) \quad (12)$$

The vapor pressure of barium is, then, not just a function of temperature. It also depends on the absolute and relative partial pressures of steam and hydrogen in the ambient atmosphere. As these additional variables range over values hypothesized to arise in reactor accidents, the rate of barium release can vary as shown in Figure 3. This dependence of release rates on gas composition and pressure is quite important for releases of more refractory radionuclides and is often absent in models used for accident analyses.

The approach of a vaporizing system to equilibrium is controlled by parallel and serial processes characterized by the effective rate coefficients, k(i,j). Many rate-controlling processes may be hypothesized. Most of these processes can be categorized in terms of the three serially-connected steps:

- condensed-phase mass transport of radionuclides to a free surface

- transport of radionuclides through a porous structure

- gas-phase mass transport of radionuclides from the fuel into the bulk gas flow.

Serially-connected, potentially rate-controlling processes alter the temperature dependences of release rates from the classic Arrhenius form as shown for a hypothetical example in Figure 4. These

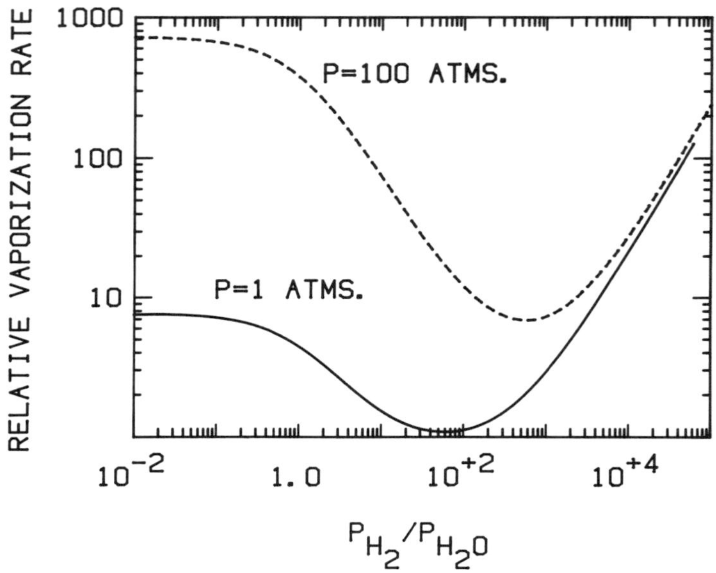

FIGURE 3. Relative vaporization rate of barium at 2200 K as a function of gas phase composition and pressure. Note that 1 atmosphere is 101325 N/m^2.

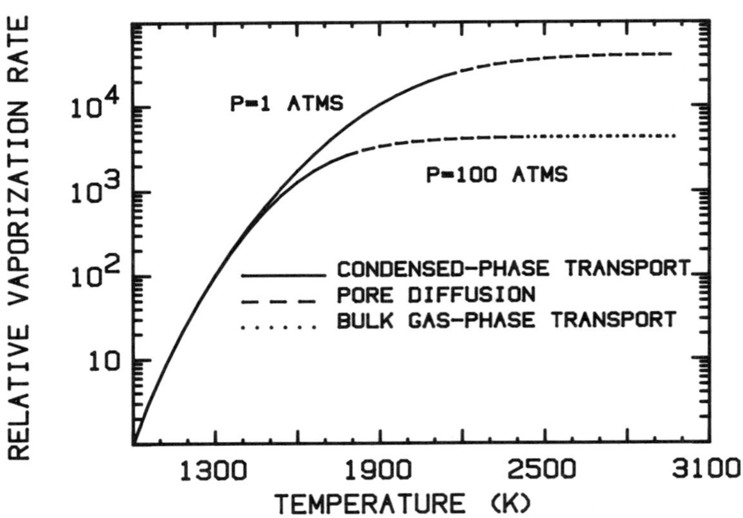

FIGURE 4. Effects of rate controlling processes on the temperature dependence of radionuclide release rates at two pressures. Curves indicate which of three mechanisms is the dominant limitation on release rates. Note that 1 atmosphere is 101325 N/m^2.

rate-controlling processes create dependencies of release on the microstructure of the fuel (and consequently fuel burnup), pressure, geometry and bulk gas flow around the fuel. These are factors which vary during accidents and among different accidents and are not universally considered in current models of radionuclide release used for accident analyses.

Radionuclide release is also dependent on the surface area available for vaporization. As fuel goes from intact rods to a coherent molten pool, the surface area per unit volume of fuel can be reduced by a factor of about 100. Release rates, too, will decrease by this amount as surface area decreases, if all other factors remain the same. Recent evidence of fuel swelling at elevated temperatures [37] suggests that as core degradation progresses, the surface area available for vaporization passes through a maximum.

The many dependencies of release over the course of an accident, make it convenient to define regimes of radionuclide release:

- release from intact rods
- release during rod or clad melting
- release from fuel that has collapsed into a rubble bed
- release from a molten fuel pool
- release when hot fuel interacts with water.

Discussions of the states of knowledge of radionuclide release in these regimes are presented below. Modern accident analyses indicate, however, that the progression of core damage may not be uniform across an entire core. Various regions of the core may simultaneously be in any of the release regimes defined above.

The discussions below are presented in terms of release into the steam and hydrogen atmosphere of the reactor coolant system during core degradation. Evidence from the damaged core at the Three Mile Island reactor (see J. Broughton and G. D. McPherson in these Proceedings) supports the contention that intact fuel will remain in the reactor vessel even if molten core debris has formed and penetrated the reactor coolant system. Air drawn into the reactor coolant system will create a radically different environment for the release of radionuclides from fuel that remains in the vessel. This new regime of release is only now being studied (see F. C. Iglesias et al. and R. Williamson and S. A. Beetham in these Proceedings) and will not be discussed here. Evidence is accumulating that patterns of release in air are radically different than release of radionuclides from fuel in steam and hydrogen.

2.1 Release from Intact Fuel Rods

By far the greatest amount of study of radionuclide release processes has been devoted to the release from intact fuel rods. Attention has focused on this stage of the core degradation because it is expected to be the period when the more volatile radionuclides--notably cesium and iodine--will be released.

A variety of models have been developed to describe radionuclide release from intact fuel rods. Before discussing the more mechanistic models, it is useful to consider the empirical CORSOR model which is widely used [6,8] for accident analyses. The model was introduced by Kress [21] and refined by Kuhlman et al. [22] as a correlation of release fraction data:

$$\frac{dF(i)}{dt} = A(i)\exp[-E(i)/RT](1-F(i)) \tag{13}$$

The principal attraction of the model is that, unlike more mechanistic models, it does treat all radionuclides of interest. This correlation of fractional release rates is readily derived from equation (1) by assuming: a) a single rate-controlling process limits the release rate, b) no changes occur in the surface-to-volume ratio of the fuel, c) ambient partial pressures of radionuclide vapors are negligibly small in comparison to equilibrium partial pressures, and d) that the vapor pressures of radionuclides in fuel depend only on temperature and are independent of pressure and gas composition. These assumptions yield a first-order kinetics rate expression with constants and geometric terms grouped into an Arrhenius rate parameter. The pre-exponential constant, A(i), is found by fitting release data. The activation energy, E(i), is found by fitting data or from the thermodynamics of unary vaporization processes such as BaO(solid) → BaO(gas) for E(Ba). The CORSOR model is intimately tied to the available experimental data base. This data base consists primarily of out-of-pile tests of irradiated fuel rods done at atmospheric pressure with flows chosen so that gas phase mass transport into the ambient atmosphere should not limit the rates of release. The data base, then, does not span the entire range of conditions that are hypothesized to arise in reactor accidents [23]. The model predictions become more suspect as conditions deviate from those studied in tests. Similarly, the first order kinetics approximation can poorly represent diffusion-controlled release processes when heating rates deviate substantially from those used to obtain data defining the kinetic parameters.

The more mechanistic models of radionuclide release from intact rods can be conveniently discussed in terms of the hypothesized controlling mechanism for release. The earliest of the mechanistic models are the so-called "Booth diffusion" models [24] in which diffusion of radionuclides through grains to the free surfaces of the fuel's pore network was assumed to control the rate of release. A variety of refinements to these models have appeared culminating in the ANS 5.4 release model [25]. For stable and long half-life isotopes, the release expression in the ANS 5.4 model is:

$$F(i) = \begin{cases} 4\sqrt{\tau/\pi} - 1.5\tau \quad \text{for } \tau < 0.1 \\ 1 - (1/15\tau) + (1/6\tau) \displaystyle\sum_{n=1}^{3} \exp\left[-n^2\pi^2\tau\right]/n^4\pi^4 \\ \qquad \text{for } \tau > 0.1 \end{cases} \tag{14}$$

where $\tau = D(i)t$ and the diffusion coefficients are:

$$D(Kr) = D(Xe) = 0.61 \exp(-25010/T)\exp(Bu/6.08) \quad (15)$$

$$D(Cs) = 0.0758 \exp(6090/T)D(Xe) \quad (16)$$

$$D(I) = 0.575 \exp(4480/T)D(Xe) \quad (17)$$

$$D(Te) = 1100 \exp(-6290/T)D(Xe) \quad (18)$$

Rest has built magnificently on the general concepts of the Booth diffusion models in the development of the FASTGRASS code [26]. He has added the idea that radionuclides can migrate through fuel grains both by atomic diffusion and by diffusion of gas bubbles nucleated in grains. Detailed modeling of the development of an interconnected pore network in the fuel has been included in the FASTGRASS code and development of this pore network, but not flow of radionuclides through the network, is considered as a rate-controlling process for radionuclide release. To account for release of radionuclides other than noble gases, fuel chemistry dictated by the uranium oxide stoichiometry has been used to define the speciation of some condensed radionuclides [27].

Modeling somewhat similar to that in the FASTGRASS code has been developed for the SINGAR model of noble gas release [28].

Models in which condensed-phase mass transport is assumed to limit the radionuclide release rates are, of course, relatively insensitive to ambient conditions such as gas flow, pressure and composition. These ambient atmospheric conditions are among the most variable features of reactor accidents. They can, especially when pressures are high and flows are low, create rate limitations quite unlike the limitations from condensed-phase mass transport. Miller [29] developed the so-called "Light Bulb" model of release based on the hypothesis that bulk gas-phase mass transport was rate limiting. Though never fully developed for accident analyses, the Light Bulb model did explain some test results.

Cubicciotti [30] has suggested a third approach in which the rates of radionuclide releases are determined by grain growth and sintering:

$$F(i) = \begin{cases} 1-[1-4(\tau_H/\pi)^{1/2}][1-4(\tau_\rho/\pi)^{1/2} + \tau_\rho] & \text{for steam conditions} \\ 1-[1+2.92\text{x}10^{-6}\exp(-32100/T)/d_o]^{-3/2} & \text{for inert conditions} \end{cases} \quad (19)$$

where $\tau_H = Dt/H^2$, $\tau_\rho = Dt/r_p{}^2$, and $D = 99\exp(-28000/T)$ cm^2/s.

Alexander has identified a rate controlling mechanism quite distinct from all cited above [31]. At high steam (or hydrogen) pressures, evaporation of the host urania lattice may expose radionuclides to free surfaces more rapidly than can diffusion. This so-called "matrix stripping" mechanism can augment releases by other processes. Somewhat similar exposure is thought to occur when fuel oxidation causes UO_2 grains to fracture and reform as U_3O_8 [32].

These models of release, each with a distinct hypothesized mechanism for controlling the rates of radionuclide release, are being amalgamated and augmented in the VICTORIA code [33]. This code conducts a detailed thermochemical analysis of the driving force for release and considers release rate limitations caused by:

- condensed phase mass transport
- pore network development
- gaseous and surface diffusion through the pore network
- gaseous diffusion in the fuel/clad gap
- reaction of radionuclides with the clad
- mass transport in the gas phase.

The model is having success in predicting releases for high pressure, strongly reducing conditions as well as for milder conditions typical of most of the available radionuclide release data base.

2.2 Release During Melting

The onset of clad and fuel melting should affect release of radionuclides though it is by no means certain how. Occlusion of the pore structure and loss of surface area might inhibit release [36]. On the other hand, foaming of the fuel and attack on urania by unoxidized clad might augment release [37]. As melting, especially of the fuel, progresses, radionuclides in metallic form can partition from the fuel into the metallic melt. This can alter the thermodynamic driving force for release of elements such as Ru, Mo, Sb and perhaps Te. Relatively little work has been done on the issue of release during melting of the clad and fuel. Observation of both cesium and iodine in melted fuel at the Three Mile Island reactor suggests that understanding of the effects of fuel and clad melting on radionuclide release may be of some importance to accident analyses. Of particular interest are the amounts of volatile radionuclides available for release during ex-vessel processes since this release will not be attenuated by transport processes in the reactor coolant system.

2.3 Release from Rubble Beds

It is often hypothesized, and evidence from the Three Mile Island reactor core indicates, that fuel may collapse within the core to form a rubble bed. Little work has been done on release from rubble beds. It is likely, however, that gas flows through the bed will be low and surface areas will be high. The release rates of radionuclides will, then, be determined largely by thermochemical equilibrium considerations.

2.4 Release from Molten Fuel Pools

Little work has been done on release from molten fuel pools within the reactor coolant system. Investigations with simulated, "fissium", melts were done as part of the SASCHA project [38]. It is suspected, however, that releases observed in these tests were limited by gas-phase mass transport peculiar to the experimental apparatus [39]. Petti has proposed recently that gas bubbles might form in molten fuel and may affect radionuclide release [40].

2.5 Release During Quenching

It was hypothesized in the Reactor Safety Study that large releases of especially tellurium and ruthenium would occur if molten reactor fuel interacted with water to cause a steam explosion [2]. These high releases were hypothesized when finely fragmented fuel was lofted into a strongly oxidizing atmosphere. These concepts have largely been abandoned in modern analyses of in-vessel release. Results of the PBF scoping test [41] do, however, show that water interactions with hot fuel can accelerate radionuclide release. The precise cause of accelerated release is not known. Possibilities include: a) high flows produced by rapid steam formation create conditions conducive to rapid release, b) fragmentation of fuel exposes radionuclides so they can vaporize, c) and oxidation of fuel by steam accelerates condensed phase mass transport of radionuclides or exposes radionuclides as fuel grains are fractured during conversion to U_4O_9.

3. TRANSPORT OF RELEASED RADIONUCLIDES

Radionuclides released from the fuel must escape the reactor coolant system before they can contribute to the radiological threat of a reactor accident. The radionuclide vapors along with massive amounts of non-radionuclide vapors from the fuel, cladding, structural materials and control rods will emerge from the core region into the much cooler regions of the balance of the reactor coolant system where vapor condensation can occur. These materials will transport through the system as a mixture of vapors and aerosols. The radionuclides may deposit on surfaces along the flowpath by:

- physical deposition of aerosols
- condensation or chemical reaction of vapors
- condensation or reaction of vapors with aerosol and subsequent deposition of the aerosols.

The physical deposition of aerosols on structural surfaces is relatively well understood. Gravitational settling, inertial deposition, thermophoresis, turbulent deposition and particle diffusion can all be important mechanisms of deposition during transport of the radionuclides. Several models are available to provide sophisticated calculations of the deposition processes as well as the growth of aerosol particles [6,42,43]. An area of residual uncertainty is whether vapor nucleation to form aerosols needs to be explicitly modeled as it is in the RAFT code [43]. Nucleation is often not explicitly modeled because coagulation of fine condensation nuclei is thought to rapidly erase any significant effects of the details of nucleation. A complete exploration of nucleation effects has not, however, been conducted.

Recent accident analyses have been based on the assumption that chemical forms of the volatile radionuclides are CsOH, CsI, Te and chemically inert aerosols [6,8]. The deposition of vapor species is described by the rate expression:

$$\frac{dN(i)}{dt} = (A/RT) \sum_{j=1}^{M(i)} k_d(i,j)[P(i,j)-P_{eq}(i,j)] \tag{20}$$

Some experimentally determined values of the deposition velocity, $k_d(i,j)$, are shown in Table 3.

The high chemical reactivities of both tellurium vapors and cesium hydroxide vapors have been noted in experimental studies of vapor transport [44-46]. Tellurium vapors react rapidly with almost any metal. Cesium hydroxide, even when simply condensed on surfaces, does not remain pure [44]. The Source Term Code Package [6] explicitly considers irreversible reactions of cesium hydroxide with surfaces.

The high chemical reactivities of cesium hydroxide and tellurium have raised questions about whether these are the real chemical forms of the radionuclides in the reactor coolant system. Questions in this area have been bolstered by results of integral tests of radionuclide release such as the STEP tests [47] and the PBF tests [41]. There is evidence tellurium may actually transport as SnTe, Ag_2Te or as a cesium-tellurium compound. Cesium molybdates, silicates, uranates, zirconates and borates have been observed deposited on flow paths.

Even when CsOH is found deposited on surfaces, it is always associated with other materials and would be in solution at elevated temperatures. The solution effects could sharply depress the vapor pressure of cesium hydroxide [54].

The stability of cesium iodide in the reactor coolant system under accident conditions has also been questioned. Several investigations have shown that cesium iodide will react with boric acid evolved from the reactor coolant or produced by steam oxidation of boron carbide control blade material [48-51]:

$$CsI + HBO_2 \rightleftharpoons CsBO_2 + HI \tag{21}$$

$$2HI \rightleftharpoons H_2 + 2I. \tag{22}$$

TABLE 3. Deposition velocities of radionuclide vapors [44,45,46,55]

Species	Measured Deposition Velocity (m/s)
CsOH (water soluble surface deposit)	$0 - 7\times10^{-4}$ $1.8\times10^{-8}\exp(9100/T)$
CsOH (insoluble, chemically reacted deposit)	$0.07\exp(-7200/T)$ $0.01\exp(-4270/T)$
CsI	3×10^{-7}
Te	0.001 - 0.1

Also, cesium iodide can partially dissociate in high temperature steam:

$$CsI + H_2O \rightleftharpoons CsOH + HI \tag{23}$$

The high reactivity of the CsOH product of this dissociation has been invoked to suggest that this equilibrium may be driven to the right and thus explain evidence of CsI decomposition in transport experiments [52].

The opportunities for chemical reactions of radionuclides are greatly enhanced by the presence of aerosols. The aerosols present far more surface area for reaction than do structures along the flow pathways in the reactor coolant system. Little attention has been given to chemical reactions of radionuclide vapors with aerosols [53] though condensation of vapors on aerosols has been found an important process [6]. Where chemical reactions are considered, their rates are usually thought to be controlled by gas phase mass transport:

$$\frac{dN(i,r_p)}{dt} = \sum_{j=1}^{M(i)} \frac{\frac{4k_m(r_p)\pi r_p^2}{RT}[P(i,j) - P_{eq}(i,j)](1+Kn)}{(1 + 1.717\ Kn + 1.333\ Kn^2)} \tag{24}$$

The non-continuum corrections in this rate expression, embodied by terms involving the Knudsen number, Kn, are important for the rates of reaction with particles of radii less than $1x10^{-6}$ m at a total pressure of $1x10^5$ N/m^2. Such fine particles can contribute a significant fraction of the surface area of a freshly formed aerosol. The importance of these finer particles decreases as the aerosol ages. The particle size at which non-continuum corrections are important decreases with increasing pressure. At pressures on the order of $1x10^7$ N/m^2, non-continuum corrections to the mass transfer rate are probably not important for aerosol particles of interest in reactor accident analyses.

Other processes that could affect the rates of vapor reactions with aerosols such as condensed phase mass transport across product layers or chemical reaction kinetics need to be investigated since they are likely to pose a greater barrier to reaction than gas-phase mass transport. Unfortunately, consideration of these other rate controlling processes requires a data base that does not now exist.

4. REVAPORIZATION

The permanence of radionuclide deposits in the reactor coolant system has become a critical issue of accident analysis. Two processes may reverse the deposition of radionuclides that occurred when released material passed through the reactor coolant system--physical re-entrainment by sudden increases in gas flow and revaporization. Re-entrainment of deposited aerosols is certainly possible especially when accelerated gas flows through the coolant system and even mechanical shocks to the structures are created by hot fuel interactions with water. It is not known, however, if entrained deposits disintegrate into particles sufficiently small to transport distances of interest for reactor accident analyses. The questions of physical re-entrainment have been given little attention. Answers to these questions may be of interest in the design of accident mitigation strategies.

Revaporization has begun to receive attention in accident analyses [10-12]. It is evident that extensive revaporization will depend on the presence of significant gas flows through the reactor coolant system, elevated temperatures in the radionuclide-bearing deposits, and vapor pressures of the radionuclides. Gas flows through the reactor coolant system appear to be inevitable. If nothing else, a natural convection flow can develop through the reactor coolant system once core debris has penetrated the reactor vessel [15]. Boil-off of residual coolant in the reactor can also provide the flow that is necessary for extensive revaporization [11]. Temperatures that can be achieved in deposited radionuclides are being debated [8,10]. It is likely that deposit temperatures will be strongly dependent on the details of the plant design and the progression of the accident. Certainly, the temperatures will depend on the extent of radionuclide release from the fuel and the nature of radionuclide deposition in the coolant system. Vapor pressures of the deposited radionuclides will depend on the chemical forms adopted by the deposited radionuclides.

Predictions of revaporization made to date have been based on the assumption of simple chemical forms for the deposited radionuclides--pure CsI and pure CsOH, for example. The observed instability of CsI under reactor accident conditions raises the question of whether there will actually be any iodine to revaporize in the reactor coolant system. The observed chemical complexity of radionuclide deposits in tests suggests revaporization rates for cesium based on the vapor pressure of CsOH may be seriously in error. Analyses considering cesium to be present as a $CsBO_2$ or mixtures of $CsBO_2$ in B_2O_3 have shown the revaporization can be retarded or even, essentially, arrested depending on the composition of the deposits [54]. Finally, it should be noted that uncertainty analyses [15] suggest Te, Ru and Ba may also be susceptible to revaporization late in reactor accidents.

5. SUMMARY

The central issues of radionuclide behavior in the reactor coolant system are radionuclide release from fuel, transport in the coolant system and revaporization. Modeling of the release of radionuclides from intact fuel rods has become quite detailed and recognizes the multiple mechanisms causing release. This modeling is, however, supported by a data base that does not span the entire spectrum of expected accident conditions. Few experimental studies have been conducted at elevated pressures and low gas flows characteristic of many important reactor accidents. Modeling of release from fuel in more advanced stages of degradation or when hot fuel interacts with water is in a more primitive state. Radionuclide transport and deposition in the reactor coolant system is now routinely considered in accident analyses. For many accidents, significant fractions of released radionuclides are predicted to be retained in the coolant system. These predictions depend on the chemical forms of the radionuclides predicted to exist under accident conditions. Chemical forms of the retained radionuclides are now only hypothesized. The actual chemical forms assumed by radionuclides deposited in the reactor coolant system will determine whether current predictions of extensive revaporization of deposited radionuclides late in an accident are realistic. Some studies of the chemical forms of radionuclides under low pressure and chemically simple circumstances are being conducted.

6. NOMENCLATURE

Symbol	Definition	SI Units
A	surface area	m^2
A(i)	pre-exponential factor for the ith radionuclide in the CORSOR release rate expression	s^{-1}
Bu	fuel burnup (GWD/metric ton uranium)	--
(c)	condensed phase	--
D	diffusion coefficient	m^2/s
d_o	initial grain size	m
E(i)	activation energy for the release of the ith radionuclide in the CORSOR release rate expression	J/mol
F(i)	fraction of the initial inventory of the ith radionuclide released	--
(g)	gas	--
H	fuel pellet height	m
Kn	Knudsen number = λ/r_p	--
k(i,j)	effective rate constant for the release of the ith radionuclide as the jth vapor species	m/s
$k_d(i,j)$	deposition velocity for the jth vapor species containing the ith radionuclide	m/s
$k_m(r_p)$	mass transport coefficient for vapor reactions with a particle of radius r_p	m/s
(l)	liquid	--
M(i)	number of vapor species bearing the ith radionuclide	--

Symbol	Definition	SI Units
$N(i)$	moles of the ith radionuclide in the fuel or deposit	mol
$N(i,r_p)$	moles of the ith radionuclide in a particle of radius r_p	mol
$P(i,j)$	ambient partial pressure of the jth vapor species containing the ith radionuclide	N/m^2
$P_{eq}(i,j)$	equilibrium partial pressure of the jth vapor over a mixture containing the ith radionuclide	N/m^2
$P_{eq}^{o}(i,j)$	equilibrium partial pressure of the jth vapor over the ith radionuclide in a pure chemical form	N/m^2
R	gas constant	J/(mol-K)
r_p	radius of a fuel pellet or aerosol particle	m
(s)	solid	--
T	absolute temperature	K
t	time	s
V	void free volume	m^3
$\gamma(i)$	activity coefficient of the ith radionuclide	--
λ	mean free path in the gas phase	m
ρ_m	molar density	mol/m^3
τ	Dt	m^2

7. REFERENCES

1. DiNunno, J. J., Anderson, F. D., Baker, R. E., and Waterfield, R. L., Calculation of Distance Factors for Power and Test Reactor Sites, TID-14844, March 1962.

2. U. S. Nuclear Regulatory Commission, The Reactor Safety Study: An Assessment of Accident Risk in U. S. Commercial Nuclear Power Plants, NUREG 75/014, WASH-1400, October 1975.

3. The American Nuclear Society and the Institute of Electrical and Electronic Engineers, PRA Procedures Guide, NUREG/CR-2300, September 1981.

4. Alpert, D. J., Chanin, D. I., and Ritchie, L. T., Relative Importance of Individual Elements to Reactor Accident Consequences Assuming Equal Release Fractions, NUREG/CR-4467, SAND85-2575, March 1986.

5. Stratton, W. R., Malinauskus, A. P., and Campbell, D. O., letter to Chairman J. Ahearne, U. S. Nuclear Regulatory Commission, August 14, 1980. Available NRC Public Documents Room.

6. Gieseke, J. A., Cybulskis, P., Jordan, H., Lee, K. W., Schumacher, P. M., Curtis, L. A., Wooton, R. O., Quayle, S. F., and Kogan, V., Source Term Code Package--A User's Guide (Mod-1), NUREG/CR-4587, BMI-2138, July 1986.

7. Silberberg, M., Mitchell, J. A., Meyer, R. O., and Ryder, C. P., Reassessment of the Technical Bases for Estimating Source Terms, NUREG/CR-0956, July 1986.

8. Fauske and Associates, Inc., Modular Accident Analysis Program User's Manual, Volumes I and II, Atomic Industrial Forum, Bethesda, MD, (undated).

9. Electric Power Research Institute, Estimation of Fission Product and Core Material Characteristics, IDCOR Technical Report 11.1, 11.4 and 11.5, Atomic Industrial Forum, Bethesda, MD, 1983.

10. Bieniarz, P. P. and Deem, R. E., Analysis of a BWR MKI Utilizing Advanced BMI-2104 Computational Techniques to Calculate Source Terms, *Source Term Evaluation for Accident Conditions*, International Atomic Energy Agency, Vienna, Austria, 1986.

11. Donahue, M., Hazzan, M., Metcalf, J., and Warman, E., Analysis of Retention/Vaporization in a BWR Mark II Power Plant, ibid.

12. Industry Degraded Core Rulemaking Project, Peach Bottom Atomic Power Station, IDCOR Task 23.1, Integrated Containment Analysis, Atomic Industrial Forum, Bethesda, MD, 1984.

13. U. S. Nuclear Regulatory Commission, Reactor Risk Reference Document, NUREG-1150 Draft for Comment, February 1987.

14. Powers, D. A. and Allen, M. D., Uncertainty Analysis of Radionuclide Release During Core Degradation, Sandia National Laboratories, Albuquerque, NM, (unpublished).

15. Powers, D. A. and Schmidt, R. C., Analysis of Revaporization of Deposited Radionuclides, Sandia National Laboratories, Albuquerque, NM, (unpublished).

16. Denning, R. S., Gieseke, J. A., Cybulskis, P., Lee, K. W., Jordan, H., Curtis, L. A., Kelly, R. F., Kogan, V., and Schumacher, P. M., Radionuclide Release Calculations for Selected Severe Accident Scenarios, Volumes 1-4, NUREG/CR-4624, BMI-2139, July 1986.

17. Leonard, M. T., Cybulskis, P., Lee, K. W., Kelly, R. F., Jordan, H., Schumacher, P. M., and Curtis, L. A., Supplemental Radionuclide Release Calculations for Selected Severe Accident Scenarios, NUREG/CR-5062, BMI-2160, February 1988.

18. Lorenz, R. A., Collins, J. L., and Malinauskus, A. P., Fission Product Source Terms for the Light Water Reactor Loss-of-Coolant Accident, *Nuclear Technology* 46 (1979) 404.

19. Tam, S. W., Blackburn, P. E., and Johnson, C. E., Effect of Core Chemistry on Fission Product Release, *Proc. International Meeting on Thermal Nuclear Reactor Safety*, NUREG/CP-0027, Volume 1, p 101, February 1983.

20. Brewer, L., Methods of Obtaining Thermodynamic Data, p 35, *Resource Papers for the Workshop on Chemical Processes and Products in Severe Nuclear Reactor Accidents*, National Academy Press, 1989.

21. U. S. Nuclear Regulatory Commission, Technical Bases for Estimating Fission Product Behavior During Reactor Accidents, NUREG-0772, June 1981.

22. Kuhlman, M. R., Lehmicke, D. J., and Meyer, R. O., CORSOR User's Manual, NUREG/CR-4173, BMI-2122, March 1985.

23. Osborne, M. F., Collins, J. L., and Lorenz, R. A., Experimental Studies of Fission Product Release from Commercial Light Water Reactor Fuel Under Accident Conditions, p 237, *Resource Papers for the Workshop on Chemical Processes and Products in Severe Nuclear Reactor Accidents*, National Academy Press, 1988.

24. Olander, D. R., Fundamental Aspects of Nuclear Reactor Fuel Elements, UC-736, TID-26711, 1975.

25. Southern Science Applications, Inc. and American Nuclear Society Working Group 5.4, Background and Derivation of ANS 5.4 Standard Fission Product Release Model, NUREG/CR-2507, January 1982.

26. Rest, J., An Improved Model for Fission Product Behavior in Nuclear Fuel Under Normal and Accident Conditions, *J. Nuclear Materials* 120 (1984) 195.

27. Rest, J., Evaluation of Volatile and Gaseous Fission Product Behavior in Water Reactor Fuel Under Normal and Severe Core Accident Conditions, *Nuclear Technology* 61 (1983) 33.

28. Brearly, I. R. and MacInes, D. A., SINGAR-A Model to Examine Fission Product Release in Accident Studies, SRD R-258, Safety and Reliability Directorate, U. K. Atomic Energy Authority, June 1983.

29. Miller, C. E., Jr., A Boundary Layer Diffusion Model of Fission Product Release From Reactor Fuels, *Nuclear Applications* 5 (1968) 198.

30. Cubicciotti, D., A Model for Release of Fission Gases and Volatile Fission Products from Irradiated UO_2 in a Steam Environment, *Nuclear Technology* 53 (1968) 5.

31. Alexander, C. A., Ogden, J. S., and Wright, R. W., Matrix Stripping and Fission Product Release at High Temperatures, *International Symposium on Severe Accidents in Nuclear Power Plants*, paper IAEA-SM-296/99, Sorrento, Italy, March 21-25, 1988.

32. Iwasaki, M., Sakurai, T., Ishikawa, N., and Kobayasha, Y., Oxidation of UO_2 Pellets in Air: Effect of Heat-Treatment of Pellet on Particle Size Distribution Produced, *J. Nuclear Science and Technology* 5 (1968) 652.

33. Camp, W. J., Release of Fission Products from Reactor Fuels During Severe Accidents: The VICTORIA Code, p 55, *Resource Papers for the Workshop on Chemical Processes and Products in Severe Nuclear Reactor Accidents*, National Academy Press, 1988.

34. Lorenz, R. A., Beahm, E. C., and Wichner, R. P., Review of Tellurium Release Rates From LWR Fuel Elements Under Accident Conditions, Paper TS-4.4, Volume 1, *Proc. International Meeting on Light Water Reactor Severe Accident Evaluation*, 1983.

35. Alexander, C. A., and Ogden, J. S., RCS Chemistry at BCL, p 19, *Resource Papers for the Workshop on Chemical Processes and Products in Severe Nuclear Reactor Accidents*, National Academy Press, 1989.

36. Rest, J., Grain-Boundary Sweeping and Liquefaction-Induced Fission-Product Behavior in Nuclear Fuel Under Severe Core Damage Conditions, *Advances in Ceramics* 17 (1986) 223.

37. Allen, M. D., Stockman, H. W., Reil, K. O., and Fisk, J. W., ACRR Source Term Experiment: ST-1, NUREG/CR-5345, SAND89-0308, forthcoming.

38. Albrecht, H., Results of the Sascha Program on Fission Product Release Under Core Melting Conditions, p 1, *Resource Papers for the Workshop on Chemical Processes and Products in Severe Nuclear Reactor Accidents*, National Academy Press, 1989.

39. Lorenz, R. A., Vaporization of Structural Materials in Severe Accidents, *Proc. International Meeting on Thermal Nuclear Reactor Safety*, NUREG/CP-0027, Volume 2, p 1090, February 1983.

40. Petti, D. A., Analysis of Fission Product Release from the TMI-2 Core, Severe Accident Research Program Semi-Annual Partners Review Meeting, Idaho Falls, ID, April 10-14, 1989.

41. Osetek, D. J., Petti, D. A., and Hagrman, D. L., Observations on Chemical Processes and Products from the Four PBF Severe Fuel Damage Tests, p 251, *Resource Papers for the Workshop on Chemical Processes and Products in Severe Nuclear Reactor Accidents*, National Academy Press, 1989.

42. Wheatley, C. J., CHARM: A Model for Aerosol Behavior in Time Varying Thermal-Hydraulic Conditions, NUREG/CR-5162, SAND88-0745, August 1988.

43. Im, K. H., Ahluwalia, R. K., and Chuang, C. F., RAFT, A Computer Model for Formation and Transport of Fission Product Aerosols in LWR Primary Systems, *Aerosol Science and Technology* 4 (1985) 125.

44. Elrick, R. M., Sallach, R. A., Ouellette, A. L., and Douglas, S. C., Reactions Between Some Cesium-Iodine Compounds and the Reactor Materials 304 Stainless Steel, Inconel 600 and Silver, Volume I Cesium Hydroxide Reactions, NUREG/CR-3197/1 of 3, SAND83-0395, June 1984.

45. Sallach, R. A., Greenholt, C. J., and Taig, A. R., Chemical Interactions of TELLURIUM Vapors with Reactor Materials, NUREG/CR-2921, SAND82-1145, March 1984.

46. Sallach, R. A., Elrick, R. M., Douglas, S. C., and Ouellette, A. L., Reactions Between Some Cesium-Iodine Compounds and the Reactor Materials 304 Stainless Steel, Inconel 600 and Silver, NUREG/CR-3197/2 of 3, SAND83-0395, August 1986.

47. Baker, L., Jr., Fink, J. K., Simms, R., Schlenger, B. J., and Herceg, J. E., Source Term Experiments Project (STEP): A Summary, EPRI NP-5753M, March 1988.

48. Elrick, R. M. and Sallach, R. A., Fission Product Chemistry in the Primary System, *Proc. International Meeting on Light Water Reactor Severe Accident Evaluation*, Paper TS 4.6, Volume 1, 1983.

49. Elrick, R. M., Sallach, R. A., Ouellette, A. L., and Douglas, S. C., Boron Carbide-Steam Reactions with Cesium Hydroxide and with Cesium Iodide at 1270K in an Inconel 600 System, NUREG/CR -4963, SAND87-1491, September 1987.

50. Beattie, I. R., Bowsher, B. R., Gilson, T. R., and Jones, P. J., The Interaction of Caesium Iodide with Boric Acid, AEEW-R 1974, June 1985.

51. Bowsher, B. R. and Dickinson, S., The Interaction of Caesium Iodide with Boric Acid: Vapour Phase and Vapour Condensed Phase Reactions, AEEW-R 2102, May 1986.

52. Elrick, R. M., Merrill, R. M., and Ouellette, A. L., The Thermal Instability of Cesium Iodide, NUREG/CR-5155, SAND88-1187, 1989.

53. *Chemical Processes and Products in Severe Nuclear Reactor Accidents-Report of a Workshop*, National Academy Press, 1988.

54. Powers, D. A and Bieniarz, P. P., Influence of Chemical Form on Cesium Revaporization from the Reactor Coolant System, *Proceeding of the Symposium on Chemical Phenomena Associated with Radioactivity Releases During Severe Nuclear Plant Accidents*, S. J. Niemczyk, editor, NUREG/CP-0078, June 1987.

55. Nichols, A. L., Reactor Coolant System (RCS) Chemistry Research in the United Kingdom, p 189, *Resource Papers for the Workshop on Chemical Processes and Products in Severe Nuclear Reactor Accidents*, National Academy Press, 1989.

The Chemistry of Water Reactor Fuel

PAUL E. POTTER
Chemistry Division
Harwell Laboratory
UKAEA, Oxon OX11 ORA, UK

INTRODUCTION

In this paper, we shall discuss features of the changes in chemical constitution which occur in fuel and fuel rods for water reactors during operation and in fault conditions. The fuel for water reactors consists of pellets of urania (UO_2) clad in Zircaloy.

An essential step in the prediction of the fate of all the radionuclides in a fault or accident is to possess a detailed knowledge of their chemical behaviour at all stages of the development of such incidents.

In this paper, we shall consider:

i) The chemical constitution of fuel during operation at temperatures corresponding to rather low ratings, together with a quite detailed discussion of the chemistry within the fuel-clad gap.
ii) The behaviour of fuel subjected to higher temperatures and ratings than those of contemporary fuel, and briefly,
iii) the changes in constitution on failure of fuel rods in fault or accident conditions.

THE FLUORITE STRUCTURED FUEL MATRIX

We firstly have to consider the behaviour of the oxide matrix of the fuel during operation. Significant factors which determine the evolution of the chemical constitution; the nature of the phases and the transport of fission product atoms and ions within the fuel matrix and to the fuel-cladding gap are the oxygen anion:metal cation ratio and the thermodynamic oxygen potential. An important requirement is to determine the relation between the increase in oxygen anion:metal atom and cation ratio and thermodynamic oxygen potential $(\bar{G}_{O_2})_0$. The value of $\bar{G}_{O_2}$ will determine the chemical constitution of the phases and pressures of the gaseous species in the fuel-clad gap of a rod. We have modelled the chemical constitution of the gap and indicated the important reactions which can act as buffers to the change in oxygen potential, and we have also used a model of the relationship between composition and $\bar{G}_{O_2}$ for the urania phase. There are several models of the urania phase which reproduce the experimental data well and are suitable for use in descriptions of fuel behaviour. A discussion of all the models for urania has been given by Potter (1987).

Lindemer and Besmann (1985) have represented all the measurements which have been made of the relationship between composition and thermodynamic oxygen potential for hypo- and hyperstoichiometric urania. These data are represented by a mutual solid solution of elements or compounds; in fact, two fluorite structured face-centred cubic components of different compositions. We describe later the use of this model for urania in calculations of the chemical constitution of the fuel-clad gap. Much more detailed models which take into account specific defect equilibria have been used. Such an application of this approach is that of Hyland (1984). Computer simulations of the ionic and electronic defects which give indications of their energies must be mentioned; Jackson et al (1986) have discussed the calculations of defect parameters in UO_2. A knowledge of the nature of the defects in the urania lattice is an essential step for the prediction of diffusion behaviour in this fluorite structured lattice and that containing fission product atoms and ions. We shall return to this topic later.

At the typical ratings of modern PWR fuel, the centre temperatures will be low, <ca.1200°C and during irradiation the urania will become a solid solution of the fission product atoms and ions; little nucleation of new phases would occur at such temperatures although the nucleation of new phases at a given temperature could be greater the higher the burn-up of the actinoid atoms. The average valency of the fission product atoms and cations will be less than four and some oxidation of the U^{4+} cations to U^{5+} cations will occur resulting in an increase of the thermodynamic oxygen potential of the fuel matrix with increasing burn-up of the actinoid atoms. If the irradiated fuel is considered as a solution of all the fission product atoms and ions then the concentration of U^{5+} cations would be greater than in a fuel matrix in which other phases have nucleated. Thus the oxygen potential for the solution phase would be greater than fuel in which other phases have nucleated. Such an increase in oxygen potential would only occur providing that there be no significant transfer of oxygen from the fuel matrix to the Zircaloy cladding. In order to allow some quantitative assessments to be made of the increase in oxygen potential with burn-up of the actinoid atoms there have been several determinations of the relationship between composition and oxygen potential of the solid solutions of the lanthanoid oxides in urania and of the urania-plutonia system. Lindemer and Brynestad (1986) have recently reviewed and presented a representation of the variation of thermodynamic oxygen potential with composition and temperature for the systems $U_{1-z}Ce_zO_{2\pm x}$ and $U_{1-z}Ln_zO_{2\pm x}$; with Ln = Y, La, Nd and Gd. In the solid solution with ceria, the cations can have valencies of three and four, whilst in the latter group the Ln cations possess a unique valency of three. The incorporation of the three-valent lanthanoid cations into the urania lattice at a given oxygen anion:metal cation ratio results in an increase in $\bar{G}_{O_2}$ of the solution with respect to urania of the same oxygen anion:uranium cation ratio. Similar behaviour occurs for the $U_{1-z}Ce_zO_{2-x}$ solid solution. The urania-plutonia solid solution is analogous to the urania-ceria solution; there is an appreciable range of composition of fluorite structured solid solution of urania and plutonia, in which the U^{4+} cations can be oxidised and the Pu^{4+} cations reduced (Markin and Street

1967); with increase of temperature the U^{4+} cations can also be reduced. The fission product rare gases Kr and Xe have been considered chemically inert, but recently MacInnes and Winter (1988) have suggested that if these fission product species reside in defect clusters they will bind oxygen and cause a depletion in the concentration of interstitial oxygen anions in the oxide lattice and thus reduce the oxygen potential of the fuel. Considerable attention has been given to the measurement of the variation of thermodynamic oxygen potential of the solid solution with composition and temperature; measurements have been essentially confined to solutions with Pu:U ratios of technological significance to their application as fuels for fast breeder reactors. Most of these measurements have again been fitted to a simple model by Besmann and Lindemer (1985, 1986).

Ball (1989) has modelled the measurements by Une and Oguma (1983) of the oxygen potential of urania to which oxides of Y, Zr, Ce, Pr and Nd were added in order to simulate burn-ups of the actinoid atoms of up to 10% in the temperature range 1000-1300°C using a thermogravimetric technique. For hyperstoichiometric solid solutions of a given oxygen anion:metal cation ratio, the oxygen potential increases linearly with simulated burn-up. In Fig.1, we show the comparison of the predictions of the model with the experimental data of Une and Oguma. We are also considering the effects of the addition of BaO and SrO; the valencies of the cations in these oxides remain invariant in the solid solution. The value of $\bar{G}O_2$ is of significance in determining the constitution of the fission product elements in the fuel-clad gap and the actual phases which would nucleate in the matrix of the fuel when subjected to higher temperatures than those characteristic of normal operation or at high burn-ups. The stoichiometry of the fuel matrix will determine, with the temperature, the concentration of the various lattice defects which influence the rate of diffusion of the many fission product atoms and ions in the fluorite lattice of the fuel.

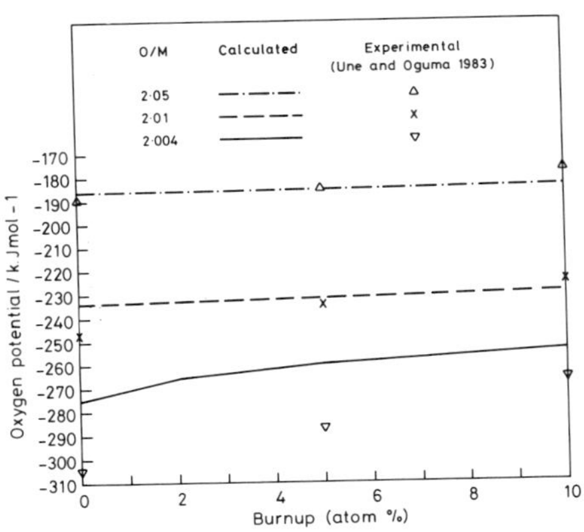

FIGURE 1. A comparison of the experimental and calculated data for $\bar{G}O_2$ of the solid solution (U, Zr, Ce, Pr, Nd, V) O_{2+x} at 1573 K.

As the oxygen potential increases within the fuel matrix even in relatively small temperature gradients with fuel centre temperatures of 1200-1400°C, some redistribution of oxygen could occur by both gaseous and solid state mechanisms (Rand and Roberts 1966, Aitken 1969, Bober and Schumacher 1978, Norris 1977). The gas phase mechanism would be predominant in hyperstoichiometric fuel and Adamson et al (1975) reported measurements on the radial distribution of oxygen in irradiated fuel rods of urania clad in Zircaloy 2. For lightly irradiated fuel of nominally stoichiometric oomposition and with an O:U ratio of 2.033 at the high rating of 525 $W.cm^{-1}$ (estimated centre temperature 1820°C), oxygen appeared to have migrated to the centre of the fuel but for the fuel of higher stoichiometry some transport of the oxygen to and reaction with the Zircaloy cladding had occurred. Examination of nominally stoichiometric urania irradiated up to 18900 and 11400 $MWdt^{-1}$ at ratings of 230 and 394 $W.cm^{-1}$ respectively, indicated that the oxygen potential in the centre of fuel had decreased for both conditions; the oxygen had probably reacted with the cladding and with caesium. Whilst these results are some of the very few for the oxygen potential of irradiated fuel, they are complicated by the effects of cooling and it is doubtful if the values of Go_2 represent those of the fuel during operation. The technique which Adamson et al (1975) used to measure oxygen potential employed an EMF solid state cell with a thoria-yttria electrolyte and Ni-NiO reference electrode. This technique has also been used to measure the oxygen potential of irradiated fast reactor oxide fuel (Ewart et al 1979, Matzke et al 1988); in these experiments small pieces of the irradiated fuel pellets were taken and measured separately whereas Adamson made measurements on an intact pellet. Further applications of this technique to determine changes in oxygen potential which occur during irradiation of fuel should be considered particularly on fuel irradiated at appropriate ratings. Leach et al (1988) indicate that PWR fuel has an increasing lattice parameter with burn-up, which might indicate a decreasing oxygen anion:metal cation ratio in the fuel matrix. The increase in lattice parameter could also result from radiation damage which can be removed by subsequent anealing; any anealing would, therefore, result in a decrease in lattice parameter. It should also be mentioned that there is evidence from measurements of lattice parameter (Walker and Mogensen 1987) that in a BWR fuel irradiated to 4.3 at.% burn-up of the actinoid atoms at a maximum power of 517 W cm^{-1} and then subjected to a transient ingress of energy (420 W cm^{-1}) that there was transport of oxygen to the centre of the fuel. Equilibration with a mixture of CO/CO_2 could also be used to measure $\bar{G}o_2$ of irradiated fuel. We shall discuss later the way in which the composition of the alloy phase Mo-Tc-Ru-Rh-Pd which can nucleate in irradiated fuel at low ratings (Kleykamp 1989) can change with oxygen potential and temperature. A knowledge of the fuel temperature or rating and burn-up at which the fuel reacts with cladding is required as such reactions could markedly influence the oxygen potential and constitution of the fuel.

THE CHEMICAL CONSTITUTION OF SOME FISSION PRODUCTS WITHIN THE GAP OF A FUEL PIN

In a PWR fuel rod, which operates at relatively low temperatures little diffusion of fission product elements to the gap would occur. Fission fragment recoil can be responsible for the inventory of fission product elements which would be present. We have already discussed aspects of the diffusion of oxygen to the cladding and in the fuel. We describe some calculations of the way in which the chemical constitution of the fuel-clad gap and the nature of the gas phase species changes with the oxygen potential of the environment. A detailed discussion of our approach and details of the calculations are given elsewhere (Ball et al 1989). We have taken a fission product inventory of 1% for a burn-up of 2.9% of the actinoid atoms. We have considered the fission product elements Cs, Te and I in the presence of the fuel matrix UO_{2+x}. In addition we have examined the presence of Mo and examined the effects of reaction with the cladding. There will be some Kr and Xe with the He in the gap. We have attempted to simulate the effect of burn-up of the actinoid atoms by modelling the phase equilibria as the oxygen potential of the system increases. The urania lattice was modelled using the representation of Lindemer and Besmann (1985) already discussed. The condensed phases which could appear in the fuel-clad gap will depend on the oxygen potential and whether Mo and Zr participate in the reaction. Particular phase assemblages will buffer the oxygen potential over a range of added amounts of oxygen and these are given in Table 1. The variation in oxygen potential (Go_2) is plotted in figures 2 and 3 against the amount of added oxygen; the plateaux are a result of the presence of these univariant phase assemblages. Fig.2 shows a case for urania with Cs, I and Te in which the plateaux relate to reactions b and c involving Cs uranates shown in detail in Table 1 with values of Go_2 for 650K. Fig.3 shows a case in which Mo and Zr have been added as additional components. These calculations have been carried using the code SOLGASMIX (Eriksson 1975) which finds the position of the minimum in total Gibbs energy of a multicomponent, multiphase system. An important aspect of the calculations of pressures of the gas phase species with oxygen potential is that the pressure of the elemental iodine species increases with increase in oxygen potential, (see Figs. 4 and 5) and even more so when Mo takes part in the reaction as was recognised earlier by Goetzmann (1982) and further discussed by Chen et al (1988). It is apparent that a considerable rise in oxygen potential is required before the iodine pressures are appropriate to that for the formation of 'ZrI'; the condition thought to be required for iodine induced stress corrosion cracking to occur. Chen et al (1988) have recently questioned the existence of this phase; the iodide with the lowest 1:Zr ratio might be an impurity stabilised Zr_6I_{12} cluster.

We expect important effects to occur when we consider the irradiation, by fission product fragments, of the gas phase species in the gap. The combined effect of all the reactions that could occur involving CsI gas is to perturb the thermal equilibrium and a new steady state is achieved in which the concentration of iodine and also tellurium are increased. The reactions considered for iodine are shown in Table 2. The kinetic simulation of the processes was simplified by assuming that all the He^+ ions recombine with electrons efficiently to form excited metastable atoms. The dose rate (D) was taken to be 21.6 Wg^{-1} (1.6 Wg^{-1} from neutron and gamma irradiation and 20 Wg^{-1} as an estimate for the contribution from fission fragment radiation). In addition the concentrations of CsI and Cs_2Te in the gaseous phase are assumed to be constant since there is, for the cases considered, a supply of condensed CsI and Cs_2Te to replenish any lost from the gaseous state by dissociation. The pressure of Cs depends on Go_2. Two examples of the effect of fission fragment irradiation on the decomposition of CsI and Cs_2Te are given in Table 3. The pressures of elemental iodine and tellurium are raised significantly by irradiation.

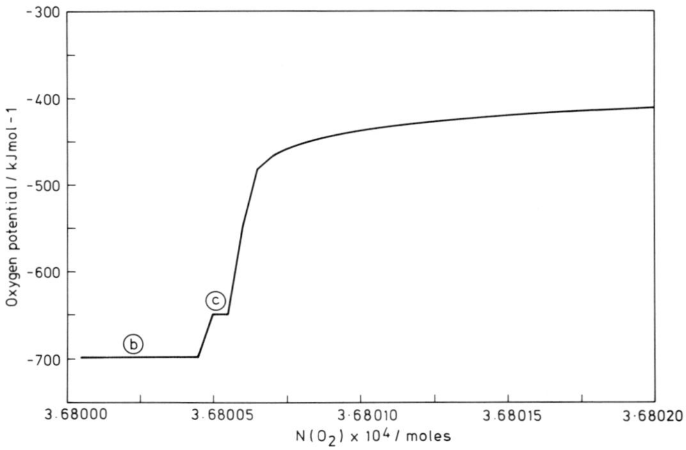

FIGURE 2. The variation of oxygen potential with the amount of added oxygen ($N(O_2)$) in the gap of a PWR rod in which the presence of Mo and Zr are excluded. The plateaux are labelled according to Table 1.

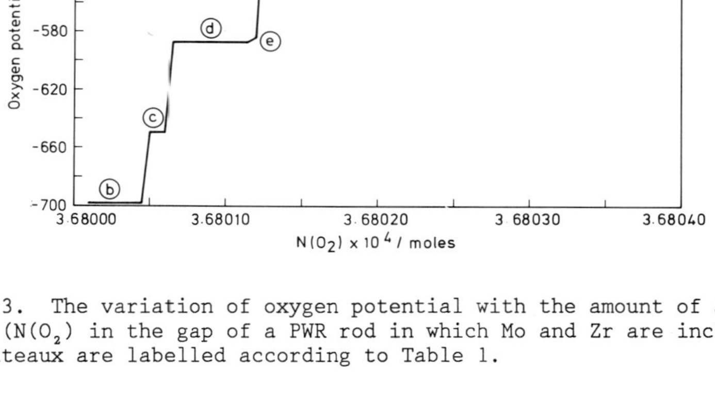

FIGURE 3. The variation of oxygen potential with the amount of added oxygen ($N(O_2)$ in the gap of a PWR rod in which Mo and Zr are included. The plateaux are labelled according to Table 1.

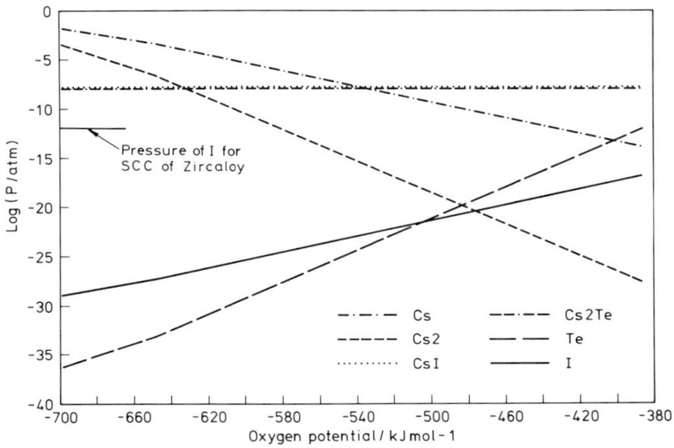

FIGURE 4. The variation of the pressures of the predominant gas phase species in the gap for the same conditions as in Fig.2.

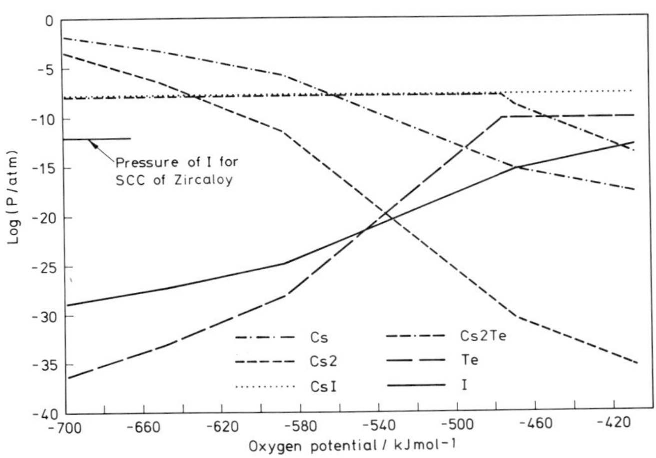

FIGURE 5. The variation of the pressures of the predominant gas phase species in the gap for the same conditions as in Fig.3.

TRANSPORT OF FISSION PRODUCTS IN FUEL

An understanding of the transport of fission product elements and ions in the lattice of the oxide fuel is required in order to predict the nucleation of new phases in bubbles and on grain boundaries. The data which are needed for the analysis of severe accidents are those from post-irradiation measurements. The kinetics of the release of fission product species in the presence of fission are considerably different from those in the absence of fission. Turnbull et al (1982) have attributed differences in the activation energy of Xe release to enhancement of the rate of diffusion to the effects of fission. Fission product species can be released from the surface of a grain or when bubbles on grain boundaries interlink. There is much interest in the behaviour of the fission gases, Kr and Xe in operational transients and this interest is being extended to other volatile fission product elements, Cs, Rb, Te and I (Matzke 1988). It has been shown that Cs, Rb, Te and I can all be found in bubbles in UO_2 and that the diffusion behaviour of these elements and ions differs from that of the inert gases. Olander (1986) and Prussin et al (1988) have recently described a model and given experimental information on the diffusion of fission product atoms and ions in the urania lattice. The model considers two release paths, the first is just the volume diffusion to a grain surface and the second is the lattice diffusion from the grain interior to the grain boundaries followed by migration in the grain boundaries to a free surface. The properties which control the release of any fission product atom or ion are its diffusion coefficient in the lattice of urania and the product of its diffusivity in the grain boundary and the distribution coefficient of the atom or ion between the lattice and grain boundary. The relative importance of these two pathways depends on these transport properties and on the grain size. The distribution coefficients for the various atoms and ions will determine the chemical speciation of the fission product elements on grain boundaries. In addition to Kr and Xe, Cs, I and Te are readily lost from fuel heated at temperatures about 1500°C (Osborne et al 1986) and so if these individual fission products collect on the grain boundaries it seems appropriate to consider their chemical constitution within the grain boundaries and the influence of this speciation on their subsequent behaviour and release from the grain boundaries to the fuel-clad gap and to the water and steam atmospheres if the cladding fails. For the release from the grains by volume diffusion directly into the gap, the chemical speciation would not change from that in the grain boundaries but directly into water or a steam atmosphere may be released in elemental rather than combined form.

The measurements of Prussin et al (1988) of the effective diffusion coefficients of several fission product atoms and ions in trace irradiated essentially stoichiometric UO_2, in the temperature range 1400-1750°C, indicated that the behaviour of the fission product atoms and ions could be classified in four groups. The results suggest a correlation between the valence state of an element in its common compounds (or its electronegativity) and the rate at which it is released from UO_2 during heating at high temperature. The elements with high electronegativities I and Te form the first group, and show high release rates. Ba and Cs form the second group; they have low electronegativities and do not dissolve significantly in the urania lattice, and show release rates which are quite large but not as large

TABLE 1. The Reactions which Buffer Oxygen Potential ($\bar{G}o_2$) within the Gap of a PWR Fuel Pin at 650K.

Reaction	$-\bar{G}o_2$ $kJ.mol^{-1}O_2$	Univariant Reaction	Other phases present
a	838	$2Cs + ZrO_2 + \frac{1}{2}O_2 \rightarrow Cs_2ZrO_3$	Cs_2Te, CsI, Mo
b	698	$4Cs + 2UO_{2+x} + [3/2 - x]O_2 \rightarrow Cs_4U_2O_7$	Cs_2Te, CsI, Mo
c	649	$Cs_4U_2O_7 + \frac{1}{2}O_2 \rightarrow 2Cs_2UO_4$	Cs_2Te, CsI, Mo
d	587	$Cs_2UO_4 + Mo + (1+x/_2)O_2 \rightarrow Cs_2MoO_4 + UO_{2+x}$	Cs_2Te, CsI
e	584	$Cs_2ZrO_3 + Mo + 3/2O_2 \rightarrow ZrO_2 + Cs_2MoO_4$	Cs_2Te, CsI
f	475	$Cs_2Te + Mo + 2O_2 \rightarrow Cs_2MoO_4 + Te$	CsI
g	469	$Mo + O_2 \rightarrow MoO_2$	CsI, Te, Cs_2MoO_4

as those in the first group. The third group consists of those fission products which remain in elemental form and are insoluble in the fuel lattice; Kr, Xe, Ru and Tc are included in this group; they have rather similar release characteristics to those fission product elements in the second group. In the fourth group are the elements of low electronegativity which can readily substitute for the U cations in the urania lattice; the lanthanoid and actinoid elements, and Zr are in this group. The elements in the fourth group were not released. Mo was difficult to classify, it gave releases considerably lower than the elemental species in the third group, but there was a finite release indicating that it did not behave as a soluble cation on those elements of the fourth group.

Grimes et al (1989) have presented calculations of the sites occupied by and solution energies of fission product atoms and ions in the urania lattice. The fission product elements Xe, Rb, Cs, Ba, Sr, Br and I have

TABLE 2. The Reactions and their Rate Constant for the He-Cs-I System considered in the Model of the Radiation Chemistry of CsI Gas.

	Reaction	Rate Constant	Reference
1	$He \rightarrow He^*$	4.15×10^{-7} Ds^{-1}	[1]
2	$He \rightarrow He^+ + e$	9.95×10^{-7} Ds^{-1}	[1]
3	$He^+ + e^- \rightarrow He^*$	10^{14} M^{-1} s^{-1}	[2]
4	$He^* + CsI \rightarrow He + Cs + I$	10^{11} M^{-1} s^{-1}	[2]
5	$He^* + I_2 \rightarrow He + I + I$	10^{11} M^{-1} s^{-1}	[2]
6	$Cs + I_2 \rightarrow CsI + I$	4.43×10^{11} M^{-1} s^{-1}	[3] [4]
7	$CsI + I \rightarrow Cs + I_2$	1.28×10^{-3} M^{-1} s^{-1}	[3] [4]
8	$Cs + I + He \rightarrow CsI + He$	3.07×10^{11} M^{-2} s^{-1}	[5] #
9	$CsI + He^* \rightarrow Cs + I + He$	2.65×10^{-11} M^{-1} s^{-1}	[2]
10	$I + I + He \rightarrow I_2 + He$	6×10^{8} M^{-2} s^{-1}	[5]
11	$I_2 + He^* \rightarrow I + I + He$	57.6 M^{-1} s^{-1}	[5]
12	$He^* + CsI \rightarrow He + Cs^+ + I^-$	10^{11} M^{-1} s^{-1}	[2]
13	$I + e \rightarrow I^-$	10^{11} M^{-1} s^{-1}	[2]
14	$CsI + e \rightarrow Cs + I^-$	10^{11} M^{-1} s^{-1}	[2]
15	$I_2 + e^- \rightarrow I + I^-$	10^{11} M^{-1} s^{-1}	[2]
16	$He^+ + Cs \rightarrow Cs^+ + He$	10^{11} M^{-1} s^{-1}	[2]
17	$He^+ + CsI \rightarrow Cs^+ + I + He$	10^{11} M^{-1} s^{-1}	[2]
18	$Cs^+ + I^- \rightarrow CsI$	10^{14} M^{-1} s^{-1}	[2]
19	$Cs^+ + e \rightarrow Cs$	10^{14} M^{-1} s^{-1}	[2]
20	$He^+ + I^- \rightarrow He + I^-$	10^{14} M^{-1} s^{-1}	[2]

[1] G values (species per 100 eV) taken from Jesse and Sadauskis (1985)
[2] Typical rate constant for this type of reaction.
[3] Konashi et al (1983).
[4] Konashi et al (1984).
[5] Husain et al (1985).
According to Husain et al (1985) reaction 8 tends to $Cs+I \rightarrow CsI$ ($k = 5.4\times10^{9}$ $M^{-1}s^{-1}$) with increase in the partial pressure of He.

been considered using lattice simulation with the Mott-Littleton method and quantum mechanical techniques. It was shown that the charge state of the dissolved fission product species will depend on the stoichiometry of the fuel; for example iodine may be stable as I^+ at U sites or I^- at O sites in UO_{2+x}, but only as I^- at O sites in hypostoichiometric UO_{2-x}. This work has recently been extended (Ball and Grimes 1989) to investigate the effect of charge state on the diffusion of Xe in UO_2.

THE CHEMICAL CONSTITUTION OF FUEL OPERATING AT HIGH RATINGS

We have already discussed the chemical constitution of the fuel-clad gap and discussed the transport of fission product atoms and ions within grains to grain boundaries of the fuel. The grain boundaries and intragranular bubbles provide sites for the nucleation of new phases. The behaviour of the fuel matrix to transients of energy and the consequent increase in temperature will depend on the rate of energy

TABLE 3. The Effect of Radiation on the Pressures of the Gaseous Species in the Gap of a PWR Fuel Pin at 650K.
Condensed Phases of CsI, Cs_2Te.

			Partial Pressures (bar)			
CsI	Cs	I	I_2	Cs_2Te	Te	Te_2
$\bar{G}o_2 = -698$ kJ.mol^{-1}						
Thermal Equilibrium						
1.8×10^{-8}	1.6×10^{-2}	1.2×10^{-29}	9.6×10^{-52}	1.3×10^{-8}	5.8×10^{-37}	1.8×10^{-57}
With Radiation						
1.8×10^{-8}	1.6×10^{-2}	1.8×10^{-15}	2.6×10^{-31}	1.3×10^{-8}	2.2×10^{-17}	3.9×10^{-35}#
1.8×10^{-8}	1.6×10^{-2}	2.5×10^{-17}	1.0×10^{-34}	1.3×10^{-8}	2.2×10^{-17}	3.9×10^{-35}Δ
$\bar{G}o_2 = -400$ kJ.mol^{-1}						
Thermal Equilibrium						
1.8×10^{-8}	4.6×10^{-14}	4.3×10^{-18}	1.2×10^{-28}	1.3×10^{-8}	6.9×10^{-14}	2.6×10^{-11}
With Radiation						
1.8×10^{-8}	4.6×10^{-14}	8.5×10^{-8}	2.6×10^{-6}	1.3×10^{-8}	1×10^{-5}	3.5×10^{-2}*

With 2nd order rate constants for reaction 8, Table 2.
Δ With 3rd order rate constants for reaction 8, Table 2
* With 2nd and 3rd order rate constants for reaction 8, Table 2

ingress. Most of our considerable knowledge of the chemical effects of burn-up derive from post-irradiation examination of oxide fuels for all reactor systems, but the greatest amount has come from an examination of fast reactor fuel - a solid solution of urania and plutonia (with Pu:(U+Pu) ratios of 0.15-0.40). The formalism for the thermodynamic treatment of the multicomponent system was developed some twenty years ago by Rand and Roberts (1966) and has been much exploited and developed since then. Kleykamp (1985) has provided an excellent review of the chemical state of the fission product elements in oxide fuels. Most of the examples presented and considered in detail are taken from studies on fast reactor fuel. These examinations have allowed the following classifications of the fission product elements (from Se to Gd):

Volatile fission product elements:
Kr, Xe Rb, Cs, Br, I, Se and Te.
Fission product elements which form metallic phases:
Mo, Tc, Ru, Rh, Pd, Ag, Cd, In, Sn and Sb.
Fission product elements which form oxide phases:
Rb, Cs, Sr, Ba, Zr, Nb, Mo.
Fission product elements which dissolve as oxides in the fuel matrix:
Sr, Zr, Nb and Y, La, Ce, Pr, Nd, Pm and Sm.

TABLE 4. The Chemical State of Fission Products in Irradiated Urania.

Fission Product Elements	Likely Chemical State
Kr, Xe	Elemental state
Y, La-Gd and actinoids	Oxides which dissolve in host matrix
Ba, Sr	Oxides which can dissolve to a limit extent in the fuel and also form separate phases. $Ba_{1-x}Sr_x[Zr_{1-w-y-z} Mo_w U_y An_z] O_3$#*
Br, I	Single phase halide solution. $Cs_{1-x}Rb_xBr_{1-y}I_y$
Rb, Cs	$Cs_{1-x}Rb_xBr_{1-y}I_y$ and compounds analogous to $(Cs_{1-x}Rb_x)_2(U_{1-y}An_y)O_4$ #
Se, Te	Single phase chalcogenide solution. $(Cs_{1-x}Rb_x)_2 Se_{1-y}Te_y$
Zr, Nb	Some dissolution in host matrix, see also Ba, Sr group
Mo, Tc, Ru, Rh, Pd	Usually single phase alloy, sometime two phase. Some Mo can oxidise to MoO_2 and also form a compound analogous to Cs_2MoO_4-$(Cs_{1-x}Rb_x)MoO_4$
Ag, Cd, In, Sn, Sb	Fission yields low; alloyed.

An represents the actinoids other than U.
* Lanthanoids can also be found in this phase.

Molybdenum can be found both in an alloy phase and an oxide phase. Some of the fission product elements are found distributed between the fuel matrix and a separate oxide phase. The likely chemical state of the fission product elements in irradiated urania is summarised in Table 4. Post irradiation examination of fast reactor fuels has shown that the predominant phases are the 'so-called' white metallic inclusions, Mo-Tc-Ru-Rh-Pd alloy and the 'so-called' grey ceramic phase which contains mainly the oxides of barium, molybdenum, uranium, plutonium, and zirconium (Kleykamp et al 1985). Much attention has been given to the analysis of the metallic alloy (Bradbury et al 1966, O'Boyle 1969, 1969, Potter 1974, Kleykamp 1989). The metallic phases mainly have an hexagonal structure (Kleykamp 1983, Bramman 1968) which can exist over a wide range of composition (Paschoal et al 1983, Haines et al 1979, Rand and Potter 1981). The main components of the alloy are Mo and Ru and these elements can migrate from hotter to colder regions of the fuel, as can Pd. If the oxygen potential is sufficiently high, the component Mo can be oxidised. Volatile gas phase species such as the oxides and Cs molybdates and ruthenates can be responsible for transport of Mo and Ru, Pd would transport in elemental form. The composition of the metallic inclusions varies markedly and depends on the fission yields, the oxygen potential of the fuel matrix, the temperature gradient within the fuel, on the rating of the fuel and the extent of burn-up.

Two phase alloys have been observed with relatively high Mo contents; and at very high burn-ups with very small Mo contents. The oxidised Mo was assumed to dissolve in the fuel matrixd (Johnson et al 1973); the ratio of Mo concentration in the alloys to that in the fuel was used as an indicator of the local oxygen potential of the fuel which is based on the equilibrium

$$\text{Mo (dissolved in alloy)} + O_2 \text{ (in fuel)} \rightleftarrows MoO_2 \text{ (dissolved in fuel)}$$

There is in fact a limited amount of MoO_2 in the fuel matrix (Giacchetti and Sari 1976). Most of the oxidised molybdenum enters the multi-component oxide precipitate - the grey phase replacing the uranium, zirconium and other ions in the lattice. Thus the composition of the metallic phase can give indications of the local oxygen potential (Kleykamp 1979) and of the temperature of the fuel during irradiation. Nucleation of these phases does occur in LWR fuel and as Kleykamp (1989) indicates the quantitative analysis of the precipitates in such fuel is difficult due to the small size of such precipitates. Kleykamp (1989) describes the post-irradiation of two LWR fuels, one at low rating (Biblis A 210 Wcm^{-1}) up to 5.9% burn-up and one at high rating (KWO 430 Wcm^{-1}) up to 4.3% burn-up. In Fig.6, taken from Kleykamp, (1989) the compositions of the alloy are in the single phase (ε, hexagonal close-packed) region except for some from the rod at higher power. Those for the rod at higher power extend into the two-phase (β (body centred cubic) + ε) region. Such analyses of the composition of the precipitates in LWR fuel should aid in determining the conditions of oxygen potential and temperature in such fuel during operational, fault and accident conditions.

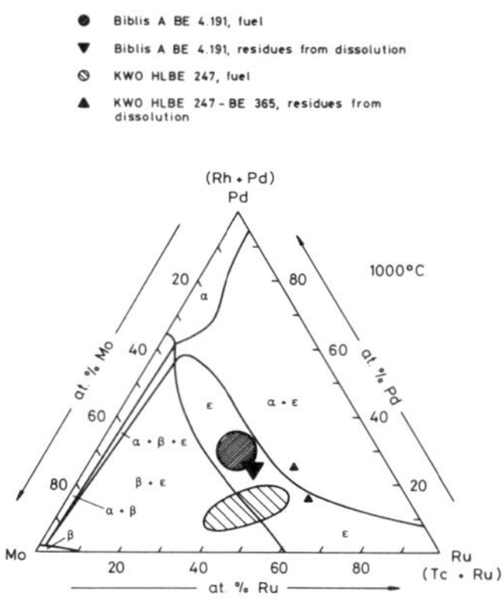

FIGURE 6. An isothermal section for the Mo-Ru-Pd system at 1000°C with the Mo-(Tc-Ru-(Rn-Pd) phases in irradiated LWR fuels.

The composition of the oxide phase, which has been observed in the matrix of fast reactor fuel has been discussed (Kleykamp 1985, Kleykamp et al 1985). The precipitates consist of phases derived from the oxides of uranium and plutonium with the fission product elements Ba, Sr, Cs, Zr and Mo together with the lanthanoids (Ln). Studies using electron probe microanalysis indicate a composition for this phase: $Ba_{1-x-y}Sr_xCs_y]$ (U, Pu, Ln, Zr, Mo)O_3. The Ln components could only be identified in fuel at high burn-ups. The amount of Mo in this phase clearly depends on the oxygen potential and the final oxygen anion metal cation ratio of the fuel; the concentration of Mo can reach a few percent at high values of $\bar{G}o_2$. These phases have the potential to nucleate on grain boundaries during the operation of LWR fuel at high ratings or at the higher temperatures found during fault and accident progressions providing that the rate of nucleation is sufficiently rapid.

A knowledge of the phase relationships essentially obtained from post-irradiation examination and laboratory measurements of simulated systems has provided much information for modelling the behaviour of fission products in operating, fault and accident conditions.

THE FAILURE OF THE FUEL RODS

If the Zircaloy cladding were to fail in fault conditions then those fission products present in the fuel-clad gap may enter the primary coolant. Much attention has been given to the fate of iodine, particularly of ^{131}I in such faults. Of particular concern has been the possible revolatilisation of iodine by radiolytic and thermal oxidation of the iodide anion in loss of coolant faults (Vikis 1989).

When the capacity to cool the reactor core is lost then the fuel will overheat and without restoration of the cooling, the fuel rods will degrade. The reaction between unoxidised Zircaloy and uranium dioxide occurs at 1300-1500°C; the minimum solidus temperatures in the U-Zr-O ternary phase diagram are at ca. 1300°C (Hofmann and Politis 1979). The reactions between the fuel matrix and Zircaloy, as already mentioned with respect to the fuel clad gap result in a drastic lowering of the oxygen potential of the fuel matrix. Detailed examination of the phase relationships of the U-Zr-O indicates that an alloy of uranium and zirconium can be formed in such reactions (Hofmann and Kerwin-Peck 1983, Olander 1983 and Yamanaka et al 1985). These reducing reactions result in the requirement to examine the constitution of the fission products before the liquidus temperature of the multicomponent system is reached. The volatile elements Cs, Rb, Br, I, Se and Te should be readily lost from the fuel under these conditions. The fate of the released fission product species in steam has to be considered in the modelling of their chemistry in other regions of the reactor system. Most of the fission product elements will remain with the core debris; the core debris will ultimately consist of two immiscible liquid phases one an oxide and the other essentially a metallic phase. The origin of the two immiscible liquids derives from the immiscibility between molten urania and stainless steel containing Mo, Tc, Ru, Rh and Pd. The extent of any liquid immiscibility in the U-Zr-O probably derives from immiscibility of uranium and urania liquids. A possible section of the U-Zr-O system at 2700°C is shown in Fig.7.

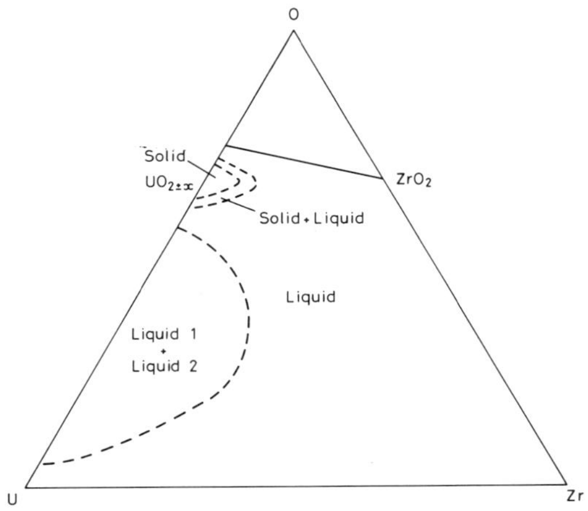

FIGURE 7. A possible isothermal section for the U-Zr-O system at 2700°C.

CONCLUSION

Studies of kinetic and thermodynamic aspects of fission product oxide fuel systems have enabled considerable understanding of the development of and changes in chemical constitution of fuel during operation and failure. The importance of having a detailed knowledge of kinetic and thermodynamic data of individual systems must be stressed in order to develop predictive capabilities to describe the behaviour of the multicomponent system - the fuel. Much understanding of the relationship between oxygen potential and the composition of the fluorite lattice is now available, and computer programs to solve multicomponent, multiphase equilibria are aiding the modelling of the evolution of chemical constitution in these systems. We must see the development of understanding of these complex problems of the chemical changes which occur during operational, fault and accident conditions as a continual interdependence of laboratory measurements of thermodynamic quantities, diffusion behaviour, radiation effects on specific systems, and examination of irradiated materials from operational and failed fuel.

ACKNOWLEDGEMENT

I am most grateful to Mr R G J Ball of Chemistry Division, Harwell Laboratory, for discussions and considerable help with the preparation of this paper. Part of this work was funded by the General Nuclear Safety Research Programme of the United Kingdom Atomic Energy Authority.

REFERENCES

Adamson, M G, Aitken, E A, Evans, S K, Davies, J H, (1975). Thermodynamics of Nuclear Materials 1974 vol.1, Vienna, IAEA p.57.

Aitken, E A, (1969). J. Nucl. Mater. 30 62.

Ball, R G J, (1989). Unpublished work, Harwell Laboratory.

Ball, R G J, Burns, W G, Henshaw, J, Mignanelli, M A, Potter, P E, (1989). J. Nucl. Mater. In press.

Ball, R G J, Grimes, R W, (1989). J. Chem. Soc., Faraday Trans 2, to be published.

Baranov, V G, Godin, Y G, (1981). Sov. Phys. Atomic Energy 51 633.

Besmann, T M, Lindemer, T B, (1985). J. Nucl. Mater. 130 489.

Besmann, T M, Lindemer, T B, (1986). J. Nucl. Mater. 137 292.

Bober, M, Schumacher, G, (1978). Advances in Nuclear Science and Technology 7 121.

Bradbury, B T, Demant, J T, Martin, P M, (1966). Report AERE R-5149.

Bramman, J L, Sharpe, R M, Thom, D, Yates, G, (1968). J. Nucl. Mater. 25 201.

Chen, T M, Kauzlarich, S M, Corbett, J D, (1988). J. Nucl. Mater. 151 225.

Eriksson, G, (1975). Chem. Scri. 8 100.

Ewart, F T, Manes, L, Matzke, Hj, Mari, C M, Toci, F, Schreiber, R, (1979). J. Nucl. Mater. 81 185.

Giachetti, G, Sari, C, (1976). Nucl. Technol. 31 62.

Goetzmann, O, (1982). J. Nucl. Mater. 107 185.

Grimes, R W, Catlow, C R A, Stoneham, A M, (1989). J. Amer. Ceram. Soc. In press.

Haines, H R, Potter, P E, Rand, M H, (1979). Thermodynamics of Nuclear Materials 1979 vol.I, IAEA, Vienna, p.471.

Hofmann, P, Politis, C, (1979). J. Nucl. Mater. 87 375.

Hofmann, P, Kerwin-Peck, D, (1983). Report KfK-3552.

Hussain, D, Plane, J M C, Xiang, C C, (1985). J. Chem. Soc., Faraday Trans 2, 81 769.

Hyland, G J, Ralph J, (1983). High Temp, High Press. 15 179.

Hyland, G J, (1984). Report EUR 9410.

Jesse, W P, Sadauskis, J, (1985). Phys. Rev. 100 1755.

Jackson, R A, Murray, A D, Harding, J H, Catlow, C R A, (1986). Phil. Mag. A 53 27.

Johnson, I, Johnson, C E, Crouthamel, C E, Seils, C A, (1973). J. Nucl. Mater. 48 21.

Kim, K, Olander, D R, (1981). J. Nucl. Mater. 102 192.

Kleykamp, H, (1979). J. Nucl. Mater. 84 109.

Kleykamp, H, (1983). In Nuclear Fuel Cycle eds. F Baumgartner et al, Verlag Chemie 2 151.

Kleykamp, H, (1985). J. Nucl. Mater. 131 221.

Kleykamp, H, (1989). J. Nucl. Mater. In press.

Kleykamp, H, Paschoal, J O, Pejsa, R, Thuemmler, F, (1985). J. Nucl. Mater. 120 426.

Konashi, K, Yato, T, Kaneko, H, (1983). J. Nucl. Mater. 116 86.

Konashi, K, Kamimura, K, Yokouchi, Y, (1984). J. Nucl. Mater. 125 244.

Leach, N A, Smith, M R, Pearce, J H, Ellis, W E and Beatham, N, (1988). IAEA Specialists' Meeting on Water Reactor Fuel Element Computer Modelling in Steady State, Transient and Accident Conditions, Preston, paper IAEA-TC-657/3.3.

Lindemer, T B, Besmann, T M, (1985). J. Nucl. Mater. 130 473.

Lindemer, T B, Brynestad, J, (1986). J. Amer. Ceram. Soc. 69 867.

MacInnes, D A, Winter, P W (1988). J. Phys. Chem. Solids 49 143.

Markin, T L, Street, R S, (1967). J. Inorg. Nucl. Chem. 29 2265.

Matzke, Hj, Ottaviani, J, Pellotiero, D, Rouault, J, (1988). J. Nucl. Mater. 160 142.

Matzke, Hj, (1989). Annales de Chemie, Science des Materiaux, Paris. In press.

Norris, D I R, (1977). J. Nucl. Mater. 68 13.

O'Boyle, D R, Brown, F L, Sanecki, J E, (1969). J. Nucl. Mater. 29 27.

Olander, D R, (1983). J. Nucl. Mater. 115 271.

Olander, D R, (1986). Advances in Ceramics, Vol.17 p.271 and Report LBL-21481.

Osborne, M F, Collins, J L, Lorenz, R A, Strain, R V, (1986). Source Term Evaluation for Accident Conditions. Vienna, IAEA p.89.

Paschoal, J O A, Kleykamp, H, Thummler, F, (1983). Z. Metallk 74 652.

Potter, P E, (1974). Behaviour and Chemical State of Irradiated Ceramic Fuels. Vienna IAEA p.115.

Potter, P E, (1987). in High Tech. Ceramics, ed. P Vincenzini, Amsterdam, Elsevier, p.2749.

Prussin S G, Olander, D R, Lau, W K, Hansson, L, (1988). J. Nucl. Mater. 154 25.

Rand, M H, Roberts, L E J, (1966). Thermodynamics vol. 1, Vienna IAEA p.3.

Rand, M H, Potter, P E, (1981). Physica B103 21.

Turnbull, J A, Friskney, C A, Findlay, J R, Johnson, F A and Walter, A J, (1982). J. Nucl. Mater. 107 168.

Une, K, Oguma, M, (1983). J. Nucl. Sci. Technol. 20 844.

Vikis, A C, (1989). Proceedings of the Second CSNI Workshop on Iodine Chemistry in Reactor Safety, Toronto, Canada, 1988, Report AECL-9923.

Walker, C T, Mogensen, M, (1987). J. Nucl. Mater. 149 121.

Wheeler, V G, Jones, I G, (1978). J. Nucl. Mater. 42 117.

Yamanaka, S, Katsura, M, Miyake, M, Imoto, S, Kawasaki, S, (1985). J. Nucl. Mater. 130 524.

Rare Earth and Actinide Release Characteristics from Solid and Molten Fuels

C. A. ALEXANDER and J. S. OGDEN
Battelle Columbus Laboratories
505 King Avenue
Columbus, Ohio 43201, USA

L. CHAN and S. B. BURSON
Nuclear Regulatory Commission
11555 Rockville Pike
Rockville, Maryland 20852, USA

ABSTRACT

A mass spectrometer with a heated inlet capable of reaching temperatures in excess of 3000K has been utilized to evaluate fission product and actinide release from both irradiated fuels and simulants. In addition to its high temperature capability, the mass spectrometer inlet can operate at a pressure of 30 bar hydrogen with up to 5 percent steam in the hydrogen gas. This inlet, as well as a conventional Knudsen cell inlet, has been employed to determine relative release rates for plutonium and several rare earths which are used as surrogates for plutonium. Release rates are shown to depend more on thermochemistry than on physical factors such as concentration or volatility of the elements. Release rates for lanthana are computed for two accident situations.

INTRODUCTION

In a severe nuclear accident, fuel temperatures can rise to levels where release of rare earths and plutonium can become significant, especially under ex-vessel releases. Not much is known of these release rates, and computation based on equilibrium thermodynamics along with estimates of solubility parameters have been used in most code development and accident simulation. Even in those cases where in-pile release rate experiments have been performed, only a very few percent of the fission product inventory is transported in these short time experiments. Predictions based either on the thermochemical estimates or on short time experiments on highly irradiated fuel are most likely to be biased because experiments have shown that a small part of the inventory releases anonalously fast and would over predict release rates.

An in-depth evaluation of lanthanum oxide properties will point out some of the shortfalls which can occur due to the complexities of the real solid solutions. In the present study, a minimum of 25 percent of the inventory was released, and all computations are based on these relatively large releases. Categorically, estimates which could be made on the first few percent would lead to over prediction of release. The release from the matrix is much slower than from grain boundaries or fission tracks or other defects which afford an easy path to the surface, as compared to bulk release from the fuel matrix.

EXPERIMENTAL

A mass spectrometer was utilized as the detector in these experiments. The spectrometer has been described previously (1)(2) and its use with a Knudsen cell(3) and with a high pressure flow inlet(4) have also been described. In the current experiments, the fuel was designated as BR-3 and had exhibited a high degree of burn-up. A total of 11 different determinations were made in this series and release rates for iodine, tellurium, cesium, xenon, strontium and barium were attained in addition to plutonium and the rare earths described herein. Results of the above species will be presented at a later date. In the experiments, a small weighed amount (1-5 mg) of silver is added to the chamber containing the fuel pellet. The silver is utilized as an internal standard for sensitivity and for relative ionization cross section determinations since the cross section for silver is well established.(5) All that is necessary in a determination is to measure an ion intensity, utilize the silver calibration to determine mass spectrometer transmission and sensitivity, and utilize published tables(6) of ionization cross sections in order to arrive at quantitative values for release rates. The same general procedure is used both in the Knudsen cell determinations and in the high pressure flow determinations.

The Lanthanum-Oxygen and the Lanthanum-Oxygen-Uranium System

While the present discussion is limited to only lanthanum among the rare earths, the other rare earth as well as plutonium would be expected to exhibit the same general characteristics. One significant characteristic of these rare earths is that the sesquioxides can exist over a homogeneity range down to the neighborhood of $R_2O_{2.8}$ or so. Concomitantly, there is a significant variation in the integral free energy of formation as shown in Figure 1 which is Figure 3 from Ackermann and Rauh.(7)

In the same paper, Ackermann and Rauh show (their Figure 2) that pLaO can vary over a factor of 400 between stoichiometric La_2O_3 and lanthana in equilibrium with molten lanthanum at a temperature of 2400K. That this behavior is not limited to lanthanum can be shown by the thermochemistry of yttria. At 1960K, the oxygen pressure in equilibrium with stoichiometric Y_2O_3 is 10^{-16} atm O_2 or higher while Y_2O_{3-x} in equilibrium with Y is 10^{-24} atm(8). In this study, cerium, dysprosium, erbium, europium, gadolinium, holmium, lutecium, praseodymium, samarium, terbium, thulium as well as lanthanum and yttrium showed the same general type of behavior, but none were as well suited as a potential crucible material for titanium melting as was yttria.

Figure 2 (9) illustrates the range of solubility of lanthana in urania which easily encompasses the entire range of rare earth fission product concentration. With the exception of the rare earths and fluorite structured dioxides, the urania lattice when predominantly in the plus four valence state tends to reject all other oxidic species.

RESULTS

In Figure 3, relative volatilities of lanthanum, cerium, europium, neodymium, praseodymium, and plutonium experimentally determined from

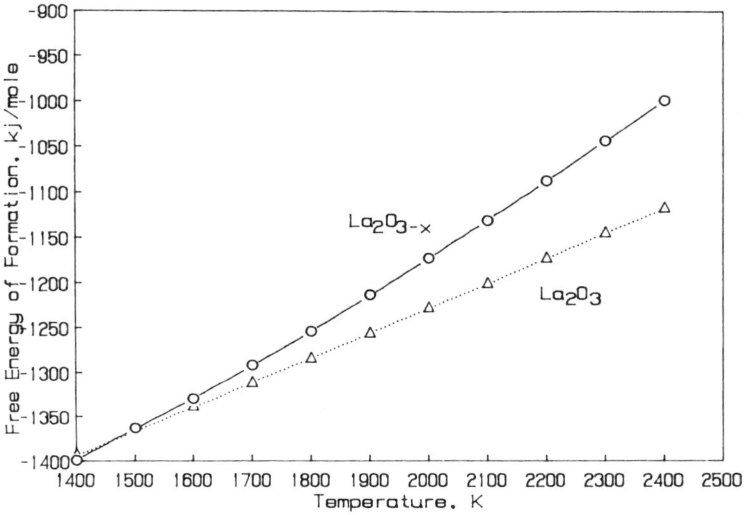

FIGURE 1. Free energies across the La_2O_3 homogeneity range.

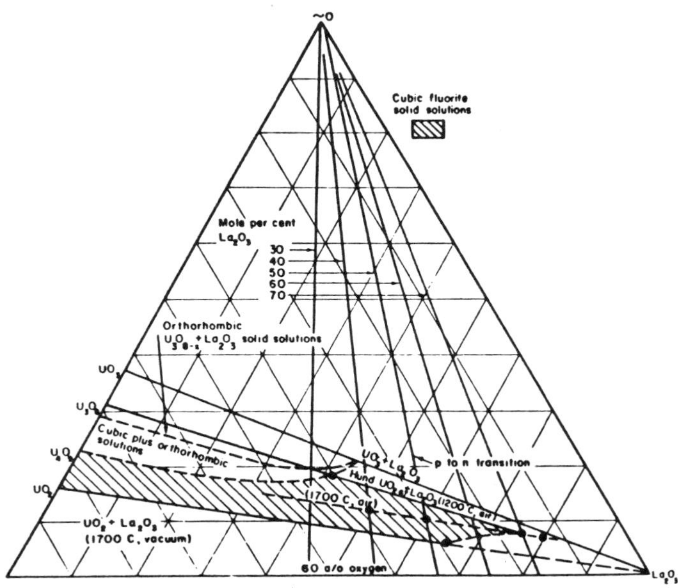

FIGURE 2. Pseudo phase diagram of the uranium-lanthanum-oxygen system.

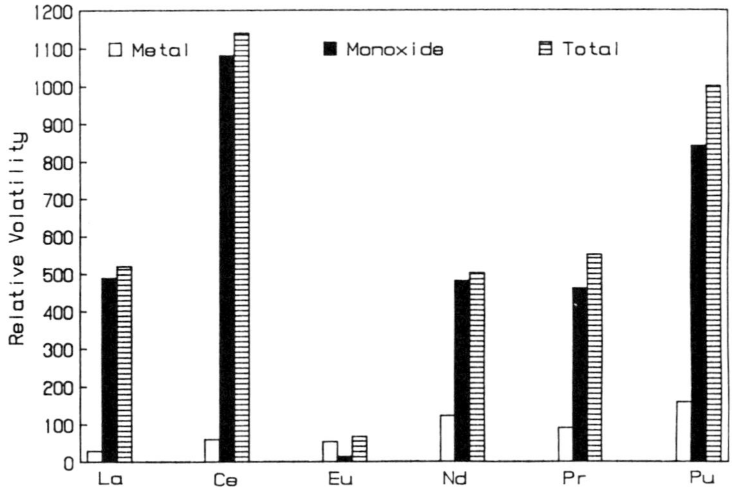

FIGURE 3. Relative volatilities at 2500K.

the BR-3 fuel are shown. The BR-3 fuel was recovered from a Belgian reactor and seemed to have an O/U ratio lower (i.e, the fuel was more reducing) than other fuels we have examined. Of the six radionuclides selected, cerium and plutonium are dominant at 2500K. Cerium would appear to be the best choice as a surrogate for plutonium.

A number of different factors might be expected to contribute to the magnitude of the release rate. We have selected concentration, volatility of the elemental form, and thermochemistry of condensed and gaseous species.

Concentration

Figure 4 shows the concentration of fission products relative to the lanthanum content. The concentrations are on a logarithmic plot, thus the amount of plutonium is 100 times that of lanthanum. As is evident cerium, europium, and plutonium show no indication that concentration plays a role in their release rate. If the operative mechanism were first order release kinetics, then $dn/dt \ \alpha$ - n. It is apparent that no generalization can be made that first order kinetics applies in all cases.

Volatility of Parent Metals

If the rate-controlling step is not driven by concentration alone, possibly the volatilization from the surface of the migrating metal atom may be a significant contributor. Figure 5 shows the vapor pressures at 2500K normalized by pm/pLa. Cerium for instance is one-tenth as volatile as lanthanum, and europium is nearly five orders of magnitude more volatile than lanthanum. Europium in the metallic state

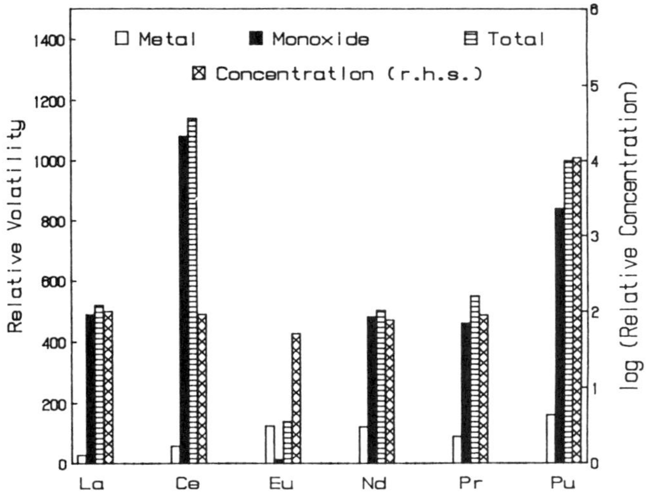

FIGURE 4. Relative volatilities at 2500K.

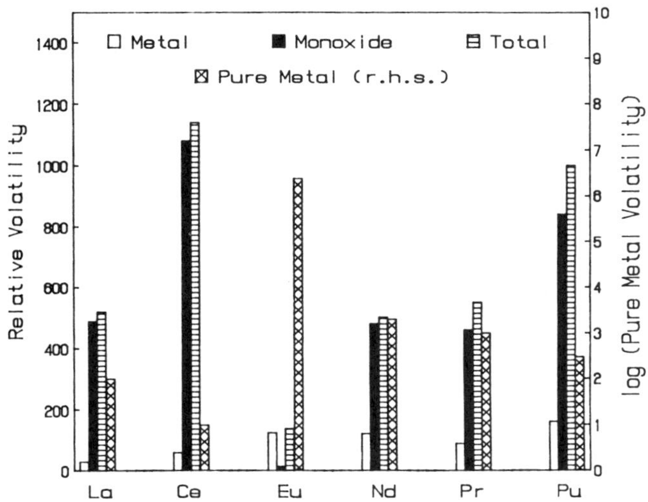

FIGURE 5. Relative volatilities at 2500K.

was not even the most volatile of metals observed, plutonium was significantly more volatile. It is apparent that while plutonium is more abundant there is no abundance-metal volatility relationship which can be applied which would not adversely affect lanthanum, cerium, neodymium, and praseodymium. We are left with the conclusion that volatility of the elemental metal plays no significant role in release rates for the rare earths.

Stability of Condensed Phase

One would intuitively think that the stability of the condensed phase would exhibit a profound effect on the release rate. In Figure 6, the release rates and the differences in standard free energy of formation of the sesquioxides with respect to lanthanum sesquioxide are shown. What is plotted is (ΔG_{2500} R_2O_3 - ΔG_{2500} La_2O_3). As can be seen, the sesquioxide of neodymium and praseodymium are more stable than La_2O_3 while the remaining sesquioxides are less stable. Cerium release correlates well, and plutonium shows a trend but neodymium and praseodymium do not, and europium is clearly counter to the supposition that the least stable in the condensed phase should have the highest release rates.

Free Energies of the Gaseous Monoxides

Since the prevalent form of volatilization (except europium) is as the gaseous monoxide, it would appear that likely the free energy of the gaseous monoxide is a significant factor. Figure 7 exhibits the volatilities and the difference of free energies of formation of gaseous monoxides relative to lanthanum (ΔG_{2500} RO - ΔG_{2500} LaO). Clearly it is obvious why europium is not present as a gaseous monoxide so the anomaly of europium is solved. The stability of the monoxide of cerium coupled with the low stability of Ce_2O_3 can explain the relatively high release rate for cerium as $CeO_{(g)}$. This trend follows throughout. If the free energy of the condensed phase is considered and then the free energy of the gaseous monoxide is considered, then it

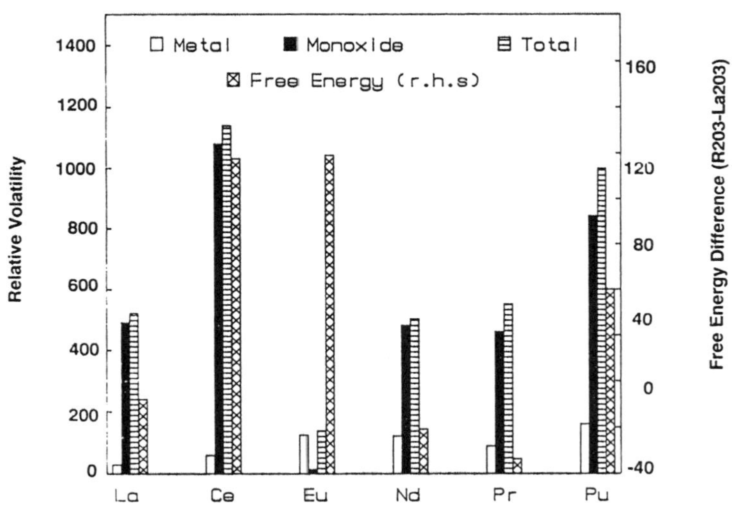

FIGURE 6. Relative volatilities at 2500K.

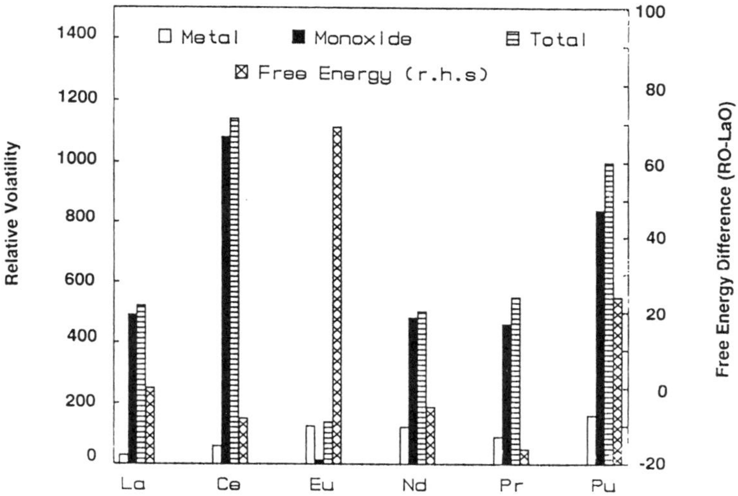

FIGURE 7. Relative volatilities at 2500K.

is possible to explain the observed variation in release rates. To most thermochemists, this conclusion is not surprising. It clearly illustrates that diffusion effects and first order kinetics--the so called mechanistic effects--are at most second order factors and that thermochemistry plays the most predominant role in determining release rates, certainly for the rare earths and plutonium.

Thermochemistry of Lanthana

Making thermochemical calculations is difficult because of the variation in free energy caused by the homogeneity range, the oxygen potential is generally not known, and the thermodynamic activity of the solute is not known. From Figure 1, at 2500K the free energy of formation of lanthana can vary between 1090 kj/mole and 960 kj/mole. For the reaction

$$La_2O_3 = 2LaO_{(g)} + 1/2\ O_{2(g)} \qquad (1)$$

$$K = \frac{(pLaO)^2\ (pO_2)^{1/2}}{(aLa_2O_3)} \qquad (2)$$

The equilibrium constant can vary between 3×10^{-9} if the free energy is as low as 960 kj/mole and 8×10^{-12} for 1090 kj/mole. The uncertainty in the value for the oxygen potential contributes at least another order of magnitude to the uncertainty in the value for pLaO as does this activity (aLa_2O_3). For the experiments performed on the BR-3 fuel at 2500K, the observables were:

$$pLaO = 2.5 \times 10^{-7}\ bar \qquad (3)$$

$pO_2 = 1.4 \times 10^{-12}$ bar (4)

The pressure of O_2 was not measured directly but obtained from the mass spectrometric measurements of the equilibrium

$$UO(g) + 1/2\ O_2(g) = UO_2(g). \quad (5)$$

Since the free energies are well known, a value for $O_2^{1/2}$ is readily obtainable from the pUO_2/pUO ratio.

If we employ the above values with a value for the equilibrium constant, we can compute the thermodynamic activity of the lanthana in the urania solid solution. With the 3×10^{-9} equilibrium constant, the thermodynamic activity is 1×10^{-8}. The molar concentration of La_2O_3 in the fuel is 3×10^{-4} moles, hence, on an ideal solution basis one would predict an activity of 3×10^{-4} while the measured values for $LaO(g)$ and O_2 would indicate a thermodynamic activity some 10^4 times lower than predictable, based on an ideal solution concept. The value for thermodynamic activity is a true representation of conditions at the fuel surface-gas interface. Care should be taken, however, in attempting to compute an activity coefficient, for undoubtedly there is a concentration gradient within the fuel and the concentration near the surface is markedly lower than the bulk concentration.

Loss Rate under Accident Conditions

In a recent publication (10) it was shown that loss rate in g/cm^2 can be related to rates determined in vacuum by the following relation:

$$\dot{m} = \frac{1.2 \times 10^{-8}\ \dot{m}_o\ T^{1.75}\ [\,1 + .05\ v]^{0.5}}{\ln\ (b/a)\ p^{0.64}} \quad (6)$$

In the above, $\dot{m}_o$ is the experimental or calculated vaporization rate in vacuum, T is temperature, kelvin, ln (b/a) is a geometrical factor which does include some gas properties. For hydrogen gas and fuel rods, 8 mm or thereabouts, ln (b/a) is of the order of unity. The velocity, v is expressed as a linear velocity in centimeters/second. The choice of units in the above expression were chosen as to be compatible with the major predictive codes.

Experimentally the observed loss rate in vacuum for ^{139}LaO at 2500K was 5.4×10^{-8} g/cm^2s. This is $\dot{m}_o$ in the above relation. If the gas is hydrogen at 7 bar and the velocity over the exposed fuel surface is 50 cm/sec then

$$\dot{m} = 2.4 \times 10^{-10}\ g/cm^2s. \quad (7)$$

To put the above in perspective, for an 8 mm diameter fuel rod a segment 4 mm in height would have 1 cm^2 surface area and would contain 1.2×10^{-3} g of ^{139}La once the fuel had undergone burn-up to 45000 MWD/T. For an incident like the Three Mile Island incident, SCDAP predictions indicate a temperature of 2500K was reached and held for about 25 minutes at a pressure of 65 bar in a hydrogen 50 percent steam atmosphere (11). The oxygen potential then was 2.5 x 10^{-5} or nearly 7 orders of magnitude higher than the experiments performed herein on the BR-3 fuel. The release rate then would be of the order of 10^{-13} g/sec cm^2. In the 25 minutes at temperature, about 1.5 x 10^{-8} g/cm^2 would have released. This is slightly more than 0.004 percent of the inventory which would release in containment.

If containment is maintained, then the release of rare earths and actinides do not reach levels of major concern.

1. Gingerich, K. A., J. Chem. Phys. **49** (1968) 14.

2. Ackermann, R. J. Rauh, E. G., Alexander, C. A., High Temp. Sci. **1** (1979) 304.

3. Alexander, C. A., Ogden, J. S., Chan, L., *Actinide Release from Irradiated Fuel at High Temperatures*, p. 77, Source Term Evaluation for Accident Conditions, IAEA, Vienna, 1986.

4. Alexander, C. A., Ogden, J. S., *Real Time Mass Spectrometric Evaluation of Fission Product Transport at Temperature and Pressure*, 2-21, Proceedings of the Symposium on Chemical Phenomena Associated with Radioactivity Releases during Severe Nuclear Plant Accidents, NUREG/CP-0078, 1986.

5. Pottie, R. F., J. Chem. Phys. **44**, (1966) 916.

6. Drowart, J., Goldfinger, P., Angew. Chem. Internat. Edit. **6** (1967) 581.

7. Ackermann, R. J., Rauh, E. G., J. Chem. Thermodynamics, **3** (1971) 445.

8. Lyons, S. R., Inouye, S., Niesz, D. E., Alexander, C. A., *Crucibles for Melting Titanium*, 239, Second International Conference on Titanium (1972).

9. Wilson, W. B., Gerds, A. F., Alexander, C. A., J. Inorg. Nucl. Chem. **20** (1961) 242.

10. Alexander, C. A. Ogden, J. S., J. High Temp. High Pres. Submitted (1988).

11. Hobbins, R. R., Langer, S., Owen, D. E., Akers, D. W., *Insights on Severe Accident Chemistry from TMI-2,* 4-1, Proceedings of the Symposium on Chemical Phenomena Associated with Radioactivity Releases During Severe Nuclear Plant Accidents, NUREG/CP-0078, 1986.

Matrix Isolation-Infrared Spectroscopic and Mass Spectrometric Studies of High Temperature Simulant Fission Products

ALAN K. BRISDON and J. STEVEN OGDEN
Department of Chemistry
University of Southampton
509 5NH UK

BRIAN R. BOWSHER and SHIRLEY DICKINSON
UKAEA Atomic Energy Establishment Winfrith
Chemistry Division
102/A50
Dorchester, Dorset DT2 8DH, UK

ABSTRACT

A novel experimental approach is described for the characterisation of fission product vapours over a wide range of temperatures and pressures. The combined techniques of matrix isolation-infrared spectroscopy and mass spectrometry are used to establish the identities of typical high temperature vapour species, and their molecular shapes and fundamental vibration frequencies. The apparatus used for these studies is described, and some recent results on simulant fission product chemistry involving the elements Cs, Te, Sn, and In are presented. The relevance of this work to the thermodynamic modelling of such systems is also discussed.

INTRODUCTION

In order to carry out detailed assessments of the consequences of hypothetical severe reactor accidents, data are required which define the quantities and physicochemical forms of the radioactive species released from the overheated core (1). In particular, the transport and attenuation of fission products in the primary circuit and containment building will depend to a large extent on the chemical species present. Under accident conditions, the fission products will be released from the UO_2 fuel in a pressurised water reactor (PWR) into an immediate environment of Zircaloy-zirconium oxide, stainless steel, Inconel, control rod debris, steam and boric acid. A realistic chemical database thus involves at least eighteen elements : U, Cs, I, Te, Ru, Mo, Zr, Sn, Fe, Cr, Ni, Mn, Ag, In, Cd, H, O and B.

There are two basic approaches to the general problem of establishing the chemical species formed in such a complex system. The first of these is essentially theoretical, and relies upon the use of existing thermodynamic data to predict equilibrium concentrations for the most likely chemical species. This approach appears very attractive because it allows predictions to be made over a wide range of pressures, temperatures and elemental compositions, as illustrated in the study by Garisto (2). However, it depends on the correct assumptions being made regarding the molecules present. If the required thermodynamic data are not available, or the system contains a significant proportion of an unsuspected species, the resulting predictions will not be reliable.

An alternative and complementary approach is to study the relevant high temperature systems in situ and to establish the identities and concentrations of the molecular species experimentally. Unfortunately, the difficulties encountered in the direct study of high temperature vapours are considerable and, despite recent advances in mass spectrometric sampling (3), most results continue to be obtained from the classical combination of vapour transport and chemical analysis of the condensate (4). However, in recent years matrix isolation has emerged as a powerful technique in the characterisation of high temperature vapours (5).

The work of these laboratories has been concerned with the characterisation of the vapour species released in vacuo from simulant fission products using molecular beam mass spectrometry and matrix isolation-infrared spectroscopy. Examples of the use of these techniques in the characterisation of molecular boric acid, and the application of the data to the thermodynamics of the caesium iodide-boric acid reaction are described in references 6 and 7. This work has recently been extended to simultaneous sampling from high temperature-high pressure systems. These latest developments are described together with results of recent studies on the characterisation of vapour species involving the elements Cs, Te, Sn, In, O and H.

EXPERIMENTAL DETAILS

The experiments described in this paper involve two methods of species characterisation - matrix isolation-infrared spectroscopy and mass spectrometry - and encompass widely different sampling conditions. Conventional mass spectrometric and matrix studies on high temperature systems have generally been confined to Knudsen-type vaporisation in vacuo, and the apparatus used in such studies is well documented (8). Sampling from a high pressure-high temperature region using a variety of small orifices and differential pumping is also a well-established technique in mass spectrometry (3), but has not previously been employed in matrix studies.

The apparatus used for this work was designed in such a way that matrix isolation and mass spectrometric studies could be made on the same vapour sample over a wide range of temperatures and pressures. Figure 1 shows how this has been achieved.

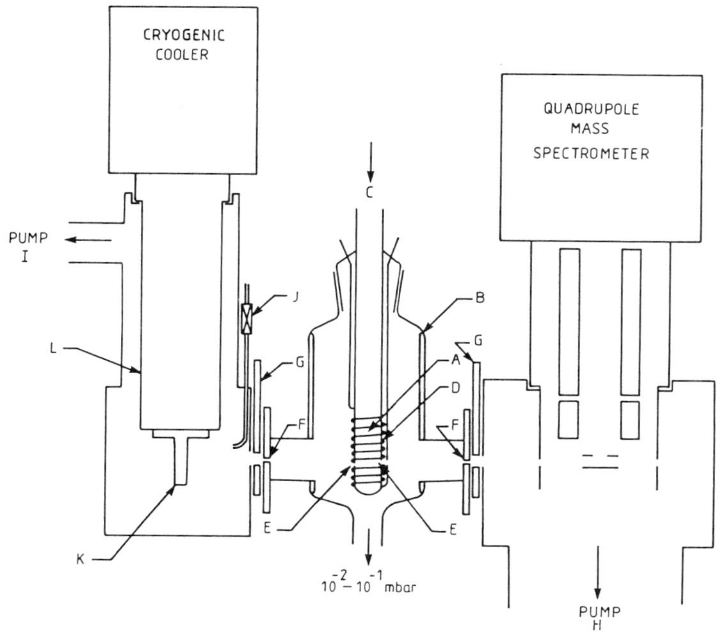

FIGURE 1. Combined matrix isolation - mass spectrometry system

The sample is contained in a Pyrex or silica holder (A) mounted in the centre of a water-cooled jacket (B). An appropriate gas is admitted to the sample via inlet (C), and sample heating is achieved by a simple resistance furnace (D). Two small holes (E, approximately 10 μm diameter) in the central sample chamber provide the source of two diametrically opposed molecular beams, which enter the differentially pumped region before passing through further pinholes (F, approximately 40 μm diameter) and sliding gate valves (G), and ultimately into the mass spectrometer and cryostat chambers. These are maintained at a pressure of typically 10^{-5} Pa using separate pumping systems (H,I) each comprising an oil diffusion pump backed by a rotary pump and incorporating a liquid nitrogen trap. The whole of the central vacuum shroud is constructed in glass, which permits visual inspection of the experiment, easy cleaning and the in situ unblocking of pinholes by a standard high fequency "Tesla" discharge coil. The matrix gas flow is controlled by a needle valve (J), and species for infrared investigation are co-condensed onto a caesium iodide or silicon window (K) held at ca. 12 K by a closed-cycle cryogenic cooler (L). A cross-beam axis mode of sampling is employed for mass spectrometric monitoring so as to minimise contamination of the quadrupole analyser. The current studies have examined the vaporisation of a variety of compounds at pressures up to 10^5 Pa, and it is intended to develop the apparatus to enable system pressures of 10^6 Pa to be studied.

Two infrared spectrometers were available for these experiments: a Perkin-Elmer 983G model and a Bruker IFS-113v. The accessible spectral range thus extends well into the far infrared. The mass spectrometers routinely used were a Vacuum Generators 12-12S quadrupole (1 - 2000 amu) interfaced with an IBM-AT compatible personal computer acting as a data station, and a lower resolution VG SXP600 quadrupole. The chemical system chosen to test the design of the apparatus was the high temperature-high pressure reaction between oxygen and metallic osmium, which is known to yield OsO_4. These initial experiments proved to be very satisfactory from the standpoint of spectral quality, and the subsequent results from systems of direct relevance to reactor chemistry are summarised below.

TELLURIUM OXIDE/WATER SYSTEM

Previous mass spectrometric studies have shown that when tellurium dioxide is heated in vacuo to 900 K, the principal vapour species released is molecular TeO_2, and that smaller amounts of TeO, Te_2O_2 and Te_2O_4 may also be present (9). This vaporisation process has also been studied by matrix isolation-infrared spectroscopy, and the TeO_2 vibrational fundamentals have been identified as 832 cm^{-1} (ν_1), 294 cm^{-1} (ν_2) and 846 cm^{-1} (ν_3). A typical nitrogen matrix infrared spectrum obtained from the vaporisation of solid TeO_2 is shown in Figure 2a.

Previous transpiration studies have shown that the volatility of tellurium oxide is enhanced in the presence of steam, and a possible explanation for this observation is the formation of a volatile oxide-hydroxide species such as $TeO(OH)_2$ (10). The TeO_2/H_2O system has therefore been studied using the pinhole sampling techniques described above in an attempt to characterise such an intermediate.

Figure 2b shows a typical nitrogen matrix-infrared spectrum obtained when steam at atmospheric pressure was passed over solid TeO_2 heated to ca. 700 K. The spectrum is dominated by an intense absorption at 1598 cm^{-1}, which is due to the bending mode in matrix isolated H_2O. However, continued heating to ca. 900 K results in the appearance of new features at 935, 846 and 717 cm^{-1} (Figure 2c). The 662 cm^{-1} band is due to CO_2, which was adsorbed on the solid TeO_2, and the feature at 846 cm^{-1} corresponds to the most intense fundamental of molecular TeO_2. The 935 cm^{-1} band, however, is new to this system, and seems to be a product of a possible TeO_2/H_2O reaction. Several studies were performed using deuterium-enriched water, which would result in a substantial frequency shift for all modes involving hydrogen atom motion. A typical spectrum obtained from such a study is shown in Figure 2d. The intense bands at 1598, 1404 and 1178 cm^{-1} correspond to the bending modes in H_2O, HDO and D_2O

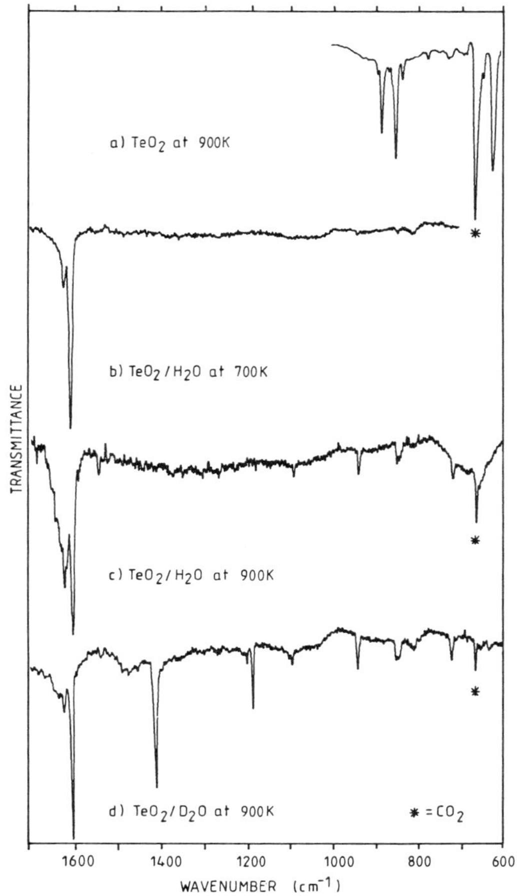

FIGURE 2. Matrix isolation-infrared spectra of TeO_2 systems

repectively, and the features at 846 and 935 cm^{-1} are again present, together with an additional absorption at ca. 840 cm^{-1}. It is believed that this absorption is the direct counterpart of the 935 cm^{-1} band, and confirms the presence of H(D) in this species. In view of its frequency, it is assigned as a Te-O-H(D) bending mode; further experiments are in progress to establish the positions of the remaining fundamentals, and to identify this new species in the vapour phase by mass spectrometry.

THE INDIUM/OXYGEN SYSTEM

Experiments on this system were carried out primarily to identify the chemical species produced when metallic indium reacts at high temperatures with an excess of oxygen at relatively high pressure. Previous studies on the vaporisation of indium oxides have established that the sub-oxide In_2O is the most important oxide species released (11). Figure 3a shows part of a typical nitrogen matrix-infrared spectrum obtained from the vaporisation of solid In_2O_3 heated in vacuo. One intense band is produced at 722 cm^{-1}, and this has been identified as the antisymmetric stretch (v_3) of molecular In_2O. Weaker features are noted at 550 and 412 cm^{-1}, and these have been assigned to the dimer In_4O_2 (12).

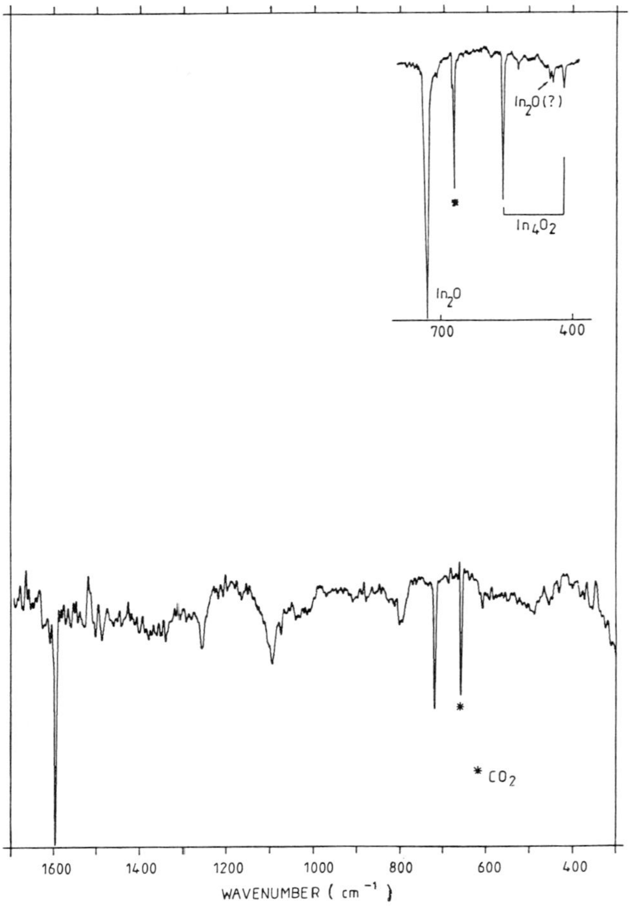

FIGURE 3. Matrix isolation-infrared spectra of indium-oxygen systems

Figure 3b shows the matrix infrared spectrum obtained using pinhole sampling of the vapour species present when indium metal is heated to ca. 1100 K in oxygen at atmospheric pressure. It is evident that under these conditions only In_2O has been isolated. Further experiments are in progress to establish the vapour species present when steam reacts with indium under these conditions, and in particular, whether volatile hydroxides might exist which could enhance the transport of indium.

CAESIUM/TELLURIUM/OXGYEN SYSTEM

As indicated above, the principal vapour species found in the tellurium-oxygen system at high temperatures is molecular TeO_2. However, in the presence of caesium, the chemistry is considerably more complex. Initial matrix isolation studies in this area have been briefly discussed elsewhere (13), but corresponding mass spectrometric data on this system have now been obtained, and the combined results are discussed below.

Typical matrix-infrared specta obtained from the vaporisation in vacuo of Cs_2CO_3/TeO_2 mixtures with different Cs/Te ratios are shown in Figure 4. For a Cs/Te atom ratio of 1:1, complex spectra are obtained (Figure 4a), with evidence for both terminal and bridging Te-O groups. This spectrum has not yet been fully interpreted, but may arise from a molecular species such as $Cs_2Te_2O_5$. Considerable simplification occurs, however, when the Cs/Te ratio is 2:1 (Figure 4b). This spectrum can be assigned to molecular

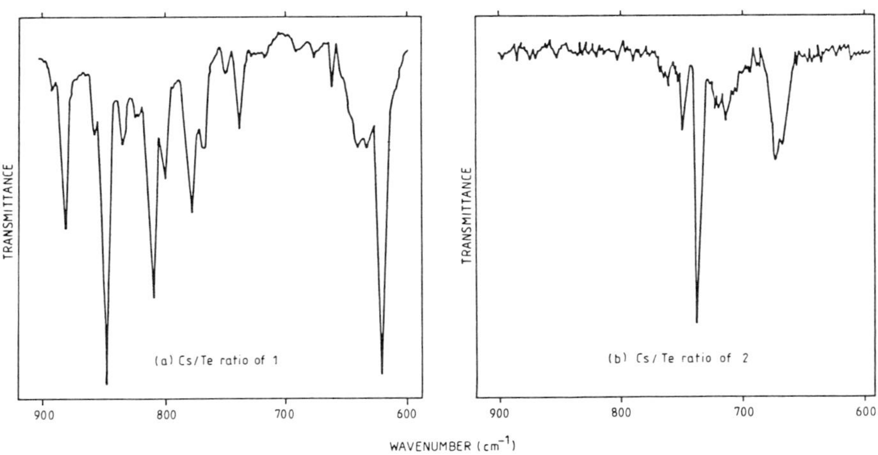

FIGURE 4. Nitrogen-matrix infrared spectra observed for Cs_2CO_3/TeO_2 mixtures with varying Cs/Te ratios

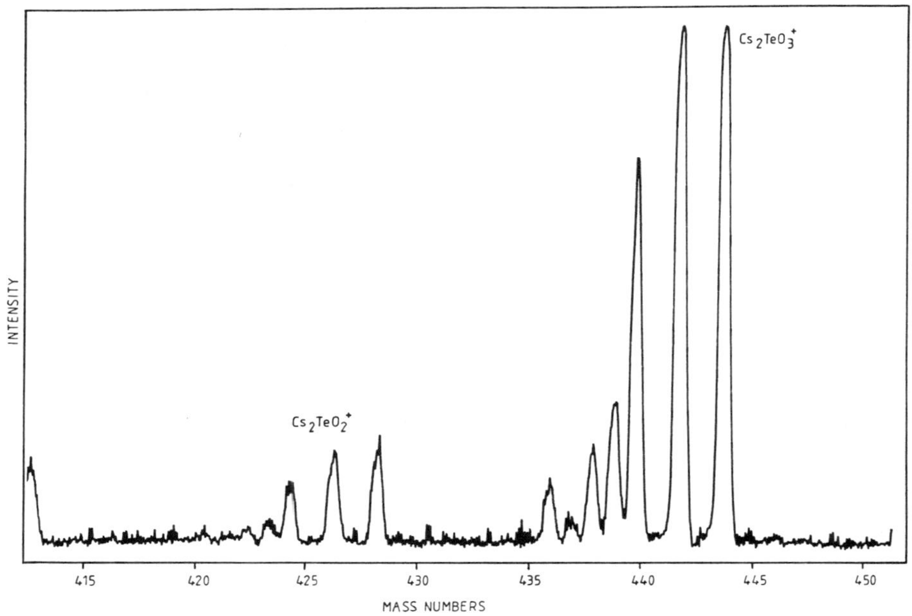

FIGURE 5. Mass spectrum observed for an equimolar Cs_2CO_3/TeO_2 mixture

Cs_2TeO_3, and O-18 enrichment studies suggest a pyramidal geometry for the central TeO_3 unit (14). These experiments provide the first vibrational characterisation of a ternary Cs-Te-O molecule, and it would appear that this species is reasonably stable at high temperatures.

The mass spectrum obtained from an equimolar mixture of Cs_2CO_3 and TeO_2 vaporised under identical conditions is reproduced in Figure 5, and shows the parent $Cs_2TeO_3^+$ ion, together with a small amount of $Cs_2TeO_2^+$ fragment. Molecular Cs_2TeO_3 thus emerges as an important high temperature species which is stable under high vacuum conditions, and which may be involved in the vapour transport of two important nuclides. Experiments are in progress to establish the coordination of the caesium atoms in this molecule, and the positions of the low-frequency caesium vibrations. It will then be possible to use these data to estimate the thermodynamic parameters of the molecule.

TIN/TELLURIUM SYSTEM

Stannous telluride is formed by reaction of tellurium with the tin component of Zircaloy cladding, and is believed to be an important compound in the transport of fission product tellurium (13). Mass spectra of the vapour above stannous telluride show intense peaks corresponding to Sn^+, Te^+ and $SnTe^+$ ions, together with weaker features due to Te_2^+, $SnTe_2^+$

and $Sn_2Te_2^+$ (15). Appearance potential measurements have shown that Sn^+ and Te^+ were fragments, whereas $SnTe^+$ and $Sn_2Te_2^+$ were parent ions. The origins of the Te_2^+ and $SnTe_2^+$ ions could not be established with certainty, although the appearance potential of $SnTe_2^+$ was consistent with its being a fragment of Sn_2Te_2. These results are in agreement with previous studies on this system (16, 17).

Infrared spectra of matrix-isolated stannous telluride vapour were recorded in both argon and nitrogen matrices over the frequency range 450 to 150 cm^{-1}, and routinely showed three features at ca. 258, 191 and 172 cm^{-1}. Figure 6 shows the spectrum from an argon matrix; the nitrogen matrix spectrum was similar but the frequencies were shifted by 1 to 2 cm^{-1}. The band at 258 cm^{-1} can be assigned to monomeric SnTe, since the observed frequency is in good agreement with the reported gas phase value of 259.5 cm^{-1} (18). The intensities of the two lower-frequency bands were found to increase when lower matrix gas flows or higher vaporisation temperatures were used (Figure 6b) indicating that these bands arose from polymeric species in the matrix. Furthermore, the intensity ratio of the two lower-frequency bands was the same in all experiments, although their intensity relative to the monomer peak varied widely according to the spray-on conditions. This indicates that the two bands arose from a single species, most probably the dimer Sn_2Te_2. Although the amount of dimer formation in the matrix is expected to be larger than in the equilibrium vapour mixture, particularly at low matrix gas/vapour ratios, the mass spectrometric studies show that this dimer is present in measureable quantities in the vapour phase.

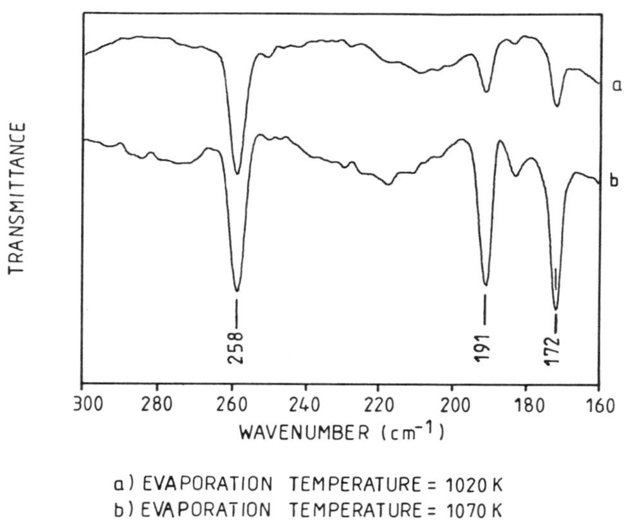

FIGURE 6. Matrix isolation-infrared spectrum of tin telluride

Matrix isolation spectra of a number of analogous Group IV-Group VI systems have been reported (eg. SiO(19), GeO(20), SnO(21), PbO(22, 23), SiS(24), GeS, GeSe, SnSe and GeTe(25)), and a general feature of these systems is that the infrared spectra characteristically show monomer/dimer features similar to those observed for the Sn/Te system. In particular, it has been postulated that the dimer molecules have planar ring structures (D_{2h} symmetry), a conclusion which is consistent with the observation of two infrared-active stretching frequencies with corresponding Raman studies in the case of Pb_2O_2 (23). On the basis of this planar ring structure, expressions for the five in-plane modes of Sn_2Te_2 have been derived in terms of the force constants and molecular geometry. A Sn-Te-Sn angle of 96° was calculated based on the observed frequencies for the B_{2u} and B_{3u} modes, and the stretching constant was calculated as 120 N m^{-1}; the bending force constant and interaction terms were arbitrarily set at 5 N m^{-1}. The frequencies of the remaining in-plane modes were thus estimated to be 185, 170 and 53 cm^{-1}. The out-of-plane B_{1u} mode, although infrared active, is predicted to occur well below the frequency range of these experiments (approximately 20 cm^{-1}). However, despite these assumptions regarding the unobserved fundamentals, it is possible to estimate the thermodynamic properties of Sn_2Te_2 with some confidence by considering an appropriate uncertainty range (see below).

INDIUM TELLURIDE SYSTEM

The mass spectrum obtained when In_2Te_3 was heated to 1100 K shows peaks due to In^+, Te^+, $InTe^+$, Te_2^+ and In_2Te^+. In_2O^+ was also observed, and this may have arisen from oxide impurities in the sample, or by reaction with the silica sample holder. These results are in agreement with previous studies of this compound (26-29). There is some disagreement in the literature as to the source of these ions. Colin and Drowart (26) assigned them all as parent ions, with In^+ and Te^+ also arising from fragmentation; Belousov et al (27) considered only In^+ and In_2^+ to arise solely from fragmentation, whereas Srinivasa and Edwards (28) and Berger et al (29) concluded that only In_2Te^+ and Te_2^+ were parent species. Not surprisingly, there are also discrepancies in the proposed reaction schemes for In_2Te_3 vaporisation, although it is agreed that in the early stages only Te_2 vapour is formed. Indium-containing species become more significant as the composition of the solid becomes more indium-rich, eventually reaching InTe which vaporises congrently as a solid (29) but not as a liquid (30). A simplified reaction scheme is therefore:

$$In_2Te_3(s) \rightarrow 2InTe(s) + 1/2\ Te_2(g)$$
$$2InTe(s) \rightarrow In_2Te(g) + 1/2\ Te_2(g)$$

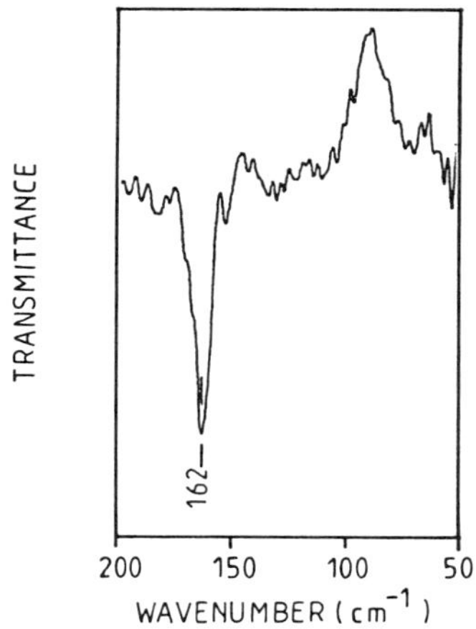

FIGURE 7. Matrix isolation-infrared spectrum of In_2Te

The infrared spectrum of matrix-isolated In_2Te is shown in Figure 7. The spectrum consists of a single band at 162 cm^{-1} (in a N_2 matrix) or 164 cm^{-1} (in an Ar matrix). In some experiments, a weak band was observed at about 130 cm^{-1}, but as this did not appear under all conditions it was assumed to be due to trace polymeric species. The position of the main band is close to that observed at 155 cm^{-1} for In_2Te in a xenon matrix (31) and was assigned to the asymmetric stretching frequency. By analogy with In_2O and Ga_2O (12), In_2Te is expected to be a non-linear molecule (C_{2v} symmetry) and would therefore have three infrared-active vibrations. Assuming the same relationship between the frequencies of the symmetric and antisymmetric vibrations as is observed for In_2O and Ga_2O - $\nu_1^2 = 0.34\ \nu_3^2$ - the symmetric stretching frequency for In_2Te was estimated to be 95 cm^{-1}. The lack of any observed band within 30 cm^{-1} of this frequency indicates that the bond angle is fairly large, and that this mode is only weakly active in the infrared.

THERMODYNAMIC MODELLING

Knowledge of the molecular structures and vibration frequencies leads in principle to the estimation of thermodynamic functions such as absolute entropy and specific heat capacity, and the recent mass spectrometric and matrix isolation-infrared studies on boric acid have already prompted a reassessment of thermodynamic equilibria involving molecular H_3BO_3 (7). The studies described above may be similarly used in thermodynamic modelling of systems containing Sn/Te and In/Te, and the results of this work are summarised below.

The only additional information required for the calculation of thermodynamic functions of SnTe is the bond length, and this was estimated to be 0.26 nm (32). The calculated thermodynamic functions are listed in Table 1. These results were found to be fairly insensitive to errors in the estimated bond length, with variations in the entropy terms of about 1.5% resulting from changes of 0.05 nm.

TABLE 1. Selected Thermodynamic Functions for SnTe

Temp (K)	H_T-H_{298} (J/mol)	Cp_T (J/mol/K)	S_T (J/mol/K)	$-(G_T-H_{298})/T$ (J/mol/K)
298	0	36.4	263.4	263.4
500	7435	37.1	282.5	267.6
1000	26049	37.3	308.3	282.2
500	44725	37.4	323.4	293.6

Estimates of the four remaining vibrational frequencies are required for Sn_2Te_2, in addition to the molecular geometry. A planar ring structure with alternating tin and tellurium atoms was assumed, with bond lengths of 0.3 nm (from ionic radii) and Sn-Te-Sn angles of 96°. The unobserved frequencies were estimated as 185, 170, 53 and 20 cm^{-1} as described above. Results of the calculations are given in Table 2. A limited study was also carried out to assess the effects of varying the force constants, bond length and out-of-plane bending frequency; the entropy and free energy functions were found to change by typically 2 to 4% for quite large variations in the estimated parameters.

TABLE 2. Selected Thermodynamic Functions for Sn_2Te_2

Temp (K)	H_T-H_{298} (J/mol)	Cp_T (J/mol/K)	S_T (J/mol/K)	$-(G_T-H_{298})/T$ (J/mol/K)
298	0	81.1	399.8	399.8
500	16544	82.4	442.2	409.1
1000	57927	83.0	499.5	441.6
1500	99435	83.1	533.2	466.9

The observation of a single infrared band for the In_2Te molecule suggests a fairly wide bond angle, and a value of 150° was assumed. The bond length was taken to be 0.27 nm (the In-Te distance in solid In_2Te_3). The frequencies of the symmetric stretch was estimated as 95 cm^{-1}, based on studies of analogous molecules, and the low frequency bending mode was estimated at 30 cm^{-1}. Results of the calculations are given in Table 3. Again, variations of 2 to 4% in the calculated entropy and free energy functions were observed when the estimated input parameters were varied over a reasonable range.

TABLE 3. Selected Thermodynamic Functions for In_2Te

Temp (K)	H_T-H_{298} (J/mol)	Cp_T (J/mol/K)	S_T (J/mol/K)	$-(G_T-H_{298})/T$ (J/mol/K)
298	0	57.6	348.2	348.2
500	11687	58.0	378.1	354.7
1000	40736	58.2	418.4	377.6
1500	69819	58.2	441.9	395.4

CONCLUSIONS

A combined matrix isolation-infrared and mass spectrometric technique has been developed for the characterisation of vapours at high temperatures and pressures. This combination provides information on the chemical identities of vapours, and their thermodynamic functions.

A number of compounds of relevance to severe accident analyses have been studied using this approach. These studies have demonstrated the potential importance of mixed fission product-reactor materials compounds in fission product transport, and have also illustrated the enhanced transport of specific compounds in steam.

ACKNOWLEDGEMENTS

The authors wish to thank C A Alexander (Battelle-Columbus Laboratories), A L Nichols (AEE Winfrith) and R A Gomme (University of Southampton) for their involvement in this work. These studies were funded by the UK Department of Energy, the Central Electricity Generating Board, the Commission of European Communities (Joint Research Centre, Ispra) and the AEA Underlying Research Programme.

REFERENCES

1 Gieseke, J.A., Cybulskis, P., Denning, R.S., Kuhlman, M.R., Lee, K.W. and Chen, H., BMI-2104, Vols I-VI, 1984.

2 Garisto, F., AECL 9552, 1988.

3 Stearns, C.A., Kohl, F.J., Fryburg, G.L. and Miller, R.A., NBS Special Publication No 561, 303, 1979.

4 Bowsher, B.R., Dickinson, S., Nichols, A.L., Ogden, J.S. and Potter, P.E., Proc. ANS Topical Meeting on Fission Product Behaviour and Source Term Research, Snowbird, July 1984, NP-4113-SR, 27-1, 1985.

5 Weltner, W., Adv. High Temp. Chem., 2, 85, 1969; Snelson, A. in "Vibrational Spectroscopy of Trapped Species", editor H.E. Hallam, John Wiley, London, 1973.

6 Ogden, J.S. and Young, N.A., J. Chem. Soc. Dalton Trans., 1645, 1988.

7 Bowsher, B.R., Dickinson, S., Ogden, J.S. and Young, N.A., Thermochim. Acta, 141, 125, 1989.

8 Hastie, J.W., editor "Characterisation of High Temperature Vapours and Gases" NBS Special Publication No. 561, 1979.

9 Muenow, D.W., Hastie, J.W., Hauge, R., Bautista, R. and Margrave, J.L., Trans. Faraday Soc., 65, 3210, 1969.

10 Malinauskas, A.P., Gooch, J.W. and Redman, J.D., Nucl. Appl. Tech., 8, 52, 1970.

11 Burns, R.P., J. Chem. Phys., 44, 3307, 1966.

12 Hinchcliffe, A.J. and Ogden, J.S., J. Phys. Chem., 75, 3908, 1971.

13 Bowsher, B.R., Dickinson, S., Gomme, R.A., Jenkins, R.A., Nichols, A.L. and Ogden, J.S., Proc. Workshop on Chemical Reactivity of Oxide Fuel and Fission Product Release, CEGB Berkeley Nuclear Laboratories, April 1987, Vol 2, 455, 1988.

14 Gomme, R.A., Ogden, J.S. and Bowsher, B.R., UK Atomic Energy Authority Report AEEW-R 2204, 1986.

15 Dickinson, S., UK Atomic Energy Authority Report AEEW-R 2227, 1987.

16 Lyubimov, A.P. and Bespal'tseva, I.I., Izv. Akad. Nauk. SSSR, Neorg. Mater., 5, 1289, 1969.

17 Colin, R. and Drowart, J., Trans. Farad. Soc., 60, 673, 1964.

18. Barrow, R.F. and Vago, E.E., Proc. Phys. Soc., 56, 78, 1944.

19. Anderson, J.S. and Ogden, J.S., J. Chem. Phys., 51, 4189, 1969.

20. Ogden, J.S. and Ricks, M.J., J. Chem. Phys., 52, 352, 1970.

21. Ogden, J.S. and Ricks, M.J., J. Chem. Phys., 53, 896, 1970.

22. Ogden, J.S. and Ricks, M.J., J. Chem. Phys., 56, 1658, 1972.

23. Khanna, R.J. and Park, Y.J., Spectrochim. Acta, 42A, 603, 1986.

24. Atkins, R.M. and Timms, P.L., Spectrochim. Acta, 33A, 853, 1977.

25. Marino, C.P., Guerin, J.D. and Nixon, E.R., J. Mol. Spectrosc., 51, 160, 1974.

26. Colin, R. and Drowart, J., Trans. Farad. Soc., 64, 2611, 1968.

27. Belousov, V.I., Vendrikh, N.F., Novoshilov, A.F. and Pashinkin, A.S., Izv. Akad. Nauk. SSSR, Neorg. Mater., 17, 1190, 1981.

28. Srinivasa, R.S. and Edwards, J.G., Monat. fur Chemie, 117, 695, 1986.

29. Berger, L.I., Strel'chenko, S.S., Bonder, S.A., Molodyk, A.D., Balanevskaya, A.E. and Lebedev, V.V., Izv. Akad. Nauk. SSSR., Neorg. Mater., 5, 872, 1969.

30. Srinivasa, R.S. and Edwards, J.G., J. Electrochem. Soc., 131, 2954, 1984.

31. Perov, P.A., Maltsev, A.A. and Shevelkov, V.F., Vestn. Mosk. Univ. Ser. 2: Khim., 16, 109, 1975.

32. Hirayama, C., Ichikawa, Y. and DeRoo, A.M., J. Phys. Chem., 67, 1039, 1963.

Fuel Behavior and Fission Product Release under Realistic Hydrogen Conditions with Comparisons between Heva 06 Test Results and Vulcain Computations

J. M. DUMAS, G. LHIAUBET, G. LEMAROIS, and G. DUCROS
Commissariat a L'Energie Atomique
CEN/FAR B.P. No. 6
F-92265 Fontenay aux Roses
Cedex, France

I - INTRODUCTION

Because of the importance of source term predictions for LWR beyond design basis safety analyses , the CEA / IPSN has developed an extensive programme to characterize the source of fission products (F.P.) released from overheated fuel and their evolution in the primary coolant system .

The VULCAIN code is a module of the ESCADRE system which is the French LWR accident analysis computer code package (1) . It describes the thermal behaviour of the core during its dewatering up to its loss of geometry , the release kinetics of F.P. and structure materials , and the thermal hydraulics in the primary coolant system , with a view to providing the main parameters governing the vapor and aerosol transport mecanisms.

The HEVA programme , a series of 8 experiments performed to date , is aimed at characterizing the fission product source associated with postulated severe accident conditions (2) . These tests , conducted at Grenoble by the department of metallurgy , were carried out in the HEVA loop with spent fuel reirradiated in the SILOE or MELUSINE reactors just prior to the experiment .

During each HEVA test transient , numerous measurements are recorded to provide key results on the following aspects :

- F.P. release kinetics as a function of temperature , carrier gas composition and cladding oxidation state ,
- cladding oxidation kinetics ,
- deposition of the F.P. and structure material vapors ,
- characteristics of the fission product aerosol source as a function of fluid temperature and composition ,
- F.P. and control rod material chemistry ,
- Ag , In and Cd control rod vaporization and interaction of these components with the F.P. (HEVA 07 and HEVA 08 tests) .

An analysis of core degradation computations performed with the VULCAIN code for the AF scenario (WASH 1400 terminology) was used to define the objectives and operating parameters of the HEVA 06 test . This experiment was designed to study F.P. release and behaviour at low pressure under realistic hydrogen conditions liable to occur in a steam-starved zone of a PWR core during its degradation .

The purpose of this paper is to show how the VULCAIN code was used to define the HEVA 06 test with realistically simulated reactor accident conditions and , reciprocally , how large benefits may be achieved by performing such an experiment for the prediction of volatile F.P. release .

II - VULCAIN RESULTS AND HEVA 06 TEST DEFINITION

II - 1 The AF accident sequence

The VULCAIN code was used to study a large family of core melt accident scenarios involving core degradation at an instant when the primary cooling system is at low pressure . Among these scenarios , the AF accident sequence reported here , is initiated by a large break LOCA (double ended guillotine failure of the cold leg) . In this hypothetical severe PWR accident , the containment spray and the auxiliary feedwater system are not in operation .

The unavoidable core dewatering , which appears one day after reactor shutdown , is due to the failure of the safety injection system in recirculation mode (deterioration of the low pressure injection pumps by recycling hot water from the containment sump) .

Particular attention has been focussed on this type of scenario because the release of F.P. occurs a short time before the containment venting system is used to avoid exceeding the design basis pressure of the reactor building .

II - 2 Analysis of F.P. release during the AF core melt sequence

II - 2 - a Gas composition at the core exit

The F.P. release starts 70 minutes after the beginning of core dewatering . It was primarely analysed on the basis of mean gas composition at core outlet . The time origin has been shifted and is taken as corresponding to the beginning of core uncovering . As seen from figure 1 , the release transient can be divided into 3 time periods :

- during the first period (70 - 100 minutes) the F.P. are transferred to the core exit in a carrier gas mainly composed of steam ,
- the fluid is then increasingly enriched with hydrogen because the steam starvation occurs in the upper part of the core . The carrier gas becomes pure hydrogen in the time period 100 - 130 minutes ,
- the steam finally reappears and becomes dominant at the core outlet in the third period of time before the core slump .

The second period has particularly engaged our attention with regard to definition of the HEVA 06 test . A substantial release of volatile F.P. is predicted in the upper part of the core during this period . Moreover the F.P. are transported by a flowing gas (figure 2) mainly composed of hydrogen (45 % in mass) , rare gas (25 %) and cesium (12 %) .

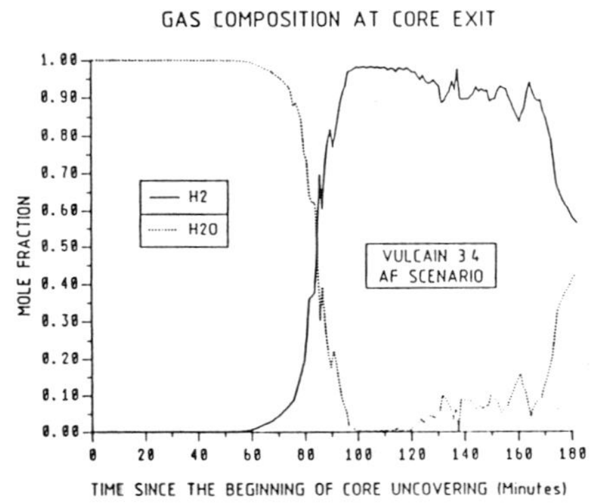

FIGURE 1

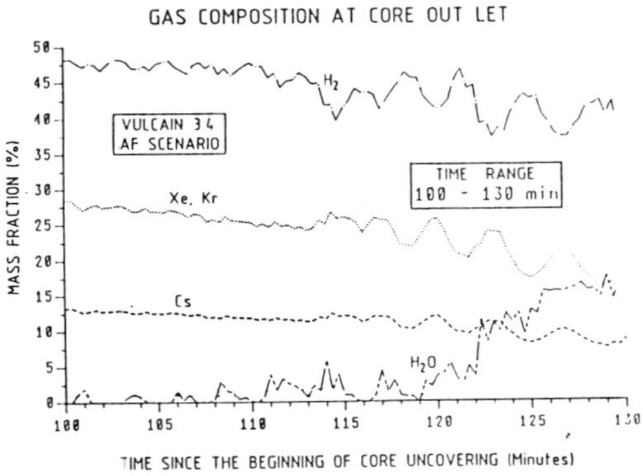

FIGURE 2

II - 2 - b F.P. release into the core cooling channels

In the VULCAIN computations , a standard 900 MWe PWR core is usually divided into 5 radial rings (i.e. mean fluid channels) and 24 equally spaced axial meshes to represent the discrepancies along the different channels from the centre to the edge of the core . Cross-flow models were applied to the fluid distribution through the core .

The gas composition at the exit of each mean channel was analysed in the time range 100 - 130 minutes . Some differences were observed between the 5 radial rings , especially the outer ring , which represents the core edge where the residual power is lower . Channel n° 4 has been taken as representative of the mean gas composition at the core outlet . The results concerning the fluid characteristics at the exit of ring n° 4 are reported in table 1 and compared with the mean parameters of the gas entering the upper plenum .

	CHANNEL N° 4		AVERAGE AT CORE EXIT	
elements	mass flowrates (g/s)	- mass fraction (%)	mass flowrates (g/s)	- mass fraction (%)
Xe,Kr	23.	26.	62.	22.
I	0.85	1.0	2.3	0.8
Cs	10.5	11.5	30.	10.5
Rb	1.25	1.5	3.7	1.3
Mo,Tc	1.0	1.6	4.5	1.8
Sr	0.3	0.4	1.2	0.45
H2	40.	48.	138.	44.

Table 1 : mean composition and flows in the time range 100 - 130 minutes after the beginning of core uncovering

The release of cesium (representative of volatiles) was therefore analysed all along the cooling channel n° 4 . As illustrated in figure 3 , Cs is increasingly escaped from the fuel in the lower axial meshes rather than the higher ones . This effect is explained by a temperature escalation due to cladding oxidation which is more and more violent as the core becomes more deeply uncovered . Figure 4 demonstrates that , in the upper part of the core (mesh n° 18) , approximately half of the cesium inventory is released in a fluid mainly composed of hydrogen .

In conclusion , experiments involving F.P. releases in a non-oxidizing environment , such as the HEVA 06 test , are particularly interesting to perform since no related data are available from current literature and , since they are in addition representative of degraded core conditions , and also because a large F.P. release is predicted in the upper part of the core when the steam starvation occurs .

II - 3 Cladding oxidation state and heat-up rate of fuel

The oxidation state of the zircaloy cladding , to be reproduced in the HEVA 06 test , was derived from VULCAIN computations carried out by using the CATHCART - PAWEL oxidation kinetics law . During the steam starvation phase , various oxidation states can be observed in the upper part of the core . Nevertheless 2/3 of the cladding is oxidized on an average when the volatiles escape from the same axial location .

This result was therefore taken into account as a limit to conduct the oxidation phase of the fuel sample by following the production of hydrogen throughout the experiment . A temperature of 1300 °C was selected to control the preoxidation of the clad with a rate high enough to avoid slow kinetics (break-away problems) but not too high to prevent any significant escape of F.P. from the fuel .

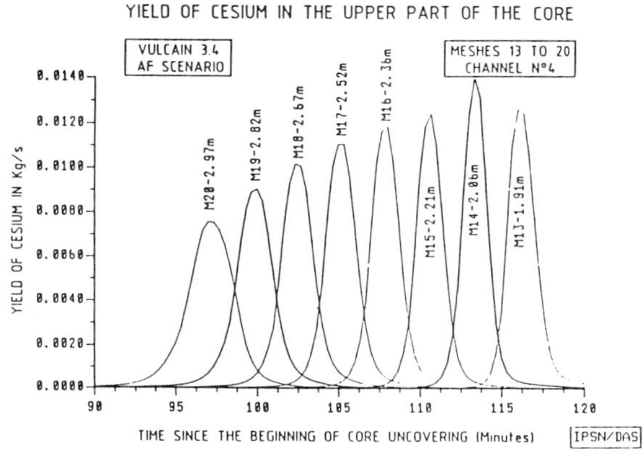

FIGURE 3

II - 4 H2 mass flowrate estimated for the HEVA 06 test

The mass flowrate of hydrogen circulating through the furnace during the temperature plateau at 2100 °C has been computed in such a way as to respect a Cs/H2 mass ratio of 0.25 deduced from the VULCAIN results reported in table 1 . This is truly important in order to reproduce the realistic conditions of concentration (i.e. partial pressure) of elements involved in the formation (i.e. thermodynamic equilibriums) of the F.P. chemical species . Precalculations of the F.P. release rate for the HEVA 06 test were performed with the EMIS module (part of VULCAIN which can be run as a stand-alone code) to obtain a mean cesium production of 0.05 mg/s during the reducing phase , which involves to adjust the H2 mass flowrate to 0.2 mg/s . The EMIS module will be detailed in section IV - 1 , in which precalculated and observed releases are compared .

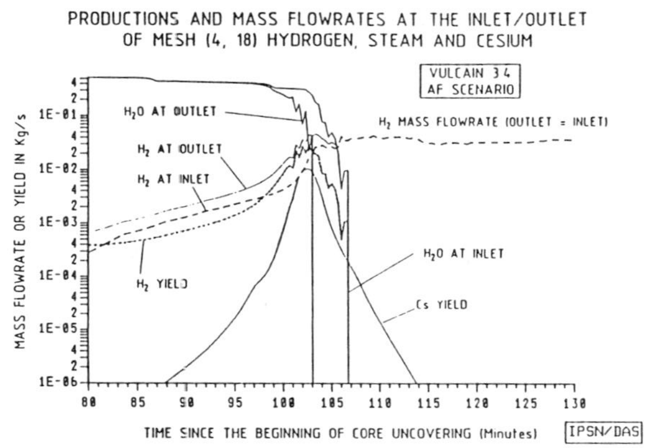

FIGURE 4

III - HEVA 06 EXPERIMENT RESULTS

III - 1 The HEVA experimental facility

III - 1 - a Description of the HEVA cell

The accident sequence takes place in a hot cell . The experimental circuit shown in figure 5 is composed of the following elements :

- a vertical high frequency induction furnace used to heat a fuel pin segment ,
- a main circuit of gas flowing around the fuel rod segment (the mass flowrate and the composition of the fluid (H2O and H2) preheated at 800 °C before entering the HF induction furnace , are continuously monitored) ,
- a secondary circuit , lightly pressurized with helium , protects the graphite susceptor from oxidation ,
- a cascade impactor , composed of 6 plates and 2 ball diffusion beds , is located 40 cm above the fuel sample (it is designed to trap selectively aerosols sorted by size in a typical range of cut-off diameter from 1 to 20 um for the impactor plates and from 0.02 to 1 um for the 2 granular beds . The remaining particles that exit the impactor , are captured in an poral filter . The temperature of this aerosol collector is maintained at 250 °C throughout the sequence) ,
- a condenser to ensure watertightness , a dryer and finally a charcoal trap to capture the active gases (essentially xenon and krypton) .

III - 1 - b Test hardware and instrumentation

During the experiment , a data acquisition system records the information provided by pressure transducers , flowmeters and temperature sensors located all along the experimental loop . The fuel segment temperature is measured by a Re sheathed / HfO2 insulated W - Re thermocouple . Three additional chromel - alumel thermocouples situated at the inlet and exit of the furnace may be used to control the fuel temperature in case of failure of the W - Re thermocouple .

The following parameters are also processed by the data logger :

- the fraction of the cladding oxidized , deduced from a differential on-line measurement of the inlet (H2) and outlet (H2 + He) mass flowrates ,
- and the current emitted by the ionization chamber in direct view of the impactor .

Three gamma-spectrometry examination stations provide the information listed below :

- the release kinetics of F.P. from degraded fuel continuously processed during the transient (straight sighting to the lower pellet of the fuel rod segment composed of 3 highly irradiated fuel pellets surrounded by 2 depleted uranium oxide blocks) ,
- just after the test , the activity distribution of the impactor before dismantling ,
- and the activity of the different components of the test facility before and after the sequence in order to draw the full balance with respect to the initial F.P. inventory (inclusive release of F.P. and distribution all along the circuit) .

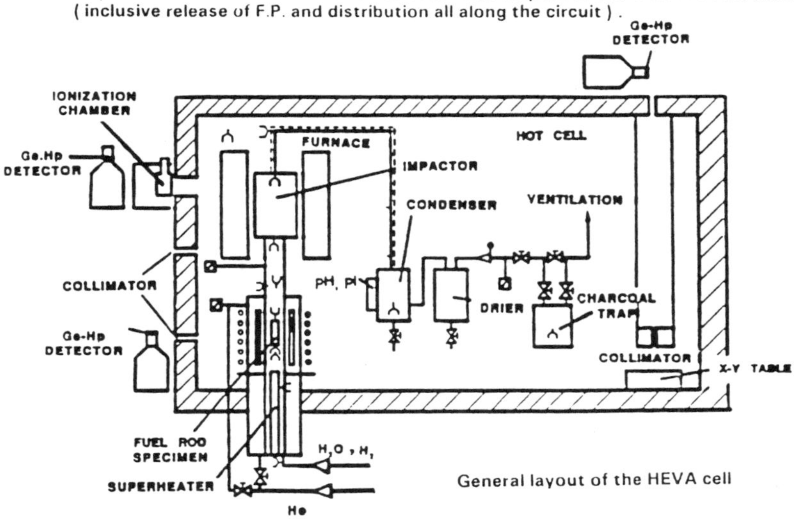

General layout of the HEVA cell

FIGURE 5

III - 2 The experimental sequence

The main parameters of each HEVA test are summarized in table 2 . Experiment n° 6 was performed with an 8 cm fuel segment cut from a rod irradiated in a commercial PWR (burn-up of 36550 MWd/tU) . The fuel sample was reirradiated in an experimental reactor at Grenoble in order to recreate the short-lived fission products liable to be present in the fuel after the reactor scram (especially iodine , tellurium , barium ...) . Fuel reirradiation is one of the most original and important technical aspects developed in the HEVA programme .

PARAMETER		TEST NUMBER							
		1	2	3	4	5	6	7	8
Fuel burn-up (GMd/tU)		19.4	19.4	28	36	36	36	0	36
Reirradiation		no	no	no	yes	yes	yes	no	yes
Gas injection	H2	0	0	0.03	0.03	0.03	0.012	0.03	0.03
(g/min at 800 °C)	H2O	6	1.8	2.2	1.8	1.5	0	1.5	1.5
Heat-up rate (°C/s)		1	1	1	1	1	2	1	1
Temperature (°C)		1630	1870	1800	2000	1800	2100	1800	1800
Hold time (min)		15	15	30	7	96	30	30	12
Impactor temperature		no	no	800	600	250	250	250	250
Control rod aerosols		no	no	no	no	no	no	yes	yes
Date of test		1983	1983	Apr-86	Dec-86	Jul-87	Mar-88	Jul-88	Mar-89

Table 2 : HEVA test matrix - main parameters

The HEVA 06 experimental sequence can be described as follows :

- The fuel temperature was increased to 1300 °C and controlled at this level during 50 minutes to reach a 2/3 cladding oxidation state (H2O and H2 mass flowrates of 25 mg/s and 0.2 mg/s respectively) .
- After that , the steam was progressively replaced by helium and a heat-up rate of 1.5 °C/s was induced to reach a fuel temperature of 2100 °C .
- Then the furnace power was adjusted to maintain the fuel at this temperature for half an hour in an environment mainly composed of hydrogen . A few minutes after the beginning of the temperature plateau , the W - Re thermocouple failed , but after that instant , all the test control parameters remained steady .
- Finally , natural cooling was operated by switching off the furnace power supply .

At the exit of the furnace , the hydrogen is diluted in a helium flow which has been adjusted to obtain an efficient and selective collection of aerosols , given the design characteristics of the impactor .

III - 3 Experimental results

As seen in figure 6 , the post-test metallographic examination (cross-sectional view of pellets) of the fuel sample highlights severe degradation of the fuel rod segment with , in particular ,

- fuel-cladding interaction with the formation of low melting point eutectics between UO_2 and the unoxidized zircaloy ,
- flow of molten cladding down to the fuel cracks ,
- fuel swelling of 9.4 % on an average (radial expansion) ,
- some blisters of the oxide layer caused by the expanding pool of molten zircaloy .

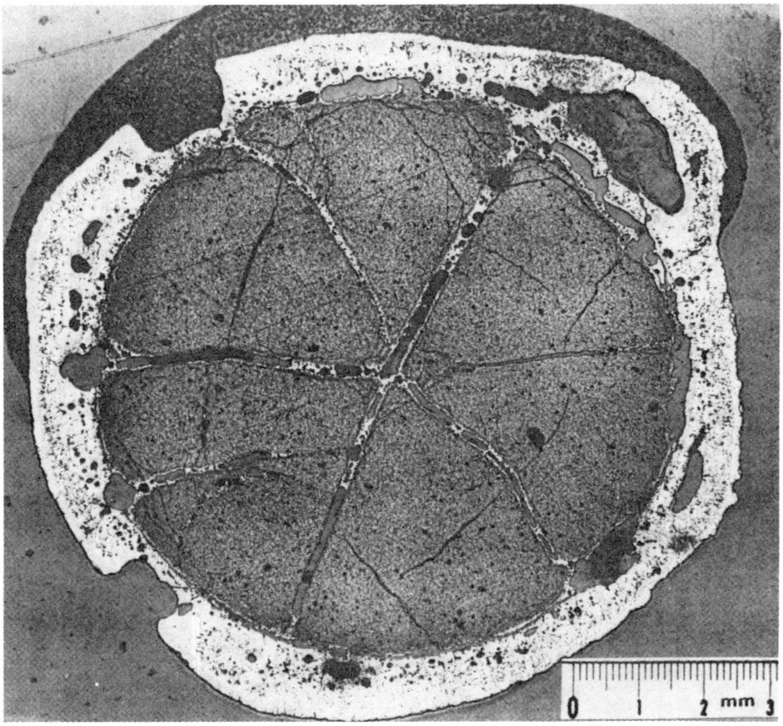

FIGURE 6 Post-test appearance of fuel after the HEVA 06 test (cross-sectional view of the lower pellet)

The on-line gamma spectrometry performed by straight sighting of the lower fuel pellet , provides the F.P. release kinetics . Figures 7 to 10 show the results obtained for I131 , Te132 , Cs137 and Ba140 . The experimental values , reported on each figure , represent a mean F.P. release from the integral fuel rod segment . The F.P. release from the lower pellet was locally higher .

At the end of the temperature plateau , the released fractions are approximately 30 % for iodine , cesium and barium and only 10 % for tellurium .

The Xenon isotopes , especially Xe135 , were followed throughout the transient . Release of this rare gas is thereby evaluated from the cold trap activity measured after the sequence . The total releases of iodine and xenon are fairly similar .

Another fission product of major interest , because chemically close to barium and radiologically important , is strontium . Its release has been deduced from the escape of Y91m which is the decay product of Sr91 . Y91m was correctely followed throughout the test . It seems to confirm a release of strontium similar to that of barium .

Molybdenum and europium would have slightly escaped from the fuel pin , approximately 5 % (value too low to be accurately quantified) .

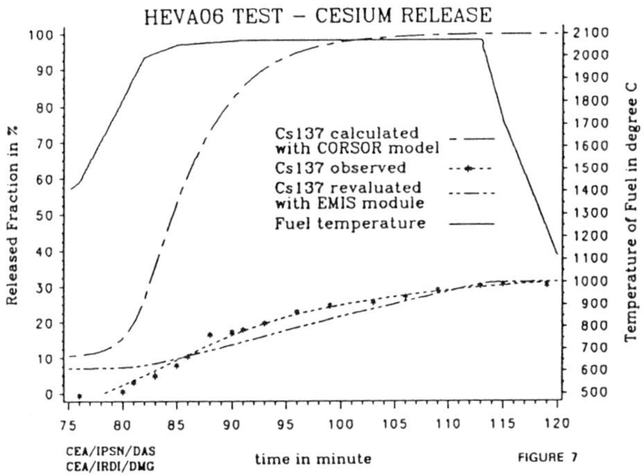

FIGURE 7

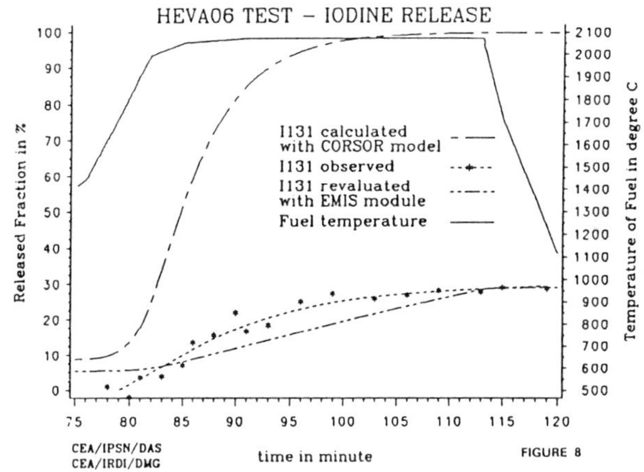

FIGURE 8

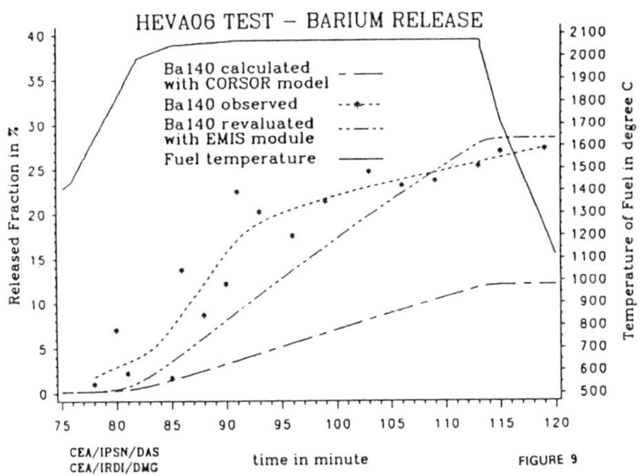

FIGURE 9

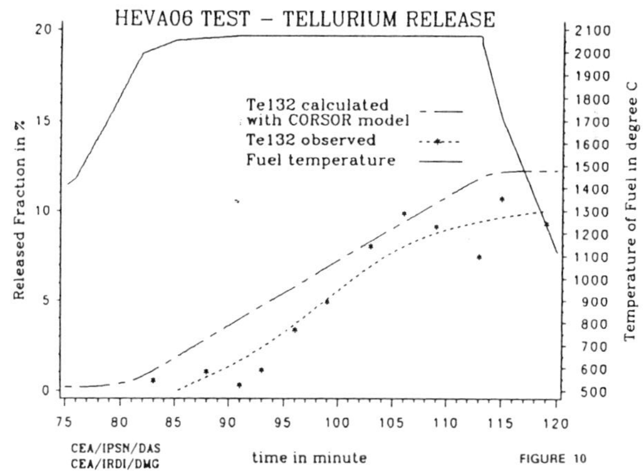

FIGURE 10

A complementary gamma tomography examination of the lower fuel pellet was made after the sequence to provide the radial distribution of F.P. . This type of examination was difficult to interpret because of a limited set of activity measurements (projections) and also owing to the severely degraded appearance of the fuel as seen in figure 6 . Nevertheless this kind of investigation clearly demonstrates that significant quantities of tellurium and barium were trapped in the cladding . Radial distributions of I131 and Te132 activity are compared in figures 11 and 12 .

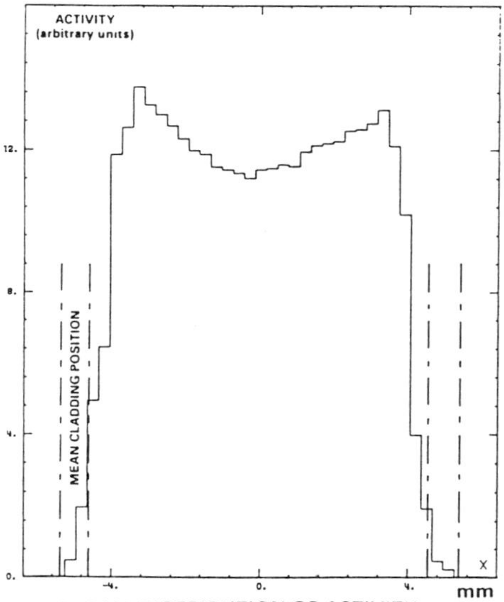

RADIAL DISTRIBUTION OF ACTIVITY

HEVA 6 * I 131

FIGURE 11

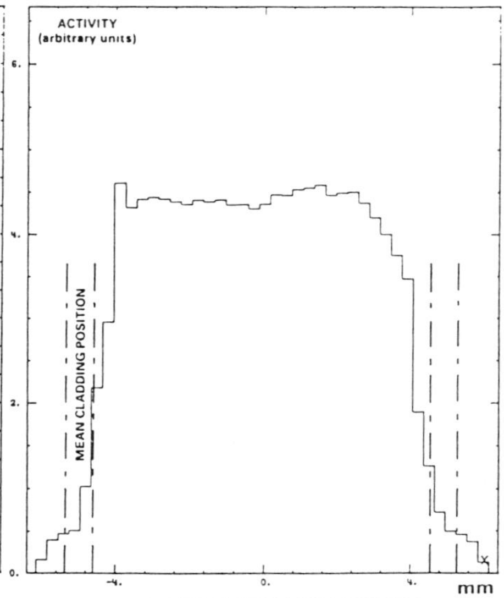

RADIAL DISTRIBUTION OF ACTIVITY

HEVA 6 * TE 132

FIGURE 12

IV - COMPARISONS OF CALCULATED AND OBSERVED F.P. RELEASES - DISCUSSION

IV - 1 Precalculated , observed and revaluated F.P. releases

As mentioned in section III . 3 , the EMIS module (subroutine of the VULCAIN code which computes the escape of F.P. from the core) has been used to precalculate the release of F.P. throughout the fuel temperature transient defined for the HEVA 06 test . This module contains the first empirical model and data base incorporated in the CORSOR code (3) using a time-dependent fractional release rate k in the following equation of first order kinetics :

$$dF / dt = k (1 - F) \quad (1)$$

which , at a constant temperature and for a time t , would result in a total fractional release of

$$F = 1 - \exp (- kt) . \quad (2)$$

The release curves , plotted on figures 7 to 10 , may reflect marked differences between escapes of cesium and iodine as predicted with the CORSOR model (100 % of initial inventory released) and as observed (30 %) . The integral released fraction precalculated for Ba is approximately three times lower than the measured one . Even if the release kinetics of tellurium seem to be slightly delayed , the total released fraction computed by using the CORSOR model is in good agreement with the observed fraction . Comparisons between release rates calculated with the 2 different CORSOR models and those observed in the HEVA 04 test (conducted in oxidizing conditions with a fuel temperature plateau of 2000 °C) and in the HEVA 06 test , are provided in table 3 .

elements	CORSOR model (2100 °C)	CORSOR - M model (2100 °C)	HEVA 04 result (2000 °C)	HEVA 06 result (2100 °C)
Cs	0.24	0.27	4.2 E-2	9.5 E-3
I	0.24	0.27	3.7 E-2	9.0 E-3
Ba	4.7 E-3	1.7 E-4	4.0 E-3	1.0 E-2
Te	4.2 E-3	4.2 E-3	4.1 E-2	3.4 E-3

Table 3 : release rate coefficients dF / dt (in fraction per minute)

Since considerable discrepancies have been revealed for Cs , I (release rates 25 times greater than those observed in the HEVA 06 test) and Ba (at least two times lower) , the release kinetics of these elements have been revaluated by introducing , in the EMIS module , the mean release rates measured in the HEVA 06 test during the temperature plateau at 2100 °C . The shape of revaluated curves demonstrates that the F.P. escape models cannot be simply described by first order kinetics . The observed decline in the F.P. escape velocities (i.e. slope of curves) would suggest that the effective release rates are also a function of the time-dependent F.P. inventory liable to be released and would decrease as the fission products are depleted .

IV - 2 Discussion

The effect of a reducing environment on F.P. release has been definitely demonstrated . Typical F.P. behaviour is actually observed in the HEVA 06 test :

- lower release rates for I , Cs , Xe , Kr , Te , Mo ,
- higher for Ba , Sr , Eu ,

than those obtained from the previous tests conducted in oxidizing conditions (4) . Recent experiments performed at SNL (ACRR - ST1 and ST2) with such reducing conditions also exhibit similar indications (5) . However , the interpretation of HEVA 06 results is somewhat complicated by the execution of a presequence in a steam environment at 1300 °C which leads to a partial oxidation of the clad before reaching the plateau of 2100 °C under hydrogen conditions .

I , Cs , Xe

In comparison with the experiments conducted under steam conditions (4) , the release velocities are perceptibly lower , which confirms the influence of UO_2 stoichiometry on the mobility of these elements . Under the hydrogen and high temperature effects , UO_{2+x} is assumed to be particularly reduced and leads to an equilibrium state estimated at $UO_{1.83}$. The oxidation of UO_2 into UO_{2+x} is usually assumed to increase the F.P. release . The change into UO_{2-x} , if it is confirmed in this test , would probably not enhance a symmetrical effect on the acceleration of the release .

Te

The low escape observed , at 2100 °C , in comparison with the release of volatiles and the significant amounts of tellurium trapped in the cladding , confirm the zircaloy affinity for this element . Sn is not detected in the impactor , as was found by SEM/EDAX examinations , bound to Te , in the preceding experiments performed under steam oxidation conditions . These observations are correlated with the low release of Te and the partial oxidation of the fuel cladding .

Mo

Unlike tests involving oxidizing conditions , Mo did not significantly escape from the fuel . In reference 4 , the high release observed is attributed to the formation of the volatile form $MoO_2(OH)_2$. In steam-starved environments , the volatility of the associated is not high enough to produce a large release of Mo at a temperature level of 2100 °C .

Ba , Sr

The high release of Ba and Sr (in relation with the observed Y release) , are probably due to their reduction into metallic forms which have a high vapor pressure at 2100 °C . This reduction can be induced by interactions between H_2 and the F.P. located in the open porosities of the fuel but also by straight contact of the F.P. oxide forms with the highly reducing equilibrium phases such as U/UO_2 and Zr/ZrO_2 shown in figure 13 , which also illustrates the oxide-to-reduce transition as a function of temperature for Ba/BaO and Sr/SrO . As a matter of fact , the above-mentioned equilibrium states were probably involved in the hydrogen reducing phase of the HEVA 06 test . If we assume that , the transitions for Ba and Sr are far above the expected range of oxygen potentials (U/UO_2 , Zr/ZrO_2) . Hence , these elements would exit as metals .

V - CONCLUSIONS

In the analysis of core melt accidents occurring at low primary pressure , the VULCAIN code predicts major steam starvation in the upper part of the core at a time when large amounts of fission products are released . The results of computations indicate that the F.P. may thus escape into a strongly reducing environment .

This kind of phenomenon may play an important role in determining the chemical behaviour of F.P. and therefore influence the characteristics of the F.P. source .

Since experimental data on the F.P. release and behaviour in reducing conditions were particularly deficient , a significant effort has been devoted to the definition of a HEVA test involving a release of F.P. in hydrogen with realistic reactor accident conditions .

From the results of HEVA06 test , we can draw the following conclusions :

- under high reducing conditions and at a temperature of 2100 °C , the measured releases of volatiles are really lower than those predicted with current models ,
- barium and strontium escapes were underpredicted by using the CORSOR models .

The results of F.P. release kinetics , revaluated with the EMIS module by introducing in the data base the new release rates measured at 2100 °C during the HEVA 06 test , seem to be really inconsistent with the assumptions currently made that the rate of release of each element is proportional to the quantity remaining in the fuel and that the constant of proportionality depends only on fuel temperature .

In conclusion , best understanding and improved modeling of F.P. release in severe fuel damage conditions are quite needful , and will allow best-estimate calculations of realistic F.P. source terms .

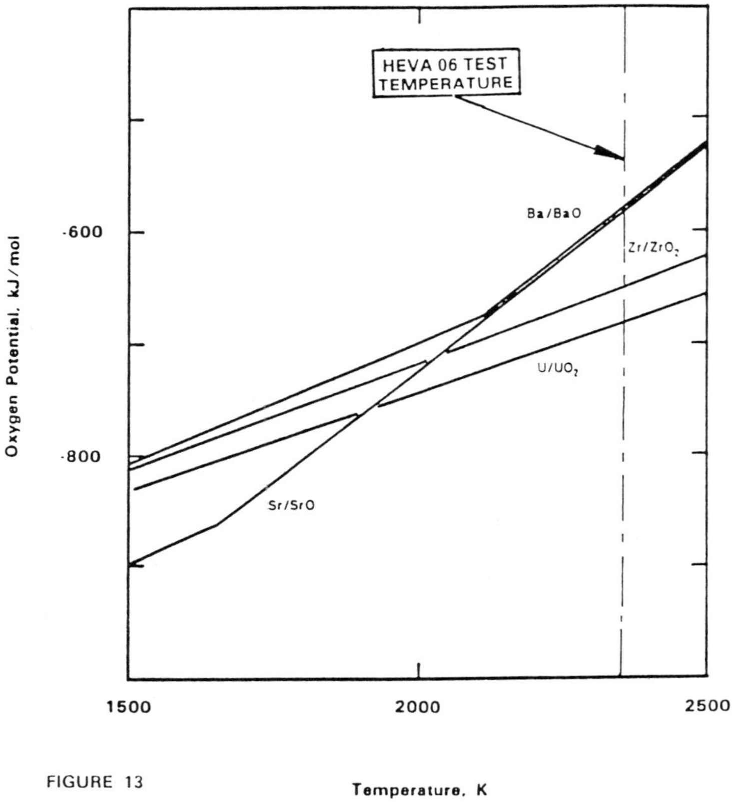

FIGURE 13

REFERENCES

1 . J. Dufresne , C. Cenerino , J.M. Dumas , J. Fermandjian , J . Gauvain , J.P. L'Heriteau , and G. Lhiaubet (CEA / IPSN - Fontenay Aux Roses) , Presentation and Application of the ESCADRE System , IAEA meeting , Sorrento - Italy , march 21-25 , 1988 .

2 . A. Mathiot , G. Le Marois , and R. Warlop (CEA / IRDI / DMG - Grenoble - France) , Analysis of fission product release during a simulated severe fuel damage accident : The HEVA progamme , IAEA technical comittee on water fuel behaviour and fission product release in off-normal and accident conditions , Vienna - Austria , november 10-13 , 1986 .

3 . M.R. Kuhlman , D.J. Lehmicke , and R.O. Meyer , CORSOR User's manual , NUREG / CR-4173 , BMI-2122 , january 1986 .

4 . G. Le Marois and R. Warlop (CEA / IRDI / DMG - Grenoble - France) ,Source term experiment for fission product transport assessment : The HEVA progam , international ANS / ENS conference , Avignon - France , october 2-7 , 1988 .

5 . M.D. Allen , F.W. Stockman , K.O. REIL , A.J. Grimley , and W.J. Camp , ACRR fission product release tests : ST-1 and ST-2 , international ANS / ENS conference , Avignon - France , october 2-7 , 1988 .

Measured Release Kinetics of Iodine and Cesium from UO_2 at High Temperatures under Reactor Accident Conditions

C. E. L. HUNT, F. C. IGLESIAS, and D. S. COX
Atomic Energy of Canada Limited
Chalk River Nuclear Laboratories
Chalk River, Ontario, K0J 1J0 Canada

1. INTRODUCTION

Among the many radioactive fission products which may be released from reactor fuel during an accident sequence, iodine and cesium are commonly the ones of most immediate concern to the general public. The quantities released and the rates at which they are released depend, in a complex way, on two primary variables, temperature and the chemical environment. For different accident sequences temperatures may range from close to operating values up to the melting point of UO_2 and the chemical environment may vary from strongly reducing to highly oxidizing conditions. The work reported here covers a temperature range from 500° to 1700°C under oxidizing conditions in either steam or air.

2. EXPERIMENTAL TECHNIQUES

Oxidation of UO_2 in both steam and air have been studied in the laboratory using unirradiated samples. Confirmatory tests were then done on irradiated UO_2 in a "hot cell" facility. Fission product releases from the irradiated UO_2 were measured using the arrangement shown schematically on Figure 1. The atmosphere at the sample was controlled by using flowing argon, steam, or air. A sensor, located downstream from the sample, gave a continuous record of the oxygen partial pressure ($P(O_2)$). The fission products remaining in the sample were continuously monitored with a gamma-ray spectrometer. Gamma ray counting times were typically 150 to 300 seconds. Longer counting times decreased the statistical scatter of the fission product releases at the expense of resolution in time. This caused some difficulty in interpretation during periods of rapid changes in release rate. Deposition rods, located in the thermal gradient of the furnace downstream from the sample, were measured after the test to determine fission product deposition patterns as a function of temperature.

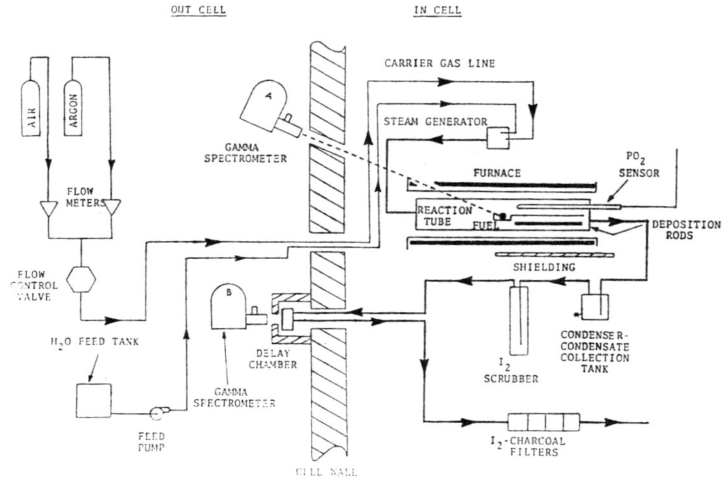

FIGURE 1. Schematic of experimental arrangement.

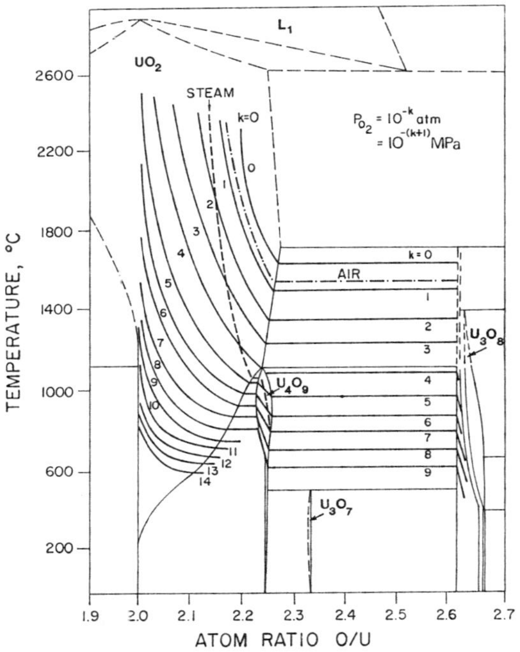

FIGURE 2. UO_2-O_2 phase diagram with oxygen partial pressure isobars and phase stabilities in air and steam.

3. UO_2 OXIDATION

The equilibrium composition of reaction products can be predicted using the well established thermodynamic properties of UO_2 and its higher oxides when the temperature and oxygen potential of the gaseous environment is known. Sample composition is a function of time and must be determined experimentally. Such kinetic information is indispensable for the understanding of fission product releases, since the timing of oxidation processes can be related to experimentally observed release events, thereby suggesting release mechanisms.

The oxygen/uranium phase diagram (1-4) shows that UO_2 can be oxidized to several higher oxides. Figure 2, derived from references 1-7, shows how phase stabilities are determined by the oxygen potential of the gas phase. At one atmosphere of air, the phase limit is defined by U_3O_8 except at high temperatures (T > 1550°C) where hyperstoichiometric UO_2 is the highest oxide. In steam at atmospheric pressure, UO_{2+x} is the highest oxide above 1050°C, while U_4O_9 is formed at lower temperatures.

In air at 400-600°C, UO_2 oxidizes rapidly to a fine U_3O_8 powder. Above 800°C, the rate of oxidation initially decreases with time in a diffusion controlled process until "breakaway" occurs when the reaction rate increases due to cracking of the previously protective oxide layer. At higher temperatures the extent of breakaway behaviour is reduced because of enhanced U_3O_8 plasticity. The kinetics of isothermal oxidation of UO_2 in air at 900 to 1200°C are presented in Figure 3.

The morphology of U_3O_8 varies significantly with temperature. Oxidation at 500°C is very rapid because the oxidation products are non-adherent. The reaction is initiated by attack at grain boundaries. Volume expansion produces a fine powder of angular fragments of U_3O_8. At 900 to 1100°C, long columnar grains grew parallel to the direction of oxygen flux. At higher temperatures, columnar grain structure is observed only after short oxidation periods, before it is mostly eliminated by plastic deformation which occurs extensively during U_3O_8 growth (8,9).

UO_2 oxidation kinetics in steam have been reported previously. Bittel et al. (10) attribute the kinetics to oxygen diffusion in UO_{2+x}. Our results show close agreement with those of Bittel, however, the rate controlling process has been interpreted to be reaction at the gas/oxide interface, a mechanism proposed by Carter et al. (11). Bittel's data is also consistent with this interpretation.

In Figure 4, for tests at 1200 and 1350°C, the weight change measurements are coincident with changes in the O/U ratios except for the last data point at 1350°C. This data point suggests weight loss at constant composition. These weight losses became more rapid with increasing temperature. The scanning electron microscope (SEM) showed a highly convoluted topography, typical of volatilization. This occurred most rapidly at grain boundaries, undercutting grains and producing tunnels. The kinetics, as well as tests in dry air, support the hypothesis that the volatile species is UO_3 gas (8).

4. IODINE AND CESIUM RELEASES

Isothermal tests for temperatures between 500 and 1700°C were performed in an air environment and similar tests, for temperatures between 1000 and

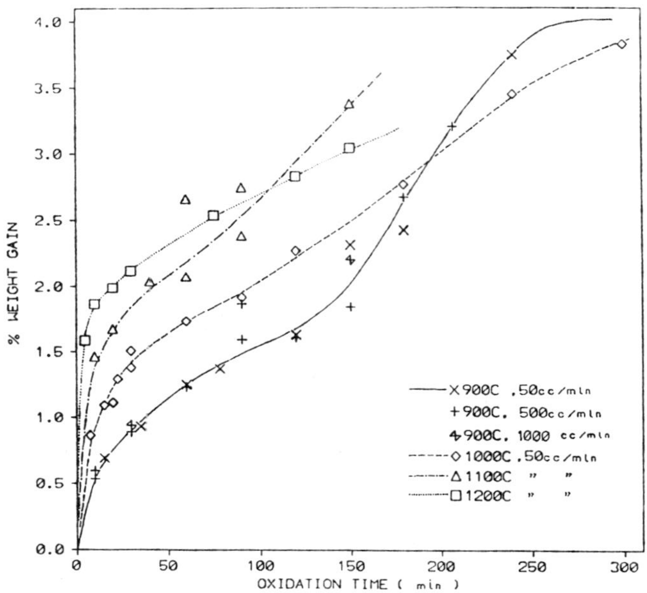

FIGURE 3. Kinetics of UO_2 oxidation in air at 900-1200°C (cylindrical specimen, 12.1 mm diameter, 5 mm long).

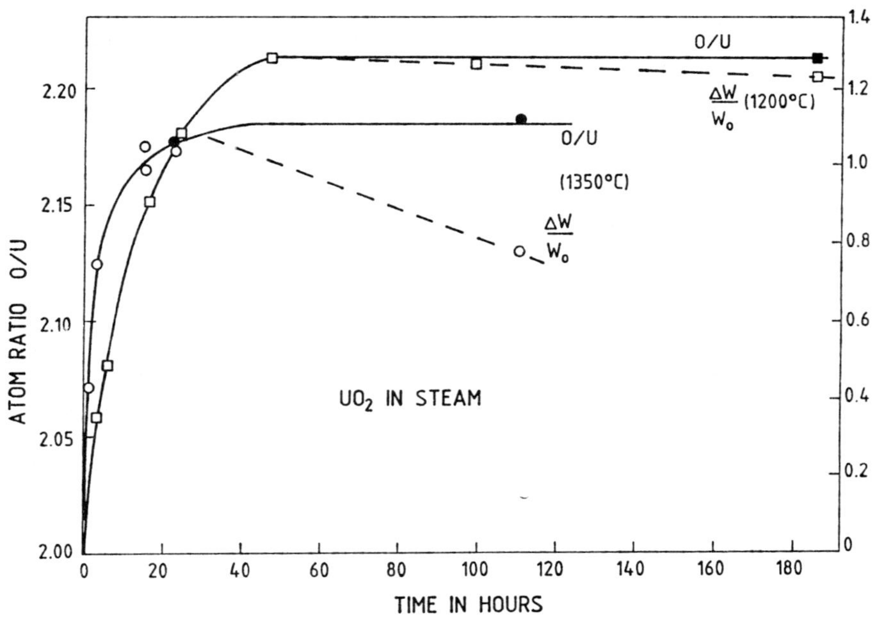

FIGURE 4. Oxidation kinetics of UO_2 in steam at 1200 and 1350°C. Solid line represents O/U and dashed line represents weight changes.

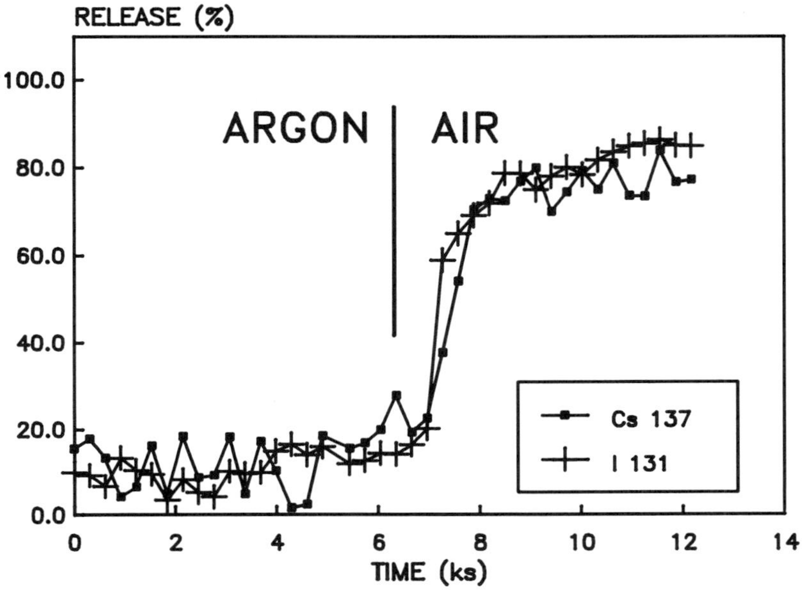

FIGURE 5. Release of iodine and cesium from UO_2 oxidized in air at 1100°C.

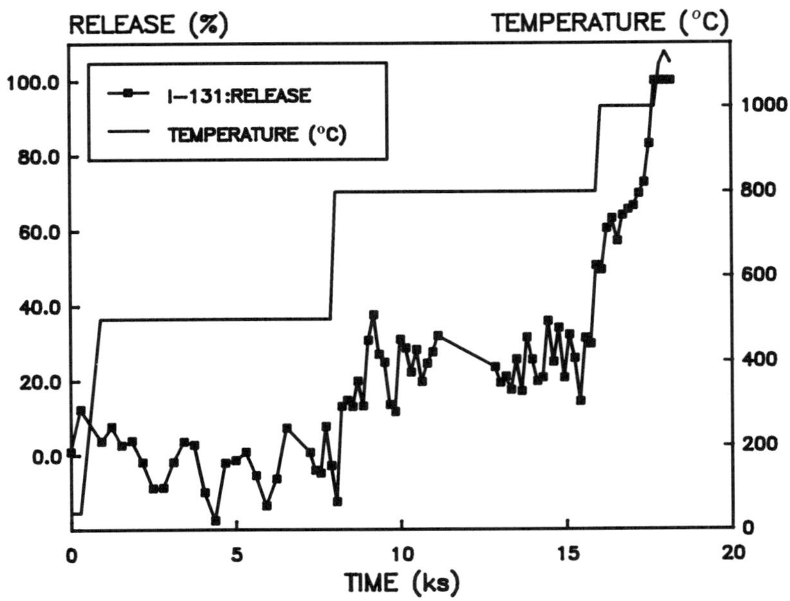

FIGURE 6. Release of iodine from UO_2 oxidized to U_3O_8 at 500°C and heated to successively higher temperatures.

1700°C, were performed in steam. UO_2 sample burnups ranged from 6000 to 30000 MWd/teU.

Iodine release was measured in air at temperatures as low as 500°C. Cesium releases were not detected below 750°C. At higher temperatures the release percentages of iodine and cesium were similar. Figure 5 is an example of a test at 1100°C in air. While the sample was maintained in an argon atmosphere the release rates of both elements were low, about 0.05%/min. When the sample started to oxidize with the addition of air, the release rate increased dramatically and when oxidation of the sample was completed, the release rates again dropped to a low value.

Figure 6 shows the effect of a step-wise increase in temperature on the iodine release. The sample was oxidized in air and fully converted to U_3O_8 at 500°C. An increase in temperature to 800°C resulted in a release of up to about 30% of the iodine inventory. This temperature was maintained for about 6000 s with almost no further release. Another temperature increase, to 1000°C, resulted in more iodine release which again showed signs of decreasing rapidly after the temperature was stabilized. A further temperature increase to 1100°C released the remainder of the iodine inventory from the sample.

Figure 7 summarizes iodine releases in air for tests with oxidation times of about 3 h and Figure 8 shows the equivalent cases for the release of cesium (12). Unlike iodine, no release of cesium is observed below a temperature of about 750°C. Otherwise, the behaviour is similar. Figure 5 shows how the cesium parallels the iodine release at the higher temperatures. Similar behaviour has been established at temperatures as high as 1700°C.

In spite of the fact that cesium and iodine are released almost simultaneously, they do not appear to be transported in air as CsI. Figure 9 shows the deposition patterns on a stainless steel rod in the thermal gradient of the furnace. During an air oxidation test at 1100°C, Cs and I had deposition peaks at different temperatures but with considerable overlap (Figure 9a). In a similar test, again done at 1100°C (Figure 9b), the system was flooded with stable iodine both before and during the sample oxidation. In this case, no radioactive iodine was deposited although the cesium deposition peak remained unchanged. Thus the iodine and cesium were not transported as CsI.

Figure 10 shows the typical cesium release behaviour in steam. This is compared with measured average sample stoichiometry. The first 4000 s of the test was done in argon, which contained an oxygen impurity partial pressure of 10^{-5} atm. When steam was introduced the oxidation rate increased as did the cesium release rate. When the sample reached equilibrium stoichiometry, the cesium release rate dropped almost to zero. Similar behaviour has been observed for iodine.

5. DISCUSSION

In all the isothermal experiments in air (13), except at 1600°C or higher, the I and Cs release occurs in two phases (for example Figure 5). In the first phase a large fraction of the inventory is released in a short time; in the second phase the remainder of the inventory is released by a slower process. Since fission product release is related to oxidation of the fuel, this release behaviour suggests that the oxidation may also occur in two stages. However, the oxidation experiments on unirradiated UO_2,

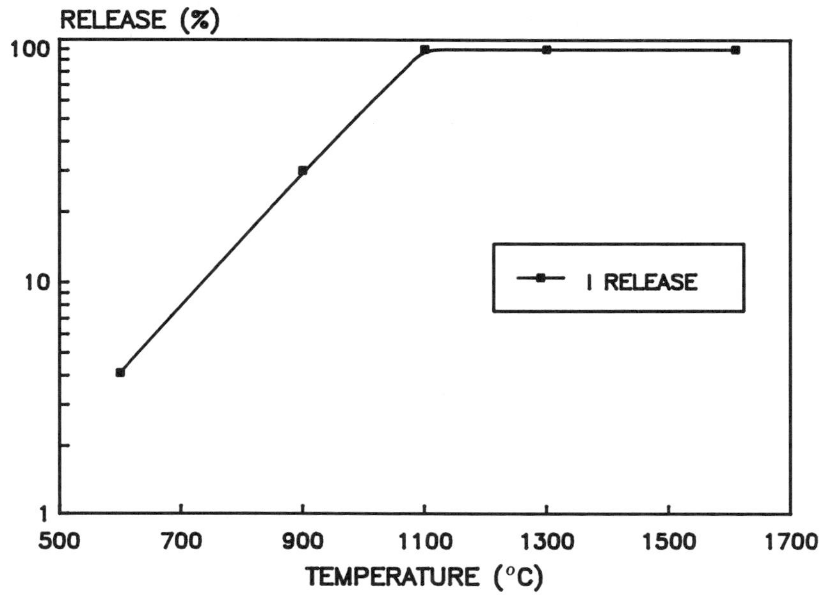

FIGURE 7. Release of iodine from UO_2 oxidized in air as a function of temperature.

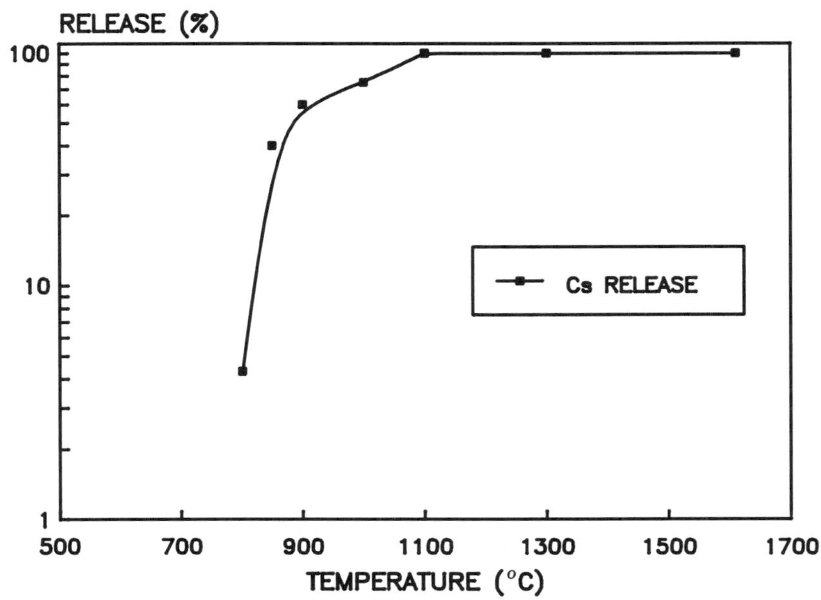

FIGURE 8. Release of cesium and iodine released from UO_2 oxidized in air at 1100°C.

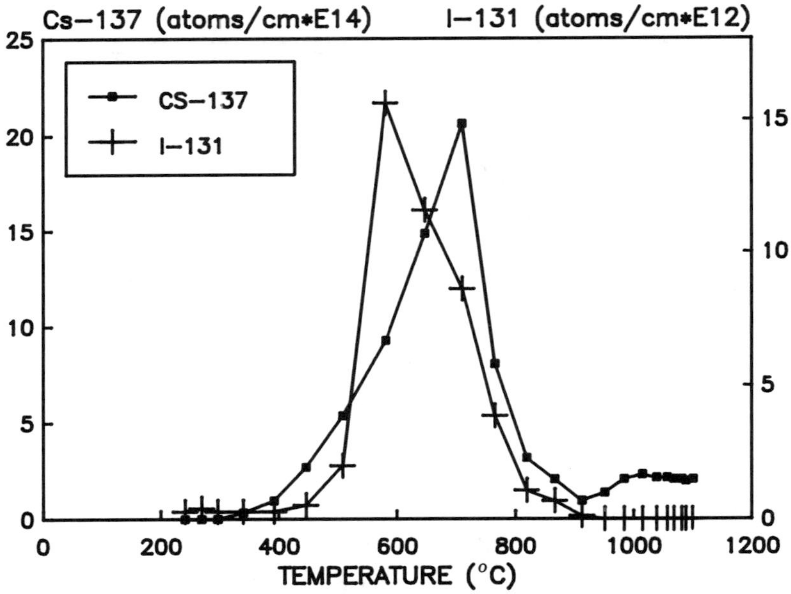

(a)

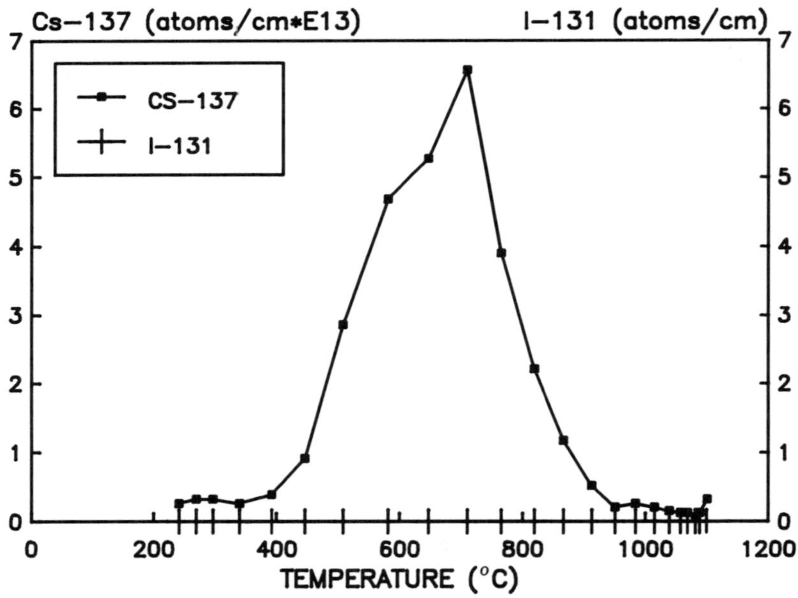

(b)

FIGURE 9. Deposition of cesium and iodine released from UO_2 oxidized in air at 1100°C.

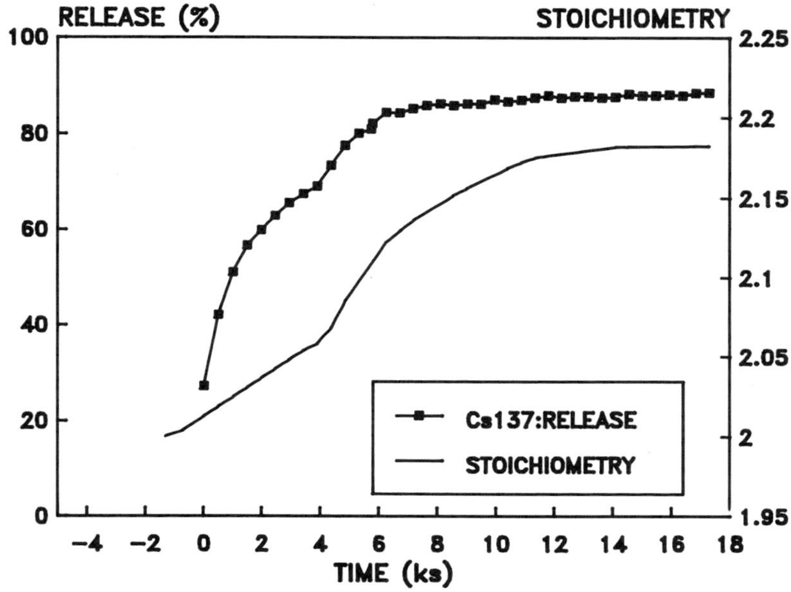

FIGURE 10. Release of cesium and change in O/U for a UO_2 sample oxidized in steam at 1400°C.

Figure 3, showed no abrupt changes in weight gain as a function of time or temperature. Furthermore, calculations based on unirradiated UO_2 kinetics and sample geometry of the time needed for complete sample oxidation, overestimate the observed oxidation times for irradiated specimens. Consequently, the oxidation of irradiated samples exhibit faster air oxidation kinetics than the unirradiated specimens. It is believed that the existence of an interconnected network of grain boundary tunnels formed during irradiation may partially explain this difference in oxidation behaviour since such tunnels would provide easy access for oxygen to the interior of the sample.

The observation that samples oxidized at temperatures of 1600°C or higher, as reported in reference (13), do not follow similar release characteristics shows that the formation of U_3O_8 (see Figure 2) is also responsible for the lower temperature behaviour.

As shown in Figures 6, 7 and 8, temperature is an important factor in determining the fraction of iodine and cesium released for temperatures below 1100°C. However, the chemical conditions influence the kinetics. In Figure 6 the U_3O_8 formed at 500°C was reheated to successively higher temperatures with accompanying step increases in the iodine released. The final temperature increase to 1100°C resulted in the almost immediate release of all the remaining iodine. Figure 5, on the other hand, shows the results of oxidizing UO_2 at 1100°C. There was a rapid release of both the iodine and cesium during the oxidation process but the step release only reached about 80%. Release continued at a much slower rate and about 1-1/4 hours later releases had reached about 95%. The study of the I and

Cs release characteristics from U_3O_8 subjected to heating is meaningful only for temperatures up to 1100°C because by this temperature 100% of the inventory is released.

Oxidation of UO_2 in air at temperatures up to 1550°C (the stability limit of U_3O_8 (Figure 2)) exhibits the characteristics shown in Figure 5. For higher temperatures, or lower oxygen partial pressure (see below), the step release is replaced by a gradual increase of the release rate when the environment is changed to oxidizing (13).

As in the air oxidation case, I and Cs release appears to be intimately associated with the dynamics of the steam oxidation process. The characteristics of steam oxidation of UO_2 (to UO_{2+x}) is represented by Figure 10. In the first part of this test the oxygen partial pressure was 10^{-5} atm. At this oxygen potential, the specimen would reach an equilibrium stoichiometry of $UO_{2+0.1}$. When steam was introduced the oxygen partial pressure increased to 4×10^{-3} atm, increasing the oxidation rate to satisfy an equilibrium stoichiometry of $UO_{2+0.18}$. It is clear from the figure that a large release took place during the initial phase of the test. When steam was introduced there was an additional release related to the increasing oxidation rate. Consequently, most of the release was associated with small changes in UO_2 stoichiometry. A similar sensitivity may be true for release during air oxidation but the rapid air oxidation kinetics prevent its observation.

6. CONCLUSIONS

1. In air, cesium and iodine are transported as separate species, not as CsI.
2. Irradiated UO_2 oxidizes in air faster than unirradiated UO_2, probably because of a network of interlinked grain boundary tunnels and U_3O_8 formation.
3. The fraction of iodine or cesium released from U_3O_8 depends primarily on the maximum temperature reached.
4. At temperatures up to about 1550°C, the stability limit for U_3O_8, oxidation of UO_2 in air releases most of the inventory in a short time period. The remaining inventory is released by a slower process.
5. At low oxygen partial pressures (steam) most of the iodine and cesium release is associated with small changes in the stoichiometry of the fuel sample.

7. ACKNOWLEDGEMENT

The research program described in this paper was sponsored by the Candu Owners Group (COG). The authors gratefully acknowledge the contribution of R.D. MacDonald in the preparation of this paper.

8. REFERENCES

1. L.M. Kovga, Dokl. Akad. Nauk SSSR 194 (1) (1970) 98; Dokl Chem. (Engl. Transl.) 194 (1) (1970) 632.

2. A.B. Johnson, E.R. Gilbert, R.J. Guenther, PNL-4189-UC-85, 1982 August.

3. M.J. Bannister, J. Nucl. Mater., 24 (3) (1967) 340.

4. F. Gronvold, J. Inorg. Nucl. Chem. 1 (6) (1955) 357.

5. K. Naito, K. Kamegashira, "High Temperature Chemistry of Ceramic Nuclear Fuels with Emphasis on Nonstoichiometry", Adv. Nucl. Sci. Technol., 9 (1976) 99.

6. A.M. Anthony, R. Kiyoura, T. Sata, J. Nucl. Mater., 10 (1963) 8.

7. S. Aronson, J. Belle, J. Chem. Phys., 29 (1958) 151.

8. D.S. Cox, F.C. Iglesias, C.E.L. Hunt, N.A. Keller, R.D. Barrand, J.R. Mitchell, R.F. O'Connor, Proc. Symposium on Chemical Phenomena Associated with Radioactivity Releases During Severe Nuclear Plant Accidents, Anaheim, CA., 1986 Sept. 4-12, NUREG/CP-0078.

9. D.S. Cox, F.C. Iglesias, C.E.L. Hunt, R.F. O'Connor, International Conference on CANDU Fuel, 1986 Oct. 6-8, Chalk River Nuclear Laboratories, ISBN 0-919784-13-5.

10. J.T. Bittel, L.H. Sjodahl, J.F. White, J. Amer. Ceramic Soc., 52 (1969) 446.

11. R.E. Carter, K.W. Lay, J. Nucl. Mater., 36 (1970) 77.

12. C.E.L. Hunt, F.C. Iglesias, D.S. Cox, N.A. Keller, R.D. Barrand, J.R. Mitchell, R.F. O'Connor, International Conference on CANDU Fuel, 1986 Oct. 6-8, Chalk River Nuclear Laboratories, ISBN 0-919784-13-5.

13. F.C. Iglesias, C.E.L. Hunt, D.S. Cox, N.A. Keller, R.D. Barrand, R.F. O'Connor, J.R. Mitchell, Workshop on Chemical Reactivity of Oxide Fuel on Fission Product Release, 1987 April 7-9, CEGB Berkeley Nuclear Laboratories.

Fission Product Release during the Air Oxidation of Irradiated Uranium Dioxide

R. WILLIAMSON and S. A. BEETHAM
UKAEA, Harwell Laboratory
Oxfordshire, OX11 ORA, UK

ABSTRACT

The ability to predict accurately the fractional release of fission products from nuclear reactor fuel during severe faults is important when assessing safety implications for the nuclear industry. Of the postulated faults at temperatures below 1000°C, the most severe are those involving air atmospheres. Under these conditions oxidation is rapid, the cubic UO_2 lattice is converted to orthorhombic U_3O_8, resulting in the loss of structural integrity of the fuel and enhanced fission product release.

Apparatus has been constructed at the Harwell Laboratory capable of heating fuel in a variety of gaseous environments at temperatures up to 1600°C. The furnace and associated glassware can accommodate approximately 100g of highly irradiated fuel, either as bare fragments or clad pellets. Volatile fission products are collected on thermal gradient tubes, particulate filters and traps which are subsequently quantitatively analysed by gamma spectroscopy and beta counting techniques.

Fractional releases of fission products from Advanced Gas-cooled Reactor fuel, air oxidised to U_3O_8, in the temperature range 300-600°C have been determined. Current studies are investigating the release kinetics under the same conditions, with particular emphasis on the transport and plate-out characteristics of volatile ruthenium species.

INTRODUCTION

During recent years there has been a substantial increase in research activities focussed on understanding the oxidation and fission product release behavicur of irradiated uranium dioxide (UO_2). The main impetus has been the need for reliable information for fuel safety assessments. Of the postulated faults in most reactor systems at temperatures below 1000°C the most severe are those involving environments containing air. For example, at various stages along the Advanced Gas-cooled Reactor (AGR) fuel handling route the fuel pins may come into contact with air atmospheres. If the cladding has failed in reactor, or is subsequently ruptured after discharge, the ingress of air will oxidise the fuel. At temperatures ≥ 250°C oxidation is rapid, the cubic UO_2 lattice is converted to orthorhombic U_3O_8 resulting in the loss of structural integrity of the fuel with potential release of fuel particulate , as well as some of the fission product inventory. Also the volume expansion caused by fuel oxidation will result in further rupturing of the cladding.

The present guidelines for calculating the release of fission products and fuel particulate from UO_2 in failed AGR pins, in air, at temperatures of 200-800°C have had to be based broadly on data from experiments conducted in the early 1960's on fuel not representative of contemporary manufacture (1) or not irradiated under AGR conditions (2). In 1983/84 a General Nuclear Safety Research Programme (GNSR) was initiated with the objective of rectifying this unsatisfactory situation by expanding the limited data base to support fuel safety assessments envisaged for all relevant reactor systems. Safety studies modelling the environmental impact of the failure of irradiated fuel require realistic estimates to be made of the radioactivity escaping from the defective fuel to provide the source term. If the data required are not available, pessimistic assumptions have to be made, which may lead to over estimates of the consequences and of the measures necessary to deal with them. This affects the design as well as the capital and operating costs of reactor systems. Experimental data on source terms under controlled conditions are vital for the development of safety cases in key areas of nuclear technology. Although the GNSR programme to date has focussed primarily on the AGR fuel handling route conditions, clearly the data are generic in their direct application to fuel faults in the Pressurised Water Reactor (PWR) and to some extent also in the Magnox (in the case of partially oxidised uranium fuel), and the Fast Breeder Reactors.

The programme was conceived as falling into two broad areas;

(i) To determine the oxidation kinetics of unclad, irradiated UO_2 and quantify the associated fractional release of fission products during air oxidation in the temperature range 250-1000°C.

(ii) To determine the resulting fission product aerosol and U_3O_8 particulate size distributions, thus enabling an overall assessment of the radiological implications of air oxidation of irradiated UO_2 to be formulated.

This paper summarises key data collected on the fractional releases of fission products from AGR fuel, oxidised in air, in the temperature range 300-600°C. Particular emphasis has been placed on the release, transport and plate-out characteristics of volatile ruthenium species.

EXPERIMENTAL DETAILS AND TECHNIQUES

A schematic diagram of the apparatus is shown in figure 1.

The samples of irradiated UO_2 are contained within a platinum wire basket which is located at the bottom of a vertically held silica reaction flask, as shown in figure 2.

The reaction flask is lined with a close fitting thermal gradient tube (TGT). The TGT provides an inert deposition surface for volatilised fission products, whilst the applied temperature gradient separates the various elements and compounds according to their respective plate-out temperatures. A filter housing caps the reaction flask and collects any remaining gas-borne particulate or aerosol for analysis.

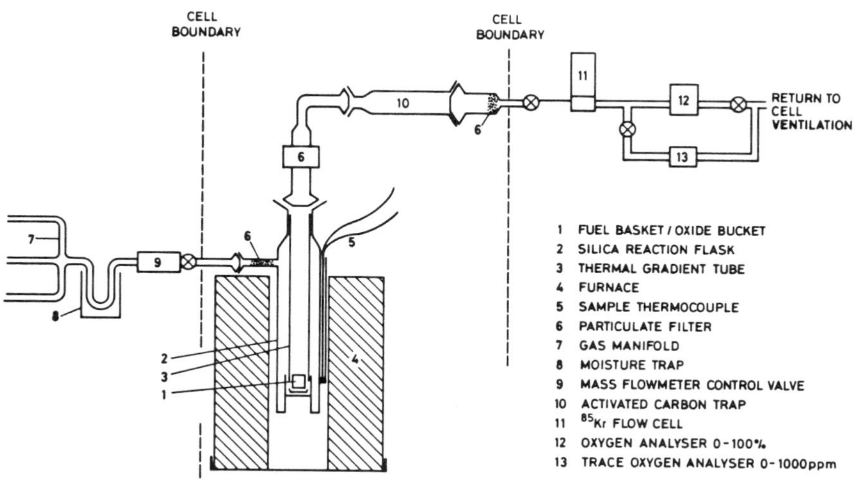

FIGURE 1. A schematic diagram of the oxidation apparatus.

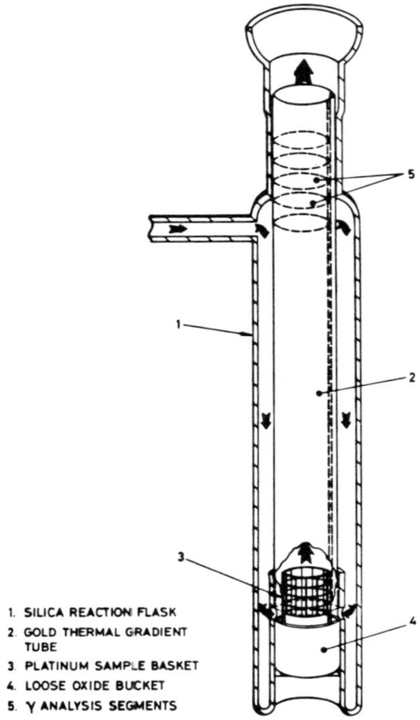

FIGURE 2. Diagram of the reaction flask, showing the location of the TGT and the fuel basket.

The exhaust gases from the flask were passed through a Nuclear Enterprise NE802 plastic scintillator flow cell optically coupled to a DM1-2 photomultiplier tube assembly. The flow cell enabled both the time dependence of the release of ^{85}Kr and its cumulative values to be measured. The counting system was calibrated before each experiment using injections of ^{85}Kr of known activity (±3%). The gas was subsequently routed through one of two oxygen detection transducers before being vented.

Using ^{85}Kr injections at septum points at various locations in the gas supply system, it was estimated that gas arising from the sample was detected in the flow cell in less than 150 seconds.

The assembled reaction flask was positioned within a vertically mounted resistance furnace with the vertical position of the furnace being adjustable using a motorised platform. The reaction flasks were fitted with external thermocouple pockets terminated in close proximity to the sample basket. Temperature calibrations of the furnace, sample basket and TGT's were undertaken in duplicate non-active apparatus, with typical accuracies of ±10°C.

Bottled gases were used for all experiments. Gas mixtures and flow rates through the reaction flask can be varied as required. Flow rates of 250 ml min^{-1} were used for all the experiments reported here except for one experiment at 100 ml min^{-1}. Two methods of raising the sample to temperature were employed. Formerly samples were raised to temperature from ambient and later cooled in flowing air. However this introduces multi-temperature oxidation during the early and latter stages of the experiment. The preferred method employed on more recent experiments involves raising the sample to temperature in $Ar/5\%H_2$, then switching to air for the duration of the isothermal heating period, then cooled to ambient in high purity argon. This procedure ensures that the fuel is held close to stoichiometry during the heating cycle and is not reduced during the cooling cycle.

The deposited activity on the TGT was subsequently quantified in two stages. Initially the tube was sectioned and monitored for gamma activity, thus providing a relative deposition profile. Alternate segments were then dissolved in aqua regia and aliquots of the solution presented to the gamma spectrometer in a standard geometry. These latter measurements enabled the initial deposition profile to be scaled and the fractional release of gamma species to be accurately determined.

Deposited strontium-90 activity was also measured, for a few samples, using precipitation techniques.

The fuel used for all the experiments reported here was discharged from Hinkley 'B' AGR in June 1983, with a burn-up of 18.64 GWd/teU. The fission product inventory of representative samples of the fuel was determined by standard radiochemical techniques (3), and measured fractional releases are with reference to the measured inventory.

RESULTS

In excess of twenty experiments have been conducted investigating fission product release behaviour from irradiated UO_2, in the temperature range 300-600°C. Eleven well characterised experiments are presented here for detailed assessment. The experimental conditions are summarised in Table 1.

TABLE 1. Experimental conditions.

Sample	Mass UO_2 (g)	Temp (°C)	Hold Time (mins)	Atmosphere Sequence heat	hold	cool
4BF/2	9.264	300	4320	air	air	air
2BF/1	5.576	400	22	Ar/H_2	air	Ar
2BF/2	4.348	400	90	Ar/H_2	air	Ar
4AF	8.295	400	630	air	air	air
3AF	10.009	400	630	air	air	air
1AF/A	4.543	400	4320	Ar/H_2	air	Ar
3BF/1	5.700	500	12	Ar/H_2	air	Ar
3AF/2	4.729	500	15	Ar/H_2	air	Ar
4BF/1	7.567	500	180	air	air	air
1AF/B	4.543	600	600	Ar/H_2	air	Ar
1AF/C	4.543	600	4920	Ar/H_2	air	Ar

This matrix provides a selection of experiments where samples were only partially oxidised to U_3O_8, and experiments where the U_3O_8 product was heated for extended periods. Eight of the eleven samples achieved weight gains equal to, or slightly in excess (103-105%) of the weight gain associated with total conversion to U_3O_8. Samples 2BF/1, 3BF/1 and 3AF/2 increased in weight corresponding to 47, 85 and 86% conversion to U_3O_8, respectively. Therefore this selection provides a data set where the release kinetics of volatile species can be assessed for UO_2 oxidation at relatively low temperatures. The release fractions, expressed as percentages of the original inventory, are given in Table 2. As the fuel was long cooled radio-iodine was not detected. However, neutron activation of the solutions of the dissolved TGT segments did not reveal iodine species.

TABLE 2. Percentage releases of volatile ^{85}Kr, ^{106}Ru, ^{137}Cs.

Sample	Temp °C	Hold Time (mins)	% Release ^{85}Kr	^{106}Ru	^{137}Cs
4BF/2	300	4320	7.61	4.16×10^{-3}	NV
2BF/1	400	22	3.8	1.86×10^{-5}	NV
2BF/2	400	90	9.18	1.07×10^{-3}	NV
4AF	400	630	7.35	2.27×10^{-2}	NV
3AF	400	630	5.54	1.56×10^{-2}	NV
1AF/A	400	4320	10.69	2.12×10^{-1}	NV
3BF/1	500	12	5.52	3.13×10^{-4}	NV
3AF/2	500	15	6.21	5.76×10^{-4}	NV
4BF/1	500	180	7.65	3.76×10^{-2}	NV
1AF/B	600	600	3.23	$3.03\text{-}6.06 \times 10^{-1}$	4.99×10^{-3}
1AF/C	600	4920	6.51	$4.35\text{-}8.71 \times 10^{-1}$	1.34×10^{-2}

(a) NV — Denotes that the detected ^{137}Cs was only associated with fuel particulate

(b) ^{106}Ru, ^{137}Cs — Percentage releases are totals which combine activity measured on the thermal gradient tubes, glassware and filters.

It should be noted that sample 1AF was heated three times; initially the sample was heated for 72 hours at 400°C and then cooled to ambient. The U_3O_8 product was subsequently re-heated twice at 600°C, for 10 hours and 72 hours respectively. For each heating a new TGT was employed.

Figure 3 illustrates the predominant observed beta/gamma activity on the TGT associated with the single experiment at 300°C.

The detection of ^{90}Sr and ^{137}Cs is considered to be associated with fuel particulate. The mean activity ratio of strontium and caesium on the TGT, and that measured in the unoxidised source fuel, is identical within the accuracy of detection, 1.76 and 1.7 respectively. The increase in activity of these radionuclides at each end of the TGT probably results from contamination of the lower end due to its close proximity to the fuel basket, and the upper end due to handling with contaminated forceps. However fuel particulate contamination in the central region of the TGT is unlikely to have been caused by handling techniques, but may be associated with small fuel particles becoming suspended in the gas stream and attaching to the tube inner surface. The ^{106}Ru profile has been corrected for the background ^{106}Ru activity associated with the fuel particulate contamination, and the pure volatile profile is shown separately. This correction for background activity due to fuel particulate has been applied to all subsequent figures and the releases in Table 2, which quote only volatile release levels.

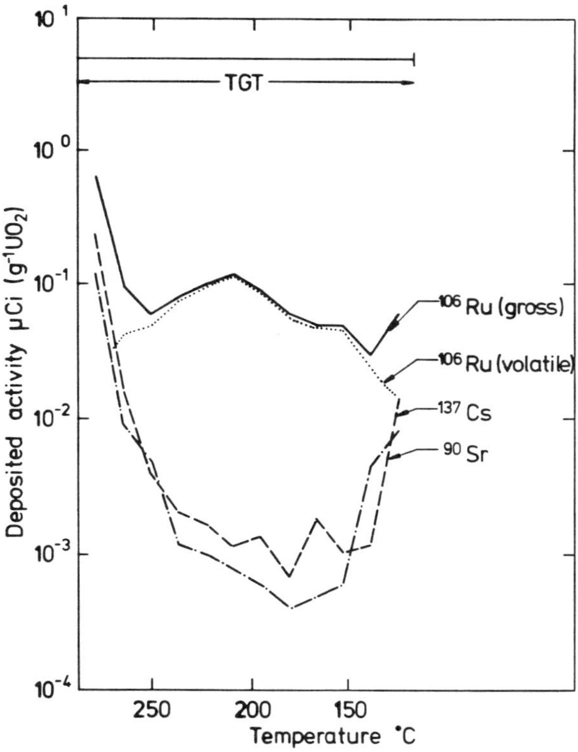

FIGURE 3. Deposited activity on the TGT resulting from air oxidation of sample 4BF/2 at 300°C.

Figure 4 is a composite of the plate-out profiles obtained for volatile ^{106}Ru at fuel oxidation temperatures between 300-600°C, for various hold times. Typically the ruthenium activity on the TGT represents greater than 70% of the total volatile fraction, indicating that the tube functions as an effective trap, with a peak deposition of ^{106}Ru occurring between 370-225°C.

A volatile form of caesium was noted from sample 1AF , experiment B and C at 600°C, with the release fraction increasing with hold time. Deposition profiles obtained for caesium are similar to those reported previously (4) by Lorenz et. al., 1980.

Releases of ^{85}Kr were in the range 3.8 - 10.69% and are generally independent of the temperature of oxidation or hold time, although a slow diffusive release (f_r = $9.04 \times 10^{-4} hr^{-1}$) was observed on continued heating of U_3O_8 at 600°C.

The releases of ^{106}Ru as a function of time are plotted in Figure 5. It will be noted that the release rates at 400 and 500°C are approximately linear. However for releases at 400°C the release behaviour is best represented by the function;

$$R_{400} = 4.91 \times 10^{-5} t \, (1 - e^{-0.00147t}) \qquad (1)$$

where R = the percentage of ^{106}Ru inventory released

t = hold time at temperature (mins)

The exponential term corrects for the lower releases observed prior to complete oxidation of the fuel. Table 3 summarises the fit of equation 1 to the experimental data.

TABLE 3. Comparison of the observed and calculated ^{106}Ru releases associated with heating UO_2/U_3O_8 at 400°C.

Sample	Temp °C	Hold Time (mins)	Release Observed %	Release Predicted %
2BF/1	400	22	1.86×10^{-5}	3.4×10^{-5}
2BF/2	400	90	1.07×10^{-3}	5.47×10^{-4}
4AF	400	630	2.27×10^{-2}	1.87×10^{-2}
3AF	400	630	1.56×10^{-2}	1.87×10^{-2}
1AF/A	400	4320	2.12×10^{-1}	2.12×10^{-1}

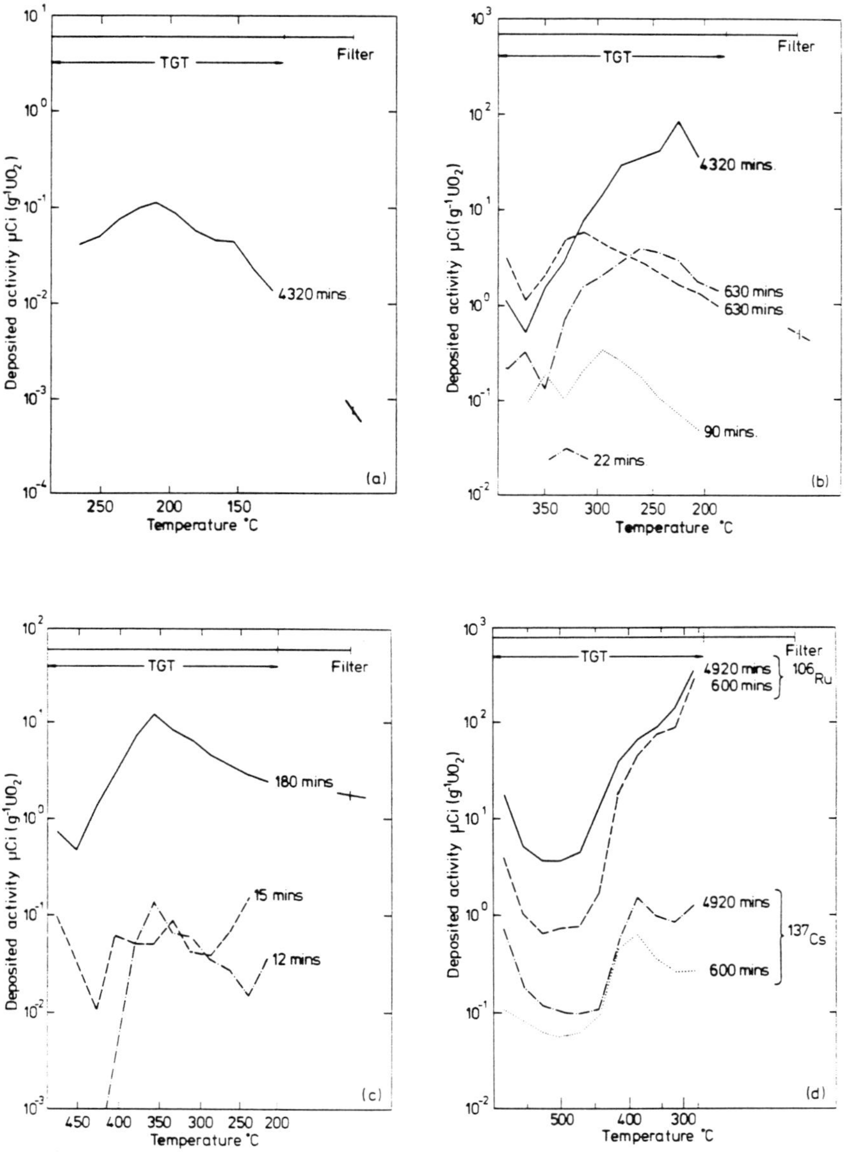

FIGURE 4. Composite of the plate-out profiles obtained for volatile ^{106}Ru and ^{137}Cs at oxidation temperatures of (a) 300°C (b) 400°C (c) 500°C and (d) 600°C.

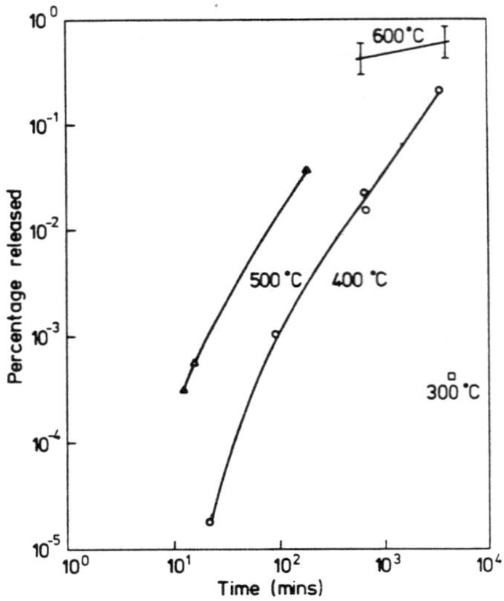

FIGURE 5. Fractional releases of ^{106}Ru as a function of time.

DISCUSSION

The operating history of the fuel used for these studies has been reported previously (3). Typically the mean centre temperature during operation was approximately 950°C. At these temperatures it may be assumed that the fission gas inventory is largely retained on intergranular sites, as single interstitial atoms, on neutral tri-vacancy sites, or in small bubbles. Therefore in the absence of a significant contribution of gas atom diffusion, due to the relatively low temperatures, it is proposed that lattice restructuring during oxidation is the primary mechanism liberating intergranular fission gas. This is supported by the observation that the release fractions of ^{85}Kr are not sensitive to the temperature of oxidation, or the duration at temperature.

The release fractions of both ruthenium and caesium species have been observed to increase with continued heating of UO_2/U_3O_8. Deposition profiles of some of the ruthenium profiles are consistent with depletion of a gas borne species, where plate-out commences when the partial pressure in the gas phase is below the saturated vapour pressure of the solid element. Previous studies (5,6) of the ruthenium-oxygen system have shown that ruthenium dioxide, RuO_2 is the only solid oxide in the system, and is formed by heating ruthenium metal in air or oxygen. The rate of formation is significant above 500°C. When RuO_2 is heated to 1000°C in an inert gas, decomposition to the elements occurs. However, when ruthenium dioxide was heated in oxygen, higher oxides were formed and analysis of the effect of oxygen pressure on the total vapour pressure in the system as a function of temperature indicated that both RuO_3 and RuO_4 were present, formed thus:

$$RuO_2(s) + \tfrac{1}{2}O_2(g) \rightleftharpoons RuO_3(g) \quad (2)$$
$$RuO_2(s) + O_2(g) \rightleftharpoons RuO_4(g) \quad (3)$$

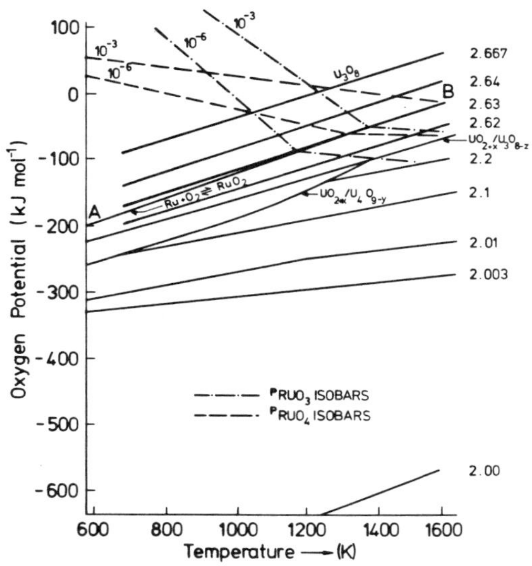

FIGURE 6. Oxygen potential diagram for the U-Ru-0 system.

At temperatures below 900°C the primary reaction is formation of the tetroxide but at temperatures above 1000°C the trioxide is the dominant species. The recommended thermodynamic data for Ru, O_2, RuO_2, RuO_3 and RuO_4 have been used (7) to calculate partial pressures of the gaseous oxides RuO_3 and RuO_4. These values have been overlaid on the oxygen potential lines associated with various O/U values in Figure 6.

These calculations indicate that even allowing for uncertainties in the thermodynamic data for phases in the $UO_{2.61}$ - U_3O_8 region, ruthenium dioxide cannot be formed unless the U_4O_9/U_3O_{8-z} or UO_{2+x}/U_3O_8 phase boundary is crossed and orthorhombic U_3O_8 is formed. It would be incorrect to infer from this that volatile ruthenium cannot be formed at O/U ratios less than 2.63 because below line AB where ruthenium metal is the stable phase, RuO_3 and RuO_4 can be produced. The experimental data reported here supports this thermodynamic assessment.

CONCLUSIONS

1. The complete oxidation of UO_2 to U_3O_8 in the temperature range 300-600°C is accompanied with the release of 3.8-10.69% of the ^{85}Kr inventory. The releases are considered to be associated with lattice restructuring and the fractional releases are not sensitive to the temperature of oxidation or the duration at temperature after complete conversion to U_3O_8. However a slow diffusive release was noted on continued heating of U_3O_8 at 600°C.

2. Releases of volatile ruthenium and caesium species are reported, with maximum values obtained at 600°C. The release fraction of both ^{106}Ru and ^{137}Cs was observed to increase with continued heating of

the oxidation product U_3O_8. The release kinetics of volatile ruthenium species are approximately linear when UO_2/U_3O_8 is heated at 400 and 500°C and the release behaviour at 400°C can be represented by the expression:

$$R_{400} = 4.91x10^{-5}t-(1-e^{-0.00147t})$$

where R = percentage of ^{106}Ru inventory released

t = hold time at temperature (mins)

However, releases of ruthenium at 600°C were not observed to vary strongly with time at temperature.

3. Thermodynamic considerations indicate that in the temperature range 300-600°C the species responsible for ruthenium transport is RuO_4, although deposition is probably via decomposition to RuO_2. Migration of RuO_2, via re-volatilisation, along the thermal gradient tube was not conclusively observed.

REFERENCES

1. Peakall K.A. and Antill J.E.
Oxidation of Uranium Dioxide in Air and 350-1000°C, J. Nucl. Mat, 2,194, 1960.

2. Parker G.W., Creek G.E., Barton C.J., Martin W.T. and Lorenz R.A., Out of Pile Studies of Fission Product Release from Overheated Reactor Fuels at ORNL, 1955-1965, ORNL 3981, 1967.

3. Williamson R. and Beetham S.A.,
Fission Product and U_3O_8 Particulate Emission Arising from the Oxidation of Irradiated Uranium Dioxide: Preliminary Results. Workshop on Chemical Reactivity of Oxide Fuel and Fission Product Release, CEGB BNL, Glos. England, 7-9 April 1987.

4. Lorenz R.A., Collins J.L., Malinauskas A.P., Kirkland O.L., and Towns R.L.
Fission Product Release from Highly Irradiated LWR Fuel, NUREG/CR-0722, 1980.

5. Rard J.A., Chemistry and Thermodynamics of Ruthenium and Some of its Inorganic Compounds and Aqueous Species, Chem. Revs.,85(1), p1-39,1985

6. Bell W.E., and Tagami M.,
High Temperature Chemistry of the Ruthenium-Oxygen System
J. Phys. Chem., 67(3), p2432-2436, 1963.

7. Whittingham A.C. Unpublished Calculations
CEGB BNL, Glos. England, 1988

Ruthenium Release Kinetics from Uranium Oxides

F. C. IGLESIAS and C. E. L. HUNT
Atomic Energy of Canada Limited
Chalk River Nuclear Laboratories
Chalk River, Ontario, K0J 1J0 Canada

F. GARISTO
Atomic Energy of Canada Limited
Whiteshell Nuclear Research Establishment
Pinawa, Manitoba, R0E 1L0 Canada

D. S. COX
Atomic Energy of Canada Limited
Chalk River Nuclear Laboratories
Chalk River, Ontario, K0J 1J0 Canada

INTRODUCTION

The Chernobyl reactor accident showed that large quantities of ruthenium can be released into the atmosphere when the reactor core is exposed to air (1). In cases where fuel is exposed to steam, the formation of volatile ruthenium oxides is also thermodynamically possible (2).

The importance of the study of ruthenium release behaviour in accidents is twofold. First, the toxicity of the volatile Ru oxides (e.g., RuO_3 and RuO_4) and the radiation hazards of the two radioactive Ru isotopes (Ru-103 and Ru-106) must be considered in accident-evaluation studies (3); and second, the Ru release characteristics in oxidizing environments are so specific that understanding the kinetic behaviour of ruthenium releases will be most useful for post-accident management analysis (4).

In steam and air environments, since uranium oxides are thermodynamically more stable, ruthenium oxides should only form in an oxidizing atmosphere after some local oxidation of UO_2 has occurred.

In this paper, a summary of ruthenium behaviour, measured in experiments at the Chalk River Nuclear Laboratories (CRNL), is presented. The release characteristics and kinetics of ruthenium isotopes, their dependence on oxygen partial pressure, and the comparison with the release of other fission products are discussed.

EXPERIMENTAL DETAILS

UO_2 samples were obtained from CANDU fuel irradiated in the NRU reactor to burnups in the range 257-522 MW.h/kg U. The fuel elements were cut

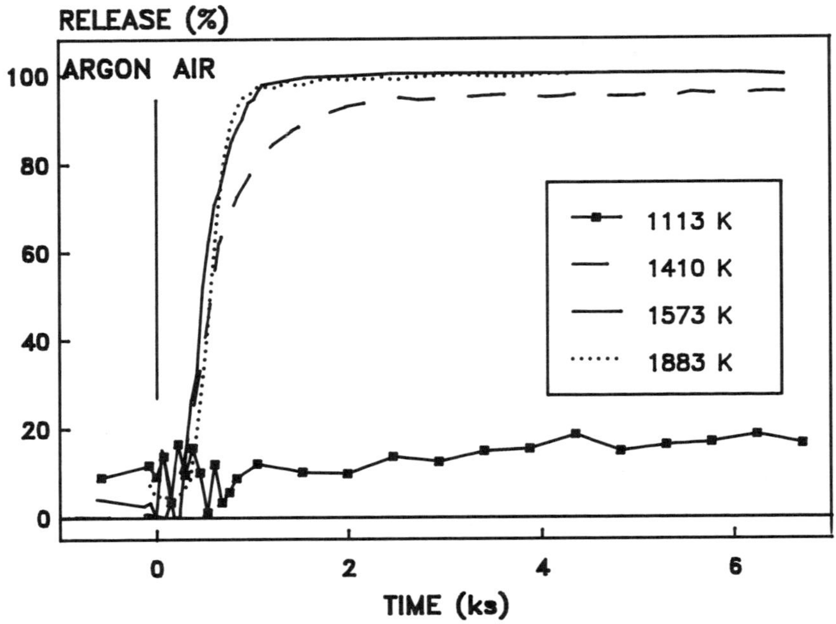

FIGURE 1. Release of ruthenium from UO_2 in air at four temperatures.

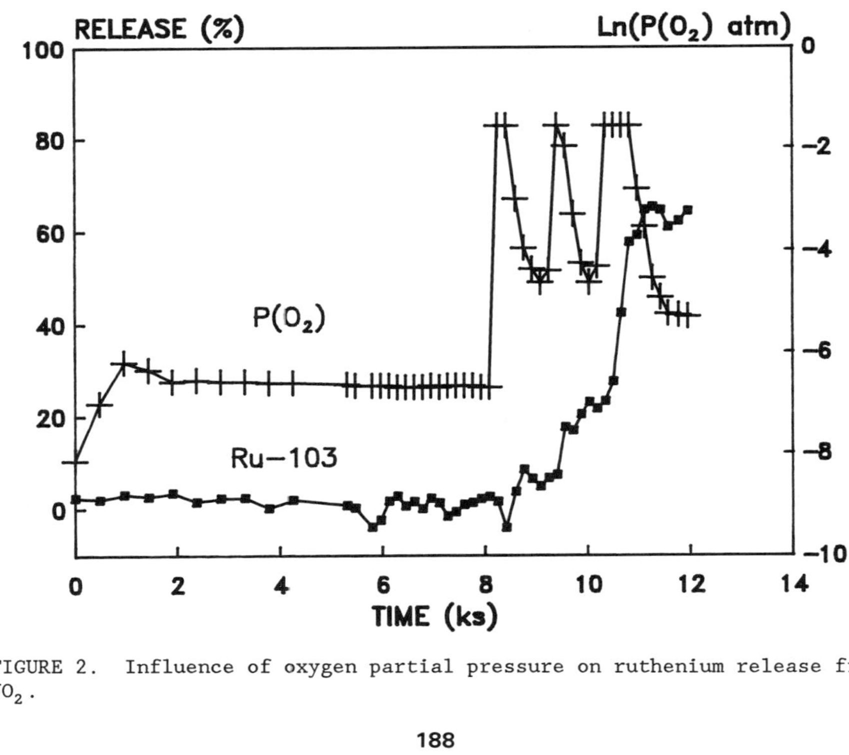

FIGURE 2. Influence of oxygen partial pressure on ruthenium release from UO_2.

open in the Universal Cell facility at CRNL and fragments of UO_2 collected. Typical fragment sizes were in the range 0.1 to 0.6 g.

Sample activity was measured by gamma-ray spectrometry using a through-wall collimator installed in one of the walls in the Universal Cell. This collimator was aligned with the centerline of the test furnace and permitted the gamma-ray spectrometer to view the fuel specimens through the side of the furnace. Thus, we were able to directly monitor the release kinetics of fission products from the fuel specimen, rather than measure the combined effects of release and transport of fission products to the cooler regions of the furnace.

For all tests, the furnace was first stabilized at the experimental temperature with an atmosphere of flowing argon (oxygen partial pressure $P(O_2)=10^{-5}$ atm). The irradiated UO_2 specimen was loaded into a boat attached to a push rod, and the end of the reaction tube sealed. The specimen was introduced into the hot zone of the furnace to give a ramp rate of about 0.5 K/s. Only when the sample reached the centre of the furnace was it in the field of view of the spectrometer. Specimen temperatures were monitored using a Pt-Pt 13% Rh thermocouple located 2 mm above the UO_2 fragments. The oxidizing atmosphere, air or steam, was introduced after the specimen reached the test temperature.

Oxygen sensors, measuring the oxygen partial pressure, were used in most of the steam tests. Steam was generated by feeding a specified amount of water through a metering pump to a stainless-steel tank maintained at 473 K. By passing the argon carrier gas through the tank, steam was carried through to the reaction tube. The experimental arrangement has been described in detail elsewhere (5).

EXPERIMENTAL RESULTS AND DISCUSSION

UO_2 is readily converted to higher oxides. In air at temperatures below about 1820 K, irradiated UO_2 samples are converted to U_3O_8. At temperatures above 1820 K, the highest oxide that can be formed is UO_{2+x}. In steam, UO_{2+x} is the highest oxide which can be formed at temperatures above 1320 K. A detailed description of the oxidation process in air and steam has been published previously (6,7).

The release of ruthenium from UO_2 has been measured in both air and steam. Release rates were estimated by calculating the derivative of the curves of release versus time, similar to those presented in Figures 1 and 6. In general, the release rates decrease with time. For tests with air (fast releases), we have calculated the initial release rates because they are less affected by inventory depletion. In the steam tests (slow releases), both an initial release rate and average release rate over the duration of the experiment were estimated. Table 1 lists details of samples, test parameters, and calculated release rates.

Releases in Air

In an air atmosphere ($P(O_2)=0.21$ atm), ruthenium releases have been measured at temperatures above 970 K. Figure 1 shows typical results. For this figure the time origin was selected as the time when the environment was switched from argon to air. For all curves it is apparent that there was an incubation time, between air reaching the

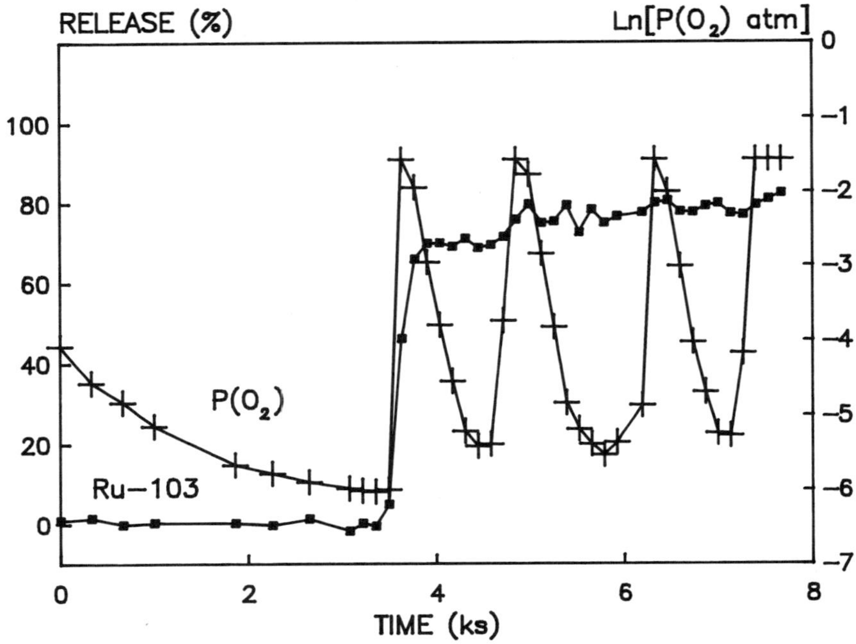

FIGURE 3. Influence of oxygen partial pressure on ruthenium release from U_3O_8.

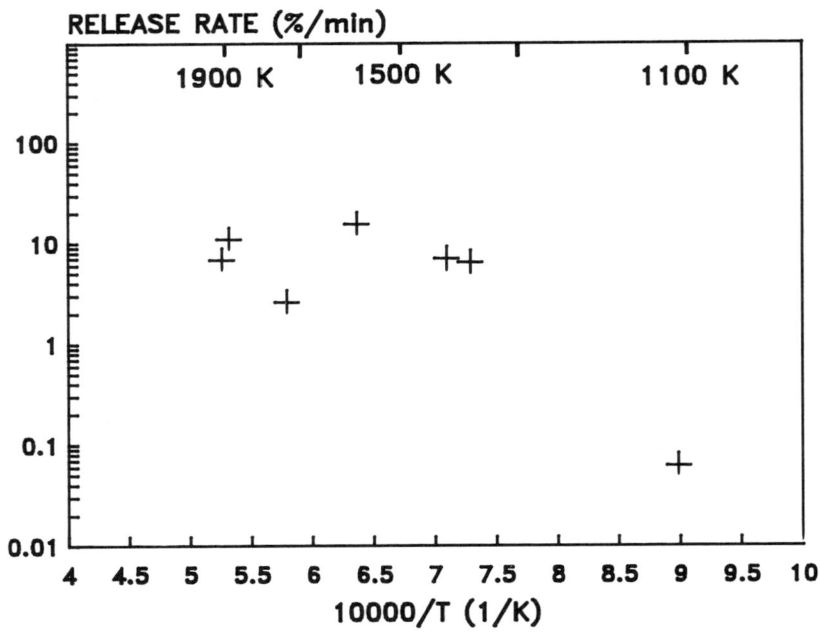

FIGURE 4. Temperature dependence of ruthenium release rate in air.

TABLE 1. Ru release rates and test parameters.

TEMP K	BU MW.h/kgU	SAMPLE WT g	NO. OF FRAGMENTS	ENVIRONMENT	RELEASE RATE %/min Initial	Average
1113	492	0.23	1	air	0.062	-
1373	522	0.44	1	air	6.48	-
1410	257	0.106	1	air	7.04	-
1573	522	-	1	air	15.6	-
1728	492	0.23	1	air	2.6	-
1883	492	0.69	4	air	10.94	-
1903	492	0.16	1	air	6.82	-
1673	512	3.28	20	steam	0.015	0.005
1682	512	2.655	22	steam	0.027	0.013
1773	512	2.81	20	steam	0.092	0.044
1923	492	0.96	2	steam	0.128	0.050
1941	512	2.65	22	steam	0.12	0.093
1963	512	2.83	21	steam	0.147	0.081

samples and the start of releases. To investigate the nature of this incubation time and the nature of the released species, tests were done with variable oxygen partial pressure. Figure 2 shows a test at 1673 K in an argon atmosphere. For 8 ks in argon ($P(O_2) = 10^{-5}$ atm) no ruthenium release was detected. Then, short pulses of air were introduced. During the initial air pulse, shown by the $P(O_2)$ curve, there was no ruthenium release. Subsequent pulses released some Ru but release stopped when the air was cut off. For the test shown in Figure 3, on the other hand, the fuel sample was fully oxidized to U_3O_8 at 770 K before the anneal at 1673 K. In this case, the first air pulse resulted in a large release of Ru without any incubation time, about 70% of the total inventory. Subsequent pulses resulted in small additional releases.

The existence of the incubation period (Figure 1) and the evidence presented in Figure 2 suggest that the ruthenium species being released are oxides. Figure 3 shows that no incubation period was required when the UO_2 sample had been previously oxidized to U_3O_8, but additional oxygen was required to convert the ruthenium species to a volatile form. The incubation time is interpreted as the time required for the more thermodynamically favorable conversion of UO_2 to higher oxides. Once the oxygen potential of the UO_{2+x} phase is near to the value for metallic Ru, oxidation of the Ru will occur.

Figure 4 shows an Arrhenius plot of the derived release rates of ruthenium in air. It is important to remember that in this figure all tests were performed at constant $P(O_2) = 0.21$ atm. These results, and the release curves presented in Figure 1 show that the release rates in air are very sensitive to temperatures below 1400 K. Contrarily, at temperatures higher than 1400 K, the release rates barely change. The reason for this transition is being investigated.

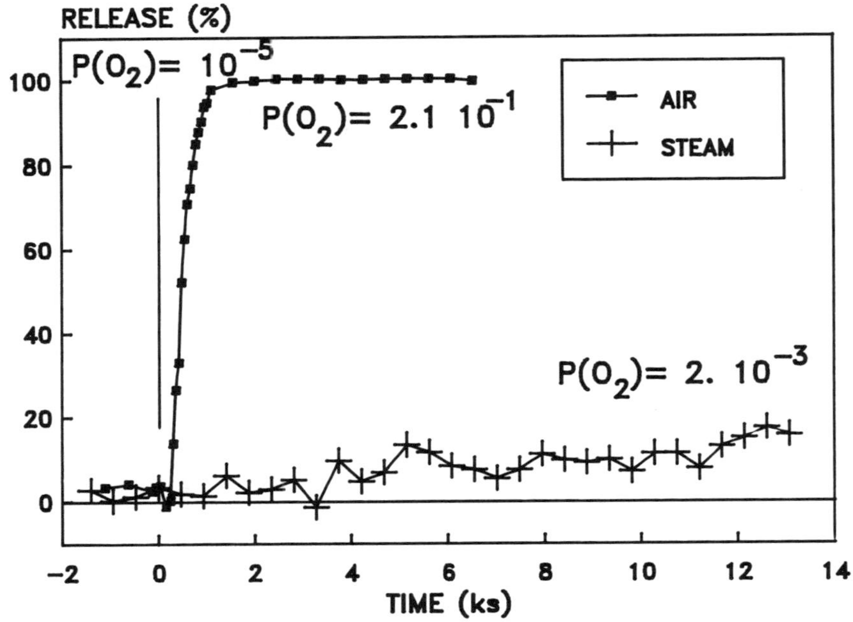

FIGURE 5. Comparison of the release of ruthenium from UO_2 in air and steam at similar temperatures.

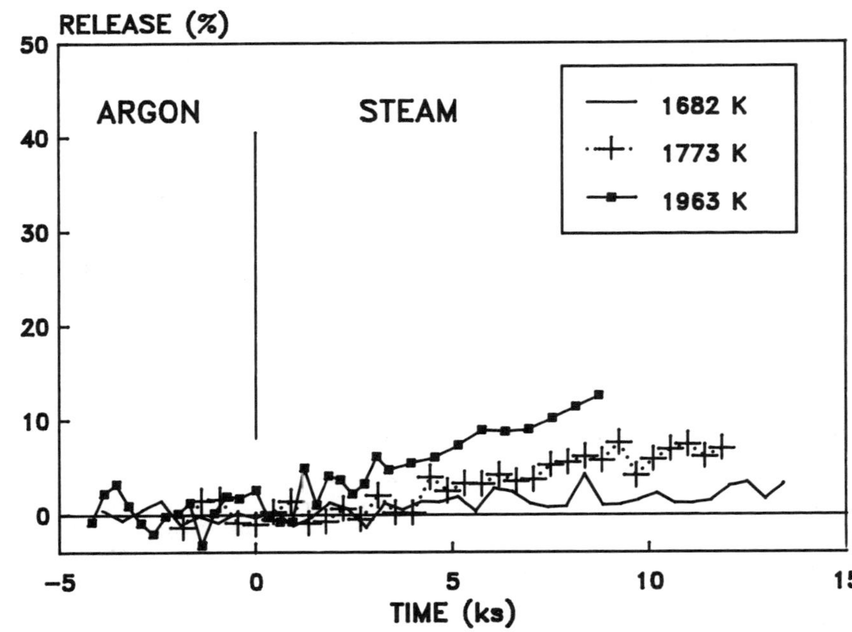

FIGURE 6. Release of ruthenium from UO_2 in steam at three temperatures.

Releases in Steam

During our experiments we compared the release rates in air with the release rates in steam. Figure 5 presents the results from two samples tested at similar temperatures. One was oxidized in air (at 1883 K) and the other in steam (at 1923 K). Neither sample released measurable amounts of ruthenium during the argon phase of the tests ($P(O_2)= 10^{-5}$ atm). During the oxidative phase, the release rates in steam were typically about 2 orders of magnitude less than air. This difference was typical of tests at all temperatures, which showed the strong dependence of the Ru release rate on oxygen partial pressure.

Release curves for three different temperatures during UO_2 oxidation in steam are shown in Figure 6. There is a dramatic difference in kinetics and release rates between these results and the air tests shown in Figure 1, but the characteristic incubation time is still present.

The temperature dependence of the initial release rates for samples annealed in steam is given in the Arrhenius plot in Figure 7. As the oxygen partial pressure in equilibrium with steam is a function of temperature, this figure shows the combined effect of temperature and oxygen partial pressure. The release rates are very sensitive to temperature up to about 1800 K, but they appear to be weakly dependent above this temperature.

Figure 8 presents both the percentage release of ruthenium and the UO_{2+x} stoichiometry during the test in steam at 1963 K. The specimen was initially annealed in an argon atmosphere ($P(O_2)= 10^{-5}$ atm) for about 5 ks, showing a steady change in stoichiometry with no associated ruthenium release. Introduction of steam raised the oxygen partial pressure to about $3\ 10^{-3}$ atm with the resulting increase in sample oxidation rate. Ruthenium release did not start until the sample stoichiometry was close to its equilibrium value and, hence, oxygen was available for the less thermodynamically favorable reaction of oxidizing ruthenium. These results strongly support the conclusions drawn from Figures 2 and 3.

In Figures 9 and 10, the release kinetics of ruthenium are compared with the more commonly studied release kinetics of cesium. Figure 9 shows results for an air environment, and Figure 10 for steam. The two tests were performed at similar temperatures. The cesium releases were similar in both tests, reaching a value of 80% in about 10 ks. It is interesting to observe that ruthenium releases, in contrast with cesium releases, were faster for the air test and slower for the steam test. Test temperature appeared to be the controlling parameter for the release of cesium (7) and oxygen partial pressure for ruthenium.

CONCLUSIONS

1. Ruthenium contained in irradiated UO_2 fuel must be oxidized before it can be released at temperatures below 2000 K.

2. Ruthenium release rates in air are very sensitive to temperature below 1400 K but are less sensitive above this temperature.

3. The releases of ruthenium are strongly dependent on the oxygen partial pressure.

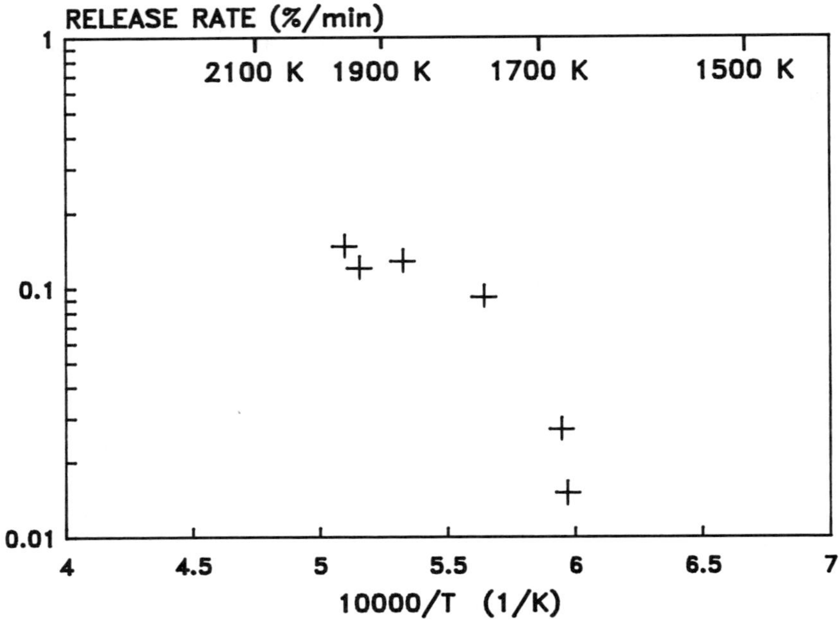

FIGURE 7. Temperature dependence of initial ruthenium release rates in steam.

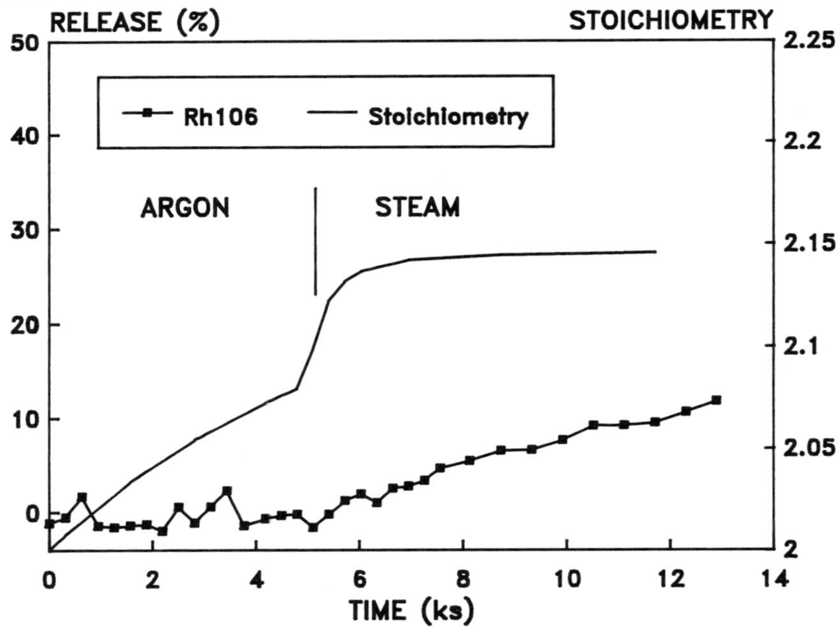

FIGURE 8. Evolution of ruthenium release and sample stoichiometry at 1963 K.

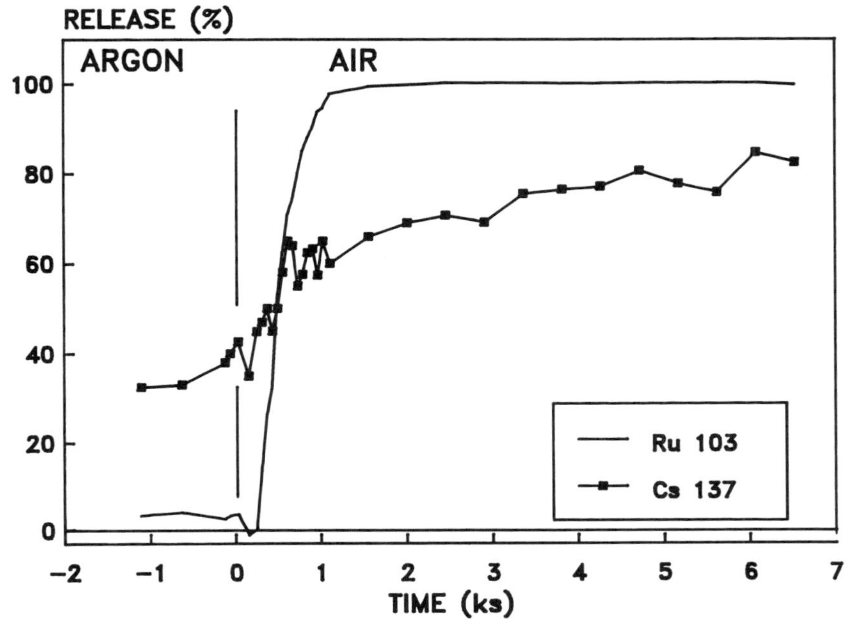

FIGURE 9. Comparison between ruthenium and cesium release kinetics in air at 1883 K.

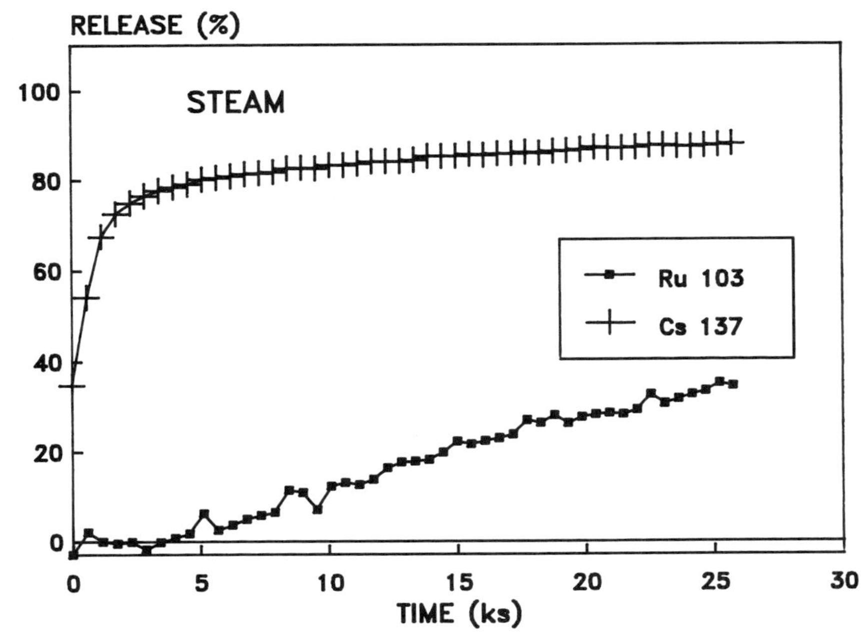

FIGURE 10. Comparison between ruthenium and cesium release kinetics in steam at 1941 K.

4. At low oxygen partial pressures (i.e., in steam), the release rates are very sensitive to temperature up to 1800 K, but weakly dependent above this temperature.

5. In steam, ruthenium release does not occur until the UO_2 specimen is close to equilibrium stoichiometry.

6. At high oxygen partial pressures (i.e., air) ruthenium releases are faster than those of Cs. At low oxygen partial pressures (i.e., steam) ruthenium releases are much slower than Cs releases.

ACKNOWLEDGEMENT

This work was funded jointly by AECL and Canadian utilities, which own CANDU reactors, under the CANDU Owners Group development program.

REFERENCES

1. Demin, V.F., Khodakovsky, I.L., USNRC Workshop on the Chemical Reactions and Processes in Severe Reactor Accidents, November 1987.

2. Garisto, F., 192nd ACS National Meeting, Anaheim, CA., 7-12 September 1986, NUREG/CP-0078, 3-107.

3. Eichholz, G.G., Prog. Nucl. Energy 2, 29 (1978).

4. Hunt, C.E.L., Iglesias, F.C., Cox, D.S., International Conference on Thermal Reactor Safety, Avignon, France, 2-7 October 1988.

5. Iglesias, F.C., Hunt, C.E.L., Cox, D.S., Keller, N.A., Barrand, R.D., O'Connor, R.F., Mitchell, J.R., Workshop on Chemical Reactivity of Oxide Fuel and Fission Product Release, Berkeley, UK, 7-9 April 1987.

6. Cox, D.S., Iglesias, F.C., Hunt, C.E.L., O'Connor, R.F., International Conference on CANDU Fuel, Chalk River, Canada, 6-8 October 1986.

7. Hunt, C.E.L., Iglesias, F.C., Cox, D.S., This Symposium Paper 2-6, Dubrovnik, Yugoslavia, 22-26 May, 1989

Fuel Behavior and Fission Product Release under HTGR Accident Conditions

KOUSAKU FUKUDA, KIMIO HAYASHI and KOREYUKI SHIBA
Japan Atomic Energy Research Institute
Department of Fuels and Materials Research
Tokai Research Establishment
Japan

1. INTRODUCTION

In early 1989 a final decision was made over construction of a 30 MWth HTGR called as High Temperature Engineering Test Reactor, HTTR, in Japan in order to utilize it for high temperature gas engineering tests and various nuclear material tests. The HTTR fuel having been developed at Japan Atomic Energy Research Institute, JAERI, is a "pin-in-block" type fuel element which is composed of a hexagonal graphite block with dimension of 580 mm in length and 360 mm in face-to-face distance and about 30 of the fuel rods inserted into the coolant channels drilled in the block. The TRISO coated fuel particles for HTTR are incorporated with graphite powder and phenol resin into the fuel compacts, 19 of which are encased into a graphite sleeve as a fuel rod.

With respect to the HTTR safety design, following three types of the accident are predicted as phenomena at extreme temperature ; core heat-up by loss of forced coolant circulation, coolant channel blockage and rapid reactivity insertion. The design estimates that the HTTR fuel is heated at about 1500°C under the core heat-up condition, it is exposed at about 1750°C by irradiation under coolant channel blockage and the rapid reactivity insertion causes fuel temperature rise over maximum 1680°C from the normally operated temperature.

It is necessary for the HTTR licensing to prove the fuel stability under these predicted accidents related to the high temperature events, and therefore, the release of the fission products and the fuel failure have been investigated in the irradiation - and the heating experiments simulating these conditions at JAERI. This report describes the HTTR fuel behavior at extreme temperature made clear in these experiments.

2. EXPERIMENT

Experiments on the fuel behavior under the accidents in the HTTR design are summarized in Table 1. For investigating the fission product release from the coated particles under the core heat-up conditions the out-of-pile heating at high temperature ranging from 1400°C to 2200°C was undertaken on the TRISO coated fuel particles irradiated to 1 %FIMA at 1300°C in an in-pile gas loop, OGL-1. Single coated particle taken out from the fuel compacts in the post-irradiation examination was used in each run of heating which was conducted isothermally or isochronally by a furnace with a graphite heater.

The accident of the coolant channel blockage was simulated by the high temperature irradiation in Japan Material Test Reactor,JMTR, where two capsules and OGL-1 were used. In the capsule irradiation experiments the fuel compacts were irradiated at normal temperature, followed by rapid increase of fuel temperature and keeping at elevated temperature for a short time of several ten hours. The OGL-1, on the other hand, could irradiate the fuel rod containing 20 fuel compacts(a total of about $3x10^5$ to $4x10^5$ coated particles were involved) at temperature ranging from normal 1350°C to accident or abnormal transient temperature of 1500°C under pressurized helium gas circulation in JMTR. During irradiation the fractional releases of the fission gas from the fuel element were determined by measuring the fission gas concentration in the primary loop of OGL-1.

TABLE 1. Summary of experiments simulating the accidents related to extreme temperature.

Phenomena	Experiment	
	Method	Conditions
Core heat-up	Out-of-pile heating for fission product release measurement	1400°C - 2200°C
Channel blockage	OGL-1 irradiation Capsule irradiation	1100°C - 1500°C 1720°C & 1900°C in maximum
Rapid reactivity insertion	NSRR irradiation	higher than 2800°C

For the simulation of the rapid reactivity insertion the reduced size fuel compacts were exposed to pulsatile irradiation with deposition energy from 260 J/gUO_2 to 2300 J/gUO_2 in Nuclear Safety Research Reactor, NSRR, where fuel temperature at center of the coated particles exceeded fairly UO_2 melting point of 2800°C in the severest case.

3. RESULTS AND DISCUSSION

3.1 Fission product release in the out-of-pile experiment

The results of the metallic fission product release from single coated particle by isochronal heating are depicted in FIGURE 1 as a typical example, where temperature at each step was kept for 2 hours and the initial contamination contribution for the release was not excluded. This heating was repeated 4 times using a different coated particle in each run, yielding similar results with it. The releases of the fission products except ^{137}Cs and ^{134}Cs from 4 coated particles were less than the detectable limit below 1900°C , but the fractional release of ^{155}Eu was extensively large comparing those of the other fission products above 1900°C. The fractional releases of cesium isotopes increased abruptly above 2100°C. The release of ^{144}Ce being largely miscible with UO_2 (1) increased steadily above 2000°C.

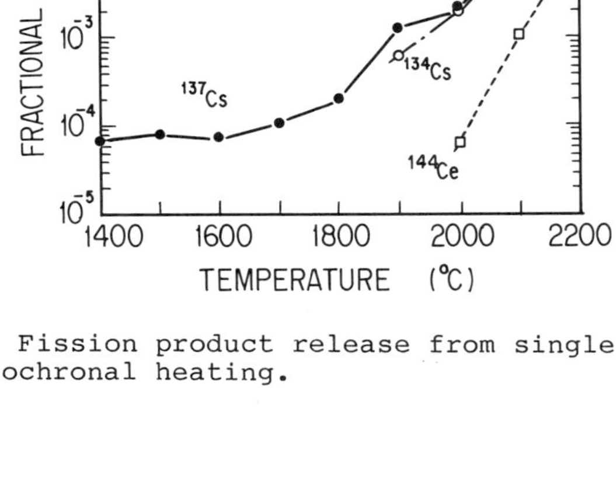

FIGURE 1. Fission product release from single coated particle by isochronal heating.

Concerning the release behavior of the most important nuclide during the accident, ^{137}Cs release was measured in the isothermal heating in the temperature range between 1600°C and 2200°C, using different single coated particle in each heating run. The results are summarized in FIGURE 2. Within heating duration of about 1000 h, a difference was recognized in the release behavior; below 2000°C the release rate was moderate, indicating the diffusion controlled release, whereas the release rate above 2100°C was fairly large. This release behavior agreed qualitatively with the isochronal release mentioned above and also with the US results obtained in CHST[2] where TRISO $(4Th/U)C_2$ particles after irradiation were ramp-heated in the temperature range between 1300°C and 2400°C. The increase of the release rate above 2100°C was influenced by change of the SiC properties at high temperature as suggested by W. Schenk et al[3]; degradation of SiC layer was characterized by corrosion by fission product above 1600°C and decomposition itself at 2200°C[3]. By comparison of thinning rate of SiC layer, the decomposition was predominant above 2100°C. Since burnup of the present coated particles was not so large, 1 %FIMA, corrosion by the fission product, palladium, was not relatively severe in these results. The present results proved that the decomposition influenced seriously on the ^{137}Cs release above 2100°C and it was possible for the coated particles to retain cesium below this temperature for about 1000 h at least.

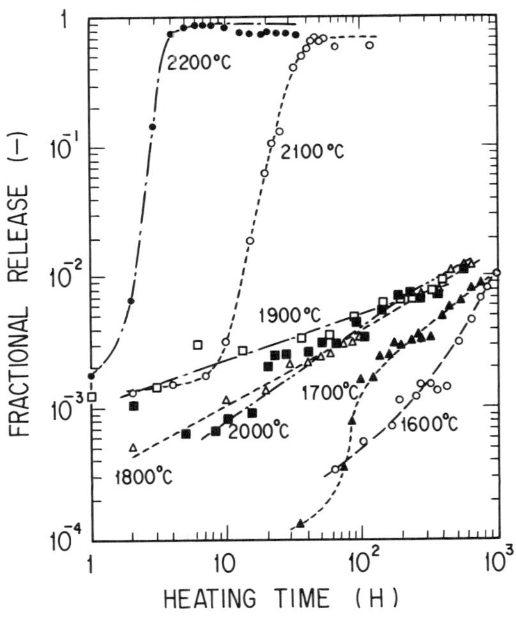

FIGURE 2. Cesium 137 release from single coated particle under isothermal heating.

3.2 Fission gas release under irradiation

The release of the fission gas measured by OGL-1 in the temperature range between 1100°C and 1500°C is shown in FIGURE 3, where the results from two experiments are included. Level of the fractional release, R/B in this temperature range was related deeply with the initial failure fraction of the coated particles in the compacts. Increase of R/B with temperature was not so abrupt that it was indicated that the remarkable failure caused by irradiation did not occur, thus proving that the heat-up by channel blockage would not give significant influence on the fuel integrity as far as temperature rise did not exceed over 1500°C.

In the case of the irradiation experiments by the closed capsule, the failure fraction increased about two order of magnitude in maximum from the initial value by irradiation at 1720°C with burnup of 4 %FIMA. Failure fraction in this case still remained less than 1 %, indicating that the severest channel blockage could keep the requirement in the design of HTTR (less than 1 %).

3.3 Rapid reactivity insertion

The instantaneous reactivity insertion would not be considered in any HTGR designs, but an event due to the reactivity insertion by drawing out the control rods

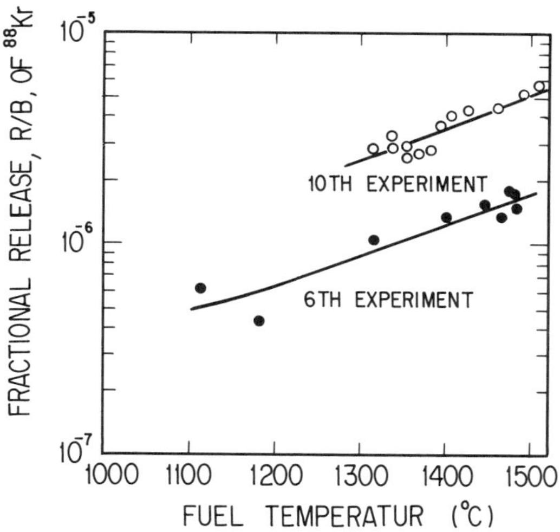

FIGURE 3. Fractional release of ^{88}Kr from OGL-1 fuel element by an in-pile gas loop irradiation.

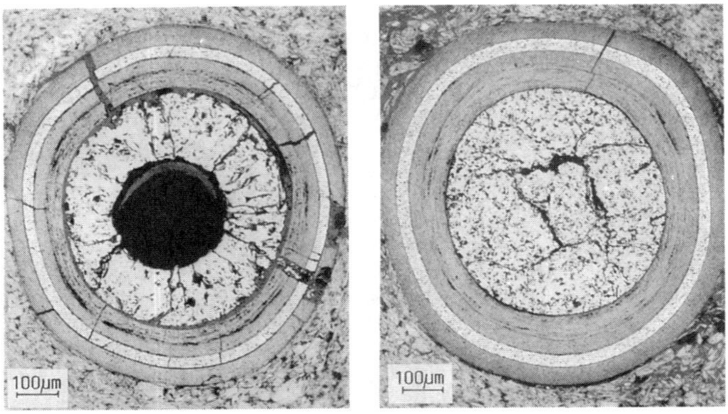

Energy deposition
2300 J/gUO_2 1050 J/gUO_2

FIGURE 4. Cross-section of coated particles irradiated in NSRR.

rapidly was predicted in the HTTR design, where energy deposited on the fuel was $1.26x10^4$ J/gUO_2 and the duration was about 8 seconds. The NSRR operation, however, could not insert the reactivity on the HTTR fuel with fully simulating the design mode, although this will be possible after 1989, and so the HTTR fuel was irradiated by the NSRR possibly operating mode, the instantaneous reactivity insertion of which pulse width(1/2 peak power) was 4.4 msec. This irradiation, however, gave us interesting information. The ceramography on the irradiated coated particles as seen in FIGURE 4 revealed that the center region of the UO_2 kernels in all of the coated particles irradiated with energy deposition of 2300 J/gUO_2 vaporized and a central void was formed there. According to observation by EPMA, it was evident that uranium vapor leaked through the cracks in the coating layers induced by high inner gas-pressure and it deposited near the boundary between the coated particles and the graphite matrix. Whereas the coated particles irradiated with the energy less than 2300 J/gUO_2 remained almost unchanged except for small cracking in the coating layers as seen in FIGURE 4.

As for appearance of the fuel compacts, only small cracks appeared in the surface even in the severest case where 2300 J/gUO_2 was deposited, and the remarkable change was not seen in the other fuel compacts deposited with lower energy. Comparing with the case of a simulation test on PWR fuel pin (4), it was evident that the HTTR fuel was by far stable under such event. In the case of PWR pins where 0.1MPa He was filled these were damaged depending on the energy deposition as follows; small cracks were introduced in the cladding by the energy deposition around 1000 J/gUO_2, the pins failed above 1400 J/gUO_2 and burst above 1500 J/gUO_2. For the HTTR fuel a relation between the failure fraction of the coated particles and the energy deposition is displayed in FIGURE 5. As seen in this figure the failure fraction of the coated particles was about 1 % by the energy deposition of 1000 J/gUO_2 and attained almost 100 % with the energy above 1500 J/gUO_2. This fuel, however, could retained the evaporated U in the graphite matrix and furthermore, the fuel compacts could keep its shape without catastrophic failure as mentioned above, even if the instantaneous reactivity with energy deposition of 2300 J/gUO_2 was inserted. This fact might ensure also that the HTTR fuel element containing the fuel rods would remain intact by the event with energy deposition less than 2300 J/gUO_2. These results were related to the adiabatic system due to the instantaneous event, but the heat flow from the coated particles would suppress keen temperature rise in the mode predicted in the HTTR design because the duration is by far long, although the energy deposition is larger than the present case. The HTTR design predicts maximum temperature by this event as 1680°C, but this estimation includes several uncertainty

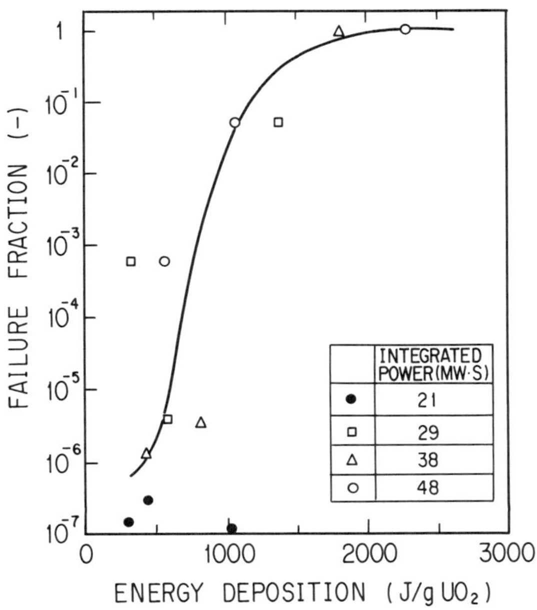

FIGURE 5. Relation between failure fraction of the coated particles and energy deposition by NSRR irradiation.

such as gap width between the buffer layer and the IPyC layer. Therefore, planned are the further experiments where the irradiated coated particles will be employed to the NSRR irradiation with longer duration.

4. SUMMARY

In the out-of-pile heating experiments the releases of metallic fission products were measured for simulating the core heat-up accident. By heating single coated particle it was evident that it would be possible for the particles to retain the fission product at 2000°C or less for 1000 h at least.

For simulation of the channel blockage it was proved that the fission gas release did not increase anomalously under temperature rise up to 1500°C from the normal operating conditions. In the irradiation experiment where the coated particle fuels were irradiated over 1700°C, although the failure fraction of the coated particles increased two orders of magnitude comparing with the initial fraction, that figure met the HTTR design requirement.

Instantaneous reactivity insertion was applied on the HTTR fuel and the fuel compacts exhibited relatively good performance by irradiation with the energy deposition of 2300 J/gUO_2, although all of the coated particles in the fuel compacts were damaged.

REFERENCES

1. Kleykamp, H., The chemical state of the fission products in oxide fuels, J. Nucl. Mater., vol. 131, pp221-246, 1985.

2. Quarterly Progress Report for Period Ending, November 30, 1977, GA-A 14744, 1977.

3. Schenk, W., Pitzer, D. and Nabielek, H., Spaltproduktfreisetzungsverlauf von Kugelbrennelementen bei Stoerfalltemperaturen, Juel-2091, 1986.

4. Fujishiro, T., A study on gap heat transfer of LWR fuel rods under reactivity initiated accident conditions, JAERI-M 84-063, 1984.

Experiments on In-Vessel Core-Melt Progression

R. W. WRIGHT
Office of Research
U.S. Nuclear Regulatory Commission
Washington, D.C. 20555, USA

Abstract: The four Severe Fuel Damage tests in the PBF test reactor have provided the basic integral information base on the early phase of in-vessel core-melt progression. This has been supplemented by results from the ACRR, NRU, and Phebus test reactors and by out-of-pile integral experiment data and fundamental materials interaction data from KfK. Nearly all the available information on the late phase of core-melt progression comes from the TMI-2 core examination. Summaries of the results and phenomenological interpretation of these experiments are provided.

Introduction

In-vessel core-melt progression concerns the state of the LWR reactor core from core uncovery up to reactor-vessel melt through in unrecovered accidents, or through temperature stabilization in accidents recovered by core reflooding. Melt progression determines: (1) the in-vessel hydrogen generation; (2) the in-vessel fission- product and aerosol release, transport, and attenuation; (3) the mode of vessel failure; and (4) the characteristics of the ejected debris that provide the initial conditions for assessing the magnitude of the ex-vessel core-melt threat to containment integrity. Sensitivity studies have shown that the greatest uncertainties in assessing the probability of early containment failure and severe-accident risk come from the uncertainties in in-vessel core-melt progression (Ref. 1).

Early-Phase Melt Progression: Generic and PWR Specific

Considerable information is available on the early phase of core melt progression that extends up to the onset of substantial relocation of ceramic core material. The principal integral information base comes from the four large integral Severe Fuel Damage (SFD) tests in the PBF test reactor that have now been completed (Ref. 2). These PWR geometry tests used one-meter-long fuel bundles with 32 or 28 rods of either trace irradiated or high-burnup fuel (36,000 Mwd/TU) The test fuel bundles were exposed to a fission power transient during coolant boildown at about 7 MPa pressure. As shown in Table 1, the PBF test conditions, one of the two high-burnup fuel tests had four PWR Ag-In-Cd control rods within zircaloy guide tubes. The initial scoping test had a steam flow rate one order of magnitude higher than a prototypic boildown, as well as water reflooding to terminate the test. (The other tests used gas for bundle cooldown.) On-line thermocouple measurements were made of temperatures in the bundle and in the gas. Extensive Post-Irradiation Examination (PIE) of the test-fuel bundle gave end-state information on materials interactions and peak temperatures and on fission-product behavior. On-line measurements were made of fission-product release (gamma spectroscopy) and, in the SFD 1-4 test with control rods, of the aerosol release. Sampling measurements were also made of fission-product release and of fission-product and aerosol deposition. The fission-product and aerosol behavior are not treated in this paper.

Table 1. PBF Severe Fuel Damage Test Conditions

Test	Burnup	Control Rods	Steam Rate	Termination
SFD-ST	Trace	No	High	Water
SFD 1-1	Trace	No	Boiloff	Steam
SFD 1-3	High	No	Boiloff	Argon
SFD 1-4	High	Yes	Boiloff	Argon

When peak temperatures in the uncovered cladding reached about 1500K, steam oxidation of the uncovered cladding started to become significant, and the heating rate increased above the "prototypic" initial rate during boiloff of about 0.5 K/sec. Initially the oxidation transient was rate-limited by diffusion through the ZrO_2 surface sheath that builds up on the zircaloy cladding, a process that is strongly temperature dependent. At about 1700K (boiloff conditions) the rate of oxidation heating increased rapidly to tens of K/sec., limited only by conversion of the entire steam flow into hydrogen (steam starvation). According to a combination of cladding temperature measurements by thermocouples (that failed above about 2100K), Post-Irradiation Examination (PIE) of the test bundle, and test analysis with the SCDAP mechanistic fuel damage code (Ref. 3), local temperatures during the rapid oxidation transient reached the range of 2800K to 3000K. During this temperature transient, molten unoxidized metallic zircaloy relocated downward toward the water surface to form a large but not complete metallic blockage across the lower fuel-rod stubs. The relocated metallic zircaloy continued to oxidize, and most of the hydrogen (85% in the SFD 1-1 test) was generated after the onset of molten zircaloy relocation. As indicated by the pseudo-binary phase diagram for UO_2 and oxygen-stabilized alpha zirconium (Ref. 4), molten zircaloy near its melting point of about 2200K can dissolve about 50% of its mass of solid UO_2 in a liquifaction process, a fraction that increases to about 85% as the temperature reaches the 2700K monotectic temperature. Considerably more UO_2 can be dissolved in molten zircaloy, however, by an inhomogeneous slurry-formation process (Ref. 4), and above the monotectic temperature, the equilibrium mass ratio of soluable UO_2 to molten zircaloy reaches about seven. This dissolution process occured in the PBF tests, as shown by the UO_2 content of the metallic (metallo-ceramic) partial blockage at the lower end of the fuel bundles and by the missing UO_2 mass in the declad fuel pellets. The control rods in the SFD 1-4 test failed near the stainless steel 1720K melting point, which is about 300K below that of zircaloy, and the control-rod material relocated downward and froze to form the lower part of the metallic blockage with the higher-melting zircaloy above.

Neutron radiographs of the post-test fuel bundles of the four PBF tests are shown in Figure 1 (from Ref. 2). The radiographs show the large partial blockage of relocated zircaloy and dissolved UO_2 (and control-rod materials in the SFD 1-4 test) across the fuel-rod stubs at the bottom of the bundle. The radiographs also show a progression in the four tests from the relatively intact rod geometry in the SFD-ST high-steam-flow fresh-fuel test that had high in-place oxidation of the cladding to the more broken up but not collapsed array of declad ceramic fuel rods and pellets in the SFD 1-4 test that had high burn-up fuel and Ag-In-Cd control rods. The SFD 1-4 test reached higher temperatures than the other high-burnup-fuel test, SFD 1-3. (In SFD 1-3 there was an inadvertent partial flow diversion from the bundle

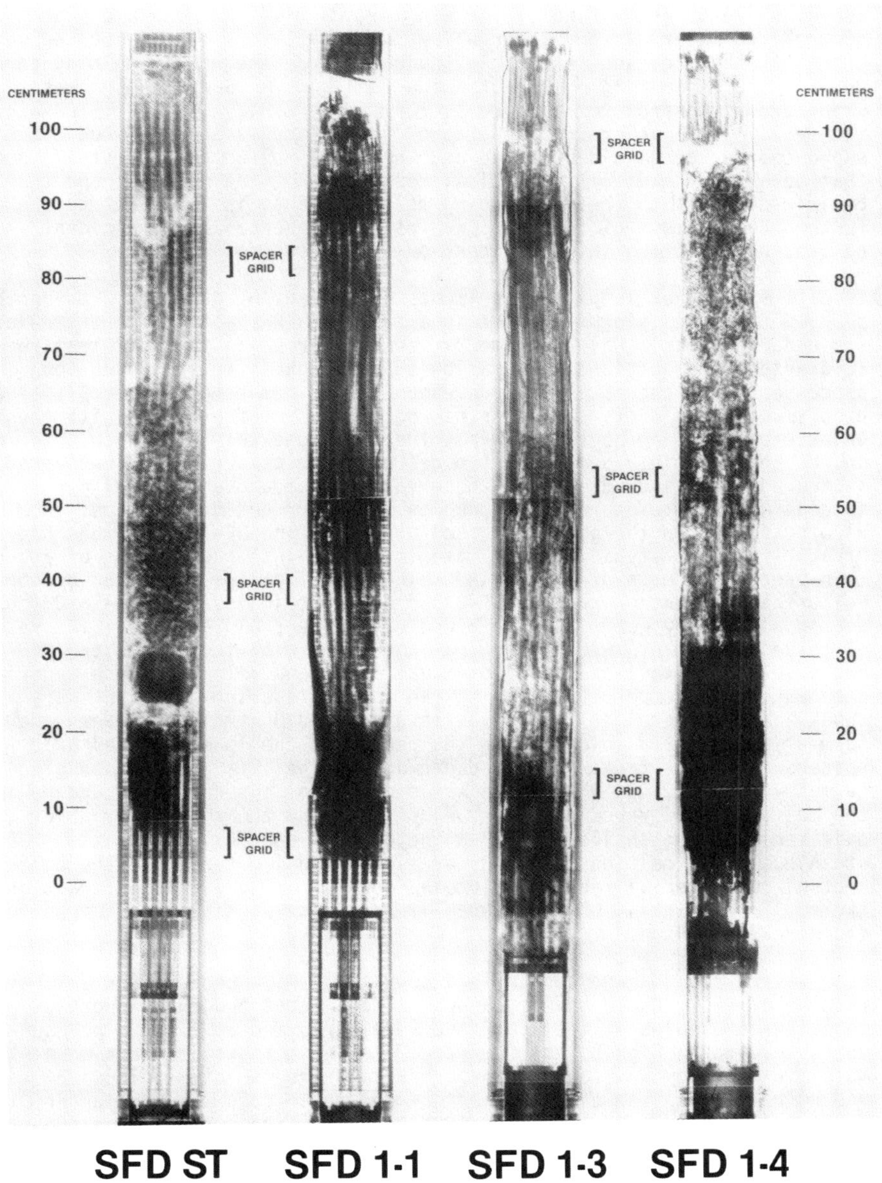

Figure 1. Post-Test Neutron Radiographs of the PBF Severe Fuel Damage Test Bundles.

during the high-temperature transient.) The upper part of the SFD 1-4 blockage contains the zircaloy with dissolved UO_2, and the lower part contains the lower melting silver, indium, cadmium, and stainless steel control rod materials and their alloys with zircaloy. The relocated metallic debris in the blockage is clearly separated from the higher melting ceramic UO_2 and ZrO_2 remaining in the standing fuel rods. (Some cadmium was vaporized at control rod failure to form an aerosol.) In the TMI-2 accident, the metallo-ceramic crust or "hardpan" formed across the fuel-rod stubs and supported the ceramic debris bed and the large molten ceramic (fuel) pool that developed later. The physical separation of the metallic and the ceramic debris with their differing melting points is important to the further development of the melt progression sequence and to accident consequences, as the TMI-2 accident demonstrated.

Results from other in-pile experiments have been in agreement with and have supplemented the information obtained from the PBF tests. Three smaller-scale damaged-fuel (DF) experiments were performed in the ACRR test reactor to provide more detailed information on early-phase melt progression processes for mechanistic code development and assessment (Ref. 5). These experiments used fifty-cm long nine-rod bundles of fresh-fuel with constant inlet steam flow at abcut 0.7 MPa. Their purpose was to furnish information on the behavior in a section (or node) near the center of the uncovered core. Instrumentation included visual observation from above, optical pyrometry, special high-temperature thermocouples, and hydrogen diagnostics. PIE was also performed on the post-test fuel bundles. The DF-1 experiment had a greater-than prototypic heating rate to investigate the effects of a greater-than-prototypic mass of molten and relocated zircaloy. DF-2 had prototypic boiloff and oxidation rates. DF-3 had a PWR stainless-clad Ag-In-Cd control rod in a zircaloy guide tube in place of the central fuel rod. The results of these experiments were in general agreement with those from the PBF tests but gave more detail on some individual phenomena. In all these experiments, the peak local temperatures during the rapid oxidation transient were about 2600K. Molten metallic zircaloy with dissolved UO_2 relocated downward to form a large partial blockage in the region of the lower inconel grid spacers, leaving stacks of eroded declad fuel pellets in their original rod-geometry columns. In DF-2 there was significant internal flow channeling of the molten zircaloy between the fuel pellets and the relatively thick ZrO_2 oxidized cladding shells. In DF-3 the control rod failed near the 1720K stainless-steel melting point, generating a cadmium aerosol. The other control rod material rapidly relocated downward with little interaction with the fuel rods. In all of these experiments, a dense tin aerosol appeared when the zircaloy melted.

Four severe-fuel-damage coolant-boildown tests have been performed in the NRU reactor at AECL, Chalk River, Canada. These tests had full-length 3.7 meter-long 12-rod bundles of test fuel and were performed at 1.4 MPa pressure. Two tests included a high burnup fuel rod. In these tests, relocation of the metallic zircaloy was found to be a repetive process of melting, relocation, and refreezing as the oxidation burn front moved down the bundle (Ref. 6). The test transient was continued for 30 minutes and for 60 minutes in two of the tests. In both cases hydrogen generation continued throughout the tests oxidizing nearly all of the uncovered zircaloy cladding in the latter test. The relocation of molten zircaloy did not produce a complete flow blockage to cut off steam flow and hydrogen generation as sometimes hypothesized.

Experiments on the early stages of fuel damage have been performed in the French Phebus test reactor (Ref. 7). Fuel damage experiments with pulse heating in a gas environment have also been performed in the Japanese NSRR test reactor (Ref. 8).

Two experiments were performed in ACRR on fission-product release from high burnup fuel in a reducing hydrogen environment and at two different pressures, 0.19 MPa and 1.8 MPa respectively, (Ref. 9). Each of the four test fuel rods in the bundle had a 15 cm long fresh-fuel "pre- heater" section and a 15 cm test section of high burnup fuel (45,000 Mwd/TU). The test fuel bundles were held on temperature plateaus of about 2500K for about 18 min. The primary experimental results on fission-product release are not reported here. Post-test x-radiographs and sectioning of the test bundles showed that, after the zircaloy clad had melted and relocated downward, the high-temperature high-burnup fuel had swelled without melting to close the flow space between the rods. The fresh pre-heater fuel rods had undergone no such swelling, even though their upper ends reached the temperature of the high-burnup fuel. PIE of the high burnup fuel rod from the steam-starved hydrogen environment near the upper end of the test fuel bundle in one NRU test also showed a similar swelling (Ref. 10). A model has been developed for this large swelling without melting of high-burnup fuel in a reducing environment.

Much of the fundamental information base on the governing phenomena in severe fuel damage and early-phase core melt progression has been obtained from out-of-pile integral experiments and basic materials interactions experiments and analysis at KfK (Refs. 4, 11). The integral experiments in the CORA and the earlier NEILS test facilities have used electricity-heated fuel rods that consist of a central tungsten heater rod, annular UO_2 pellets, and zircaloy cladding. The objective of these ongoing experiments is to obtain information on the integral meltdown behavior of fuel rod bundles, metallic melt relocation and blockage formation, and reflood quench behavior. The effects on meltdown behavior of various bundle components has been studied, including the effects of stainless-steel-clad Ag-In-Cd and B_4C neutron absorber rods and inconel and zircaloy spacer grids.

These experiments at KfK have covered much of the range of early phase melt progression phenomena. They have provided basic information on the oxidation of zircaloy cladding and cladding failure, and on the processes of metallic melt relocation and blockage formation. By visual observation, melt relocation was found to occur mostly by non-coherent, non-coplanar rivulet flow. Large but not complete metallic blockages were formed between the lower fuel rods by frozen relocated metallic melt. There was significant dissolution of the solid UO_2 by molten zircaloy and UO_2 relocation with the molten zircaloy. It was shown that eutectic alloying interactions of the absorber materials, stainless steel, and inconel with the zircaloy can significantly lower the temperature for metallic melt formation. Grid spacers were found to be a collection point for blockage formation, even through the grid spacers disappeared in melting. The addition to the bundle of a PWR Ag-In-Cd control rod in a zircaloy guide tube greatly increased the quantity of relocated metallic melt. The results of the out-of-pile experiments in CORA and NEILS and the fundamental separate- effects experiments and analysis on materials interactions at KfK are in good agreement with the integral results of the in-pile tests in PBF, ACRR, and NRU, and with the results of the TMI-2 core examination. The KfK work supplements these results. (The early fundamental KfK research and NEILS experiments preceded the post-TMI-2 in-pile tests.)

Early-Phase Melt Progression: BWR Specific

The ACRR DF-4 experiment was performed to provide information on the effects of the B_4C control blades and the channel boxes on early-phase core-melt progression in BWRs (Ref. 12). The experiment also provided a check on the hypothesis that a complete blockage of the steam flow within intact channel boxes would occur near the melting threshold of the zircaloy fuel-rod cladding, cutting off all further hydrogen generation. The DF-4 experiment represented a section of a BWR core including a control-blade tip, channel-box walls, and 14 fresh fuel rods. For experimental convenience, the channel box in DF-4 enclosed the control blade, with the fuel outside. This reversed the actual BWR core configuration,but has no significance. The 50cm-long test section was supplied with prototypic boiloff steam flow, with the proper flow split between the gap around the control blade and the fueled sections. Instrumentation, including visual observation, was the same as in the other ACRR DF experiments.

When the control blade reached about 1550K, 170K below stainless melting, the control blade failed from the eutectic interaction of boron and stainless steel. This occured before the rapid oxidation temperature transient. The control-blade melt attacked and failed the molten zircaloy channel box walls, opening up the geometry like a PWR. The control blade material relocated down to the bottom of the test section, an effect that could produce recriticality problems if a damaged BWR core were to be reflooded with unborated water. The control blade material along with the unoxidized metallic zircaloy and alloyed material formed a massive but not complete metallic blockage at the colder bottom of the test section. Oxidation of the relocated metal continued for about ten minutes until the experiment was terminated. These results disagree with the hypothesized cut off of hydrogen generation by metallic blockage formation in intact channel boxes mentioned above. They are also disagree with the modeling in the BWRSAR accident analysis code (Ref. 13), that assumes that the channel box and control-blade melt (including eutectic alloys) drains out of the core to collect in layers on the core plate in the sequence of the material melting (and eutectic alloying) temperatures. (BWRSAR uses a simplified candling model for the molten cladding relocation.) The alternative to this metallic melt drainage from the core is freezing of the relocating metallic melt to form a large metallic blockage across the fuel rod stubs as occured in the PBF, ACRR, and CORA tests. This is the metallic blockage process that initiated formation and later drainage of the large in-core pool of high temperature ceramic melt in the TMI-2 accident as shown in Figure 2. It should be noted, however, that the TMI-2 accident (with water in the lower core), the DF-4 experiment, and other experiments in which this lower metallic blockage have occurred all had high axial temperature gradients where the metal froze to form the blockage. In depressurized BWR accidents, the blowdown lowers the water level below the core plate, so that the steaming rate and the axial temper- ature gradient in the core are both reduced. No data currently exist on metallic relocation and blockage formation under these conditions, and conclusions for these conditions might be different. The SCDAP and MELPROG codes (Refs. 3 and 14) treat metallic melt drainage and blockage formation mechanisticly, but experimental data to validate the models are few. The author considers this question of metallic melt drainage or blockage formation to be the current major uncertainty regarding in-vessel core-melt progression.

Late Phase Melt Progression

The primary information base on the late phase of in-vessel core-melt progression (that starts with the onset of substantial ceramic material relocation) comes from the TMI-2 core examination (Ref. 15). The TMI-2 accident was successfully recovered by reflooding the core and had a complex accident sequence. Nevertheless the basic melt-progression phenomenology involved in TMI-2 is nearly all applicable to unrecovered PWR accidents as well as to PWR accidents that are recovered by core reflooding. Figure 2 (from Ref. 15) shows a schematic drawing of the end-state configuration of the core as determined by the TMI-2 core examination, and will be referred to frequently

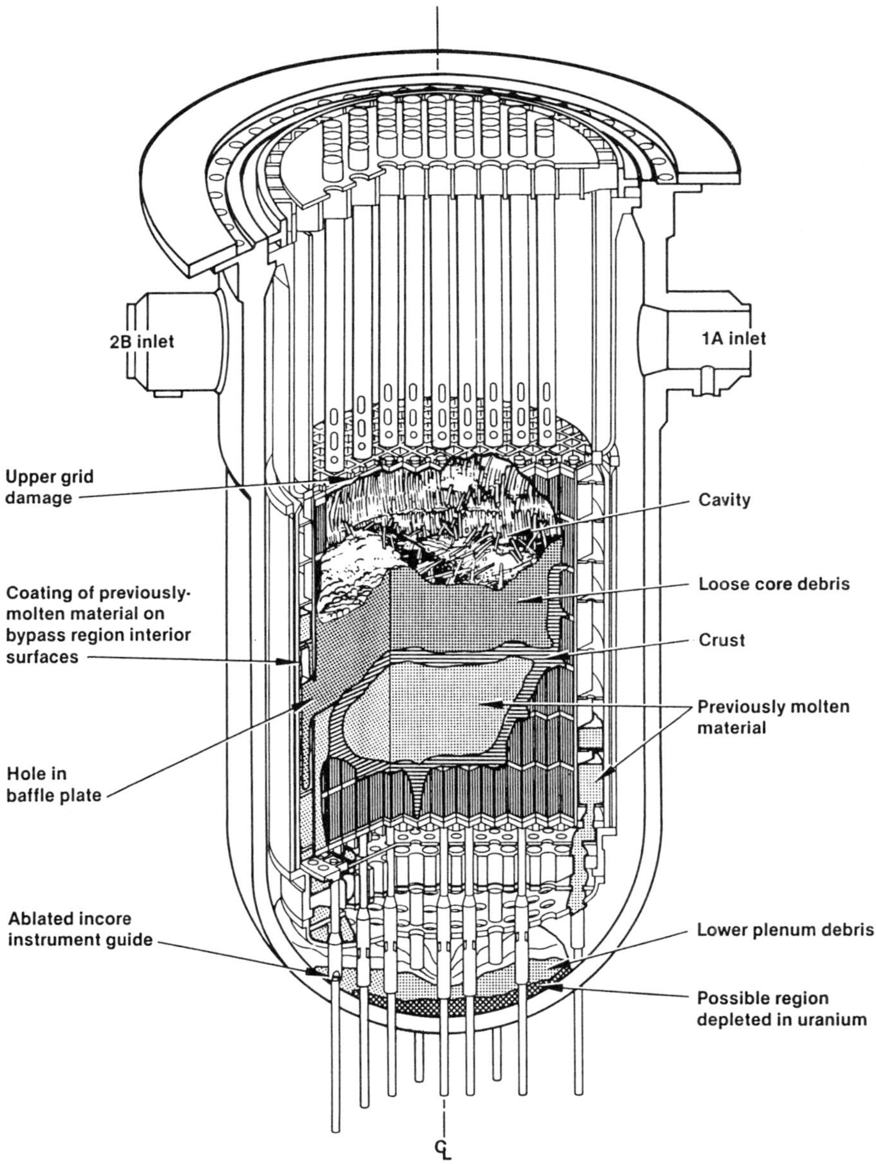

Figure 2. TMI-2 Core End State Configuration.

As shown by the TMI-2 core examination, the early phase of melt progression was consistent with the behavior observed in the PBF tests and the other experiments previously discussed. A large metallic blockage or crust was formed across the fuel-rods in the lower portion of the core above the water surface. This crust consisted of control-rod materials, zircaloy with dissolved UO_2, and alloys of these materials. Metallographic examination of debris from the upper portions of the TMI-2 core and the PBF test fuel bundles showed that local temperatures during the rapid oxidation temperature transient reached the 3000K UO_2 melting point, but that large-scale ceramic melting did not occur (Refs. 2 and 15).

At 174 minutes into the accident there was a brief flow transient when one of the coolant pumps was started. Considering NSRR results (Ref. 8), it is thought that mechanical agitation from the pump water shattered the declad hot ceramic fuel pellets that were still largely in free-standing rod geometry at that time in a configuration similar to the post-test PBF fuel bundles (Ref. 15). The shattered pellets collapsed to form a largely ceramic particulate debris bed supported by the lower metallic crust across the lower fuel-rod stubs. This collapse formed a large central void in the upper region of the core, and constituted the beginning of late-phase core-melt progression. Core heatup from fission-product decay and continued oxidation of relocated metallic zircaloy continued, and a mostly ceramic (UO_2, ZrO_2) melt pool grew within the debris bed. The ceramic pool temperature must have exceeded the 2800K UO_2-ZrO_2 eutectic temperature. A thin, mostly ceramic upper crust formed at the top of the ceramic pool, as shown in Figure 2.

At about 200 minutes the high pressure injection (HPI) pump was started to reflood the core. Reflooding did not quench the entire core, and, according to analysis, the molten ceramic pool in the debris bed continued to grow (Ref. 16). At 226 minutes the ceramic pool melted through the crust at one side, as shown in Figure 2. This side melt out was caused by natural circulation in the internally-heated molten ceramic pool. About 20% of the core mass drained from the side of the pool down through the edge of the core and into the lower plenum. At this time the water level was above the top of the pool, and analysis indicates that the pool contained over 50% of the core mass. Readings from in-core self-powered neutron detectors (SPNDs) indicate that melt drainage into the lower plenum took only about one minute. The thin upper ceramic crust and the debris bed above the crust collapsed down to the height of the melt-through opening as the ceramic melt drained from the pool. According to analysis, even though the partially drained pool was under water, it continued to grow for another 20 minutes after the melt relocation, when temperature stabilization finally occured (Ref. 16).

There was no large energetic thermal explosion (steam explosion) when the 20 metric tons of ceramic melt drained into the lower plenum water. This is thought to be because the vessel pressure at that time (12 MPa) was too high for steam explosion triggering. As the melt stream broke up and was cooled, however, there was non-explosive rapid steam generation (or a steam spike) of 2 MPa above the ambiant 12 MPa pressure, and this lasted for about 15 minutes. There is little direct information on the mean diameter (surface area) of the fragmented melt during this process because of later agglomeration of much of the debris in the lower plenum.

A most significant result of the TMI-2 accident is that the vessel lower head did not fail when the 20 metric tons of ceramic melt drained into the lower plenum water. The pressure load on the lower head was then 12 MPa (14 MPa during the 15-minute steam spike). Both penetration failure and gross creep-rupture failure modes have been examined, but the analysis is unclear as to which would have occured first. Examination of metallographic samples from inside the lower head should provide significant information on the margin to failure at TMI-2 and on failure mechanisms.

Fast-reactor safety experiments have shown that frozen ceramic surface crusts provide effective insulation for metal structures during transient ceramic pours, and models for this process have been developed (Ref. 17). The relative lack of thermal attack by the ceramic melt on the core support plate and other lower plenum structure during melt drainage is consistent with these results. Fast reactor safety experiments have also been performed on the melting behavior and thermal characteristics of dry beds of particulate UO_2 debris, both with and without particulate stainless steel in the bed (Ref. 18). The results are applicable to LWR late-phase melt progression modeling.

Summary Conclusions

The experiments that have been performed, the TMI-2 accident, and corollary analysis have provided considerable information on the governing phenomena involved in in-vessel core-melt progression, particularly in the early phase. Three separate and distinct major material relocation processes occur during melt progression. These involve metallic melts, ceramic melts, and solid ceramic debris, and they occur at different local temperatures and times during accident sequences. For a mechanistic treatment of melt progression, it is necessary to treat these separate processes separately, as is done in the mechanistic MELPROG, SCDAP, and CORMLT codes (Refs. 3, 13, and 18), and not to lump them into a fictitious "core slump" of a hypothetical "corium" mixture. A major current question, particularly for BWRs, is whether core blockages are formed by the freezing of relocated metallic melt (as happened in the TMI-2 accident), or whether metallic melts (and later ceramic melts) drain out of the core when formed, as currently modeled in the BWRSAR code (Ref. 13). This question has a major effect upon the progression of the accident and upon its consequences.

References:

1. Lipinski, R.J., Mast, P.K., Powers, D.A., and Walker, J.V.,"Uncertainty in Radionuclide Release Under Specific LWR Accident Conditions (QUEST Study)," Sandia National Laboratories, SAND 84-0410, Vols. 1-4, 1985.

2. Osetek, D.J., "Results of the Four PBF Severe Fuel Damage Tests," Proc. Fifteenth Water Reactor Safety Information Meeting, Gaithersburg, Maryland, NUREG/CP- 0090, October 1987.

3. Allison, C.M., Berna, G.A., Wagner, R.J., Chang, J.C., Siefken, L.J., and Hagerman, D.L., "Capabilities of the Integrated SCDAP/RELAP5/ TRAP-MELT Severe Accident Analysis Computer Code," Proc. IAEA International Symposium on Source Term Evaluation for Accident Conditions, Columbus, Ohio, October 1985.

4. Hofmann, P., Hagen, S., Schanz, G. and Skokan, A., "Reactor Core Materials Interactions at Very High Temperatures," Nuclear Technology, forthcoming.

5. Reil, K.O., Marshall, A.C., Gauntt, R.O., Ostensen, R.W., Pickard, P.S., Fryer, C.P., and Stalker, K.T., "Results of the ACRR DFR Experiments," Proc. ANS/ENS Topical Meeting on Thermal Reactor Safety, San Diego, California, February 1986.

6. Lombardo, N.J. and Lanning, D.D., "Data Report FLHT Experiment 2," Pacific Northwest Laboratories, PNL-6551, April 1988.

7. Gronnier, M.C., Fabrega, M.S., Scott de Martinville, E., and Geoffroy, R., "In-Pile Investigations on the Behavior of PWR-type Fuel Bundles in Severe Accident Conditions Beyond the Design Criteria," Proc. International ENS/ANS Conference on Thermal Reactor Safety, 'NUCSAFE 88', Avignon, France, Vol. 5, October 1988.

8. Saito, S., and Shiozawa, S., "Severe Fuel Damage in Steam and Helium Environments Observed in In-Reactor Experiments," Proc. Fifth International Conference on Thermal Nuclear Reactor Safety, Karlsruhe, FRG, KfK-3380, Vol. 2, p. 938, December 1984.

9. Allen, M.D., Stockman, H.W., Reil, K.O., Grimley, A.J., and Camp, W.J., "ACRR Fission Product Release Tests ST-1 and ST-2," Proc. International ANS/ENS Conference on Thermal Reactor Safety, 'NUCSAFE 88', Avignon, France, Vol. 5, October 1988.

10. Feherenbach, P., unpublished.

11. Hagen, S., Sepold, L., Hofmann, P., and Schanz, G., "Out-of-Pile Experiments on Severe Fuel Damage Behavior of LWR Fuel Elements," Proc. IAEA/NEA Symposium on Severe Accidents in Nuclear Power Plants, Sorrento, Italy, Vol. 2, p. 275, March 1988.

12. Gauntt, R.O., Gasser, R.D., Fryer, C.P., and Walker, J.V., "Results and Phenomena Observed from the DF-4 BWR Control Blade/Channel Box Damage Test," Proc. International ANS/ENS Conference on Thermal Reactor Safety, 'NUCSAFE 88', Avignon, France, Vol. 5, October 1988.

13. Hyman, C.R., and Ott, L.J., "Effects of Improved Modeling of Best-Estimate BWR Severe-Accident Analysis," Proc. Twelfth Water Reactor Safety Research Information Meeting, Gaithersburg, Maryland, NUREG/CP-0058, October 1984.

14. Dosanj, S.S., Tomkins, J.L., Young, M.F., Dearing, J.F., Smith, R.C., Kelly, J.F., Grimley, A.T., Wheatley, C.J., Maudlin, P.J., Heames, T.J., and Camp, W.J., "MELPROG-PWR/MOD1: A Two- Dimensional, Mechanistic Code for Analysis of Reactor Core Melt Progression and Vessel Attack under Severe Accident Conditions," NUREG/CR-5193, March 1989.

15. Broughton, J.M., Kuan, P., Petti, D.A., and Tolman, E.L., "A Scenario of the TMI-2 Accident," Nuclear Technology, forthcoming.

16. Kuan, P., "Damaged TMI-2 Core Consolidated Region Crust Behavior," Nuclear Technology, forthcoming.

17. Pilch, M. and Mast, P.K., "PLUGM: A Coupled Thermal-Hydraulic Computer Model for Freezing Melt Flow in a Channel," NUREG/CR-3190, April 1984.

18. Hitchcock, J.T., and Fryer, C.P., "The Postirradiation Examination of the DC Melt Dynamics Experiments," NUREG/CR-4625, March 1988.

19. Denny, V.E., Mertol, A., Sehgal, B.R., and Merillo, M., "PWR Core Degradation Modeling in the CORMLT Code," Proc. International ENS/ANS Conference on Thermal Reactor Safety;, 'NUCSAFE 88,' Avignon, France, Vol. 3, p. 917, October 1988.

Fission Product Release as a Function of Chemistry and Fuel Morphology

R. R. HOBBINS, D. J. OSETEK, D. A. PETTI, and D. L. HAGRMAN
Idaho National Engineering Laboratory
EG&G Idaho, Inc.
P.O. Box 1625
Idaho Falls, Idaho 83415, USA

INTRODUCTION

Analysis of the consequences of severe reactor accidents requires knowledge of the location and chemical form of fission products throughout the accident sequence. Two factors that strongly influence the location and chemical form of fission products are the chemistry within the core and the morphology of the fuel or fuel-bearing debris. This paper reviews the current understanding of the these factors garnered from integral and separate effect experiments and the TMI-2 accident, and provides perspective on the significance of contributing phenomena for the analysis of severe accidents, particularly during the in-vessel phase.

Information has been obtained recently on phenomena affecting the release of fission products from fuel and the reactor vessel during the in-vessel melt progression phase of a severe accident. The influence of a number of these phenomena will be reviewed, including fuel chemistry, H_2/H_2O ratio, fuel liquefaction, molten pools, and debris beds.

FUEL CHEMISTRY

The chemistry of uranium dioxide fuel changes as a function of burnup due to the production of fission products and the concomitant increase in the oxygen-to-metal (O/M) ratio. However, the O/M ratio is only a weak function of burnup indicating that burnup has a relatively small influence on the oxidation state of fission products under accident conditions. The main effect of burnup is on the concentration of fission products, the distribution of fission products within the fuel, and the fuel structure itself. The uranium dioxide undergoes restructuring as a function of burnup resulting in the formation of fission gas bubbles on grain boundaries that tend to interlink to form tunnels along grain edges at burnups above $\sim$ 5,000 MWd/MtU. This restructuring is enhanced by irradiation at higher powers and elevated fuel temperatures. During irradiation, fission gases and vapors migrate to the grain boundaries and, if tunnels have developed, the gas bubbles can be released fairly easily to the gap outside the fuel pellets. Grain boundary tunnels also provide a pathway for volatile fission product release during fuel heatup under

---

This work was sponsored by the U.S. Nuclear Regulatory Commission under DOE Contract No. DE-AC07-76ID01570.

accident conditions. The GRASS family of computer codes[1] explicitly models these changes in fuel morphology as a function of burnup. However, it has been found that simple Booth diffusion modeling gives results that are in good agreement with in- and out-of-pile experimental data[2,3] by varying the effective diffusion distance. Burnup can enhance release by a factor of as much as thirty at a burnup of 47,000 MWd/MtU[4].

The increase in fission product inventory with higher burnup affects the chemical form of fission products in the gas phase in the reactor vessel following release from the fuel rod. Calculations have shown a strong dependence of the chemical form of iodine on the concentrations of iodine and cesium in the gas phase relative to hydrogen and steam[5]. With trace-irradiated fuel, the H_2/I ratio is large and hydrogen iodide tends to be favored as the preferred iodine chemical form for gas phase transport at high temperatures, whereas with highly irradiated fuel, the H_2/I ratio is reduced and cesium iodide is favored. Such differences in chemical form can lead to differences in the transport of fission products within the reactor coolant system after release from the fuel.

HYDROGEN TO STEAM RATIC

After fuel rod failure, the oxygen potential of the fuel is determined by the H_2/H_2O ratio in the gas space of the damaged reactor core. Chemical equilibrium is a reasonable assumption at the high temperatures experienced by core materials during severe accidents. Free energies of formation of oxides of various core materials are displayed in Figure 1. Also shown in the figure is the oxygen potential corresponding to H_2/H_2O ratios from ten to one-tenth. This range brackets many of the conditions anticipated in the reactor vessel during core melt progression in a severe accident. Toward the high oxygen potential end of this range, oxides can be formed by the fission products molybdenum and antimony. Tin from the zircaloy cladding, indium from silver-indium-cadmium control rods, and iron from structural stainless steel can also be oxidized. At oxygen potentials lower in the range, representative of steam-starved conditions, these materials would be expected to exist as metals. The volatility of these oxides is very different than the volatility of the corresponding metals. Fission product cesium is stable as an oxide at low temperatures (<800 K) but should exist as elemental vapor at the fuel temperatures expected under accident conditions (>2000 K).

In the example of the TMI-2 accident, there is evidence from the temperature of control rod drive lead screw material in the upper plenum and the nature and thickness of the oxide on the surface of this material that the H_2/H_2O ratio was unity or less over most of the time that high temperature effluent was exiting the core[6]. Reference to Figure 1 indicates that gas with H_2/H_2O ratio in the range 1.0 to 0.1 is capable of oxidizing UO_2 to $UO_{2.05}$. As indicated in a previous review[4], such oxidation is capable of increasing volatile fission product release rates by factors of four or more and is the basis for models of enhanced volatile fission product release in solid UO_2. Some evidence of localized fuel oxidation has been observed in the examination of materials from the TMI-2 core[6] and from the steam-rich test Severe Fuel Damage-Scoping Test (SFD-ST)[7] in the Power Burst Facility (PBF), but evidence of widespread fuel oxidation in TMI-2 or in the integral in-pile tests conducted to date (most of which have been steam limited) has not been found.

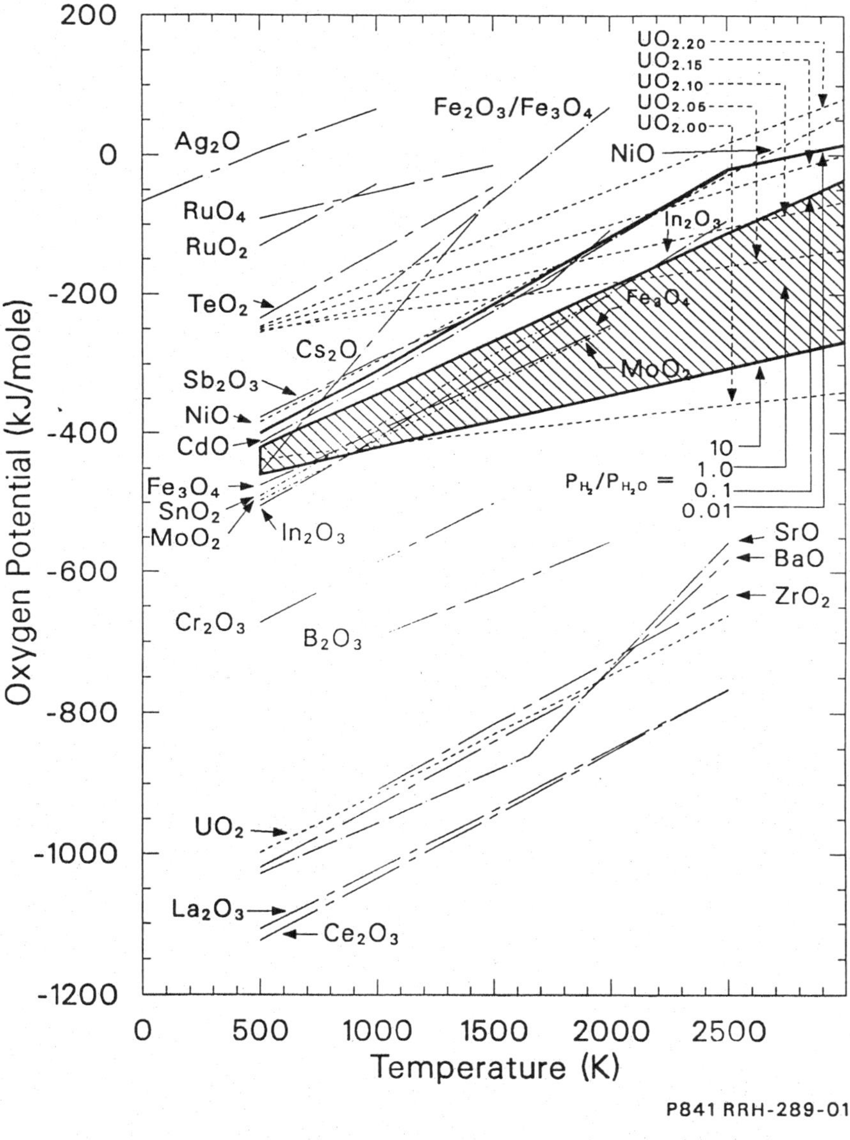

FIGURE 1. Free energies of formation of core material oxides.

A strong effect of the oxidation of zircaloy cladding on the release of tellurium has been observed in small scale laboratory tests at Oak Ridge National Laboratory[8] and confirmed in larger scale in-pile SFD tests in the PBF[9]. The tellurium released from the fuel apparently reacts with zirconium and tin in the fuel rod cladding and is released when the cladding is oxidized enough to concentrate the tin in a thin unoxidized layer. The release data from TMI-2 are consistent with this theory[6]. In both SFD-ST and in the TMI-2 accident, the tellurium that exited the core transported far downstream suggesting an unreactive chemical form such as SnTe or CsTe. SnTe has been found to be evolved from zircaloy containing small amounts of tellurium which was then oxidized [10].

Very small releases of barium and strontium are expected when steam is present because these materials are expected to be in the form of low volatility oxides in the fuel. Indeed, the releases of barium measured in the SFD tests in PBF are one percent or less[9] and the release of strontium from the fuel in the TMI-2 accident is also small[6]. Recent in-pile measurements of fission product release from highly irradiated fuel in a reducing atmosphere (hydrogen/inert gas mixture) in the ST-1 and ST-2 experiments in the Annular Core Research Reactor (ACRR) at Sandia National Laboratories indicate barium and strontium releases of several percent[11]. These results are consistent with those of the out-of-pile SASCHA experiments[12] and can be explained as the reduction of the oxides to more volatile elemental barium and strontium.

According to Figure 1, ruthenium should exist in its elemental form under severe accident conditions and very small releases are expected based on the low volatility of the metal. It is interesting to note that large releases from the fuel, but negligible releases from the core, have been reported for ruthenium in the TMI-2 accident[6]. Examination of debris removed from the TMI-2 core indicates that ruthenium is retained in molten metallic materials in agreement with earlier observations from out-of-pile meltdown studies[6].

The foregoing discussion of the effects of steam oxidation demonstrates that the H_2/H_2O ratio in the gas surrounding the degrading fuel and the extent of zircaloy and UO_2 oxidation are important factors in determining the release of fission products from fuel, their volatility and their chemical forms in the gas and condensed phases. Results of equilibrium thermodynamic calculations of chemical species for vapor transport of fission products under postulated severe accident conditions are found to be dependent on fission product concentrations, system temperature and pressure, H_2/H_2O ratio, and the variety of species considered in the calculations[13]. This realization leads to the consideration of the chemical effects of a variety of materials in LWR cores on the distribution of fission products within a reactor vessel undergoing a severe accident.

CONTROL MATERIAL CHEMISTRY

Boric acid in the reactor coolant and in the emergency coolant is used as a reactivity control material in PWRs, and boron carbide encased in stainless steel control blades is used in BWRs. Boric acid vapor reacts with cesium iodide in either the vapor or condensed states to form cesium borate and hydrogen iodide [14]. Boric acid vapor also reacts with cesium hydroxide in either the vapor or condensed phases to form cesium borate [14]. Separate effects tests have demonstrated that hydrogen iodide can be formed by reaction of boron carbide in steam with cesium iodide [15].

The ability of boron compounds to interact with fission product cesium and iodine depends on the reactor type and the accident sequence. The in-pile integral-effects Test DF-4 has shown that boron carbide in a BWR control blade may be prevented from interaction with steam and cesium iodide by preferential interactions with the stainless steel cladding of the blade [16]. This in-pile result confirms earlier results from small-scale out-of-pile laboratory experiments that showed boron carbide undergoes interaction with steel to form boron and carbon eutectics with iron and nickel [17].

In a PWR, the availability of boric acid vapor to interact with cesium iodide and cesium hydroxide depends on the accident scenario. Large break loss-of-coolant accidents without emergency core cooling tend to result in nearly complete loss of water and thus boric acid before core heatup and fission product release, whereas small break loss-of-coolant accidents and station blackout entail slow boiloff of the coolant such that considerable steaming and boric acid vaporization could be present during fission product release from the fuel. Accumulator injection in a small-break loss-of-coolant accident would provide an additional source of boric acid.

Thermodynamic studies indicate that silver vapor from silver-indium-cadmium control rods can form silver iodide under severe accident conditions in PWRs. The formation of silver iodide is favored by high silver-to-iodine ratios in the vapor[18]. High pressure accident scenarios result in control rod failure by melting of the stainless steel cladding and runoff of much of the control alloy with relatively little vaporization of silver[19]. In this case, the formation of silver iodide is not favored. In accident scenarios at low system pressure (such as a large break loss-of-coolant accident) the control rods fail by bursting and silver is sprayed throughout the core, tending to enhance vaporization[19]. These conditions tend to maximize the vaporization of silver and the formation of silver iodide.

FUEL LIQUEFACTION

Fuel liquefaction, the creation of liquids containing UO_2 below the melting point of UO_2 (3100 K), can occur at 2200 K by dissolution with molten zircaloy, or, perhaps, at even lower temperatures by dissolution in eutectic mixtures of zircaloy with transition metals and/or silver[20]. The dissolution rate of UO_2 in liquid zircaloy has been measured[21] but dissolution in eutectic mixtures at temperatures below 2200 K has not yet been studied. The dissolution process is complicated and involves the diffusion of oxygen out of the fuel and into the metallic melt. Oxygen diffusion takes place most rapidly along grain boundaries and the formation of molten uranium metal in the grain boundaries occurs just ahead of the broad moving boundary between the UO_2 and the molten metal. However, to model this process it is important to realize that the rate of dissolution is associated with the broad moving boundary and that the grain boundaries are not filled with liquid metal at great distances beyond the uniform moving boundary.

Fuel liquefaction destroys the crystal structure of the UO_2 so that the release of fission products is now governed by the migration of atoms and bubbles in a liquid. This is a much faster process than diffusion in a solid and fission product release rates should be enhanced. This effect has been measured in out-of-pile tests[22] using small chunks of irradiated fuel, but has not been measurable in in-pile integral effects tests such as the PBF-SFD test series, possibly because of the smearing in time and space of the liquefaction process inherent in an intrgral experiment.

MOLTEN POOLS

Molten pools of primarily $(U,Zr)O_2$ ceramic formed in the TMI-2 accident[23] and in severe fuel damage tests, both in-pile[3,7,24,25,26] and out-of-pile[20]. Analysis of fission product release from the molten ceramic pool in the TMI-2 accident indicates that fission product chemical form and concentration dominate the release of medium and low volatile fission products and that bubble dynamics dominates the release of fission gases and volatile fission products[2]. Fission product chemical forms are influenced by the oxygen potential of the large molten pool. The presence of iron oxides in the melt in TMI-2 establishes a lower limit on the oxygen potential of about -150 kJ/mole at 2800 K. At this oxygen potential, rare earth elements such as europium and cerium would exist as oxides (Eu_2O_3 and Ce_2O_3 or CeO_2), strontium would exist as SrO, and ruthenium and antimony would be present as metals immisible in the ceramic melt. Low releases are calculated for these materials primarily because of the low concentration and volatility of these species in the melt and the low surface-to-volume ratio of the molten region. Virtually all of the fission gases and volatile fission products should be released from the melt due to bubble coalesence and buoyancy. However, cesium has been found in the molten debris transported to the lower plenum of the TMI-2 reactor vessel[6]. The retention of cesium in a previously molten ceramic is unexpected and has implications for fission product chemical forms and for accident analyses that are discussed later. Evidence for cesium retention in previously molten ceramics has also been found in the SFD experiments, but the results are ambiguous[3,7,26]. The retention of the less volatile fission products in the melt that relocated to the lower plenum is discussed below.

DEBRIS BEDS

Two debris beds were formed in the TMI-2 accident, one on top of the molten pool, and one in the lower plenum. In the TMI-2 accident[23], the PBF SFD-ST[7], and the LOFT FP2 test[24], the upper debris bed is thought to have been produced when coolant was introduced into the hot core/bundle causing thermal shock and fragmentation of the oxidized fuel rods in these steam-rich transients. In less steam-rich transients, such as Test SFD 1-4[3], an upper debris bed was formed by the melting (probably aided somewhat by eutectic reactions with structural and control materials) and relocation of the zircaloy fuel rod cladding leaving the fuel pellets without mechanical restraint.

The upper debris bed is found to consist principally of fuel fragments with little zircaloy cladding or control materials. The mass mean diameter of the particles in the upper debris bed in the TMI-2 core is about 1 mm. This is two orders of magnitude larger than fuel grain size. As a consequence, fragmentation into a debris bed should have no significant affect on fission product release from solid fuel.

The molten debris in the pool in the TMI-2 accident failed the upper crust between the pool and the upper debris bed along one side near an upper corner by thermal ablation causing $\sim$20 metric tons of molten material to relocate into the lower plenum in a period of one to two minutes[23]. This liquid stream interacted with the coolant in the lower plenum to produce the debris in the lower plenum. This debris has been observed via video probe to vary from small particles about 1 mm in diameter to large chunks several centimeters in size to regions of consolidated debris forming walls in the debris bed.

Samples of large chunks of lower plenum debris contain primarily previously molten ceramic $(U,Zr)O_2$ with transition metal oxides in the grain boundaries. Fission product retention in the debris in the lower plenum and also in the upper debris bed is presented in Table 1 where it can be seen that the retentions of cesium and iodine in the upper debris bed are both about 20 percent, but the retentions in the lower head debris are quite different, about 3 percent for iodine and 13 percent for cesium.

The similar retentions of iodine and cesium in the upper debris are in agreement with the vast number of observations of similar release rates of these fission products from solid fuel. The high retention of cesium in the previously molten ceramic debris in the lower plenum is unexpected, based on analyses of the release of volatile fission products from melts discussed earlier in this paper, and suggests that some of the cesium is in an involatile chemical form stable at very high temperatures.

Some indication of possible chemical associations of cesium in the lower plenum debris has been obtained through elemental analysis of cores removed from debris particles and micro gamma scanning across surfaces sectioned from particles[27]. Results of this work suggest that cesium may be combined with transition metal oxides in the grain boundaries of the solidified ceramic in forms such as chromates.

Other fission product retention data of interest in Table 1 are the small retentions of antimony and ruthenium in both debris beds. Antimony is generally thought to be in the mid-volatility group and ruthenium to be of low volatility. Neither of these fission products were released from the reactor vessel in large amounts but they have been found to be associated with previously molten transition metals in the debris. These materials have migrated out of the ceramic phases to be sequestered in molten metallic phases where they would have low volatilities due to solution or alloying. The fission products strontium and cerium remain with the ceramic phase, cerium as an oxide (Ce_2O_3 or CeO_2) and strontium as an oxide (SrO) or uranate ($SrUO_4$). Only about 6 percent of the fission product tellurium was released from the core and much of the tellurium remaining in the core appears to be associated with metallic debris in the crusts surrounding the ceramic pool.

SUMMARY AND CONCLUSIONS

The locations of fission products and their chemical forms at the time containment integrity is lost are important in the analysis of the consequences of an unmitigated reactor accident. Fission products that are released from the core during the heatup and the in-vessel core melt progression phases of an accident likely will be deposited on upper plenum and reactor coolant system piping surfaces from which they may slowly revaporize depending on the accident scenario. Revaporization is likely to be an important contributor to the source term from containment only in the case of a relatively late containment failure (well after reactor pressure vessel failure). It is important to consider physical and chemical interactions between fission product vapors and aerosols when analyzing the behavior of fission products in the reactor coolant system.

TABLE 1. Fission Product Retention in TMI-2 Debris

	Percent of inventory retained			
	Lower Plenum		Upper Debris	
Radionuclide	Average	Range	Average	Range
I-129	3	0-25	23	10-29
Cs-137	13	0-34	19	5-39
Sb-125	6	0-35	47	37-78
Ru-106	6	0-17	49	31-77
Sr-90	107	68-178	87	74-96
Ce-144	96	78-142	99	78-113

If containment integrity is lost before reactor pressure vessel failure, deposition of fission products released from the core during heatup and core melt progression is an important retention mechanism. However, in this case as well as the case in which containment integrity is lost shortly after reactor pressure vessel failure, fission product releases due to ex-vessel processes (such as high pressure melt ejection and core-concrete interaction) will probably dominate the source term. For such cases, the fission product inventory and the chemical forms of the fission products in the molten debris that exits the reactor pressure vessel are controlling parameters for the source term. Analyses of ex-vessel fission product releases need to account for the small fraction of the core inventory of iodine and a larger fraction of the core inventory of cesium that may still be present in ceramic debris. It is also important to recognize that large fractions of the core inventory of tellurium, antimony, and ruthenium can be retained in metallic debris.

The results of the experimental work to date on core melt progression indicate that core melt progression phenomena influence the nature, timing, amounts, and chemical forms of fission product released during the in-vessel phase of severe accidents. This brief survey of results indicates the importance of careful modeling of core melt progression phenomena in severe accident source term analyses.

REFERENCES

1. Rest, J., An Improved Model for Fission Product Behavior in Nuclear Fuel Under Normal and Transient Accident Conditions, *J. Nuclear Materials*, vol. 120, pp. 195-212, 1984.

2. Petti, D. A., Adams, J. P., Anderson, J. L., and Hobbins, R. R., Analyses of Fission Product Release Behavior During the TMI-2 Accident, *Nuclear Technology*, forthcoming.

3. Petti, D. A., Martinson, Z. R., Hobbins, R. R., Allison, C. M., Carlson, E. R., Hagrman, D. L., Cheng, T. C., Hartwell, J. K., Vinjamuri, K., and Siefken, L. J., PBF Severe Fuel Damage Test 1-4 Test Results Report, EG&G Idaho Rept., NUREG/CR-5163, EGG-2542, February 1989.

4. Hobbins, R. R., Osetek, D. J., and Hagrman, D. L., In-Vessel Release of Radionuclides and Generation of Aerosols, *Proceedings IAEA Symposium on Source Term Evaluation for Accident Conditions, Columbus, Ohio*, pp. 45-63, 1985.

5. Cronenberg, A. W., and Osetek, D. J., Reaction Kinetics of Iodine and Cesium in Steam/Hydrogen Mixtures, *Nuclear Technology*, vol. 81, pp. 347-359, 1988.

6. Hobbins, R. R., Cronenberg, A. W., Langer, S., Owen, D. E., and Akers, D. W., Insights on Severe Accident Chemistry from TMI-2, *Proc.Am. Chemical Society Symposium on Chemical Phenomena Associated with Radioactivity Released During Severe Nuclear Plant Accidents, Anaheim, California*, pp. 4-1 to 4-20, 1986, Rept. NUREG/CP-0078, June 1987.

7. Knipe, A. D., Ploger, S. A., and Osetek, D. J., PBF Severe Fuel Damage Scoping Test - Test Results Report, EG&G Idaho Rept. NUREG/CR-4683, EG&G-2413, March 1986.

8. Collins, J. L., Osborne, M. F., and Lorenz, R. A., Fission Product Tellurium Release Behavior Under Severe Light Water Reactor Accident Conditions, *Nuclear Technology*, vol. 77, pp. 18-31, 1987.

9. Osetek, D. J., Results of the Four PBF Severe Fuel Damage Tests, *Trans. Fifteenth Water Reactor Safety Information Meeting, Gaithersburg, Maryland*, 1987, Rept. NUREG/CP-0090, pp. 20-15 to 20-16.

10. Alexander, C. A., and Ogden, J. S., Mass Spectrometric and Kinetic Evaluation of Zircaloy-Tellurium Interactions, *J. Nuclear Materials*, forthcoming.

11. Allen, M. D., Stockman, H. W., Reil, K. O., Grimley, A. J., and Camp, W. J., ACRR Fission Product Release Tests ST-1 and ST-2, Proc. International ENS/ANS Conference on Thermal Reactor Safety, Avignon, France, vol. 5, 1988.

12. Albrecht, H. and Wild, H., Review of the Main Results of the SASCHA Program on Fission Product Release Under Core Melting Conditions, *Proc. ANS Topical Meeting on Fission Product Behavior and Source Term Research, Snowbird, Utah*, pp. 3-1 to 3-14, 1984, EPRI Rept. NP-4113-SR, July 1985.

13. Cubicciotti, D., and Sehgal, B. R., Vapor Transport of Fission Products in Postulated Severe Light Water Reactor Accidents, *Nuclear Technology, 65*, pp. 266-291, 1984.

14. Bowsher, B. R., and Dickinson, S., The Interaction of Cesium Iodide with Boric Acid: Vapor Phase and Vapor-Condensed Phase Reactions, Rept. AEEW-R 2102, Winfrith, UKAEA, May 1986.

15. Elrick, R. M., and Sallach, R. A., Fission Product Chemistry in the Primary System, *Proc. ANS International Meeting on Light Water Reactor Severe Accident Evaluation, Cambridge, Massachusetts*, vol.1, pp. 4.6-1 to 4.6-5, 1983.

16. Gauntt, R. O., Gasser, R. D., and Ott, L. J., The DF-4 BWR Blade/Channel Box Fuel Damage Experiment, SNL Rept. NUREG/CR-4671, forthcoming.

17. Hagen, S., and Hofmann, P., Physical and Chemical Behavior of LWR Fuel Elements up to Very High Temperatures, Karlsruhe, FRG Rept. KfK-4104, 1987.

18. Hahn, R., and Ache, H. J., Thermodynamic Calculations on the Behavior of Gaseous Iodine Species Following a Hypothetical Severe Light Water Reactor Accident, *Nuclear Technology*, vol. 67, pp. 407-410, 1984.

19. Petti, D. A., Silver-Indium-Cadmium Control Rod Behavior in Severe Reactor Accidents, *Nuclear Technology*, vol. 84, pp. 128-151, 1989.

20. Hagen, S., Hofmann, P., Schanz, G., and Sepold, L., Results of CORA Experiments on Severe Fuel Damage with and without Absorber Materials, *Proc. 26th ASME/AIChE/ANS National Heat Transfer Conference, Philadelphia*, 1989, forthcoming.

21. Hofmann, P., Uetsuka, H., Wilhelm, A. N., and Garcia, E. A., Dissolution of Solid UO_2 by Molten Zircaloy and its Modelling, IAEA International Symposium on Severe Accidents in Nuclear Power Plants, Sorrento, Italy, 1988.

22. Kaspar, G., Haas, W., and Bleier, A., Untersuchung der Spaltproduktexhalation aus abgebranntem Brennstoff unter Anwesenheit von Zircaloy und Sauerstoff, KWU Report R 917/86/021, November 1986.

23. Broughton, J. M., Kuan, P., Petti, D. A., and Tolman, E. L., A Scenario of the TMI-2 Accident, *Nuclear Technology*, forthcoming.

24. McPherson, G. D., and Hicks, D., LOFT Goes out with a Severe Accident, *Nuclear Engineering International*, p. 24, April 1986.

25. Reil, K. O., Marshall, A. C., Gauntt, R. O., Ostensen, R. W., Picard, P. S., Fryer, C., and Stalker, K. T., Results of the ACRR-DF Experiments, *Proc. of the International ANS/ENS Topical Meeting on Thermal Reactor Safety, San Diego, California*, vol. 4, pp. XIX.8-1 to XIX.8-8, 1986.

26. Martinson, Z. R., Petti, D. A., and Cook, B. A., Volume 1: PBF Severe Fuel Damage Test 1-1 Test Results Report, EG&G Idaho Rept. NUREG/CR-4684, EGG-2463, Volume 1, October 1986.

27. Akers, D. W., Hartwell, J. K., and Hobbins, R. R., Gamma-Ray Emission Tomographic Examination of TMI-2 Core Samples, *Nuclear Technology*, forthcoming.

Fission Product Release from Fuel Elements under Stationary and Accident Conditions

ANDRZEJ L. STRUPCZEWSKI
Institute of Atomic Energy
PL-05-400 Otwock-Swierk
Poland

INTRODUCTION

Fuel pellet material is the first and very important barrier in the system of barriers (fuel material, cladding, pressure boundary of the reactor cooling system, containment) preventing fission product release outside the nuclear power plant. Under stationary conditions UO_2 pellets hold more than 99 % of the activity of fission products, and under accident conditions, even after cladding burst and significant fuel heating, most of solid fission products are kept in the fuel until very high temperatures are reached.
In the Reactor Safety Study |1| the release of fission products under accident conditions was evaluated in two stages: first, the activity of fission products accumulated in the pellet-to-cladding gap was assumed to be released, partially or wholly, depending on the nuclide, then - if the accident proceeded to fuel melting - the release of fission products from molten fuel was defined by certain escape coefficients generally reflecting a large conservatism necessary in view of the limited experimental data base available at the time.
A similar method was approved for use in CMEA countries for VVER safety analysis |2|. However, with the development of codes for time related analysis of fuel heating during severe accidents the need arose for a realistic approach to the process of fission product release in function of fuel temperature and accident time. Much progress in the mechanistic description of gaseous and volatile fission product migration in fuel has been achieved, as shown e.g. in the work of J.Rest |3|. On the other hand, the methods used in complex reactor accident analyses have been mostly based on macroscopic approach proposed by NRC |4|, in which the fuel rod is treated as a whole and the average rod cross-section temperature is used to determine release rate coefficients for various fission products.
This approach was used in CORSOR code applied e.g. by Battelle Memorial Institute |5| and validated recently by Lorenz |6|. The release rate coefficients for volatile and solid fission products were studied by Andriesse |7| who provided an excellent theoretical basis for the observed characteristics of solid fission product migration in high temperature regions of UO_2 fuel.
In VVER fuel analyses performed in Poland a different approach has been used |8|, based upon studies of steady state fission product releases made by Luzanova and her co-workers from the Kurchatov's Institute of Atomic Energy in Moscow |9|. This has the advantage of considering the influence of structural changes in UO_2 pellets in various temperature zones on

fission product release and of relating the released fractions of various nuclides to their decay constants. Since the theory of Luzanova was limited to short and medium-lived gaseous and volatile fission products, the theory developed by Andriesse |7| was used to evaluate solid fission product releases under steady state and accident conditions. Moreover, the contribution of fission product release from intergranular bubbles and pores under accident conditions has been introduced into the calculation code independently from the volume diffusion processes. This seems to give better agreement with experimentally measured data and to provide sufficient conservatism in safety analyses made for short-time transients in which the fuel does not melt or is subject to limited melting only.

FISSION PRODUCT RELEASE FROM FUEL UNDER STATIONARY CONDITIONS

Fuel Structure Transformations in High Temperatures

During UO_2 irradiation the structure of fuel matrix changes depending on fuel burnup and its operating temperature. Postirradiation studies of VVER fuel elements irradiated to high burnups at power rating up to 52 kW|m have shown |9| that the volume of irradiated fuel consisted of three zones:

- outer zone of unchanged primary structure, in which the temperature did not exceed 1300^oC,
- zone of equiaxed grain growth situated in the fuel region of temperatures ranging from 1300^oC up to 1700^oC,
- zone of columnar grains situated in the central part of the pellet, where the temperature exceeded 1700^oC.

Other studies |4| have shown that starting at 1300^oC the formation, swelling and coalescence of bubbles of fission products occurs. Ronchi et al. observe, that in-pile bubble mobility of fission products is well described by theoretical models assuming that the bubbles migrate by a surface diffusion mechanism |12|. Since the surface diffusion enthalpy in UO_2 is high (approximately 120 kcal|mol) the onset of bubble migration and coalescence in a fuel rod corresponds to a threshold temperature which for slow operational transients is of the order of 1400^oC |12|.
The expanded bubbles work to mechanically separate the grains allowing escape of gas and vapour. In still higher temperatures the bubbles link together forming high porosity central zone of fuel in which many tunnels reach the voids, opening regularly under the influence of excessive pressure of gases and letting fission products flow out into the central hole in the fuel rod. Thus the increase of fuel temperature brings about not only the acceleration of fission product diffusion in the UO_2 grains, but also provides new release routes, thus significantly increasing fission product release ratios.

Fission Product Release from the Low Temperature Zone

In the low temperature zone the process of fission product release occurs mostly due to kinetic energy of fission fragments. A fraction of fission product nuclei born in the fuel layer neighbouring to the surface leaks out. It is proportional to the mean free path of fission product nuclei i in fuel ϱ_i and is a function of fuel pellet geometry as shown by the formula

$$\frac{\text{direct release rate}}{\text{birth rate}} = \left(\frac{R}{B}\right)_o = \frac{\varrho_i S_f k_f}{4 V_f} \tag{1}$$

where S_f - fuel geometrical surface, m^2

V_f - fuel volume, m^3

k_f - the ratio of the overall fuel surface to S_f.

For a VVER fuel element with cylindrical pellets of outer diametr d and inner diameter d_o the formula (1) will be reduced to

$$\left(\frac{R}{B}\right)_o = \frac{\rho_i k_f}{d - d_o} \qquad (2)$$

In the effect of local heating of fuel material due to the dissipation of kinetic energy of fission fragments travelling across fuel grains there occurs temporary evaporation of fuel particles and fission products accumulated in the fuel matrix, followed by their solidification in slightly different locations. This so called "knock-out effect" adds to the direct release of fission products described above. Luzanova and Slaviagin |9| have proposed the following formula for knock-out effect release of fission products of decay constants $10^{-6}\ s^{-1} < \lambda < 10^{-3}\ s^{-1}$

$$\left(\frac{R}{B}\right)_o + \left(\frac{R}{B}\right)_1 = \frac{k_f \rho_i S_f}{4 V_f} \left(1 + 2\frac{\bar{\rho}\, \Delta V_1\, F\, \overline{K_z^2}}{\lambda \rho_i}\right) \qquad (3)$$

where: $\bar{\rho}$ - average value of mean free paths of fission product nuclei in fuel, m

ΔV_1 - fuel volume evaporating due to the transfer of kinetic energy of the fission fragment, m^3

F - average fission density in fuel element, fissions/(m^3s),

$\overline{K_z^2}$ - average of square values of axial heat flux distribution coefficients for the fuel element.

For UO_2 density equal to about 95% of the theoretically possible maximum density the value of $k_f(\Delta V_1) = 10^{-25} m^3$ |9|. Then for VVER fuel pellets of average fission density $F = 8.45\ 10^{10}$ fissions /(m^3s), taking $\overline{K_z^2} = 1.3$ one can obtain for fission products of atomic mass number around 130 ($\rho_i = 5\ \mu m$)

$$\left(\frac{R}{B}\right)_o + \left(\frac{R}{B}\right)_1 = 7\ 10^{-4} \left(1 + \frac{2.6\ 10^{-6}}{\lambda} \right) \qquad (4)$$

which shows that the knock-out effect provides a significant contribution (more than 25%) only for radionuclides of $\lambda < 10^{-5} s^{-1}$ |11|. The release from fuel of long lived nuclides like ^{131}I equals to about 0.25% of their birth rate, which is in good agreement with the results of experimental investigations of both VVER and PWR fuel elements which were operated in temperatures below 1300°C |9, 13|.

Gaseous and Volatile Fission Product Release from Zones of Intermediate Temperatures, 1300-1700°C and High Temperatures, Above 1700°C.

Under irradiation, inside fuel grains a restructuring takes place characterized by the presence of a large number of gas microbubbles |average diameter 2 nm, density up to $10^{24}\ m^{-3}$ |11|. At such high concentrations of microbubbles the probability of fission product nuclei reaching the grain boundary purely through diffusion without being captured by microbubbles is very small, as the lifetime of a FP nucleus in the UO_2 matrix before capture by a microbubble is about 1500 s at 1300°C and decreases with increasing temperature |11|, while the halflives of all important fission products are much longer. The release of FP may be therefore assumed to occur in two steps, namely:

- diffusion of fission products via atomic and gas bubble mobility from the grains to the grain faces,
- release of noble gases and volatile fission products (VFP) from grain faces through a network of interconnected tunnels created due to fission product gas swelling and fuel structure changes described above.

The leakage of FP due to the migration of gas microbubbles may be evaluated assuming that all generated nuclides are contained in the microbubbles. As the velocity of microbubble displacement in the grains measured directly by Backer |11| is $V_m = 3 \; 10^{-13}$ m/s, the leakage of nuclides due to it may be described by the formula

$$\left(\frac{R}{B}\right)_2 = \frac{V_m \, S_g}{V_g} \tag{5}$$

where: S_g, V_g - geometrical surface and volume of UO_2 grains.

An analysis provided by Luzanova et al |11| proved that this mechanism of FP transport by microbubbles is not sufficient to describe FP release, since the maximum relative release rate from grains provided solely by microbubble movements would not exceed 4%, while measured values for ^{131}I may come close to 100%. Therefore a more complicated mechanism must be assumed, including multiple captures and dissolutions of FP nuclei, going on in a cyclic way till the nuclei leave the UO_2 grains or decay. Then the leakage of FP due to atomic diffusion may be evaluated assuming that microbubbles inside grains constitute immobile traps for migrating radionuclides of constant volumetric concentration m(at|m^3) and equal diameters d_p. Then the general equation of continuity of nuclides migrating in UO_2 lattice may be written as

$$\frac{dn}{dt} + \mathrm{div}\,(\bar{v}\, n) = -\lambda n - k \pi d_p^2 \, m |\bar{v}| n + b\, n_1 + \alpha_k F \tag{6}$$

where: n - volumetric concentration of FP nuclei in fuel, atoms/m^3

$\bar{v}$ - velocity vector of FP displacement

k - proportionality constant

b - probability of dissolving of the microbubble, proportional to the flux of fission fragments in fuel and to the geometrical cross-section of the microbubble,

α_k - cumulative generation of the nuclide per fission of ^{235}U

Denoting the reciprocal of mean free path of the fission product nucleus during its lifetime from the point of birth to the point of decay as ω, the value of which is defined in |9|, Lazunova and Slaviagin have shown that the leakage from the fuel material to intergranular pores can be described by the formula

$$\left(\frac{R}{B}\right)_2 = \frac{1 - \exp(-\omega H_o)}{\omega H_o} \tag{7}$$

where H_o is the radial size of the grain.

The build-up of gaseous fission products pressure in the pores leads to periodical opening of immobile pores remaining in the vicinity of fuel pellet surface |9| and expulsion of accumulated gases. For pellet surface temperature around 1700°C the average pore collapse time is $\tau = 3.5\ 10^{5}$ s |11|. The relative release rate of fission products from the pore may be described |9| as equal to $[1 - \exp(-\lambda\tau)]/\lambda\tau$. Assuming that the fraction of fuel neighbouring to the periodically opening pores is δ_2 and the release from the fuel grains to the pores is $(R/B)_2$, the fraction of FP released outside fuel pellets will be

$$\left(\frac{R}{B}\right)_3 = \left(\frac{R}{B}\right)_2 \delta_2 \frac{1 - \exp(-\lambda\tau)}{\lambda\tau} \tag{8}$$

However, the analysis of experimental data performed in |9| has shown that for nuclides of decay constants $3\ 10^{-5}\,s^{-1} < \lambda < 10^{-4}\,s^{-1}$ the leakage rate is a weak function of λ , while according to the above given formula it should be approximately proportional to $\frac{1}{\lambda^2}$. Luzanova and Slaviagin postulated therefore existence of a fuel volume which releases fission products of $\lambda < 10^{-3}\ s^{-1}$ practically without delay |9|. Such a leakage may occur in the fuel layer neighbouring to the central hole and characterized by high temperature in this volume enhancing fast release of fission products both in the way of diffusion and of microbubble movement. Finally, the formula proposed in |9| and used in REFRAC calculation code developed in Poland |10| is as follows:

$$\left(\frac{R}{B}\right)_3 = \delta_2 \left(\frac{R}{B}\right)_2 \frac{1 - \exp(-\lambda\tau)}{\lambda\tau} + \delta_3\, \eta\,(10^{-3} - \lambda) + \left(\frac{R}{B}\right)_0 + \left(\frac{R}{B}\right)_1 \tag{9}$$

where $\eta(10^{-3} - \lambda)$ - Heavyside function, equal to 1 for $\lambda \leq 10^{-3}\,s^{-1}$ and equal to 0 for $\lambda > 10^{-3}\,s^{-1}$.

The coefficients δ_2 and δ_3 depend on the fuel structure in the zone around the surface of fuel surrounding the central hole in fuel pellets. For the zone of equiaxed grain growth $\delta_2 = 0.02$ and $\delta_3 = 7\ 10^{-4}$ while for the zone of columnar grains $\delta_2 = 0.06$ and $\delta_3 = 1.5\ 10^{-3}$.

The formula |9| is valid for noble gases and VFPs having no long-lived precursors in the schemes of radioactive decay. In a general case the precursors should be considered. Similarly, the parameters of fuel operation such as temperature fields and power distribution should be taken into account by integration of the corresponding expressions.

Migration of Solid Fission Products

In low temperature area the release of solid fission products occurs due to the same mechanisms of direct release and knock-out effect as in the case of noble gases and VFP. In high temperature zone, however, the situation is more complicated and the formula |9| is insufficient to describe the release of solid fission products. It has been therefore conservatively assumed |8| that once a nucleus of a solid fission product i is released from a UO_2 grain to the interconnected intergranular tunnels it is ready to be expelled from the gas bubble together with noble gases and halogens, so that its release characteristics in this second stage are the same as those of iodine. The difference lies in the first stage: the diffusion of various nuclides through UO_2 lattice. Taking the release of iodine ^{131}I as the reference point the following simplified formula has been used in REFRAC code to describe the escape of solid fission products from UO_2 grains:

$$\left(\frac{R}{B}\right)_{2,i} = \left(\frac{R}{B}\right)_{2,{}^{131}I} \frac{\int_{T1}^{T2} D(T)_i \, dT}{\int_{T1}^{T2} D(T)_{{}^{131}I} \, dT} \qquad (10)$$

where $D(T)_i$ - diffusion coefficient for species i at temperature T in UO_2 grains, taken from the work of Andriesse |7|, T1 and T2 - lower and upper temperature limit in each zone of fuel pellet structure considered.

FISSION PRODUCT RELEASE FROM FUEL PELLETS UNDER ACCIDENT CONDITIONS

Principal Mechanisms Controlling the Rate of Fission Product Release from Fuel Under Accident Conditions.

There are five principal mechanisms which control the rate of release of fission products from LWR fuel under accident conditions: 1) Burst release, 2) Diffusional release of the pellet-to-cladding gap inventory, 3) Grain boundary release, 4) Diffusion from UO_2 grains and 5) Release from molten material. Each mechanism becomes dominant at a succedingly higher temperature.

Burst release. It is the release of noble gases accumulated in the pellet-to-cladding gap, in the fuel plenum and in the open voids in the fuel pellets, occuring at the moment of cladding rupture. This rupture takes place when the cladding temperature reaches the range of 750-1100°C, depending on the pressure inside the cladding. The noble gases released from the gap carry with them a small fraction of cesium and iodine being initially in the gap space. The experiments of Lorenz et al. |14| have shown that for the typical PWR LOCA conditions these fractions are below 1% of inventory in the gap.

Diffusional release of the pellet-to-cladding gap inventory. This release of iodine and cesium remaining in the gap space after burst is a slow process and as seen from Lorenz's experiments - the activities of volatile fission products released in this way are smaller than those released initially during the burst, unless the temperature of the gap raises considerably and is held for a long time.
In the case of severe accident studies one of the possible scenarios assumes that the operation of passive ECCS provides core reflooding and the fuel becomes waterlogged, but then, due to the failure of active ECCS, the fuel dries out and becomes overheated. In such a case the water accumulated in the gap would be forcibly evaporated and the resulting steam would leave the gap under high pressure through the rupture point. For this scenario it has been assumed |10| that the volatile and particulate fission products which had been released to the gap would be carried out with the outcoming steam. The release from the gap would be thus nearly complete before fuel reached higher temperatures involving the next mechanism of fission product release.

Thus in the case of LOCA the correlations of Lorenz et al. |14| are assumed to be correct, while in the case of severe accidents a complete release of gap inventory is assumed.

Grain boundary release. According to NRC |4|, at 1350°C noble fission gases, iodine and cesium previously accumulated at the grain boundaries are released, the released fraction of initial fuel inventory of stable isotopes of the above elements reaching up to 20% for high burnup fuel rods. The experiments with VVER fuel elements irradiated in the MR reactor have shown similar effects. E.g. after abrupt cooling of a fuel rod operated in very high temperatures, with the zones of equiaxed grain growth and of columnar grains reaching for about 25% of fuel volume each, the fraction |9| of ^{131}I released was equal to about 30% of full ^{131}I inventory in the fuel.

The experiments of Lorenz et al. |15| with short-time heating of H.B. Robinson fuel up to 1400, 1445 and 1600°C are an excellent example of the importance of grain boundary release. Although in the NRC report on fission product releases |4| they are given as examples of data for the evaluation of overall fractional release rate coefficient and therefore fall high compared to other, longterm heating studies, they should be rather used to determine the grain boundary release rates. This has been done in a recent study of fission product release from VVER fuel under severe accident scenario |10|.

The results of Lorenz et al. |15| show the following general features:
- The fraction of noble gases and VFP released during heating grows with temperature from about 0.3% of total inventory of these fission products at 1300°C through 7% at 1440°C up to 25% at 1610°C. Further heating does not change this fraction, the release continues in the way of intragranular diffusion.
- The release from intergranular pores and tunnels proceeds quickly, e.g. in the temperature of 1440°C over 50% of the activity to be released escapes within 0.4 min. Prolongation of heating time above 2-3 min.does not produce any significant increase of this fraction, except for the slow increase corresponding to intragranular diffusion.

In view of the scarcity of data a precise correlation describing grain boundary release rate can not be formulated yet. However, it is possible to provide an approximate expression based on Lorenz's results |15|. In REFRAC code the following formula has been used to determine the activity of nuclide i released from the pores in time interval Δt_j at temperature T:

$$\Delta A_i(T,t_j) = \left\{ A_{fuel}\ f_p \left[\frac{T(t_j) - T_b}{T_e - T_b} \right]^2 - \sum_{1}^{j-1} \Delta A_i \right\} \left(1-\exp(-k_p \Delta t)\right) \quad (11)$$

where: A_{fuel} - initial inventory of nuclide i in fuel,

f_p - fraction of fuel inventory occuring in intergranular pores,

T_b and T_e - temperatures at which the grain boundary activity release begins and ends, respectively,

k_p - grain boundary release rate coefficient,

A good agreement with Lorenz's results is reached assuming

$f_p = 0.25$, $T_b = 1250°C$, $T_e = 1600°C$, and $k_p = 2\ min^{-1}$ |10|.

Diffusion from UO_2 grains. In the REFRAC code the approach formulated by NRC |4| was adopted, i.e. the total fraction F of inventory of a nuclide present in fuel at the beginning of a time step Δt released during an isothermal test is calculated from the formula

$$F = 1 - \exp(-k(T)\Delta t) \qquad (12)$$

where k(T) - fractional release rate coefficient at temperature T, used in REFRAC |10|.

The values of k(T) were taken from the work of Andriesse |7| whose theory on solid FP propagation in UO_2 matrix seems to provide the best explanation of observed phenomena.

Release from molten material. For time dependant descriptions of FP releases in severe accident studies the approach proposed by NRC |4| and the data determined in SASCHA experiments |16| have been adopted in REFRAC code |10| Formerly, for accidents leading to fuel melting the fractional release coefficients determined by CMEA expert committee |2| were used.

Process of Fuel Heating

In order to provide a realistic evaluation of FP releases during severe accidents it is necessary to be able to calculate the course of fuel heating. This heating is due to two processes, namely residual heat generation and Zr-steam reaction. The latter process has been studied by many authors. In REFRAC code the rate of Zr oxidation is defined on the basis of the results of Leistikow et al. |17| which in temperatures up to 1500°C lie close to most other contemporary results and in higher temperatures approach the results of Baker and Just.

A difficult problem in the case of severe fuel damage consists in determining the behaviour of partly oxidized Zr cladding after the melting point of Zr has been reached. In REFRAC code the criterion of Tanabe and Matsumoto |18| is adopted. If the local fraction of ZrO_2 in the cladding thickness is smaller than the critical value f_c, then it is assumed that the oxide layer will break at the moment of Zr melting and the molten metal will flow down along the outer cladding surface. If $f(ZrO_2)$ is larger than f_c, then the ceramic layer is strong enough to withstand the pressure of molten Zr and prevents its outflow. If there is another oxide layer of thickness larger than f_c on the inside surface of the clad, then the molten material remains in place and provides further heating due to Zr-steam reaction.

An example of the former case is presented in Fig. 1, showing the course of fuel heating in a VVER-440 reactor after LB LOCA with simultaneous total blackout, including emergency Diesel power supply. The accident sequence was assumed to be terminated at a moment of partial core melting (at t = 2200 s) by restoring the power to ECCS Diesels.
In such a sequence, corresponding to sequence AB' according to Rasmussen's nomenclature, the time - release relationship is important for good evaluation of accident results.

Fission Product Releases from VVER Fuel in a Severe Accident of AB' Type.

Fig. 2 shows the releases of iodine ^{131}I from VVER fuel elements of high ratings in the case of the above mentioned AB' accident. The overall release for the whole core at the moment of accident termination reaches about 55% of initial ^{131}I inventory in fuel. The releases of solid FPs are much smaller, the fraction of released Ba reaching some 0.2% and the fraction of released Sr remaining below 0.03%.

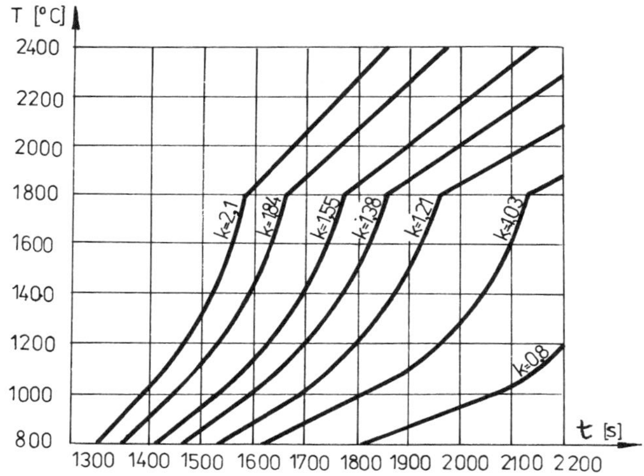

FIGURE 1. Heating of VVER-440 fuel rod sections of various relative power ratings k_i in an AB' type accident. |10|

$$k_i = \frac{\text{power rating in fuel section i}}{\text{average power rating in the core}}$$

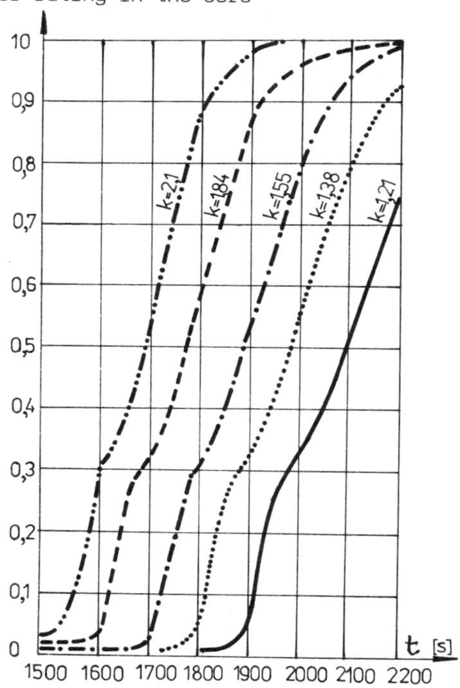

FIGURE 2. Iodine release from VVER-440 fuel rod sections of various relative power ratings k_i in an AB' type accident. |10|

These results might be compared with the fraction of released FPs which would be obtained for a 15% core melt according to the CMEA guidelines |2|, namely for ^{131}I - 15%, for Ba - 0,75% and for Sr 0,75%.

It is clear, that the presented approach, besides giving insight into the time sequence of fuel melting and FP release, makes it also possible to obtain results which reflect better the expected FP releases in case of severe accidents with partial core melting.

LITERATURE

1. Rasmussen N. et al.: Reactor Safety Study. WASH-1400, 1975, NRC, Washington D.C.
2. CMEA Guidelines on Evaluation of Radiation Effects of LOCA in NPP with VVERs, ORB No 15, Moscow 1983 (in Russian).
3. Rest J.:Evaluation of Volatile and Gaseous Fission Product Behaviour in Water Reactor Fuel Under Normal and Severe Core Accident Conditions, Nuclear Technology, Vol.61, Apr. 1983, p. 33-48.
4. NRC: Technical Bases for Estimating Fission Product Behavior During LWR Accidents. NUREG -0772, 1981, Washington D.C.
5. Gieseke J.A.,Cybulskis P.,Denning R.S.,Kuhlman M.R.,Lee K.W.,Chen H. - Radionuclide Release Under Specific LWR Accident Conditions, BMI-2104 Battelle Columbus Laboratories, 1985.
6. Lorenz R.A.: Status of Validation of the CORSOR Computer Code used for the Accident Source Term Reassessment Study, in: ORNL|TM-8842, 1985.
7. Andriesse C.D.: Melting nuclear fuel. Kema Scientific and Technical Reports. 1985, Vol. 3, No 7, p. 87-107.
8. Strupczewski A.: The release of fission products from WWER-440 reactors under DBA conditions-calculation methods and results for Żarnowiec NPP In: Proc. of Technical Committee Meeting on Water Reactor Fuel Behaviour and Fission Products Release in off Normal and Accident Conditions, IAEA, Vienna 10-13 Nov. 1986, IAEA-TC-579|43.
9. Luzanova L.M.,Slaviagin P.D.: Fission Product Release from Defective Fuel Rods of Sintered UO_2, IAE-3723|4, 1984 (in Russian).
10. Strupczewski A.: Evaluation of Containment in Żarnowiec NPP, IAE Report 19|E-BP|88.
11. Luzanova L.M.,Miglo W.N.,Slaviagin P.D.: Release of Fission Products into Primary Coolant of Operating VVER Reactor, in: Atomnye Elektricheskije Stancyi, 1987, No 9, p. 124-131. (in Russian).
12. Ronchi C.,Coqnerelle M.,Sakellaridis J.,van de Laar J.: Analysis of the Fission Product Behavior Under Normal and Off-Normal Reactor Conditions, IAEA-TC-579|34.
13. Gittus J.H.: PWR Degraded Core Analysis, ND-R-610 (S), April 1982.
14. Lorenz R.A.,Collins J.L.,Malinauskas A.P.: Fission Product Source Terms in the LWR LOCA. Nuclear Technology, Vol.46, Dec. 1979, p. 404-410.
15. Lorenz R.A.,Collins J.L.,Malinauskas A.P.,Osborne M.F. and Towns R.L.: Fission Product Release from Highly Irradiated Fuel Heated to 1300-1600°C in Steam, NUREG|CR 1386, 1980.
16. Albrecht H.,Wild H.: Behavior of I, Cs, Te, Ba, Ag, In and Cd During Release from Overheated PWR Cores, Proc. of Int. Meeting on LWR Severe Accident Evaluation 1983, Cambridge.
17. Leistikow S.,Schanz G.,Berg H.V,Aly A.E.: Comprehensive Presentation of Extended Zircaloy-4 Steam Oxidation Results. In: Proc.of OECD-NEA-CSNI| IAEA Specialists Meeting on Water Reactor Safety and Fission Product Release in off-Normal and Accident Conditions, 20 May 1983, Risö, Denmark, IWGFPT|16, IAEA 1983.
18. Tanabe F.,Matsumo K.: Development and Verification of Models for Cladding Oxidation in Early Stage of Severe Accident, Intern.Symp.Sorrento, 1988

Source Term of Iodine and Noble Gas Fission Products in the Fuel-to-Sheath Gap of Intact Operating Nuclear Fuel Elements

B. J. LEWIS
Royal Military College of Canada
Department of Chemistry and Chemical Engineering
Kingston, Ontario K7K 5L0, Canada

C. E. L. HUNT, F. C. IGLESIAS, L. D. MACDONALD, and C. CHE
Atomic Energy of Canada Limited Research Company
Chalk River Nuclear Laboratories
Chalk River, Ontario, K0J 1J0 Canada

1. INTRODUCTION

The extent to which fission products are released from reactor fuel and escape from the primary system is determined primarily by their volatility and chemical behaviour. The gap release component consists of that part of the fission-product inventory collected in the fuel voids and fuel-sheath gap during normal operation that is free to escape as a gas or vapour from the fuel element in the event of a sheath failure. Of the fission products that reach the gap, the inert noble gases, Kr and Xe, migrate freely to the rupture point, while the volatile halogens, I and Br, may undergo chemical reactions with the cladding and other fission products, such as those between iodine and zirconium or cesium, and therefore may be prevented from escaping from the fuel element.

In order to study such short-lived fission-product release behaviour in intact operating CANDU*-type fuel, several experiments at Chalk River [1-9] were performed using a sweep gas technique. In these tests, the fuel-to-sheath gap was continually purged with an inert carrier gas-flow transporting the active species which were released from the UO_2 fuel to a gamma-ray spectrometer. This technique was used to determine the physical mechanisms of gas release for unoxidized fuel through the

---

* CANada Deuterium Uranium

measurement of an instantaneous release rate. However, from a safety analysis perspective, it is also important to present the data as an integrated source term in the fuel-to-sheath gap which is immediately available for release after sheath failure. In this study, a methodology is developed in which the sweep gas results are related to a free gap activity during steady reactor operation and for reactor shutdown and startup. These results are then compared to predictions of the ANS 5.4 standard fission-product release model [10].

2. EXPERIMENTAL

2.1 Experiment Description

The fuel, sheath and irradiation conditions in these experiments are briefly summarized in Table 1. The experiments covered a variety of operating conditions, with fuel linear ratings ranging from 17-65 kW/m, to a maximum burnup of 220 MW.h/kgU.

The physical setup of the sweep gas system is shown in Figure 1. The fission products were swept from the fuel-to-sheath gap with an inert carrier gas (He-2% H_2) at a system pressure of 3 MPa. Activity concentrations C_m (Bq/m^3) of the various isotopes were calculated

TABLE 1. Summary of Sweep Gas Experiments at Chalk River*

Characterisitic	FIO-122	FIO-124	FIO-133	FIO-134	FIO-138	FIO-141
Fuel	UO_2	UO_2	UO_2	UO_2	UO_2	UO_2
Enrichment (wt% U-235 in U)	5.02	4.5	1.38	5.9	1.38	5.9
Density (Mg/m^3)	10.71	10.65	10.64	10.64	10.69	10.65
Grain Size (μm)	10	10	10	10	10	10
Pellet Diameter (mm)	12.15	12.15	18.06	11.71	18.06	11.71
Fuel Stack Length (mm)	477	477	378	477	379	477
Sheath	Zircaloy-4	Zircaloy-4	Zircaloy-4	Zircaloy-4	Zircaloy-4	Zircaloy-4
Sheath Outside Diameter (mm)	13.11	13.11	19.85	13.06	19.86	13.08
Wall Thickness (mm)	0.43	0.43	0.81	0.63	0.80	0.63
NRX Test Loop	X-1	X-1	X-4	X-4	X-2	X-4
Coolant	Pressurized Water	Pressurized Water	Pressurized Water	Pressurized Water	Pressurized Water	Pressurized Water
Coolant Pressure (MPa)	8.5	8.5	8.5	8.5	9.6	8.5
Coolant Flow (kg/s)	1.1	1.1	0.24	0.24	0.9	0.24
Inlet Temperature (°C)	240	240	260-275	260	275-290	260
Linear Power Range (kW/m)	39-45	55-60	55-65	60-62	53-60	17-57
End-of-Life Burnup (MW·h/kgU)	70	50	50	220	50	50
Test Conditions	Normal	Normal	Normal and Dryout	Normal and Oxidizing	Normal Blowdown	Normal, Solid vs Annular Pellets
Temperature** (°C, max.)	1370	1770	1760	2100	1900	1700
Reference	1,6	1,6	3,4,7	2	7,9	5

* Taken from Ref. 8.
** Estimated temperatures for normal operation.

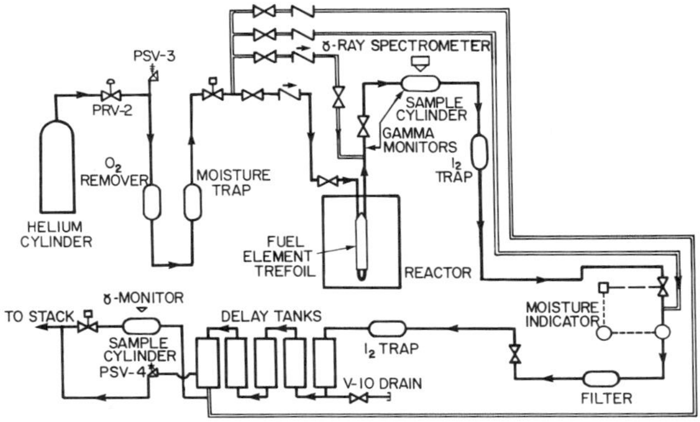

FIGURE 1. Schematic diagram of sweep gas system. (Taken from Ref. 8).

directly from the measured count-rate data at the spectrometer. The isotopic release rates R (atoms/s) from the fuel were then evaluated as a function of time from the measured downstream concentrations by the expression [11]

$$R(t-T_r) = (V/\lambda)\ \exp(\lambda^* T_r)\ C_m(t) \qquad (1)$$

where: T_r = transit time from the fuel to the spectrometer
V = volumetric flow rate (m^3/s)
$\lambda^* = \lambda + f\sigma\phi$; λ = loss-rate constant for radioactive decay, f = fraction of transit time in the neutron flux, σ = microscopic absorption cross section, ϕ = neutron flux.

2.2 Experimental Results

Figure 2 shows the best-fit results of the Chalk River tests [1,2,3,5,6] for normal operation under steady-state reactor conditions. Here the release-rate to birth-rate ratio (R/B) for the noble gases is plotted against the decay constant. This ratio generally exhibits a behaviour close to $\lambda^{-1/2}$ which is indicative of a diffusional-controlled release of fission gas through the UO_2 matrix [12]. None of the fission-product iodine, however, was directly measured at the spectrometer during any phase of normal (intact) operation and it was deduced that this species was adsorbed on internal fuel or sheath surfaces. Iodine releases were calculated from the behaviour of the Xe-133 and Xe-135, by continued sweeping during reactor shutdowns. Such analysis indicated that iodine had a diffusion behaviour in the fuel matrix similar to that of noble gas, where the isotopes I-133 and I-135 were observed to lie on the respective curves in Fig. 2 [1].

Figures 3a and 3b show the behaviour of the short-lived krypton and xenon gases during a normal reactor shutdown and startup in the FIO-133 test from a linear power of 55 kW/m at a burnup of 27 MW·h/kgU. When the release rate was normalized by the fission yield, Y, a pulse release behaviour of $R_t/Y \propto \lambda^{-3/2}$ was reported for both the shutdown and startup transients [4,8]. This behaviour is consistent with a

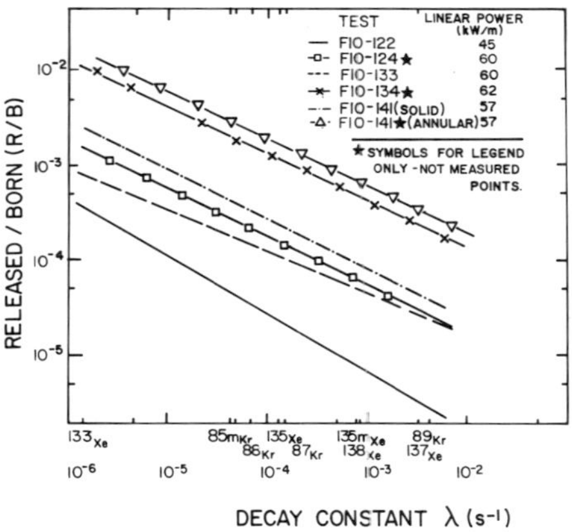

FIGURE 2. R/B versus λ for Chalk River data. (Taken from Ref. 8).

mechanism whereby fission products initially diffuse during steady operation to isolated sites within the fuel where an equilibrium inventory (q) is established as a result of decay:

$$\frac{dq}{dt} = R_s - \lambda q = 0, \qquad (2)$$

The steady-state source term (R_s), from the above discussion, is given by

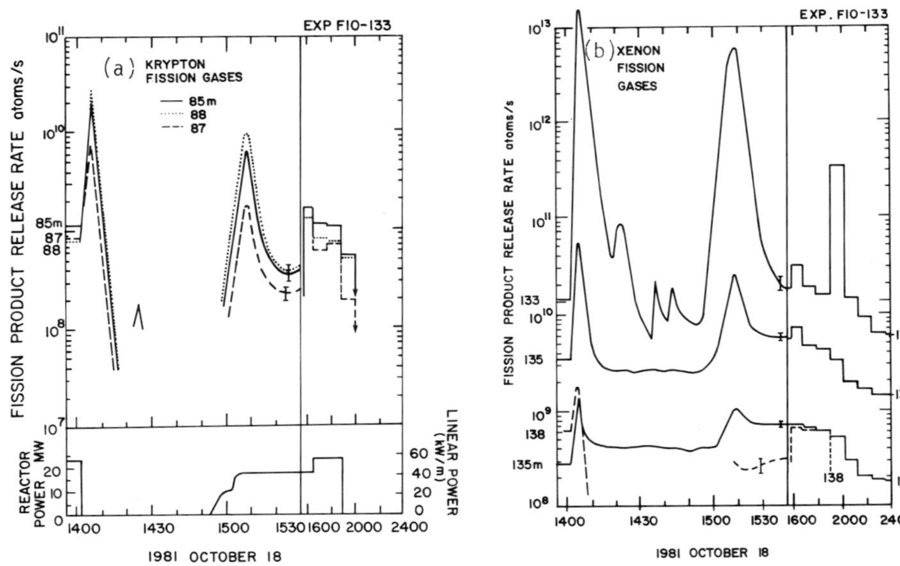

FIGURE 3. Release rates of noble gas fission products for a reactor shutdown and startup in experiment FIO-133. (Taken from Ref. 4).

$$R_s/Y = k\lambda^{-1/2}. \quad (3)$$

During the transient, fission gases are vented to the sweep gas as a first-order rate process when cracks in the fuel due to thermal shock intersect these locations. Hence, using Eqs. (2) and (3):

$$R_t/Y \propto q/Y = k\lambda^{-3/2}. \quad (4)$$

3. MODEL DEVELOPMENT

Given the rate of release in Figs. 2 and 3, an activity inventory in the gap can be calculated as a function of time. In an intact element, the following source and loss terms for the mass balance of fission products must be considered. As previously described, the source is given by the measured rate of release (R). However, fission products are also lost as a result of radioactive decay and through neutron transmutation during reactor operation (transmutation is only important for the isotope Xe-135). The rate of change with respect to time of a given isotope in the gap (N_g) is:

$$\frac{dN_g(t)}{dt} = R(t) - \lambda^* N_g , \quad (5)$$

where $\lambda^* = \lambda + \sigma\phi$ and $\sigma\phi$ is the rate constant for neutron transmutation.

In the case of equilibrium conditions (such that $dN_g/dt=0$), the gap activity is proportional to the release rate,

$$\lambda N_g = (\lambda/\lambda^*)\ R . \quad (6)$$

With a reactor shutdown occurring at t=0, followed by startup at $t=t_o$, the time-dependent solution of Eq. (5) is given by the integral expression:

$$\lambda N_g(t) = [\lambda N_{go} + \lambda \int_0^t R(\tau) \exp\{\lambda\tau\}\, d\tau] \exp\{-\lambda t\} , \qquad 0 \le t < t_o \quad (7a)$$

$$= \{ [\lambda N_{go} + \lambda \int_0^{t_o} R(\tau) \exp\{\lambda\tau\}\, d\tau] \exp\{(\lambda^*-\lambda)t_o\} + \lambda \int_{t_o}^{t} R(\tau) \exp\{\lambda^*\tau\}\, d\tau\} \exp\{-\lambda^* t\} , \quad t \ge t_o \quad (7b)$$

where λN_{go} is the initial activity at the start of the transient. For all isotopes other than Xe-135, it is assumed that $\lambda^*=\lambda$. In this case, Eq. (7a) is applicable for all $t \ge 0$.

4. FUEL-TO-SHEATH GAP ACTIVITY

4.1 Noble Gas Activity

The steady-state activity of noble gas in the fuel-to-sheath gap for the various experiments can be evaluated immediately using Eq. (6) and the data in Fig. 2. For a given isotope, the gap activity in Bq is

$$\lambda N_g = (R/B)\ B\ \xi\ H \qquad (8)$$

where: B = [= FY] birth rate (atoms/s)
F = fission rate (fissions/s) = $P \times 3.121 \times 10^{13}$
Y = cumulative fission yield (atoms/fission) (see Table 2)
P = fuel-element power rating (kW)
ξ = (λ/λ^*) = 0.106 for Xe-135, or 1 for all other isotopes
H = precursor correction factor (see Table 2).

The factor H must be applied in Eq. (8) since the (R/B) data have been corrected for precursor effects [1,13]. The above expression for F is based on a value of 200 MeV/fission.

The gap activity as a result of reactor shutdown and startup in the FIO-133 test can be evaluated with the use of Eq. (7) and Fig. 3. An equilibrium activity is assumed at the start of the transient where λN_{go} is given by Eq. (6). The release rates are numerically integrated using an extended trapezoidal rule [14] with interpolated values for constant time intervals. Results of this calculation for the krypton and xenon fission gases are shown in Figs. 4a and 4b, respectively. The gap activity for Xe-137 has the greatest uncertainty because its half-life is shorter than the transport time T_r [see Eq. (1)]. The activities of fission gases with short-lived precursors are seen to decay at their natural rate following the initial reactor shutdown. Only those isotopes with relatively long-lived iodine precursors (Xe-133, Xe-135 and Xe-135m) display an increase in the integrated release. This behaviour is due to the decay of fission-product iodine adsorbed on internal element surfaces (see discussion below). It is also interesting to note that the gap activities did not significantly increase as a result of pellet cracking during reactor shutdown or startup. In fact, these activities are within a factor of 4 of the

TABLE 2. Isotopic Data for Noble Gas and Iodine

Isotope	Decay Constant λ (s^{-1})	Fission Yield Y (a/fission)	Precursor Correction Factor H
Kr-85m	4.30E-05	1.30E-02	1.019
Kr-87	1.52E-04	2.52E-02	1.016
Kr-88	6.78E-05	3.55E-02	1.002
Xe-133	1.53E-06	6.70E-02	1.205
Xe-135m	7.56E-04	1.10E-02	6.373
Xe-135	2.12E-05	6.54E-02	3.353*
Xe-137	3.02E-03	6.06E-02	1.072
Xe-138	8.18E-04	6.42E-02	1.003
I-131	9.98E-07	2.88E-02	1.0
I-132	8.37E-05	4.30E-02	11.7
I-133	9.26E-06	6.70E-02	1.1
I-134	2.20E-04	7.73E-02	2.1
I-135	2.91E-05	6.30E-02	1.0

* Value for an effective decay constant $\lambda^* = 2 \times 10^{-4}\ s^{-1}$.

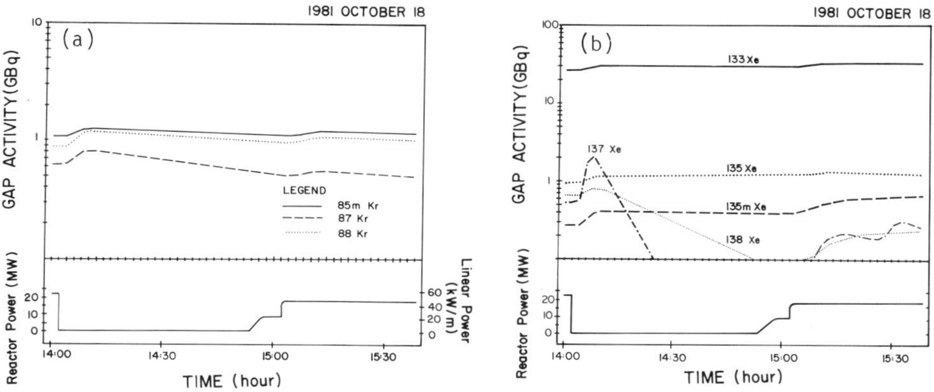

FIGURE 4. Activity inventories in the fuel-to-sheath gap of (a) krypton and (b) xenon fission products for a reactor shutdown and startup transient in experiment FIO-133.

pre-transient values even though the release rates increased by a factor of 10^3 for short periods of time (compare Figs. 3 and 4). The shortest-lived isotopes generally exhibited the greatest increases in activity.

4.2 Iodine Activity

As discussed previously, the iodine species have a diffusion behaviour in the UO_2 fuel matrix similar to that of noble gas. Hence, steady-state activities can be obtained directly from Fig. 2 using the methodology developed in Section 4.1.

Iodine that is released into the gap is chemically adsorbed onto the UO_2 or inner sheath surfaces. Inspection of Fig. 3 suggests that the constant releases for Xe-133, Xe-135 and Xe-135m (just prior to startup) result mainly from the decay of the deposited iodine precursors which have access to the sweep gas, rather than from direct mechanisms of lattice diffusion, which should be negligible at fuel temperatures existing during shutdown, and pellet cracking. The activity of I-133 and I-135 can therefore be determined from the measured release-rate data of Xe-133, Xe-135 and Xe-135m (Fig. 3b) by the relation:

$$(\lambda N)_i = R_{xe}/f_{xe}, \tag{9}$$

where f_{xe} is the isotopic branching fraction (= 1 for Xe-133, 0.835 for Xe-135, and 0.165 for Xe-135m). Indirectly measured values for these isotopes are shown in Fig. 5.

Assuming similar release processes, the release behaviour of iodine can also be modelled from the noble gas kinetic data. Iodine release rates can be predicted from Eqs. (3) and (4):

$$R_i(t) = \{\lambda_i/\lambda_n\}^{-m} (Y_i/Y_n) (H_i/H_n) R_n(t) \tag{10}$$

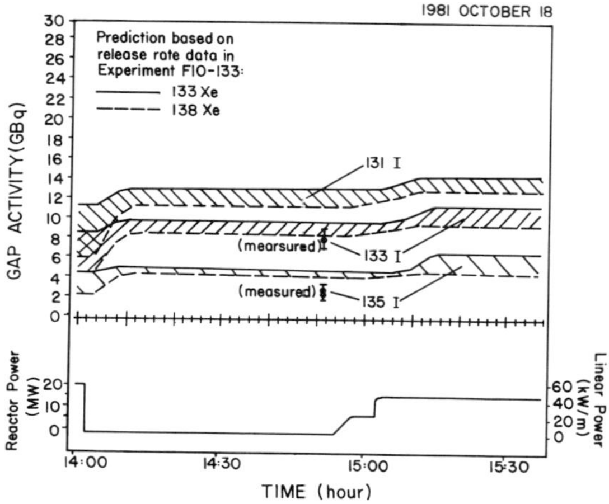

FIGURE 5. Comparison of predicted and measured iodine activity in the fuel-to-sheath gap for a reactor shutdown and startup transient in experiment FIO-133. The model predictions are based on Eq. (10) and the release-rate data in Fig. 3b.

where m equals 1/2 for steady-state operation and 3/2 for transient operation, and the subscripts i and n refer to iodine and noble gas, respectively. Releases for iodine can therefore be deduced with the isotopic information in Table 2, and the release-rate data for Xe-133 (5.25 d half-life) and Xe-138 (14.13 m half-life) in Fig. 3b. With a knowledge of the time-dependent release rate, a gap activity can then be estimated using Eq. (7) for the nuclides of interest (e.g. I-131, I-133 and I-135). Agreement between the two estimates in Fig. 5 for a given iodine nuclide, based on a long and short half-life isotope, indicate that the release behaviour is reasonably modelled. As seen in Fig. 5, indirectly measured activities of I-133 and I-135 are in good agreement with the model predictions.

5. COMPARISON WITH FISSION-PRODUCT RELEASE MODELS

The sweep gas data can be used to assess the various fission-product release models used in the ELESIM2 fuel-performance code [15]. The ANS 5.4 algorithm [10] is based on a simple Booth diffusion-type model where the physical processes of atomic diffusion, gas-bubble nucleation, bubble migration, bubble coalescence and irradiation resolution are modelled with the use of an empirical diffusion coefficient. The FREEDOM code [16] is a more physically based model detailing the transient, microstructure-dependent fission gas release and swelling behaviour for both the radioactive and stable species.

Figure 6a shows a comparison of the measured versus predicted fractional release* for noble gas, as a function of decay constant, for experiment FIO-133 for the steady-state situation prior to reactor

---

* Defined as the ratio of the gap inventory to total inventory in the fuel at time t.

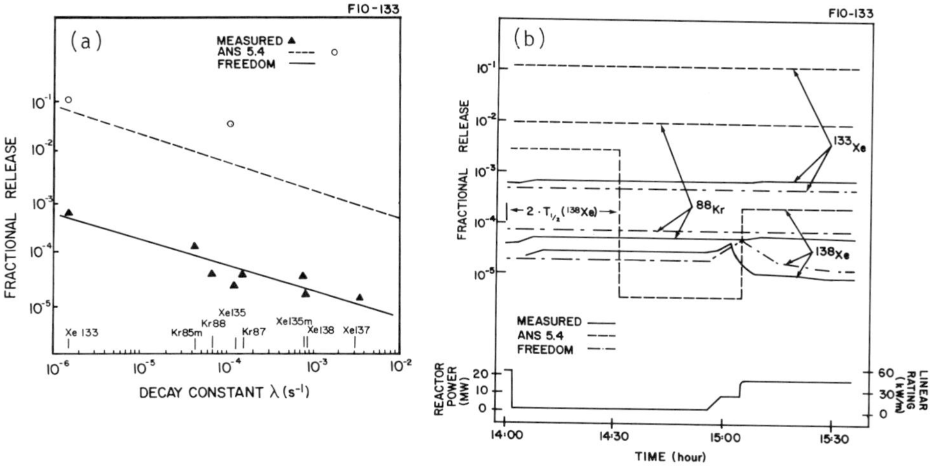

FIGURE 6. Comparison of ANS 5.4 and FREEDOM model predictions with release data from experiment FIO-133 for (a) steady-state conditions and (b) a reactor shutdown and startup transient.

shutdown in Fig. 4. A $\lambda^{-1/2}$ behaviour is observed both experimentally and theoretically since diffusion is the dominant release process. For this data sample, at relatively high power (55 kW/m) and low burnup (27 MW·h/kgU), the release is overpredicted by the ANS 5.4 model by two orders of magnitude, but is predicted well with FREEDOM. Figure 6b shows a comparison of the fractional release as a function of time during reactor shutdown and startup. Again, the ANS 5.4 model consistently overpredicts the gas release for the longer-lived isotopes, Xe-133 and Kr-88; however, an underprediction occurs after shutdown for the short-lived Xe-138. This underprediction arises from the low temperature-release calculation in the ANS 5.4 model, which uses the peak temperature reached during the last two half-lives in the irradiation history. The release fractions also remain constant over the transient as the algorithm cannot mechanistically model burst releases due to pellet cracking. In comparison, FREEDOM is in good agreement with experiment and is even capable of predicting the reactor startup trend for Xe-138 using a simplistic, radial pellet-cracking model for fuel-power increases.

6. CONCLUSIONS

1. A methodology has been developed to calculate the gap activity of fission gas in intact fuel elements during both steady-state and transient reactor operation. This methodology has been applied to experimental results from sweep gas tests performed at Chalk River. A model has also been developed to estimate the gap activity of the short-lived iodine species based on the kinetic release behaviour of noble gas. This model is in good agreement with indirectly measured values of I-133 and I-135.

2. For experiment FIO-133, at a high power of 55 kW/m and burnup of 27 MW·h/kgU, the ANS 5.4 model overpredicted the steady-state and transient release of the longer-lived isotopes by several orders of magnitude while the more physically based FREEDOM model was able to predict the release behaviour within a factor of 2.

7. ACKNOWLEDGEMENTS

The authors would like to acknowledge the pioneering experimental work and analysis of I.J. Hastings, J.J. Lipsett and C.E.L. Hunt. The authors would also like to thank P.J. Fehrenbach, J.M. Bunge and J.H.K. Lau (Ontario Hydro) for helpful discussions and assistance in the preparation of this paper.

This study was funded under COG Work Package No. 494.

8. REFERENCES

1. Hastings, I.J., Hunt, C.E.L., Lipsett, J.J. and MacDonald, R.D., Tests to Determine the Release of Short-Lived Fission Products from UO_2 Fuel Operating at Linear Powers of 45 and 60 kW/m: Methods and Results, AECL-7920, September 1985.

2. Hunt, C.E.L., Lipsett, J.J. and Hastings, I.J., Release of Short-Lived Fission Products from Operating Fuel under Oxidizing Conditions, AECL-8564, September 1984.

3. Hastings, I.J., Hunt, C.E.L., Lipsett, J.J. and Gray, R.G., Transient Fission Product Release during Dryout in Operating UO_2 Fuel, AECL-7832, December 1982.

4. Lipsett, J.J., Hunt, C.E.L. and Hastings, I.J., Transient Fission Product Release within Operating UO_2 Fuel Elements during Power Cycles, AECL-8069, May 1983.

5. Hastings, I.J., Hunt, C.E.L. and Delaney, R.D., Short-Lived Fission Product Release from the Surface and Centre of Operating UO_2 Fuel, AECL-8353, July 1984.

6. Hastings, I.J., Hunt, C.E.L., Lipsett, J.J. and MacDonald, R.D., Behaviour of Short-Lived Fission Products within Operating UO_2 Fuel Elements, Res Mechanica, vol. 6, pp. 167-175, 1983.

7. Fehrenbach, P.J., Hastings, I.J., Walsworth, J.A., Spencer, R.C., Lipsett, J.J., Hunt, C.E.L. and Delaney, R.D., Zircaloy-Sheathed UO_2 Fuel Performance during In-Reactor LOCA Transients, AECL-8569, October 1984.

8. Hastings, I.J., Hunt, C.E.L. and Lipsett, J.J., Release of Short-Lived Fission Products from UO_2 Fuel: Effects of Operating Conditions, J. Nucl. Mater, vol. 130, pp. 407-417, 1985.

9. Lipsett, J.J., Hastings, I.J., Hunt, C.E.L., Fehrenbach, P.J., Delaney, R.D., Larson, L.L. and Spencer, R.C., Short-Lived Fission-Product Release within a UO_2 Fuel Element during In-Reactor LOCA Transients, in Advances in Ceramics, vol. 17, ed. I.J. Hastings, pp. 173-189, American Ceramic Society, Columbus, Ohio, 1985.

10. "Method for Calculating the Release of Fission Products from Oxide Fuel", background and derivation of ANS 5.4 Standard Fission Product Release Model, American Nuclear Society, NUREG/CR-2507, U.S. Nuclear Regulatory Commission (1982).

11. Lewis, B.J. and Duncan, D.B., Determination of Release Rate from Measured Activity of Depositing Fission Products, ISBN 0-919784-13-5, Proc. International Conference on CANDU Fuel, Chalk River, Canada, pp. 460-469, 1986.

12. Booth, A.H., A Suggested Method for Calculating the Diffusion of Radioactive Rare Gas Fission Products from UO_2 Fuel Elements and a Discussion of Proposed In-Reactor Experiments that may be Used to Test its Validity, AECL-700, September 1957.

13. Lewis, B.J., Phillips, C.R. and Notley, M.J., A Model for the Release of Radioactive Krypton, Xenon, and Iodine from Defective UO_2 Fuel Elements, Nucl Technol., vol. 73, pp. 72-83, 1986.

14. Press, W.H., Flannery, B.P., Teukolsky, S.A. and Vetterling, W.T., Numerical Recipes: The Art of Scientific Computing, pp. 107-108, Cambridge University Press, Cambridge, 1987.

15. Notley, M.J.F., ELESIM: A Computer Code for Predicting the Performance of Nuclear Fuel Elements, Nucl. Technol., vol. 44, pp. 445-450, 1979.

16. Macdonald, L.D., Duncan, D.B., Lewis, B.J. and Iglesias, F.C., FREEDOM: A Transient Fission-Product Release Model for Radioactive and Stable Species, AECL-9810, Proc. of the IAEA Technical Committee Meeting on Water Reactor Fuel Element Computer Modelling in Steady-State, Transient and Accident Conditions, Preston, England, September 18-22, 1988.

Correlation of Recent Fission Product Release Data

T. S. KRESS and R. A. LORENZ
Oak Ridge National Laboratory
Oak Ridge, Tennessee 37831, USA

T. NAKAMURA
Japan Atomic Energy Research Institute
Department of Fuels and Materials Research
Tokai Research Establishment
Japan

M. F. OSBORNE
Oak Ridge National Laboratory
Oak Ridge, Tennessee 37831, USA

INTRODUCTION

For the calculation of source terms associated with severe accidents, it is necessary to model the release of fission products from fuel as it heats and melts. Perhaps the most definitive model for fission product release is that of the FASTGRASS computer code [1] developed at Argonne National Laboratory. FASTGRASS contains mechanistic models for such phenomena as: diffusive transfer of species within grains to internal bubbles and to grain boundaries; diffusive transfer of bubbles to grain boundaries; diffusion of fission product species along grain boundaries and through interlinked cracks; growth of the grain size with irradiation and with time at temperature, and the effects of such grain growth on the transport. There is persuasive evidence that these processes, as well as additional chemical and gas phase mass transport processes, are important in the release of fission products from fuel. Nevertheless, it has been found convenient to have simplified fission product release correlations that may not be as definitive as models like FASTGRASS but which attempt in some simple way to capture the essence of the mechanisms. One of the most widely used such correlation is called CORSOR-M [2] which is the present fission product/aerosol release model used in the NRC Source Term Code Package. The CORSOR type of modelling had its formal beginning as an expedient in the NUREG-0772 exercise [3] (actually it was patterned after the technique being used at KfK to correlate their SASCHA experimental data [4]). CORSOR has been criticized as having too much uncertainty in the calculated releases and as not accurately reproducing some experimental data (notable PBF, ACRR, and the recent ORNL HI and VI series of tests). For example, Figure 1 shows cesium release data from some recent ORNL tests [5] that indicate release coefficients as much as an order of magnitude lower than CORSOR and that these coefficients systematically decrease with time. It is currently believed that these discrepancies between CORSOR and the more recent data have resulted because of the better time resolution of the more recent data compared to the data base that went into the CORSOR correlation. The release rate coefficients for CORSOR and CORSOR-M had been derived on the basis of

---

This research was sponsored by the Office of Nuclear Regulatory Research, U.S. Nuclear Regulatory Commission under Interagency Agreement DOE 1886-8012-7B with the U.S. Department of Energy under contract DE-AC05-84OR21400 with the Martin Marietta Energy Systems, Inc.

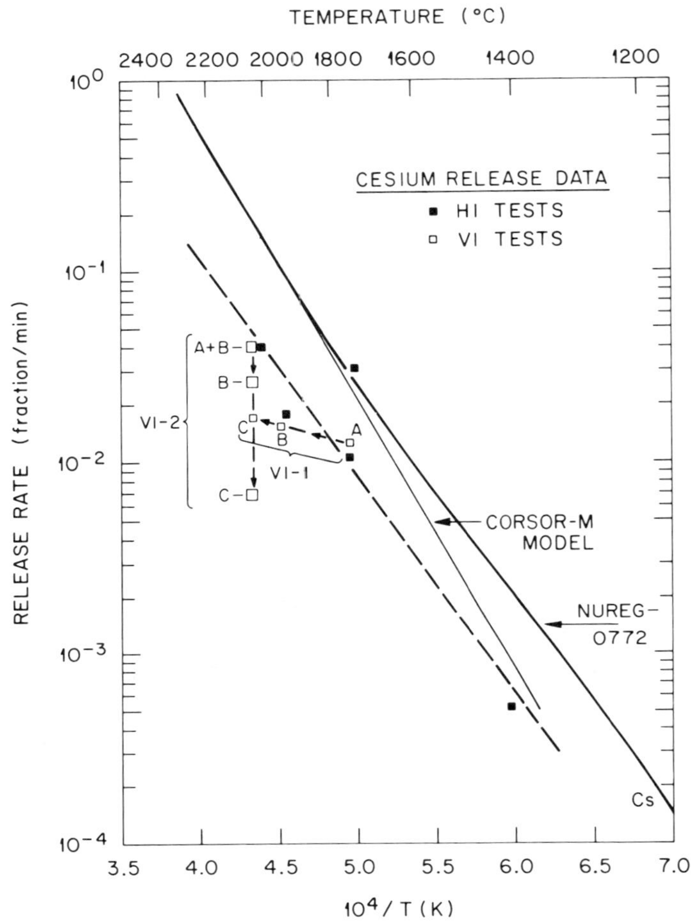

FIGURE 1. Comparison of release rate coefficients from tests VI-2 and VI-1 with earlier work.

relatively short term constant temperature release experiments and on simulated fuel experiments conducted prior to 1983. Both of these conditions tend to bias the results to the high side.

The systematic decrease in the release rate coefficients with time at constant temperature as observed in the ORNL experiment results shown in Figure 1 indicates that the first order rate equation that is the basis for CORSOR may be an inadequate correlational model. (Note that the A, B, and C designations on Figure 1 indicate successively later times in the indicated experiment.) Recently some success has been had in better correlating both the ORNL and the PBF data with a "diffusive-like" correlational model [6]. This is not too surprising because the mechanisms (believed realistic) in FASTGRASS are generally diffusive in nature, and even pure surface evaporation can be described by equations of similar form.

A reasonable choice, then, for a simple correlational model is one that has a diffusive-like character to it. The question then arises, how best can the existing data base that tends to be oriented toward release rate coefficients be utilized in such a model? A technique for doing this was developed by the authors for use in connection with the NUREG-1150 risk uncertainty exercise sponsored by NRC at Sandia National Laboratories and is presented below.

SIMPLIFIED CORRELATIONAL MODEL

A "diffusive-like" correlational model was developed for fission product release from heating fuel:

$$\frac{\partial C}{\partial t} = D \frac{1}{r} \frac{\partial^2}{\partial r^2} (r\ C)$$

where $D = D_o \exp(-Q/RT)$.

It is not implied that this is in any way a true mechanistic description of the governing phenomena - just that it appears to be a convenient empirical correlational scheme. It is intended that the above equation be coupled with the thermal hydraulic calculation of the fuel temperature and be solved in finite difference form using some effective diffusion length parameter. The correlational parameters to be empirically based, then are:

1. an effective spherical length parameter, a, that could be viewed as being related to the mean grain radius,
2. an effective "diffusion coefficient frequency factor," D_o, and
3. an effective "activation energy," Q.

In the ORNL Hot Cell Fission Product Release tests using real discharged fuel, the Cs generally dominates the released activity from the test fuel specimens [7]. It has been possible, then, to use the gamma measurements made downstream of the furnace for each of the tests to get continuous measurements of the Cs release. The correlational model proposed above was "best fit" to a significant number of the recent ORNL tests to obtain for cesium release from high burnup fuel:

$$a = 6 \text{ microns},$$
$$D_o = 0.0000763 \text{ cm}^2/\text{s}, \text{ and}$$
$$Q = 74300 \text{ cal/mole}.$$

Typical comparisons of the predicted and experimental release patterns using these parameters are shown in Figure 2 (the present model is referred to as the "Booth" model on Figure 2).

If sufficient time-resolved data existed, it would be possible to similarly establish a set of such parameters for each fission product elemental group (and other released material). Unfortunately such a data base for the various fission product and inert aerosol groups does not presently exist. Even if only total (integral) release data were available from a significant number of release experiments conducted under different temperature conditions, then it would be possible to fit these via regression techniques to determine appropriate values for the parameters. While this is the present approach of the ORNL research, as of now the quantity of high temperature data is still insufficient to fully

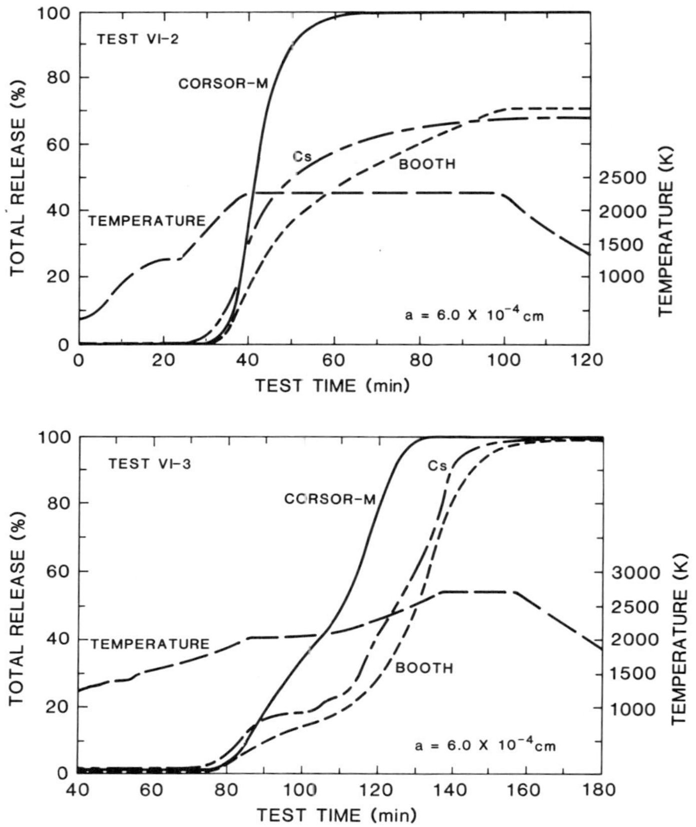

FIGURE 2. Typical comparisons of ORNL Fission Product Cesium Release Data with a "Diffusion" model (Booth).

implement it. There is at present insufficient data to develop values for D_0 and Q for each elemental group. Consequently, we are proposing an alternate approach that utilizes past fission product release experience for application with the proposed simplified "diffusion" correlation.

Lorenz [8] has observed that there is a remarkable consistency in the fission product and structural material release data when reported in terms of a release rate coefficient (for small releases, this is essentially the integral release fraction for fuel heated to very high temperatures divided by the release time at temperature). Figures 3 and 4, taken from Reference 8, illustrate this consistency. If any particular tests' results for the measured release rate coefficients are placed on a logarithmic scale ordinate in such a way that they form a straight line, then the resulting positions of the constituents on the linear abscissa defines what Lorenz called a "relative volatility" for each. Lorenz observed that the release rate coefficient data from other tests at markedly different conditions, when plotted on these same scales, also resulted in straight lines. The large difference in slopes for the relative volatility of species in the SASCHA tests (Fig. 3), and other tests

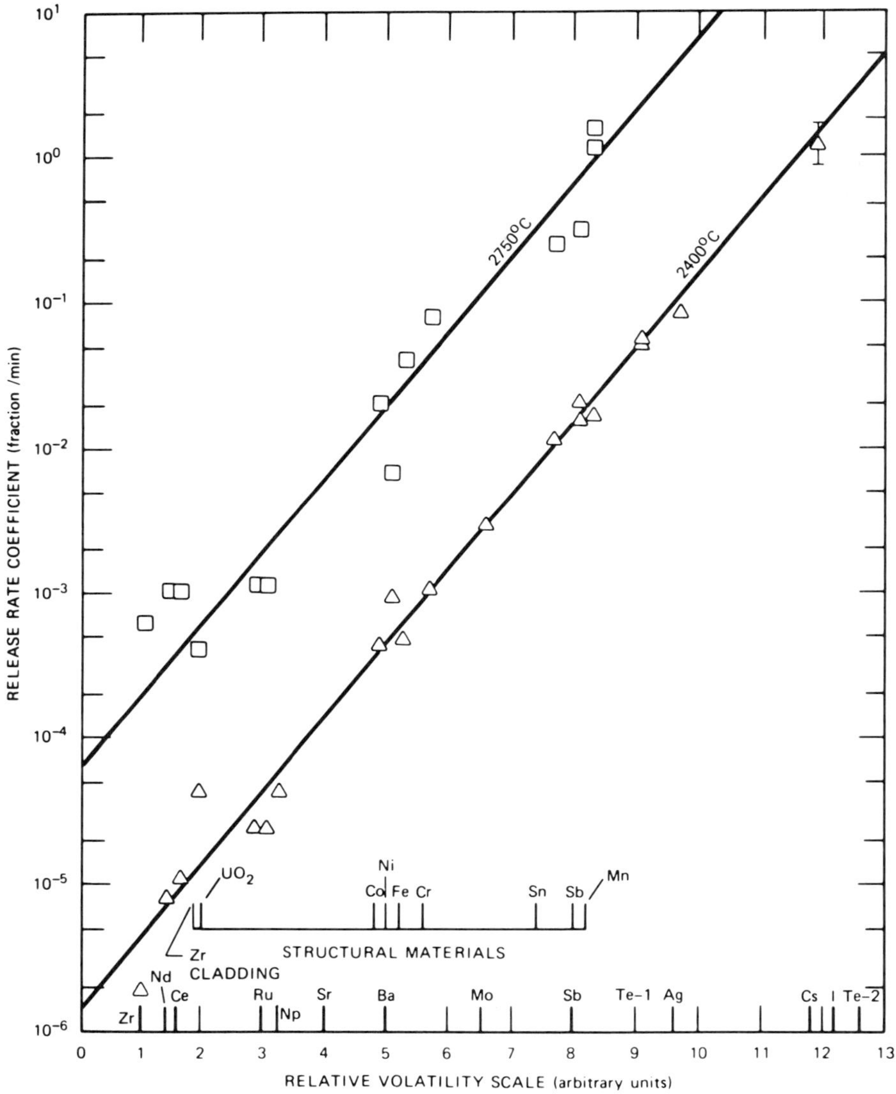

FIGURE 3. Release rate of species in SASCHA tests in air at 2400 and 2750°C.

(Fig. 4) appears to be related to the size of the fuel specimen and the geometry of the tests. The relative release rates for structural material components used in the SASCHA tests were found to be directly proportional to their partial pressures [8].

If the straight line relationship can be considered to be a universally applicable principle (to be proven), then we have a powerful tool,

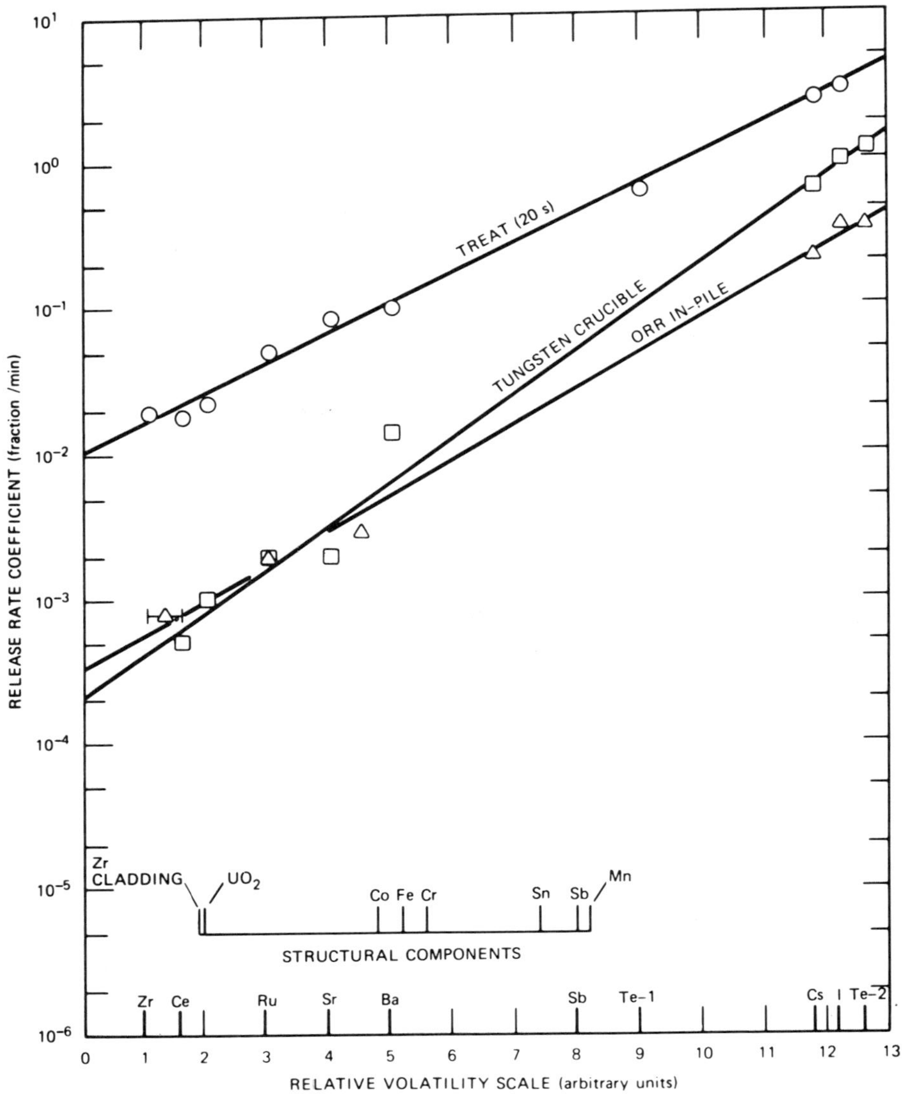

FIGURE 4. Release of fission products from molten UO_2.

indeed. The "relative volatility" relationship permits a semi-quantitative comparison of release results even though the exact time-temperature history may be unknown (as is the case for the TREAT transient heating test data shown in Fig. 4). Because of the way the release rate coefficients were derived for Figs. 3 and 4, it would be expected that plotting the integral release values on the relative volatility scale would also give a straight line. Therefore, we need only to know the release values for two elements (to define the straight line) then we can interpolate to find the release values for all the other constituents knowing only their "relative volatility" values.

We already have one good standard in the D_o, Q values for Cs — so we need only to look for one more. Available data from the most recent HI and VI series of tests conducted in the ORNL Fission Product Release Program have been examined to see if they are adequate for developing values for a second standard. Sufficient data exist to develop D_o and Q values for Cs, Sb, and Kr. However, some judicial selection was necessary. The "relative volatility" value for Kr is too near that of Cs for it to be a good second standard. Unfortunately, the "relative volatility" value for Sb on Figures 4 and 5 primarily come from SASCHA data which were developed under conditions for which considerable unoxidized metallic phases (primarily Zr) were present during the release periods. It is known from recent ORNL results that the presence of unoxidized Zr hinders the release of Sb just as it does Te in the CORSOR-M model. The release rate for Sb when the cladding becomes fully oxidized is much higher, probably around 11 on the relative volatility scale of Figs. 3 and 4: much too close to Cs to be a good second standard. Therefore, to use Sb as the second standard in the proposed technique, the D_o and Q values would have to be developed from tests that reflect the unoxidized Zr conditions under which the position of Sb on the "relative volatility" scale was determined. At this time, it appears that only two of the ORNL tests satisfy this criteria. It is recognized that the integral release values from only two tests of different fuel types do not represent an adequate data base on which to place much confidence in the parameters for the second standard for the proposed model. Nevertheless, we have fit the model to these two tests for purposes of illustrating the technique and for providing interim preliminary standards. The values will need to be refined as additional high temperature data become available.

Therefore, the preliminary values for the use of Sb as the second standard are:

$$D_o = 2.81 \times 10^{-6} \text{ cm}^2/\text{s}$$

$$Q = 82{,}500 \text{ cal/mole.}$$

The simplified fission product release algorithm, then, is as follows:

1. For the given temperature transient, input an average value for the grain radius (6 microns is recommended), or some correlation for grain size growth, and solve the basic model equation for two "standard" species to obtain their fractional release values. Here, for illustration, we recommend Cs and Sb, for which:

Species group	D_o (cm²/s)	Q (cal/mole)
Cs	7.63×10^{-5}	74300
Sb	2.81×10^{-6}	82500

2. Use the "relative volatility", RV, scale as taken from Figure 4 and shown in Table 1 (note that the arbitrary scale originally selected on Figure 4 has been adjusted in Table 1 to give a value of 1.00 for the "relative volatility" for Cs).

TABLE 1. Relative volatility values

Species group (i)	RV(i)
Fission Products	
NG	1.1
Te	1.07 (oxidized cladding)
I	1.03
Cs	1.00
Sb	0.68 (unoxidized cladding)
Ba	0.42
Sr	0.34
Ru	0.25
La	0.14
Ce	0.085
Structural Materials	
Mn	0.69
Sb	0.68
Sn	0.63
Cr	0.47
Fe	0.44
Co	0.41
UO_2	0.17

3. With the release value for Cs, (RCs), and for Sb, (RSb), calculated in (1) and the RV(i) values listed in (2), use the following interpolation formula to calculate the fractional release R(i) for any other species group:

$$R(i) = \frac{RCs}{\exp\left[\frac{\ln(RCs/RSb)\ RV(Cs)}{RV(Cs) - RV(Sb)}\right]} \exp\left[\frac{\ln(RCs/RSb)\ RV(i)}{RV(cs) - RV(Sb)}\right]$$

It is noted here that, in addition to Sb, certain other species, (notably Te and possibly Ru) are held up by unoxidized Zr clad. The above release calculation for such species represents only the release from the fuel to the clad. At some higher clad oxidation level these will then get re-released from the clad to the gas space. In addition, special provisions may have to be made to include the effects of highly oxidizing conditions on other species (particularly Ru and Mo) as well as for the effects of burnup and for conditions in which the resistance in the gas phase controls.

To fully implement this proposed correlation, additional data are needed to establish a better second "standard" and, perhaps, additional refinement of the RV scale. Because of the data sources, the suggested technique is most appropriate for high temperature release.

REFERENCES

1. Rest, J., An Improved Model for Fission Product Behavior in Nuclear Fuel Under Normal and Accident Conditions, *J. Nucl. Mater.*, 120, 195 (1984).

2. Kuhlman, M. R. (BCL), Lehmicke, D. J. (BCL), and Meyer, R. O. (NRC), "CORSOR User's Manual," NUREG/CR-4173, BMI-2122, Battelle's Columbus Laboratories, March 1985.

3. Technical Bases of Estimating Fission Product Behavior During LWR Accidents, NUREG-0772, U.S. Nuclear Regulatory Commission (June 1981).

4. Albrecht, H., and Wild, H., Review of the Main Results of the SASCHA Program on Fission Product Release Under Core Melting Conditions, *Proc. ANS Topl. Mtg. Fission Product Behavior and Source Term Research*, Snowbird, Utah, July 15-19, 1984, Electric Power Research Institute (1985).

5. Osborne, M. F., Collins, J. L., and Lorenz, R. A., Experimental Studies of Fission Product Release from Commercial LWR Fuel Under Accident Conditions, *Nucl. Technol.* 78(2), 157–69 (August 1987).

6. Nakamura, T., and Lorenz, R. A., Effective Diffusion Coefficients Calculated From ORNL Fission Product Release Tests, ORNL Technical Report (in progress).

7. Collins, J. L., Osborne, M. F., Lorenz, R. A., and Malinauskas, A. P., Fission Product Iodine and Cesium Release Behavior Under Severe LWR Accident Conditions, *Nucl. Technol.* 81(1), 78–94 (1987).

8. Lorenz, R. A., The Vaporization of Structural Materials in Severe Accidents, *Proceedings of the International Meeting on Thermal Nuclear Reactor Safety*, NUREG/CP-0027, Vol. 2, August 29 - Sept. 2, 1982.

A Thermodynamic/Mass-Transport Model for the Release of Ruthenium from Irradiated Fuel

FRANK GARISTO
Atomic Energy of Canada Limited
Whiteshell Nuclear Research Establishment
Pinawa, Manitoba, ROE 1L0 Canada

F. C. IGLESIAS and C. E. L. HUNT
Atomic Energy of Canada Limited
Chalk River Nuclear Laboratories
Chalk River, Ontario, K0J 1J0 Canada

1. INTRODUCTION

Some postulated nuclear reactor accidents lead to fuel failures and hence release of fission products into the primary heat transport system (PHTS). To determine the consequences of such accidents, it is important to understand the behaviour of fission products both in the PHTS and in the reactor containment building.

At high temperatures (> 1100 K), chemical reactions are generally fast and chemical equilibrium should be rapidly attained after the fission products are released into the PHTS. For these cases, chemical equilibrium calculations can be used to determine the chemical speciation of the fission products [1,2]. The behaviour of the fission products can then be analyzed in terms of the physical and chemical properties of the predominant fission product chemical species identified in the equilibrium calculations.

Ruthenium metal has a high boiling point and is nonvolatile under reducing conditions. However, under oxidizing conditions ruthenium can form volatile oxides at relatively low temperatures and, hence, could escape from failed fuel and enter the containment building. The ruthenium radioisotope Ru-106 presents a potentially significant health risk [3] if it is released outside the reactor containment building. Consequently, it is important to understand the behaviour of ruthenium during a nuclear reactor accident.

In this paper, we first briefly review the thermodynamic behaviour of ruthenium at high temperatures. The qualitative behaviour of ruthenium, predicted using thermodynamic calculations, is then compared with experimental results from the Chalk River Nuclear Laboratories (CRNL). Finally, a simple thermodynamic/mass-transport model is proposed to explain the release behaviour of ruthenium in a steam atmosphere.

2. THERMODYNAMIC BEHAVIOUR OF RUTHENIUM AT HIGH TEMPERATURES

In this section we describe the thermodynamic behaviour of ruthenium under various conditions. The ruthenium species included in our chemical equilibrium calculations are Ru(s), RuO_2(s), Ru, RuO, RuO_2, RuO_3 and RuO_4. (The letter s indicates that the species is a solid. All the other ruthenium species are gases.)

The results of our equilibrium calculations are presented in the form of ruthenium speciation versus temperature diagrams (see Figure 1). These diagrams give the abundance of a particular ruthenium species as the fraction of ruthenium in the system. The vapour pressures, $p(RuO_n)$, of the ruthenium gases are related to the fractions, f, obtained from the speciation diagrams by the equation

$$p(RuO_n) = f(RuO_n) \times p \times (N_{Ru}/N_T) \quad (1)$$

where p is the system pressure, N_{Ru} is the total amount of ruthenium in the system, and N_T is the total number of moles of gas in the system.

The chemical speciation of ruthenium in a steam atmosphere is shown in Figure 1. Ru(s) is the predominant ruthenium species at all temperatures. RuO_2 is the most abundant gaseous ruthenium species at high temperatures. However, even at 2000 K, only 0.3 percent of the ruthenium is in the gas phase. This corresponds to a vapour pressure of about 10^{-4} atm (1 atm = 101.325 kPa).

The behaviour of ruthenium in steam-hydrogen mixtures was also studied. The presence of hydrogen reduces the volatility of ruthenium. For example, for the release of 0.01 moles of ruthenium into a 50% steam-hydrogen mixture (0.15 moles of steam), the total vapour pressure of the ruthenium gases is only about 10^{-8} atm at 2000 K. This is 10^4 times smaller than for the case shown in Figure 1. RuO is the predominant ruthenium gaseous species in a 50% steam-hydrogen atmosphere.

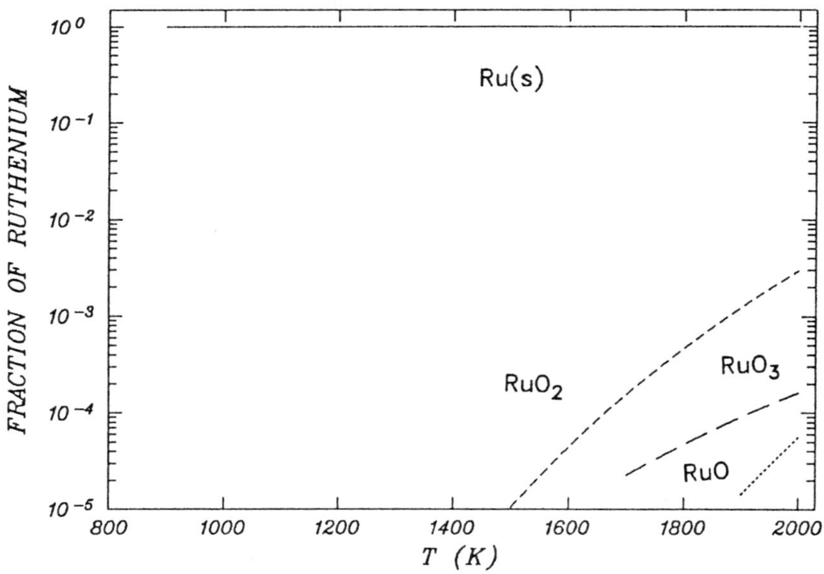

FIGURE 1. Ruthenium speciation diagram for a system containing 0.01 moles of ruthenium and 0.3 moles of steam at 1 atm pressure.

The chemical speciation of ruthenium in a 50% steam-air mixture is shown in Figure 2. The presence of oxygen significantly increases the volatility of ruthenium. Under these conditions $RuO_2(s)$ is the stable ruthenium solid at low temperatures. However, at high temperatures, $RuO_2(s)$ decomposes to Ru(s) even in the presence of air,

$$RuO_2(s) \rightarrow Ru(s) + O_2. \tag{2}$$

Reaction 2 proceeds to the right if the oxygen partial pressure, $p(O_2)$, is less than the dissociation pressure, p_{dis}, of $RuO_2(s)$. The dissociation pressure (in atm) can be calculated using the equation

$$\ln(p_{dis}) = 39.577 - 2.5093 \ln(T) - 38083.1/T \tag{3}$$

The dissociation pressure exceeds 0.21 atm for temperatures above 1695 K. The decomposition of $RuO_2(s)$ leads to a discontinuity in the $p(RuO_n)$ versus 1/T curves. This discontinuity is also apparent in Figure 2.

In irradiated fuel, ruthenium is not found as Ru(s) but is alloyed with other metals [4]. The major components of this metallic alloy are ruthenium and molybdenum, and the minor components are technetium, palladium and rhenium. Does the chemical form of ruthenium in the fuel affect the volatility of ruthenium?

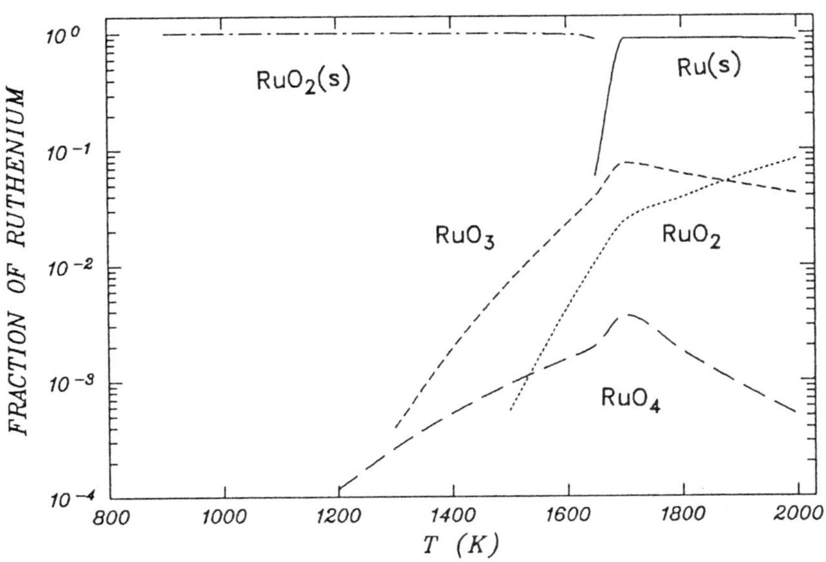

FIGURE 2. Ruthenium speciation diagram for the release of 0.01 moles of ruthenium into a 50% steam-air atmosphere (0.15 moles of steam) at 1 atm pressure. At about 1630 K $RuO_2(s)$ is unstable and decomposes to Ru(s).

For simplicity, we will only consider a Ru/Mo alloy. Molybdenum has a higher affinity for oxygen than ruthenium and is more volatile than ruthenium under oxidizing conditions. For example, a pure steam atmosphere is oxidizing with respect to formation of $MoO_2(s)$, but reducing with respect to formation of $RuO_2(s)$. Thus, in a steam atmosphere, the molybdenum in the inclusions should be oxidized to $MoO_2(s)$, leaving behind Ru(s). Consequently, for the cases of interest in this paper, the ruthenium in irradiated fuel should have approximately the same volatility as Ru(s).

3. RUTHENIUM RELEASES IN THE CRNL EXPERIMENTS

Fission product releases from irradiated UO_2 fuel were measured under various high-temperature conditions at CRNL. In these experiments sample activity was measured using a through-wall collimator which permitted a gamma-ray spectrometer to view the fuel specimens through the side of the furnace. Thus, the release kinetics of fission products from the fuel could be monitored directly.

The release of ruthenium from UO_2 was measured in both air and steam. (Ruthenium was not released when the specimen was in an inert argon atmosphere.) Release rates were estimated from the curves of release versus time, similar to those presented in Figures 3 and 4. For the experiments in air, the initial release rates were calculated since these give a clearer indication of the temperature dependence of the release. For the steam experiments, an average release rate over the duration of the experiment was calculated. Table 1 lists details of samples, test parameters, and calculated release rates.

TABLE 1

RUTHENIUM RELEASE RATES AND TEST PARAMETERS

TEMP (K)	BURNUP (MWh/kg U)	SAMPLE WT (g)	ENVIRONMENT	RELEASE RATE (%/min)
1113	492	0.23	AIR	0.062
1373	522	0.44	AIR	6.48
1573	522	NA	AIR	15.5
1728	492	0.23	AIR	2.6
1883	492	0.69	AIR	10.94
1903	492	0.16	AIR	6.82
1673	492	0.29	STEAM	0.008
1673	512	3.29	STEAM	0.005
1682	512	2.65	STEAM	0.013
1773*	512	2.81	STEAM	0.044
1873*	492	0.96	STEAM	0.050
1941	512	2.65	STEAM	0.093
1963*	512	2.84	STEAM	0.081

*The uncertainty in these temperatures is ± 50 K.

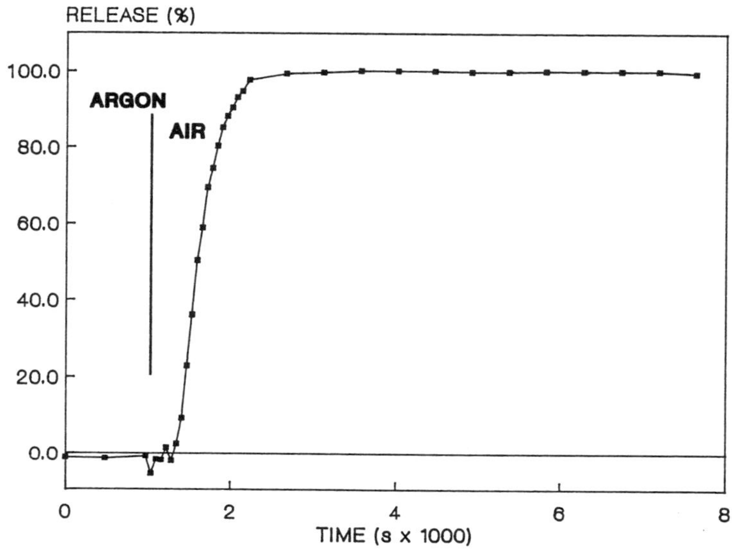

FIGURE 3. Measured releases of Ru-103 versus time from a UO_2 fuel sample heated in air at 1883 K.

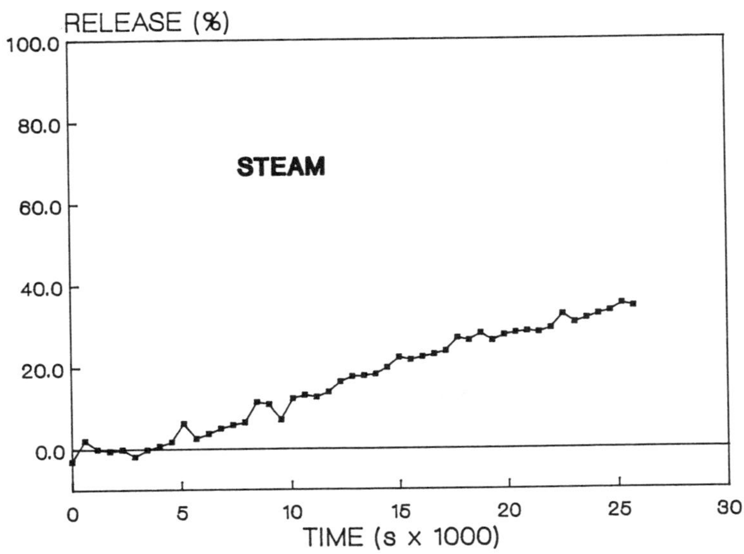

FIGURE 4. Measured releases of Ru-103 versus time from a UO_2 fuel sample heated in steam at 1941 K.

3.1 Releases in Air

Ruthenium releases in air were measured at temperatures above 970 K. Typical results are shown in Figure 3. It is apparent that there is an incubation time between when air reaches the sample and when the release starts. This incubation time is interpreted as the time required for preferential oxidation of the fuel. After the oxygen potential of the UO_{2+x} phase exceeds the corresponding value for Ru(s), oxidation of ruthenium can occur. This interpretation is supported by release tests on fuel samples that were pre-oxidized to U_3O_8 at 770 K. In these latter tests large releases of ruthenium were observed as soon as air reached the fuel sample.

Figure 5 shows an Arrhenius plot of the ruthenium release rates in air and the calculated total vapour pressures of the gaseous ruthenium species. As expected, this latter curve shows a discontinuity at 1690 K, the temperature at which the stable solid changes from RuO_2(s) to Ru(s). The temperature dependence of the release rate data is consistent with the temperature dependence of the calculated vapour pressure.

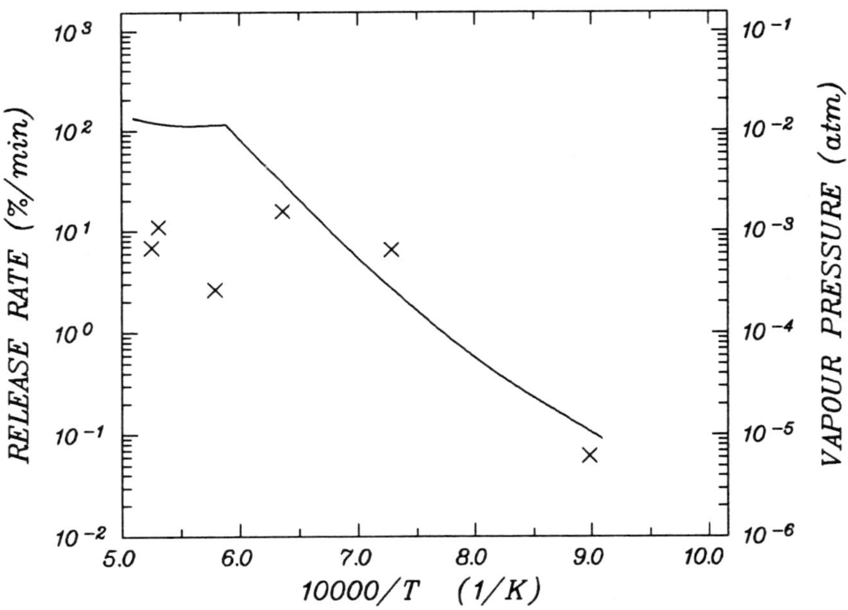

FIGURE 5. Arrhenius plot of ruthenium release rates in air. The total vapour pressure of all the ruthenium gases has also been included for comparison.

3.2 Releases in Steam

The results from a typical test in steam are illustrated in Figure 4. As in the air tests, there is an incubation time between when steam reaches the fuel and the release of ruthenium. However, the release rates in steam are typically two orders of magnitude less than the release rates in air. This strong dependence of the ruthenium release rate on the oxygen partial pressure parallels the strong dependence of the ruthenium volatility on the oxygen partial pressure predicted by the thermodynamic calculations.

Figure 6 shows an Arrhenius plot of the ruthenium release rates in steam and the total vapour pressure of the ruthenium gases in equilibrium with Ru(s) in a steam atmosphere. The temperature dependence of the total vapour pressure shows fair agreement with the temperature dependence of the release rate data.

For both air and steam environments, the slopes of the total vapour pressure curves and the release rate data agree reasonably well. This agreement, to some degree, validates our thermodynamic model and suggests that thermodynamic equilibrium is maintained between the ruthenium gases and the ruthenium solid in the fuel, at least in the boundary layer near the fuel surface.

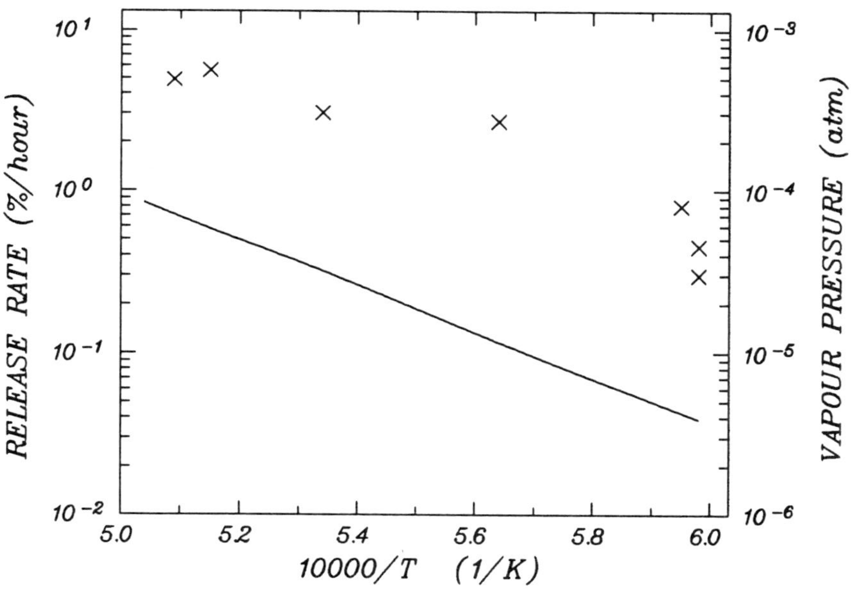

FIGURE 6. Arrhenius plot of ruthenium release rates in steam. The total vapour pressure of all the ruthenium gases has also been included for comparison.

4. MODEL FOR RUTHENIUM RELEASES IN STEAM

In postulated reactor accidents in which the failed fuel is in contact with a large volume of air and/or steam, chemical equilibrium between the fuel and the gas phase is attained only after very long times, i.e., times greater than the duration of the accident. For these accident scenarios mass transport effects are important in the calculation of fission product releases.

Nevertheless, even under the aforementioned conditions, thermodynamic calculations can be useful in predicting the behaviour of fission products. For example, if it is assumed that equilibrium is maintained between the ruthenium in the failed fuel and the gases in the boundary layer surrounding the fuel, then the rate of release of ruthenium, k, can be estimated using [5]

$$k = k_m \, S \, p_{eq} \tag{4}$$

where k_m is a mass transfer coefficient [5], S is the surface area of the failed fuel and p_{eq} is the total vapour pressure of all the ruthenium gases in the boundary layer. Thermodynamic data can be used to determine the vapour pressure of the gaseous oxides RuO_n above either $RuO_2(s)$ or Ru(s) and, hence, the ruthenium release rate.

The ruthenium release rates in steam (see Figure 6) are approximately proportional to the total vapour pressure, p_0, of the ruthenium gases in equilibrium with ruthenium metal in a steam atmosphere, i.e.,

$$k(\%\cdot h^{-1}) \simeq (1.24 \pm 0.60) \times 10^5 \, p_0 \tag{5}$$

where p_0 is in atmosphere units. Can we derive an expression for the proportionality constant in Equation 5?

In the Chalk River experiments, fuel samples are placed in a furnace tube and steam is passed over the fuel. For very low steam flow velocities, v, the atmosphere above the fuel should be saturated with ruthenium gases, i.e., the total pressure of the ruthenium gases in the tube is equal to p_0. In this case the ruthenium release rate is given by

$$k(mol\cdot s^{-1}) = vC_0\pi a^2 = vp_0\pi a^2/RT \tag{6}$$

where C_0 is the concentration of the ruthenium gases above the fuel, a is the radius of the furnace tube, R is the gas constant, and T is the temperature in Kelvin. This release rate can be converted into the measured quantity, given the number of moles of ruthenium in the fuel sample. Fuel samples with a burnup of 512 MWh/kg U contain ~ 0.93×10^{-5} moles of ruthenium per gram of fuel; therefore, for the larger fuel samples,

$$k(\%\cdot h^{-1}) \simeq (1.7 \times 10^8) vp_0\pi a^2/T \tag{7}$$

Using a = 2.8 cm and average values for v ($3.0\ cm\cdot s^{-1}$) and T (1800 K), we obtain $k(\%\cdot h^{-1}) = (6.9 \pm 1.0) \times 10^6 p_0$. This estimated ruthenium release rate is ~ 60 times greater than the experimental value. Therefore, we conclude that the atmosphere above the fuel is not saturated with ruthenium vapours.

Let us now assume that only the gas boundary layer around the fuel is saturated with ruthenium vapours, i.e., the gas concentration at the fuel surface is equal to C_0. For the conditions in our experiments (low flow velocities and small fuel samples sizes) the empirical correlation [5] relating k_m (through the Sherwood number) to the Reynolds and Schmidt numbers is not applicable. Instead we can obtain a crude estimate of the steady-state ruthenium release rate by multiplying the right-hand-side of Equation 7 by the factor $S/\pi a^2$, where S is the "effective" surface area of the fuel sample. S is not easily determined. Assuming that the ruthenium is released predominantly from the top of the fuel samples, we estimate that $S = 1.3 \pm 0.5\ cm^2$ for the larger fuel samples. Therefore, for this ruthenium release model we obtain

$$k(\%\cdot h^{-1}) = (3.7 \pm 1.8) \times 10^5\ p_0 \qquad (8)$$

The theoretical result in Equation 8 agrees reasonably well with the experimental results in Equation 5. We are currently studying alternate ruthenium release models and carrying out further experiments to ensure that the agreement between the theoretical and the experimental results is not fortuitous.

5. SUMMARY

Thermodynamic calculations show that ruthenium can be volatilized at fairly low temperatures under oxidizing conditions. However, during postulated reactor accidents, the ruthenium release rate determines the amount of ruthenium that escapes from failed fuel. Both the experimental and theoretical results suggest that the rate of ruthenium release is proportional to the vapour pressure of the ruthenium gases in the boundary layer surrounding the fuel.

ACKNOWLEDGEMENTS

This work was funded jointly by AECL and the electrical utilities under the COG-CANDEV development program.

REFERENCES

1. Garisto, F., Thermodynamics of Iodine, Cesium and Tellurium in the Primary Heat-Transport System Under Accident Conditions, Atomic Energy of Canada Limited Report, AECL-7782 (1982).

2. Garisto, F., Thermodynamic Behaviour of Ruthenium at High Temperatures, Atomic Energy of Canada Limited Report, AECL-9552 (1988).

3. Eichholz, G.G., Hazards and Control of Ruthenium in the Nuclear Fuel Cycle, Prog. Nucl. Energy 2, 29 (1978).

4. Kleykamp, H., The Chemical State of the Fission Products in Oxide Fuels, J. Nucl. Mater. 131, 221 (1985).

5. Szekely, J., Evans, J.W., and Sohn, H.Y., Gas-Solid Reactions, Academic Press, New York, 1976.

Influence of VVER-1000 Reactor Failure Model on Kinetics of Fission Product Release in Severe Accidents

B. E. VOLKOV, G. A. VOLKOV, N. S. FILL, A. S. SIDOROV, and V. P. SPASSKOV
OKB Gidropress
Podolsk, USSR

ABSTRACT

The necessity of constructing the scenario of the core melting accident on the basis of design criteria is shown. An example of the accident scenario effect on the reactor fuel melt relocation and the radioactivity release time history is given.

The code complex for analyzing reactor system parameters during core melting accidents being at present under development at the organization of the chief designer of VVER type reactors is briefly described.

INTRODUCTION

The development of the code complex to analyse severe core melting accidents is currently underway at the organization of the chief designer of VVER type reactors.

The code complex is intended to support safety measures for the new generation of the reactor system design considering postulated core melting accidents.

The phenomena in a pressurized water reactor during core melting are known to be studied unsufficiently at present. This fact requires to make conservative design decisions. The searching of the most unfavourable scenario of the accident under consideration from the viewpoint of the radioactivity release is one of the problems arising.

ACCIDENT SCENARIO SEARCHING

The typical cases of core melting accidents are the cases of the primary system initially leaky and of the primary system initially tight. The scenarios of the accident for these cases can be built using the criterion of the most rapid radioactivity release from the reactor as the basis.

The largest portion of fission products is known to release during fuel melting. This means that in the case if the primary system is initially tight the most rapid radioactivity release from the vessel can occur as result of the most rapid meltdown of the reactor vessel bottom.

On the other hand, the most rapid fission product release from the reactor occurs if the fuel melt is kept within the core as long as possible for the case if the primary system is initially leaky. This is equivalent to the maximal extension of the fuel melt surface.

It is necessary to construct the event tree taking account of all the possible events and factors to analyze the core components behaviour during the accident. Some of the events and factors within the reactor are as follows:

1. The preliminary core cooling;
2. The remaining water in the reactor;
3. The coolant circulation in the reactor and the primary system;
4. The heat generation axial profile;
5. The location of the heat generation maximum;
6. The fuel melting temperature;
7. The sequence of the fuel and clad melting;
8. The functioning of the upper core grid;
9. The fuel pellet integrity and the retaining of the fuel pellets;
10. The functioning of the core lower grid;
11. The fuel element material relocation before its melting;
12. The core structure position;

Let us carry out the qualitative analysis of the influence of the listed events and factors on the process of core destruction during the accident under consideration from the viewpoint of the two criteria.

The first criterion is the fuel melt reaching and melting down the reactor vessel bottom as rapidly as possible.

The second criterion is the most rapid fuel melt surface extension and the most rapid fission product release from the reactor through the reactor or reactor system leak.

The preliminary core cooling.

It is obvious that the absence of the preliminary core cooling down can be assumed from the viewpoint of both the criteria.

The remaining water in the reactor.

The absence of the remaining water in the reactor can be assumed on the sufficient basis from the viewpoint of the melting through the reactor vessel bottom as rapidly as possible.

On the other hand, the cladding oxidation due to the metal-water reaction would enhance to the more rapid fuel uncovery. As a result of this the space between fuel elements can be filled with fuel pellet wrechage. This factor leads to the extention of the fuel melt surface.

The coolant circulation in the reactor and the primary system

The absence of the coolant circulation can be assumed from the viewpoint of both the criteria.

The heat generation axial profile

The fuel melting starts earlier if the heat generation profile is cosine rather than plain. So the cosine profile can be assumed from the viewpoint of the fuel melt reaching the reactor vessel bottom as rapidly as possible.

The cosine profile can also be assumed from the viewpoint of the fuel melt surface maximal extension because the fuel melting occurs in this case more gradually that enhances to the extension of the fuel melt surface.

The location of the heat generation maximum

The lower portion of the fuel elements melts earlier if the heat generation maximum is at a lower location and there can be less favourable conditions for filling the space between fuel elements with the fuel element material. Therefore, the lower location of the heat generation maximum can be assumed from the viewpoint of the first criterion.

The upper location of the heat generation maximum can be assumed from the viewpoint of the second criterion because the path of the fuel melt relocation becomes longer in this case and better conditions for the extension of the fuel melt surface are created.

The fuel melting temperature

The fuel melting starts earlier at the minimal fuel melting temperature. Therefore, the minimal fuel melting temperature can be assumed from the viewpoint of the second criterion.

On the other hand, the melting of the reactor vessel bottom occurs more quickly with the more hot fuel melt on it.

It can be shown that the decrease of the duration of the vessel melting through as a result of the greater fuel melt temperature would be more than the increase of the fuel melting duration. Thus the greater fuel melting temperature can be assumed from the viewpoint of the first criterion.

The sequence of the fuel and clad melting

The space between the fuel elements would be filled with the fuel wreckage as a result of fuel scattering out of the cladding earlier than the fuel melting occurs if the cladding melts earlier than the fuel (assuming the loss of the fuel pellets integrity and the fuel pellets being not held by the cladding). That enhances to the surface extension of the fuel melt arising later.

The fuel melt reaches the bottom more quickly if the clad melting occurs immediately after the fuel melting.
Thus, the earlier fuel melting can be assumed from the viewpoint of the first criterion and the earlier clad melting can be assumed from the viewpoint of the second one.

The functioning of the upper core grid

The scattering down of the fuel pellet wreckage can take place if the upper core grid holds fuel elements parts remaining above the place of the clad melting (assuming the loss of the fuel pellet integrity and the cladding not holding the fuel pellets). In this case the space between the fuel elements becomes filled with the fuel pellet wreckage. The wreckage can make it difficult to the fuel melt to reach the reactor bottom thus enhancing to the surface extention of the fuel melt flowing through the core space between the fuel elements.

If the core upper grid cannot hold the fuel elements parts remaining above the place of cladding melting those parts can fall down onto the lower parts of the fuel elements. In this case the scattering down of the fuel pellets can be less than in previous case. The fuel will come out of the fuel elements mainly because of its melting when the fuel pellets are assumed to keep their integrity and to be held by the cladding. And if assumed so the fuel wreckage scattering either will not occur when the upper grid holds the fuel elements.

Thus, from the viewpoint of the first criterion it is no matter how the upper grid functionsif the fuel pellets are assumed to be held by the cladding.

From the viewpoint of the second criterion the upper grid should be assumed to hold the claddings (supposing the fuel pellets do not keep integrity and the claddings do not hold the fuel pellets).

The fuel pellet integrity and the retaining of the fuel pellet

The fuel pellet integrity and the retaining of the fuel pellet within the cladding prevent the filling of the space between the fuel elements with fuel wreckage thus facilitating the fuel melt falling down onto the reactor bottom.

The fuel pellets destroying and scattering out of the cladding lead to the filling of the space between the fuel elements with the fuel wreckage which enhances to the surface extension of the fuel melt.

The functioning of the core lower grid

The core lower grid destroying and not holding the fuel elements lead to the blockage of the lower part of the core and the space under the core. That makes it difficult for the fuel melt to reach the vessel bottom.

There are less obstacles for the fuel melt to reach the vessel bottom if the lower grid functions properly and holds the fuel elements.

Thus, from the viewpoint of the first criterion the **proper** functioning of the lower grid and the holding of the fuel elements with it are to be assumed. The lower grid destruction can be assumed from the viewpoint of the second criterion.

The fuel element material relocation before its melting

The continuous removal of the unmolten fuel element material from the space between the fuel elements can be assumed from the viewpoint of the first criterion.

The filling up of the space between the fuel elements with the destroyed fuel elements material can be assumed from the viewpoint of the second criterion.

The core structure position

The core structure drop onto the reactor bottom moves the fuel melt closer to the bottom and enhances to a faster melting down of the bottom.

Thus, the core structre drop onto the reactor bottom can be assumed from the viewpoint of the first criterion, and the core steady position can be assumed from the viewpoint of the second one.

The scenario of the core destruction

The analysis performed shows that from the viewpoint of meeting the first criterion the process of core destroying can be presented as follows. After the maximal heat-powered fuel melting achieved the whole fuel melts. The clad in the corresponding place melts after the fuel melting and the fuel and clad melt relocates to the reactor bottom. Most of the core being melt reaches the reactor bottom after the average heat-powered fuel melts.

From the viewpoint of the second criterion the process of core destroying can be presented as follows. The fuel wreckage scattering down into the space between fuel elements occurs in the melting place of the cladding according to the maximal heat rate. The amount of the fuel wreckage can achieve 50% of the total fuel amount in the core if the cladding melting occurs in the middle of the core . Then the fuel melt flows throughout the fuel wreckage blocking the fuel melt reaching the vessel bottom. Most of the core being melt reaches the vessel bottom after the fuel with minimal heat rate melts.

The core melt reaches the reactor bottom and the reactor bottom melts down earlier in the first case. Only a portion of the fission products releasing from the core can enter the containment as a result of possible operation of the pressurizer safety valves.

The fuel melt surface is well extended in the second case. The strong fission product release into the reactor volume and from there through the reactor system leak into the containment takes already place up to the time of melting of the average heat-powered fuel.

The second scenario case seems to be considered as one of the limiting cases to specify the requirements to the fission product catching systems within the containment.

The analysis performed can serve as a qualitative basis for the computer aided searching of the core melting scenario. Some general ideas about the required codes can be seen from the analysis.

CODE COMPLEX FOR ANALYSIS OF CORE MELTING ACCIDENTS

The code complex fcr analyzing reactor system parameters during core melting accidents is currently developed at the organization of the chief designer of VVER type reactors. A short description of the constituting codes is given below.

The code for evaluating the core state

The code is intended for evaluation of heating up, melting, geometrical changes, the substance and phase compound of the core during core melting. The modelling of the intact fuel, of fuel elements with injured clads, of fuel wreckage, of liquid and resolidified fuel melt and the account of possible gas volumesare provided.

The code for evaluating the reactor vessel state

The code is intended for modelling temperature fields of the vessel during the accident under consideration. The core melt can be placedwithin the core structure, on the lower core grid and on the reactor vessel bottom. The account of the water available in reactor and/or of the water supplied into the reactor is provided.

The code for evaluating temperature fields in a concrete well

The code is intended for evaluating temperature fields of the reactor concrete well. The possibility of the account of the heat removal through a passive heat exchanger is provided.

The code for evaluating core melt and water interaction

The code is intended for evaluating parameters of the medium in the primary system and reactor, dynamical forces upon the reactor internals and vessel during of the process of water coming onto the core melt and during the process of core melt entering the water.

The code for evaluation of reactor vessel melting

The code is intended for modelling the interaction of core melt with the reactor bottom. The account of core melt convection, heat generation and temperature distribution, the determination of structural and physical properties of the core melt, of the components and effects arising during the process, of the parameters of the components leaving reactor through the melted bottom are provided.

The code for evaluation of the process in a dry catcher

The code is intended for modelling the process of interaction of the medium coming through the melted reactor bottom and the structure of a dry catcher, for evaluating parameters of the medium entering the containment.

The code for evaluating the process in a wet catcher

The code is intended for modelling the process of interaction of the medium coming through the melted reactor bottom with the structure and components of a wet catcher, for evaluating parameters of the medium entering the containment. The possibility is provided of the account of two initial catcher states with water and without it.

The code for evaluating lomg-term cooling of the catcher

The code is intended for evaluating the process of long-term cooling of the catcher with core components within it and for evaluating the parameters of the catcher medium entering the containment. The possibility of the account of fuel melt boiling at appropriate conditions in the catcher is provided.

The code for evaluation of the catcher heat removal passive system

The code is intended for calculating the thermal and hydraulic performance and the heat exchange area of the catcher heat removal passive system for the accident considered.

The code for evaluating generation of hydrogen

The code is intended for evaluating generation of hydrogen and conditions of its burning and explosion.

The structure of the code complex

The possibility of the code complex performance with an adjustable structure depending on the problem solved is provided. The input and output data of the constituting codes are mutually agreed and are to be sent across automatically.

CONCLUSIONS

Two scenario cases of the core destruction precess during the severe core melting accident are constructed on the basis of a qualitative analysis meeting two different criteria which are the fastest reaching reactor bottom by the core melt and the maximal fission product release through the reactor system leak directly into the containment. The influence of the accident scenario on the fission product release from the reactor is shown.

The examples given can present an idea of the requirements to the codes for searching scenarios of the accident on the basis of design criteria.

The basic codes constituting the code complex for the analysis of the core melting accidents being developed at the organization of the chief designer of VVER type reactors are briefly described

3. TRANSPORT PHENOMENA IN THE REACTOR COOLANT SYSTEM

LECTURE

Phenomena Occurring in the Reactor Coolant System during Severe Core Damage Accidents

A. P. MALINAUSKAS
NRC Programs Office
Oak Ridge National Laboratory
P.O. Box 2008, MS 6135
Oak Ridge, Tennessee 37831-6135, USA

INTRODUCTION

The reactor coolant system (RCS) of a nuclear power plant consists of the reactor pressure vessel and the piping and associated components that are required for the continuous circulation of the coolant which is used to maintain thermal equilibrium throughout the system. In the event of an accident, the RCS also serves as one of several barriers to the escape of radiotoxic material into the biosphere.

The physical and chemical processes occurring within the RCS during normal operation of the reactor are relatively uncomplicated and are reasonably well understood. When the flow of coolant is properly adjusted, the thermal energy resulting from nuclear fission (or, in the shutdown mode, from radioactive decay processes) and secondary inputs, such as pumps, are exactly balanced by thermal losses through the RCS boundaries and to the various heat sinks that are employed to effect the conversion of heat to electrical energy. Because all of the heat and mass fluxes remain sensibly constant with time, mathematical descriptions of the thermophysical processes are relatively straightforward, even for boiling water reactor (BWR) systems. Although the coolant in a BWR does undergo phase changes, the phase boundaries remain well-defined and time-invariant.

Moreover, much of the focus in normal operation is on the performance of the fuel and the characteristics of the coolant; transport processes involving other species, such as fission products (from leaking fuel rods) or oxidized debris ("crud"), are of secondary importance and their effect on the thermalhydraulics is negligible.

In contrast to normal operating conditions, severe core damage accidents are characterized by significant temporal and spatial variations in heat and mass fluxes, and by eventual geometrical changes within the RCS. Furthermore, the difficulties in describing the system in the severe accident mode are compounded by the occurrence of chemical reactions. These reactions can influence both the thermal and the mass transport behavior of the system. In addition, behavior of the reactor vessel

---

[1]Research sponsored by the Office of Nuclear Regulatory Research, U.S. Nuclear Regulatory Commission under Interagency Agreement DOE 1886-8012-7B with the U.S. Department of Energy under contract DE-AC05-840R21400 with the Martin Marietta Energy Systems, Inc.

internals and of materials released from the core region (especially the radioactive fission products) in the course of the accident likewise become of concern to the analyst.

SEVERE CORE DAMAGE ACCIDENTS

Severe core damage accidents are believed to result typically from the failure of an engineered safety feature (ESF) to respond to a reactor upset condition, either because of failure of the ESF to perform its intended function, or because of its deliberate disablement. As a consequence, a mismatch occurs between the thermal power production of the core and the rate at which the thermal energy is transported by the coolant from the core of the reactor to the appropriate heat sinks. This mismatch may be sudden (such as in the case of a reactivity insertion accident), or it may proceed over a prolonged period of time (a station blackout sequence, for example).

Several stages of a severe core damage accident which involve the RCS can be identified. The first of these is the power-coolant mismatch, i.e., the inability of the coolant to adequately remove the heat being generated within the core region. This results in the heatup of the coolant and its consequent loss through either a breach in the RCS or relief or safety valves. The second stage involves uncovery of the core, heatup of the uncovered portions of the fuel elements, the onset of steam-cladding chemical reaction, and the eventual loss of fuel cladding integrity. It is thus at this point that fission products begin to escape from the fuel rods. The initial releases, however, primarily involve only that fraction of the fission product inventory that is associated with the interconnected void volumes within the rods, that is, the so-called "gap inventory." Hence, the initial releases consist of a small amount of fuel fines and their associated fission products, and small fractions of the inventories of the noble gas and volatile fission products (primarily nuclides of cesium and iodine). Should the accident be terminated at this stage (a "controlled loss-of-coolant accident"), the amount of radioactivity introduced into the environment would be inconsequential [1].

The third stage of a severe core damage accident is characterized by the appearance of molten material and the relocation of core components into the bottom head of the reactor pressure vessel. It also includes interaction of the molten core debris with residual water in the bottom head. Major fractions of the inventories of the risk dominant radionuclides (^{131}I, ^{134}Cs, ^{137}Cs) are introduced into the RCS at this time, and a dense aerosol (comprised mostly of control rod and structural material) is formed.

The fourth stage involves failure of the reactor pressure vessel and the introduction of molten material into the reactor containment. This was previously regarded as the terminus for considerations of phenomena occurring within the RCS. More recently, however, it has been realized that decay heat from radioactive materials that had become deposited on the RCS surfaces during the earlier stages of the accident could heat these surfaces to sufficiently high temperatures that revaporization of some of the deposited material could occur. This, then, is an additional stage of a severe core damage accident that involves the RCS. Moreover, since this stage represents the delayed release of radioactive material from the RCS, its neglect may lead to non-conservative estimates of the consequences of several accident scenarios.

SEVERE CORE DAMAGE ACCIDENT PHENOMENA

The degree of mathematical complexity that is required to describe the physicochemical processes occurring within the RCS, even under steady-state conditions, is determined by the temporal and spatial scales that are of interest, and by the accuracy of the results that are desired. Thus, for example, estimates of the average core exit temperature of the coolant that are adequate for most purposes can be had using only the power rating of the core, the flow rate of the coolant through the core, its core inlet temperature, and the specification of a small number of thermophysical properties of the coolant and RCS structural materials.

The problem would become considerably more complex, however, if the complete specification of the temperature distribution throughout the core were desired. Moreover, the calculational method would have to be extended to include an interdependence between fission product transport within the uranium dioxide fuel matrix and the thermal performance of the fuel elements, since the thermal performance of the fuel rods degrades with time due to the release of fission products into the fuel-cladding gap region. Furthermore, the resultant decrease in gap conductance increases the operating temperature of the fuel, and this promotes additional transport of fission products into the gap region.

In slowly developing accidents (i.e., those not associated with reactivity transients), a mathematical description of the thermophysical processes in the early stages presents no additional complexities relative to their description under normal reactor operation. In fact, if decay heat represents the major mode of energy production (the reactor has been tripped), some simplification results, since the temperature gradients within individual fuel pellets are reduced significantly. Later into the accident, however, the appearance of a second coolant phase in the covered portion of the core does result in considerable complication. As core uncovery occurs, the complexities multiply. One source is due to the initiation of steam-cladding chemical reaction. This reaction has a two-fold effect; first, it introduces a new and significant source of thermal energy, and second, the hydrogen produced in the course of the reaction alters the thermophysical properties of the coolant.

Other complications arise when the fuel rod claddings are breached and fission products are released, and again when reactor vessel temperatures become sufficiently high that structural and control rod materials begin to relocate. This alters the problem from one involving purely thermalhydraulics considerations to one concerning the interdependence of thermalhydraulics and materials behavior.

The initial release of fission products from the fuel rods is almost inconsequential. As the accident progresses, however, their further release and subsequent transport, which are determined by thermalhydraulics factors, begin to influence the thermalhydraulics. An example of this is evident from a consideration of the data presented in Table 1. At reactor shutdown, elements which form volatile or moderately volatile species under severe accident conditions (Kr, Xe, I, Br, Cs, Rb, Te, Se) constitute more than a third of the total decay heat generation of the core. These species are essentially completely released into the RCS as the fuel forms low temperature melting eutectics with other core components or attains temperatures in excess of about 2000°C [2]. One hour into the accident (after reactor shutdown), the radionuclides in volatile form still contribute about a third of the total decay heat generation, and the total power production due to decay heat decreases to less than a quarter of its value at shutdown. Thus, the thermalhydraulics models must account both for a decrease and a redistribution of the thermal power sources over time.

As temperatures in the core continue to rise, less volatile fission product species, such as those which include the nuclides of Ba, Sr, Sb, Ru, etc., likewise begin to be released in significant amounts and thus further deplete the thermal energy production of the (now molten) fuel. These releases are accompanied by the formation, by both mechanical and thermal processes, of a dense aerosol. In the case of most pressurized water reactor (PWR) systems, major components of the aerosolized mass are the constituents of the control rods, i.e., Cd, In, and (possibly) Ag. In addition, for both BWR and PWR systems, a large fraction of the aerosol is comprised of structural components (Fe, Cr, Mn, Sn) [3]. These aerosols present new surfaces for fission product interaction and provide additional modes of transport and deposition of radioactive material within the RCS. Throughout the accident, the coolant and the hydrogen produced from the steam-metal reactions represent the primary motive fluids for aerosol and fission product transport.

Because of the wide variations in temperature that occur within the reactor pressure vessel in the course of a severe core damage accident, materials behavior is influenced by a number of competing processes. Molten species can flow downward in the vessel and resolidify in cooler regions; moreover, the process can be repeated many times, so long as parts of the vessel remain covered with water. Some of the molten material is unoxidized metal (Zr, Fe, Cr, Ni), which can act as a trap for some of the fission products (Te, Sb, Ru), and thus delay their release until the metals become oxidized.

Materials that become airborne may nucleate to form aerosols; a fraction of these aerosols will fall or be driven back into the core region, where they may revolatilize to repeat the release process. The remainder of the aerosol mass may impact stationary surfaces or other particles in the course of their transit through the RCS. Other airborne material may become attached to aerosol particles as a result of physical or chemical processes, become sorbed by stationary surfaces, undergo chemical reaction, or simply be transported by the motive fluid through the RCS.

The chemical processes are no less complex. The presence of metallic zirconium during the early stages of a severe core damage accident, and of hydrogen in the later stages, does result in generally chemically reducing conditions for many species of interest. However, the elevated temperatures involved and the large gradations in temperature throughout the RCS permit the possibility (from a thermodynamics standpoint) of many types of chemical reactions. For example, thermodynamics considerations of the gaseous phase chemical reaction,

$$CsI + H_2O = CsOH + 1/2\ H_2 + I,$$

indicates that CsI could decompose in a steam-hydrogen environment at temperatures characteristic of the core region in a severe accident, provided that there is a preponderance of steam relative to iodine. But as the temperature of the system is reduced, as would occur ex-core, the reaction can be reversed; re-formation of CsI could then occur [4]. In the interim, however, the CsOH may react with other species (such as boron oxide neutron poisons or with RCS metal components) and thus may no longer be available for CsI formation.

The number of chemical reactions that can be identified for the highly complex chemical system provided by the contents of the reactor pressure vessel in a severe core damage accident is prohibitively large. Moreover, most considerations of the chemical aspects tacitly assume rapid kinetics and no limitations on transport, and the latter has recently been shown to have a significant effect on even vaporization processes [5]. Thus, a rigorous, complete mathematical formulation of the chemical processes appears to be intractable; the use of carefully selected experiments, designed to identify those interactions that have major impacts on the consequences of severe accidents, seems to be the more appropriate solution to this problem.

The appearance of molten materials adds another degree of complication, viz., a time dependence of the geometrical parameters. This becomes particularly bothersome when it includes the heat source (i.e., the fuel). And as the molten material falls into residual water in the lower head of the reactor pressure vessel, a new set of phenomena must be considered. These phenomena include core debris-coolant interactions, the possibility and consequences of steam explosions, the release of mechanically-generated aerosols, and an enhancement in the processes causing release of materials from the RCS.

Even prior to this phase, in a PWR, natural circulation patterns are predicted to develop within the primary circuit [6]; these can have a significant effect on the course of the accident, since they provide a process for the loss of RCS integrity at an elevated location relative to the lower head of the pressure vessel. If such a breach in the RCS were to occur, it would preclude the possibility of core melt ejection under conditions of high RCS pressure.

Natural circulation patterns can lead to highly localized heating in the upper regions of the core; this is obvious from examination of the upper gridwork of the Three Mile Island Unit 2 Reactor.

Once the debris falls into the lower plenum of the reactor pressure vessel, and the water that had been present there is removed through vaporization and chemical reaction, ablation of the pressure vessel occurs. This would eventually result in failure of the pressure vessel due to the combined effects of the weight of the debris and the melting of the vessel metal. Alternately, loss of pressure vessel integrity could result from failure of one or more penetrations in the lower head. The manner in which the vessel fails, the nature of the debris, and the mode in which the debris is ejected from the vessel can significantly impact events which occur afterward outside the RCS (and are thus not considered in this paper).

Phenomena occurring within the RCS even after the pressure vessel has been breached can influence the consequences of the accident if they result in additional releases of radiotoxic materials after the reactor containment has been breached. Paramount among these phenomena is the revaporization of material that had become deposited on the RCS surfaces in the earlier phases of the accident. This can occur if the decay heat from the deposited radioactive material is sufficient to elevate the temperatures of the areas of deposition to the point at which significant re-vaporization results. However, the extent to which re-vaporization can contribute to overall accident consequences will depend upon the magnitude of the motive forces at the time re-vaporization occurs. For example, the phenomenon would probably have little impact if gaseous diffusion or (for aerosols) Brownian motion or thermophoresis were the primary modes of transport (as would be the case if the only breach in the pressure vessel were failure of the bottom head). If, on the other hand, the RCS were breached in such a manner as to promote a chimney effect, the re-volatilized material could be more efficiently released into the containment.

Re-volatilization also serves to illustrate how an apparently "conservative" approach can in fact underestimate the consequences of a severe accident. For example, at first glance it would appear to be conservative to ignore deposition in the RCS altogether. While this would certainly maximize the amount of radioactive material introduced into the reactor containment, it would also maximize those removal processes whose rates vary directly with concentration, and would preclude re-vaporization from RCS surfaces from ever occurring. Thus, should the accident involve a breach of the reactor containment at a time late into the accident, this apparently "conservative" approach would overestimate removal processes occurring within the containment prior to its loss of integrity, and underestimate radioactive material concentrations within the breached containment afterwards.

It has been tacitly assumed throughout this paper that intervention of the accident does not occur, i.e., that none of the systems, such as coolant injection systems, which are designed specifically to mitigate severe core damage, could perform their intended functions. Considerations of the operation of these systems, or of other mitigative actions, such as the deliberate depressurization of the RCS, provides additional levels of complexity, and their impact is not always clear even in a global sense.

CONCLUDING REMARKS

It is obvious that sophisticated computational programs are necessary in order to describe the course of a severe core damage accident in sufficient detail that reasonably accurate assessments of its radiological consequences can be made. Consequently, a number of computer codes have been developed to describe the various physicochemical processes that occur in a severe accident. Several of these, and their limitations, are described elsewhere in these Proceedings and will not be discussed in this paper. We conclude, however, with some general observations concerning these computer codes.

First, none of the codes has been developed solely on the basis of "first principles;" nor does such an approach appear to be practical or warranted. As a consequence, however, it has become necessary to rely upon experimental data to guide code development and to obtain values for input parameters. Moreover, some of the input data (constants in constitutive relations, for example) are devoid of physical meaning, whereas others, because they lack an adequate experience base, are left as parameters to be adjusted at the discretion of the code user.

The validity of the results of most of the severe accident computer codes therefore rests primarily with the experience and capabilities of the code user, and not necessarily with the codes themselves; the codes lack rigidity. Code flexibility is of course desirable while the code is in an evolutionary stage and while sensitivity studies are of interest, but the codes should become less flexible in their employment of user-adjusted parameters as the computer programs mature.

A related concern involves code "completeness." Since the codes are semi-empirical in nature, elements of the computer program necessarily reflect only the experimental data base available. In many cases this data base is limited, and extrapolation beyond the base of experience becomes necessary. At the other extreme is the tendency to include all possible physical and chemical processes in developing the computational model, without regard to their significance. This approach, while more rigorous mathematically, can complicate the problem unnecessarily. In both cases, an acceptable solution appears to be an appeal to experiment, i.e., the conduct of carefully selected integral-type experiments, which could test the extrapolation assumptions on the one hand, and the significance of selected phenomena on the other.

Lastly, a major criticism is that virtually all of the computer codes lack a pedigree. As the computer programs were being developed, they were distributed, at various stages in their development, to potential users. In many cases these users developed the codes further, often in parallel and independent of each other. There has thus become a proliferation of modified versions of certain computer codes, with little or no indication that modifications had been made. This renders code verification exercises as suspect.

Some recognition of the problem has occurred, and attempts toward rectification are underway. One of these has been the formation of the International Code Assessment and Applications Program (ICAP), whose purpose is to provide independent assessment (and, hopefully, provide a pedigree as well) of three major thermalhydraulics computer codes: RELAP5, TRAC-PWR, and TRAC-BWR [7]. Such activities should be extended to all of the computational programs currently being used to assess the course and consequences of severe core damage accidents.

TABLE 1. Decay Heat Generation in a 1000MW(e) PWR[a]

	Decay Heat Generation (MW)	
Element	At Shutdown	1 Hour Post-shutdown
Kr	6.50	1.13
Xe	7.65	0.53
I	15.75	7.96
Br	4.92	0.13
Cs	13.87	1.69
Rb	12.87	1.05
Te	6.92	1.79
Se	1.71	0.01
Sb	6.38	0.76
Ba	8.25	1.27
Sr	11.00	2.02
Actinides	9.20	5.30
All others	91.98	19.50
Total	197.0	43.14

[a]Calculated using ORIGEN2 [8, 9].

REFERENCES

1. Lorenz, R. A., Collins, J. L., and Malinauskas, A. P., *Fission Product Source Terms for the Light Water Reactor Loss-of-Coolant Accident*, Nuclear Technology 46, 404, 1979.

2. Silberberg, M., Mitchell, J. A., Meyer, R. O., and Ryder, C. P., Reassessment of the Technical Bases for Estimating Source Terms, U.S. Nuclear Regulatory Commission Report NUREG-0956, 1986.

3. Butland, A. T. D. and Kuhlman, M. R., Factors Affecting Primary System Radionuclide Retention in an LWR Severe Accident, *Proc. Symposium on Source Term Evaluation for Accident Conditions, Columbus, Ohio*, October 28 - November 1, 1985. International Atomic Energy Agency, Vienna, pp. 197-225, 1986.

4. Technical Bases for Estimating Fission Product Behavior During LWR Accidents, U.S. Nuclear Regulatory Commission Report NUREG-0772, 1981.

5. Clough, P. N., Starkie, H. C., and Taig, A. R., Vaporization and Transport of Structural Constituents from LWR Degraded Cores, *Proc. Symposium on Source Term Evaluation for Accident Conditions, Columbus, Ohio*, October 28 - November 1, 1985. International Atomic Energy Agency, Vienna, pp. 177-193, 1986.

6. Sehgal, B. R., Denny, V. E., Stewart, W. A., and Chen, B. C.-J., Effects of Natural Convection Flows on PWR High Pressure Severe Accidents, *Proc. Symposium on Source Term Evaluation for Accident Conditions, Columbus, Ohio*, October 28 - November 1, 1985. International Atomic Energy Agency, Vienna, pp. 293-305, 1986.

7. Ting, P., Hanson, R., and Jenks, R., International Code Assessment and Applications Program. Annual Report, U.S. Nuclear Regulatory Commission Report NUREG-1270, Vol 1, 1987.

8. Croff, A. G., ORIGEN2 - A Revised and Updated Version of the Oak Ridge Isotope Generation and Depletion Code, Oak Ridge National Laboratory Report ORNL-5621, 1980.

9. Croff, A. G., *ORIGEN2 - A Versatile Computer Code for Calculating the Nuclide Compositions and Characteristics of Nuclear Materials*, Nuclear Technology 62, 335, 1983.

The Effects of Phenomena not Modeled by STCP on Fission Product Transport in a PWR System Accident

PETER P. BIENIARZ
Risk Management Associates, Inc.
2309 Dietz Farm Rd., NW
Albuquerque, New Mexico 87107, USA

ABSTRACT

The Source Term Code Package has been in use for close to three years in applications in the US and abroad. However, until the release of MELCOR, there is no readily available means of evaluating the influence of some phenomena, such as natural circulation, heat losses/gains through insulation and fission product behavior after vessel failure, not modeled in the STCP. This paper examines the effects of these phenomena by recalculating the Zion TMLB' accident sequence as described by the sample problem supplied with the STCP. A modified version of the STCP was used to model the missing phenomenology and the results of these analyses are presented.

1.0 INTRODUCTION

Following the accident at TMI-2, the USNRC undertook a major effort to determine why the radiological release from the accident was so much smaller than had been expected. The resulting BMI-2104, (Ref. 1), analyses represented the first attempt at a realistic calculation of radioactive releases from severe accidents.

Subsequent research on the part of the USNRC has resulted in a calculational suite of computer codes known as the Source Term Code Package (STCP). The STCP is a major advance over BMI-2104 in that it incorporates phenomenology previously not explicitly treated. Nevertheless, limitations still exist as this paper demonstrates. The code suite that Risk Management Associates, Inc. (RMA) utilizes to perform source term analyses also had its basis in the BMI-2104 codes but many phenomenological models beyond those presently incorporated in the STCP have been incorporated.

The STCP was released for general usage in late 1986, and the USNRC feels it is an acceptable analytical code suite to perform all Integrated Plant Evaluation, (IPE), analyses. An examination, (Ref. 2), of the STCP has revealed several phenomenological limitations that can cave a significant impact on calculated "back-end" analyses and the source term results. Three specific limitations are discussed in this paper. The first limitation is the adiabatic treatment of the reactor system throughout the analyses. The second limitation is the lack of explicit treatment of natural circulation within the reactor vessel, and the third limitation is the inability of the STCP to perform a post-vessel failure analysis of the deposited fission products on the reactor system internals. The code suite used by RMA to perform source term analyses models this phenomenology and others not present in the STCP. Calculations have been performed to assess the effect of these limitations in the STCP and the results for one accident sequence are presented here.

2.0 MCT2 PHENOMENOLOGY

The advanced code developed and utilized by RMA to analyze fission product transport is MCT2. The primary purpose for MCT2 is to provide a means of developing "back-end" and source term calculations for IPE's and PRA's. A complete description of MCT2 and its models is described in Ref. 2 and will not be repeated here other than to briefly describe the two modifications dealing with insulation heat losses/gains and natural circulation.

Two types of insulation are modeled, aside from the adiabatic treatment of the system. These are the bulk type and the reflective mirror type insulation. Energy transfer is allowed through the insulation to and from the containment, depending on the conditions in the reactor system and the containment. Thus during late phases of core-concrete interaction when the temperature in the containment can be higher than in the reactor system, (especially the BWR MK1 containment) heat transfer and heating up of the reactor system by the containment could occur and result in revolatilization of the deposited materials in the reactor system.

Modeling of the natural circulation phenomenon in MCT2 is a one-dimensional approximation necessitated by the lumped parameter modeling that is characteristic of the BMI-2104 and the STCP code suites. The model applies the Bernoulli equation to describe the flow field between two adjacent control volumes. The driving mechanism that sustains the flow is the buoyancy difference resulting from the temperature differences between the two volumes. The Bernoulli equation together with the buoyancy term and the ideal gas law yields the natural circulation velocity. An assumption is made that there is no pressure difference between the two volumes so that the flow of gases between the two volumes must be balanced. This approach tends to somewhat overpredict the heat transfer but the induced overprediction is well within the overall uncertainty of the calculational methods.

3.0 ANALYSIS

Four sets of analyses were performed and are presented here. The first set, shown in Figure 1, summarizes the results of the straight STCP analysis of the TMLB' accident sequence at Zion in terms of the suspended and deposited fractions of the CsI species. The second set is an analysis performed by MCT2 during which the adiabatic treatment of the reactor system has been replaced by explicitly accounting for heat transfer through insulation to the containment. As in the STCP case, natural circulation has not been accounted for. The results of this analysis are shown on Figure 2. The adiabatic treatment of the reactor system would be expected to result in a lower retention of the species CsI because of the somewhat higher structure temperatures one would expect with adiabatic treatment. Comparison of the CsI suspended and deposited fractions in both cases clearly confirms this effect.

Figures 2 and 3 also show the results of the Zion TMLB' analysis without and with inclusion of the natural convection phenomenon. Clearly, the effect of recirculation into the core region and the mixing in upper plenum is to establish a mixed gas temperature in the reactor vessel. It should be pointed out that for this accident sequence, the through-flow out of the core during the majority of the accident was much lower than the natural convection flow. The gas temperatures in Figure 2 are typical of a through-flow where there is no recirculating flow into the core region. Thus, the temperatures in the reactor system tend to be lower further away from the core. If, however, natural convection is allowed, the recirculating flow back into the core continues to bring energy out of the core region into the upper plenum, the outlet plenum, and the hot leg region, thus increasing the gas temperatures in those regions. The intermixing between the volumes assures that the temperature differences between them are low, as shown in Figure 3. Since there is no natural convection between the surge line and the hot leg due to geometry restrictions, there is no mixing between these volumes, although mixing does exist between the surge line and the pressurizer volumes. It should be mentioned here that the analysis did not consider the flow splitting between the hot leg and the steam generator, although this is an additional capability of the MCT2 code.

Because of the higher gas temperatures in the system, the natural circulation case results in a more severe temperature gradient in the heat structures, as shown in Figures 2 and 3. The higher structure temperatures imply that the volatiles condensed on the structures will more readily revolatilize. This is readily shown by comparing Figures 2a and 3 for CsI behavior. Here, the case without natural circulation shows high retention of the CsI species. The case with

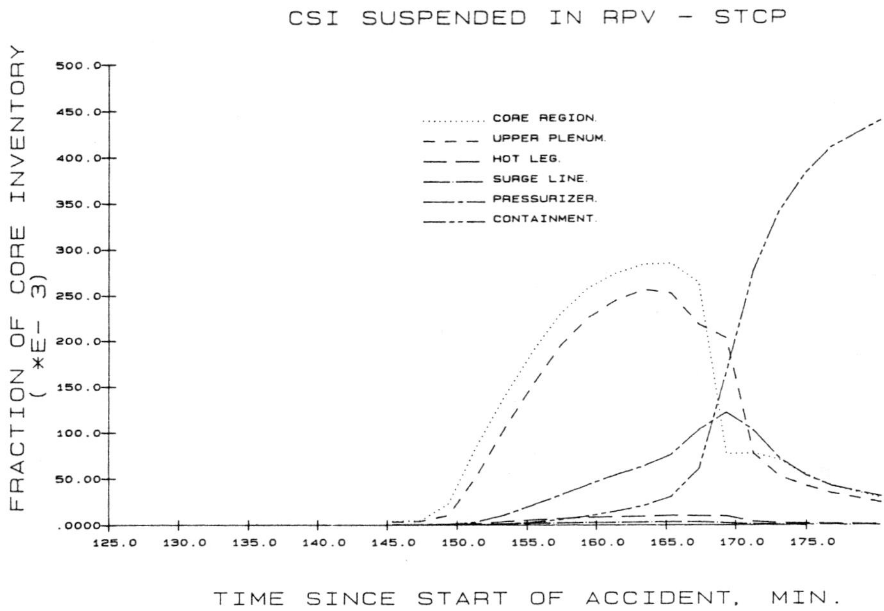

FIGURE 1 — Zion TMLB' STCP Analysis

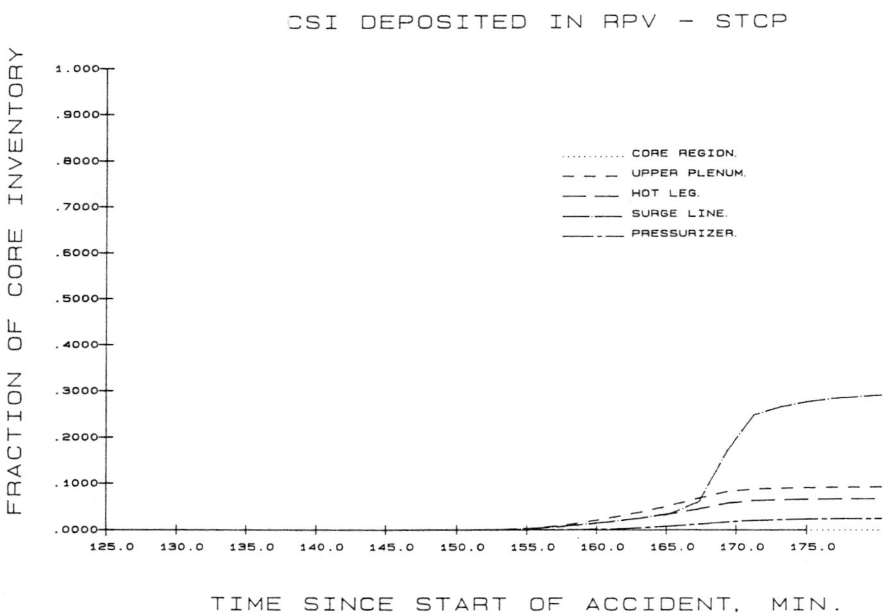

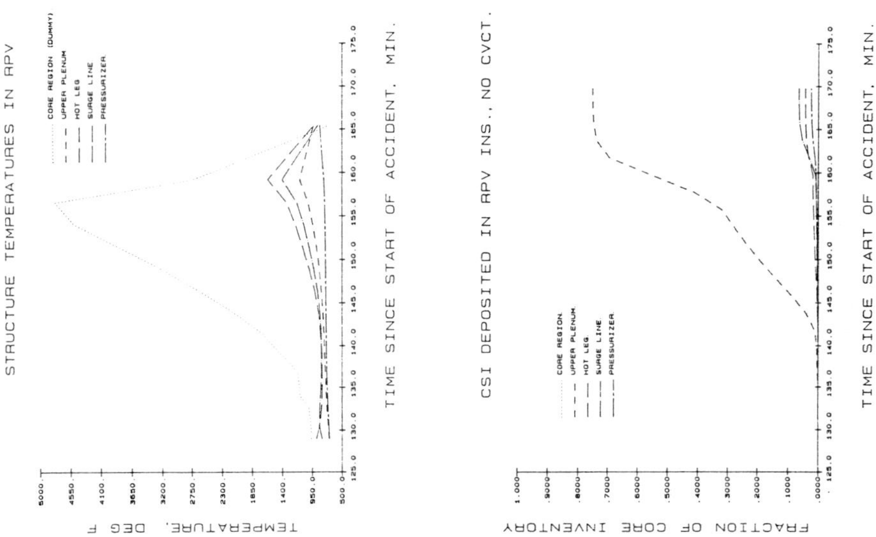

Figure 2—Zion TMLB'MCT2 Analysis--No Natural Cirulation
Heat Transfer Through Insulation is on

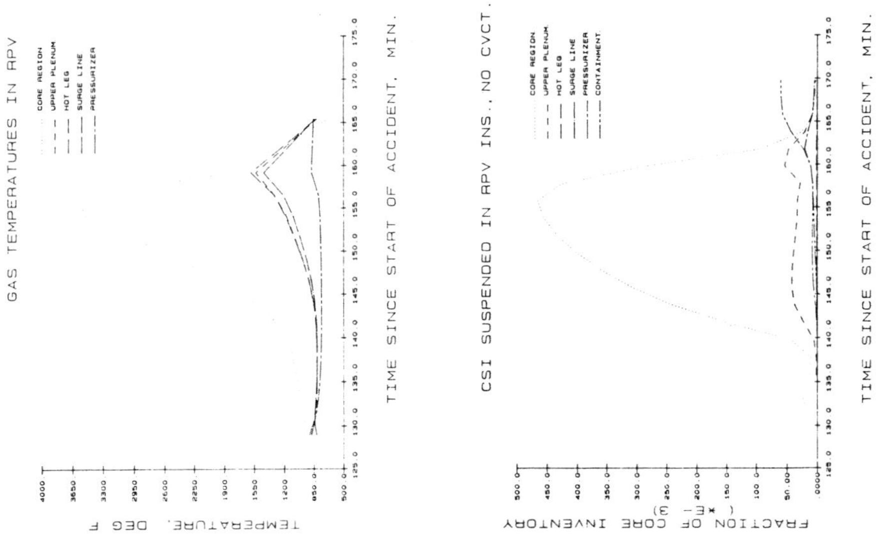

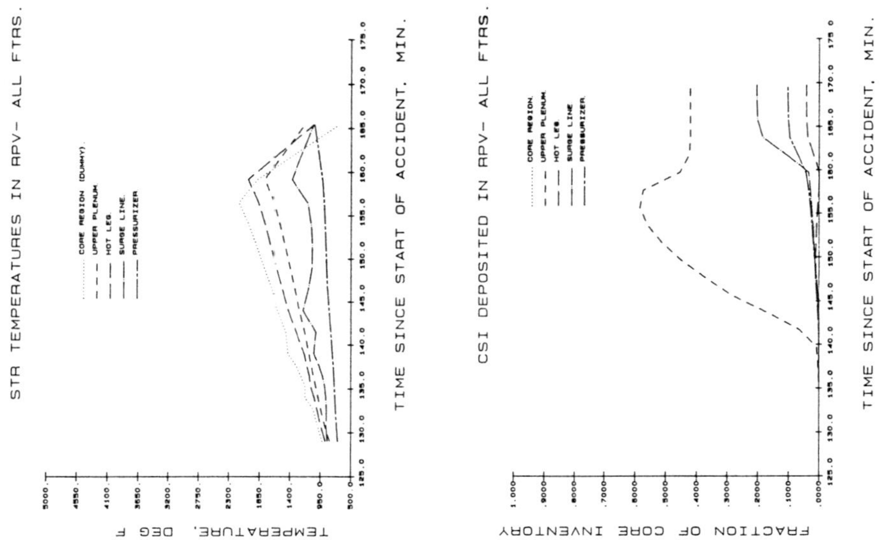

Figure 3— Zion TMLB'MCT2 Analysis--Natural Circulation and Heat Transfer Through Insulation is on

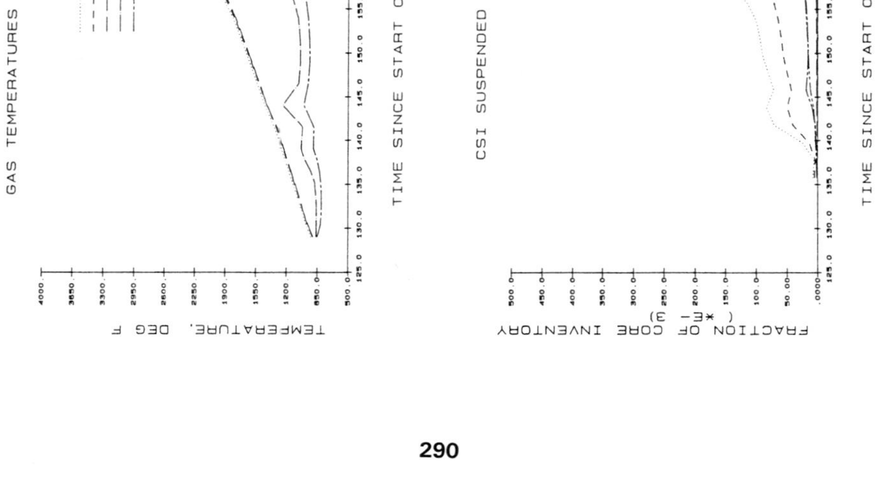

natural circulation results in a significant drop in retention.

Figure 4 shows the effect of carrying the in-vessel source term analyses beyond the vessel failure point. This particular case is an AE case for a BWR. Figure 4 shows that at the time of vessel failure the retention factors of the CsI and CsOH species is high. Here it is seen that the reactor system is very effective in retaining nearly all of the released volatiles. If one were to stop the calculations at this time, a conclusion regarding the release of CsI and CsOH might be made which is not true at all, as illustrated by the next two plots of Figures. Here, it is observed that after the vessel fails, the deposited radionuclides heat the structures on which they are deposited and eventually revolatilize. The liberated vapors travel to the subsequent volumes by natural circulation as is evidenced by the peaks and valleys in the retention plots of Figure 4. Finally, once the materials condense in the upper head and that structure heats up, the revolatilized materials are pushed out of the reactor system through the downcomer region, due to gas heat up and the corresponding pressure increase. The downcomer region remains relatively cool throughout the analyses and commences to heat up only after appreciable amounts of fission products are deposited. Eventually the retention factor of the reactor system becomes that of the downcomer, and as the downcomer heats up it is possible that further revolatilization could occur.

The significance of the post-vessel failure analysis is closely tied to the time of containment failure. Thus, if the containment were to fail at the time of maximum revolatilization in the reactor system, the calculated source term could be greatly affected. The post-vessel failure analysis also depends on the modeling of the natural circulation processes. Clearly, if no natural circulation were modeled in the case presented above, each volume would retain its inventory of volatiles and the calculated retention by the reactor system would be close to 0.

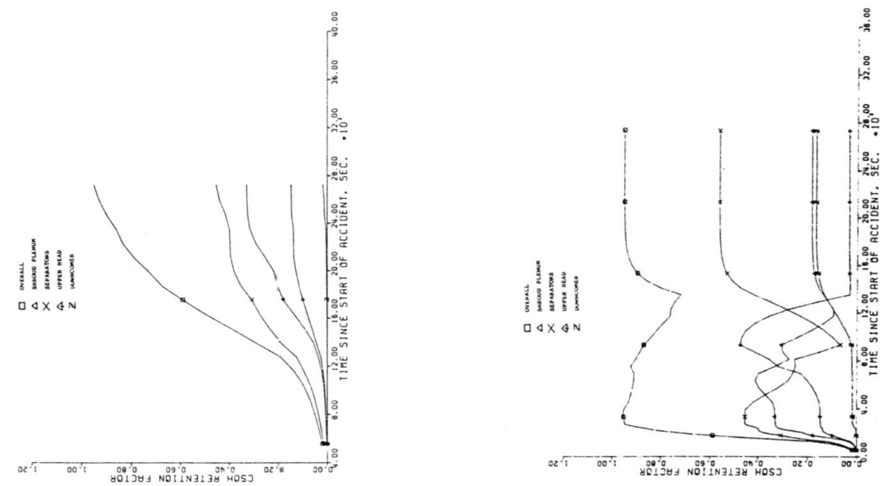

Figure 4—Illustration of the Importnace of Treating Post-Vessel Failure Period, a BWR Analysis

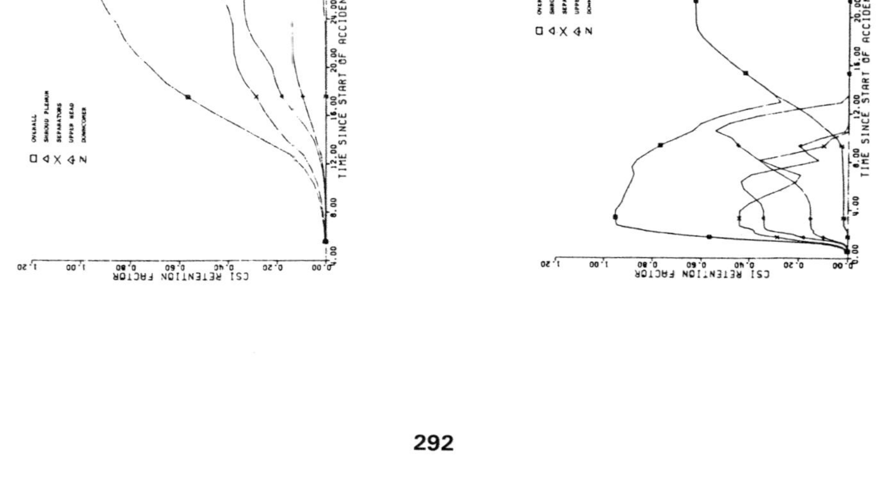

4. CONCLUSIONS

The STP is a major improvement in source term methodology. However, it is important to qualify its use. As was shown the STCP can be applied reasonably well to these accident sequences which would negate the existence of natural circulation, i.e., sequences which have high rates of through-flow during the fission product release period. Such sequences, however, are not the risk dominant sequences which have high rates of through-flow during the fission product release period. Such sequences, however, are not the risk dominant sequences usually found in IPE or PRA analyses. For sequences such as ATWS and TQUV sequences for a BWR or a TMLB sequence for a PWR, significant periods of low through-flow would exist, especially during the period of time when the core is completely uncovered, where natural circulation flow would play an important role in redistributing the fission products to the cooler volumes. In addition, it is rare that any accident sequence would result in an almost complete non-retention of volatiles by the reactor system prior to containment failure. Therefore it is important to analyze the behavior of the in-vessel fission products during the post-vessel failure period with respect to the availability of the volatile vapors at the time of containment failure.

Analyses described here, and others performed using the advanced RMA code suite, lead to the conclusion that it is necessary to include the natural circulation and post-vessel failure analysis of the in-vessel deposited fission products to properly analyze the revaporization phenomena. It is felt that the validity of any source term results could be challenged without inclusion of these processes. If the STCP methodology is to be used to perform "back-end" and source term analysis for IPE's and PRA's, potential users of the STCP should be made aware of these limitations and the possible misleading conclusions to which these limitations may lead.

5.0 REFERENCES

(1) Gieseke, J.A., Cybulskis, P., Denning, R.S., Kuhlman, M.R., Lee, K.W., and Chen, H., "Radionuclide Release Under Specific LWR Accident Conditions," Battelle Columbus Laboratories, BMI-2104 Draft Reports, Volumes I-VII (July, 1983 - February, 1985).

(2) Comments on NUREG-1150, Reactor Risk Reference Document submitted by New York power Authority, James A. Fitzpatrick Nuclear Power Plant, Docket No. 50-333-DPR 59, Indian Point 3 Nuclear Power Plant Docket No. 50-286-DPR-64.

Thermalydraulic Processes in the Reactor Coolant System of a BWR under Severe Accident Conditions

STEPHEN A. HODGE
Boiling Water Reactor Severe Accident Technology (BWRSAT) Program[1]
Oak Ridge National Laboratory
P.O. Box 2008, MS 8057
Oak Ridge, Tennessee 37831-8057, USA

INTRODUCTION

Boiling water reactors (BWRs) incorporate many unique structural features that make their expected response under severe accident conditions very different from that predicted in the case of pressurized water reactor accident sequences [1]. Automatic main steam isolation valve (MSIV) closure as the vessel water level approaches the top of the core would cause reactor vessel isolation while automatic recirculation pump trip would limit the in-vessel flows to those characteristic of natural circulation (as disturbed by vessel relief valve actuation). This paper provides a brief discussion of the BWR control blade, channel box, core plate, control rod guide tube, and reactor vessel safety relief valve (SRV) configuration and the effects of these structural components upon thermalhydraulic processes within the reactor vessel under severe accident conditions. The dominant BWR severe accident sequences as determined by probabilistic risk assessment are briefly described and the expected timing of events for the unmitigated short-term station blackout severe accident sequence at the Peach Bottom Atomic Power Station is presented.

IN-VESSEL STRUCTURES

The internal structure of the BWR reactor vessel is shown schematically in Figure 1. Forced circulation flow during power operation is downward through the jet pumps into the lower plenum surrounding the control rod guide tubes and then upward through the core. As indicated, the normal reactor vessel water level is near the top of the steam separators. Steam generated in the core region passes from the separators through the dryers into the upper vessel head, where it enters one of the four main steam lines. It should be noted, however, that a bypass path around the dryers is opened whenever the vessel water level falls below the bottom of the dryer skirt. Thus, approximately 10% of the flow rising from the damaged core region under severe accident conditions would be expected to bypass the dryers.

[1]Research was sponsored by the Office of Nuclear Regulatory Research, U.S. Nuclear Regulatory Commission under Interagency Agreement DOE 1886-8045-2B with the U.S. Department of Energy under contract DE-AC05-84OR21400 with the Martin Marietta Energy Systems, Inc.

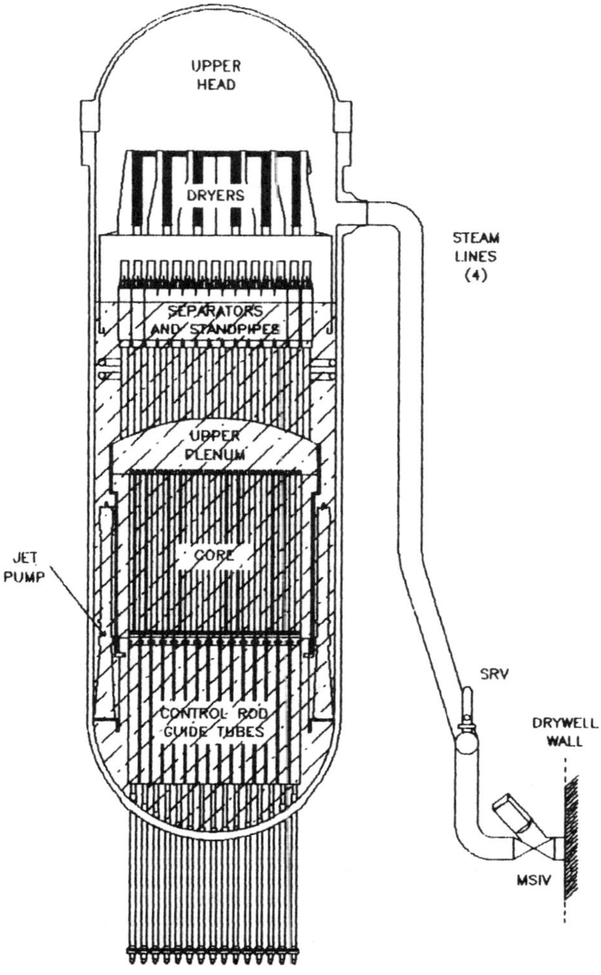

FIGURE 1. The major internal components of the BWR reactor vessel are the control rod guide tubes, the core, the upper plenum, the standpipes and steam separators, and the dryers. Shading indicates the volume of liquid water under normal operating conditions.

SAFETY RELIEF VALVES

The BWR reactor vessel is protected from overpressure by multiple (eleven at Peach Bottom) safety relief valves (SRVs) mounted on the horizontal piping runs of the main steam lines between the vessel and the inboard MSIVs as indicated on Fig. 1. The SRV tailpipes (not shown) pass from the lower drywell into the wetwell and terminate in discharge devices (T-quenchers) well underwater in the pressure suppression pool. The SRVs actuate automatically to release steam from the reactor vessel whenever the vessel pressure reaches their setpoint (about 7.58×10^6 N/m^2), then reseat after a blowdown of about 52×10^4 N/m^2. Alternatively, as long as normal or battery power remains available, the SRVs can be remote-manually actuated by the control room operators as necessary to control the reactor vessel pressure within any desired lower band. The steam released from the reactor vessel via the SRVs is condensed in the pressure suppression pool and any fission products carried with the steam under severe accident conditions are subject to retention by the pool. BWRs have an Automatic Depressurization System (ADS) designed to rapidly depressurize the reactor vessel under accident conditions should the vessel water level approach the top of the core (due to failure of the high-pressure injection systems) while the low-pressure injection systems are running and capable of injection into a depressurized vessel. ADS actuation under such conditions would rapidly permit the low-pressure injection systems to flood the vessel and terminate the accident sequence without core damage. However, for accident sequences that involve loss of the low pressure injection systems as well as the high pressure systems (such as Short-Term Station Blackout), automatic vessel depressurization would not occur. Under these circumstances, the BWR Owners Group Emergency Procedures Guidelines (EPGs) direct the control room operators to manually actuate the ADS when the reactor vessel water level has fallen to about one-third core height [2].

The effect of manual ADS actuation in accordance with the EPGs is shown in Figure 2. It is important to recognize that for the case without reactor vessel injection available, the EPG instructions with regard to manual ADS actuation have been carefully crafted by the BWR Owners Group to cause the control room operators to depressurize the reactor vessel precisely at a time when most of the core is uncovered and at elevated

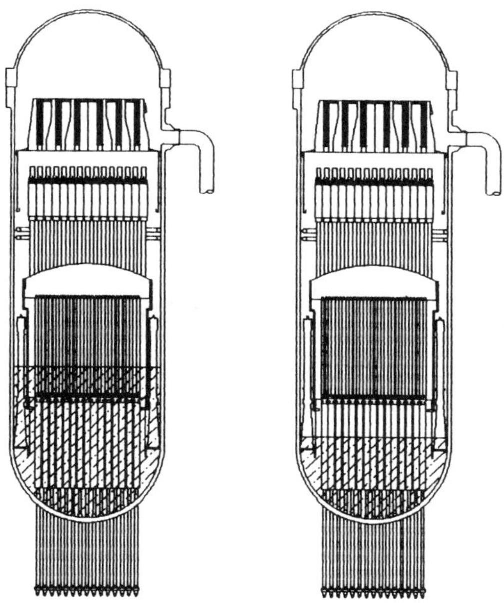

FIGURE 2. Flashing attendant to manual ADS actuation with vessel water level at 1/3 core height lowers the water level into the lower plenum, but does not uncover the jet pump exits.

temperatures, but none of the core is at temperatures sufficient for runaway zirconium oxidation. Thus, the effect of the sudden depressurization is to cool the previously uncovered regions of the core. For Short-Term Station Blackout, calculations indicate that the hottest regions of the uncovered core are cooled by about 300 K and that about 30 minutes of time are gained by this maneuver, before core temperatures again approach runaway zirconium oxidation levels [3].

CONTROL BLADES

Multiple control blades (185 at Peach Bottom) are employed in the BWR design. The control blades are driven hydraulically upward into the core region from the reactor vessel lower plenum, where they are housed in control rod guide tubes that occupy approximately two-thirds of the lower plenum volume beneath the core. Each control rod guide tube supports four fuel assemblies in the core region. Each fuel assembly is comprised of 62 fuel rods and two water rods, and is cooled by an upward channeled flow that is guided by a square canister or channel box as shown on the right side of Figure 3. When the control blades are inserted into the core, they occupy the interstitial region between the channel boxes, as shown in the center of Figure 3.

The control blade neutron poison is B_4C powder, stored within the neutron absorber rods located within the control blade sheaths as indicated on the left side of Figure 3. The presence of B_4C powder within the core region has two important ramifications with respect to the BWR response under severe accident conditions. First, the stainless steel sheath and absorber rod walls have a lower melting temperature than the Zircaloy of the channel box walls and fuel cladding. Thus, severe accident calculations predict control blade structural failure and relocation while the fuel assemblies remain standing in the uncovered region of the core [3]. This early relocation of the control blade was also observed in the DF-4 experiment in the ACRR at Sandia National Laboratories [4] and clearly raises the question of recriticality should reactor vessel water injection capability be restored after partial core degradation has

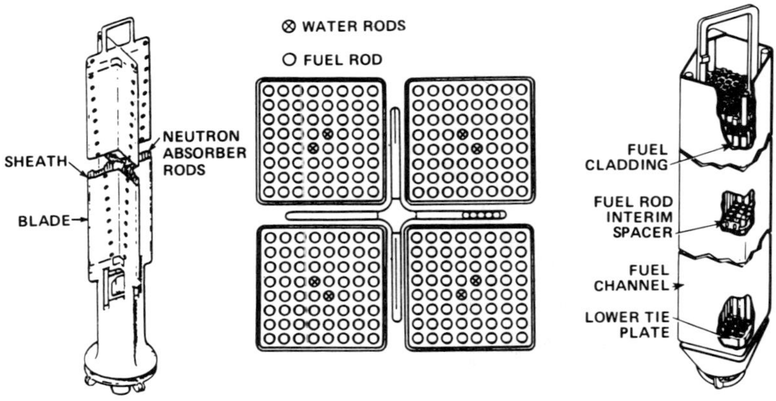

FIGURE 3. Arrangement of control blades and channel box fuel assemblies within the BWR core.

occurred. The effect is aggravated by the tendency of the B_4C powder to form a lower-melting-temperature mixture with the surrounding stainless steel, causing an even earlier blade relocation than that which is predicted by consideration of the melting temperature of stainless steel (1672 K).

The second important ramification of the presence of B_4C powder in the BWR core is that the reaction of B_4C with steam produces, among other gases, methane [5]. To recognize the significance of this, it should be recalled that under accident conditions, reactor vessel gases would be passed to the pressure suppression pool via the SRVs. By this means, volatile fission products in particulate form released from the fuel after cladding failure would be carried with the flow of steam into an underwater release in the pressure suppression pool, where the steam would be condensed. The effectiveness of the pool in retaining CsI is well known [6], but the addition of methane from the B_4C-steam reaction would induce transformation of some of the iodine into the much more volatile methyl iodine form. Therefore, it is most important that the reaction of B_4C powder with steam be considered in any BWR severe accident thermal hydraulic calculation that is to provide boundary conditions for a fission product transport analysis.

CORE DEGRADATION AND RELOCATION INTO THE LOWER PLENUM

The chief result of the manual ADS actuation as directed by the EPGs for BWR accident sequences without reactor vessel injection available is that structural deformation and downward relocation of molten control blade, channel box, candling clad, and fuel (in that order) would occur within a totally dry region above the core plate [3]. What subsequently happens to the core plate is a matter of great importance in predicting the progression of a BWR severe accident.

As indicated in Figure 4, the BWR core plate separates the core region from the very large lower plenum, where the control rod guide tubes that hold the withdrawn control blades during power operation are located. It should be noted that the core plate does not support the core, but merely ensures proper lateral alignment of the upper portions of the control rod guide tubes. Clearly, if the relocating molten material falling on the dry core plate causes local core plate failure, then the molten debris would fall into the lower plenum and be quenched, forming an underwater debris bed (there is more than enough water in the BWR lower plenum to quench an entire molten core). On the other hand, if the core plate remains intact, then the BWR debris bed would be expected to form above the core plate, in a fashion similar to that observed in the Three Mile Island accident (PWR). Thus, it is most important to consider the fate of the core plate in BWR severe accident analyses.

DOMINANT BWR SEVERE ACCIDENT SEQUENCES

The dominant BWR severe accident sequences leading to core melt are Station Blackout and Anticipated Transient Without Scram (ATWS). For Peach Bottom, the recently completed Probabilistic Risk Assessment (PRA) performed for the U.S. Nuclear Regulatory Commission (NRC) by the Accident Sequences Evaluation Program (ASEP) assigns 60% of the risk to Station Blackout, 31% to ATWS, and 9% collectively to all other possible accident sequences [7]. However, there are two forms of Station Blackout [8].

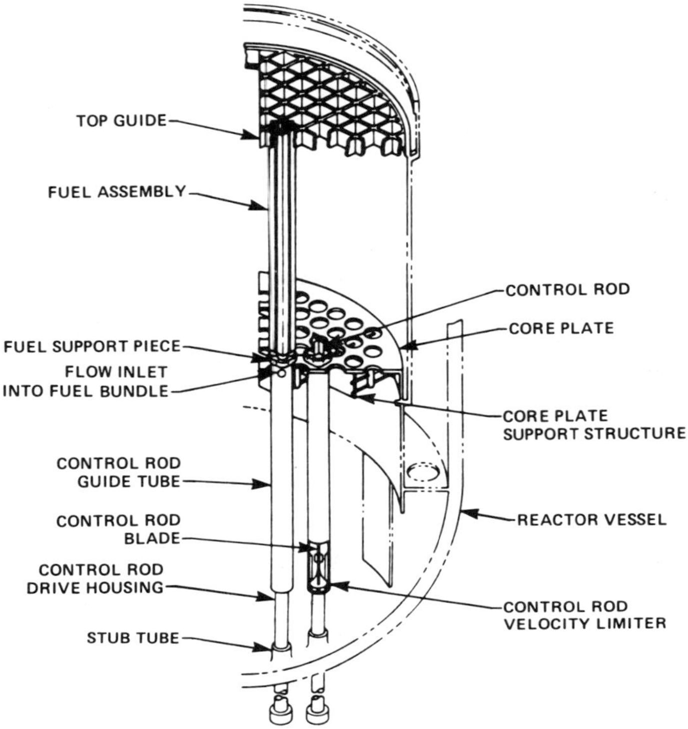

FIGURE 4. The BWR core plate separates the core region from the reactor vessel lower plenum. The core fuel assemblies are supported by the fuel support pieces, the control rod guide tubes, the control rod drive housings, and the stub tube welds.

Long-Term Station Blackout implies maintenance of reactor vessel injection capability by means of steam turbine-driven systems until exhaustion of the plant batteries (6-8 hours after scram), whereas Short-Term Station Blackout implies independent failure of the steam turbine-driven injection systems and provides the most rapid progression of events to core degradation, reactor vessel failure, and fission product escape from primary containment. It should also be noted that the reactor vessel would be depressurized during the periods of core degradation and bottom head failure in Short-Term Station Blackout whereas battery failure would preclude manual SRV actuation during these periods for the long-term case.

Severe accident analysis codes sponsored by the NRC for the purpose of providing specific models for representation of the sequence of events within a BWR reactor vessel under severe accident conditions include the APRIL code developed under joint sponsorship with ESEERCO at Rensselaer Polytechnic Institute [9], the BWRSAR code developed by the BWRSAT Program at Oak Ridge National Laboratory [10], and the BWR portion of the MELCOR code currently under development at Sandia National Laboratories [11]. Close coordination has been maintained over the years between these BWR code development activities and it is intended that all of the capabilities of APRIL and BWRSAR will be available in MELCOR when it is released for general use and becomes the NRC flagship code for general severe accident applications (both BWR and PWR). In the meantime, the BWRSAR code is being extensively applied in BWR station blackout studies. In the next section, results of recent BWRSAR calculations of the short-term station blackout accident sequence are discussed.

SHORT-TERM STATION BLACKOUT

The recent ASEP results assign 30% of the total risk of Peach Bottom core melt to the short-term station blackout accident sequence [7]. The estimated timing of events for this relatively fast-moving accident sequence as recently calculated with the BWRSAR code at Oak Ridge National Laboratory [3] is provided in Table 1. Since, by definition, all forms of reactor vessel injection are lost at the inception of the Short-Term Station Blackout, this calculation produces the shortest time to core uncovering of any accident sequence not involving ATWS or LOCA. Control room operator manual actuation of the ADS is assumed to be in accordance with the EPGs and occurs about 80 minutes after scram. It should be noted that core plate dryout is predicted to occur almost immediately after ADS actuation; however, because of the concomitant steam cooling of the uncovered regions of the core, the onset of debris relocation is delayed for some 3000 s. Since the core plate is dry when debris relocation does occur, the relocating molten material immediately induces local core plate failure.

TABLE 1. Timing of events for Peach Bottom short-term station blackout

Event	Time after scram, s
Swollen level below top of core	2 412
ADS system actuation	4 800
Core plate dryout	4 854
Debris relocation begins	7 944
First local core plate failure	7 962
Central fuel column collapse	13 368
Lower plenum dryout	15 294
Bottom head penetration failure	15 300

The establishment of natural circulation loops within the BWR reactor vessel after the vessel water level falls below the core plate is prevented by the action of the SRVs. In Short-Term Station Blackout, the valves associated with the ADS (5 at Peach Bottom) are opened when the system is manually actuated; subsequently the prominent streamlines within the reactor vessel are directed toward the main steam lines and release to the pressure suppression pool via the SRVs.

The very pronounced falloff of the radial power profile in the outer regions of the BWR core also has significant effect in differentiating the events of a BWR severe accident sequence from those that might occur in a PWR under similar circumstances. Indeed, in the current calculation, the outer 20% of the Peach Bottom core is predicted to be undamaged at the time that collapse of the central fuel column occurs. Because of the reduced power at the core periphery, events in the outer regions of the BWR core would lag far behind the situation in the inner core. This time lag in the onset of fuel degradation is sufficient that collapse of the outer regions is predicted by BWRSAR only after the central core has relocated into the vessel lower plenum, boiled off the water there, and caused thermally-induced failure of the control rod guide tubes that support the outer region fuel assemblies.

The initial release of fission products from BWR fuel under severe accident conditions would occur at higher cladding temperatures than for PWR fuel under similar conditions [12]. This is because the pressure differential across the cladding would be much less for the BWR. The internal BWR fuel rod pressure is generated by the heatup and expansion of the initial helium fill gas (30.4×10^4 N/m^2 at room temperature for an 8×8 P fuel rod) plus the fission gas released to the rod plenum and void spaces during the previous periods of normal reactor operation. If the reactor vessel remains pressurized as in Long-Term Station Blackout, then the internal fuel rod pressure would never exceed the vessel pressure and cladding failure would not occur until cladding temperature had increased to the range of 1450-1550 K, so that temperature-induced embrittlement causes failure by means of cracking or fragmentation.

For Short-Term Station Blackout, however, the reactor vessel would be depressurized by actuation of the ADS and the internal fuel rod pressure would exceed the reactor vessel pressure by about 106×10^4 N/m^2 at the time of cladding failure. With this pressure differential, fission product release from fuel would be expected to begin within the core whenever the maximum cladding temperature reached 1280 K. This occurs at time 5 994 s for the accident sequence described by Table 1.

The pathway from the reactor vessel for fission products released from fuel involves passage through the upper plenum and the separators and standpipes, whose location within the reactor vessel is identified in Figure 1. Most of the escaping fission products would also pass through the dryer assembly, although there is a small bypass path around the dryers at the low water levels associated with severe accidents, and approximately 10 percent of the flow would follow this bypass to the main steam lines.

Since the upper reactor vessel structures have large surface areas, it is important to consider the potential that they offer for fission product deposition. This depends to a large extent on their temperatures during the periods that fission products are being carried through these structures as part of the gas streams flowing toward the main steam lines and the open SRVs. Experience with BWR severe accident calculations at Oak Ridge has shown that the general range of upper structure temperatures is accident sequence-dependent. For Short Term Station Blackout, these temperatures are predicted to remain relatively low, increasing from approximately 775 K to 1425 K during the period of rapid release of volatile fission products, which occurs between time 7 500 and 11 700 s for the accident sequence described by Table 1.

SUMMARY

It has been possible in this short paper to identify and briefly discuss only the major unique characteristics of BWR severe accident response that would affect considerations of in-vessel fission product transport. BWRs are very different from PWRs and the important differences in structure, modes of pressure control, and emergency operating procedures must be considered in performing severe accident and fission product transport analyses. In particular, it is absolutely necessary that the effects of SRV actuation upon the heatup of the core and upon the establishment of flow patterns within the reactor vessel be included in any BWR severe accident analysis that is intended to be realistic. Furthermore, for fission product transport calculations, the reaction of the B_4C control rod powder with steam and the attendant production of methane must also be considered.

Of the dominant BWR accident sequences leading to core melt, Short-Term Station Blackout involves the shortest period of time between reactor shutdown and the onset of core degradation and fission product release from fuel. For this reason, much of the NRC-sponsored severe accident analytical effort for BWRs has been focused upon this important accident sequence. The BWR-specific models developed at Rensselaer Polytechnic Institute [9] and Oak Ridge National Laboratory [10] as necessary for use in ongoing severe accident studies have all been provided to the formal code development programs at Sandia National Laboratories. In the meantime, the best available estimate of the sequence of events for Station Blackout is believed to be achieved by application of the BWRSAR code at Oak Ridge.

REFERENCES

1. Greene, S. R., Realistic Simulation of Severe Accidents in BWRs — Computer Modeling Requirements, NUREG/CR-2940, ORNL/TM-8517, April 1984.

2. The General Electric Company and the BWR Owners Group, EMERGENCY PROCEDURES GUIDELINES, BWR 1 through 6, Revision 3, December 1982.

3. Hodge, S. A., Hyman, C. R., and Ott, L. J., Primary Containment Response Calculations for Unmitigated Short-Term Station Blackout at Peach Bottom, NUREG/CR-5317, ORNL/TM-11057, Forthcoming.

4. Gauntt, R. O., and Gasser, R. D., The DF-4 BWR Control Blade/Channel Box Experiment, NUREG/CR-4671, SAND86-1443, Forthcoming.

5. Beahm, E. C., Wichner, R. P., and Weber, C. F., Chemical Factors Affecting Fission Product Transport in BWR Severe Accidents, *Proc. Twelfth Water Reactor Safety Research Information Meeting, Gaithersburg*, Vol. 3, pp. 109-121, 1983.

6. Wichner, R. P., Weber, C. F., Hodge, S. A., Beahm, E. C., and Wright, A. L., Noble Gas, Iodine, and Cesium Transport in a Postulated Loss of Decay Heat Removal Accident at Browns Ferry, NUREG/CR-3617, ORNL/TM-9028, August 1984.

7. Kolaczkowski, A. M., Cramond, W. R., Daniel, S. L., Maloney, K. J., Sype, P. T., and Wheeler, T. A., Analysis of Core Damage Frequency: Peach Bottom, Unit 2 Internal Events, NUREG/CR-4550 Rev. 1, Vol. 4, SAND 86-2084, Forthcoming.

8. Greene, S. R., Harrington, R. M., Hodge, S. A., Hyman, C. R., Ott, L. J., Sozer, A., Station Blackout at Browns Ferry Unit One, NUREG/CR-4571, ORNL/TM-10011, Forthcoming.

9. Kim, S. H., Kim, D. H., Koh, B. R., Pessanha, J., SiAhmed, El-k, Podowski, M. Z., Lahey, R. T., The Development of APRIL.MOD2 — A Computer Code for Core Meltdown Accident Analysis of Boiling Water Nuclear Reactors, NUREG/CR-5157, ORNL/Sub/81-9089/3, July 1988.

10. Hodge, S. A., and Ott, L. J., Boiling Water Reactor Severe Accident Response (BWRSAR) Code Description and Assessment, NUREG/CR—5318, ORNL/TM 11058, Forthcoming.

11. Haskin, F. E. and Webb, S. W., Development and Status of MELCOR Code, *Proc. Fourteenth Water Reactor Safety Research Information Meeting, Gaithersburg*, 1986.

12. Wichner, R. P., Davis, W., Weber, C. F., Hodge, S. A., Lorenz, R. A., Mitchell, A. D., Station Blackout at Browns Ferry Unit One — Iodine and Noble Gas Distribution and Release, NUREG/CR-2182, Vol. 2, ORNL/NUREG/TM-455/V2, pp. 29-35, August 1982.

Fission Product Transport during In-Vessel Progression of PWR Severe Accidents

B. R. SEHGAL
Electric Power Research Institute
P.O. Box 10412
Palo Alto, California 94303, USA

A. T. WASSEL and D. C. BUGBY
Science Applications International Corporation

R. L. RITZMAN
Electric Power Research Institute
P.O. Box 10412
Palo Alto, California 94303, USA

ABSTRACT

Fission product transport processes in the primary coolant system (PCS) of a PWR depend on the prevailing flow and temperature fields. In turn, the self-heating fission products on deposition and revaporization, may affect the prevailing flow and temperature distributions. Thus, thermal hydraulic and fission product transport processes are coupled and should be so modeled.

Simultaneous treatment of the fission product transport, thermal hydraulics, and material transport processes has been modeled in the EPRI-sponsored SIAM code, which couples the codes CORMLT, PSAAC, and RAFT which, respectively, treat core degradation, PCS thermal hydraulics, and PCS aerosol transport. Calculation proceeds in a coupled manner at each time step, simultaneously in these three codes. The time dependent fission product release to the containment is calculated. Validation of parts of the SIAM code has been performed by comparison to measured aerosol transport data.

I. INTRODUCTION AND BACKGROUND

It has become apparent that during the in-vessel progression of PWR severe accidents, fission product release and transport and the thermal-hydraulic phenomena provide feedback to the other.

The magnitude and rate of release of the fission product and structural material species from the overheated core region to the reactor pressure vessel (RPV) upper plenum and the PCS depends not only on the prevailing temperatures in the core[1], but also on the accessibility of the release locations to the in-vessel gas flows. The core melting process leads to substantial reduction of flow areas due to candling and solidification of melt material from the control and the fuel rods. Large planar blockages could be formed which alter the flow pattern and restrict the access of the gas flow to the interior regions where the fission product release may be occurring. The release may be restricted until either the access of gas flows is re-established or the fission product vapors diffuse to regions where gas flows are not obstructed.

The temperature distributions in the PWR PCS for the risk dominant scenarios TMLB' (station blackout) and the S_2D (small break) are determined primarily by the natural convection flow patterns[2] that get established during the early part of these accidents.

Revolatilization of the fission products deposited on the PCS surfaces will occur when the local temperatures exceed the saturation temperatures of the deposited compounds. Fission products will move as in a cascade, progressively from the hotter regions to the colder regions and could be retained in the PCS for a long time after the initiation of the core melt. The revolatilization release may be of special concern if it occurs after the postulated containment failure in a particular scenario.

The earliest work, in coupling thermal-hydraulics and fission product transport processes was incorporated in the MCT code[3], in which the thermal-hydraulic distributions calculated in the MERGE code[4] were passed to the TRAP-MELT code[5] at each time step. The core fission product vapor source at each time step was calculated with the MARCH[6] and CORSOR[7] codes. The information on the deposited aerosol fission products was transmitted back to the MERGE code from the TRAP-MELT code and the local structure temperatures were recalculated taking account of the fission product self-heating. The flow patterns in the MERGE and TRAP-MELT codes, however, were not changed. These code combinations did not model the multi-dimensional natural convection flow fields for the TMLB' and S_2D accidents. The results of calculations with the MCT code for the PWR scenarios showed[8] strong influence of the fission product deposition and revolatilization on the rate of release of the individual fission product species from the PCS to the containment.

The MAAP[9] code developed under U.S. industry sponsorship incorporated an integrated analysis of thermal-hydraulics and aerosol transport processes. The fission product vapor and core material sources generated during core heat up and degradation are converted to an aerosol source, whose transport and deposition is treated through a correlation approach[10]. Fission product self heating and revolatilization are treated. The MAAP code also models the natural convection flow fields in PWR vessel and piping. The effects of flow splits and recirculation on the transport and deposition of fission product vapors and aerosols, however, is not modeled.

In this paper, we will describe the models incorporated in the EPRI-sponsored code SIAM[11]. The SIAM code combines the codes CORMLT[12], PSAAC[13], and RAFT[14], which individually model, respectively, the in-vessel heat up and melt progression, the PCS thermal-hydraulics, and the vapor and aerosol transport in the PCS. Models incorporated in these stand-alone codes will be described very briefly. Earlier versions of the SIAM code have been described previously (15,16). The recent progress made and the results obtained will be described in this paper.

II. BRIEF DESCRIPTION OF CORMLT, PSAAC, AND RAFT CODES

II.A. CORMLT Code

The CORMLT code models the heatup, liquefaction, and slumping of the core and vessel internals for PWRs from the time of core uncovery to vessel failure. The models in the code have been described earlier (17,18,19). Briefly, for the high pressure accidents, the code models the natural convection flows in the vessel with axi-symmetric recirculating flow loops between the core and upper plenum, between the upper plenum and the upper head; (the flow loop between the upper plenum and the Westinghouse steam generators is modeled in the PSAAC code(20)). The code calculates heatup and liquefaction in sequence of (a) Ag-In-Cd control rods, (b) the vessel upper internals, (c) the Zircaloy clad on its oxidation, and (d) the fuel material. The stainless steel jacket is failed at a user-specified temperature and the control rod melt is ejected into the surrounding channels. Candling of the control rod melt is treated and the candling flows progress towards lower elevations, where on energy exchange with the substrate, freezing may occur and blockages may be formed.

The stainless steel components in the upper plenum melt next (due to heat transported by the natural convection flows to the upper plenum) and follow the control rod material towards lower elevations where they also may freeze. The (UO_2-Zr) eutectic melt material candles following the control rods and the upper plenum internals and also may freeze at lower elevations. The Zircaloy clad liquefaction reaction kinetics in the CORMLT code are taken from Hofmann-Kerwin Peck(21). If the candling flows completely fill the flow area, complete blockages may be formed. The blockages change the natural convection flow fields. The candling flows can be modeled as either film or rivulet type; the latter travelling much longer distances than the former.

The release of the radionuclides from the overheated core is modeled as per the BMI-2104 prescription. The fission products are transported only as vapors in the RPV; the aerosolization is modeled only in the RAFT code.

II.B PSAAC CODE

The Primary System Accident Analysis Code (PSAAC), is a modular code which predicts time dependent thermal hydraulics in the PCS during postulated degraded core accidents. The code models the PCS as a network of control volumes with flows of a mixture of steam, hydrogen, and fission product vapors. The code calculates, as a function of time, (a) the system pressure, (b) the mass flow rates into and out of each control volume, (c) temperatures of the gas phase and of each of the structures, (d) heat and mass transfer coefficients, (e) thermal and concentration boundary layer thicknesses, (f) water levels, and (g) mechanical integrity of PCS boundaries.

The PSAAC code includes an extensive nodalization scheme for the various structures within the PCS. The energy equations for each structure account for the convection and radiative heat transfer between the gas and structures. The gas phase mathematical formulation allows for a compressible binary mixture of steam and hydrogen, or a mixture of homogenous two-phase flow (steam and water) and hydrogen, and predicts the temporal and spatial mass fractions of steam and hydrogen within the PCS. The heat transfer between the gas phase and structures is by radiation and either forced or natural convection.

The convective flow patterns currently treated in the PSAAC code include a once-through flow, a stratified counter-current flow in the hot legs and steam generator tubes, a pulsating flow due to actuation of the safety/relief valves, and a well mixed upper-plenum flow due to the strong mixing which results from the buoyancy-driven flows.

The PSAAC code has been employed in a preliminary analysis of the TMI-2 accident[22] and for calculation of the thermal-hydraulic conditions that may occur in the PWR PCS during postulated accidents.

II.C. THE RAFT CODE

The RAFT (Reactor Aerosol Formation and Transport) code predicts the time dependent behavior of fission product vapors and aerosols within the PCS during LWR degraded core accidents. A unique feature of the RAFT model is the inclusion of the homogeneous nucleation mechanism which provides the capability to predict the particle size spectrum from the inception stage, without making ad-hoc assumptions concerning initial size and number density. Since both homogeneous and heterogeneous condensation mechanisms are simulated in the model, the particle formation calculations are sensitive to reactor accident conditions such as pressure, temperature gradients and the release rate of fission products. Other processes modeled are: particle agglomeration, particle deposition on surfaces, slip between the particles and gas, gas phase equilibrium chemistry and vapor condensation on solid structures and aerosols. RAFT requires, as input, time histories of pressure and gas and tube surface temperatures along the PCS. It also requires the time histories of mass flow rates of steam, hydrogen, fission products, control rods and other core material vapors available at top of core.

The RAFT code incorporates the necessary thermophysical property and transport data and a chemistry package to treat a flow system composed of steam, hydrogen, 13 fission product elements, control rod and structural materials and a generalized inert gas. The elements treated include Cs, I, Te, Ag, In, Cd, Sn, Sr, Mo, Ru, Ba, Zr, and Mn. Depending on the local gas temperature and pressure along the flow, these various elements, steam, and hydrogen will chemically equilibrate to form various combinations of molecular species and elements. The principal fission product species for most accident scenarios, however, are CsI, CsOH, and Te. The gas phase chemistry package in RAFT performs these computations by either Gibbs free energy minimization or by algebraic elimination based on the ordering of the various chemical reactions as a function of their rates.

The homogeneous nucleation rates in RAFT are computed by classical nucleation theory. An option is available, however, for computing non-classical nucleation rates. Heterogeneous growth of aerosols is modeled via the fundamental theory for heat and mass transfer to a droplet. The agglomeration mechanisms included in RAFT consist of Brownian motion, turbulence, and gravitation. The particle removal mechanisms include Brownian diffusion, thermophoresis, eddy impaction, inertial deposition on bends, and sedimentation.

Factors that favor homogeneous over heterogeneous condensation promote the formation of a large number of submicron particles and vice-versa. Parametric calculations indicate that the evolved particle number density is inversely proportional to the initial concentration of the fission product vapors (i.e., the release rate). They also show that a low system pressure and fast cooling rate lead to a large number density of particles and small average size.

Implementation of the RAFT code has yielded some important results for reactor accident analysis. Under nominal accident conditions, the particle size distribution depends mainly on the homogeneous and heterogeneous condensation processes, and the competition between these mechanisms, while agglomeration contributes only secondarily. Furthermore, the mass of homogeneously-nucleated material is quite small compared to that which condenses on pre-existing nuclei via heterogeneous condensation RAFT has been validated against experimental data obtained at Marviken facility in Sweden and Hot Tube apparatus at Argonne National Laboratory.

III. INTEGRATED ANALYSIS METHODOLOGY

The integration of the thermal hydraulic and aerosol processes has been achieved by combining the CORMLT, PSAAC and the RAFT codes. The objective is to obtain a code package which will simulate (a) effects of flow patterns on fission product transport, (b) effects of temperature and flow distributions on aerosolization, particle size distribution and deposition, (c) effects of fission product deposition on gas and local structural heating, and (d) effects of structural heating on the flow patterns. In this respect, a key requirement was the incorporation of aerosol transport modeling, which would account for the buoyancy-driven flow loops, the flow branching into various loops and the periodic opening and closing of valves.

The approach employed is to use the PSAAC code as the driver code and employ either the semi-coupled or the fully coupled procedures indicated in Fig 1. In both of these procedures there is complete coupling between the CORMLT and the PSAAC codes; with PSAAC providing pressure and gas mass flow rates at the inlet and outlet nozzle for each loop, and CORMLT providing the steam, hydrogen and core material vapor flow rates at the outlet nozzles in each loop. The PSAAC code provides the thermal hydraulic data for the in-vessel and ex-vessel parts of the PCS to the RAFT code. When the data on the deposited and the suspended fission product spatial distributions, developed in the RAFT code, is transferred to the PSAAC code for recalculation of the thermal hydraulic conditions, the coupling of the thermal hydraulic and the aerosol codes is termed complete; otherwise the coupling is termed semi-complete. In the latter case the structure and gas temperatures are not modified for the fission product self-heating effects. It should be noted the data transfer occurs at each time step.

There were several issues in the use by the RAFT code of the thermal hydraulic data generated by the codes CORMLT and PSAAC. These arose due to the different mesh and numeric structure employed by RAFT and necessitated either processing of the output from the CORMLT and PSAAC codes, before it could be used by the RAFT code, or a modification of the RAFT code. We will describe these procedures and modifications briefly in the following paragraphs.

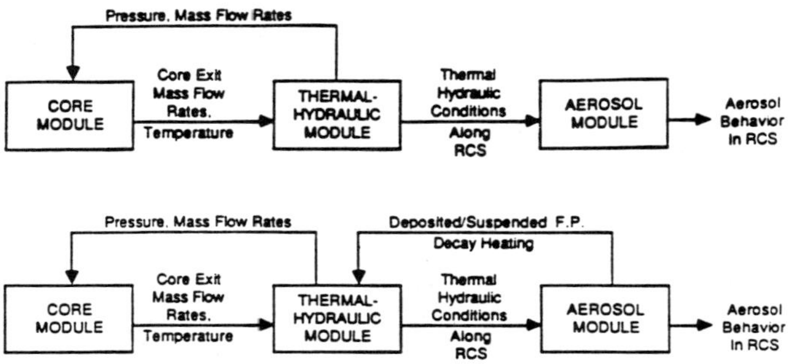

Fig. 1: Semi Coupled and Fully Coupled Procedures

III.A. PCS TEMPERATURE DISTRIBUTION

The CORMLT and PSAAC codes employ control volumes and calculate an average temperature at the center of each control volume, whereas RAFT is a differential code and varies the key variables continuously along a (one dimensional) coordinate. Thus the coarse temperature distributions obtained in the CORMLT and PSAAC codes have to be smoothed. A method was developed to accomplish this. It is based on weighted linear interpolation using the calculated temperatures. The length of each control volume (taken as volume/cross section area) was used as the weight for its temperature to obtain the interface temperature between two control volumes. Within each control volume the temperature was assumed to vary linearly from one interface to the other. This way, a smooth and continous gas temperature distribution was made available to the RAFT code.

III B. MULTIPLE STRUCTURE CAPABILITY

The RAFT code as supplied, incorporates only one structure per unit length of the PCS, whereas there may be several structures within a particular control volume of the CORMLT/PSAAC codes. It is important to determine the fission product deposition on each structure for obtaining the structure temperature history and its failure probability during the course of the accident. In addition, it is also important to obtain the fission product revaporization rates and determine the release from the PCS to the containment. For these reasons the RAFT code was expanded to include more than one structure in each component of the RAFT calculation.

The procedure followed for this extension of the code is to introduce the requisite data for the structures in a RAFT component and modify the code to include the necessary equations for the deposition rates of vapor and aerosols in the gas onto those structures. Deposition rates calculated for each structure are integrated to obtain the accumulation of the deposited vapor and aerosols on each structure as a function of time. These are then employed to determine the structure temperatures as a function of time.

We must note that this additional capability introduced in RAFT does not provide the ability to model strutures at different orientations to the flow and does not distinguish between structures of different shape, which may have somewhat different deposition rates/unit surface area. Secondly, the mass transfer coefficients are taken to be the same for each structure at a given axial flow path location in the RAFT component, when, in fact, they may vary due to the different hydraulic diameters/unit surface area for the individual structures. Thirdly, the effects of having a mixture of vapor species on the driving pressure for deposition are ignored and each species is treated as separate. This approximate modeling may result in somewhat lower deposition rates and somewhat higher revaporization rates than they should be.

III C. FISSION PRODUCT HEATING

The decay energy produced by the deposited fission products was incorporated in the gas phase and the structure energy equations, solved in the CORMLT/PSAAC codes. For the gas phase, the right hand side of the energy equation was augmented by two additional energy sources in each control volume representing, (a) the heat absorbed by the gas from the fission products suspended in it, and (b) the heat absorbed by the gas from the fission products deposited on structures.

The energy equation for a structure in control volume is also modified to account for the fission product decay heat generation. The two terms added to the energy equation are (a) the decay heat absorbed by the structure from the deposited fission products, and (b) the decay heat absorbed by the structure from the fission products suspended in the gas. The additional terms added to both the gas phase and the structure energy equations contain absorption coefficients for the gamma and beta rays emitted by the fission products. These absorption coefficients are determined for the gas mixture and for each structure, (while taking account of its geometrical configuration) in the control volume.

III D. MULTI-STREAM CAPABILITY

The RAFT code as supplied, incorporates a single flow stream along which the aerosol and vapor transport calculations are performed. In the modeling of accident sequences, it became apparent that more than one flow stream would be involved. For example, the flow from the upper plenum carrying vapor aerosols splits into flow streams to intact and broken loops; or the natural circulation flow patterns in the TMLB' accident, carry vapors and aerosols around the steam generators and also to the surge line and the pressurizer. For such diverse requirements, the RAFT code was expanded to treat systems characterized by one independent main flow stream with several dependent side streams, which branch from and merge with the main stream. This multi-stream capability of RAFT is a major advancement over the single stream RAFT code. This capability, however, had to be adapted to model two independent flow paths as would occur e.g., in one stream going to the broken loop and the other to the unbroken loop. A typical PWR loop flow system, illustrated in Fig. 2, with a main flow M_{HN} and loop flow M_L is modeled for the RAFT calculation as shown in Fig. 3. A fictitious upper plenum extension of one meter is added, from which a main stream flow of M_{HN} and side stream flow of M_L are extracted. A fictitious main stream extension of one meter is added at the other end where the side stream flow joins the main stream flow. The side stream conditions are used as input to the CORMLT code for the loop flow. This way, two independent streams are modeled in RAFT, and the CORMLT code receives the proper vapor conditions in the flow entering back into the upper plenum. The conditions in the fictitious extensions of the upper plenum and the main stream are specified such that no appreciable nucleation or deposition of vapor occurs.

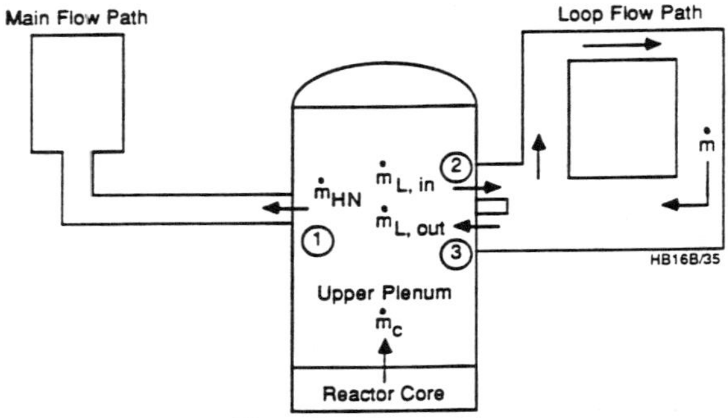

Fig. 2: Typical PWR Loop Flow Configuration

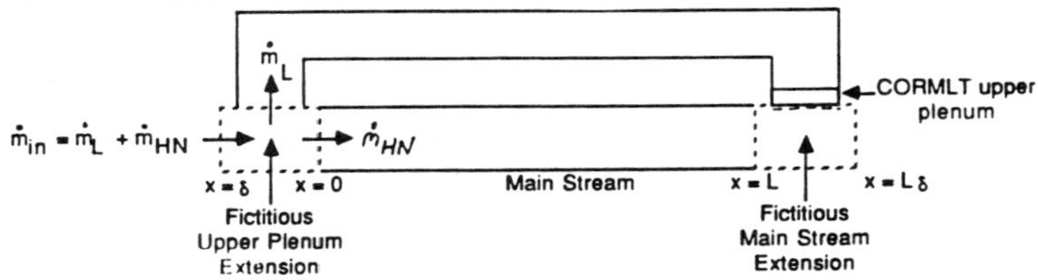

Fig. 3: Flow Systems Approximation for SIAM

IV. SOME OBSERVATIONS FROM PRELIMINARY SIAM APPLICATIONS

The SIAM code has been used for analysis of risk-dominant PWR accident scenarios, e.g., station blackout (TMLB') and small break (S_2D) LOCA for both Westinghouse and Babcock and Wilcox plants. Sensitivity calculations were also performed to determine the dependence of calculated results on variations of parameters, e.g., fission product mass release rates, heat and mass transfer coefficients, structure parameters, etc. In the following paragraphs, we will describe some observations derived from the limited SIAM applications made so far. These observations should be considered tentative at this time.

The SIAM calculations employed the fully coupled procedure in which the CORMLT and the PSAAC codes were coupled at the hot leg nozzle. For the Westinghouse station blackout scenario, the natural convection flows through the hot leg and steam generator were modeled as the secondary loop flow connected to the main flow path through the pressurizer and the relief valves.

The results from the calculations indicate that the fission products released from the reactor core can substantially influence PCS thermal-hydraulic response via the decay heating. Deposited and suspended fission products were found to increase the PCS gas and structure temperatures by as much as 400K, depending upon the fission product mass released into and deposited/suspended within the PCS, and upon the prevailing convection and/or radiative heat transfer rates from gases to structures. In case, the core region fission product mass release rates are low, the decay heating has little impact on the PCS thermal-hydraulics. Similar conclusion is reached for those cases in which the gas velocities are high (as in S_2D accident scenario), even though the fission product mass release rate is high. The most interesting result of the coupled code calculations, is that even with significant changes in the thermal-hydraulics, due to fission product heating, the effect of the changed thermal-hydraulics on the aerosol size distributions and depositions appears to be small. Further experience with SIAM code applications is needed before definitive conclusions with respect to the mode of coupling can be delineated. Furthermore, additional validation of the code against existing data remains to be performed.

REFERENCES

1) Nuclear Regulatory Commission, "Technical Basis for Estimating Fission Product Behavior During LWR Accidents", NUREG-0772 (June 1981).
2) B. R. Sehgal, V. E. Denny, W. A. Stewart, and B. C-J Chen, "EFfects of Natural Convection Flows on PWR System Temperatures During Severe Accidents", ASME/AIChE/ANS National Heat Transfer Conference, Denver, Colorado (August 1985).
3) Risk Management Associates and New York Power Authority, "Source Term Safety Assessment of Radio Nuclide Release Under Severe Accident Conditions at Indian Point 3 Nuclear Power Plant", Vol. 1 and 2, pub. by New York Power Authority (July 10, 1984); and, P. P. Bieniarz and R. E. Deem, "An Analysis of a BWR MkI Utilizing Advanced BMI-2104 Computational Techniques to Calculate Source Terms", IAEA-SM-281/48, IAEA Symposium on Source Term, Columbus, Ohio (October 28-November 1, 1985).
4) R. G. Freeman-Kelly and R. G. Jung, "A User's Guide for MERGE", Battelle Columbus Laboratories, NUREG-4172, BMI-2121 (March 1985).
5) H. Jordan and M. R. Kuhlman, "TRAP-MELT User's Manual", NUREG/CR-4205, BMI-2124 (February 1977).
6) R. Wooten and P. Cybulskis, "March 2 Code Description and User's Manual", Battelle Columbus Laboratory Report (1983).
7) M. R. Kuhlman, D. J. Lehmicke, and R. O. Meyer, "CORSOR User's Manual", NUREG/CR-4173, BMI-2122 (March 1985).
8) P. P. Bieniarz, R. Deem, J. Favora, S. S. Iyer, and S. Masciulli, "Development of Realistic Source Terms for Indian Point 3", Proceedings of ANS Topical Meeting on Fission Product Behavior and Source Term Research, Snowbird, Utah, EPRI NP-4113-SR (July 15-19, 1984).
9) Fauske & Associates, Inc., Technical Report 16.2-3, "MAAP Modular Accident Analysis Program User's Manual", Vol. 1 and 2 IDCOR Program Report Atomic Industrial Forum, Washington D.C. (January 1985).

10) P. G. Ellison, M. Epstein, R. E. Henry, "Aerosol Modeling in LWR Integrated Severe Accident Analysis", Proceedings of the International ANS/ENS Topical Meeting on Thermal Reactor Safety, San Diego, California (February 2-6, 1986).
11. D. C. Bugby, A. T. Wassel, J. L. Farr, Jr.,"Severe Accident Integrated Accident Methodology: SIAM Code", EPRI NP report to be published (1989).
12. V. E. Denny, "The CORMLT Code for the Analysis of Degraded Core Accidents", EPRI NP-37667-CCM (1984).
13. A. T. Wassel, et al., "Thermal-Hydraulic Modeling of the Primary Coolant System of Light Water Reactors During Severely Degraded Core Accidents", EPRI NP-3563 (1984).
14. K. H. Im, R. K. Ahluwalia, and C. F. Chuang, "RAFT: A Computer Model for Formation and Transport of Fission Product Aerosols in LWR Primary Systems", Aerosol Sci. and Tech., Vol. 4, pp.125-140 (1985); and, K. H. Im, R. F. Ahluwalia, and H. C. Lin, "The RAFT Computer Code for Calculating Aerosol Formation and Transport in Severe LWR Accidents", EPRI NP-5287-CCM (July 1987).
15. B. R. Sehgal and R. L. Ritzman, "Integrated Analysis of Fission Product Transport in Reactor Coolant Systems", Proceedings of the Symposium on Chemical phenomena Associated With Radioactivity Release During Severe Nuclear Power Plant Accidents, American Chemical Society, Anaheim, California (September 1986).
16. D. C. Bugby, J. L. Farr, Jr., A. T. Wassel, and B. R. Sehgal, "Coupled Thermal-Hydraulic and Fission Product Transport Analysis of Reactor Coolant Systems During Degraded Core Accidents", Proceedings of the National Heat Transfer Joint Conference, Pittsburgh, Pennsylvania (August 9-12, 1987).
17. V. E. Denny, "The Role of Natural Circulation in Severe Accident Analysis", Proceedings of the Topical Meeting on Reactor Physics and Safety, Saratoga Springs, New York, Vol.2, pp.699-719, NUREG/CP-0080 (September 1986).
18. V. E. Denny, A. Mertol, and B. R. Sehgal, "CORMLT Modeling of Severe Fuel Damage in Postulated Accidents", National Heat Transfer Conference, Pittsburgh, Pennsylvania (August 1987).
19. B. R. Sehgal, V. E. Denny, M. Merilo, and A. Mertol, "PWR Core Degradation Modeling in the CORMLT Code", Proceedings of the International ANS/ENS Conference on Thermal Reactor Safety, Avignon, France (October 2-7, 1988).
20. A. T. Wassel, M. S. Hoseyni, J. L. Farr, Jr., and B. R. Sehgal, "Thermal-Hydraulics of the Primary Coolant System of Light Water Reactors During Severely Degraded Core Accidents", Proceedings of the Third International Meeting on Reactor Thermal Hydraulics, Newport, Rhode Island (October 15-19, 1985).
21. P. Hofmann and D. K. Kerwin-Peck, "UO/Zircaloy-4 Chemical Interactions from 1000 to 1700° C Under Isothermal and Transient Temperature Conditions", J. Nuclear Materials, Vol. 124, pp.80-105 (1984).
22. S. M. Ghiaasiaan, A. T. Wassel, M. S. Hoseyni, and B. R. Sehgal, "Thermal-Hydraulics and Heat-Up of Light Water Reactor Cores During Severe Accidents", ASME/AIChE/ANS National Heat Transfer Conference, Denver, Colorado (August 1985).

SOPHIE: A Computer Code for the Prediction of the Transport and Deposition of Fission Product Vapors in a PWR Primary Circuit during a Severe Core Damage Accident

C. LEUTHROT
Commissariat à l'Energie Atomique (CEA)
Départment de Transfert d'Energie (ETD)
Service d'étude des coeurs et systèmes (ETD)
Centre de'Etudes Nucléaires de Cadarache (CNRC)
13108 Saint Paul Lez Durance
Cedex, France

1. INTRODUCTION

The ESCADRE code system (Reference 1) developed at the CEA/IPSN accurately simulates a severe accident in a pressurized water reactor, in order to study its radiological consequences.

The transport and deposition of fission products in the primary circuit are analyzed using two codes from this system (Figure 1).

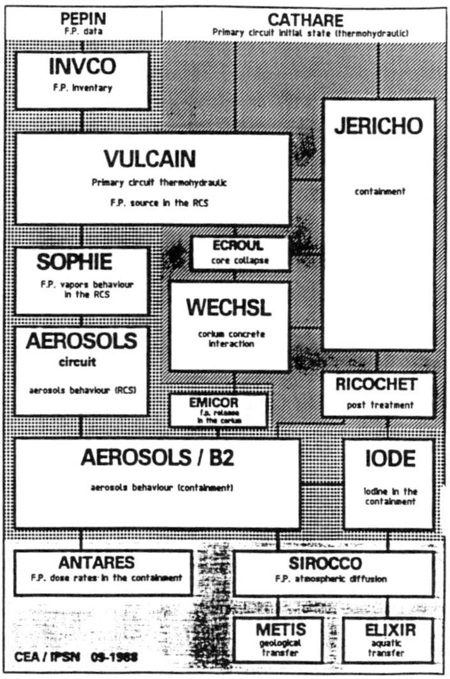

ESCADRE system
Analysis codes system for PWR accident

FIGURE 1.

SOPHIE: This code processes the transport and deposition of F.P. (fission product) vapors in the section of the circuit where the gas temperature is higher than the transition temperature, and provides the F.P. mass available between the vapor phase and the aerosol phase, together with the vapor source for a possible salting-out.

AEROSOLS: This code processes aerosol transport and deposition in the section of the circuit where the gas temperature is lower than the transition temperature.

The code uses the SOPHIE outputs as input data.

The choice of this method of analysis was based on our current knowledge of the phenomenology of aerosol/vapor transfers, and particularly on the following observations:

- Parametric studies concerning the influence of the median radius of the aerosol source in the vessel, have shown that atmospheric release is not greatly affected by this parameter.

- The mechanisms which form aerosols from vapor condensation (transition from vapor phase to aerosol phase) are virtually instantaneous. The corresponding procedures are highly complex and presuppose a precise knowledge of temperature, vapor concentration and the number of condensation nuclei. Given that this knowledge cannot be obtained, it is impossible to obtain a more accurate idea of aerosol size than that obtained by the experimental correlation drawn from representative studies of fission product release in PWR irradiated fuels. (Reference 2).

- The HEVA experimental studies confirm that, below a certain temperature (transition temperature), almost all the mass of the F.P and the structural materials in suspension in the gas, will be in the aerosol phase.

- Sensitivity studies show that in the reactor containment, and for a given aerosol mass concentration, particle size rapidly becomes similar, regardless of the initial value.

Consequently, precise knowledge of the original particle size is not essential.

The SOPHIE code, presented in this paper, was therefore developed with this in view. It is used to calculate the transport, deposition, and possible revaporization of volatile fission products in the vapor phase (iodine, cesium and tellurium), which are released from the core during the accident and propagated inside the primary circuit.

2. PHENOMENON MODELLING

Two types of physical models are used in the SOPHIE code:

a) Calculation of the transport of vapors in the fluid (composed of a mixture of water vapor and hydrogen); this calculation is based on the mass transfer/heat transfer analogy and uses the correlations deduced from this modelling.

b) Calculation of the parameters for vapor deposition on the internal structures and the walls of the primary circuit; the following mechanisms are taken into account:

- condensation and evaporation of the chemical species of the fission products studied according to wall temperature and the concentration of these species.
- F.P. adsorption and desorption via physical mechanisms (reversible), or chemical mechanisms (partially or totally irreversible).

2.1. Mass Transfer Calculation

Thermodynamic properties of the carrier fluid. Knowledge of thermodynamic properties of the carrier fluid (H_2/H_2O mixture) is required for the correlations used to calculate the vapor transfer from the fluid to the circuit walls. These correlations are based on the mass transfer/heat transfer analogy.
The properties of the carrier fluid are calculated from status equations. The following parameters are calculated:

- density of mixture
- enthalpy
- molecular diffusivity
- dynamic viscosity
- thermal conductivity
- heat capacity

The calculation can be carried out for mixtures with a hydrogen content of between 10^{-5} and 0.999.

Mass transfer coefficient. For molecules transported in a circuit by a flowing fluid, the current towards the wall is given by:

$$J = h\ (C_B - C_W)$$

where: C_B = F.P. vapor concentration in the bulk.
C_W = F.P. vapor concentration in the fluid at the wall.

A mass transfer coefficient h is defined in this way.

h is calculated using standard formulations from the mass transfer/heat transfer analogy for all possible hydraulic conditions during the accident:

- natural convection
- forced convection in laminar and turbulent flow.

2.2. Deposition Calculation

Adsorption of F.P. on the walls. Whatever type of force it is submitted to, an atom located at the surface of a solid differs from an internal atom, in that the resultant of the forces applied to it is not equal to zero. Any process that decreases free surface energy and therefore, the surface tension of the solid, occurs spontaneously. Adsorption of a gas by a solid is therefore a spontaneous process that tends to minimize the surface tension of the solid.

The adsorption phenomenon will display different characteristics, according to the magnitude of the energies involved and can therefore be classified in one of two categories: physical adsorption or chemical adsorption.

Most of the theories developed:

- either involve a general study of the adsorption - desorption phenomenon at equilibrium, i.e. assume that these mechanisms are extremely rapid, compared to mass transfer,

- or involve the assumption that the surface is a perfect receptor that cannot be saturated, and on which desorption is always zero. These hypotheses lead to the concept of "deposition rate".

Neither of these two approaches is appropriate for the problem under consideration. We have therefore used more sophisticated modelling, although it is probably even more simplistic as regards the phenomena studied.

The most interesting models (since they are the closest to physical reality), are BRUNAUER-EMMET-TELLER's model for physical multilayer adsorption, and FREUNDLICH's empirical equation for chemisorption.

LANGMUIR's model is simply drawn from the BRUNAUER-EMMET-TELLER theory (deposits below a single layer) but its simplicity, and the fact that it is suitable for both physical and chemical adsorption, has led us to study it in detail and use it in our modelling.

LANGMUIR's isotherm. In his theory, Langmuir assumed that the surface of the adsorbent contained a determined and constant number of adsorption sites n/n_∞. Maximum simplification of mathematical processing is obtained if the surface is also considered to be homogenous, i.e. that all the sites are identical and that the adsorbate is adsorbed in a non-dissociative manner. In this case, the adsorption rate at a given instant t, when the cover rate has the value $\frac{n}{n_\infty}$, is proportional:

- to the number of molecular collisions on the wall.

According to the kinetic gas theory, the number of collisions on the wall is:

$$v = \frac{1}{4} n v_m = \frac{P}{\sqrt{2 \pi m k t}}$$

where P = partial pressure of adsorbate
m = molecular mass
k = Boltzmann's constant

- to the surface not yet covered by the adsorbate: $1 - \frac{n}{n\infty}$

- to a sticking probability correctly represented by Arrhenius' theory.

$$Ce^{-\frac{E_a}{RT}}$$

where E_a is the adsorption activation energy

$$\frac{dn}{dt} = \frac{P}{\sqrt{2 \pi m k T}} C \left(1 - \frac{n}{n\infty}\right) e^{\frac{-E_a}{RT}}$$

Under the same conditions, desorption rate is proportional to the covered fraction of the surface and also to a temperature factor including E_d, the desorption activation energy.

$$-\frac{dn}{dt} = K \frac{n}{n\infty} e^{-\frac{E_d}{RT}}$$

where the value K is represented by a number of molecules /cm^2s

Expressed more simply, the adsorption current $\vec{J_a}$ towards the wall is given by

$$\vec{J_a} = K_1 C_f$$

where C_f is the concentration in the fluid at the wall (at/m³)
K_1 is the absorption rate (m/s) where

$$K_1 = K_{10}\, e^{-\frac{Q_a}{RT}}$$

Q_a = adsorption activation energy

Likewise, the desorption current can be written as:

$$\vec{J_d} = K_2\, C_d$$

where C_d = surface deposit at the wall (at/m^2)
K_2 = desorption rate (s^{-1})

and
$$K_2 = K_{20}\, e^{-\frac{Q_d}{RT}}$$

Q_d = desorption activation energy

<u>Multilayer physical adsorption - chemisorption</u>. The modelling described in Chapter 3.2, derived from LANGMUIR's model, only applies to single-layer physical adsorption, which is probably not the case for volatile fission product depositions on the steel parts of the primary circuit.

In the SOPHIE code, a second adsorption/desorption mechanism is used in series with the first, in order to refine modelling and consider chemisorption or diffusion of the species absorbed in the metallic surfaces (i.e. nonlinear phenomena). This second mechanism has two rates, K_3 and K_4, that differ from rates K_1 and K_2. The diagram displaying the exchanges between the various layers is presented in Figure 2.

The adsorption current for the second mechanism is written as:

$$\vec{J}_{a2} = K_3 C_1$$

where C_1 = surface concentration (at/m^2) deposited by the first adsorption mechanism.

The desorption current is written as:

$$\vec{J}_{d2} = K_4 C_2$$

where C_2 = surface concentration (at/m^2) deposited by the second mechanism.

This pair of adsorption/desorption mechanisms should make it possible to simulate complex deposition phenomena. An experimental program is being defined at the CEA to determine the characteristic coefficients of this modelling.

Condensation - Evaporation. Condensation of F.P. vapors at the wall occurs when the pressure of this vapor in the carrier fluid is greater than the saturation vapor pressure at wall temperature. Inversely, a condensed vapor evaporates when the fluid pressure falls below the saturation vapor pressure, at the temperature of the wall on which it is condensed.

This mechanism is taken into account in SOPHIE, by calculating the saturation vapor pressure as a function of temperature, for the chemical species being studied. (Figure 2).

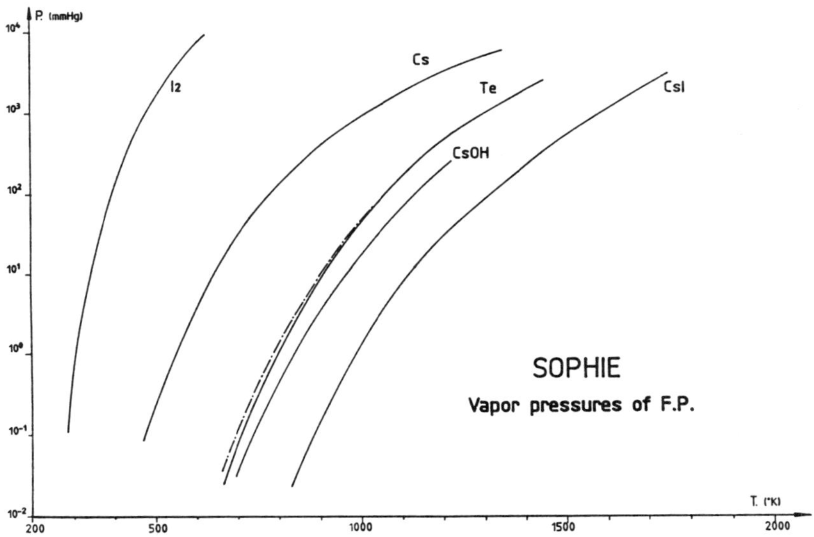

FIGURE 2.

This calculation is carried out from thermodynamic data gathered for the various chemical species considered.

The condensation/evaporation current of a chemical species towards the wall is given by:

$$J_c^{\rightarrow} = h\ (C_g - C_{SAT})$$

where C_g = vapor concentration in the fluid (at/m^3)
C_{SAT} = saturation vapor concentration for the chemical species considered at wall temperature (at/m^3)
h = mass transfer coefficient in the mesh.

3. TRANSPORT AND DEPOSITION MODELLING OF VAPORS IN A FLOWING GASEOUS FLUID

The model used is described in Figure 3. It takes into consideration the various types of modelling for physical and physicochemical phenomena described in the preceding chapters.

The matter transported through the carrier fluid is assumed to be the combined result of axial convection and radial diffusion, turbulent or otherwise.

As they reach the surface of the pipe, the molecules interact with the wall material, by condensation, physical adsorption, or chemisorption.

SOPHIE

F.P. TRANSPORT AND DEPOSITION MODELS

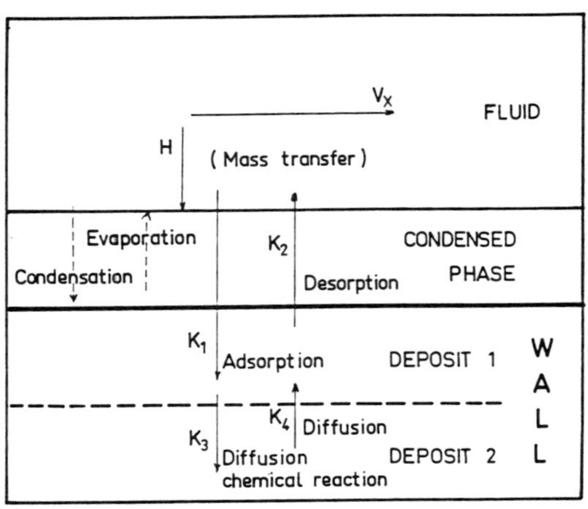

FIGURE 3.

3.1. Transport and Deposition Equations

The calculation is carried out in a mesh of given geometry, in which all the parameters are constant.

Adsorption/desorption and condensation are processed separately.
Each mesh can contain two surfaces with different temperatures.
However, in order to write the equation more simply, the case of a single wall is presented here.

Let an isothermic mesh have a volume V and a deposition surface S. Inside this mesh, there is a vapor source σ for the fission product considered.

Mass balance in the fluid. The mass conservation equation for the mesh considered is written as.

$$\frac{\delta C_g}{\delta t} + v \frac{\delta C_g}{\delta x} + \lambda C_g + \frac{hS}{V} (C_g - C_1) - \frac{\sigma}{V} = 0 \tag{1}$$

The mass flow rate between the fluid volume and the boundary layer is in the form h $(C_g - C_1)$. C_1 is adjusted so that the flux entering and leaving the surface of the mesh remains equal. This equality is illustrated by:

$$h (C_g - C_1) = \underbrace{k_1 C_1 - k_2 C_{D1}}_{\text{adsorption/desorption}} + \underbrace{h' (C_g - C_{SAT})}_{\text{condensation/evaporation}} \tag{2}$$

in the absence of evaporation/condensation, i.e. if

$C_g < C_{SAT}$ and $C_C = 0$ then $h' = 0$

if $C_g \geq C_{SAT}$ or $C_C \neq 0$ $h' = h$

C_1 is drawn from (2), and the mass conservation equation (1) in the mesh becomes:

- if $C_g < C_{SAT}$ and $C_C = 0$

$$\frac{\delta C_g}{\delta t} + v \frac{\delta C_g}{\delta x} + \left[\lambda + \frac{hs}{V} (1 - \frac{h}{k_1 + h})\right] C_g - \frac{hS}{V} \frac{k_2}{k_1 + h} C_{D1} - \frac{\sigma}{V} = 0 \tag{3}$$

- if $Cg \geq C_{SAT}$ or $C_C \neq 0$

$$\frac{\delta C_g}{\delta t} + v \frac{\delta C_g}{\delta x} + \left[\lambda + \frac{hS}{V}\right] C_g - \frac{hS}{V} \frac{k_2}{k_1 + h} C_{D1} - \frac{h^2 S}{V} x \frac{1}{k_1 + h} C_{SAT} - \frac{\sigma}{V} = 0 \tag{4}$$

Mass balance in adsorption layer 1. The mass conservation equation for adsorption layer 1 in the mesh is written as:

$$\frac{\delta C_{D1}}{\delta t} + (\lambda + k_2 + k_3)\, C_{D1} - k_1 C_1 - k_4 C_{D2} = 0 \tag{5}$$

When (2) is taken into account, this equation becomes:

- if $C_g < C_{SAT}$ and $C_C = 0$

$$\frac{\delta C_{D1}}{\delta t} + (\lambda + k_2 + k_3 - \frac{k_1\, k_2}{h + k_1})\, C_{D1} - \frac{k_1\, h}{h + k_1}\, C_g - k_4\, C_{D2} = 0 \tag{6}$$

- if $C_g \geq C_{SAT}$ or $C_C \neq 0$

$$\frac{\delta C_1}{\delta t} + (\lambda + k_2 + k_3 - \frac{k_1\, k_2}{h + k_1})\, C_{D1} - \frac{k_1\, h}{h + k_1}\, C_{SAT} - k_4\, C_{D2} = 0 \tag{7}$$

Mass balance in adsorption layer 2. The mass conservation equation for adsorption layer 2 in the mesh is written as:

$$\frac{\delta C_{D2}}{\delta t} + (\lambda + k_4)\, C_{D2} - k_3\, C_{D1} = 0 \tag{8}$$

Mass balance for the condensation layer.

$$\frac{\delta C_C}{\delta t} + \lambda\, C_C - h\,(C_g - C_{SAT}) = 0 \tag{9}$$

4. AF TEST CASE

The primary circuit grid used in the calculation is shown in Figure 4. It takes the following into account:

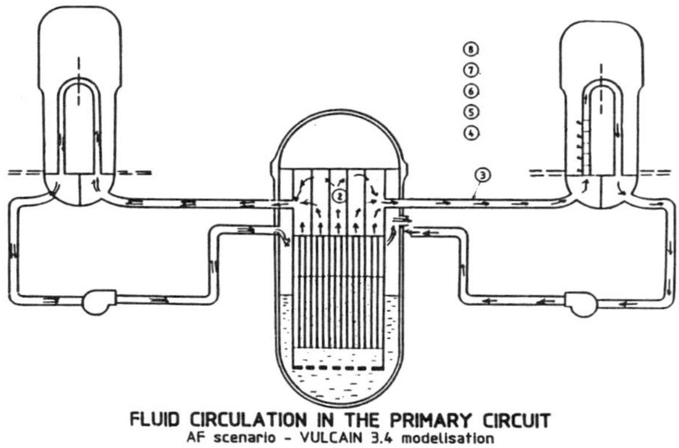

FIGURE 4

- upper plenum
- hot leg
- 80 cm of SG tubes, for which the fluid temperature is greater than 500°C (Figure 5)

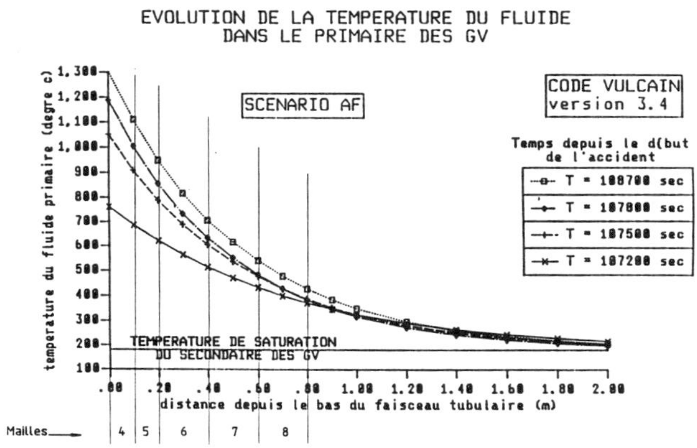

FIGURE 5.

Of all the results given by the code, this report gives the concentrations of two species (I_2 and CsI) in the fluid at the exit of the last 3 meshes (fission product source used in the AEROSOL code - Figures 6 and 7). The deposits in the primary circuit are given 130 minutes after dewatering the core (Figure 8). These calculations were carried out using adsorption rate values available in the literature.

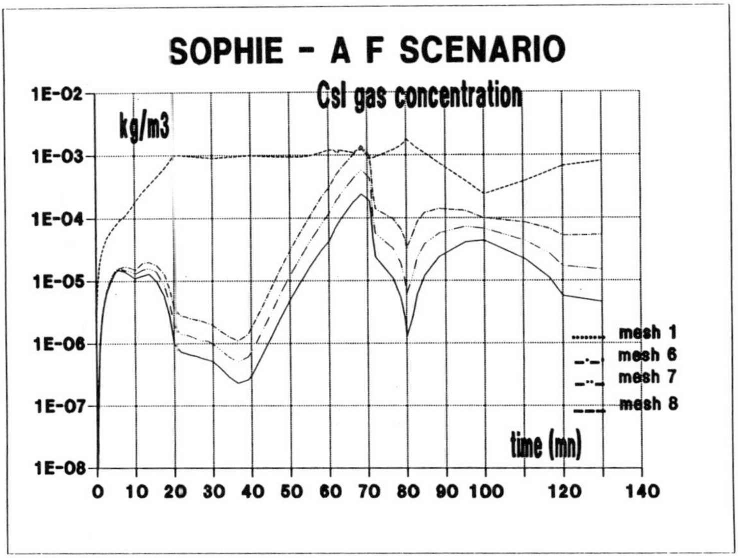

FIGURE 6.

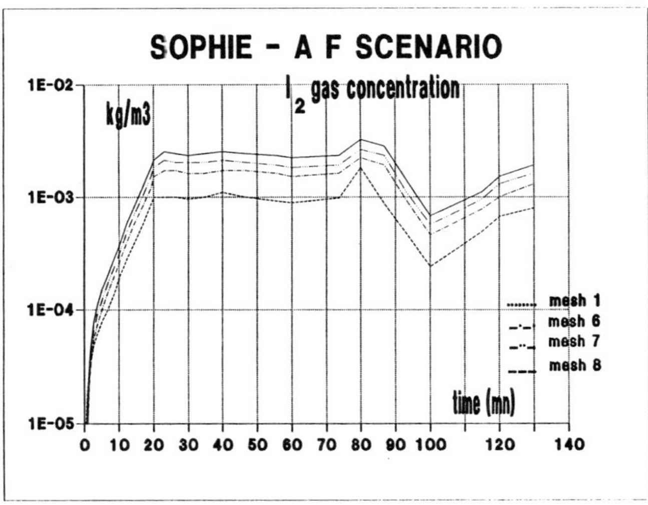

FIGURE 7.

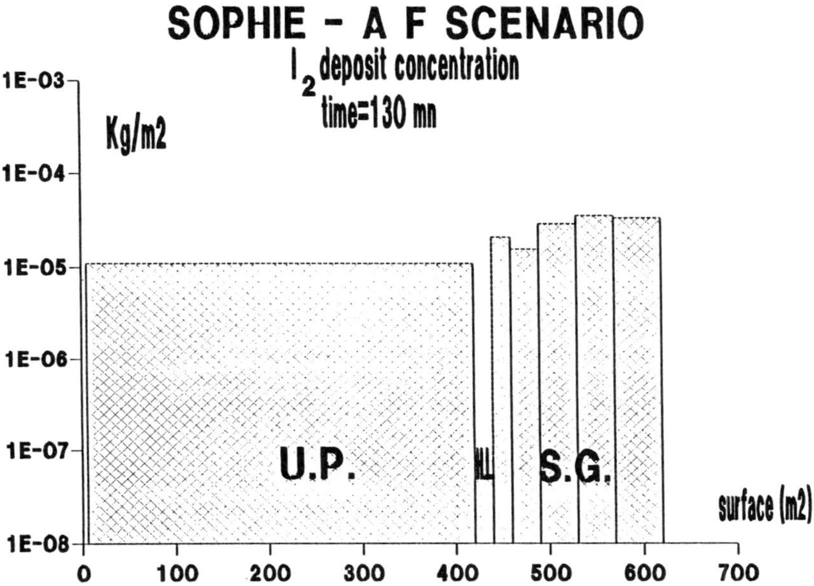

FIGURE 8.

5. CONCLUSION

The tests carried out with the Sophie code demonstrate that it exactly meets the requirements expressed by the CEA/IPSN for the development of the ESCADRE accident code system, and that it can be used to calculate the transport and deposition of fission products which exist in vapor form in the primary circuit of a water pressurized reactor. The data currently available concerning adsorption isotherms is, however, inadequate, and an experimental program is in progress at the CEA in order to provide data which will eventually be integrated into the code.

REFERENCES

1. DUFRESNE, J., CENERINO, C., DUMAS, J.M., FERMANDJIAN, J., GAUVAIN, J, L'HERITEAU, J.R. and LHIAUBET, G., Presentation and Application of the ESCADRE System (IPSN/CEA/FRANCE) IAEl *Meeting SORRENTO, 21-25 March 1988.*

2. LE MAROIS, G., WARLOP, R., Source Term Experiment for Fission Product Transport Assessment : the HEVA Program. *International Conference ENS/ANS Avignon 2-7 october 1988.*

Fission Product Transport in the Reactor Coolant System for a Spectrum of Interfacing System LOCA Scenarios

EDWARD P. WARMAN, JAMES METCALF, ROBERT HESSIAN, JR., and MICHAEL DONAHUE
Stone & Webster Engineering Corporation
Boston, Massachusetts, USA

INTRODUCTION AND SUMMARY

One of the most important potential severe accident sequences for any pressurized water reactor (PWR) is a loss of coolant accident (LOCA), or V-sequence, in one of the interfacing systems. As initially described in the Reactor Safety Study (WASH-1400) (1), interfacing system LOCAs involved the failure of check valves in emergency core cooling systems (ECCS), but could also involve the residual heat removal (RHR) systems. The check valves protect the low-pressure portions of these systems from the high pressures of the reactor coolant system (RCS) to which they are connected to provide cold leg injection. A consequent break in the low-pressure piping outside the containment may result in core damage and a direct pathway for fission products to be transported from the core, through the RCS and ECCS or RHR to the auxiliary building, from which they can escape to the environment.

This paper addresses the retention and transport of fission products (specifically, CsI) in the RCS in V-sequence scenarios. It summarizes some of the major differences between models resulting from the latest version of the Industry Degraded Core Rulemaking (IDCOR) MAAP Computer Program (2), MAAP 3.0B, developed for the Electric Power Research Institute (EPRI) (3), and the U.S. Nuclear Regulatory Commission (NRC) Source Term Code Package (STCP) (4). Discussed in greater detail are the differences in: 1) fission product transport and retention in small, medium, and large ECCS pipe breaks, as well as the effect of ECCS and auxiliary feedwater (AFW) system operation; and 2) fission product retention in the various regions of the RCS as calculated by MAAP 3.0B and the STCP (4).

Intermediate results are presented based on work performed by Stone & Webster Engineering Corporation (Stone & Webster) for a program sponsored by EPRI (5). The overall objective of the EPRI project is to analyze retention in various auxiliary building configurations for various interfacing system LOCAs. Results are presented for 4 of the 20 MAAP analyses prepared for various V-sequence configurations, with the data limited to those portions of the analyses which address fission product retention in the RCS.

Discussion of fission product retention in the various regions of the RCS incorporates data from one of the above-mentioned MAAP analyses, as well as a published NRC-sponsored analysis of a V-sequence that used the STCP (6). Unfortunately, the STCP analysis, performed by Brookhaven National Laboratory (BNL), used the Surry Plant data whereas the MAAP analyses used Zion Plant data; however, the effect of these differences on RCS retention are believed to be small in comparison with the effects of the differences in modeling between the MAAP and STCP methods.

The differences in CsI retention in the various regions of the RCS are so great between the MAAP and STCP models as to indicate that considerable work is required to improve the understanding of the processes and bring the two methodologies into closer agreement.

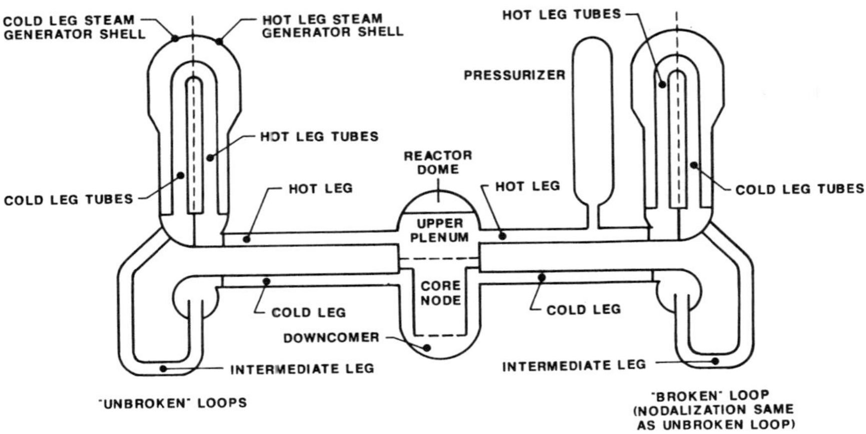

FIGURE 1. PWR primary system nodalization in MAAP 3.0B

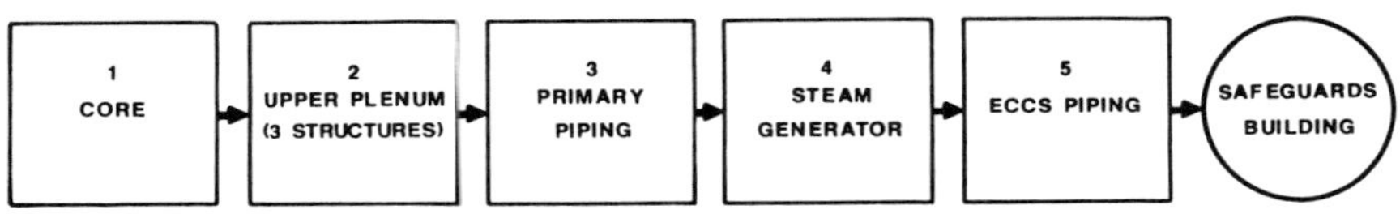

FIGURE 2. Schematic of TRAPMERGE control volumes in STCP

MODELING OF REACTOR COOLANT SYSTEM IN MAAP AND STCP

Figure 1, which is adapted from Figure 7–6 of the IDCOR Technical Summary Report (7), depicts the primary system nodalization used in the PWR version of MAAP. As can be seen from this figure, both the broken loop (i.e., the loop in which the break occurs in the RCS piping, or in the interfacing system piping in the case of the STCP model) and the unbroken loops (three in the case of the four-loop Zion Plant) are included in the MAAP model. Recirculation is modeled and fission product retention in the intact loops has been found to be significant. It should be noted that MAAP does not currently include interfacing system piping.

Figure 2 shows the nodalization (called control volumes in this figure) used for fission product transport analysis in the STCP. The STCP does include interfacing system piping (labeled ECCS piping in this figure); however, the TRAPMERGE portion of the STCP does not include the intact loops. It is a serial-type model in which flow is unidirectional from node to node, thus there is only one pathway for fission products, i.e., from the core to the environment on a node-to-node basis.

Table 1 lists the comparable MAAP and STCP nodalizations. The middle column comprises the grouping of the MAAP nodes selected for comparison with the STCP nodes. The data included in Figures 3–6 refer to these groups of nodes when addressing MAAP results; in fact, these data represent the sum of the CsI retained in the nodes represented by each group.

TABLE 1. System Nodalization in MAAP and STCP Analyses of V–sequence

MAAP nodalization	Grouping of output for comparison	STCP[1] nodalization
Core	Core	Core
Upper plenum Pressure vessel dome	Upper plenun & dome	Upper plenum
Downcomer (in pressure vessel)	Downcomer	(not modeled)[2]
Broken loop hot leg Pressurizer	Broken loop hot leg & pressurizer	Hot leg
Broken loop hot leg stm. gen. tubes Broken loop cold leg stm. gen. tubes	Steam generator	Steam generator
Broken loop intermediate leg Broken loop cold leg	Broken loop cold leg	(not modeled)[2] [3]
Unbroken loops hot legs Unbroken loops hot leg stm. gen. tubes Unbroken loops cold leg stm. gen. tubes Unbroken loops intermediate legs Unbroken loops cold legs	Intact loops	(not modeled)[2]
ECCS piping (not modeled)	(not modeled)[4]	ECCS piping

[1]TRAPMERGE nodalizations used for fission product transport analysis.
[2]Serial nature of the STCP is limited to single node–to–node pathway to the environment.
[3]TRAPMERGE nodalization excludes the piping between the steam generator and the ECCS cold leg injection point.
[4]Not currently included in MAAP.

In addition to nodalization differences between MAAP and STCP, several other important differences in modeling and assumptions exist that substantively affect the analysis of fission product transport and retention in the RCS. Some of the major differences are briefly summarized below.

Revaporization of fission products in the RCS has been found to be a very significant phenomenon. Volatile fission products, e.g., iodine, cesium and tellurium, are released from the core in vapor form and some are transported in the RCS as vapor, while a significant fraction quickly condenses on aerosols or homogeneously nucleates to form aerosol particles.

Deposition on RCS metal surfaces occurs primarily as a result of vapor condensation, aerosol deposition, and inertial impaction. As the deposition layer increases, the volatile fission products heat up because of their decay heat and the other heat sources and "revaporize," and are transported from node to node. The rate of revaporization is closely tied to the gas and heat sink temperatures in each node. In the STCP, the revaporization analysis stops at the time of failure of the lower head of the pressure vessel, because the thermal hydraulic data needed to continue the analysis, which are obtained from the MARCH program portion of the STCP up to the time of lower head failure, are no longer available. In contrast, the MAAP program includes revaporization throughout all phases of the accident scenario. Post-vessel failure revaporization has been found to be significant in MAAP analyses of V-sequences, as discussed below.

The core slumping models differ beween MAAP and STCP. Although both MAAP and STCP estimate that core collapse occurs about 50 minutes after core uncovery, the subsequent behavior is quite different in the V-sequence for which detailed comparisons are presented. In MAAP the lower pressure vessel head is assumed to fail at one of the numerous penetrations within about 3 minutes of core collapse. In contrast, the STCP model predicts that it takes about 15 minutes to boil off the water in the lower head and more than one hour for the vessel head to fail. During this entire time period copious amounts of hydrogen are generated in the STCP model due to the zirconium metal-water reaction in the lower head. The core exit gas temperature exceeds 2000 K during much of that time. This has the effect of raising the temperature in the downstream nodes in the STCP model with the concomitant release of previously deposited fission products. The amount of hydrogen produced in the lower head and remaining portions of the core in MAAP is a fraction of that in STCP and the core exit gas temperatures are far lower in MAAP, by many hundreds of degrees.

The resultant effects include a totally different characteristic behavior of fission products retained in the RCS nodes and in the ECCS piping node in STCP, than is predicted by the MAAP model. Comparisons of fission product retention as a function of time, as calculated by MAAP and STCP, are presented later.

FISSION PRODUCT RETENTION IN THE RCS FOR VARIOUS MAAP V-SEQUENCE ANALYSES

On behalf of EPRI, Stone & Webster performed a parametric study of fission product transport and retention in auxiliary buildings (5). The objective of that study, which included plant visits to operating nuclear-power plants, was to identify a representative combination of parameters to characterize most potential V-sequence scenarios. Sixteen PWR plant types were visited, including Westinghouse, Combustion Engineering, and Babcock and Wilcox reactors in combination with auxiliary buildings designed by all major architect-engineering firms, such as Stone & Webster, Bechtel, etc.

Following the plant surveys, MAAP 3.0B analyses were performed for 20 selected configurations and assumptions, such as pipe break size, etc. Although the objective of the EPRI-sponsored study was the study of retention in auxiliary buildings during interfacing system LOCAs, this paper focuses on the intermediate results from that study, i.e., retention in the RCS. Table 2 lists the four cases selected for comparisons.

Table 2 shows that a large interfacing system LOCA pipe break, i.e., 10 in. (15.4 cm) in diameter (Run B), results in earlier core uncovery and failure of the reactor pressure vessel bottom head, approximately 22 minutes earlier than a medium break, i.e., 6 in. (15.24 cm) in diameter (Run A). When the ECCS operates following the break, as is the case in Run K, core uncovery does not occur until 2.7 hr and bottom head failure occurs at about 4 hr.

Fission product transport in the RCS is characterized in this paper by the transport of the fraction of the core inventory of iodine (i.e., as CsI) released into the RCS. Figure 3 depicts the fraction of CsI in the RCS for the four runs selected for comparison. Retention exceeding 75 percent is observed early in Run K, the case where the ECCS operates to delay core uncovery.

TABLE 2. MAAP V-sequence Analyses Selected for Comparison

Run	Pipe Break Size[1]	RCS Interfacing Sys. Connection	ECCS Available[2]	AFW Available[2]	Event Timing (hr): ECCS Failure	Uncovery	Bottom Head Failure
A	Medium	Cold Leg	No	No	0.00	0.81	1.66
B	Large	Cold Leg	No	No	0.00	0.67	1.31
D	Small	Hot Leg	No	Yes	0.00	1.39	2.59
K	Medium	Cold Leg	Yes	No	1.24	2.78	4.06

[1]Small break - 2 in. (5.08 cm) diameter, 0.02 ft^2 (0.002 m^2) break area
Medium break - 6 in. (15.24 cm) diameter, 0.20 ft^2 (0.018 m^2) break area
Large break - 10 in. (25.4 cm) diameter, 0.54 ft^2 (0.051 m^2) break area
[2]Unavailability indicates that the harsh environment created by the ECCS piping break flow is assumed to have affected the availability of these systems to operate to provide water to the core.

However, subsequent revaporization results in permanent retention of slightly more than 40 percent of the available CsI after 48 hr. Run D, in which a small break occurs with AFW functioning, also results in slightly more than 40 percent retention of CsI in the RCS, although the initial retention is not as high as for Run K, i.e., about 55 percent initial peak retention.

In both Run A and Run B, neither the ECCS nor AFW function and much earlier core uncovery and release of fission products to and deposition in the RCS occurs. The first peak shown in Figure 3 (at 1.3 hr) is for Run B, which includes a large ECCS piping break, and an earlier core uncovery time than Run A. Initially, approximately 50 percent of the available CsI is retained in the RCS, prior to bottom head failure in Run B. For Run A, which includes a medium ECCS piping break, the peak RCS retention is 44 percent of the available CsI. However, both Run A and Run B result in similar permanent retention in the RCS of 20 to 22 percent, after two days. No subsequent revaporization is anticipated based on the system temperatures, etc.

To illustrate the distribution within the various groups of nodes listed in Table 1 for one of the cases studied, Figure 4 depicts the retention of CsI for each group of nodes as a function of time for Run A, i.e., a medium ECCS piping break, with both ECCS flow and AFW flow unavailable. A substantial portion (32 percent) of the available CsI is initially deposited in the three intact loops in this four-loop plant. However, by roughly 3 hr after bottom head failure, about 20 percent remains in the intact loops. During the decrease in retention in the intact loops, there is an increase in the upper plenum and dome and the downcomer regions, totaling about 18 percent. (Note: The total retention is shown in Figure 3).

Migration of CsI between the broken loop cold leg, the intact loops, and the downcomer and upper plenum and dome is observed over the first few hours following release from the core, which occurs about 1.5 hr into the accident. However, deposition in the broken loop hot leg and pressurizer, which is only about 3 percent, occurs prior to vessel failure and continues after bottom head failure at 1.66 hr but does not decrease once deposited.

The data in Figure 4 are plotted out over time to the point that the retention is essentially stabilized, at 28 hr, with approximately 20 percent of the core inventory of iodine retained in the RCS.

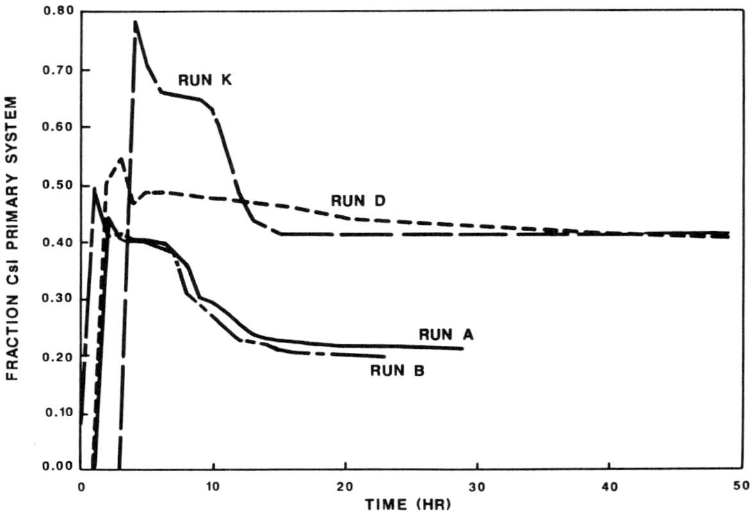

FIGURE 3. CsI retention in primary system in MAAP V–sequence analyses

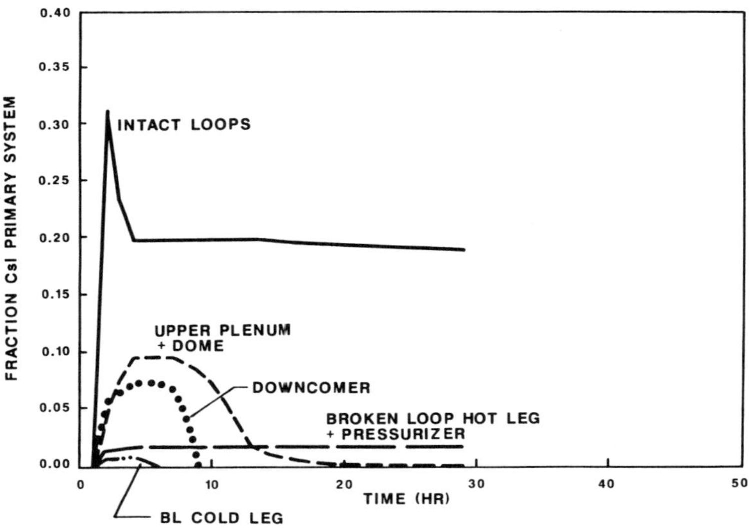

FIGURE 4. CsI retention in groups of nodes for MAAP V–sequence Run A

COMPARISON OF V–SEQUENCE FISSION PRODUCT RETENTION AS CALCULATED BY MAAP AND STCP

An STCP analysis of a V–sequence for the Zion plant was not available for comparison with the results of the MAAP analyses discussed earlier. However, BNL presented details of a V–sequence analysis for Surry in NUREG/CR–4629 entitled, "Independent Verification of Radionuclide Release Calculations for Selected Accident Scenarios" (6). In that report, BNL compared the Surry V–sequence analyses as reported in the initial BMI–2104 Volume I (8) with the updated analysis in BMI–2104 Volume V (9), which was published in 1984. The BNL analysis was published in 1986 and is believed to be more representative of the latest STCP versions.

In both the NUREG/CR–4629 STCP analyses and the MAAP analysis Run A, a medium 6 in. (15.24 cm) diameter, 0.20 ft^2 (0.018 m^2) ECCS piping break is assumed to occur with no ECCS or AFW flow to the core. The BMI–2104 and BNL analyses are incorrect in that intermediate and cold leg piping are omitted in the Surry model. Nonetheless, these differences are not expected to appreciably affect the results, and the STCP model of the RCS for the three-loop Westinghouse Surry plant would not be substantially different from an STCP model of the four-loop Westinghouse Zion plant, because only the broken loop is modeled in STCP.

The respective event timing for the MAAP Run A for Zion and the STCP V–sequence analysis for Surry are as follows:

EVENT TIMING (minutes)

EVENT	MAAP Zion Run A	STCP Surry V–Sequence
ECCS Failure	0	0
Core Uncovery	49	38
Core Collapse	97	76
Vessel Dryout	--	88
Bottom Head Failure	100	165

The data for MAAP Run A from Figure 4 have been replotted in Figure 5 for the first 5–hr time period.

Comparable data for the STCP V–sequence analysis of Surry are shown in Figure 6, as taken from NUREG/CR–4629. Although both analyses resulted in 40 percent retention in the RCS at 5 hr in the respective analyses, that apparent agreement masks the fact that the two analyses are very different. (Note: Data for times greater than 5 hr are presented in Figure 4 for the MAAP analysis.)

The major differences between the MAAP analysis results in Figure 5 and the STCP analysis results in Figure 6 are discussed briefly below:

1. Most of the CsI retention in the MAAP analysis is in the intact loops, which are not included in the STCP model.

2. Most of the retention in the STCP analysis initially is in the ECCS piping, which is not included in the MAAP model. However, if the STCP analysis is correct, the retention in the ECCS piping drops to zero by 2 hr into the accident, thus its effect is to delay release into the auxiliary building, in the STCP analysis.

   Note: ECCS piping typically is about 200 ft (60 m) between the RCS and postulated break location and involves four or more 90° pipe bends.

3. Approximately 20 percent of the available CsI is retained in the broken loop's steam generator in the STCP analysis and essentially none is retained there in the MAAP analysis.

4. CsI retention in the upper plenum was about 14 percent of the core inventory when bottom head failure occurred, at 2.75 hr in the STCP analysis. In the MAAP analyses, approximately 9 percent of the available CsI was retained in that region at about 4 hr and remained there until it subsequently revaporized between 5 and 15 hr leaving essentially no CsI in that region (see Figure 4).

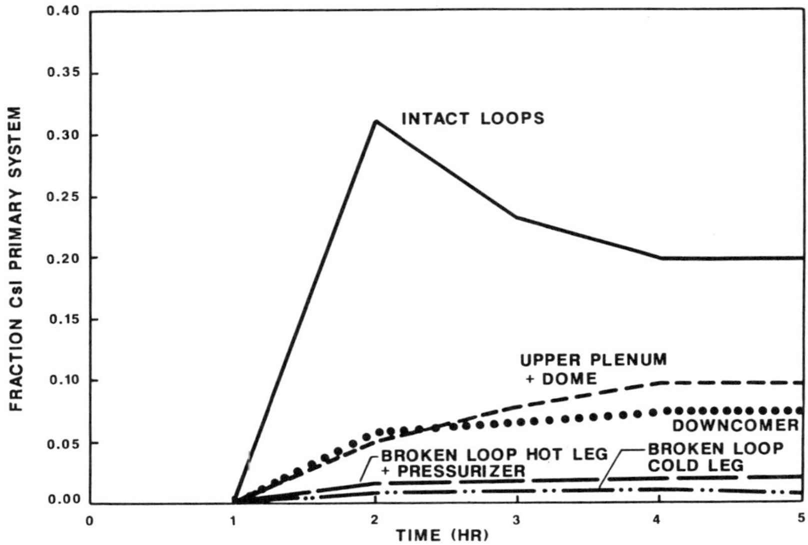

FIGURE 5. Initial 5-hour CsI retention in groups of nodes for MAAP V-sequence Run A

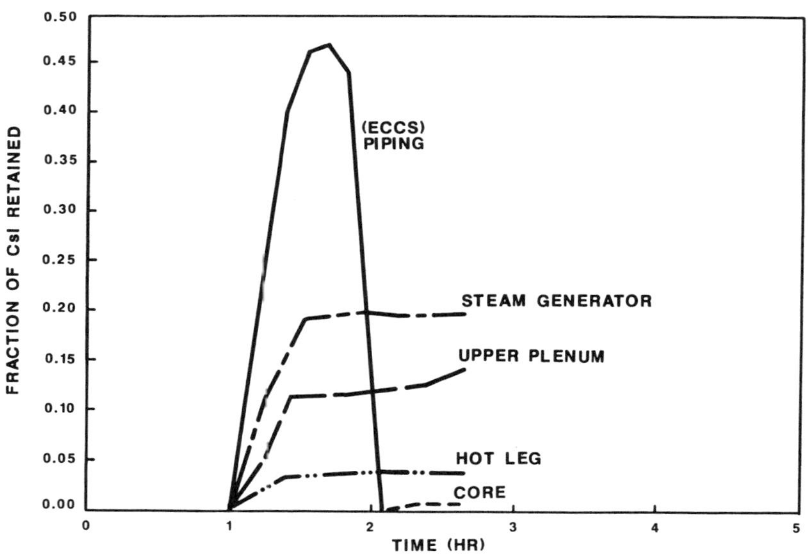

FIGURE 6. CsI retention in STCP analysis of V-sequence for Surry

5. In the MAAP analysis, retention in the downcomer within the reactor pressure vessel rose to about 7.5 percent within 5 hr, as shown in Figure 5, but subsequently revaporized and reduced to zero by about 9 hr, as shown in Figure 4. As noted earlier, that region does not exist in the STCP model as recirculation flow back through the reactor vessel is not modeled.

6. Retention in the hot leg and pressurizer is comparable in both analyses, i.e., approximately 3 percent in MAAP and 4 percent in STCP. Another 2 percent is retained in the cold leg in MAAP, which is not modeled in STCP.

It may be concluded that fission product transport and retention in the RCS differ substantially between MAAP and STCP.

The interdependence of thermal hydraulic and fission product transport processes are much more nearly correct as modeled in MAAP because it is an integral program with recirculatory flow and no disruption of the analysis at the time of bottom head failure. In addition, MAAP includes a much finer nodalization, with a much larger number of heat sinks.

However, a number of the results of STCP fission product behavior, such as are shown in Figure 6, are thermal–hydraulic–driven by such conditions as the long lower head water boiloff period with its associated hydrogen production and extremely high core exit gas temperatures. If those processes are real and are modeled correctly in STCP, then those portions of the analyses may be more nearly correctly modeled in STCP than in MAAP. However, those modeling differences may not affect the retention in the intact loops which MAAP calculates to be very substantial and which STCP does not even model.

One should be cautious in interpreting the above comparisons as generally applicable beyond V–sequence analyses. There is evidence indicating that much closer agreement exists between MAAP and STCP for PWR high pressure sequences such as TMLB', where the flows are different and post–vessel failure revaporization is not calculated to occur in MAAP, even though it is not analyzed in STCP.

Some of the differences noted between the MAAP and STCP analyses derive from different assumptions, e.g., those related to core melt progression and slumping into the lower head of the pressure vessel. Until further research is completed, it may be appropriate to bracket analyses by quantifying the fission product transport under both sets of assumptions. However, other differences derive from more fundamental modeling considerations, e.g., the very limited once–through serial nature of the nodalization in STCP in contrast with the recirculatory nodalization including the broken loop and the intact loops as modeled in MAAP. One approach to improving the understanding of the modeling and phenomenological differences involved is to perform comparable analyses with the NRC's recently developed MELCOR program (10). A three–way comparison among MAAP, STCP, and MELCOR for the same sequence at the same plant would be a major step toward bringing the analysis of fission product transport in the RCS into closer agreement than currently exists.

REFERENCES

1. Rasmussen, N. et al, Reactor Safety Study, U.S. Nuclear Regulatory Commission (NRC) WASH–1400 (NUREG–75/014), October 1975.

2. MAAP–3.0 Modular Accident Analysis Program User's Manual, Fauske & Associates Inc. Industry Degraded Core Rulemaking (IDCOR) Technical Report 16.2–3, IDCOR Program, February 1987.

3. Henry, R. et al, Modular Accident Analysis Program, (MAAP) Version 3.0B, Developed by Fauske & Associates, Inc. under EPRI RP–2726.

4. Silberg, M. et al, Reassessment of the Technical Bases for Estimating Source Terms, NRC NUREG-0956, July 1986.

5. *Evaluation of the Consequences of Containment Bypass Sequences*, EPRI Research Project No. 2726 (to be published in 1989).

6. Lee, M. et al, Independent Verification of Radionuclide Release Calculations for Selected Accident Scenarios, NUREG/CR-4627, Brookhaven National Laboratory, July 1986.

7. IDCOR Technical Summary Report - Nuclear Power Plant Response to Severe Accidents, Technology for Energy Corp., IDCOR Program, Atomic Industrial Forum, November 1984.

8. Gieseke, J. et al, Radionuclide Release Under Specific LWR Accident Conditions, Battelle Columbus Laboratories BMI-2104 Vol. 1, July 1983.

9. Gieseke, J. et al, Radionuclide Release Under Specific LWR Accident Conditions; Volume V - PWR Large, Dry Containment Design (Surry Plant Recalculations), Battelle Columbus Laboratories BMI-2104 Volume V, July 1984.

10. Alpert, D. J. et al, MELCOR Accident Consequence Calculation Code System, Sandia National Laboratories, NUREG/CR-4691 (in press).

Review of the Major Predicted Phenomena during FP Transport and Deposition in the RCS and Containment Building under Severe Accident Conditions

A. MARKOVINA and P. FASOLI-STELLA
Commission of the European Communities
Institute of Safety Technology-Ispra Site
21020 Ispra (Varese), Italy

A. MAILLIAT
IPSN-SEMAR
Cadarache, France

1. INTRODUCTION

The CEC (Commission of the European Communities) has promoted through the Shared Cost Action programme a collaboration among several European organizations to review the main results of national Source Term evaluations. The ojective of this action was to contribute to a coherent definition of a programme for the Phebus F.P. (Fission Product) project, an integral in pile experiment on FP/aerosol transport in the reactor cooling system (RCS) and in the containment building (CB) to be realized in Cadarache (F) as a joint CEA/CEC effort. The programme intends to investigate relevant phenomena which take place in the main reactor components under sufficiently prototypical conditions, rather than to quantify the FP/aerosol retention during severe accidents.

Therefore the main intent of this collaboration was to compare the phenomena and the corresponding range of parameters, as calculated by the codes available at national level, to assess the similarities and discrepancies of the various analyses and possibly to derive a common strategy for the definition of the test matrix.

Relevant information on the accident sequences considered in the national studies and the status of the analytical tools being utilized came also from this collaboration.

Ten organizations grouped in 5 national teams have participated in this exercise:

1.	GRS Garching IKE Stuttgart SIEMENS/KWU Erlangen	FRG
2.	AEE-Winfrith SRD-Culcheth	UK
3.	ENEL-Milan Univ. of PISA	Italy
4.	CIEMAT-Madrid Univ. of MADRID	Spain
5.	CEA-IPSN	France

This paper presents a summary of the comparison and analysis of these studies. The data presented in this paper will be analysed and compared without reference to the authors, and the five analytical groups will be simply identified by a number without any prefixed order.

2. GENERAL BACKGROUND

This study was not directed to any specific LWR plant design: the reactor system modelled was a "large modern PWR" and there was no direct connection with safety assessments which may be made for a real plant.

In general there is no substantial difference in the core and primary circuit design of the Nuclear Power Plants taken into consideration, while differences may be found in the design of the containment system.

Each group has been requested to assess the accident sequences which present the highest interest from the point of view of the risk to the population and of the relevant phenomena involved.

Sixteen accident sequences (Table 1) belonging to the basic three families: Large Break LOCA (AB), Small Break LOCA (S_2D) and Transient (TMLB') have been analysed.

In Table 2, for each basic family of accident sequences, the codes utilized by the five groups of analysts are indicated.

Concerning the core behaviour the most widely used code is BOIL from MARCH, with exception of the German S_2D calculations based on national codes MELSIM and LUECKE and the French calculations made with VULCAIN, which is part of the French system of codes ESCADRE.

Fission product and aerosol release from the core were computed by the US code CORSOR or by the national codes VULCAIN (F), FIPREM (FRG), FAEREL (UK). The thermalhydraulic response and the FP/aerosol transport in the reactor coolant system have been predicted by the Source Term Code Package (STCP) modules, namely MERGE and TRAPMELT, sometimes in national versions (SPRITE-TRAPMELT-UK, TRAPMELT/ENEL).

The containment thermalhydraulic response was mainly computed with the STCP modules, namely MACE from MARCH, with two exceptions: Germany (ASTRO-WAVCO-RALOC) and France (JERICHO).

The aerosol behaviour in the containment was computed with three different codes, namely NAUA (STCP), AEROSIM-M (UK) and AEROSOL B1 (France). Aerosol generation from corium-concrete interaction was obtained from VANESA (STCP).

TABLE 1. Accident Sequences Analyzed

Group of analysts	Number of sequences	Sequences
I	2	S_2B cold leg / TMLB
II	1	TMLB
III	3	AB hot leg / S_2D / TMLB
IV	3	AB hot leg / S_2D / TMLB
V	7	AB hot leg / AF cold leg / S_2D ALB cold leg / TCLD / 2 x TMLB'

TABLE 3. Timing of Key Events (time in minutes)

Sequence	Events \ Group	I	II	III	IV	V
AB	Core uncovery			14	42	5
	Start melt	NA	NA	38.6 }47.5	66 }18	10 }33
	Core slump			86.1	84	43
	Bottom head fail.			175.5	150	75
S_2D	Core uncovery	27.9		34.3		31
	Start melt	51.7 }33.2*	NA	61 }42.2	NA	155 }40
	Core slump	221.4		103.2		195
	Bottom head fail.	243.4		128.9		262
TMLB'	Core uncovery	119.8	140.1	149		47.5
	Start melt	154.7 }31.1	165.3 }20.5	188.4 }31.6		75 }59
	Core slump	185.8	185.8	220		134
	Bottom head fail.	195.3	191.5	246.3		136.5

NA Not available
* Accumulator discharge is assumed. A second core melt onset is delayed with respect to the first one by 136.5 m

TABLE 2. Codes Utilized in the Analysis of AB, S_2D and TLMB' Sequences

Sequence	Group	Core and RCS			Containment			
		Thermal-hydraulics	FP/aerosol release	FP/aerosol deposition	Thermal-hydraulics	Core/concrete interaction	FP/aerosol from C.C.I.	FP/aerosol deposition
AB	I							
	II							
	III	MARCH2/SPRITE	MATREL	TRAPMELT2	MARCH2		hand calc.	AEROSIM-M
	IV	MARCH/MELSIM2	ORIGEN/QUELLE	TRAPMELT	RALOC	WECHSEL	QUELLE	NAUA/IMPAIRG
	ALB	MARCH3	CORSOR		MARCH3	CORCON		
	V AB	BOILK/MERGE*	PEPIN/BOILK	TRAPMELT*	JERICHO	**		AEROSOL B1
	AF	VULCAIN	PEPIN/VULCAIN		JERICHO	WECHSEL		AEROSOL B1
S_2D	I	MARCH/MERGE	ORIGEN/CORSOR	TRAPMELT2 ENEL	MARCH3	CORCON		NAUA
	II					NO		
	III	MARCH2/SPRITE	MATREL/FAEREL	TRAPMELT2	MARCH2		VANESA	AEROSIM-M
	IV	MELSIM/LUECKE/RELAP	FIPREM	TRAPMELT	ASTRO/WAVCO			
	V	BOILK/MERGE*	PEPIN/BOILK	TRAPMELT*	JERICHO	**		AEROSOL B1
TMLB'	I	MARCH/MERGE	ORIGEN/CORSOR	TRAPMELT2 ENEL	MARCH3	CORCON		NAUA
	II	MARCH3/MERGE	ORIGEN/CORSOR	TRAPMELT3	MARCH3	CORCON	VANESA	NAUA
	III	MARCH2/SPRITE	MATREL/FAEREL	TRAPMELT UK	MARCH2			AEROSIM-M
	IV				ASTRO/WAVCO			
	TLCD	BOILK/MERGE*	PEPIN/BOILK	TRAPMELT*	JERICHO	**		AEROLOL B1
	V 2xTLMB'	MARCH3/MERGE	CORSOR	TRAPMELT3	MARCH3	CORCON		

* published data ** from published results

3. CORE AND RCS THERMOHYDRAULICS

In Table 3 the timing of some key events for the five groups of analysts is compared considering 3 basic sequences, namely AB, S_2D and TMLB'. As can be expected, there are differences in the predicted times, while the duration of the core melting, that is the phase from the melt onset up to core slumping is predicted to range between 20 m and 1 h, without a strong difference between the three analyzed sequences. The average expected time interval between core uncovery and core slump is slightly over one hour.

In order to characterize the thermohydraulic conditions during the time interval in which the most significant release of FPs from the core takes place, a time window is considered in which Cs is emitted from 5% to 90% of the total inventory. If the core collapse in the vessel bottom head occurs before 90% of Cs is released, the corresponding time is considered the end of the time window.

Table 4 describes the core response during the time window.

The duration of Cs release (column 4) never exceeds 2500 s, with the only exception of the AF sequence, where a long phase of core cooling by recirculation pump injection allows the decay heat to decrease considerably before core uncovery, approximately by a factor of two.

The two core average temperatures (column 2) correspond to 5% and 90% of Cs release respectively, which seems to indicate that the fuel temperature is still well below the UO_2 melting point, even though some quantity of liquid compounds may exist. The temperature ramp rate indicates a rather uniform behaviour of the three accident sequences. The two values given in column 3 correspond respectively to the maximum and minimum slope during the time window. The typical value can be placed in the range 0.5 to 1.5 K/sec.

Approximate values for the predicted mass flow rates are given in column 5. The flow composition during the time window is not uniform. In general, as illustrated in Fig.1, three different flow conditions can be identified for all the analyzed sequences:

1. A steam prevailing atmosphere with a mass flow rate decreasing from 10 kg/sec to low values.
2. An hydrogen rich atmosphere with a mass flow rate typically in the range 0,1 - 0,3 kg/sec.
3. A steam atmosphere with very high mass flow rates generated by the core collapse in the vessel bottom head.

The two values of mass flow rates given in Table 4 correspond to the first two phases just mentioned. For all the sequences more than 50% of noble gases and Cs is released during the second phase (in reducing atmosphere), and approximately 80% or more of noble gases and Cs before core collapse.

TABLE 4. Characteristic Parameters of AB, S_2D and TMLB' Sequences before Core Collapse

Sequence \ Column		1	2	3	4	5	6	7
		Pressure range (MPa)	Core maximum average temp. or max. temp. (K)	Temp slope (K/s)	FP release duration 5-90% based on Cs (s)	Flow rate range (kg/s)	FP fraction release before core collapse based on Cs (%)	Oxidation before core collapse (%)
AH	hot leg	0.14	2200	2.2 / 0.6	1200	7.5 → 0 / 0.1	≳90	60
AB	hot leg	0.35-0.3	1600-2760	1.4 / 0.3	1320	5 → 0 / 0.3-0.1	100	46
AF	cold leg	0.36	1400-2000	0.6 / 0.05	7200	6 → 0 / 0.1	88	93
ALB	cold leg	0.2	1400-2300	1.1 / 0.2	1450	16.1 → 0 / 0.1	≳80	35
AB	hot leg	low press.	NA	NA	1200	3 → 0 / 0.1	99	26
S_2D		8.2-4.3	2100	1.4 / 0.6	1000 ***	15 / 0.4	Cs: 36% *	23 **
S_2D		8.6-1.7	1300-2740	1.5 / 0.3	1560	<1 / 0.4-0.1	100	54 **
S_LD		7.9-1.9					≳98	59
S_LCD		NA	NA	NA	1800	NA	95	67
TMLB'		16.5	1600-2760		2000	1 → 0 / 0.2	95	NA
TMLB'		17.45	1300-2400	0.25 / 0.1	2400	1 → 0 / 0.05	≳90	26
TMLB'		17.4	2200	~0.8	NA	15 → 0 / 0.12		22
TLCD		16	NA	NA	2400	NA	100	NA
TMLB' β/δ		8 - 15.5	1400-2300	1.2 / 0.5	1300	14 → 0 / 0.3	77	45

* FP release not modeled during reaction; ** accumulator not activated; ***smaller time frame for release; (1) only partial core collapse predicted

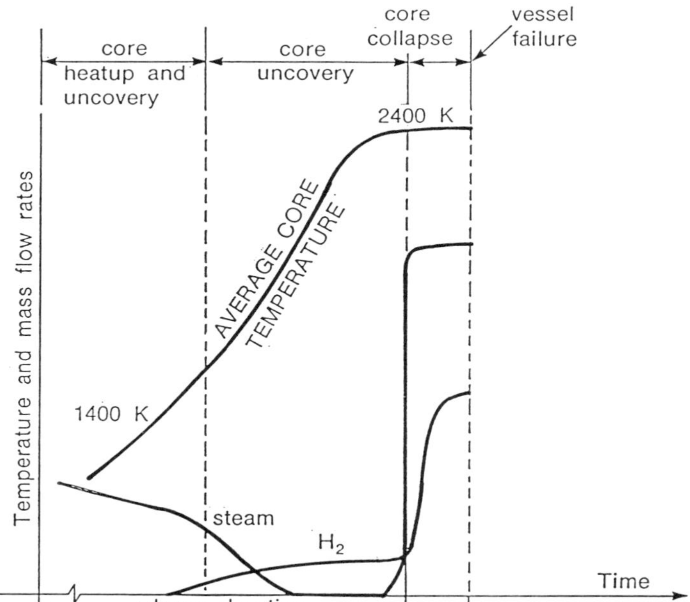

FP release	release duration ~1 H		
core chemistry	oxidizing conditions	reducing conditions	oxid. cond.
core temperature	low	medium-high	high
steam flow	medium (x . 1 kg/s)	low or none	high 10 kg/s
H_2 flow	low	low	high

FIGURE 1. Typical Thermohydraulic Conditions in the Reactor Core

4. CONTAINMENT THERMOHYDRAULICS

In Table 5 the main parameters characterizing the containments are presented, all referred to 0.1 MPa and 300°K. Large differences can be observed between containment volumes and available surfaces for heat transfer, as well as large differences for the sedimentation area.

In Table 6 the reactor power and RCS water versus containment volume are given.

The overall containment thermohydraulic behaviour can be described considering three distinct phases as shown in figure 2.

The first one, the short term phase, starts with the beginning of core uncovery and ends just before the vessel bottom head failure. Its duration is of the order of one hour.

The second phase, the intermediate term phase, begins with the vessel failure and ends when the reactor cavity is dryed out. Its duration ranges between one and three hours.

The third phase, the long term phase, concerns the containment behaviour for several days.

It can be seen in Table 7 that, with the only exception of the AF sequence, the first period of FP emission never exceeds 1 hour; for the second period of emission the results are more scattered, but all within a time interval of 1 hour. Therefore the total FP emission time does not exceed two hours.

The containment pressure never exceeds 0.5 MPa for any sequence and its variation from the time of core uncovery (first pressure column) to the bottom head dry-out time (second pressure column) is less than 0.05 MPa. The atmosphere temperature never exceeds 400 K with a variation in this time interval of ~30 K. The main parameters characterizing the intermediate range phase are the coolant inventory of the primary system and of the accumulator (and eventually the ECCS) which are discharged into the lower cavity and interact with the molten core material. It is clear that for low pressure sequences all the coolant has been already released through the RCS break and this phase practically does not exist. For high pressure sequences (e.g. TMLB') and intermediate pressure sequences (e.g. S_2D), whose pressure remains above the accumulator discharge set point, this phase becomes significant.

The data shown in Table 8 are affected by some inconsistency as some computations were done assuming the immediate onset of the corium-concrete interaction while the others assuming the interaction starting upon cavity dryout. Nevertheless the maximum duration predicted for this phase is 3 hours, while the pressure increase is within 0.3 MPa.

TABLE 5. Main Characteristics of the Reactor Containment Building

Group	Volume (m^3)	Heat transfer area (m^2)	Sedimentation area (m^2)	Volume / HT area (m)	Volume / sed.area (m)	Initial mass of air (P=0.1 MPa T=300 K) (ton)
I	73000	19750	1550	3.7	47	86.1
II	56917	28058	1256	2	45	66.7
III	85500 91100	41827	1350	2 - 2.2	63 - 67	100.9 - 107.3
IV	68000	39700	2400	1.7	28	80.0
V	49400	50000	1060	1	47	58

* minimum and maximum values

TABLE 6. Energy and Coolant Mass per Unit Containment Volume

Group	Core nominal thermal power (MW)	Thermal power / cont. volume (kW/m^3)	RCS water / cont. volume (kg/m^3)
I	2775	38	3.4
II	2696	47	4.4
III	3411	40	4.18
IV	3733	55	3.9 *
V	2785	56	5.7

* Without accumulator discharge

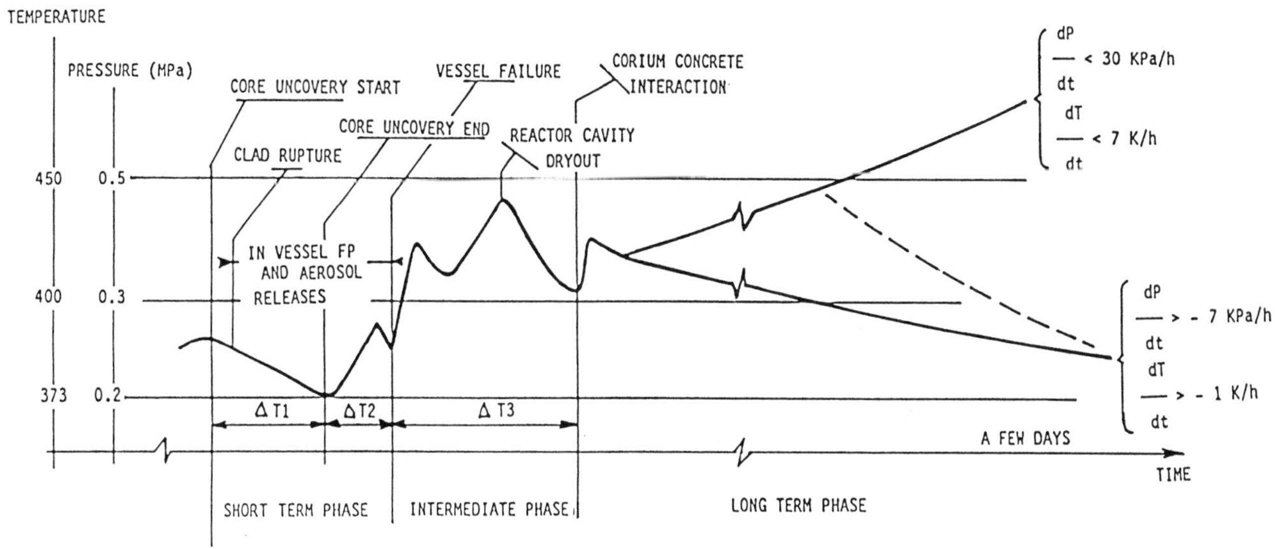

FIGURE 2. Overall Thermohydraulic Behaviour of the Containment Building

TABLE 7. Thermohydraulic Parameters of the Short Term Phase for AB, S_2D and TMLB' Sequences

Sequence	Duration (s)		Atmosphere pressure (MPa)		Atmosphere temperature (K)	
	Δt1	Δt2	Core uncovery	Bottom head dry-out	Core uncovery	Bottom head dry-out
AB	1200	260	0.35	0.31	400	416
AB	1320	2000	0.34	0.36	398	401
AB	1200	120	0.14	0.16	365	375
ALB	1450	730	0.2	0.26	377	365
AF	7200	11700	0.45	0.48	410	443
S_2CD	1800	250	0.2	0.2	371	370
S_2D	-	-	0.19	0.28	365	387
S_2D	1000	1100	0.165	0.18	362	354
S_2D	1560	100	NA	NA	NA	NA
TMLB'	2400	360	0.21	0.267	372	383
TMLB'	1680	264	0.26	0.28	385	388
TMLB'	2000	300	0.1	0.23	320	326
TLCD	2400	150	0.26	0.22	382	371
TMLB'	1300	120	0.27	0.37	388	400

TABLE 8. Thermohydraulic Parameters of the Intermediate Phase (for TMLB' Sequences)

Sequence	Duration (min.)	Over pressure (MPa)	Temperature increase (K)
S_2D	17	0.235	120
TLMB'	190	0.2	373
TMLB'	70	0.2	26
TLMB'	190	0.15	30
TLCD	NA	0.3	44

The long-term phase is dominated by the corium-concrete interaction and may last for a few days.

In three cases siliceous concrete and in two others limestone concrete were considered. The initial corium temperature ranged from 1650 K to 3030 K.

As shown in Table 9 large differences exist in the computed duration of this phase. In average terms the containment initial pressure is 0.3 MPa and the temperature is 380 K.

5. FP/AEROSOL TRANSPORT AND DEPOSITION

The main objectives of the analysis of FP/aerosol transport are to identify the phenomena prevailing in specific time intervals and in specific system components, the retentions expected in the reactor coolant system and the locations where the largest deposits may occur.

5.1 Reactor Cooling System

In Tables 10, 11 and 12 the predicted retentions for the three families of accident sequences (AB, S_2D, TMLB') are provided. They are given for CsI, CsOH, Te and aerosols, and are expressed as percent of the total core emission at the time of core collapse. It can be noted that there is practically no retention of CsI and CsOH in RCS for large break LOCA sequences.

For small break LOCA sequences it can be said that roughly 60-85% of core release is retained in RCS. In this case steam generator is predicted to be the location of largest deposits.

For transient initiated sequences the deposition in RCS is predicted to be in the range 70-95%. For Cs compounds and Te the upper plenum appears as the most effective trapping location. Aerosols present also an important deposition in the upper plenum, but the reactor core is also effective.

The major mechanisms contributing to the FP/aerosol retention in RCS are shown in Tables 13 and 14.

For the small break LOCA transients, chemisorption is identified as the major mechanism for CsOH and Te retention in the upper plenum. Aerosol behaviour is predicted to be dominated by thermophoresis in one case and by settling in another one. Turbulent deposition and settling are the main mechanism predicted in the steam generator.

TABLE 9. Thermohydraulic Parameters of the Long Term Phase

Sequences	Duration (hours)	Pressure (MPa)		Temperature (K)	
		Initial	Final	Initial	Final
AH	48	0.12	0.1	361	373
AB	47	0.367	0.469	400	406
S_2D	32	0.18	0.145	354	342
TMLB'	46	0.25	0.338	375	398
S_2D	20	0.3	0.38	392	414
TMLB'	20	0.28	0.4	385	411
TMLB	5	0.474	0.595	412	417
AB ε *	168	0.27	0.5	381	404
S_2CD ε *	168	0.2	0.57	370	404
TLCD ε *	168	0.53	0.57	416	404
AB β *	11	0.27	0.2	381	355
S_2CD β *	8	0.2	0.16	370	362
TLCD β *	10	0.53	0.2	416	357
ALB	31	0.21	0.22	380	370
AF	28	0.485	0.27	443	510

* Conditions of containment isolation/failure according to WASH-1400 nomenclature

TABLE 10. Deposition in RCS at Core Collapse Expressed as Percent of Total Core Emission for AB Sequences

AB	Group	CsI	CsOH	Te	Aerosols
Total deposition as	III	0	0.4	30.1	6.6
% of core emission	V	3	4	70	48
Core	III	0	0	0	30
	V	NA	NA	63	79
Upper plenum	III	0	100	100	70
	V	NA	NA	37	21

TABLE 11. Deposition in RCS at Core Collapse Expressed as percent of Total Core Emission for S_2D Sequences

S_2D	Group	CsI	CsOH	Te	Aerosols
Total deposition as	I **	21.8	23.95	38.64	25.83
% of core emission	III	62.4	62.7	82.8	61.7
	V	74	76	90	73
Core	I	0.2	1	2	7.4
	III	0	0	0	3.1
	V	NA	0	NA	40
Upper plenum	I	0.1	7	17.6	8.4
	III	0	2.4	61.2	20.4
	IV	0	NA	NA	NA
	V	39	50	97	31.5
Hot leg	I	0.05	0.6	3.8	3.3
	III	0	0.1	5.2	5.3
	IV	0/3.3 *	NA	NA	NA
	V	16	14.5	NA	8
Steam generator	I	99.65	91.4	76.2	80.9
	III	100	97.5	33.6	71.2
	IV	30/15 *	NA	NA	NA
	V	43	35.5	NA	20.5

* with/without resuspension. Calculations terminated due to hot leg failure approx. 250 s prior to core collapse
** core quenching occurs during FP release

TABLE 12. Deposition in RCS at Core Collapse Expressed as Percent of Total Core Emission for TMLB' Sequences

TMLB'	Group	CsI	CsOH	Te	Aerosols
Total deposition as % of core emission	I	86.5	85.2	73	83.3
	II	69.5	72.6	39.3	71.6
	III	69.3	82	84.1	96.1
	V	85	85	85.4	93
Core	I	1.1	10.8	16.5	23.5
	II	0	0	0	0
	III	0	0	0	73.6
	V	4	4	19.5	33
Upper plenum	I	76	66.7	62.7	49
	II	76.1	77	56.6	63.6
	III	53.8	72.8	96.9	22.6
	V	80.5	82	77	51.7
Hot leg	I	8.2	7.7	3	6.3
	II	7.7	7.4	10.3	7.2
	III	11.5	7.4	1.8	1.2
	V	*	*	*	*
Surge line	I	9.3	8.8	6.6	14.4
	II	14.3	13	18.8	27.5
	III	13.3	7.2	0.65	0.8
	V	**	**	**	**
Pressurizer	I	5.4	6	11.2	6.8
	II	1.9	2.6	14.3	1.7
	III	21.4	12.6	0.65	1.8
	V	15.5	14	3.5	14

* FP/aerosol retention included in upper plenum retention

** FP/aerosol retention included in pressurizer retention

TABLE 13. Main Retention Mechanisms in Upper Plenum and Steam Generator at Core Collapse for Small Break LOCA Sequences

Groups / Upper plenum	I		III		IV
CsI	No retention (deposit revaporized)		No retention		No retention (deposit revap.)
CsOH	Sorption	100%	Sorption	100%	
Te	Sorption	100%	Sorption	100%	
Aerosols	Thermophoresis Settling (AMMD 1-3μm)	87% 7%	Settling Thermophor.		
Steam generator					
CsI	Turbul. deposition Settling	52% 48%	Settling		Turb. deposition Settling
CsOH	Turbul. deposition Settling	52% 48%	Settling		
Te	Turbul. deposition Settling Sorption	50% 26% 24%	Settling Sorption	40% 60%	
Aerosols	Turbul. deposition Settling (AMMD 3-6μm)	62% 38%	Settling	100%	

TABLE 14. Main Retention Mechanisms in the Upper Plenum at Core Collapse for Transient Initiated Sequences

Groups	I		II		III	
CsI	Settling Thermophoresis	99% 1%	Settling	100%	Settling Vapor cond.	92% 8%
CsOH	Settling Condens. + sorpt.	93% 7%	Settling	≥99%	Settling Sorption Vapor cond.	79% 18% 2%
Te	Sorption	100%	Settling Sorption	68% 32%	Sorption	100%
Aerosols	Settling (AMMD 1-6μm)	≥99%	Settling	100%	Settling (AMMD 1.8-8.5 μm)	≥99%

For transient initiated sequences agglomeration and gravitational sedimentation are dominating, because of the small velocity associated to high density of the carrier gas. Te retention is still dominated by chemisorption.

5.2 Containment Building

As not all the analysis groups have taken into account the gases, FPs and aerosols emitted from corium concrete interaction, the comparison of results is here limited to the time interval preceeding the onset of interaction.

A typical description of phenomena is given in table 15 for the TMLB' sequences.

Before core collapse, steam condensation on aerosol particles takes place and wall condensation is low. Therefore settling is more effective than diffusiophoresis. After core collapse, a large amount of hot steam flows to the containment and produces an increase of atmosphere temperature. In these conditions the steam condensed on particles will evaporate again and the overall particle size will decrease, thus reducing gravitational settling. At the same time wall condensation will increase, favouring the enhancement of diffusiophoresis.

If an intermediate phase is present between the short- and long-term phases, the accumulator water, discharged on the corium in the cavity, will evaporate and will produce roughly the same effect described previously.

In order to compare the results of the various calculations the main parameters characterizing sedimentation and diffusiophoresis were reduced with respect to the total emitted mass (note that these quantities are not adimensional parameters and therefore cannot be used for dimensional analyses). In Table 16 the mass settled, plated on walls and suspended are expressed as percent of the mass emitted and are also referred to the unit of settling surface, lateral wall surface and containment volume, so to take account of the differences in containment geometries.

It can be seen that the reduced parameters for settling range between 1.7 and 7.7 after six hours. These values can be considered as relatively homogenous, considering that the emitted masses vary by a factor of 20.

For diffusiophoresis the reduced parameters range between 4.2 and 18. The largest value is found for a containment with significantly lower area available for diffusiophoresis.

TABLE 15. Mechanisms Dominating Aerosol Transport and Retention Before the Onset of Corium-Concrete Interaction in a TMLB' Sequence

Phenomena	Parameters	Range of param.	Relat. contr.
A) 158.6-185.5 min			
Steam cond. on and evap. from particles	Steam cond. evap. rate on bulk	100-0 g/s 0-54 g/s	
Gravitat. settling	AMMD	27.28-2.4 μm	95.4%
Diffusiophoresis	St.cond.rate on walls	1E4-4E3 g/s	4.6%
Agglomeration	Particles concentrat.	5E5-6E9 part/m^3	
B) 185.5-191.5 min			
Steam evap. from part.	St.evap.rate on bulk	0-684 g/s	
Diffusiophoresis	St.cond.rate on walls	1E4-4E3 g/s	80.6%
Agglomeration	Particles concentrat.	6E9-1E12 part/m^3	
Gravitat. settling	AMMD	0.634-0.35 μm	19.4%
C) 191.5-193.5 min (Puff release)			
Diffusiophoresis	St.cond.rate on walls	8E3-1E5 g/s	85.4%
Agglomeration	Particles concentrat.	~1E12 part/m^3	
Gravitat. settling	AMMD	0.634-0.6 μm	14.6%
D) 193.5-334.5 min			
Agglomeration	Particles concentrat.	1E12-1E11 part/m^3	
Diffusiophoresis	St.cond.rate on walls	2.5E3-5E4 g/s	59.4%
Gravitat. settling	AMMD	1.2-0.6 μm	40.6%

TABLE 16. Characteristic Values of Aerosol Depletion Six Hours after Start of Release

Groups		m_T (kg)	m_T/V (g/m^3)	m_S/m_T (%)	$\frac{m_S}{m_T \cdot S_F}$ ($10^4/m^2$)	m_D/m_T (%)	$\frac{m_D}{m_T \cdot S}$ ($10^6/m^2$)	m_a/m_T (%)	$\frac{m_a}{m_T \cdot V}$ ($10^6/m^3$)
I	S_2D	743	10.2	47	3.7	15	7.7	37	5.2
I	TMLB	75	1	32	2.5	36	18	32	4.4
II	TMLB	120	2.1	21	1.7	37	13	38	6.6
III	TMLB	303	3.5	88.7	6.6	10.7	4.2	0.6	0.07
IV	AB	1300	19	46	1.9	18	4.6	9	1.2
V	AF	640	12.8	76	7.7	NA	NA	19	3.9

m_T = total aerosol mass emitted
m_S = settled mass by sedimentation
m_D = settled mass by diffusiophoresis and diffusion
m_a = suspended mass
S = sedimentation area
S_F = diffusiophoresic area
V = containment volume

For the suspended mass, neglecting the value of 0.07 which was obtained considering available for sedimentation all the horizontal surfaces (and corresponding therefore to the highest sedimentation value), the reduced parameters are within a relatively narrow range.

6. CONCLUSIONS

1) For sixteen accident sequences the Source Term has been analysed by 5 analytical groups referring to "modern large PWRs" of different designs and utilizing different sets of codes.

2) If core uncovery and heat-up is not delayed by emergency cooling systems, the core collapse into the reactor vessel bottom bead is predicted to occur after about one hour from the core uncovery.
The predicted fuel temperature ranges between 1400 and 2400°K during uncovery with a typical ramp rate of $\leq$1.5°K/sec.

3) During core heat-up more than 90% of Cs and I are anticipated to be released from the fuel. The flow composition in this time interval passes rapidly from oxydizing to reducing conditions, until core slumping takes place producing a rapid surge of steam.

4. The aerosol retention predicted in RCS is roughly 60% for low pressure sequences, increasing to 70-90% for high pressure sequences.

5) Gravitational settling and diffusiophoresis appear to be the two main depletion processes in the containment building during in vessel FP/aerosol emission.

The comparison of the thermohydraulic and FP/aerosol transport calculations has shown that, beyond largely expected differences due to plant design, detailes of the accident sequences analyzed and codes utilized, some overall simularities may be found. On this basis a coherent definition of the Phebus F.P. experimental programme will be undertaken and in particular the relevant matrix of tests will be identified.

Representation of the Fission Products Behavior during Severe Accidents into a Global Experiment: Dimensioning of the Phebus F.P. Experimental Set-Up

E. SCOTT DE MARTINVILLE and C. HUEBER
Centre d'Etudes Nucleaires de Cadarache
13108 Saint Paul Lez Durance
Cedex, France

I INTRODUCTION

The general objective of the PHEBUS F.P.programme /1/ is to provide data for the code systems validation, in conditions suitable to involve the coupling of the different phenomena that influence the F.P. behavior up to the release to the atmosphere.
On one hand, knowing the type of the sequences for which it is desired to improve the code systems, a first description of the phenomena and the range of parameters to be respected can be established. A set of scenarios to check the dimensioning of the set up can be described in terms of boundary conditions, once scaling laws have been defined.
On the other hand a preliminary design of the set up has to be proposed for verification calculations. This design has to take into account the major phenomena to be represented, to integrate them with the constraints of the facility, and to propose scaling laws in order to perform the code verification in the right physical domain for the parameters. This is the subject of this paper.
A general consideration is that the core, reactor cooling system (RCS) and containment involve so many phenomena that it is not possible to obtain the similitude on all of them for the fission product behavior. The number of phenomena will be chosen in correspondance with the number of parameters to be fixed for the experimental set up. The scaling criteria for dimensioning and representativity problem will be successively discussed hereafter for each part of the experimental set up : fuel bundle, circuits and containment.

II. REACTOR CORE REPRESENTATION

As it is whished to represent the F.P. behavior in the RCS and containment, the criteria of the PHEBUS fuel bundle representativity are the composition of the gaseous flow issued from a reactor core, and its temperature and pressure. The use of real irradiated fuel is the main feature that ensures the representation of the reactor. As the temperature level is the main parameter governing the F.P. release, the fuel temperature map and temperature history will have to be similar to the reactor one in order to respect the composition of the mixture including F.P. and carrier gas. But this similarity needs to be achieved only during the F.P. release period, which allows some freedom for a preconditioning phase up to the temperature of F.P. release.

II.1. F.P. inventory and reirradiation policy

The inventory is obtained primarily through the irradiation of the fuel which should lie between 20 and 30 GWd/t. But this fuel will be disactivated for some years before the experiment. It remains that the chemical composition is not far from a reactor one, but short life fission products will have disappeared. Simple calculations have shown that some days to some week for the reirradiation would be sufficient to build up a large proportion of the activity inventory (Fig 1). A 15 days reirradiation was chosen to ensure mainly the F.P. activity with respect to possible radiolysis effect in the sump water.

Considering a 900 MW nuclear plant for which the mean burn up is around 15 GWd/t (between 10 and 20 GWd/t) and comparing it with a 20 rods fuel bundle of one meter high at average burn up between 20 and 30 GWd/t, an inventory ratio (Ir) can be defined to compare the quantity of available F.P. between the reactor and the experiment. For our case, this inventory ratio is defined as :

$$Ir = \frac{41500 \text{ pins X } 3{,}66 \text{ m X } 15 \text{ (GWd/t)}}{20 \text{ pins X } 1 \text{ m X } 25 \text{ (GWd/t)}} = 4500$$

depending on the respective burn up this number can range between 2500 and 7500.
It will have a major importance in the dimensioning of the set up.

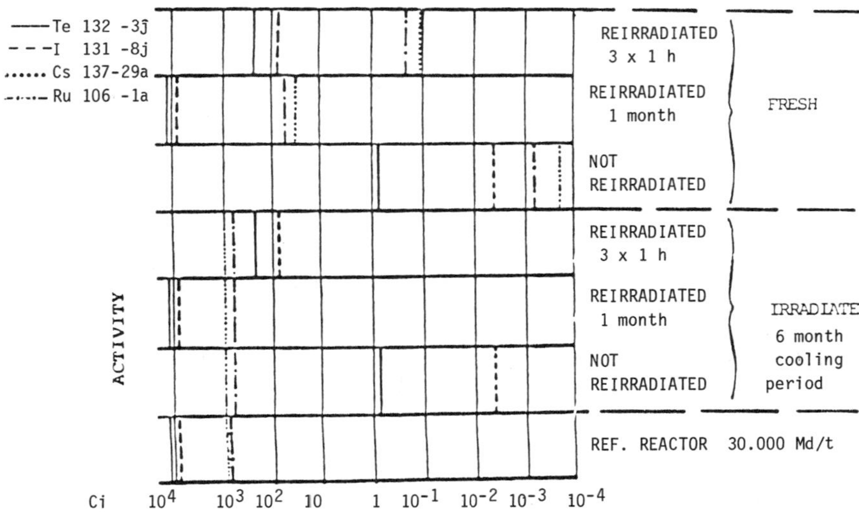

FIGURE 1. Activity of the fuel used in PHEBUS as function of reirradiation.

II.2 F.P. release and fuel bundle degradation

With respect to the composition of the carrier gas charged with F.P. this implies a given space-time temperature for the fuel, but also a fixed inlet carrier gas mass flow which is transformed by zircaloy oxidation when crossing the bundle. This implies that the power delivered to the fuel should be determined also in space and time, in order to produce the desired temperature map history for the chosen carrier gas flow. This consideration has to be modified due to two aspect :

- there is some freedom to preoxidize the bundle during the first part of the test (below F.P. release temperature), which gives some freedom on the inlet carrier gas mass flow rate and composition.

- The temperature level necessary for the F.P. release, will result in a high degradation of the fuel bundle associated with a possible relocation of the fuel. May be this phenomenology is quite comparable to the reactor case, also considering the presence of a control rod in the tested bundle, but a major difference between the reactor and the test, consists in the fact that the power to heat up the fuel relocates with the fuel in the reactor case, whereas the power distribution in the fuel is determined by the driver core in the case of the experimental set up. The possibility to improve the prediction of the fuel degradation in severe accidents will be given by the SFD program which is presently underway.

III. REPRESENTATION OF THE RCS.

III.1.Criteria for scaling

The PHEBUS F.P. design includes constant characteristics for the fuel and containment but it is wished to represent the main flow paths that can be used by the fluid from the core to the containment or to the atmosphere. A simplified approach, by scaling down volumes, areas and structure masses shows that pipes diameter, length or thickness are not compatible with the constraints of the set up. Moreover, the representation of the circuits needs to be performed on well defined phenomena. Aerosols deposit may happen through turbulent diffusion and impaction, thermophoresis, diffusiophoresis and settling. Therefore the choice of representativity criteria may be very difficult.In that quantity of phenomena, the ones that were chosen for scaling down were :

- fluid space time temperature for aerosols formation,
- mean residence time of fluid in the circuit for aerosols settling
- vapor condensation on walls.

These criteria may be written in analytical form as :

$$tr/trp = 1$$

$$\frac{\partial T_f}{\partial x} \Big/ \left(\frac{\partial T_f}{\partial x} \right)_p = 1$$

$$\frac{\partial T_s}{\partial t} \Big/ \left(\frac{\partial T_s}{\partial t} \right)_p = 1$$

$$\frac{1}{M} \frac{dM}{dt} \Big/ \left(\frac{1}{M} \frac{dM}{dt} \right)_p = 1$$

Letters without subscript refer to the reactor.
III.2. Analytical expression of scaling criteria

Simplified balance equations can be established for the heat exchange between the fluid and the pipe wall :

$$H \cdot A \cdot (T_f - T_s) = S \cdot Cs \cdot \frac{\partial Ts}{\partial t} \quad (1)$$

for the heating of the pipe, neglecting the heat losses of the pipe. The temperature change of the fluid is then :

$$H \cdot A \cdot (T_f - T_s) = Q_f \cdot C_f \cdot \frac{\partial Tf}{\partial x} \quad (2)$$

In the case of turbulent flow, the Nusselt number is :

$$Nu = 0.02 \cdot Re^{0.8} \cdot Pr^{0.4} \quad (3)$$

The F.P. condensation through turbulent deposition of a given gaseous species on the wall may be expressed by :

$$\frac{dM}{dt} = K.A.\ (Y - YS) \quad (4)$$

the concentration Y in the fluid is usually much bigger that the equilibrium concentration at the wall temperature YS, so that the relative variation of the concentration may be expressed as :

$$\frac{1}{M} \frac{dM}{dt} = K \cdot A \ / \ QV \quad (5)$$

K, the mass transfer coefficient may be expressed with the help of the diffusion coefficient F of the considered species in the fluid and of the Sherwood number SH as :

$$K = F \cdot SH \ / \ D \quad (6)$$

In turbulent deposition the Sherwood number is :

$$SH = 0.023\ RE^{0.83} \cdot S_C^{0.33} \quad (7)$$

If we define the following dimensionless parameters to compare the reactor case to the PHEBUS one :

α =	Q / Qp	mass flow rate ratio
β =	D / Dp	diameter ratio
γ =	S / Sp	structure mass ratio
ϵ =	L / Lp	length ratio

It is easy to show that :

$$RE / RE_p = \alpha / \beta \quad (8)$$

$$SH / SH_p = \alpha^{0.83} / \beta^{0.83} \quad (9)$$

$$K / Kp = \alpha^{0.83} / \beta^{1.83} \quad (10)$$

$$H / H_p = \frac{\alpha^{0.8}}{\beta^{1.8}} \quad (11)$$

so that finally the scaling criteria may be expressed as :

$$tr / tr\ p = \beta^2 \cdot \epsilon / \alpha \qquad \text{for the ratio of residence times} \quad (12)$$

$$\frac{\partial T_f}{\partial x} \Big/ \left(\frac{\partial T_f}{\partial x}\right)_p = \epsilon / \alpha^{0.2} \cdot \beta^{0.8} \quad \text{for the fluid cooling} \qquad (13)$$

$$\frac{\partial T_s}{\partial t} \Big/ \left(\frac{\partial T_s}{\partial t}\right)_p = \alpha^{0.8} . \epsilon / \beta^{0.8} . \gamma \quad \text{for the wall heating} \qquad (14)$$

$$-\frac{1}{M} \cdot \frac{dM}{dt} \Big/ \left(\frac{1}{M} \frac{dM}{dt}\right)_p = \epsilon \Big/ \left(\alpha^{0.17} . \beta^{0.83}\right) \quad \text{for the condensation on the walls} \qquad (15)$$

It is possible to eliminate γ , with the additional condition that the stress should be equal in the reactor and PHEBUS pipes : $\gamma = \beta^2$

III.3. Discussion on the scaling

A first comparison between the different similitude conditions shows that the fluid temperature evolution (equation 13) is very similar to that concerning the F.P. condensation on walls (equation 15). These two conditions will be assumed to be identical for the rest of the analysis. It is possible to make a graphical representation of the different criteria with α and β as parameters (Fig 2). On that Figure 3 the constraints of the experimental set up are represented by :

$\beta \geqslant 10$ PHEBUS diameter smaller than 74. mm
$\alpha \geqslant 2000$ This comes from the inventory effect to keep the different species concentration.

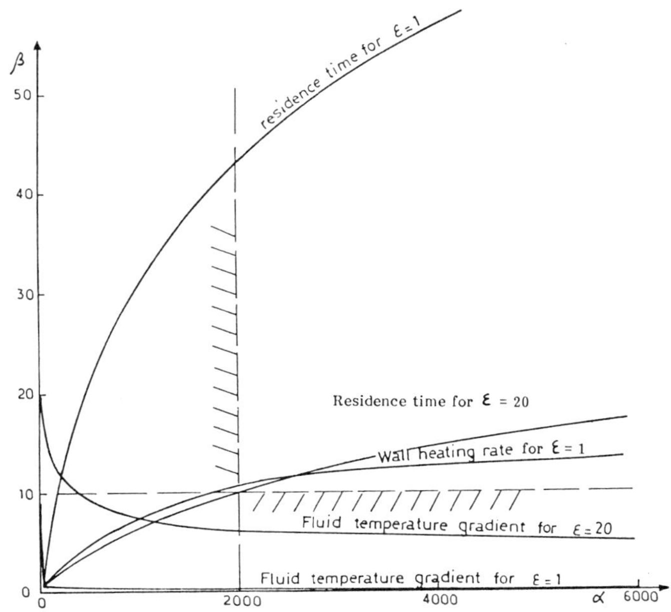

FIGURE 2. Representation of the scaling criteria for the RCS in relation with the experimental constraints.

On that diagramme, it can be seen that for ϵ increased, that is reducing the length of the pipes in PHEBUS, it is possible to find a solution to minimize the representativity problems. This can be done on portions of the circuits but not everywhere.
On the contrary, some portions of the circuits cannot be shorten because of experimental constraints : the circuit has to cross the driver core pool.
In that case, the fig. 2 shows that there is no hope to find a similitude on the whole length of the pipes.
For the case of long pipes a compromise was therefore proposed including a short length of representative pipe and a length for which the depletion of F.P. has to be minimized. For these pipes which are desired to be neutral with regard to F.P. behavior it is proposed to fulfill two conditions :

- control the wall temperature to avoid F.P. condensation
- minimize the residence time in these pipes to decrease the aerosol settling.

A preliminary design of the set up that could be used for low pressure accident with a leak in cold leg, is represented on Figure 3 and Table 1
According to the preceding analysis and to the constraints of the set up the length L1, L3 and L6 should represent respectively the upper plenum, hot leg and cold leg, whereas the pipe length L2, L4 and L7 should be heaten to their fluid inlet temperature and have smaller diameter in order to avoid a too high F.P. retention.
As regard the steam generator, an easy representation can be found by keeping the S.G. tubes geometry but decreasing there number by the inventory ratio factor.

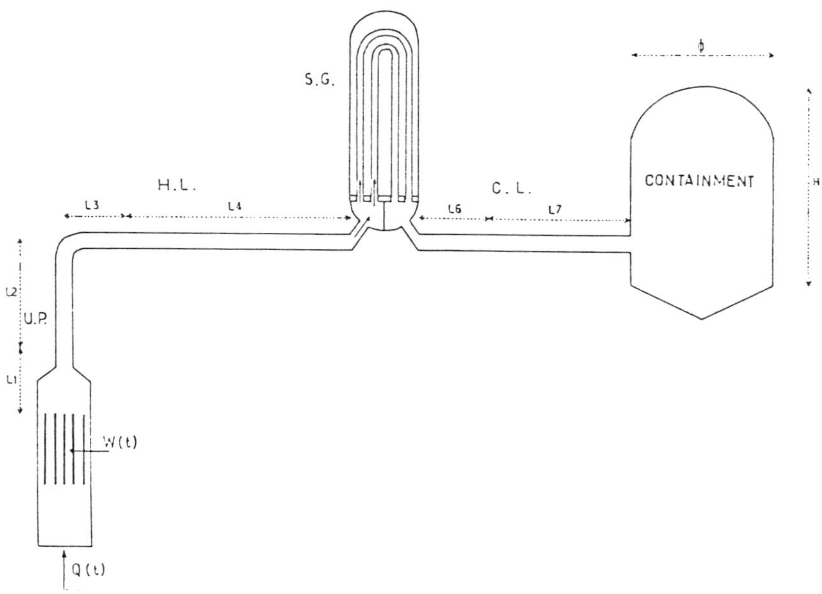

FIGURE 3. Experimental lay out for PHEBUS F.P. low pressure scenario with leak in cold leg.

TABLE 1. Dimensions for PHEBUS P.F. low pressure scenario with cold leg break

Item	Identi-fication	Length m m	Range m m	Internal diameter mm	Range mm	Thickness mm	Range mm	Boundary conditions type	Range on T in k)
Upper Plenum	L1 L2	1 2.5		52	< 52			Heat losses Temperature control	Safety B.C 520.1000
Hot Leg	L3 L4 L3 + L4	0.3 10.7 11	0.12 0.12 11 - 12	66 15	< 100 < 100	6 1.5	1 - 10 1 - 10	Heat losses compensated Temperature controlled	520 - 530 520 - 1000
Steam Generator	2 tubes	20	0 - 20	20		1.25		Heat losses compensated Secondary conditions for T	Wet or dry secondary side
Cold Leg	L6 L7 L6 + L7	1 2 3	0 - 5 0 - 5	66 20	< 100 < 100	6 1.5	1 - 10 1 - 10	Heat losses compensated Temperature controlled	520 - 530 520 - 1000

This geometry was submitted to the verification calculations and it was found that the dimensioning was correct, except for the diameter of long narrow pipes that would produce too much aerosol depletion through turbulent impact on the walls. A mitigation of that drawback was found by increasing these pipes diameters.

IV. CONTAINMENT REPRESENTATION

IV.1. Scaling criteria

The list of phenomena to be considered in the containment representation includes many items on thermalhydraulics, aerosols behavior and chemistry. If it is wished to design a volume for such a purpose with a few numbers, i.e. diameter, height, wall thickness, it is desirable to limit the number of phenomena considered in the preliminary dimensioning. It was decided that radio-chemistry of the sump and chemistry on walls and in the atmosphere of the volumes would be ignored in the preliminary design, whereas aerosol physics would be the main item.
The considered quantitaty are therefore aerosol concentration, diffusiophoresis and settling of aerosols.
Making the hypothesis that the other conditions like thermahydraulics are similar to the reactor ones, the ratios to be considered are then :

$1 = (Vc/Q_f) / (Vc/Q_f)_p$ for aerosol concentation (16)
$1 = (Vc/Aw) / (Vc/A_w)_p$ for aerosol diffusiophoresis (17)
$1 = (Vc / Ah) / (Vc/Ah)_p$ for aerosol settling. (18)
If agglomeration is not considered to influence the settling.

IV.2. Application of the criteria

Due to the inventory of F.P. in the fuel, the condition (16) shows that the PHEBUS containment volume should be in between 15 and 25 m^3.
For an inventory ratio of Ir = 7000,and if the containment is considered as a cylinder, conditions (16) and (18) would yield a diameter of 0.8 m and a height of 38 m for the representation of a 900 MW PWR.
This results in a very large wall area (= 98 m^2) that would amplify

50 times the desired diffusiophoresis. If conditions (16) and (17) are considered, it results in a large diameter cylinder of a small height in which the aerosol settling would be very much enhanced. These considerations show that there may be an intermediate dimensioning, fulfilling the aerosol concentration condition, that would increase in the same factor the diffusiophoresis and the settling.
Let us then consider the following conditions :

$$Vc/Vcp = Ir \quad \text{which comes from (16)} \tag{19}$$

$$(Ah/Aw) / (Ah/Aw)_p = 1 \quad \text{which would ensure} \tag{20}$$

to have the same proportion of diffusiophoresis and settling as in the reactor case, then (20) is equivalent to :

$$\left(\frac{Ah}{Aw}\right)_p = \left(\frac{D}{H}\right)_p = \frac{D}{H} \tag{21}$$

The right proportion of diffusiophoresis to settling is obtained if the PHEBUS containment has the same proportion as the reactor one.

In that case, using (19) and (21) it comes :

$$(A_f/Vc)_p / (A_f/Vc) = (Aw/Vc)_p / (Aw/Vc) = Ir^{1/3} \tag{22}$$

which means that if the volume is decreased with the ratio Ir with respect to the reactor, the same proportion of the containment would give way to the same proportion in diffusiophoresis and settling, but both would be amplified with the factor $Ir^{1/3}$ with respect to the reactor.

4.3. Testing of the proposed design

Calculations were performed with the AEROSOL code on simplified conditions. The reference reactor case is a 50000 m^3 containment similar to the one presented in /4/, with fixed thermahydraulic conditons.
This calculation is compared with two other cases having an inventory ratio of 2000 :

- the case referred to as PHEBUS 1, has a 25 m^3, the same H/D and the amount of aerosol source is devided by 2000. In that case the thermal boundary conditions are deduced from the reactor one with a time history accelerated with a factor 12,6 ($2000^{1/3}$).

- The PHEBUS 2 case has the same design as PHEBUS 1 but the boundary conditions and the source term injection are on the same time schedule as in the reactor.
  It was shown that if agglomeration of the aerosols is not considered, the reduced time scale PHEBUS 1 exhibits the same aerosol concentration as in the reactor case at a later time.
  If the agglomeration is considered (Fig 4) the settling is not accelerated as much as it should in the PHEBUS case.

Both these calculations show that if the aerosols injection time is short with respect to the containment depletion processes, then the maximum aerosol concentratio should not be affected. Therefore, the possible acceleration of long term processes in the PHEBUS containment does not involve to accelerate the test in the fuel bundle and the circuits.
With regard to the PHEBUS containment, the possibility to control the

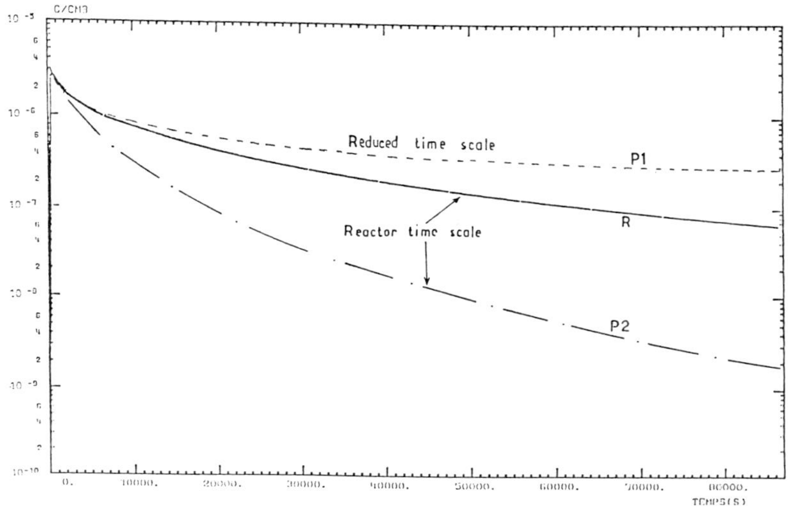

FIGURE 4. Aerosol concentration comparison between PHEBUS and reactor.

thermalhydraulic and chemical processes with a coherent speed up with regard to the aerosol physics is considered through calculations that are presented elsewhere /2/

V. CONCLUSION

The dimensioning of the PHEBUS F.P. as been studied as a global experiment one, in order to keep the representativity for the main phenomena.

The use of real fuel irradiated fuel gives the opportunity to study the complex coupling that may arise between the thermalhydraulics the aerosols physics and the wide spectrum of chemical reaction of the fission products. The first dimensioning has been obtained on the basis of a limited amount of simple scaling criteria. Some compromises were proposed on the basis of the analytical study presented here, but much more detailed verification and improvement should be obtained through calculations that take into account a much larger number of phenomena

NOTATION

Letters

A	Surface area	m^2
C	Specific heat	J/(K.kg)
D	Pipe diameter	m
F	Diffusion coefficient	m^2/s
H	Heat transfer coefficient	$W/(m^2.K)$
Ir	F.P. inventory ratio	
K	Mass transfer coefficient	m/s
L	Pipe length	m
M	F.P. Mass	kg

P	Pressure	Pa
Q	Fluid mass flow rate	kg/s
QV	Fluid volumetric flow	m^3/s
S	Pipe mass	kg
T	Temperature	k
t	time	s
tr	residence time of the fluid in the pipe	s
V	containment volume	m^3
x	**absice** on the pipe	m
Y	concentration	
α	mass flow rate ratio	
β	diameter ratio	
γ	structure mass ratio	
ϵ	length ratio	
λ	thermal conductivity	W/(m.K)
μ	dynamic viscosity	kg/(m.s)
ρ	fluid density	kg/m^3
σ	stress	Pa/M^2
PR	Prandtl number	$C. \mu/\lambda$
RE	Reynolds number	UD/μ
SC	Schmidt number	$\mu/\rho F$
SH	Sherwood number	K.L/D

Subscript

f	fluid
s	structure
p	in PHEBUS
c	containment
w	wall
h	horizontal

R E F E R E N C E S

1/ J. BUSSAC, H. HOLTBECKER, The PHEBUS fission products in pile test programme, NUCSAFE 88 - CONFERENCE, AVIGNON 2 - 7 OCTOBER 1988

2/ A. MARKOVINA, P. FASOLI-STELLA, A. MAILLIAT, Review of the major predicted phenomena during F.P. transport and deposition in the RCS and containment building under severe accident conditions, ICMH Conference, DUBROVNIK 1989

3/ C. GONNIER, S. FABREGA, E. SCOTT DE MARTINVILLE, R. GEOFFROY, In pile investigations at the PHEBUS facility on the behavior of PWR-type fuel bundles in severe accident conditions beyond the design criteria, NUCSAFE 88 - CONFERENCE, AVIGNON 2 - 7 OCTOBER 1988

4/ J. GAUVAIN, G. LHIAUBET, Influence of structural insoluble aerosols in severe accident Scenarios with and without steam condensation on to soluble aerosols fission products. Point NEA/CEC Workshop on water cooled reactor code evaluation and uncertainty assessment. BRUXELLES 9 - 11 SEPTEMBER 1987

LECTURE

Modeling Aerosol Behavior in Reactor Cooling Systems

B. H. MCDONALD
Containment Analysis Branch
Whiteshell Nuclear Research Establishment
Atomic Energy of Canada Ltd.
Pinawa, Manitoba, ROE 1LO Canada

INTRODUCTION

This paper presents an overview of some of the areas of concern in using computer codes to model fission-product aerosol behaviour in the reactor cooling system (RCS) of a water-cooled nuclear reactor during a loss-of-coolant accident. The basic physical processes that require modelling include: fission product release and aerosol formation in the reactor core, aerosol transport and deposition in the reactor core and throughout the rest of the RCS, and the interaction between aerosol transport processes and the thermalhydraulics. In addition to these basic physical processes, chemical reactions can have a large influence on the nature of the aerosol and its behaviour in the RCS. Here, however, the focus is on the physics and the implications of numerical methods used in the computer codes to model aerosol behaviour in the RCS.

Calculated fission product releases from the RCS are subject to many potential errors. Numerical methods used in computer codes, which use algebra to estimate solutions to integro-differential equations, introduce errors of truncation, round-off, and "numerical diffusion." The integro-differential equations contain models for aerosol particle generation, agglomeration, and deposition, which are based either on validated physical (or chemical) models, or on postulated models for which adequate data for do not yet exist to completely validate the models. These models are often non-linear, which means that the integrated effect of the models working together within the framework of a numerical procedure could produce unexpected (and possibly incredible) results.

Aerosol and thermalhydraulic processes, even without the complications of chemistry, constitute a large-scale, non-linear, highly interactive, extremely complex set of phenomena [1-3]. This paper presents a case study, based on the development at Atomic Energy of Canada Limited (AECL) of a prototype coupled thermalhydraulics/aerosol code package, CATHENA/PACE, to illustrate one approach to building an RCS aerosol behaviour code, and the variation in results that might be expected from different aerosol model assumptions. PACE (Program for Aerosol Code Evaluation) is an aerosol behaviour code developed to study the containment aerosol codes used in a 1985 OECD/GREST code comparison exercise [4,5], and is being extended to include published RCS models from other state-of-the- art codes. CATHENA (Canadian Algorithm for THErmalhydraulic Network Analysis) is the AECL code specially developed to model CANDU RCS thermalhydraulics [6]. The desirability of coupled thermalhydraulics/aerosol code packages has been recognized previously [7].

Because a review such as this, and a generic type of model such as the coupled CATHENA/PACE algorithm, should be based on published material from the open literature, this paper necessarily avoids direct reference to current work that is as yet unpublished. Readers should bear in mind that many of the topics being discussed in other presentations at this seminar will have received less-than-up-to-date treatment here.

The paper begins by outlining the CATHENA/PACE strategy and rationale for the coupled thermalhydraulics/aerosol equations. The PACE code, which contains selectable aerosol physics models from other codes, is described briefly, followed by a brief summary of the CATHENA thermalhydraulics code. Then the basic aerosol models are described, including the governing equation for both single-component and multicomponent aerosols, the basic physical properties calculated from local thermalhydraulic conditions, and aerosol particle agglomeration and removal/deposition mechanisms with special models for handling pipe bends. Models used for RCS aerosol sources are indicated, based on material release rates and vapour condensation rates, and on aerosol resuspension; the special considerations needed due to a dependence on thermalhydraulic conditions are noted.

The numerical methods and strategies in CATHENA/PACE are described briefly and several code calculations based on an S_2D test case are presented to illustrate the effects of different model and numerical parameters. In particular, the effect of a simple model for bend deposition is described based on centripetal force.

THE CATHENA/PACE PROTOTYPE CODE FOR COMPARING RCS AEROSOL MODELS

To facilitate comparing the effects of the various aerosol model assumptions, a simple finite-difference code has been built, and permits aerosol physics models to be emulated from other state-of-the-art aerosol codes (e.g., HAA4, REMOVAL, NAUA, AEROSIM, CONTAIN/MAEROS, AEROSOLS/B1, RAFT, VICTORIA, TRAP-MELT) by switch selection. This code is called PACE [4,5], and in its basic form requires thermalhydraulic conditions to be provided as input data. While all integrations are done numerically, an equivalent log-normal distribution at each time step may be selected. Thus, virtually all of the major possibilities are available. By making appropriate selections, this model should be able to emulate the performance of any of the codes included, provided, of course, that the models have been faithfully transported. The numerical method chosen was the Method of Weighted Residuals [8] in a finite-difference formulation with parameters chosen to match AEROSIM [5]. Semi-implicit backward Euler integration was used to advance the solution in time, with the automatic time-step control used in the thermalhydraulics code CATHENA [6]. The documentation available for the particular codes emulated in PACE was followed in modelling all the variations for agglomeration, removal and condensation. Thus, by specifying the code to be emulated as input data, the appropriate models are selected automatically. Many variations can be emulated by over-ride selection of parameters such as those in the collision efficiency model, wet or dry aerosols, or the Stephan flow model. CONTAIN/MAEROS models [9] were used in cases where the available documentation is unclear about model selection.

The PACE code itself is strictly an aerosol behaviour model and does not incorporate any aerosol-thermalhydraulics interaction. Thus it cannot really emulate the performance of CONTAIN (for example) which does include a full-containment thermalhydraulics package. PACE simply distributes condensing vapour over the aerosol particles, and the method used is that reported for AEROSIM [4].

CATHENA [6] was developed to specifically calculate the thermalhydraulic response of the CANDU RCS during accident conditions. CATHENA uses a node-link representation for the RCS and has a two-fluid capability and a sophisticated heat transfer package to calculate CANDU fuel temperature transients during accident conditions.

PACE has been coupled to CATHENA to provide an integrated analysis. At each time step (controlled by CATHENA), thermalhydraulic data are passed to PACE subroutines to calculate the local aerosol physics response, and PACE returns any specific results to CATHENA where they will have an impact on the thermalhydraulic calculations for the next time step. The CATHENA/PACE code strategy is shown in Figure 1.

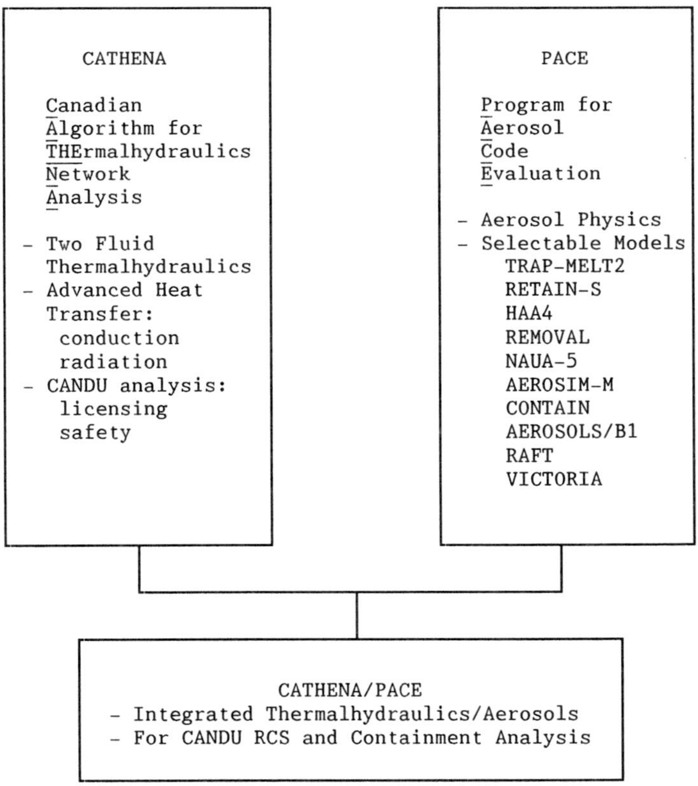

FIGURE 1. The integrated CATHENA/PACE Code for coupled thermalhydraulics and aerosol physics analysis.

GOVERNING AEROSOL EQUATIONS

To begin with, let us look at the basic equation that must be solved in obtaining a solution to RCS aerosol transport and deposition processes. In general, the core and the RCS are divided into connected control volumes within which the aerosol and carrier gas atmosphere are assumed to be well-mixed. This means that there are no point-to-point differences within the control volume in carrier gas properties or aerosol particle sizes or concentrations. All gradient information is lost. (Some of it must be empirically reconstructed to prescribe the particle removal rates.) All new aerosol particles introduced into the control volume are assumed to be distributed instantly and uniformly throughout. In some cases, the well-mixed assumption could be considered to be a gross over-simplification of the real problem. The major advantage of the well-mixed aerosol assumption is that it reduces an extremely complex physical problem to a fairly simple one.

These connected control volumes may then be considered to represent actual RCS circuit components, and gas-flow velocity data are used to describe the transport of aerosols from one volume to the next (the Eulerian approach); alternately, these volumes may be considered to be packets of gas and particles that "go with the flow" (the Lagrangian approach). The Eulerian approach is generally simpler and faster but less accurate than the Lagrangian. (The reader should note that there are always tradeoffs between cost and accuracy when using computer models.)

For a single aerosol species, a single equation, of integro-differential form, is used to describe the problem [4]:

$$\frac{\partial}{\partial t} C(m,x,t) + \frac{\partial}{\partial x} [v(m,x,t)\, C(m,x,t)] =$$

$$\frac{1}{2} \int_0^m d\mu\ \phi(\mu,m-\mu)\, C(\mu,x,t)\, C(m-\mu,x,t) - C(m,x,t) \int_0^\infty d\mu\ \phi(\mu,m)\, C(\mu,x,t)$$

$$- \frac{\partial}{\partial m} [\xi(m,x,t)\, C(m,x,t)] - R(m,x,t)\, C(m,x,t) + S(m,x,t) \qquad (1)$$

where C(m,x,t) is the number of particles of mass m at distance x in the RCS at time t. The first term on the left of this equation prescribes the time rate of change of C(m,x,t) and the second term on the left prescribes the gradient of the flux of C(m,x,t) along the RCS in the x direction, where v(m,x,t) is the aerosol particle velocity.

The first two terms on the right, the integrals, describe the process where small particles agglomerate to form larger ones. In the first term, all particles combining to form new particles of mass m are added. The agglomeration kernel, $\phi(\mu,m-\mu)$, prescribes the rate at which particles of mass μ and $m-\mu$ combine to produce particles of mass m. The coefficient 1/2 is needed because the integral counts each agglomeration twice. In the second term, all particles of mass m combining with all other particles through agglomeration are subtracted. The third term describes the growth of particles due to condensation of steam. The condensation rate is $\xi(m,x,t)$. In the fourth term, the removal rate R(m,x,t) describes the rate at which particles of mass m are removed from the atmosphere by deposition on structural surfaces or by leakage to the environment. In the last term, new particles of mass m are injected into the atmosphere at the rate S(m,x,t).

To emulate those aerosol codes that have a true multicomponent capability, an equation like (1) must be specified for each component. If $q_k(m,x,t)$ is the mass of component k in particles of mass m at distance x along the RCS at time t, the multicomponent aerosol population balance equation analogous to equation (1) is:

$$\frac{\partial}{\partial t} q_k(m,x,t) + \frac{\partial}{\partial x} [v(m,x,t)\, q_k(m,x,t)] =$$
$$\int_0^m d\mu\, \phi(\mu,m-\mu)\, q_k(\mu,x,t)\, C(m-\mu,x,t) - q_k(m,x,t) \int_0^\infty d\mu\, \phi(\mu,m)\, C(\mu,x,t)$$
$$+ \delta_{ik} \xi_k(m,x,t)\, C(m,x,t) - \frac{\partial}{\partial m} [\xi_k(m,x,t)\, q_k(m,x,t)]$$
$$- R(m,x,t)\, q_k(m,x,t) + S_k(m,x,t) \qquad (2)$$

Here $C(m,x,t)$ remains the total number of aerosol particles of mass m. If there are L species,

$$\sum_{k=1}^{L} q_k(m,x,t) = m\, C(m,x,t) \qquad (3)$$

All terms are equivalent to those in Eq. (1), except the term δ_{ik} $\xi_k(m,x,t)\, C(m,x,t)$. The Kronecker delta function δ_{ik} is unity when $k = i$, defined as the condensing (or evaporating) component, and is zero for other components. This term corresponds to the change in the airborne mass of this component. In other words, it defines the condensation of vapour of component k on aerosol particles of mass m, and is separate from the familiar differential term that corresponds to the shifting of mass up (or down) the mass range due to condensation (or evaporation) [4]. (The reader may have noticed the similarity in form between the $\partial/\partial m$ and the $\partial/\partial x$ terms in these equations and wondered if the Eulerian/Lagrangian approach considerations might apply to condensation modelling as they do to physical aerosol transport. They do.) PACE incorporates the multicomponent equation (2).

The solution of the equation involves specifying rules for calculating physical parameters at any location and time in the RCS (or alternatively, specifying the input data). Aerosol and gas source rates, aerosol particle agglomeration and removal rates, vapour condensation rates, and gas flow velocities are needed. Then a numerical method is prescribed to replace the governing integro-differential equation with an algebraic equation at a number of locations in the RCS. In these equations, algebraic coefficients are based on the physical parameters. These algebraic equations are solved in the computer code, producing an estimate for the changes in aerosol concentrations (the C or q values) over the time step and thus advancing the solution to the new time.

BASIC PHYSICAL PROPERTIES / AGGLOMERATION / DEPOSITION

The basic rules for calculating the agglomeration kernel (ϕ) and the removal rate (R) are the same for the RCS as they are for containment studies. Therefore the aerosol physics packages found in the containment codes and previously incorporated in PACE [4,5] remain effective and permit NAUA physics, for example, to be used for RCS studies. Particular variations found in the documentation of the RAFT [10,11], TRAP-MELT [12], and VICTORIA [13] codes for the basic physical properties, for agglomeration, and for the principal deposition mechanisms have been added to PACE and are selectable in the same way as the others.

DEPOSITION AT PIPE BENDS

Aerosol deposition calculated using the basic code rules for the removal rate (R) may underestimate the effects at bends in pipes, and additional modelling is required to apply the codes to aerosol deposition at bends in pipes. Special models for turbulent deposition and bend deposition are normally incorporated and some of these are summarized here.

Turbulent deposition in AEROSIM [14] and HAA4 [15] is modelled using correlations based on the Liu-Agarwal test data in RAFT [10] with a simplified Im-Chung correlation [16], and in the TRAP-MELT codes using either the Friedlander-Johnstone model [17] with the Battelle modification [18] or the Wood model [19].

To handle special effects at bends, both AEROSIM [14] and TRAP-MELT2.2 [20] use turbulent deposition models with modifications to include bend pressure loss in the calculation, HAA4 [15] uses a special transport efficiency model, and RAFT [10] incorporates a similar model where bend deposition is assumed to be due only to inertial impaction on the outer walls of the pipe bends. (That is, these codes use what is essentially a Stokes' stopping-distance modelling strategy.)

SOURCES - FISSION PRODUCT RELEASE AND AEROSOL FORMATION

There are essentially three sources of aerosol material: micron-sized fragments, seed particles formed by core material vapour condensation, and materials previously deposited that are resuspended.

The release of fission products as vapour from fuel is related directly to the temperature of the fuel, and considerable experimental evidence exists to support the fission product release models used for LWR reactors. Aerosol particles are produced not only by condensation of material vapours, including structural materials, but also directly by fragmentation and powdering. In RAFT, for example, primary aerosol particles are produced when the material vapours assumed to be released from overheated fuel [10] condense. The correct estimation of local release of material vapours, coupled to a correct estimation of gas flow and temperature to obtain the condensation, and correct estimate of the numbers and size distributions of particles produced directly by fragmentation and powdering, are all required to obtain a correct estimate of the aerosol source at a given location in the reactor core at a given time.

In most codes, there remains the option of providing aerosol source rates as input data, and for the examples here, aerosol sources are specified directly using PACE emulations of the other codes.

CONDENSATION

For containment applications, the condensation of steam to produce water droplets or wet fission product aerosol particles is of prime importance in water-cooled reactors [21]. As well, steam condensation on structural surfaces in containment leads to enhanced aerosol deposition through diffusiophoresis. Condensation has an equally important role to play in the RCS: structural materials and fission products are vaporized at extremely high core temperatures, and these vapours will condense in regions of the core and the RCS where the temperatures are lower. The physics is the same, although the temperatures are higher and the condensing vapours are different from those encountered in containment, where steam is the principal condensing vapour. Accordingly, condensation models (or input data) are needed to obtain the correct physics. The homogeneous nucleation models found in RAFT [10] provide one example, and such models are presently being incorporated in the PACE package to enhance its capabilities. Local condensation depends on local thermalhydraulic conditions and can have a feedback effect on the thermalhydraulics by inducing pressure fluctuations. Intimate coupling of the thermalhydraulics with the aerosol physics may be as important for RCS studies as it is proving to be in containment analysis [22,23].

NUMERICAL METHODS

The solution of the particle-size distribution equation for the RCS may be thought of as a three-step process. The equation itself is an integro-differential equation in time, space, and particle size or mass. The first step is to convert the equation to a system of partial differential equations in space and time using numerical methods to transform the particle size integro-differential terms to algebraic terms. Codes of the HAA and RETAIN families use a method of moments technique to accomplish this, AEROSOLS/B1 uses a finite-element technique, and the others use finite differences, although in some cases the actual differences are largely semantic. The PACE code uses a first-order Galerkin method with particle size regions (or "bins") and an option to use a log-normal particle size distribution function. The second step in the solution procedure is to convert the equations to a system of ordinary differential equations in time-only dependent variables. These variables may be viewed as parameters from which the size distribution may be constructed through specified functional relationships. The partial-differential terms in the space coordinate are recast as time-dependent algebraic functions of the variables by applying another numerical procedure, usually finite differences. The third and last step in the process is to integrate the the system of ordinary differential equations to produce the time history of these variables.

The first step (replacing the integro-differential particle size terms with algebraic analogues) has been described previously [5]. Some work has been reported [24] for a moving-grid approach (or in effect a method of characteristics) to handle the vapour condensation terms. While attractive in terms of reduced numerical diffusion, methods of characteristics are usually time-step limited and can consume a lot of computing effort.

The second step (replacing the first-order partial differential system in space and time with a system of ordinary differential equations in time) reduces to the solution of the mass balance equation in a flowing system. In PACE, the existing method for doing this in CATHENA was implemented [6]. It consists of an upwind differencing scheme based on donor-cell technology. Another approach, based on an application of the method of characteristics with a marching technique, may be found in RAFT [10].

The last step (integrating the system of ordinary differential equations in time-dependent variables) was also discussed previously [5]. PACE uses the CATHENA method: semi-implicit backward Euler integration with dynamic time-step control. The method in RAFT is bound up with the marching technique.

These three steps may be performed sequentially in a once-through algorithm, in an algorithm with some iteration, or with a fully iterative scheme. Generally speaking, a fully iterative scheme is preferred for accuracy, as it ensures that all the variables are updated together over the time-step. Otherwise, a time lag is introduced for certain parameters, and this could affect accuracy. Iterative methods tend to be more expensive (for example, this is why CATHENA uses a semi-implicit method rather than a fully-implicit method) and we must ask if the additional accuracy is worth the additional expense. There is one school of thought that suggests that all the equations, that is both aerosol transport and thermalhydraulic equations, should be solved simultaneously to provide maximum accuracy. While this approach could be extremely useful if it produces a reference algorithm, it could prove to be too expensive to use in routine safety analyses.

Readers are reminded that computer codes deal only with algebraic expressions and only produce numbers. These numbers are usually coefficients in the algebraic approximations of the integro-differential equations for which a solution is sought. Working backwards, using, for example, Taylor-series expansions, the actual integro-differential equation being approximated may be deduced. This equation usually turns out to be the equation sought plus some higher-order terms that are often diffusive in nature (diffusive terms are helpful in that they help stabilize an algorithm and may help to permit use of crude nodalization without risk of code failure) and tend to smear out, or average, physical phenomena. There are always tradeoffs between speed and accuracy when dealing with computer code algorithms, and this application is certainly no exception.

CATHENA/PACE EXAMPLES - AN S_2D TEST CASE

In this section, several CATHENA/PACE code calculations on an S_2D accident scenario will be presented to illustrate some of the differences that can occur using different physical correlations and numerical methods. The test case parameters and the schematic are shown in Figure 2, and all cases are run to about 4300 s. To facilitate comparisons, distributions of system aerosol mass (deposited on the metalwork, released to containment, and airborne) are presented as percentages of the source aerosol masses prescribed for all cases as input data. The CATHENA/PACE nodalization for all cases is shown in Figure 2. Time-step size was held constant at 50 s for all but one test run. For all but one run (and HAA4 log-normal cases), five bins were used for the aerosol particle size distribution. Five variations in basic aerosol code physics are employed: PACE/NAUA is the default setting in the code, and cases are shown with HAA4 physics (and a forced log-normal particle size distribution), RAFT, AEROSIM, and TRAP-MELT aerosol physics. CATHENA-run parameters were left unchanged between runs. CATHENA/PACE is still under development and errors in representing physical models used in the other aerosol codes may exist; the results are simply to illustrate that CATHENA/PACE may be used to evaluate the sensitivity of predictions to the physics models selected.

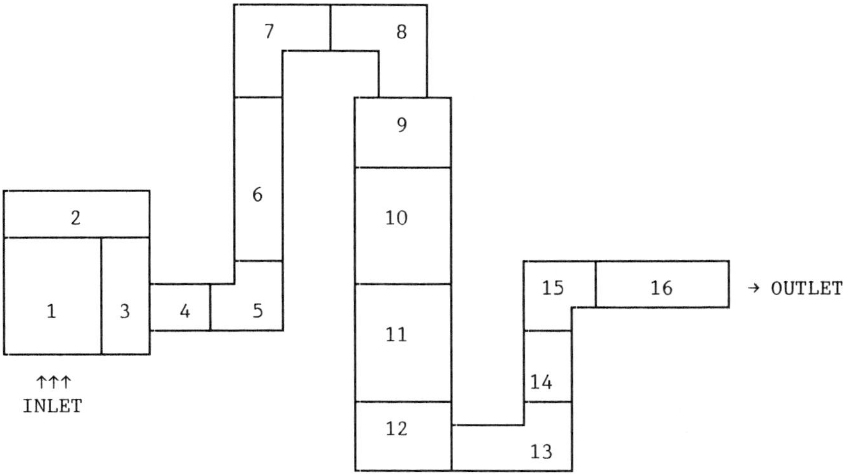

Volume #	Volume (m**3)	Length (m)	Cumulative Length (m)	X-Sect. (m**2)	Perimeter (m)	Structure
1	22.41	2.74	2.74	8.17	51.55	Upper Plenum
2	12.50	1.53	4.27	8.17	22.05	Upper Plenum
3	3.18	1.74	6.01	1.83	22.26	Exit Plenum
4	1.35	2.05	8.06	0.66	2.88	Hot Leg
5	1.63	2.50	10.56	0.65	2.86	Hot Leg
6	7.61	11.70	22.26	0.65	2.86	Hot Leg
7	1.63	2.50	24.76	0.65	2.86	Hot Leg
8	1.63	2.50	27.26	0.65	2.86	Hot Leg
9	7.45	1.02	28.28	7.30	9.58	S.G. Inlet
10	20.63	8.49	36.77	2.43	688.00	S.G. Tubes
11	20.63	8.49	45.26	2.43	688.00	S.G. Tubes
12	7.45	1.02	46.28	7.30	9.58	S.G. Exit
13	0.40	1.00	47.28	0.40	2.24	Cold Leg
14	0.72	1.80	49.08	0.40	2.24	Cold Leg
15	0.40	1.00	50.08	0.40	2.24	Cold Leg
16	1.00	2.50	52.58	0.40	2.24	Cold Leg

FIGURE 2. The S_2D reactor case parameters and schematic

Base Case Comparisons

For base case comparison, code emulations were run with CATHENA/PACE and the five aerosol code physics models without any special bend deposition models operational. The results are shown in Figure 3. Significant differences are apparent: the HAA4 emulation shows up to 53% of the aerosol mass deposited on the metalwork at 4300 s, whereas the AEROSIM and TRAP-MELT emulations show only about 8%. This difference may be an artifact of the log-normal distribution in HAA4 and has been noticed before [25]. It is interesting to note that the aerosol released to containment by 4300 s varies only from about 20% to 38% for all cases in the scenario.

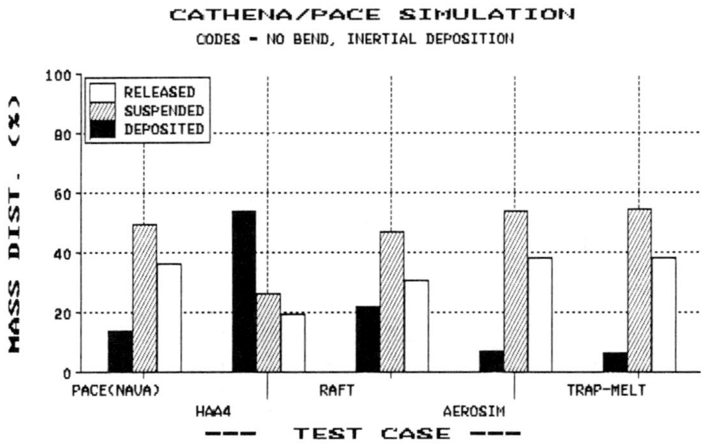

(a) Overall Aerosol Distributions at 4300 s into the S_2D Transient.

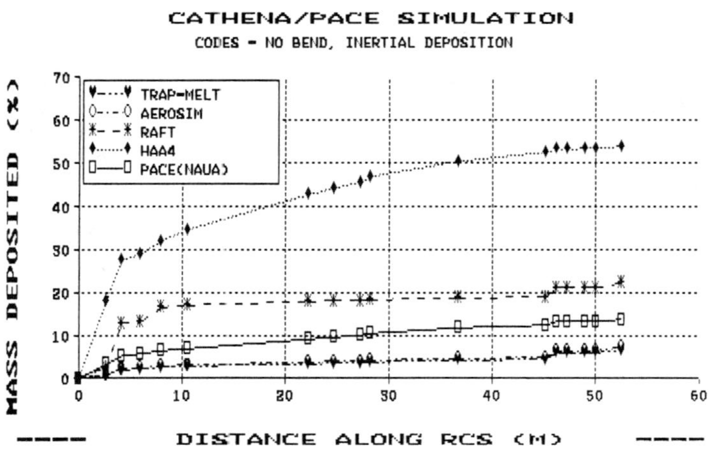

(b) Distribution of Deposited Aerosol in the RCS at 4300 s.

Figure 3. Aerosol Distributions for CATHENA/PACE S_2D Simulations of Codes Without any Special Treatment of Inertial Deposition or of Deposition at Bends in Pipes.

Base Case Comparisons With Special Bends Treatment

The same cases were run with inertial deposition models active, and the results are shown in Figure 4. It is somewhat surprising that these results are effectively the same as the base case runs without inertial deposition. A double check of the computer code revealed no obvious errors, and a brief analysis of the detailed code output revealed that inertial deposition did not appear dominant at any time in this scenario. It would appear that published models that describe the enhancement of aerosol deposition at bends in pipes may sometimes make little difference.

A Special Enhanced Bend Deposition Model

In order to investigate in a little more detail the effect of deposition at pipe bends, a special model was constructed for the test using the classical centripetal acceleration equation with Stokes' law to determine whether or not particles would hit the pipe surface while "turning the corner." (The gravitational settling velocity for a particle was modified by replacing the gravitational acceleration factor with the quantity v^2/r, where, as before, v is the particle velocity and r is the radius of the bend.) Three variations were tested. For the results labelled "WITH ID & BE", both inertial deposition and this Stokes' stopping distance model are employed. For the results labelled "REDUCED X-C", both models are used, but with a higher flow velocity obtained by artificially reducing the flow area at bends by a factor of the square root of ten to observe the effect on the Stokes' law model. For the results labelled "REDUCED RADIUS", both models were used but with the bend radius artificially reduced by a factor of ten to observe this effect on the Stokes' stopping distance model. All of these variations should not be viewed as anything more than sensitivity studies.

Figure 5 shows the results of the comparison with NAUA physics. The only significant variation observed is with increased flow through the first bend in the hot leg. The comparison of the results with HAA4 physics is shown in Figure 6. No significant variation is observed; the bend deposition models appear to have no effect. Figure 7 compares the results with RAFT physics. Minor variations are observed in deposition at the first and second bends in the hot leg, with reduced radius having the most noticeable effect. Figure 8 compares the results of the comparison with AEROSIM physics and Figure 9 compares them with TRAP-MELT physics. No significant differences are observed.

The results of this exercise are not conclusive. It would appear that more work is needed to be done to sort out the modelling of aerosol particle deposition at bends in pipes, at least considering the published models. Perhaps some light will be shed on the problem in other papers representing new work at this seminar.

Perturbation Studies

To observe any effects on solutions due to variations in numerical methods parameters, Figure 10 presents the results of two additional tests. In each case, CATHENA/PACE with NAUA aerosol physics is used. For the case labelled "REDUCED TIME STEPS", the time steps were reduced to ten seconds from fifty, and no significant difference was observed.

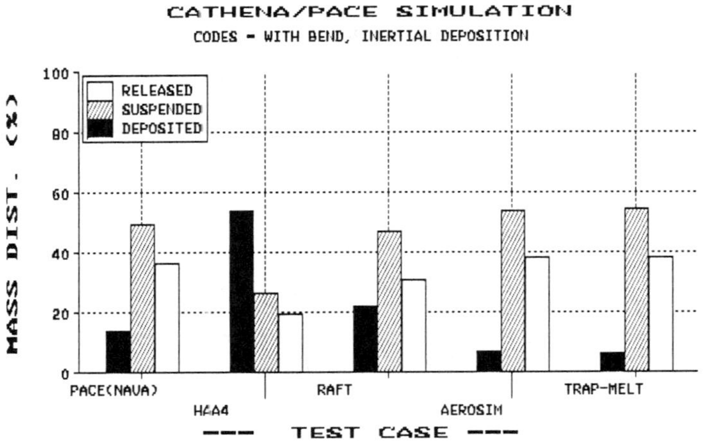

(a) Overall Aerosol Distributions at 4300 s into the S_2D Transient.

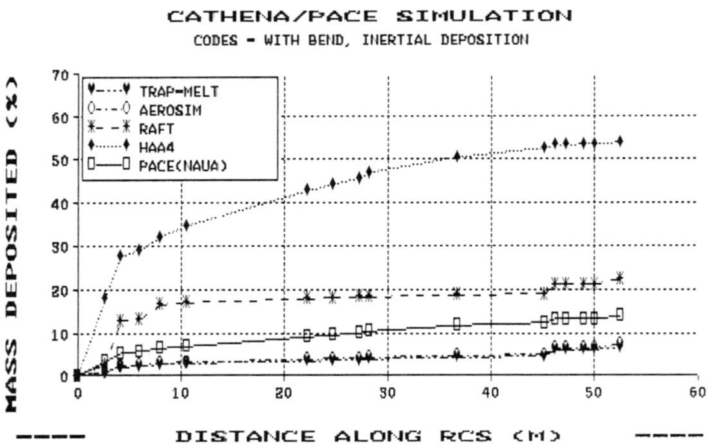

(b) Distribution of Deposited Aerosol in the RCS at 4300 s.

Figure 4. Aerosol Distributions for CATHENA/PACE S_2D Simulations of Codes With Prescribed Treatment of Inertial Deposition and of Deposition at Pipe Bends.

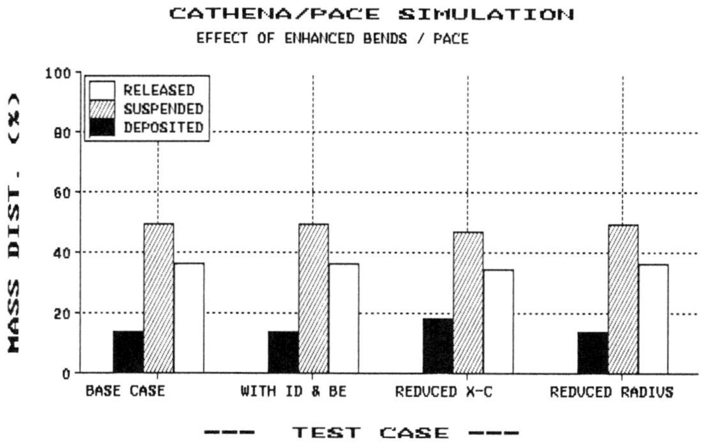

(a) Overall Aerosol Distributions at 4300 s into the S_2D Transient.

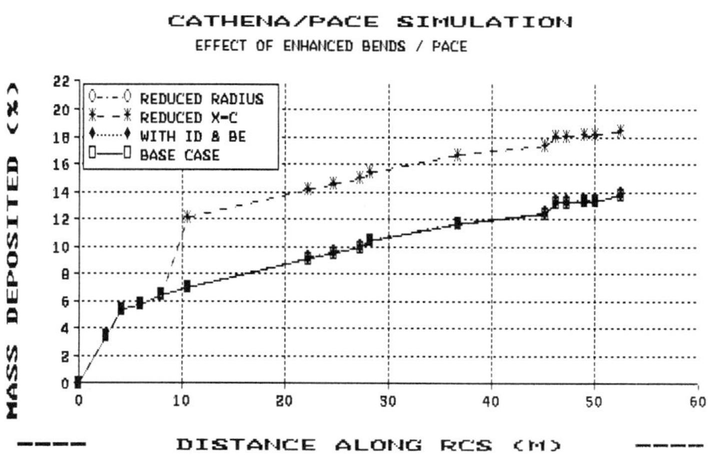

(b) Distribution of Deposited Aerosol in the RCS at 4300 s.

Figure 5. Aerosol Distributions for CATHENA/PACE Simulations of S_2D (with PACE/NAUA physics) Showing the Effect of Enhanced Special Bend Deposition.

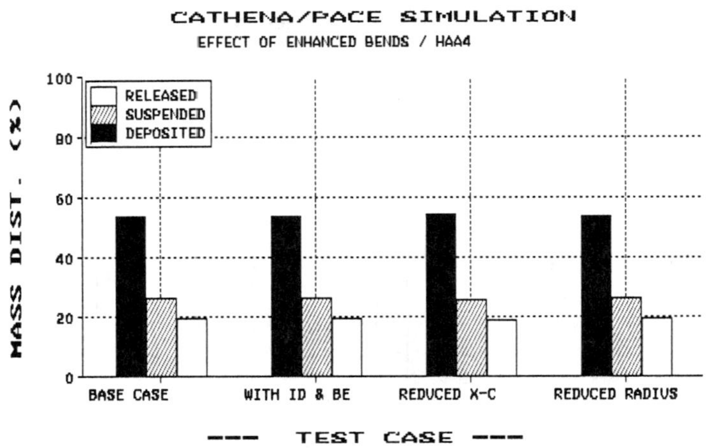

(a) Overall Aerosol Distributions at 4300 s into the S_2D Transient.

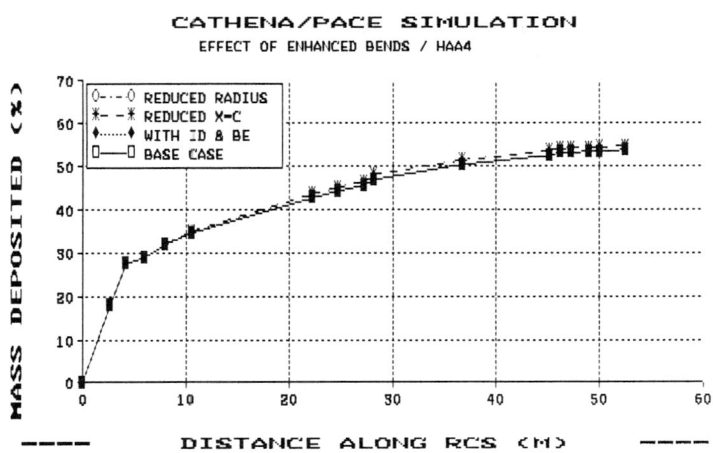

(b) Distribution of Deposited Aerosol in the RCS at 4300 s.

Figure 6. Aerosol Distributions for CATHENA/PACE Simulations of S_2D (with HAA4 physics) Showing the Effect of Enhanced Special Bend Deposition.

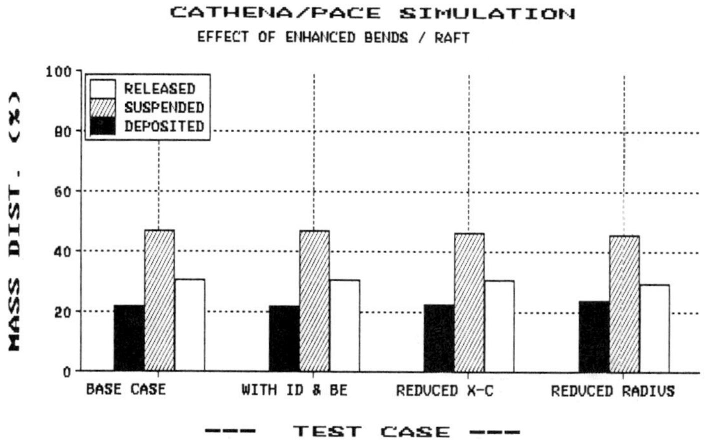

(a) Overall Aerosol Distributions at 4300 s into the S_2D Transient.

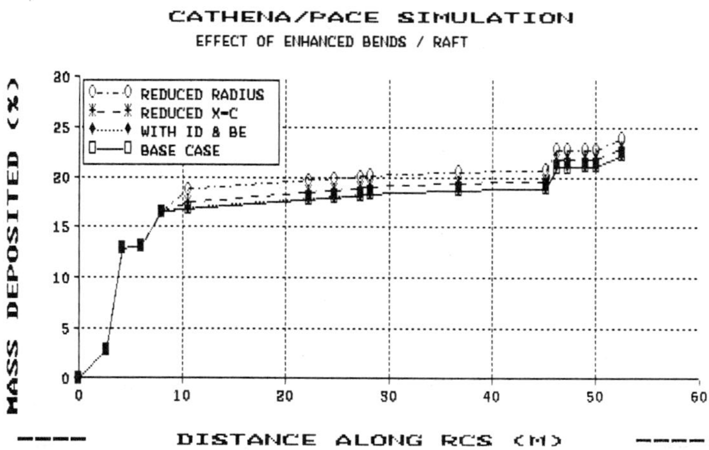

(b) Distribution of Deposited Aerosol in the RCS at 4300 s.

Figure 7. Aerosol Distributions for CATHENA/PACE Simulations of S_2D (with RAFT physics) Showing the Effect of Enhanced Special Bend Deposition.

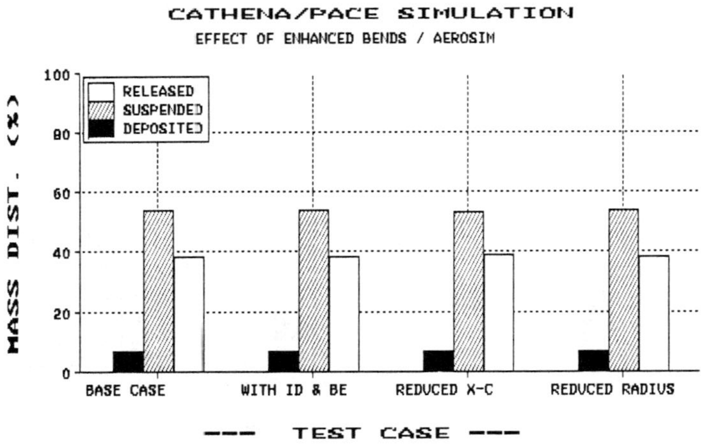

(a) Overall Aerosol Distributions at 4300 s into the S_2D Transient.

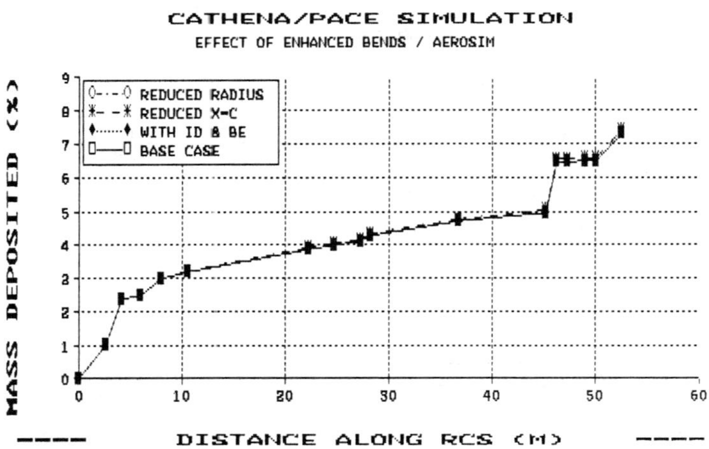

(b) Distribution of Deposited Aerosol in the RCS at 4300 s.

Figure 8. Aerosol Distributions for CATHENA/PACE Simulations of S_2D (with AEROSIM-M physics) Showing the Effect of Enhanced Special Bend Deposition.

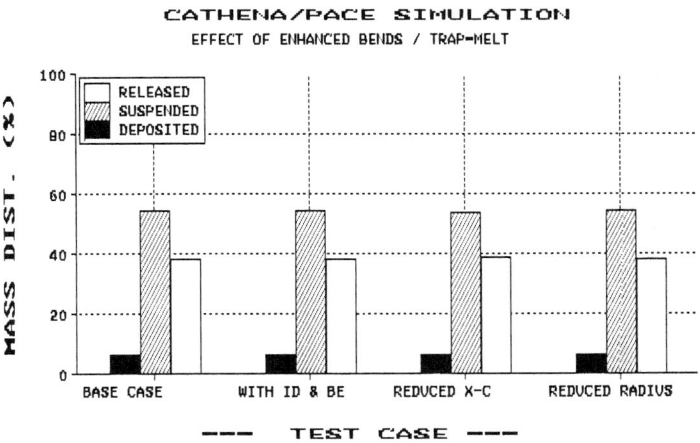

(a) Overall Aerosol Distributions at 4300 s into the S_2D Transient.

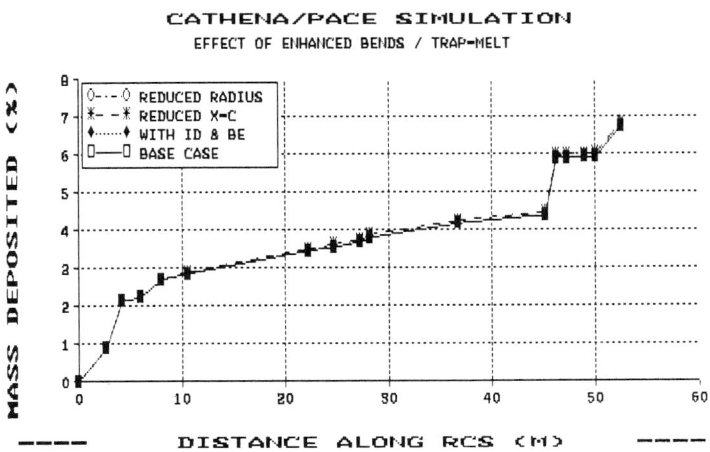

(b) Distribution of Deposited Aerosol in the RCS at 4300 s.

Figure 9. Aerosol Distributions for CATHENA/PACE Simulations of S_2D (with TRAP-MELT physics) Showing the Effect of Enhanced Special Bend Deposition.

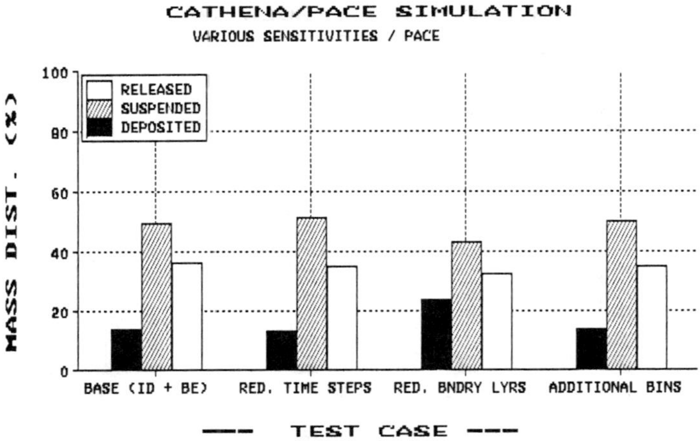

(a) Overall Aerosol Distributions at 4300 s into the S_2D Transient.

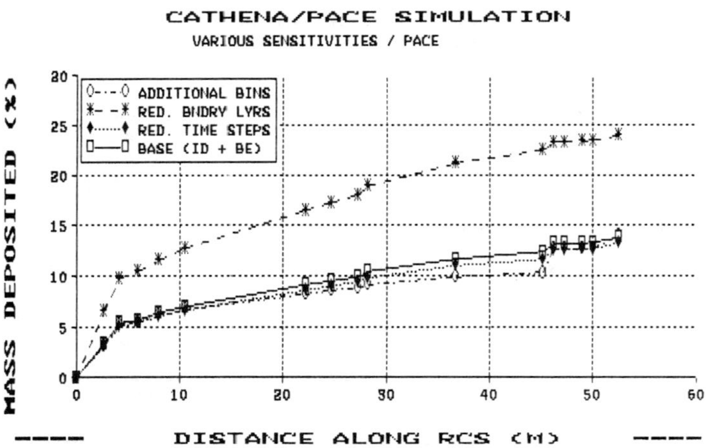

(b) Distribution of Deposited Aerosol in the RCS at 4300 s.

Figure 10. Aerosol Distributions for CATHENA/PACE Simulations of S_2D (with PACE/NAUA physics) for Various Perturbations from the Base Case with Inertial Deposition and Special Bend Deposition.

For the case labelled "ADDITIONAL BINS", the number of aerosol size bins was increased from five to ten, and only a slight difference was observed through the steam generator. (A case run with twice as many segments along the RCS demonstrated no significant difference from the base case.) For comparison purposes, a third test result is presented: for the case labelled "REDUCED BOUNDARY LAYERS", nominal values of the diffusion, thermophoretic and diffusiophoretic boundary layer thicknesses were reduced from 2 mm to 0.5 mm, which resulted in twice as much deposition on the metalwork. Note however, that the percentage of aerosol released is relatively unchanged at 4300 s, varying only from 35 to 32; the increase in deposited mass is reflected in reduced airborne mass in the piping.

It is noteworthy that no variation tried in the numerical parameters had as much effect as changing the boundary layer thicknesses. It will be interesting to observe how the modelling of aerosol processes evolves as comparisons between codes and actual test data [26] find their way into the literature.

CONCLUDING REMARKS

The development of the CATHENA/PACE coupled thermalhydraulics/aerosol code is an ongoing activity at AECL that will result in the availability of software to analyze fission product release and transport experiments and to perform CANDU safety analyses for both RCS and containment phenomena. As well as putting in place numerical procedures that will be both sufficiently accurate and efficient, it is extremely important to have in place adequate aerosol behaviour models. By incorporating other code models in the PACE aerosol package, and by continuing the development by updating existing models and introducing new models as further research work is published, the PACE package should provide the arena wherein models specifically suitable for CANDU scenarios may be defined. At this stage of its development, the CATHENA/PACE package provides a useful vehicle for illustrating the effects of sensitivity in modelling RCS fission product transport. The results I have presented here lead me to make several comments. Published models for enhancing aerosol deposition at bends in pipes may not be adequate, and more work may need to be done and reported. Results seem to be less sensitive to a variation in parameters in the numerical methods than they are to variations in boundary-layer thicknesses; more work may need to be done and published on basic aerosol deposition phenomena. Condensation phenomena may require closer coupling between thermalhydraulics and aerosol physics in the RCS than has been seen to date. Analysts should bear in mind that all numerical procedures introduce errors, and sometimes a numerical quirk, be it a predetermined shape function for the aerosol particle size distribution, or unseen numerical diffusion, can dominate the physical processes modelled and produce code results that are, at the least, misleading.

On the bright side, there is a lot of work currently under way at various Laboratories worldwide, and as this work matures and is published, the state of the art will advance rapidly.

REFERENCES

1. Kress, T.S, "Review of Areas that may Require Simultaneous Coupled Solution of the Thermal Hydraulics and Fission Product/Aerosol Behaviour Equations for Source Term Determination," In Proceedings of CSNI Specialist Meeting on Nuclear Accidents in Reactor Safety, Karlsruhe, September 1984.

2. Butland, A.T.D., Johns, N.A., Williams, D.A., "Validation of Models Used to Predict Primary System Thermal-Hydraulics and Radionuclide Transport and Deposition in a PWR Severe Accident," In Proceedings of an IAEA Symposium on Source Term Evaluation for Accident Conditions, pp. 247-265, Columbus Ohio, 1985 October 28 - November 1.

3. McDonald, B.H., Wren D.J, "Thermalhydraulics and Activity Transport," In ICHMT Int. Seminar on Fission Product Transport Processes in Reactor Systems, Dubrovnik, May 22-26, 1989.

4. McDonald, B.H., "Assessing Physical Models Used in Nuclear Aerosol Transport Models," In Proceedings of a CEC/OECD Workshop on Water-Cooled Reactor Aerosol Code Evaluation and Uncertainty Assessment, pp. 206-226, Brussels, September 9-11 1987, available as EUR-11351-EN or AECL-9098, 1987.

5. McDonald, B.H., "Assessing Numerical Methods Used in Nuclear Aerosol Transport Codes," In Proceedings of a CEC/OECD Workshop on Water-Cooled Reactor Aerosol Code Evaluation and Uncertainty Assessment, pp. 185-204, Brussels, September 9-11 1987, available as EUR-11351-EN or AECL-9099, 1987.

6. Richards, D.J., Hanna, B.N., Hobson, N., Ardron, K.H., "CATHENA: A two-fluid code for CANDU™ LOCA Analysis," In Third International Topical Meeting on Reactor Thermal Hydraulics, Newport, Rhode Island, USA, Oct 15-18, 1985, PP. 7.E-1, 7.E-14.

7. Allison, G.A., Berna, G.A., Wagner, R.J., Cheng, T.C., Siefken, L.J., Hagrman, D.L., "Capabilities of the Integrated SCDAP/RELAP5/TRAP-MELT Severe Accident Analysis Computer Code," In Proceedings of an IAEA Symposium on Source Term Evaluation for Accident Conditions, pp. 137-144, Columbus Ohio, October 28 - November 1, 1985.

8. Finlayson, B.A., "The Method of Weighted Residuals and Variational Principles with Application in Fluid Mechanics, Heat and Mass Transfer," Academic Press, New York, 1972.

9. F. Gelbard, "MAEROS User Manual," Sandia National Laboratories Report, NUREG/CR-1391, SAND80-0822, December 1982.

10. Im, K.H., Ahluwalia, R.K., Chang, C.F., "RAFT: A Computer Model for Formation and Transport of Fission Product Aerosols in LWR Primary Systems," In Proceedings, American Nuclear Society Meeting on Fission Product Behavior and Source Term Research, EPRI NP-4113-SR, July 1985.

11. Ritzman, R., Rahn, F., "Recent Improvements and Applications of the RAFT Code," In Proceedings of a CEC/OECD Workshop on Water-Cooled Reactor Aerosol Code Evaluation and Uncertainty Assessment, pp. 76-87, Brussels, 1987 September 9-11, available as EUR-11351-EN.

12. Jordan, H., Kughman, M.R., "TRAPMELT2 User's Manual," NUREG/CR-4205, BMI-2124.

13. Johns, N.A., "A Comparison of TRAP-MELT and VICTORIA for an Aerosol Transport Problem," Reactor Systems Analysis Division, UKAEA Winfrith, PWR/SAWG/P(88)394, January 1988.

14. Kissane, M.P., "Using a Containment Aerosol Code in a "Lagrangian Mode" to model Aerosol Processes in Pipe Flows," In Proceedings of a CEC/OECD Workshop on Water-Cooled Reactor Aerosol Code Evaluation and Uncertainty Assessment, pp. 366-374, Brussels, September 9-11, 1987, available as EUR-11351-EN.

15. Otter, J.M., Vaughn, E.U., "HAA-4 Code Description and User Manual," AI-DOE-13528, U.S.A., December 1986.

16. Im, K.H., Chung, P.M., "Particulate Deposition from Turbulent Parallel Streams," AIChe Journal, 29, No. 3, 498-505 (1983).

17. Friedlander, S.K., Johnstone, H.F., "Deposition of Suspended Particles From Turbulent Gas Streams," Ind. Eng. Chem. 49, pp. 1151-1157, 1957.

18. Gieseke, J.A., Lee, K.W., Goldenberg, M.A., "Measurement of Aerosol Deposition Rates in Turbulent Flows," Battelle-Columbus Laboratories, January 1980. Report, NUREG/CR-1264, BMI-2041.

19. Wood, N.B., "A Simple Method for the Calculation of Turbulent Deposition to Smooth and Rough Surfaces," J. Aerosol Sci. 3, pp. 275-290, 1981.

20. Kuhlman, M.R., Kogan, V., Schumacher, P.M., "TRAP-MELT2 Code: Development and Improvement of Transport Modeling," Battelle-Columbus Laboratories, July 1986. Report, NUREG/CR-4677, BMI-2141.

21. Clement, C.F., "The Calculation of Aerosol Growth Rates in Conjunction with Heat and Mass Transfer," In Proceedings of a CEC/OECD Workshop on Water-Cooled Reactor Aerosol Code Evaluation and Uncertainty Assessment, pp. 156-169, Brussels, September 9-11, 1987 available as EUR-11351-EN.

22. Sehgal, B.R., Wassel, A.T., Ahluwalia, R.K., Im, K.H., Denny, V.E., "Coupled Thermal-Hydraulic/Aerosol Transport Analysis Capability for Severe Accidents," In Proceedings of an IAEA Symposium on Source Term Evaluation for Accident Conditions, pp. 687-704, Columbus Ohio, October 28 - November 1, 1985.

23. Abbey, F., Schikarski, W.O., "Aerosol Behaviour and Light Water Reactor Source Terms", Nuclear Technology, 81 (2), pp. 293-306, May 1988.

24. Gelbard, F., "Aerosol Growth in a Nuclear Reactor Containment Environment", Proceedings of a CEC/OECD Workshop on Water-Cooled Reactor Aerosol Code Evaluation and Uncertainty Assessment, pp. 256-266, Brussels, September 9-11, 1987, available as EUR-11351-EN.

25. Haggblom, H., "Use of RETAIN-S for Calculation of Fission Product Transport," In Proceedings of an IAEA Symposium on Source Term Evaluation for Accident Conditions, pp. 307-320, Columbus Ohio, October 28 - November 1, 1985.

26. Wilson, J.H., Wright, A.L., Arwood, P.C., "LACE Aerosol Code Comparison Activities at Oak Ridge National Laboratory," In Proceedings of a CEC/OECD Workshop on Water-Cooled Reactor Aerosol Code Evaluation and Uncertainty Assessment, pp. 8-20, Brussels, September 9-11, 1987, available as EUR-11351-EN.

Limitations of Current Primary Circuit Aerosol Transport Codes

D. A. WILLIAMS
UKAEA, Winfrith
Dorchester, Dorset, DT2 8DH, UK

ABSTRACT

A great deal of effort has been expended in order to generate a database which can be used to validate primary circuit aerosol transport codes and to highlight their limitations, however certain aspects are only poorly represented. Several missing phenomena and uncertainties arising from particular modelling aspects have been identified. Most of the individual models for aerosol deposition and agglomeration contained in the transport codes are supported by small-scale, single-effects experiments. Large-scale experiments have confirmed that the methodology used is basically sound but that several phenomena not modelled in the codes may be important under certain accident conditions.

Typically, missing phenomena include aerosol resuspension, the behaviour of thick deposits of aerosols, deposition in bends in pipework, deposition between adjacent volumes, chemical interactions between vapours and aerosols, and possible interactions between different mechanisms. Uncertainties in the results of calculations performed using these codes also arise from limitations in the databases supporting each individual model, the paucity of suitable thermodynamic data to identify the chemical species being transported, modelling compromises made in order to simplify calculations. Uncertainties in the thermal hydraulics codes may play a significant role in limiting the accuracy of the transport calculations, and in particular the modelling of aerosol nucleation is sensitive to predictions of temperature and temperature gradient. In addition, the accuracy of aerosol codes may be dependent on the numerical schemes incorporated into each code, and exhaustive testing and comparisons are required to ensure that numerical effects do not dominate the physics of each particular calculational problem.

A significant amount of work remains to be done to address the outstanding problems and many of the items discussed are under investigation at present. Some consideration is given to the priorities to be attached to each of these items. Developments of thermal hydraulic modelling may also be required if this proves to be the major source of uncertainty. Care must be taken to ensure that the codes, and all their inherent assumptions and simplifications, are not taken for granted and become accepted as "tablets of stone"; new numerical methods and relevant data should be exploited wherever this proves to be of benefit to the accuracy and reliability of the codes.

1 INTRODUCTION

Computer codes designed to model the transport of aerosols and fission products through the primary circuit of a nuclear reactor undergoing a severe accident have now been under development for up to a decade. Many of the models used in these codes have been adapted from other areas of science, and care must be taken to ensure that they are also relevant under conditions typical of hypothetical severe accidents.

In general, most of the individual models for aerosol deposition and agglomeration contained in the transport codes are supported by a database generated from small-scale, single-effects experiments. However, extrapolation to severe accident conditions is necessary and several large-scale, integrated experimental programmes have been conducted for this purpose (eg Marviken-V [1], LACE [2]). A great deal of effort has been invested in the generation of such databases which can be used to validate the transport codes and also to investigate their limitations, although certain aspects (eg chemical interactions, aerosol nucleation) are only poorly represented. These large-scale experiments generally confirm that the methodology used in many of the codes is basically sound but that several of the phenomena not modelled in the codes may be important under certain conditions.

2 NUCLEATION

The nucleation of supersaturated vapour species to form aerosols is determined by three competing mechanisms: homogeneous, ionic, and heterogeneous nucleation. It is to be expected that under reactor accident conditions ionic nucleation will dominate homogeneous nucleation, and competition between ionic and heterogeneous nucleation will determine the ultimate size of the nuclei at the point at which vapour condensation ceases. However, nucleation is strongly dependent on the temperature and temperature gradients in the gas stream, and it may not be possible to predict these parameters with sufficient accuracy to support a credible nucleation calculation.

A further limitation in the nucleation modelling is the probability that the gas flow in the region of the supersaturated vapour may not be entirely aerosol-free as particles may already have formed via nucleation or physical means elsewhere in the degrading core. Nucleation of further aerosol particles may be suppressed in this instance, and vapour condensation onto the existing aerosols would take place to relieve the supersaturation. One possible pre-existing aerosol species is boric oxide which could arise from the boron present in the reactor coolant. During a blowdown the surface of the coolant will be boiling violently and this may give rise to boric oxide particles as ejected water droplets are entrained and evaporated in the steam flow. The existence of these aerosols may have a profound effect on fission product transport, and under certain conditions may dominate the release to the containment.

The chemical forms of the fission product vapours will play an important part in the nucleation process, and will determine the temperature at which supersaturation will occur and also the surface tension of the material forming the nuclei. Nucleation in multi-component systems is not understood at present, and the paucity of chemical thermodynamic data (see Section 5) for predicting the species present further limits the accuracy of nucleation predictions. Complications may also arise from chemical reactions which result in vapour species being converted to solid species on a very short timescale.

Given these uncertainties and limitations the mechanistic calculation of nucleation rates and sizes of nuclei may be no better than the approach adopted by many codes which require the user to specify these parameters. It is to be hoped that the agglomeration of the nuclei will be sufficiently rapid that the final particle size will be independent of the size and concentration of the nuclei, and studies need to be undertaken to assess this point.

3 AGGLOMERATION

The agglomeration of aerosol particles has been traditionally modelled in transport codes as being the result of three independent mechanisms. These mechanisms are gravitational sedimentation, Brownian agglomeration, and turbulent agglomeration. No account is taken of possible interactions between these mechanisms, and the agglomeration rates due to the gravitational and turbulent components are usually summed in quadrature and added to the Brownian component to give the total rate. Mathematical arguments have been used [3] to justify this summation technique, and usually one mechanism dominates over the others. However, under conditions where two mechanisms result in similar agglomeration rates this assumption could be an important limitation requiring investigation.

A further limitation in the scope of aerosol transport calculations would arise if not all mechanisms resulting in particle agglomeration were considered in the codes. An obvious mechanism worthy of evaluation in this respect is sonic agglomeration [4]. Sonic agglomeration arises from the carrier gas turbulence generated by an intense acoustic field, and may therefore be considered as an increase in the turbulent agglomeration rate. The questions which must then be answered are what are the sources of noise and what is the level of noise generated around the reactor coolant circuit during a severe accident, and information on these subjects does not appear to be readily available.

One basic assumption inherent in all current transport codes is that particle agglomerates may only grow with time, and this growth arises from further agglomeration and vapour condensation. The possibility that agglomerates may become fragmented due to turbulence, acoustic noise, and collisions with other agglomerates etc is not considered. The energy required to fragment an agglomerate would be determined by the chemical species comprising the particle (eg the stickiness of the particle), and the size and shape of the agglomerate. The issue of stickiness is also of relevance to the calculation of agglomeration rates where it is assumed that the particles brought into contact by the agglomeration mechanisms invariably stick together with 100% efficiency.

Particle collision shape factors, as applied to agglomeration modelling, are at best only poorly understood and really only serve as scaling factors by which calculations may be adjusted to fit experimental data. Since the collision shape factors may only be obtained by back-fitting experimental data the task of validating agglomeration models is made more difficult, and this is a further limitation on the capabilities of the aerosol transport codes.

In order to simplify the calculations, some assumptions must be made regarding the aerosol concentrations within the volumes of interest, and the most common of these is the well-mixed volume approach. Agglomeration rates depend on the square of the aerosol concentration and it is therefore evident that the assumption of a homogeneous aerosol concentration could result in a significant uncertainty. For high Reynolds number flows the turbulent nature of the fluid should ensure that the assumption is valid; however, care must be exercised with calculations for laminar flow conditions and under conditions where deposition rates are large. Lagrangian-based numerical schemes do not suffer from this limitation but have other problems (eg inability to model aerosols settling against the direction of fluid flow) which off-set their advantage.

4 DEPOSITION

Aerosol deposition within the primary circuit may also be brought about by one, or any combination, of several mechanisms, each of which is modelled as being wholly independent of the other mechanisms. The most important mechanisms common to most codes are gravitational sedimentation, thermophoresis, and deposition from laminar or turbulent flow. These mechanisms are all supported by extensive databases arising mainly from non-nuclear sciences, and work is in hand to validate the integration of these models within the transport codes.

Experiments such as the Marviken-V and LACE programmes have been designed to provide suitable data for code assessment and validation and, as a consequence of these tests, several deposition phenomena not modelled within the codes have been identified as being of possible importance under certain severe accident conditions. One such mechanism is the deposition of aerosols in bends in pipes, and the development of models to predict this phenomenon is in hand at several institutions. There is no model capable of reproducing all the bend deposition data currently available, and this is probably due to the over-simplified treatment of the fluid flow patterns within the bends which are assumed to be stochastic.

Early analysis of the Marviken-V [5] and ORNL-ATT data [6] indicated possible problems in modelling thermophoresis with transport codes such as TRAP-MELT 2 [7] which seriously underpredicted the measured wall deposition rates. Investigation of the cause of this discrepancy indicated that the estimation of the heat transfer boundary layer was limited to the region of turbulent forced convection, and this gave rise to errors at low Reynolds numbers. The solution to this particular problem [8] was relatively straightforward but highlights the gross simplifications in the fluid flow physics adopted by the transport codes. Similarly, assessment of the VICTORIA code [9] for turbulent pipe flow indicated that at Reynolds numbers of the order of 10^5 there is an uncertainty of up to 50% in the predicted turbulent inertial deposition rate due to the choice of correlation used in the calculation [10]. This limitation arises from the extrapolation of correlations to points beyond the range of data from which they were produced and requires a re-evaluation of the data and possible extension of the database to resolve the choice of correlation.

In all current codes it is assumed that an aerosol sticks to a deposition surface with 100% efficiency on arrival and, equally importantly, stays fixed to that surface regardless of subsequent changes in thermal hydraulics etc. There are ample data available to demonstrate that both these premises are false; particles may rebound from the deposition surface [11] or be resuspended from the surface at a later time [12]. Both of these processes will be governed by thermal hydraulics and aerosol material properties (and hence chemistry), and may also be affected by vibrations of the deposition surfaces which is a problem analogous to the sonic agglomeration question discussed in the previous section. Resuspension of aerosols is an important consideration in source term analysis since it has the potential for the production of significant sources of radioactivity to the containment building at a late time during the accident, and this could alter the source term in the event of containment leakage or failure. Revaporisation of deposited material poses a similar threat, and this is addressed in Section 5.

Models for the bouncing and resuspension of single aerosols have been developed [11, 12], and work is in hand to adapt these models for use in primary circuit aerosol transport codes [13]. However, it would be naive to suggest that this is a simple step to take. Visual evidence from the Marviken-V clean-up operations suggest that resuspension of thick layers of aerosol deposits had taken place. Sheets of material had been stripped from some of the pipe and vessel walls leaving sharp transition regions between areas devoid of deposits and adjacent areas with thick deposits. It is possible that the single-particle resuspension theory may not lend itself to being extended to the thick deposits likely to occur under severe accident conditions, and an alternative model may be required. It is also important to understand the fragmentation of a sheet of deposited aerosols once it is resuspended from the surface. The resultant aerosol particle size distribution and concentration will determine the subsequent behaviour of this material.

With few exceptions all plant calculations performed to date have assumed that the aerosol particles are spherical. Whilst this assumption may be true for some of the primary particles generated during core degradation, it will not be true of agglomerates if the steam in the primary circuit remains superheated. Early analysis of Marviken data [14] has clearly demonstrated the importance of shape factors in determining particle transport behaviour, and the equivalent data for nuclear aerosols are extremely sparse. A further complication arises from the agglomeration processes which may, under certain conditions, result in size dependent shape factors, and this is overlooked by all the codes currently available.

The assumption that deposition mechanisms are independent of one another is an over-simplification, and data are now available from experiments investigating the interaction between sedimentation and turbulence [15], and also between thermophoresis and turbulence [16]. In both cases the interaction increases with increasing Reynolds number, and this may produce errors in the results of calculations performed for accident sequences such as the V-sequence during which high flow rates are possible.

5 CHEMISTRY

The importance of chemistry effects on the prediction of fission product and aerosol behaviour within the primary circuit has been largely overlooked until quite recently, and transport codes have adopted an extremely simple-minded approach to chemical interactions. Most code developers have made assumptions, based on thermodynamic calculations [17], as to the likely chemical species which would be transported during a severe accident, and the volatile species were usually identified as caesium iodide, caesium hydroxide, and elemental tellurium. Other fission product species and chemical interactions between species were largely ignored, and this could have significant implications for deposition, resuspension, and revaporisation predictions.

Laboratory studies [18] have demonstrated the possible chemical complexity of fission product transport under severe accident conditions, and the effect of varying the chemical form of iodine and tellurium species transported has been found to be an important consideration [19]. The most recent of the aerosol transport codes (eg RAFT [20], VICTORIA) are designed to address these issues and include chemical equilibrium thermodynamic calculations in order to predict the chemical species present within the primary system as a function of both time and location. Both these codes are currently under development and the number of species modelled increases with time; VICTORIA is capable of handling ~ 100 species at present (based on the elements Cs, I, Te, O, H, Ba, Sr, Zr, Sn, U, Kr, Xe, Fe, Mo), and RAFT considers ~ 80 species (based on the elements Cs, I, Te, O, H, Ba, Sr, Zr, Sn, Kr/Xe, Mn, Ru, Mo, Ag, In, Cd). The species predictions may then be used as input to nucleation calculations to model the production of aerosols as the vapours are transported through the circuit. Experimental studies have shown that the list of species included in these codes is by no means comprehensive; for example the VICTORIA code does not presently include compounds containing control rod materials and neither code includes boron compounds [21], or oxides, hydroxides and ternary oxides of elements such as ruthenium, barium, strontium, etc. The relevant thermodynamic data for some of these species are not known and require study. The present database may also require extending to more relevant conditions for some species (ie high temperature).

Scoping calculations [19] have indicated that the neglect of chemical interactions between vapours and aerosols is a serious limitation on code capabilities, and interaction with aerosols could significantly reduce the source of radioactivity into the containment for a TMLB' sequence [22]. Current versions of both RAFT and VICTORIA are limited in the complexity of their calculations of chemical interactions between vapours and aerosols. The chemical composition of aerosol particles is assumed to be independent of size, and each particle is assumed to be homogeneous. This assumption is not supported by experimental data, and consequently the modelling of chemical interactions neglects any effects arising from non-homogeneities in the aerosol composition (eg vapour species x may only interact with aerosol species y which is only to be found in size range z). The next stage in the development of VICTORIA will be to extend the modelling to include multi-component aerosols, and this should also include a more realistic approach to vapour-aerosol interactions.

Certain transport codes (eg TRAP-MELT 2) included crude models which attempted to make allowance for interactions between vapour species and structure surfaces, and this was achieved by the use of temperature independent irreversible vapour deposition velocities. Surface passivation by oxidation or other chemical interactions is not included in these correlations and this further limits their usefulness, particularly in the case of tellurium. An improvement on this approach was the derivation of Arrhenius-type correlations based on extensive experimental data [23]; application of these correlations to accident sequences [21] demonstrated the sensitivity of primary system retention to the interaction between vapours and structures. The modelling capabilities of the VICTORIA code are also an important step towards a realistic description of this process, and further developments are planned. VICTORIA considers interactions between vapour species contained within a boundary layer adjacent to a structure and condensed species within a surface layer on the structure. Vapour species concentrations are replenished by diffusion from the bulk gas, and the diffusion process is mass-transfer limited. An equilibrium chemical thermodynamic calculation is performed for these vapour and condensed species. The model presently ignores the effects of diffusion through the surface layers of the structure but can predict the re-evolution of vapour species due to further reactions as well as revaporisation. Revaporised material then becomes available for further aerosol nucleation, interaction with aerosols, and interaction with structures.

No allowance is made in any of the codes for chemical effects due to the presence of liquid water, and this could have an effect on subsequent transport behaviour of soluble vapour species. Since it may not always be the case that the path to the break will be dry, care must be exercised in applying codes which have only been developed for dry conditions.

6 NUMERICAL STRUCTURE

Although the numerical structures of many of the aerosol transport codes have been tried and tested over a number of years it is quite possible that they have yet to be exercised over the full range of conditions for which they are expected to be required. It is also possible that problems arising from the numerical methods have been overlooked even for those conditions wehre they have been used. A recent UK study [22] using TRAP-MELT 2 indicated numerical problems associated with the aerosol number density which must also have been present in similar calculations performed in other countries. Users must be wary of this possibility, particularly when attempting to apply the codes to new scenarios.

Code comparison exercises [24] and exchanges at various specialist analysis meetings [25, 26] have demonstrated that two users with the same code can obtain differing results, and this can be as much dependent on the nodalisation adopted by each user as on the choice of input data. Care must be exercised in setting up nodalisation schemes and, where possible, sensitivity studies should be undertaken to evaluate the likelihood of discrepancies in the results. Analyses of the aerosol behaviour within the Marviken-V pressuriser vessel were performed by sub-dividing this vessel into three well-mixed volumes. However, before the advent of the VICTORIA code, no code had the ability to model aerosols settling against the fluid flow from one of these pseudo-volumes. It has been shown [27] that this choice of nodalisation scheme, together with the missing settling capability, resulted in all codes under-predicting the pressuriser deposition, aerosol particle size and over-predicting the deposition downstream of this vessel.

7 PRIORITIES

The assignment of priorities for future work to reduce or eliminate the code limitations discussed in the earlier sections is a difficult task which invites political, as well as technical, attention. Nevertheless, it is an important task which should focus effort on the areas of major difficulty, and also promote discussion amongst those requiring the results of such aerosol transport codes. The work to be undertaken falls into two main interdependent categories, namely chemistry and aerosol physics.

It is debatable whether aerosol resuspension or chemical interactions between vapours, aerosols and structures is the most important problem to be solved and it would be satisfying to see effort applied to both issues simultaneously. Coupled to the issue of chemical interactions is the question of the adequacy of the available thermodynamic and kinetic databases (ie species and range of data), and further effort will be required before the number of species considered in the transport calculations can be reliably increased. In general these items probably represent a greater code limitation than the sum of all other topics discussed in this paper, although this will obviously depend on the particular conditions of each transport scenario.

The remaining limitations and uncertainties appear to arise from inadequacies in the current state-of-the-art of aerosol physics, although chemistry will play a part in defining aerosol material properties such as stickiness. None of these "second order" issues would appear to be significantly more important than the rest, and it is unlikely that, with the exception of the time-evolution of particle shape factor, and the development of nucleation theory, any would take an inordinate level of effort to solve. An assessment of the potential for generation of boric oxide aerosols during a reactor blowdown should be undertaken, and this could take the form of a series of small-scale experiments where pools of borated water of various concentrations could be violently boiled under simulated accident conditions. Small-scale single effects experiments can be, and in some cases (eg interaction between thermophoresis and turbulence) are being used to address the issues of the additivity of deposition and agglomeration rates. The existing turbulent deposition models should be re-evaluated against the supporting database to resolve the large discrepancy at high Reynolds numbers, and it is possible that the existing data will be adequate for this purpose thus minimising the effort required. Current models which describe the behaviour of aerosols in bends in pipes are not based on any understanding of the physics involved. It would be sensible to perform detailed thermal- hydraulic calculations for fluid flow around bends so that the systematic flow regimes are understood before the aerosol physics is addressed. There is ample scope for small-scale experiments to investigate these phenomena under a wide range of flow conditions. Similarly, detailed thermal-hydraulic calculations could shed light on some of the uncertainties in the nucleation models.

8 CONCLUSIONS

Code assessments and validation efforts which utilised data from both large and small scale experiments have indicated that the methodology of most codes is basically sound but that several phenomena ignored by the codes may be important under certain accident conditions. A significant amount of work remains to be done, particularly in addressing aerosol resuspension and various chemical issues, and many of the items discussed in this paper are being investigated at present.

Care must be taken to ensure that the codes, and all their inherent assumptions and simplifications, are not taken for granted and become accepted as "tablets of stone". The continued validity of these assumptions should be questioned regularly. Users must be aware of limitations arising from some of these assumptions (eg well-mixed volumes). New numerical methods and supporting chemical and aerosol data should be exploited wherever this proves to be of benefit to the accuracy and reliability of the codes, and developments in computer hardware are also important in this respect.

9 ACKNOWLEDGEMENT

This work has been part funded by the Department of Energy and the CEGB.

10 REFERENCES

1 Collén, J., Unneberg, H., Mecham, D., Proc ANS Topical Meeting, Snowbird, Utah, pp 20-1/20-9 , July 1984.

2 Rahn, F.J., Collén, J., Wright, A.L., Nucl. Technol., 81, 158, 1988.

3 Saffman, P.G., Turner, J.S., J. Fluid Mech., 1, 16, 1956.

4 Malherbe, C., Boulaud, D., Boutier, A., Lefevre, J., Aerosol Sci. Tech., 9, 93, October 1988.

5 Williams, D.A., Butland, A.T.D., UK Atomic Energy

Authority Report, AEEW-M 2147, October 1984.

6 Wright, A.L., Proc. 2nd Int. Aerosol Conf., Berlin, September 1986.

7 Jordan, H., Kuhlman, M.R., NUREG/CR-4205, May 1985.

8 Williams, D.A., UK Atomic Energy Authority Report, AEEW-M 2298, April 1986.

9 Camp, W.J., Eur. Appl. Res. Report - Nucl. Sci. Technol, 5, 1605, 1984.

10 Williams, D.A., Chown, N.M., UK Atomic Energy Authority Report, forthcoming publication.

11 Reed, J., J. Phys. D, 18, pp 2329-2337, 1985.

12 Reeks, M.W., Reed, J., Hall, D., J. Phys. D, 21, pp 574-589, 1988.

13 Wright, A.L., forthcoming publication.

14 Williams, D.A., UK Atomic Energy Authority Report, AEEW-M 2371, December 1986.

15 Reeks, M.W., Oxford Conf. on Mass Transfer, 1988.

16 Owen, I., El-Kady, A.A., Cleaver, J.W., Proc. 2nd National Heat Transfer Conference, Glasgow, 1988.

17 Potter, P.E., Rand, M.H., Alcock, C.B., J. Nucl. Mater., 130, 139, 1985.

18 Nichols, A.L., Proc. Workshop on Chem. Processes and Products in Severe Reactor Accidents, Captiva Island, Florida, December 1987.

19 Butland, A.T.D., Nichols, A.L., Williams, D.A., ENC '86 Trans., 3, 625, 1986.

20 Im, K.H., Ahluwalia, R.K., Chuang, C.F., Aerosol Sci and Tech., 4, 125, 1985.

21 Bowsher, B.R., Dickinson, S., UK Atomic Energy Authority Report, AEEW-R 2102, May 1986.

22 Chown, N.M., Williams, D.A., UK Atomic Energy Authority Report, AEEW-R 2429, March 1988.

23 Bowsher, B.R., Dickinson, S., Nichols, A.L., Seminar of the International Centre for Heat and Mass Transfer: Fission Product Transport Processes in Reactor Accidents, Dubrovnik, Yugoslavia, 22-26 May 1989.

24 Wright, A.L., Wilson, J.H., Arwood, P.C., Proc. 2nd Int. Aerosol Conf., Berlin, September 1986.

25 Williams, D.A., Johns, N.A., Butland, A.T.D., Bowsher, B.R., Nichols, A.L., UK Atomic Energy Authority Report, AEEW-R 2263, December 1985.

26 Nichols, A.L., Williams, D.A., UK Atomic Energy Authority Report, AEEW-M 2486, August 1986.

27 Williams, D.A., Johns, N.A., UK Atomic Energy Authority Report, forthcoming publication.

The Suspension of Fine Particles in a Reactor Coolant System

M. W. REEKS
CEGB, Berkeley Nuclear Laboratories
Berkeley, Gloucestershire, UK

ABSTRACT

Resuspension is an important process in determining the net deposition of an aerosol released into the coolant of a nuclear reactor. This paper describes a statistical model that has been used to predict the resuspension of aerosol particles by reactor coolant flows where the principle adhesion forces are Van der Waals forces. Particles are released from the surface when they have accumulated enough vibrational energy from the local flow to escape from the surface adhesive potential well. The form of the resuspension rate constant is very similar to that for the desorption of molecules from a surface. A brief description is given of the measurements that have been made of the aerodynamic lift force which are necessary inputs in the practical application of the model. Consideration is given to the implications of the model for:
(a) resuspension rates from rough surfaces which in the long term vary almost inversely as the time of exposure to the flow;
(b) the decay of gas-borne particles in a reactor coolant system.

INTRODUCTION

Many design basis faults and severe accidents in nuclear reactors will involve the release of radioactive aerosols. The deposition of those aerosols to surfaces within the reactor coolant system (RCS) and containment building will reduce the potential release of radioactivity to the atmosphere. It is therefore important for safety assessment to predict this deposition reliably. In general that deposition will be controlled by three processes:

1. aerosol arrival rate to the depositing surface;
2. the fraction of aerosol arriving at the surface that adheres or bounces upon impact (impact adhesion),
3. the fraction of aerosol initially adhering upon impact which is resuspended aerodynamically sometime later.

This paper is concerned with our work on the third process - resuspension. It begins with a brief description of the background to that work, highlighting the motivation and objectives. This is then followed by a description of our theoretical model for resuspension which contains several novel and interesting features. Then an account is given of the measurements that have been made of the aerodynamic forces acting on a resuspending particle which are necessary inputs in the practical application of the model. Consideration is then given to the practical implications of the model for the resuspension of particles from rough surfaces and for the decay of a gas borne aerosol in a reactor coolant, comparison being made between model predictions and those obtained from recent experiment. The paper is concluded with a brief description of our current and future work aimed at validating the quantiative predictions of the model.

PURPOSE OF WORK

Historically our work on particle resuspension is associated with a requirement to model and predict particle plateout in Civil Advanced Gas-Cooled Reactors (CAGRs). As noted earlier this plateout reduces the environmental release of a radioactive aerosol accidentally released into the reactor coolant. The principle concern has been with the release of radioactive particles following the accidental dropping of fuel in the fuel route. This aerosol may take the form of fuel element debris or more likely U308 particles resulting from the direct exposure of fuel to the CO_2 coolant. To determine the plateout of such an aerosol within the reactor circuit, a series of aerosol injection experiments were carried out in Windscale Advanced Gas Cooled Reactor (WAGR) (Garland et al., 1982) and Hinkley Point B CAGR (Garland et al., 1984) in which the gas-borne aerosol concentration was measured as a function of time. The injections were limited to a range of aerosol size and flow. An important practical objective of our work has therefore been to interpret this decay in terms of the various plateout mechanisms with a view to extrapolation to a continuous range of aerosol sizes and flows appropriate to the reactor fault. A major uncertainty in that extrapolation has been the aerosol resuspension following initial deposition. We recognised that the observed decay was attributable in part to two forms of resuspension:

1. short-term resuspension occurring in times less than a second after initial deposition; this reduces the effective deposition velocity to a surface,
2. long term resuspension occurring in times greater than a second which depends upon the history of the gas-borne concentration.

So our work on resuspension has been first to build a working model that quantifies both these forms and then secondly to compare its predictions with the results obtained from carefully controlled laboratory experiments.

THE REEKS, REED AND HALL RESUSPENSION MODEL - A BRIEF SUMMARY

This new approach to resuspension [3] illustrates a close analogy of particle motion in a turbulent flow with the behaviour of thermodynamic systems, - in particular with the desorption of molecules from a surface, and with the escape of Brownian particles from a potential well.

The model for resuspension is based upon the view that an adhering particle and its substrate are deformed elastically by their surface adhesive forces. In static equilibrium (no external forces or flow) there is a balance between the adhesive force and an elastic restoring force. However, when a particle is exposed to a turbulent flow, there is a transfer of turbulent energy to the particle, which causes the particle and substrate to deform continually about their static equilibrium configuration (changing the area of contact). Particles in contact with the substrate are confined to motion within a surface adhesive potential well derived from the dependence of adhesive and elastic restoring forces upon particle-substrate deformation (see Fig. 1) i.e., absence of slip or rolling.

In this approach a particle is released from a substrate (resuspended) when it receives enough energy to escape from the adhesive potential well.

The transfer of energy takes place through the agency of aerodynamic lift forces where:

(a) the average component modifies the shape and height of the adhesive potential well;
(b) the random fluctuating component causes the particle and surface to deform in a random oscillatory fashion from their static equilibrium deformation (point of minimum potential, A in Fig. 1).

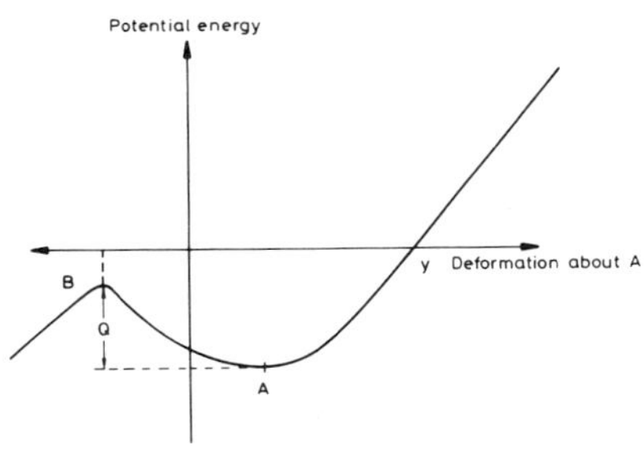

Fig.1 The surface adhesive potential energy of a sphere on a surface with a constant applied lift force. Two positions of equilibrium: (A) stable and (B) unstable.

The motion is represented by that of a very stiff lightly damped harmonic oscillator with a forcing term provided by the random fluctuations in aerodynamic lift force. Thus for a particle of mass m at time t with a deformation y about the point of minimum potential,

$$\frac{dv}{dt} + \beta v + \omega^2 y = m^{-1} f_L(t); \ \frac{dy}{dt} = v, \qquad (1)$$

where

(a) β is the fluid and mechanical damping term (particle inverse response time). The fluid damping is generally greater than its equivalent value based on Stokes drag, being dependent upon the frequency of vibration (vibrational damping). The mechanical damping corresponds to an energy loss due to elastic wave propagation in the solid substrate.

(b) ω is the natural frequency of the particle and substrate based on the harmonic approximation at the point of minimum potential. Using a model of adhesion by Johnson, Kendal and Roberts [JKR] [3], the stiffness $m\omega^2$, for zero applied load, has the form

$$m\omega^2 = \frac{9}{10}\left(6\pi\Delta\gamma K^2R^2\right)^{1/3}, \tag{2}$$

for a particle of radius R on a surface where $\Delta\gamma$ is the adhesive surface energy per unit area and K an elastic constant given by

$$K = \frac{4}{3}\left[\frac{1-\sigma_1^{\,2}}{E_1} + \frac{1-\sigma_2^{\,2}}{E_2}\right]^{-1}, \tag{3}$$

where E_i, σ_i (i=1, 2) refer to the Young's modulus and Poisson's ratio of either substrate or particulate materials. ω is typically $\sim 10^7$ Hz for a 10 μm diameter spherical particle on a flat surface.

(c) $f_L(t)$ is the fluctuating component (zero mean) of the aerodynamic lift force. The two most important statistical quantities associated with $f_L(t)$ in this model are its r.m.s., $\langle f_L^{\,2}\rangle^{1/2}$, and its energy spectrum $E_L(\upsilon)$, where υ refers to the frequency of a Fourier component of $f_L(t)$. If τ is the integral timescale of f_L, then for a stiff lightly damped oscillator

$(\omega\tau)^2 \gg 1; (\omega\tau)^2 \gg \beta\tau$.

These conditions are satisfied in most practical cases of particle resuspension.

The transfer of turbulent energy eventually maintains some constant average potential energy <PE> of particles within the surface potential well. It is considered that the balance of this energy with the height, Q, of the surface potential well is a more appropriate criterion than a force balance criterion for the initiation of resuspension. Consideration of the random motion of a particle in the well led to a simple formula for the probability per unit time of particle release from a surface (the rate constant), p, in situations where $\langle PE\rangle/Q \ll 1$. This situation refers to long-term resuspension, meaning that the average lifetime of a particle on the surface is very much greater than both the timescale, τ, of the aerodynamic lift force, and the particle response time (β^{-1}). Explicitly

$$p = \frac{\omega_0}{2\pi}\exp\left[-\frac{Q}{2\langle PE\rangle}\right], \tag{4}$$

where ω_0 is the typical frequency of vibration.

It was noted that this formula has the same form as that for the rate constant associated with the desorption of molecules from a surface and many chemical activation processes.

The value of <PE> is the sum of two contributions

(a) a contribution from the low frequency (off-resonance) components of the fluctuating lift force. For these frequencies there is a balance between the instantaneous driving force and the harmonic restoring force. This is the response typical of most of the energy spectrum of f_L

(b) a contribution from the near resonance components of f_L, ie. with frequencies close to ω. This contribution represents the most efficient way in which the accumulation of vibrational energy (and hence resuspension) can take place. The analogy is a familiar one of a vibrating spring where repeated applications of a relatively small force can in time have an equal effect on the amplitude of vibration as that of a much larger force applied at a much lower frequency.

Using the expression for <PE> involving these two contributions with that of Q involving the force of adhesion f_a we obtain for the resuspension rate constant

$$p = \frac{\omega_0}{2\pi} \exp\left[-\frac{k(f_a - \langle F_L\rangle)^2}{\langle f_L^2\rangle(1 + \eta)}\right] \tag{5}$$

where

k is a numerical constant dependent upon the shape of the potential (for the surface adhesive potential $k \sim 0.5$

η refers to the resonance contribution and given by

$$\eta = \frac{\pi\omega^2 E_L(\omega)}{2\beta\langle f_L^2\rangle} \tag{6}$$

where $E_L(\omega)$ is the energy spectrum of the fluctuating lift force at the resonant frequency ω. The dependence on β reflects the fact that η is limited by energy dissipation in the local fluid and substrate.

The typical frequency of vibration ω_0 is given by

$$\omega_0 = \omega\left[\frac{\eta + \langle \dot{f}_L^2\rangle \;/\; \langle f_L^2\rangle\;\omega^2}{\eta + 1}\right] \tag{7}$$

It is a natural consequence of the influence of 'resonant energy transfer' that our resuspension model will in certain circumstances predict significantly lower threshold velocities for detachment of a particle from a surface than that based on a force balance of aerodynamic and adhesive forces.

MEASUREMENTS OF THE AERODYNAMIC LIFT FORCE

To evaluate the resuspension rate constant we require values for

1. the mean and rms of the lift force
2. the power/energy spectrum of the fluctuating component of the lift force especially near the resonant frequency of the surface particle deformation.

At the time the resuspension model was first developed reliable values for these quantities for the range of particle sizes and reactor flows of interest did not exist. What values we currently use are based on our own experimental measurement.

Direct measurement of the lift force acting on a particle ~ several microns in size would present considerable technical difficulties. Hall [4, 5] has overcome this problem by exploiting a suitable scaling law so that in practical terms a 10 micron diameter particle in a CAGR could be modelled using a mm diameter particle in an air flow of about 2 m/s at NTP. Forces typically ~ 10^{-8}N (the weight of a grain of castor sugar) have been measured by securing the mm size particle to a wire attached to a very sensitive pressure transducer (rms noise level nominally 10^{-6}N.) Even so the electronic noise was still greater than many of the signals it was desired to resolve. Full details of the experiment and how these problems were overcome are described in [4, 5].

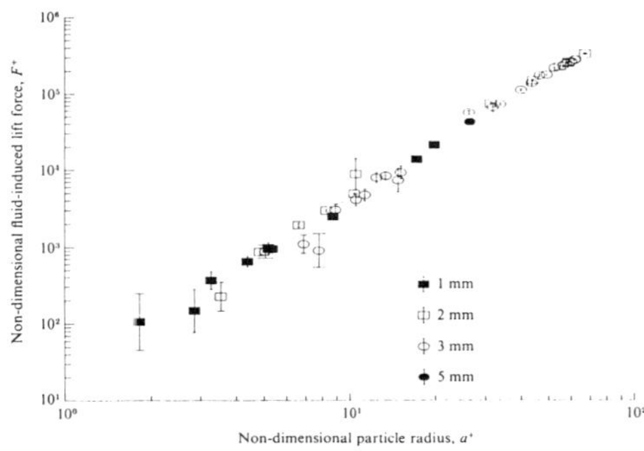

Fig.2 Lift force on particles on a smooth surface for various particle diameters (68% confidence limits).

With this device Hall has obtained values for the particle mean, rms and energy spectrum of the lift force for a variety of particle shapes and sizes on both rough and smooth surfaces. The data have been reduced to a suitable dimensionless form appropriate to a particle in a fully developed turbulent boundary layer. For a flow of density ρ, kinematic viscosity ν

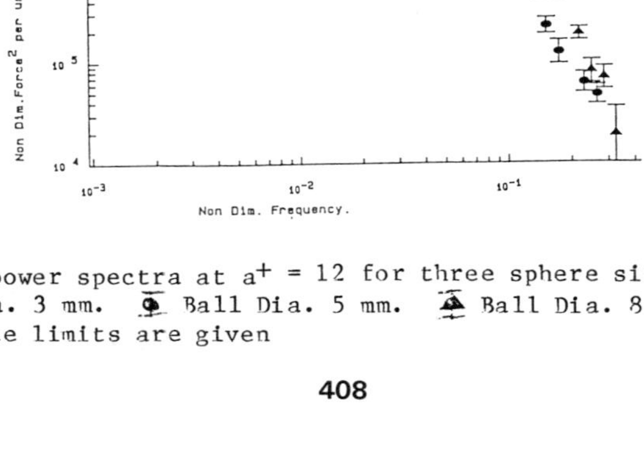

Fig. 3 The power spectra at $a^+ = 12$ for three sphere sizes.
■ Ball Dia. 3 mm. ● Ball Dia. 5 mm. ▲ Ball Dia. 8 mm.
68% Confidence limits are given

and friction velocity u_τ, Fig. 2 shows the measurements of the normalised mean lift force $F^+ = \frac{\langle F_L \rangle}{\nu^2 \rho}$ against normalised particle radius $a^+ = \frac{a u_\tau}{\nu}$ for a sphere on a smooth surface of various diameters. Fig. 3 shows the corresponding measurements of the normalised energy spectrum $E_L^+(\upsilon^+) = E_L(\upsilon) \frac{u_\tau^2}{\nu}$ against normalised frequency $\upsilon^+ = \frac{\nu \upsilon}{u_\tau^2}$. In this case the rms lift force was ~ 0.2 of the mean lift force for the range of particle sizes and flow velocities considered.

As an illustration of the application of the resuspension model we show in Fig. 4 predictions based on Hall's lift force measurements. The case is for removal in air of U308 spheres in contact with a smooth substrate. Fig. 4 in particular shows the influence of resonant energy transfer in reducing the force necessary to remove a particle from a surface. The resonant factor η is between 5-20 for the range of particle sizes (1-100 microns) considered, and is sufficient to reduce the force necessary to remove a particle by a factor of 20-30 from that in a static situation. The normalised resonant frequency ~ 0.4 which is just outside the range of Hall's measurements.

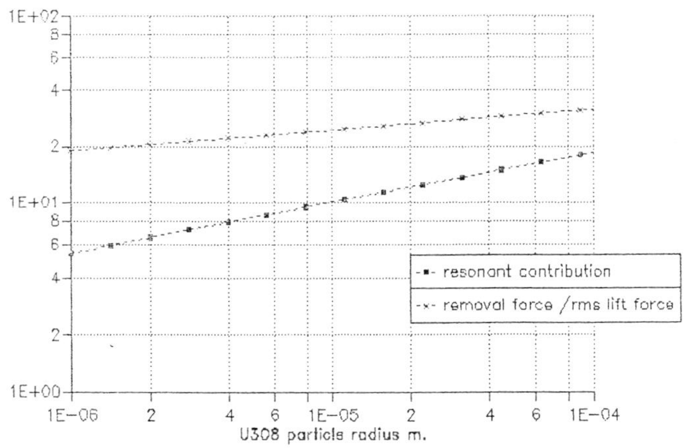

Fig.4 Rms Lift Force and Resonant Factor Predictions for 50% removal after 1s of U_3O_8 spheres in smooth contact with steel surface.

RESUSPENSION FROM ROUGH SURFACES

An important consideration appropriate to resuspension in a recirculating reactor flow is the form of the resuspension rate from a rough surface exposed to the flow. Most surfaces in practice are rough i.e. they have a surface topography which can be characterised by a spread in the height and radius of curvature of surface asperities. This in turn gives rise to a spread in surface adhesive forces. The value of the resuspension rate constant seen in equation (5) is very sensitive to the force of adhesion f_a. Thus even when surfaces are nominally smooth the spread of adhesive forces can have significant effect on the net resuspension rate.

If $\phi(f_a)$ is the distribution of adhesive forces, then the resuspension rate $\Lambda(t)$ at time t is given by

$$\Lambda(t) = \int_0^\infty p(f_a)e^{-p(f_a)t}\phi(f_a)df_a \quad (8)$$

Numerical calculations in [1] show the variation of $\Lambda(t)$ naturally dividing into two regimes:

1. A short initial response over times $\lesssim 10^{-2}$ seconds in which the resuspension rate is very high. Such resuspension is usually responsible for a significant fraction of the total resuspension from a rough surface.
2. A region of 'longer term' resuspension in which the resuspension rate is seen to vary almost inversely with the time of exposure i.e.,

$$\Lambda(t) = \zeta t^{-\varepsilon} \quad (9)$$

These calculations also show that this longer term inverse relation is maintained over a wide range of exposure times ($10^{-2}s<t<10^5s$) and in fact is very robust to wide changes in flow, particle size and surface roughness.

This longer term behaviour has since been confirmed by a number of separate and independent experiments [7,8,9,10]. Most interestingly we note that the most recently reported measurements of Wen and Kasper [10] use a similar Arrhenius activation form for the resuspension rate constant as used by Reeks et al. [3] and draw the same basic conclusions about the inverse form for the longer term resuspension rate.

MEASUREMENT OF THE DECAY IN GAS-BORNE CONCENTRATION OF SMALL PARTICLES IN A CAGR

A striking example of the influence of longer-term resuspension is seen in the decay of gas-borne particle concentration in the coolant flow of a CAGR. In such a recirculating turbulent flow, deposition and resuspension at surfaces exposed to the flow occur simultaneously. The initial resuspension increases the decay time by reducing the net deposition to a surface; the longer-term resuspension reduces the decay in a manner which depends on the history of the gas-borne concentration.

The appropriate equation for the concentration c(t) at time t is an integro-differential equation of the form [11].

$$\frac{\partial c}{\partial t} = - ac(t) + b\int_0^{t-t_i} \Lambda(t - s)c(s)ds \quad (10)$$

where a is a decay constant modelling all particle removal mechanisms including deposition and initial resuspension from surfaces plus natural leakage; b is a decay constant modelling only the deposition to surfaces exposed to the flow; and t_i is the duration time for initial resuspension. (The value of t_i is not critical so long as it is sufficient to include all initial resuspension.)

The integral term on the right-hand side of equation (10) represents the contribution to the decay rate from longer-term resuspension. Our previous considerations imply that the fractional resuspension rate, $\Lambda(t - s)$, will have the form

$$\Lambda(t - s) = \frac{\zeta}{(t - s)^{\varepsilon}} \qquad (11)$$

where $\varepsilon \simeq 1$.

Numerical solutions of equation (10) with Λ given by equation (11) behave in a manner consistent with the observed decay of gas-borne particles injected into the coolant of a CAGR [2]. The injected particles were labelled with radioactivity for identification, and measurements made of sampled particle gas-borne activity over a period of 1-200 min from the start of the injection.

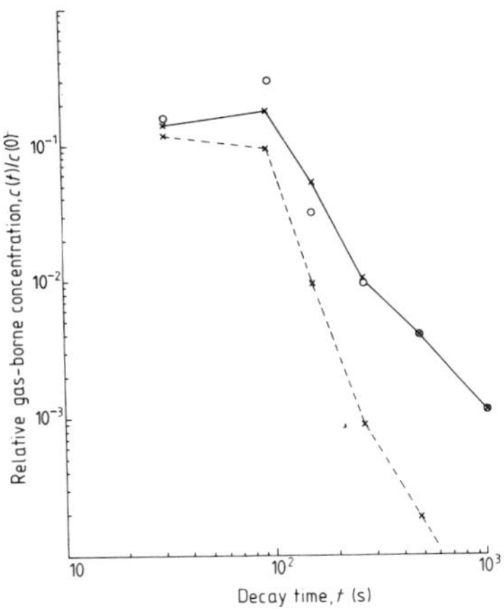

Fig.5 Injection experiment results for gas-borne concentration of 2 μm diameter ion oxide particles. (The initial rise in concentration is due to mixing.) Symbols: 0, experiment; —— x, deposition plus initial resuspension plus longer-term resuspension; ---x, deposition plus initial resuspension.

As an example of the results for the decay in sampled activity (corrected for background) for 2 μm size iron oxide injected particles are shown in figure 5. Shown also are the corresponding values obtained from the best-fit solution of equation (10) using equation (11) for $\Lambda(t - s)$ with $\varepsilon = 1.1$ (numerical solutions for $c(t)$ with ε outside the range $1 < \varepsilon \leqslant 1.1$ gave unacceptably poor fits).

For comparison we have also shown the decay arising from deposition and initial resuspension alone (i.e. no longer-term resuspension). Of the two decays, that based on the integro-differential equation (with longer-term resuspension) is clearly a much better fit to the data. It shows the marked influence of longer-term resuspension on the final stages of the decay. (Please and Wilmott [12] have shown that in the limit of $t \to \infty, c(t)$ from equation (10), arising from an instantaneous point source, behaves $\sim t^{-1}$.)

Further investigations suggest that forms of $\Lambda(t - s)$ in equation (10), other than the form proposed in equation (11), give unacceptable fits to the experimental data. Therefore, not only are these data evidence of longer-term resuspension but, more precisely, evidence of a longer-term resuspenion rate which decays almost inversely with the time of exposure on the surface.

CURRENT EXPERIMENTS

Hall is presently engaged on a resuspension experiment to confirm the more quantitative aspects of the resuspension model. It is essentially a more accurate repeat of the earlier Corn and Stein [13] experiment in which particles of a known size distribution were exposed to a range of fully developed air flows (10 - 120 m/s). The fraction resuspended after a known period was plotted as a function of flow velocity (or more appropriately the friction velocity). The resuspension depends not only upon the flow but also upon the particle surface adhesion which was measured in a separate experiment. This involved placing the particle-surface specimen in an ultra-centrifuge and measuring the fraction of particles remaining as a function of centrifuge speed. From these measurements a distribution for the adhesive forces was deduced. The experiment suffered not only from inaccuracies in the measurement but also in the analysis of the date [14]. The present experiment is designed to irradicate these deficiencies.

SUMMARY AND CONCLUSIONS

1. The work described forms part of an overall objective of predicting the plateout of an aerosol in a reactor coolant flow.

2. We have described our current statistical model for resuspension based on the escape of a particle from a surface adhesive potential; vibrational energy is accumulated via the fluctuating aerodynamic lift force until it is sufficient to detach the particle from the surface.

3. We have given a brief description of the measurements of the mean, rms and energy spectrum of the aerodynamic lift force at the surface.

4. We have considered the implications of the model for the resuspension from rough surfaces and the decay of gas-borne particles in the RCS of a CAGR.

5. Mention has been made of the current experiments to verify the quantitative aspects of the model.

ACKNOWLEDGMENT

The work reported was carried out at Berkeley Nuclear Laboratories and is published with the permission of the Central Electricity Generating Board.

REFERENCES

[1] Garland, J. A., Wells, A. C. and Highnam, E. J., UKAEA Report AERE-R10491., 1982.

[2] Garland, J. A., Wells, A. C., Hedgecock, J., Smith, I. C. and Davis, R. E., ARPWG/P(84)58, 1984.

[3] Reeks, M. W., Reed, J. and Hall, D., On the Resuspension of Small Particles by a Turbulent Flow, J. Physic. D., 21, pp 574-589., 1988.

[4] Johnson, K. L., Kendall, K. and Roberts, A. D., Proc. Roy. Soc. A 324, pp 301-13, 1971.

[5] Hall, D., Measurements of the Mean Force on a Particle near a Boundary in Turbulent Flow, J. Fluid Mech., 187, pp 451-466, 1988.

[6] Hall, D., Measurements of the Fluctuating Lift Force on a Particle at a Boundary in Turbulent Flow, submitted for publication in J. Fluid Mech., 1988.

[7] Garland, J. A., UKAEA Report AERE R9452, 1979.

[8] Braaten, D. A., Shaw, R. H. and Paw, U. K. T., Particle detachment in turbulent boundary layers, 2nd International Aerosol Conf. Berlin, Aerosols: Formation and Reactivity, Gt. Britain, Pergamon Journals Ltd., 1986.

[9] Erhert, D. L., Turbulent Aerosol Deposition and Reentrainment, Ph.D. Thesis, University of California, Berkeley, 1986.

[10] Wen, H. Y. and Kasper, G., On the Kinetic of Particle Reentrainment from Surfaces, submitted to J. Aerosol Sci., 1989.

[11] Reeks, M. W. and Hall, D., Deposition and Resuspension of Gas-Borne Particles in Recirculating Turbulent Flows, J. Fluids Eng. 10, pp 165-171, 1988.

[12] Please, C. and Wilmott, P., Math. Eng. Ind. 1, pp 21-32, 1987.

[13] Corn, M. and Stein, F., Am. Ind. Hyg. Assoc. J., 26, pp 325-336, 1965.

[14] Reeks, M. W. and Reed, J., An Interpreation of Experimental Results on Particle Re-entrainment by a Turbulent Flow using a recently developed Model for Resuspension, CEGB Rep TPRD/B/0642/N85., 1985.

Chemical Interactions with Primary Circuit Surfaces under Severe Accident Conditions

BRIAN R. BOWSHER, SHIRLEY DICKINSON, and ALAN L. NICHOLS
AEA Technology
Chemistry Division, Winfrith
Dorchester, Dorset, DT2 8DH, UK

ABSTRACT

Although it is generally accepted that fission-product attenuation in the reactor coolant system will be relatively small in comparison to attenuation in the main containment building, specific chemical reactions in the primary circuit may significantly alter the chemical forms of the radioactive release into the containment. An adequate understanding of fission-product behaviour in the primary system is required, and any important chemical changes and resuspension of material need to be identified to assist in a more confident assessment of the consequences. Fission-product vapour species can undergo physicochemical reactions with primary circuit surfaces, resulting in the initial attenuation and possible resuspension of material in different chemical forms after an ill-defined period of time.

Experiments have been undertaken to identify and characterise the important interactions between fission-product vapours and primary circuit materials. Preliminary studies have focussed on the behaviour of iodine, caesium and tellurium species. First-order rate constants have been derived for the caesium hydroxide vapour reaction with 304L stainless steel, and the reaction of elemental tellurium with Zircaloy. Mass spectrometry has also been used to determine the chemical species formed in these systems. The new compounds generated under these conditions alter the ultimate volatility and transport properties of fission-product caesium and tellurium. The implications of the caesium iodide-boric acid reaction will also be discussed in terms of the formation of hydrogen iodide in the reactor coolant system and the potential attenuation of this species by interaction with bulk surfaces.

INTRODUCTION

Under severe accident conditions the reactor core will overheat and fission product vapours will be released into the damaged coolant system. Over the past fifteen years increased efforts have been made to quantify fission-product vapour and aerosol behaviour in the primary circuit of light water reactors (LWRs) in order to determine more precisely the physical and chemical forms of the volatile species evolved from the damaged fuel. Mechanistic models have been introduced into such computer codes as TRAP-MELT2, RAFT and VICTORIA (1-3), and thermodynamic data bases have been used to predict the chemical forms of the subsequent release into the containment building and the environment (4-6). The fission products will be transported through some portion of the primary circuit, dependent upon the accident sequence, and will interact with steam, hydrogen and aerosols generated from the core, structural surfaces and pipework. Although it is accepted that attenuation in the reactor coolant system will be small in comparison to potential deposition processes in the main containment building, specific chemical reactions within the primary circuit could result in significant changes in the form of the ultimate release. Furthermore, these effects will be of increasing importance in the assessment of certain accident sequences in which the containment building is by-passed (for example, steam generator tube rupture and failure of the emergency core cooling system, accompanied by other malfunctions of the plant). A sound understanding of fission product behaviour in the primary system is needed so that important chemical changes and any delayed resuspension of radioactivity can be quantified to produce more confident estimates of the consequences of reactor accidents.

Specific fission-product vapours may undergo heterogeneous reactions with components of the primary circuit pipework and miscellaneous support structures within the vicinity of the overheated core. These reactions will be governed by first-order rate constants (k) defined by

$$J_R = k\ C_S$$

where J_R is the flux, and C_S is the concentration in the vapour phase of the reactant at the metal surface. Any decrease in the mass of vapour represents the rate at which the reactant deposits on the bulk surface (dM_D/dt), such that

$$\frac{1}{A}\ \frac{dM_D}{dt} = k\ C_S$$

for a container of surface area A and volume V. Assuming the vapour concentration at the surface is approximately the same as the bulk concentration in the gas phase ($C_S \approx C_B$), the rate of vapour deposition on the surface is matched by the rate of decrease in the mass of vapour in the bulk ($dM_D/dt = - dM_B/dt$), and thus

$$\frac{1}{A}\frac{dM_D}{dt} = -\frac{1}{A}\frac{dM_B}{dt} \approx k\, C_B$$

$$-\frac{1}{A}\frac{dM_B}{dt} \approx k\,\frac{M_B}{V}$$

This equation can be integrated to give

$$[M_B]_t = [M_B]_o \exp-(k.At/V) = [M_B]_o \exp-(V_d.At/V)$$

where $[M_B]_t$ and $[M_B]_o$ are the total masses of the vapour species in the container at times t and zero respectively, and V_d(or k (cm s^{-1})) is the surface-reaction rate constant (or vapour deposition velocity as used in the TRAP-MELT2 code (1)). This form of analysis is only valid for relatively slow reactions ($C_S \approx C_B$) in which the heterogeneous chemical reaction is the rate-controlling process.

Vapour deposition velocities have been measured for the reactions of specific iodine-, caesium- and tellurium-containing species with unoxidised and preoxidised metals as discussed in Sections 2 and 3 below. These elements are important volatile fission products, and it is generally accepted that their predominant chemical forms in the hydrogen-steam atmosphere of a severe reactor accident are caesium iodide, caesium hydroxide, elemental tellurium and hydrogen telluride (7-9). The transport properties of these species in the primary circuit have been studied in terms of their stability and rates of reaction with 304L stainless steel, Inconel 600 and Zircaloy. Some of this work has been reviewed below on the basis of bulk surface interactions, first-order rate constants, and efforts to determine the secondary chemical species formed in these systems. Emphasis has been placed on experiments in which metal samples with a significant oxide coating were used, rather than virgin materials. No consideration has been given to the possibility of further reactions in the vapour phase and competitive reactions between the bulk surfaces and aerosols for the condensing and reactive fission-product vapours, although the equivalent interactions of these vapour species with representative LWR aerosols have been considered in a companion paper (10).

CAESIUM HYDROXIDE - 304L STAINLESS STEEL INTERACTION

Caesium hydroxide undergoes chemical reaction with 304L stainless steel at temperatures above 600°C (11). The reaction products depend upon the precise experimental conditions: experiments at Sandia National Laboratories (12) support the formation of a mixed caesium-silicon oxide ($Cs_2Si_4O_9$), while equivalent studies at Winfrith (13) have not produced any significant evidence for the stabilisation of a specific chemical species of caesium. This interaction of caesium hydroxide with stainless steel is temperature dependent, and measurements at Winfrith have been made over the temperature range 500 to 1000°C. The resulting data have been expressed in terms of vapour deposition velocities and give a quantitative description of the interaction of caesium hydroxide vapour with 304L stainless steel.

The studies were carried out in a dual-wound resistance furnace (Figure 1) comprising a 25 mm diameter nickel-lined silica tube with two nichrome wire elements for independent heating of the caesium hydroxide and stainless steel deposition coupon. Experiments were conducted in various gaseous atmospheres of steam with 4% hydrogen in argon (Table 1). Water was introduced directly into the lower-temperature furnace at a controlled rate using a peristaltic pump. A nickel crucible containing approximately 100 mg of caesium hydroxide (99% purity) was placed at the centre of the lower-temperature furnace (typically at 500(±10)°C (T_v)), while a 304L stainless steel coupon (80 mm x 20 mm x 0.1 mm) was placed in the centre of the high-temperature furnace (typically at 600 to 1000°C (T_d) with an axial variation of up to 20°C).

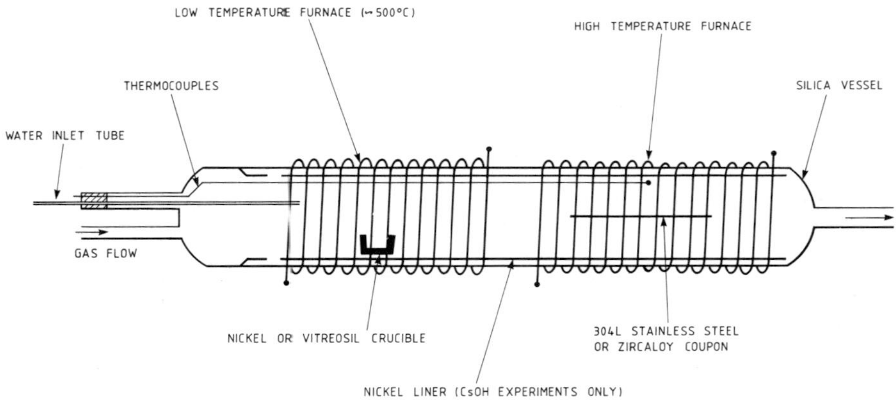

FIGURE 1. Dual-wound resistance furnace

The stainless steel coupons were removed from the furnace after each experiment, photographed and cut into sections for either microscopic or chemical analysis. Detailed surface examination of specific sections involved the use of a Jeol JSM-35CF scanning electron microscope with energy dispersive

TABLE 1. Experimental Conditions for Caesium Hydroxide - 304L Stainless Steel Studies

Carrier Gas	Flow (cm^3 min^{-1})	Time (min)	T_v (°C)	P_{CsOH} (Pa)	T_d (°C)
82.8%Ar-3.4%H_2-13.8%H_2O	75	360	500	16	570
82.8%Ar-3.4%H_2-13.8%H_2O	73	360	510	21	770
82.8%Ar-3.4%H_2-13.8%H_2O	72	360	510	21	965
61.9%Ar-2.6%H_2-35.5%H_2O	107	360	510	21	960
100% H_2O	130	360	500	16	960

TABLE 2. Vapour Deposition Velocity Data for Caesium Hydroxide - 304L Stainless Steel Interaction

Mass Deposited (μg CsOH cm^{-2})			C_B CsOH (μg cm^{-3})	$\frac{10^3}{T_d}$ (K^{-1})	Vapour Deposition Velocity (cm s^{-1})		
Soluble Component	Insoluble Component	Total			Soluble Component	Insoluble Component	Total Caesium
536	36	572	0.34	1.19	7.6×10^{-2}	5.1×10^{-3}	8.1×10^{-2}
164	167	330	0.31	0.959	2.2×10^{-2}	2.3×10^{-2}	4.5×10^{-2}
30	197	228	0.31	0.808	4.8×10^{-3}	3.1×10^{-2}	3.6×10^{-2}
8	147	155	0.23	0.811	1.3×10^{-3}	2.3×10^{-2}	2.4×10^{-2}
-	-	253	0.23	0.811	-	-	5.5×10^{-2}

spectroscopy (SEM/EDS). Previous studies had shown that the caesium was present in both water-soluble and water-insoluble forms (11), and therefore samples were leached in cold water for one hour to remove the water-soluble component for assay, before dissolving the coupon in boiling aqua regia. It was assumed that the partial pressure of caesium hydroxide in the gas stream (P_{CsOH} (Pa)) was equal to the saturated vapour pressure at the source temperature (T_V (K)), and this was calculated from the equation

$$\log_{10} P = 9.76 - \frac{6611}{T_V}$$

as determined by Elrick et al (14). The contributions of the soluble and insoluble caesium components are listed separately in Table 2, and these data were used to calculate the vapour deposition velocities and derive the temperature-dependent equations (Figure 2):

$$\ln(V_d) = \frac{(9100 \pm 5800)}{T} - (13.2 \pm 5.5) \text{ for the soluble caesium species,}$$

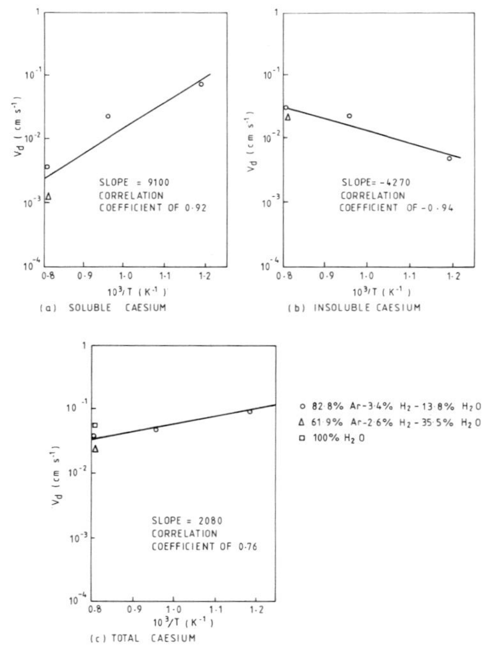

FIGURE 2. Vapour deposition velocity data for CsOH interaction with 304L stainless steel

$$\ln(V_d) = -\frac{(4270 \pm 2320)}{T} - (0.05 \pm 2.2)$$ for the insoluble caesium species.

The SEM/EDS data showed that a duplex oxide layer had formed on the surface of the 304L stainless steel. Localised concentrations of caesium were detected in deposits that also contained significant concentrations of phosphorus, manganese, copper and silicon, although no specific caesium species were identified.

The reaction of caesium hydroxide vapour with 304L stainless steel is slow compared to a gas-phase mass transfer-limited process, and the rate-determining step can be identified with the growth of the oxide at the metal surface, caesium reaction and transport within this oxide layer, diffusion of specific trace components through the metal, or a combination of these phenomena. A simple equation can be used to represent the reaction of caesium hydroxide (15)

$$CsOH_{(g)} \underset{k_2}{\overset{k_1}{\rightleftarrows}} CsX_{(c)} \underset{k_4}{\overset{k_3}{\rightleftarrows}} CsY_{(c)}$$

where $CsX_{(c)}$ is the soluble caesium species identified with the initial sorption process, and $CsY_{(c)}$ is the insoluble caesium species arising from the slow diffusion of caesium into the surface of the stainless steel. Although it is difficult to define the rate-limiting process for the overall reaction, the various studies indicate that the reaction is controlled by the diffusion of caesium cations into the metal oxide to give an ill-defined product with a Gibbs energy estimated to be 50 to 75 kJ mol^{-1} lower than caesium hydroxide (15).

Mass spectrometric experiments have been undertaken to study the release of caesium from 304L stainless steel coupons that have interacted with caesium hydroxide vapour. Samples were prepared by heating stainless steel coupons to various temperatures in a stream of caesium hydroxide vapour and argon-steam carrier gas. Each sample was then contained in a platinum foil tube within an alumina holder, which was placed inside the water-cooled vacuum jacket adjacent to the mass spectrometer (Figure 3). A Stanelco 3kW induction furnace was used to heat the sample by coupling to a tantalum susceptor around the alumina holder. The temperature was measured by an R-type thermocouple passing through the alumina holder, while the resulting vapour species were monitored using a VG-SXP 600 mass unit quadrupole spectrometer. The amount of caesium released from the coupons prepared at 580°C increased gradually with temperature over the range 500 to 800°C, but decreased at higher temperatures with no further release observed above

1100°C although heating continued up to 1500°C. Samples prepared at 800°C showed an initial caesium release when reheated to 800°C, and this emission increased with temperature up to ~ 1000°C, before decreasing as the temperature was raised further; a second release of caesium occurred between 1300 and 1500°C, accompanied by the melting and partial vaporisation of the stainless steel. The release curves for the lower-temperature samples suggested an evaporation process, whereas the second release from the higher-temperature sample did not show a smooth variation with temperature.

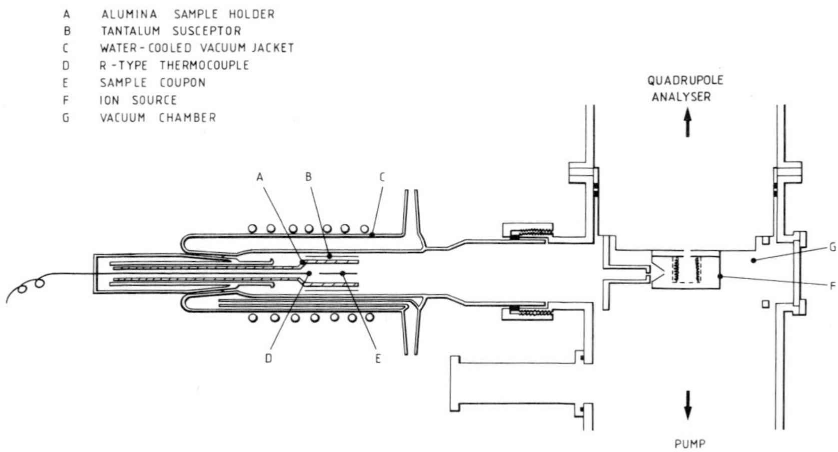

FIGURE 3. Quadrupole mass spectrometer

A very weak peak at 382 atomic mass units was observed when one of the samples prepared initially at the higher temperature was heated from 800 to 1000°C, indicating the release of molecular caesium chromate from the metal-oxide surface. This peak was very close to the limits of detection of the instrument, and was not observed in any of the other experiments. However, supporting evidence for the vaporisation of molecular caesium chromate was obtained from studies of CsOH-chromia and Cs_2CO_3-chromia mixtures. Caesium volatility was observed from both mixtures at temperatures above 700°C, and the resulting mass spectra were dominated by Cs^+, with much weaker Cs_2O^+ and $Cs_2CrO_4^+$ peaks. Appearance potential measurements showed that the parent ion was $Cs_2CrO_4^+$, while the other ions were produced by molecular fragmentation. The chemical form of the caesium released at higher temperatures has not been identified. Elemental Cs and Cs_2O species are the most likely candidates, since these species dominate the vapour above caesium oxides over a wide range of oxygen potentials (16). However, the Cs_2O species could not be detected in the mass spectrometer because of the presence of a background group around 280 mass units which would have dwarfed any weak peak from the sample.

Supporting experiments showed that the caesium release from caesium oxide-silica mixtures was negligible below 1000°C, but at higher temperatures this emission increased quite rapidly. The release of caesium began at temperatures that depended upon the Cs/Si ratio of the sample, and varied between 1100 and 1350°C for Cs/Si values in the range 0.5 to 0.1. Cs^+ was the only ion observed in these experiments, and the measured appearance potential was consistent with parent ion production.

These results suggest that two forms of caesium are present within the oxide layer, in addition to the more loosely-bound soluble material at the surface. The first species is associated with the chromium oxide component, either as a specific compound or as a less well-defined Cs-Cr-O phase (13); this caesium is released as molecular caesium chromate vapour at temperatures above 800°C. The remaining caesium is retained in a more tightly bound form and is only released near to the melting point of the stainless steel; this observation is consistent with the release of elemental caesium vapour from an ill-defined silicate species with a fairly low Cs/Si ratio, although the initial formation of other species such as ferrates or nickelates can not be discounted. Unfortunately, a further complication could occur that has not been studied in this analysis: CsOH may react with boric acid (H_3BO_3) to produce caesium borates with different deposition characteristics on primary circuit surfaces.

TELLURIUM - ZIRCALOY INTERACTION

Genco et al (17) have measured the reaction of elemental tellurium with zirconium at temperatures as low as 400°C. However, no measurable interaction was observed at 500°C by Sallach et al (18), and this was attributed to the formation of a protective film of zirconium dioxide. Alexander and Ogden (19) have shown that tellurium will react with clean zirconium/Zircaloy surfaces, and that the extent of this reaction is dictated by the temperature of the metal. Any residual oxide film protects the Zircaloy surface (98% zirconium and 1.5% tin), inhibiting the tellurium-zirconium reaction, but not the tellurium-tin reaction to form stannous telluride (SnTe).

Experiments were conducted in atmospheres of high purity argon (99.995%, ≤ 5 ppm O_2), 5% hydrogen in argon and water vapour/5% hydrogen in argon using the dual-wound resistance furnace (Figure 1). The water vapour was introduced by bubbling the carrier gas through a water bath at ambient temperature to give a typical water content of 3% v/v before passing the mixture into the furnace. A silica boat (Vitreosil, 45mm x 10mm x 7mm) containing approximately 200 mg of tellurium powder (Alfa Chemicals, 99.999%) was placed at the centre of the lower-temperature furnace (500 ± 10°C). The deposition coupon (60mm x 20mm x 0.1mm) was placed at the centre of the high-temperature furnace (between 600 and 1000°C with an axial variation of up to 25°C). Additional experiments were also performed to investigate the tellurium interaction with tin in an argon/hydrogen atmosphere; known weights of tin (BDH Analar, 99.5%) were contained in the silica boat, and these samples were heated in the dual-wound furnace.

The foils from each experiment were removed and cut into sections for either microscopic or chemical analysis. Pre-oxidation of the coupons in an autoclave resulted in lower vapour deposition velocities at temperatures below 600°C, but had little effect above 800°C. However, the presence of steam in the carrier gas reduced the tellurium reaction significantly at all temperatures (Figure 4). The interaction of tin with tellurium vapour in argon/hydrogen exhibited the lowest reaction rate. These data can be used to express the vapour deposition velocities in terms of the temperature of the coupons and the fractional degree of Zircaloy oxidation (x), assuming that the reaction is not gas-phase mass transfer-limited:

$$\ln(V_d) = -\frac{(4000 \pm 1000)}{T} - \frac{(4200 \pm 1200)}{T} x^{1/2} + (2.0 \pm 1.0)$$

The mechanism for Zircaloy oxidation has been identified as the inward movement of oxygen into the metal (20, 21). Compressional stress in the thin surface oxide will initially enhance the protective nature of the film by opposing the formation of cracks. As the oxide thickens, the compressional strain will cause a sudden collapse and breakaway of this protective film. Furthermore, hydride precipitation will lead to embrittlement of the oxide coating, possibly due to the formation of a hydride at the oxide-metal interface. Alexander and Ogden (19) have proposed that an adherent zirconium dioxide coating will completely inhibit the tellurium-zirconium reaction, but not the reaction between tellurium and tin. Thus when the zirconium in the Zircaloy is 100% oxidised, the measured vapour deposition velocities are believed to represent only the interaction of the tellurium vapour with the tin component of the alloy.

Mass spectrometric experiments have been undertaken to study the chemical form(s) of the tellurium re-released from Zircaloy coupons. These coupons were exposed to tellurium vapour at 520 and 1000°C, before being reheated to 700°C and their mass spectra monitored over the range 240 to 262 atomic mass units. The spectrum of the sample prepared at 520°C was dominated by Te_2^+, with small quantities of $SnTe^+$ also being detected. Although $SnTe^+$ peaks of similar intensity were observed from the coupon prepared at 1000°C, the Te_2^+ spectrum was much weaker and $SnTe^+$ was the dominant vapour species (Figure 5).

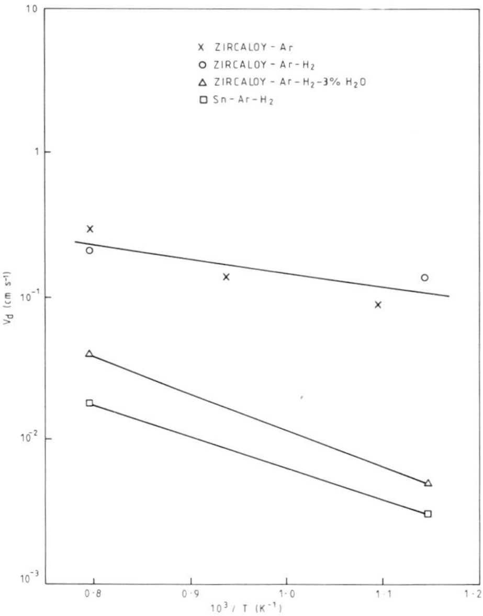

FIGURE 4. Vapour deposition velocity data for tellurium interaction with Zircaloy

It is not certain from these experiments whether the observed Te_2 arose from elemental tellurium which was weakly adsorbed onto the Zircaloy surface, or from the breakdown of tellurium-zirconium species formed within the Zircaloy. However, the larger amounts of Te_2 released from the coupon prepared at 520°C indicated that the tellurium was more strongly bound to the coupon treated at 1000°C, probably due to the formation of a stable compound of zirconium-tellurium at the higher temperature. These results showed that SnTe is formed at ~ 500°C when Zircaloy is exposed to tellurium vapour, with a zirconium-tellurium compound also being stabilised at higher temperatures (~ 1000°C). This implies that the tellurium-Zircaloy reaction is dominated by kinetic factors such as the diffusion of tellurium atoms through the Zircaloy matrix.

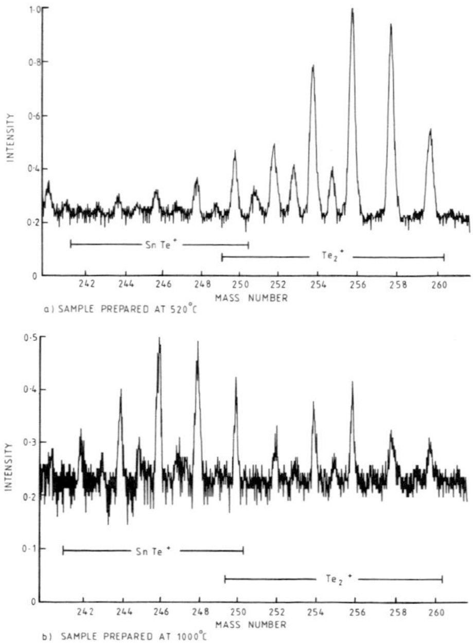

FIGURE 5. Mass spectra of vapour released from Zircaloy coupons after exposure to tellurium vapour

HYDROGEN IODIDE - 304L STAINLESS STEEL

Experimental studies by Nicolosi and Baybutt (22) have indicated that caesium iodide (CsI) vaporises congruently over the temperature range 820 to 1040°C to condense on representative metal surfaces without any detectable interaction. These observations have been supported in equivalent experiments by Sallach et al (23), and mass spectrometric analyses of CsI stability in equilibrium conditions representative of the primary circuit (24). Thermodynamic calculations have shown that CsI will be stable towards thermal dissociation in steam at temperatures up to ~ 1500°C (8)

$$CsI_{(g)} + H_2O_{(g)} \rightleftarrows CsOH_{(g)} + HI_{(g)}$$

Nevertheless, Collins et al (25, 26) and Tang et al (27) have noted the partial decomposition of this compound at low caesium and iodine concentrations with the formation of a more volatile iodine species. The stability of CsI under high-temperature steam conditions will depend on the relative concentrations of the reactants, possible interference from other ill-defined species, and mechanisms that result in the removal of one or more of the products of the equilibrium reaction.

Both PWRs and BWRs contain significant quantities of boron as either boric acid in the coolant system or boron carbide control rods. Boron carbide (B_4C) will react with steam to generate boric acid (28), and studies at Winfrith and the University of Southampton have shown that boric acid will react with CsI at relatively low temperatures to produce monomeric and polymeric hydrogen iodide, HI-water adducts and caesium borate (29-31):

$$CsI + H_3BO_3 \rightarrow CsBO_2 + HI_{(g)} + H_2O_{(g)}$$

Experiments have been undertaken in which the hydrogen iodide vapour species formed from the CsI-H_3BO_3 reaction has been passed over 304L stainless steel foils, and micrographs of the resulting surfaces are shown in Figure 6. The rectangular platelets are characteristic of hydrogen iodide attack, and EDS analyses of the deposits showed the iodine concentrations to be associated with the iron and nickel constituents of the stainless steel. Significant quantities of iron and nickel were also leached from these foils when they were washed with water, giving relative metal concentrations in solution of 57% Fe, 40% Ni and ≤ 3% Cr for

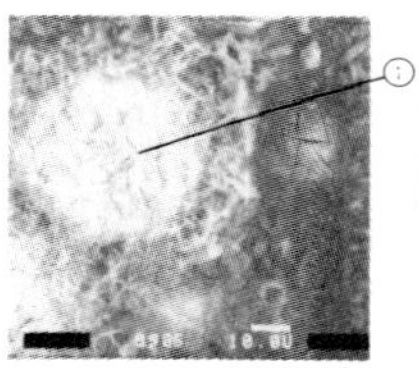

EDS ANALYSIS (ATOM PERCENTAGE)

POSITION	I	Cs	Cr	Mn	Fe	Ni
1 FIBROUS MATERIAL	8·96	0	3·4	0	69·8	17·8

(a) 3cm FROM CRUCIBLE (~680°C)

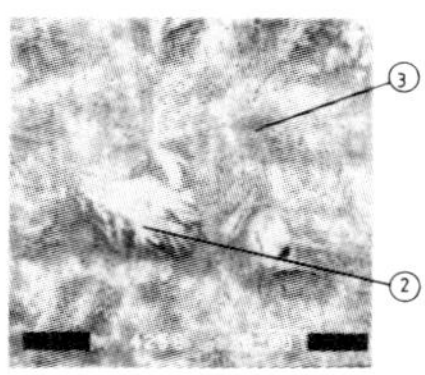

EDS ANALYSIS (ATOM PERCENTAGE)

POSITION	I	Cs	Cr	Mn	Fe	Ni
2 DEPOSIT	2·92	0·22	19·3	1·5	68·8	7·3
3 BACK-GROUND	4·58	0·28	18·3	1·6	67·6	7·6

(b) 17cm FROM CRUCIBLE (~175°C)

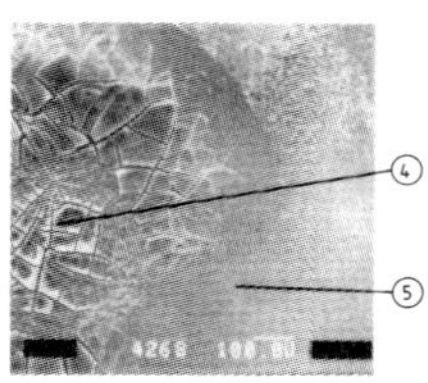

EDS ANALYSIS (ATOM PERCENTAGE)

POSITION	I	Cs	Cr	Mn	Fe	Ni
4 PLATELET	3·37	0·24	20·7	1·9	65·9	7·9
5 BACK-GROUND	0·88	0	19·9	1·7	69·7	7·7

(c) 27cm FROM CRUCIBLE (~110°C)

FIGURE 6. Micrographs and EDS data for HI interaction with 304L stainless steel (CsI + H_3BO_3 studies)

a foil held at ~ 100°C, and 30% Fe, 60% Ni and ≤ 10% Cr for a foil at ~ 500°C (compared with 304L stainless steel composition of 70% Fe, 9% Ni and 18% Cr). Experimental and thermodynamic assessments of the transition metal iodides have shown that both ferrous and chromous iodides are less stable than nickel iodide in these conditions. At 500°C nickel iodide is thermodynamically favoured, while ferrous iodide is formed preferentially at lower temperatures due to kinetic factors. Thus, the HI vapour that evolves from the interacting CsI and H_3BO_3 will react rapidly with the stainless steel constituents, although the resulting metal iodide species are susceptible to thermal decomposition.

CONCLUSIONS

There are significant uncertainties associated with the interaction of fission product vapours with primary circuit surfaces. Small-scale separate effects experiments have revealed the complexity of these surface reactions involving the more volatile species of caesium, tellurium and iodine, while demonstrating their potential to attenuate significant fractions of any CsOH and Te vapours released from the damaged fuel. Two of the more important reactions involve CsOH - 304L stainless steel and Te - Zircaloy, and these systems have been studied over the temperature range 500 to 1000°C in a variety of atmospheres, including various hydrogen and steam concentrations. Vapour deposition velocities (first-order rate constants) have been determined for these reactions as a function of temperature.

(a) The CsOH - 304L stainless steel reaction involved a physisorption process followed by the slow diffusion of caesium into the metal oxides formed in steam. Insoluble caesium species were formed as a consequence of this diffusion process, and their concentration depended upon the temperature of the coupon. Mass spectrometric studies indicated that the re-release of the insoluble caesium occurred in two stages: evaporation of a moderately volatile species (tentatively identified as Cs_2CrO_4), followed by a further release (probably elemental Cs) close to the melting point of 304L stainless steel. These results imply that caesium may be stabilised within the chromia (Cr_2O_3) and silica (SiO_2) phases of the duplex oxide layer formed on the surface of steam-oxidised stainless steel.

(b) The extent of the Te - Zircaloy reaction depended upon the degree of zirconium oxidation. A rapid reaction occurred at the metal surface in the absence of a protective oxide film; the initial formation of an oxide layer prior to an experiment inhibited the reaction, while rapid steam oxidation during an experiment would appear to result in the re-evolution of tellurium. The chemical species evolved from oxidised samples has been identified by mass spectrometry as stannous telluride (SnTe), which has different transport properties to the original fuel release of elemental tellurium (Te_2).

The chemical stability of CsI in the reactor coolant system may be in some doubt following experimental studies of the CsI - H_3BO_3 reaction, which results in the generation of HI at relatively low temperatures. This work has emphasised the importance of HI in the reducing conditions of the primary circuit, and the need to quantify the interactions of this volatile iodine species with the various structural materials and their oxides. Further studies are also required to assess the reactions of elemental Te and SnTe with 304L stainless steel and Inconel 600, CsOH with Zircaloy and Inconel 600, and H_3BO_3-$CsBO_2$ with Zircaloy, 304L stainless steel and Inconel 600.

The chemical reactions of fission-product vapours with primary circuit surfaces can be classified as short-term or partial attenuation processes in a severe reactor accident. Temperatures in the reactor coolant system may continue to rise significantly to re-release the fission products in different chemical forms. This could induce a relatively persistent radioactive emission into the main containment or the environment some considerable time after the initial release of fission products from the fuel. Nevertheless, some care is required when attempting to quantify fission-product reactions with primary circuit surfaces. These chemical processes will be in direct competition with equivalent reactions involving aerosols generated from bulk materials such as boric acid and the silver-indium-cadmium control rods. If vapour-aerosol interactions are dominant, the fission products will be preferentially transported as aerosol debris rather than being chemically trapped in the bulk structure of the primary circuit. However, the absorption of fission-product vapours onto aerosols will be limited by the rate at which their latent heats of condensation can be dissipated from the suspended particles (32). This effect may prevent significant vapour condensation onto aerosols, but will not be as important for primary circuit surfaces which constitute large heat sinks.

ACKNOWLEDGEMENTS

The authors wish to thank R J C Dunn and A F Kingsbury for the SEM/EDS studies. This work was funded by the UK Department of Energy and the Central Electricity Generating Board.

REFERENCES

1 Jordan, H. and Kuhlman, M. R., NUREG/CR-4205, BMI-2124, 1985.

2 Im, K. H., Ahluwalia, R. K. and Chuang, C. F., Aerosol Sci. Technol., 4, 125, 1985.

3 Camp. W. J., Eur. Appl. Res. Report - Nucl. Sci. Technol., 5, 1605, 1984.

4 JANAF Thermochemical Tables, Second Edition, Project Directors: Stull, D. R. and Prophet, H., US National Bureau of Standards, Washington DC, NSRDS-NBS 37, 1971 and supplements.

5 Thermodynamic Properties of Individual Substances, Editors: Glushko, V. P., Gurvich, L. V., Bergman, G. A., Veits, I. V., Medvedev, V. A., Khachkuruzov, G. A. and Yungman, V. S., Nauka, Moscow, 1978-82.

6 Cordfunke, E. H. P., Konings, R. J. M., Prins, G., Potter, P. E. and Rand, M. H., Thermochemical Data For Reactor Materials and Fission Products, Eur. Contract ETSN-0005-NL, 1988.

7 Cubicciotti, D. and Sanecki, J. E., J. Nucl. Mater., 78, 96, 1978.

8 Technical Bases for Estimating Fission Product Behaviour During LWR Accidents, Chapters 5 and 6, NUREG-0772, 1981.

9 Cubicciotti, D. and Sehgal, B. R., Nucl. Technol., 65, 266, 1984.

10 Benson, C. G., Bowsher, B. R. and Nichols, A. L., Seminar of the International Centre for Heat and Mass Transfer: Fission Product Transport Processes in Reactor Accidents, Dubrovnik, 22-26 May 1989.

11 Bowsher, B. R., Dickinson, S., Jenkins, R. A., Nichols, A. L. and Ogden, J. S., Source Term Evaluation for Accident Conditions, p.267, IAEA-SM-281/22, IAEA Vienna, 1986.

12 Elrick, R. M., Sallach, R. A., Ouellette, A. L. and Douglas, S. C., NUREG/CR-3197/1, SAND83-0395, 1984.

13 Allen, G. C., Bowsher, B. R., Dickinson, S., Fotios, G. M., Nichols, A. L. and Wild R. K., Oxid. Met., 28(1/2), 33, 1987.

14 Elrick, R. M., Powers, D. A. and Sallach, R. A., Reactor Safety Research Semiannual Report January-June 1987, Volume 37, Editor: Walker, J. V., Sandia National Laboratories Report, NUREG/CR-5039(1 of 2), p.195, 1988.

15 Taig, A. R., Bowsher, B. R., Dickinson, S., Elrick, R. M., Nichols, A. L. and Sallach, R. A., Proc. ANS Topical Mtg. on Fission Product Behaviour and Source Term Research, Snowbird, Utah, 16-19 July 1984, EPRI Report NP-4113-SR, p.38-1, 1985.

16 Lamoreaux, R. H. and Hildenbrand, D. L., J. Phys. Chem. Ref. Data, *13*, 151, 1984.

17 Genco, J. M., Berry, W. E., Rosenberg, H. S. and Morrison, D. L., BMI-1863, 1969.

18 Sallach, R. A., Greenholt, C. J. and Taig, A. R., NUREG/CR-2921, SAND82-1145, 1984.

19 Alexander, C. A. and Ogden, J. S., Battelle Columbus Laboratories, private communication, 1985.

20 Birks, N. and Meier, G. H., Introduction to High Temperature Oxidation of Metals, Edward Arnold, London, 1983.

21 Dollins, C. C. and Jurisch, M., J. Nucl. Mater., *113*, 19, 1983.

22 Nicolosi, S. L. and Baybutt, P., NUREG/CR-2713, 1982.

23 Sallach, R. A., Elrick, R. M., Douglas, S. C. and Ouellette, A. L., NUREG/CR-3197/2, SAND83-0395, 1986.

24 Alexander, C. A. and Ogden, J. S., Proc. ACS Symp. Chemical Phenomena Associated with Radioactivity Releases During Severe Nuclear Plant Accidents, 9-12 September 1986, Anaheim, California, NUREG/CP-0078, p.2-21, 1987.

25 Collins, D. A., McIntosh, A. E., Taylor, R. and Yuille, W. D., J. Nucl. Energy, Parts A/B, *20*, 97, 1966.

26 Collins, D. A., Collins, R. D., Taylor, R., McIntosh, A. E. and Roys, W. B., UK Atomic Energy Authority Report, TRG-1332(W), 1967.

27 Tang, I. N., Aronson, S., Munkelwitz, H. R., Dickey, J. M. and Growcock, F. B., BNL/NUREG-50689, 280, 1978.

28 Elrick, R. M., Sallach R. A., Ouellette, A. L. and Douglas, S. C., NUREG/CR-4963, SAND87-1491, 1987.

29 Bowsher, B. R. and Nichols, A. L., UK Atomic Energy Authority Report, AEEW - R 1973, 1985.

30 Bowsher, B. R. and Dickinson, S., UK Atomic Energy Authority Report, AEEW - R 2102, 1986.

31 Dickinson, S., Ogden, J. S. and Young, N. A., ACS Symp. Nuclear Reactor Severe Accident Chemistry, Third North American Chemical Congress, 7-10 June 1988, Toronto.

32 Ford, I. J. and Clement, C. F., UK Atomic Energy Authority Report, Theoretical Physics Division, Harwell Laboratory, TP.1307, 1988.

The Interaction of Fission Product Vapors with Aerosols in Severe Reactor Accidents

CHRISTOPHER G. BENSON, BRIAN R. BOWSHER, and ALAN L. NICHOLS
AEA Technology
Chemistry Division, Winfrith
Dorchester, Dorset, DT2 8DH, UK

ABSTRACT

Complex mechanistic codes have been developed to model the physical and chemical characteristics of the radioactive materials transported through a nuclear power plant during the course of a severe reactor accident. However, a significant source of uncertainty in such calculations is the interaction of labile fission product vapours with aerosols generated from other reactor materials within the overheated region of the core. The surface area of these aerosols could be approximately 500 times greater than the total surface area of the primary circuit, implying that fission product vapour-aerosol interactions could dominate the transport behaviour of the radioactive emission.

An induction furnace and associated equipment have been incorporated within a glove box system (Falcon) to study the transport of fission products and aerosols as they pass from the high-temperature reducing conditions of the primary circuit to the predominantly oxidising and steam-condensing environment in the containment. The furnace has been used to heat simulant and trace-irradiated UO_2 fuel samples clad in Zircaloy. The resulting airborne debris has been characterised in studies with thermal gradient tubes, filter samplers and a stainless steel containment vessel, using analytical techniques such as gamma-ray spectroscopy, aerosol particle-size analysers and scanning electron microscopy. These experiments have been conducted in the presence and absence of substantial concentrations of aerosol generated by heating silver-indium-cadmium control rod alloy clad in stainless steel. The data have been analysed in terms of the overall vapour and aerosol deposition profiles and their microscopic features.

INTRODUCTION

Mechanistic codes are being developed to determine the radioactive release from severe accidents in nuclear power plants (1). However, these programs do not include models to represent all the interactions of fission product vapours with aerosols generated from structural materials within the core. These materials include components of the control rod (an alloy of 80%Ag-15%In-5%Cd clad in stainless steel for PWRs and B_4C for BWRs), boric acid, stainless steel, Zircaloy and Inconel, and could generate an aerosol concentration in excess of 1 g m^{-3} (2, 3). The surface area of these aerosols is approximately 500 times greater than the total surface area of the primary circuit, implying that fission product vapour-aerosol interactions could dominate the transport of the fission products released from the damaged fuel (3, 4).

While uniform thermal-hydraulic conditions do not exist within the reactor core during a severe accident (for example, the peripheral fuel rods may be breached relatively late during the accident sequence, when the central rods are already molten), it has been shown that the broad time-envelope of structural aerosol release will coincide with the emission of the volatile fission products (5). The interactions may consist of physisorption and chemisorption, chemical reaction, and condensation, although vapour condensation onto aerosols is the only process currently modelled accurately in reactor accident calculations. Although the sorption phenomena are limited to only one or two molecular layers at the surface, Spence and Wright (3) have calculated that between 10 and 100 moles of vapour could be physisorbed on aerosol generated in the primary circuit. This quantity of vapour is significant with respect to the inventory of the volatile fission products (2), and implies that even this relatively weak interaction could be important in determining fission product transport.

A number of calculations have been undertaken to assess the importance of sorption of fission product vapours onto aerosols. For example, Spence and Wright (3) have estimated that the tellurium source term from the primary circuit to the containment could be increased by a factor of 370 for the AB hot leg sequence if sorption onto aerosols occurred. Furthermore, Butland et al (6) have calculated that the dominant mechanism for hydrogen iodide retention in a S_2D accident will change from reaction on the internal surfaces of the steam generator to reaction with the aerosol and subsequent deposition in the upper plenum when this latter mechanism is included in the calculations.

Fission product attenuation within the primary circuit is only one aspect of the consequences of vapour interactions with aerosols. This phenomenon may also increase the importance of related effects such as resuspension of aerosols, trapping of aerosols in liquids, and dissolution of fission products from the surface of particles (3). In particular, the interactions between fission product vapours and aerosols will define the physicochemical forms of the source term from the primary circuit to the containment. A recent sensitivity study has indicated that lack of knowledge of this phenomenon represents the major uncertainty associated with source term work (7). Given this uncertainty, there has been a surprising lack of research into this area associated with:

(i) sorption of fission product vapours onto aerosols,

(ii) chemical reactions between the fission product vapours and the aerosols.

The interaction of caesium iodide and caesium hydroxide vapours with silver, chromium (III) oxide, iron metal and iron (III) oxide have been studied in a static powder-bed geometry by Spence and Wright (3). Significant sorption was observed for most systems and isotherms were derived to describe these processes. They also reported evidence for a chemical reaction between caesium hydroxide vapour and iron and iron (III) oxide aerosols, and this system has also been studied by Bowsher and Dickinson (8). Beard et al (9) have studied the interaction of molecular iodine vapour with silver-indium-cadmium control rod aerosol deposited on 304L stainless steel foils. These experiments indicated that the reaction of iodine with the cadmium-based aerosol was approximately an order of magnitude greater than the equivalent reaction with the steel substrate. This was attributed mainly to the more extensive surface area of the aerosol deposit, although thermodynamic factors also influenced the reaction with cadmium iodide being formed even in a steam environment.

However, few experiments have been designed to study the reactions of fission product vapours with suspended aerosols in a flowing system. Preliminary separate-effects tests have been undertaken at Winfrith in which the reactions of individual simulant fission product vapours including caesium iodide, iodine, caesium hydroxide and tellurium were studied with control rod aerosol in a flowing system (10). This paper describes the development of these studies whereby the interactions of a number of fission products with structural aerosols have been examined. Simulant and trace-irradiated fuel samples clad in Zircaloy were heated up to ~ 2100°C in the absence and presence of control rod aerosol. The material released from the fuel/control rods was characterised by collecting the debris in thermal gradient tubes, a stainless steel containment vessel, and on a series of filters. These samples were analysed by various techniques including gamma-ray spectroscopy, aerosol particle-size analysis and scanning electron microscopy with energy dispersive spectroscopy. The results of these preliminary experiments provide substantial evidence for the importance of vapour-aerosol interactions in source term calculations.

EXPERIMENTAL PROCEDURE

The Falcon facility (Figure 1) has been designed to study the interaction of mixtures of fission product vapours with representative aerosols in integral experiments. Zircaloy-clad fuel pellets and control rod samples clad in 304L stainless steel were inductively heated either separately or together to temperatures of up to 2100°C. Any fission products transported through the apparatus were monitored by gamma-ray spectroscopy and aerosol particle-size analysers. Material was also collected on filter membranes and deposition coupons for further study. Two experimental arrangements have been used:

(i) thermal gradient system (TG),

(ii) containment vessel (CT).

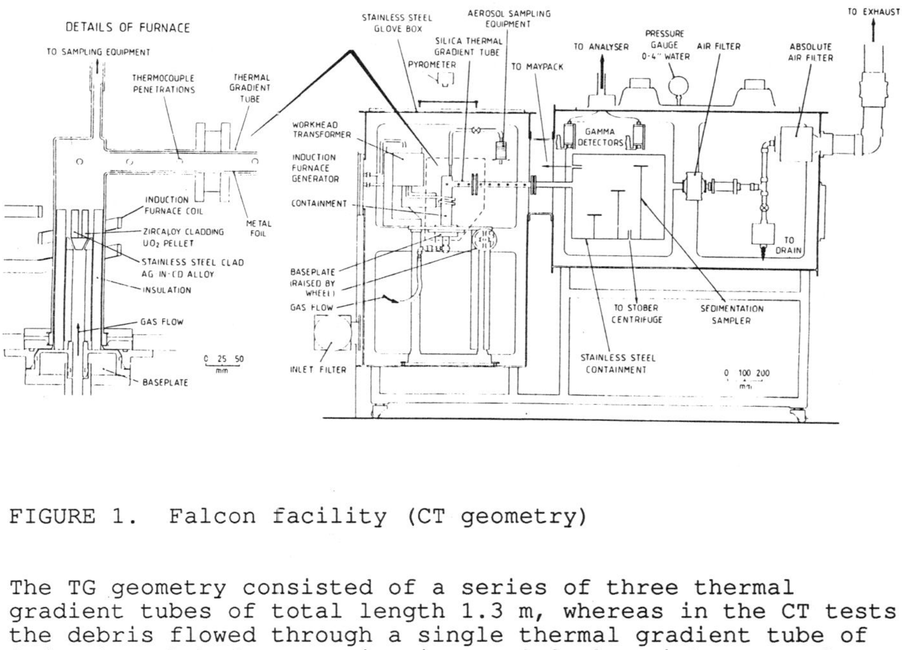

FIGURE 1. Falcon facility (CT geometry)

The TG geometry consisted of a series of three thermal gradient tubes of total length 1.3 m, whereas in the CT tests the debris flowed through a single thermal gradient tube of 0.6 m length before passing into a 0.3 m^3 stainless steel containment vessel. The thermal gradient tubes were lined with nickel or stainless steel foils which were removed after each experiment for chemical analysis. Nuclepore filter samples were obtained for periods of 5s from above the degrading fuel, and sequential sedimentation samplers were used to collect the aerosol deposits in the containment vessel. The stainless steel containment was dismantled and the interior cleaned systematically to determine the relative deposition on horizontal and vertical surfaces. Emphasis was placed on data that quantified the distribution of the fission products and the control rod elements throughout the apparatus, and the composition and morphology of the primary aerosol particles. The experiments are summarised in Table 1. The simulant fuel samples (experiments TG1 to 3 and CT1 to 5 in Table 1) consisted of trace quantities of simulant fission products representative of high burn-up fuel in a UO_2 matrix. Zircaloy-clad uranium dioxide samples were also irradiated in a low energy reactor for 3 h at 3 kW to give ~ 1 GBq of mixed fission products after a 24 h cooling period (experiments TG4 to 6 in Table 1). The control rod samples consisted of ~ 24 g of an 80%Ag-15%In-5%Cd alloy clad in 304 stainless steel and located within a Zircaloy shroud.

TABLE 1. Fission Product Vapour-Aerosol Interaction Studies

Experiment	Atmosphere	Fuel (a)	Control Rod Alloy (b)	Time (s)	Temperature (°C)
TG1	Ar	√	-	1070	1730
TG2	Ar	-	√	870	1400
TG3	Ar	√	√	605	> 2100
TG4*	Ar	√	-	1830	1680
TG5*	Ar	√	-	1920	1800
TG6*	Ar	√	√	1935	> 2100
CT1	97%Ar-3%H_2O	-	√	315	1430
CT2	77%Ar-23%H_2O	-	√	600	1430
CT3	97%Ar-3%H_2O	√	-	450	1430
CT4	77%Ar-23%H_2O	√	-	660	1930
CT5	97%Ar-3%H_2O	√	√	510	1380

(a) Simulant fuel sample: 59.94 mg CsOH, 55.80 mg Mo, 34.42 mg BaO, 21.66 mg SrO, 10.64 mg Te, 8.84 mg CsI in 18.16 g UO_2.

(b) Control rod sample (wt%): 79.83 Ag, 15.07 In, 5.06 Cd, 0.024 Cu, 0.018 Zn, 0.001 Ni, Pb, Sn and Fe.

*Trace-irradiated fuel sample.

RESULTS AND DISCUSSION

TG Experiments with Simulant Fuel (TG1, 2 and 3)

Mass balance data were obtained for all of the experiments, and the resulting percentage releases are given in Table 2. Cadmium dominated the release from the control rod alloy. Although the relative concentrations of the control rod constituents in the airborne material were in excellent agreement with previous data (11), their absolute magnitudes were lower due to significant interaction with the Zircaloy sheath. The percentage releases of caesium from the simulant fuel samples in experiments TG1 and 3 were approximately an order of magnitude lower than those of iodine, although the absolute quantities of caesium and iodine released from the simulant fuel were similar, indicating the formation of a low-volatility caesium species within the fuel matrix. The release data for experiment TG1 were in reasonable agreement with NUREG-0772 calculations (12); however, such comparisons are questionable given the nature of the simulant fuel samples.

TABLE 2. Experiments TG1, 2 and 3: Release Data

Element	% Release		
	TG1	TG2	TG3
I	6.0	-	36.0
Cs	1.4	-	4.6
Te	7.7	-	2.3
Ba	0.1	-	0.7
Sr	0.01	-	0.9
Mo	<0.01	-	0.02
U	$<10^{-3}$	-	5×10^{-4}
Cd	-	15.7	47.8
In	-	0.13	0.02
Ag	-	0.02	1×10^{-3}

The primary objective of these experiments was to monitor and understand the transport and deposition behaviour of the fission products. Detailed analytical studies were undertaken of the deposition profiles along the thermal gradient tubes, and examples of the iodine, caesium, barium and cadmium profiles are given in Figure 2. The deposition profiles of iodine and caesium in experiments TG1 and 3 were very similar, and indicated the transport and deposition of caesium iodide (Figures 2a and b). The deposition profiles for caesium and iodine in experiment TG1 were characterised by a sharp leading edge with an approximately exponential decay in concentration with decreasing temperature. Such profiles are representative of vapour transport and deposition mechanisms, showing good agreement with other thermal gradient studies involving simulant and irradiated fuel (13, 14, 15). Both barium and tellurium appeared to deposit in a relatively random manner on discrete sections of

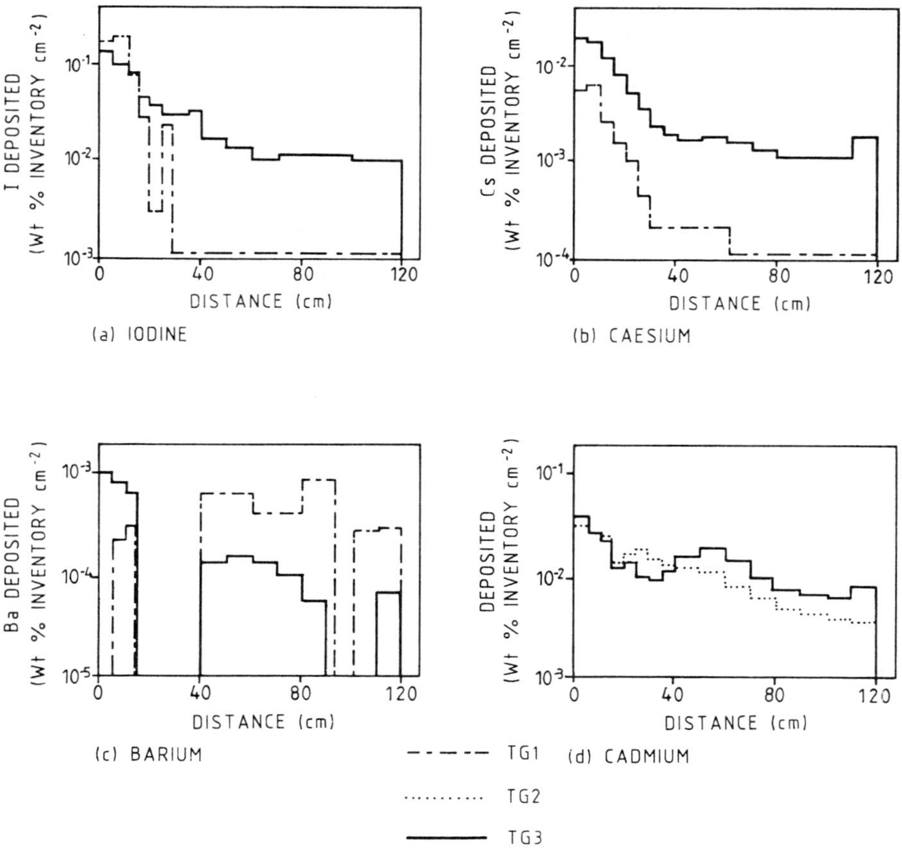

FIGURE 2. Experiments TG1 to TG3: deposition profiles along the thermal gradient tube

the thermal gradient tube (e.g Figure 2c); these data indicated that their release and transport were via an aerosol process following rupture of the Zircaloy cladding. The magnitude of the tellurium release in experiment TG3 was lower than that observed in TG1. This can be attributed to the stabilisation of tellurium as zirconium telluride within the cladding in TG3 (16), whereas in experiment TG1 a significant mechanical release of aerosol occurred due to the violent rupture of the Zircaloy cladding.

The deposition profiles of cadmium in experiments TG2 and 3 showed a gradual decrease in concentration along the thermal gradient tube (Figure 2d). Two distinct forms of cadmium particles were detected in the microscopic examination of the deposits: an underlying deposit of spherical particles (~ 0.3 μm geometric diameter) originating from the initial aerosol release (11), and a surface layer of much finer debris (< 0.1 μm geometric diameter) arising from the revaporisation and deposition of cadmium. Although the aerosol from experiment TG2 consisted predominantly of cadmium, some indium-based particles were also detected. The deposition profiles for these two elements differed significantly, indicating that the cadmium- and indium-based aerosols retained their individual identities to some degree (11).

The most important findings from these experiments arose from a comparison of the elemental deposition profiles for the simulant fission products in the absence (TG1) and presence (TG3) of aerosol. Two main effects were observed:

(i) significant quantities of the iodine and caesium release were transported further along the thermal gradient tube in the presence of aerosol (Figures 2a and b),

(ii) the barium deposition profiles were similar in both experiments (Figure 2c).

The increased mobility of iodine and caesium was attributed to their release from the fuel at approximately the same time as the cadmium-based aerosol was generated. The subsequent interaction resulted in these vapours being transported in association with the cadmium rather than as separate species. Microscopic examination of the deposits in the thermal gradient tube supported this conclusion: whereas no discrete caesium- or iodine-based particles were detected in experiment TG1 (consistent with a vapour transport and deposition process), many particles were found in experiment TG3. These particles consisted predominantly of cadmium, but a significant proportion contained both caesium and iodine in concentrations up to 16 wt%. Depth profiling studies demonstrated that the fission product simulants were found at the surface of the aerosol, indicating condensation/sorption of the caesium and iodine compounds onto the aerosol. This enhanced transport behaviour was not observed with barium since the majority of the cadmium evolved from the control rod sample before the release of any barium species (as low volatile barium oxide). Hence, relatively little interaction occurred between the barium and the cadmium, with the barium oxide transported as aerosol in both experiments.

TG Experiments with Trace-Irradiated Fuel (TG4, 5 and 6)

The irradiated fuel sample and thermal gradient tube were monitored by gamma-ray detectors during the heating phase of experiments TG4 to 6. Significant releases of radioactive material occurred, and the following radionuclides were detected on the deposition foils and filters:

I-131, I-132, I-133, Cs-137, Te-132, Ba-140, Ru-106 and La-140.

Although the presence of volatile iodine, caesium and tellurium was expected (2), the release of ruthenium and lanthanum was surprising. Ruthenium will be stabilised in the fuel as the low volatile metal, and would only be volatile as RuO_4 under highly oxidising conditions (2, 17). While lanthanum will be stabilised within the fuel as lanthanum (III) oxide, under strongly reducing conditions this compound will decompose to form the more volatile monoxide (2). However, the conditions of the experiments should not have allowed the formation of either of these more volatile species, and further assessments of these releases are in progress.

The deposition profiles for I-131 in experiments TG4 and 5 (without control rod aerosol) were similar to those observed in experiment TG1 with simulant fuel, indicative of vapour transport and deposition. Most significantly, the amount of fission product iodine transported to the second gamma-ray detector (approximately 0.8 m along the thermal gradient tube) was a factor of twenty higher in the presence of aerosol (TG6) than without aerosol. Whilst the different concentrations of fission products in the simulant and trace-irradiated fuel will affect the physicochemical behaviour of the fission products, it is evident that vapour-aerosol interactions are important for both systems.

CT Experiments with Simulant Fuel (CT1 to 5)

The containment (CT) experiments have been designed to study vapour/aerosol interactions in the high-temperature region simulating the core, through pipework representing the upper plenum and hot leg, and into a steel vessel simulating the containment. These tests were undertaken in an argon-steam atmosphere, with and without control rod aerosols. Mass balance measurements were made together with morphological studies of the aerosol deposits on coupons, Nuclepore filters and sedimentation samplers. As with the TG studies, cadmium dominated the aerosol release from the control rod experiments CT1 and 2, with the same two distinct forms of cadmium deposit. The majority of this aerosol plated out on surfaces within the upper plenum and hot leg pipework, while the remainder underwent gravitational deposition in the containment vessel.

The experiments with simulant fuel samples alone (CT3 and 4) generated a wide range of aerosol particles, including long Zircaloy shards (~ 30 μm), UO_2 particles (~ 10 μm geometric diameter), and submicron debris that contained caesium and iodine together with smaller quantities of molybdenum, tellurium and barium. The larger particles can be assumed to arise from the disruptive mechanical failure of the fuel cladding; the submicron particles were distributed throughout the apparatus (> 50% of the caesium release was found in the containment vessel after CT4), implying an aerosol formation and transport mechanism. Some of this fine debris was also associated with the Zircaloy shards, indicating that this material acted as condensation centres for the more volatile fission product vapours.

The simulant fuel rod ruptured at 1230°C (after 180s) and the control rod cladding failed at 1380°C (after 510s) in the combined experiment CT5. These temperatures were somewhat lower than those measured for the single system experiments, and may reflect the formation of a low melting point solidus as the two samples came into contact. The transport properties of the fission products were similar to the experiments with no control rod, indicating little interaction between the cadmium-based aerosol and the iodine, caesium and tellurium vapours because of the significant period of time between the failure of the two components (~ 330s). However, larger concentrations of molybdenum, barium and strontium were found in the containment vessel; this implies that the release of these simulant fission products coincided with the delayed failure of the control rod cladding, resulting in their interaction.

Separate-effects tests have been conducted to support these experiments, involving the interaction of specific simulant fission product vapours (iodine, caesium iodide, caesium hydroxide and tellurium) with cadmium aerosols in a flowing system. Significant interactions were observed in all cases: chemical reactions occurred between the cadmium aerosol and the iodine and tellurium vapours with the formation of cadmium iodide (CdI_2) and cadmium telluride (CdTe), caesium iodide had a limited chemical reaction with the cadmium aerosols, whilst caesium hydroxide vapour underwent a simple condensation/sorption process. These separate-effects studies confirm some of the trends observed in the integral experiments. Condensation/sorption of caesium iodide and caesium hydroxide vapours onto cadmium aerosols would account for the increased mobility of these fission products observed in the experiments with control rod aerosol, whilst the studies with iodine and tellurium vapours and control rod aerosols illustrate the potential for chemical reactions between vapours and aerosols in more complex systems.

APPLICATIONS TO SEVERE REACTOR ACCIDENT ANALYSES

These qualitative studies have illustrated the potential importance of interactions between fission product vapours and reactor materials aerosols, including condensation, sorption processes and chemical reactions. However, the last two mechanisms are not included in current modelling codes. Spence and Wright (3) have suggested that the sorption processes can be represented by modified Freundlich isotherms and treated in the codes to model accidents under equilibrium conditions (e.g. the TMLB sequence) by the use of expressions such as:

$$R = Ak(C_b - C^*),$$

where R is the rate of sorption onto the surface, A is the area of the surface, k is the rate constant, C_b is the bulk gas concentration and C* is the gas concentration in equilibrium with the gas already sorbed. It is possible that extensive sorption under equilibrium conditions may be better represented by BET isotherms, whilst Langmuir isotherms may reflect sorption under non-equilibrium conditions (e.g. for the AB sequence). Nonetheless, experiments are required to provide the necessary constants for inclusion in the relevant computer codes. One limitation of this approach is that it does not take into account the latent heat of vaporisation of the fission product vapours as they condense on aerosols, and further theoretical work is required to determine the importance of this phenomenon in assessing the partition of the fission product vapours between bulk surfaces and aerosols.

The inclusion of chemical reactions between vapours and aerosols in computer codes represents a relatively straightforward extension of current thermochemical modelling (although it may also be necessary to address the possibility of kinetic limitations in these processes). Thermochemical data for a number of fission product species are included in some sophisticated programs. The chemical species stabilised in the circuit throughout the accident sequence are predicted by minimising the Gibbs' energies of the fission product species (18). However, whilst significant progress has been made in collecting a reliable set of thermodynamic data of condensed phases and vapour species (19), large gaps remain in this data library. A number of molecular species still require identification and characterisation (for example gaseous hydroxides, oxyhydroxides and ternary oxides), and further work is required to obtain suitable thermochemical data on mixed fission product-reactor material compounds such as indium telluride.

CONCLUSIONS

Aerosols generated from various reactor materials during a severe accident could dominate the transport behaviour of the fission products released from the damaged core. However, apart from the simple condensation process onto existing aerosols, vapour-aerosol interactions are not accurately modelled in mechanistic codes, despite evidence from sensitivity calculations that they constitute a major uncertainty in defining the behaviour of fission products within the containment (7). These phenomena, together with the thermal-hydraulic conditions, will define the magnitude and transport characteristics of the radioactive source term released from the primary circuit to the containment.

Experiments have been initiated in the Falcon facility to study the interaction of fission product vapours from simulant and trace-irradiated fuel with control rod aerosols in a flowing system. Small segments of fuel were heated up to ~ 2100°C in the absence and presence of aerosol, and the resulting vapour-aerosol release was studied using thermal gradient tubes, a stainless steel containment vessel, and filter samplers. Measurements were made of the overall deposition profiles of the elemental components and the microscopic features of the aerosol. The following observations were made in these preliminary studies:

(i) extensive physicochemical interactions occurred between the fission product vapours and the control rod aerosol;

(ii) the stainless steel cladding of the control rod alloy failed at ~ 1400°C, resulting in the emission of a cadmium-based aerosol, in coincidence with the release from the fuel of volatile iodine, caesium and tellurium;

(iii) the deposition profiles of the volatile fission products were characteristic of vapour condensation onto bulk surfaces when there was no initial aerosol present; however, these profiles were less well-defined and there was increased mobility when a significant mass of aerosol was generated from the heated control rods;

(iv) both the iodine and tellurium fission product vapours reacted chemically with the cadmium aerosol to form cadmium iodide (CdI_2) and cadmium telluride (CdTe).

Physicochemical reactions between the fission product vapours and bulk aerosols could have a major effect on the transport characteristics of the radioactive emission from a severe reactor accident. These effects could be included in mechanistic calculations in order to assess the radiobiological consequences of a severe accident. However, further experimental studies are required to obtain thermodynamic and kinetic data for a number of important vapour-aerosol systems.

ACKNOWLEDGEMENTS

The authors wish to thank A M Beard and N A Rowe for assistance with the experimental work, R K Alty and G R Brown for the chemical analyses, R J C Dunn and A F Kingsbury for the SEM/EDS studies, and B A Bellamy for the XRD analyses. This work was funded by the UK Department of Energy, the Central Electricity Generating Board, and the Commission of European Communities, Joint Research Centre, Ispra.

REFERENCES

1 Gieseke, J. A., Cybulskis, P., Denning, R. S., Kuhlman, M. R., Lee, K. W. and Chen, H., BMI-2104, Volumes I-VI, 1984.

2 Bowsher, B. R., Prog. Nucl. Energy, 20, 199, 1987.

3 Spence, R. D. and Wright, A. L., Nucl. Technol., 77, 150, 1987.

4 Butland, A. T. D. and Kuhlman, M. R., p.197 in Source Term Evaluation for Accident Conditions, IAEA-SM-281/20, IAEA Vienna, 1986.

5 Hobbins, R. R., Osetek, D. J. and Hagrman, D. L., ACS Symp. Nuclear Reactor Severe Accident Chemistry, Third North American Chemical Congress, 7-10 June 1988, Toronto.

6 Butland, A. T. D., Nichols, A. L. and Williams, D. A., ENC 86 Trans., 3, 625, 1986.

7 Longworth, J. P., Barry, K., Chown, N., Kissane, M. P., and Williams, D. A., CEGB, NNC and UKAEA, private communication, 1988.

8 Bowsher, B. R. and Dickinson, S., UKAEA, private communication, 1985.

9 Beard, A. M., Bowsher, B. R. and Nichols, A. L., p.201 in Severe Accidents in Nuclear Power Plants, IAEA-SM-296/108, Vol 2, IAEA Vienna, 1988.

10 Beard, A. M., Bowsher, B. R., Dickinson, S. and Nichols, A. L., ACS Symp. Nuclear Reactor Severe Accident Chemistry, Third North American Chemical Congress, 7-10 June 1988, Toronto.

11 Bowsher, B. R., Jenkins, R. A., Nichols, A. L., Rowe, N. A. and Simpson, J. A. H., UK Atomic Energy Authority Report, AEEW-R 1991, 1986.

12 Technical Bases for Estimating Fission Product Behaviour During LWR Accidents, NUREG-0772, 1981.

13 Collins, D. A., Collins, R. D., Taylor, R., McIntosh, A. E. and Roys, W. B., UK Atomic Energy Authorit Report, TRG-1332(W), 1967.

14 Norwood, K. S., NUREG/CR-4105, 1985.

15 Collins, J. L., Osborne, M. F., Lorenz, R. A., Norwood, K. S., Travis, J. R. and Webster, C. S., NUREG/CR-3930, 1985.

16 Bowsher, B. R., Dickinson S., Gomme, R. A., Jenkins, R. A., Nichols, A. L. and Ogden, J. S., p.455 in Proc. Workshop on Chemical Reactivity of Oxide Fuel and Fission Product Release, CEGB Berkeley, Vol 2, 1988.

17 Garisto, F., p.3-107 in Proc. ACS Symp. Chemical Phenomena Associated with Radioactive Releases During Severe Nuclear Plant Accidents, NUREG/CP-0078, 1987.

18 Williams, D. A., Seminar of the International Centre for Heat and Mass Transfer: Fission Product Transport Processes in Reactor Accidents, Dubrovnik, 22-26 May 1989.

19 Cordfunke, E. H. P., Konings, R. J. M., Prins, G., Potter, P. E. and Rand, M. H., Thermochemical Data for Reactor Materials and Fission Products, Eur. Contract ETSN-0005-NL, 1988.

Metal Oxides and Hydroxides: Their Relevance to Vapor Transport in Severe Reactor Accidents

IAN R. BEATTIE and PETER J. JONES
Department of Chemistry
University of Southampton
509 5NH, UK

BRIAN R. BOWSHER and ALAN L. NICHOLS
UKAEA Atomic Energy Establishment Winfrith
Chemistry Division
102/A50, Dorchester, Dorset DT2 8DH, UK

PAUL E. POTTER and MALCOLM H. RAND
AEA Technology
Harwell Laboratory
Didcot, Oxon OX11 ORA, UK

ABSTRACT

A severe reactor accident in a light water reactor would result in the release of fission products and other components of the core by vaporisation into the reducing steam environment of the primary circuit. The behaviour of relevant metal/metal oxide systems in the presence of water vapour is not well understood, and much of the present knowledge stems from work undertaken some thirty to forty years ago. Water can react with elemental and oxidic species to form gaseous hydroxides and oxyhydroxides at high temperatures. A summary will be given of the chemistry of a number of relevant hydroxides, with emphasis on the need for more extensive studies to measure their thermodynamic quantities. Equipment has been developed to undertake a preliminary study of vapour transport in inert gas/steam mixtures at ambient pressure and temperatures up to 1000°C, and results will be presented on the transport of indium in steam using this apparatus.

INTRODUCTION

The magnitude of any radioactive release into the environment during a severe reactor accident would depend on the mode(s) of plant failure, the resulting thermal-hydraulic conditions, and the chemical forms of the fission product and actinide elements released from the damaged fuel and transported through the primary circuit and the containment (1, 2). The effectiveness of any natural and engineered attenuation processes can only be estimated with confidence if the chemical species and their physical forms can be defined throughout the plant. Mechanistic codes are being developed to assess the consequences of severe reactor accidents. These computer programs require thermodynamic and kinetic data to estimate the fission products and structural materials released from the core, their chemical and physical forms, the interaction of these species with each other, steam, hydrogen, and bulk materials, and the timing of their emission from the primary circuit into the containment.

Temperatures, pressures and hydrogen-steam ratios (thermodynamic oxygen potential) will determine the nature of the chemical species vaporised from the core, their reactions with bulk materials, and the extent of aerosol formation and deposition. It can be confidently postulated (3) that transient oxides, hydroxides and oxyhydroxides will be formed in the gaseous phase above the degrading core by reactions such as

$$M_{(s)} + H_2O_{(g)} \rightleftharpoons MO_{(g)} + H_{2\,(g)}$$

$$2M_{(s)} + 2H_2O_{(g)} \rightleftharpoons 2MOH_{(g)} + H_{2\,(g)}$$

$$M_{(s)} + 2H_2O_{(g)} \rightleftharpoons M(OH)_{2\,(g)} + H_{2\,(g)}$$

$$M_2O_{(s)} + H_2O_{(g)} \rightleftharpoons 2MOH_{(g)}$$

$$MO_{(s)} + H_2O_{(g)} \rightleftharpoons M(OH)_{2\,(g)}$$

$$M_2O_{3\,(s)} + 3H_2O_{(g)} \rightleftharpoons 2M(OH)_{3\,(g)}$$

$$M(OH)_{3\,(g)} \rightleftharpoons MO(OH)_{(g)} + H_2O_{(g)}$$

$$MO_{2\,(s)} + H_2O_{(g)} \rightleftharpoons MO(OH)_{2\,(g)}$$

$$MO_{3\,(s)} + H_2O_{(g)} \rightleftharpoons MO_2(OH)_{2\,(g)}$$

Only monomeric formulations are shown for the gaseous products, although polymeric species could be generated in many cases. A wide range of oxide, hydroxide and oxyhydroxide species can be postulated, and such molecules could be vaporised from the damaged fuel and other components of the reactor core.

The volatility of fission products in steam has been identified as a major source of uncertainty in defining the radioactive release to the environment (1), and experiments are required to characterise the species involved and to determine their transport behaviour. The chemistry is briefly reviewed, and relevant transpiration/transportation experiments are described to study the gaseous species involved in the release of indium from the control-rod alloy (Ag-In-Cd) of a pressurised water reactor.

VAPOUR TRANSPORT IN STEAM

The vaporisation and transport of inorganic compounds at high temperature in various environments are not well characterised. The condensed material may undergo several phase changes as volatilisation occurs, and the vapours may contain a variety of different polymeric species. Furthermore, the nature of the vaporisation process will depend on the temperature and pressure. Brewer (4) has rationalised the unexpectedly complex molecular character of high temperature vapours: if a high-temperature vapour system contains several molecular species in equilibrium with a condensed phase, the lower-concentration species may become more significant as the temperature rises. Brewer also noted that thermodynamic assessments are worthless if the major species are not identified and included in such calculations. Furthermore, high-temperature vapours, melts and solids are highly reactive so that significant interaction will occur with the surrounding environment (bulk structures, water vapour and containment atmosphere). The behaviour of high-temperature vapours is complex (3), and the lack of understanding of such systems prevents the incorporation of these species in mechanistic analyses of severe reactor accidents.

Water is an excellent solvent for inorganic materials because it can act as a complexing ligand and is a reasonably good oxidising agent. While many hydroxides decompose to oxides on heating with the evolution of water vapour, they can become a significant component of the gaseous phase at high temperatures. For example, $Be(OH)_2$ decomposes to BeO and water on heating at atmospheric pressure, but at 1300°C BeO reacts with water vapour to form the gaseous hydroxide (5, 6). Various fission products and other components of the reactor core could form gaseous hydroxides and oxyhydroxides according to the following reactions

uranium (7): $UO_{3\,(s)} + H_2O_{(g)} \xrightarrow{1050\text{-}1400°C} UO_2(OH)_{2\,(g)}$

molybdenum (8, 9): $MoO_{3\,(s)} + H_2O_{(g)} \xrightarrow{600\text{-}690°C} MoO_2(OH)_{2\,(g)}$

$$Mo_{(s)} + 4H_2O_{(g)} \xrightarrow{1200\text{-}1500°C} MoO_2(OH)_{2\,(g)} + 3H_{2\,(g)}$$

tellurium (10): $$TeO_{2\,(s)} + H_2O_{(g)} \xrightarrow{550\text{-}700°C} TeO(OH)_{2\,(g)}$$

barium (11, 12): $$BaO_{(s)} + H_2O_{(g)} \xrightarrow{1170\text{-}1510°C} Ba(OH)_{2\,(g)}$$

iron (13): $$Fe_{(s)} + 2H_2O_{(g)} \xrightarrow{1300\text{-}1460°C} Fe(OH)_{2\,(g)} + H_{2\,(g)}$$

nickel (14): $$Ni_{(s)} + 2H_2O_{(g)} \xrightarrow{1175\text{-}1420°C} Ni(OH)_{2\,(g)} + H_{2\,(g)}$$

It should be noted that the temperature ranges listed with these equations represent the conditions under which the experiments were conducted, and do not denote the limits of the reactions. The nature of the volatile species can often be predicted by considering the analogous chemistry of their fluorides, chlorides and oxyhalides. Molybdenum trioxide (MoO_3) vaporises mainly as trimer and tetramer at low temperatures, whereas the four coordinate monomer ($MoO_2(OH)_2$) would not be predicted to polymerise to the same extent by analogy with gaseous MoO_2X_2 (X = halogen). The transport of iron (III) oxide (Fe_2O_3) in steam might be expected to occur by means of the hydroxide molecule $Fe(OH)_3$ or its dimer, although $Fe(OH)_2$ and FeO(OH) could be formed under appropriate conditions. It should also be noted that an enhancement of volatility can occur when the molecules in the vapour phase are surrounded by water ligand molecules. Another important example of relevance to severe reactor accidents is the transport of caesium hydroxide in steam close to the critical temperature of this system (15).

Vapour transport in steam is an important and poorly understood area of chemistry. Considering the plethora of elements in a nuclear reactor, it is clear that a whole range of important reactions could occur, many of which are not included in current thermodynamic calculations and assessments because of a lack of suitable data.

VAPORISATION AND TRANSPORT OF INDIUM IN STEAM

Many pressurised water reactors (PWRs) include an alloy of 80% Ag, 15% In and 5% Cd in the form of control rods clad in 304L stainless steel. Cadmium accounts for 99% of the vapour/aerosol generated in argon when the material is overheated and the cladding fails at ca. 1400°C, whereas up to 40% of the airborne material is indium-based when equivalent experiments are conducted in steam (16). This change in composition can be attributed to the transient formation of volatile indium species which are subsequently stabilised at lower temperatures as indium (III) oxide, In_2O_3.

The thermodynamic properties of In_2O_3 have been studied in detail (17 - 21) and many of the results have been summarised in reference 22. Heating In_2O_3 to high temperatures causes decomposition, and the only vapour species detected have been In_2O, In and O_2 (23, 24, 25). The presence of In_2O has been detected by electron diffraction, mass spectrometry and matrix isolation-infrared, Raman and electronic spectroscopy (26-31); however, there appears to be little evidence that In_2O is stable as a solid (32, 33). $InOH_{(g)}$ is a well-known species for which spectroscopic and thermodynamic data are available (34-37). Calculations with these data suggest that both $In_2O_{(g)}$ and $InOH_{(g)}$ will be important in the vaporisation of indium and its alloys in steam. However, there has been no unambiguous identification of the major indium vapour species formed in the presence of water vapour. Thus, experiments have been undertaken to determine the role of steam in the vaporisation and transport of the indium species (38). Transpiration mass spectrometry has also been used to assist in the identification of the volatile indium species (39).

EXPERIMENTAL STUDIES

An alumina transpiration system (Figure 1) was used to study the vapour transport of indium species generated in a well-defined gas flow at high temperatures. The main furnace consisted of a nichrome wire heating element wound to give a 30 mm central section of constant temperature (± 10°C). Before each experiment, the alumina furnace was thoroughly washed with distilled water and dried for 12 h at 150°C. After each experiment, any oxide formed was reduced to the metal by passing hydrogen through the apparatus at 800°C; the indium was then dissolved in concentrated hydrochloric acid, and the alumina tube washed thoroughly with distilled water and dried to constant weight.

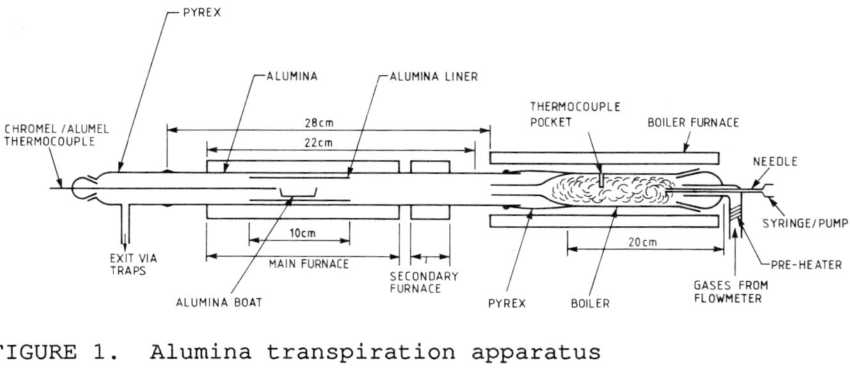

FIGURE 1. Alumina transpiration apparatus

The indium metal was supplied by Aldrich (99.999% purity) and used without further treatment, while the water was distilled and degassed; both the helium and hydrogen were research grade (British Oxygen Company), and the carbon monoxide and carbon dioxide were technical grade (The Distillers Company). Problems were encountered with impurities in the helium gas supply, and 99.999% purity material was used in the later experiments. Hydrated indium oxide was obtained by dissolving indium metal in concentrated hydrochloric acid, diluting, precipitating the indium hydroxide with ammonia, and then centrifuging and washing repeatedly with water until the washings were free of chloride; the resulting white deposit was dried in an oven at 150°C for 24 h, and an infrared spectrum of the product was in good agreement with a reference spectrum. Indium oxide (In_2O_3) was obtained by heating the hydrate to 800°C according to the method of Stubbs et al (17).

The recrystallised alumina boat (20 mm x 5 mm x 5 mm), sample and liner were weighed and the apparatus assembled. Helium was passed through the furnace at a rate of 50 cm^3 min^{-1} for at least two hours before the boiler was switched on (the total volume of the apparatus was approximately 120 cm^3 giving at least 50 volume changes). When the boiler had reached approximately 170°C, the main and secondary furnaces were switched on. The ancillary furnace was set to a temperature of approximately 350°C, and the reacting gas was passed into the system after the main furnace had reached the required temperature. This gas flow was normally stopped after 2 h, but the furnaces were maintained at temperature for a further 15 min, before being allowed to cool under a helium gas flow for 3 h. The apparatus was dismantled, and the boat and tube reweighed. Some of the important experiments undertaken during these studies are listed in Tables 1, 2 and 3 (38).

Supportive experiments were attempted using a transpiration mass spectrometer (VG 12-12S) to analyse the vapour species formed from the interaction of indium metal with steam at ~ 800°C. It should be noted that these studies were carried out on vapour-gas systems that had been expanded through an orifice into the vacuum chamber of the apparatus (39), and a significant change could alter the chemical stability of the indium vapour species. Further experiments are planned with the transpiration mass spectrometer to determine the vapour species stabilised over a wider range of steam pressures and temperatures.

RESULTS

The experimental data can be separated into a number of specific results that assist in defining the chemical form of

TABLE 1

INDIUM TRANSPIRATION STUDIES: INDIUM METAL/HELIUM, HELIUM/HYDROGEN AND HELIUM/OXYGEN

Temp (°C)	Helium Flow Rate (cm^3 min^{-1})*	Additional Gas Flow Rate (cm^3 min^{-1})	Initial Weight of In (g)	Weight Gain (+) or Loss (−) (g)	Weight Gain of Liner (g)
760	84	-	1.5839	+ 0.0013	ND
864	84	-	1.0747	+ 0.0059	ND
860	75(a)	H_2 17	0.8875	- 0.0079	0.0067
850	50(a,b)	H_2 17	0.8612	- 0.0022	zero
885	84	O_2 8	1.3537	+ 0.0083	ND
860	50(a,b,c)	O_2 8	0.9487	- 0.0005	0.0027

ND, not determined

*Half of the total gas flow passed over the sample unless stated otherwise

(a) Pure helium (99.999%)

(b) 2/3 of the total gas flow passed over the sample.

(c) 4h experiment

TABLE 2

INDIUM TRANSPIRATION STUDIES: INDIUM METAL/HELIUM/STEAM, WITH AND WITHOUT HYDROGEN OR OXYGEN

Temp (°C)	Helium Flow Rate ($cm^3\ min^{-1}$)*	Steam Flow Rate (mmole min^{-1})	Additional Gas Flow Rate ($cm^3\ min^{-1}$)	Initial Weight of In (g)	Weight Gain (+) or Loss (−) (g)	Wt Gain of Liner (g)
790	84	0.69	–	1.4091	− 0.0601	ND
857	84	0.50	–	0.7407	− 0.0729	ND
884	84	0.74	–	1.0309	− 0.1766	ND
830	75	0.79	H_2 17	1.0191	− 0.2020	0.2003
850	75	0.74	H_2 17	1.1543	− 0.2011	ND
852	67(a)	0.39	H_2 17	0.9630	− 0.0626	0.0621
855	17(a)	0.39	H_2 17	0.9874	− 0.0445	0.0468
850	34(a)	0.41	H_2 17	1.0446	− 0.0703	0.0732
859	50(a)	0.90	H_2 17	1.2096	− 0.1191	0.1303
860	34(a)	0.88	H_2 36	1.0209	− 0.1083	0.1079
861	50(a)	0.51	H_2 36	0.8767	− 0.0646	0.0600
857	34(a)	0.60	H_2 36	0.8009	− 0.0671	0.0637
861	84(a)	0.65	O_2 8	1.1649	+ 0.0045	0.0030
887	84	0.69	O_2 8	1.0318	+ 0.0075	ND

ND, not determined

*Half of the total gas flow passed over the sample unless stated otherwise

(a) Pure helium (99.999%), with 2/3 of the total gas flow passed over the sample

TABLE 3

INDIUM TRANSPIRATION STUDIES: MISCELLANEOUS EXPERIMENTS

Temp (°C)	Helium Flow Rate ($cm^3\ min^{-1}$)*	Steam Flow Rate ($mmole\ min^{-1}$)	Additional Gas Flow Rate ($cm^3\ min^{-1}$)	Initial Weight (g)	Weight Gain (+) or Loss (-) (g)	Wt Gain of Liner (g)	Remarks
840	75	-	H_2 17	In_2O_3 0.2441	- 0.0455	- 0.0002	Reduction to metal; no transport
860	67 (a)	-	-	In_2O_3 0.1464	- 0.0021	- 0.0007	No transport
860	84 (a,b)	0.79	-	In_2O_3 0.3692	- 0.0034	+ 0.0006	No transport
859	75 (a)	-	-	In_2O_3 0.1443 In 0.6068	- 0.0117	+ 0.0150	Vapour pressure (In species) ~ 50 Pa
860	84 (a,b)	-	-	In_2O_3 0.1349 In 0.6518	- 0.0133	+ 0.0120	Vapour pressure (In species) ~ 40 Pa
858	67 (a,b)	-	-	In_2O_3 (sample from In above experiment)	- 0.0098	+ 0.0174	Vapour pressure (In species) ~ 40 Pa
852	67 (a)	-	-	$In(OH)_3$ 0.1559	- 0.0309	+ 0.0007	Formation of In_2O_3; no transport
840	75	-	CO 9 (c)	In 0.9114	- 0.0249	+ 0.0225	Vapour pressure (In species) ~ 100 Pa
830	75	-	CO_2 5 (c)	In 1.0266	- 0.0046	+ 0.0016	Vapour pressure (In species) ~ 20 Pa
861	-	-	CO 34 (b, c) CO_2 34 (b, c)	In 1.0257	- 0.0053	+ 0.0083	Vapour pressure (In species) ~ 20 Pa
863	-	-	CO 66 (b, c) CO_2 6 (b, c)	In 0.7257	- 0.0067	+ 0.0104	Vapour pressure (In species) ~ 25 Pa

*Half of the total gas flow passed over the sample unless stated otherwise

(a) Pure helium (99.999%)

(b) 2/3 of the total gas flow passed over the sample

(c) Technical grade

the indium species:

(a) No significant weight change was observed in the tests with helium, helium/hydrogen and helium/oxygen gas mixtures up to ~ 850°C (< 10 mg (Table 1)). The lack of indium transport in helium/hydrogen can be explained in terms of the low vapour pressure of elemental indium at 850°C (< 1 Pa); it can also be inferred that a protective coating of indium (III) oxide was formed on the metal surface in the helium/oxygen experiments.

(b) Appreciable weight losses occurred from the sample when steam was introduced into the transpiration apparatus at ~ 850°C (up to 180 mg (Table 2)). These results proved to be somewhat erratic probably due to the formation of an oxide film on the metal surface, but the data showed increased indium transport at higher concentrations of water vapour. Indium (III) oxide was the only product detected on the liner from the steam experiments, in agreement with the studies involving control-rod alloy (16). The introduction of oxygen into the helium/steam gas flow resulted in a dramatic decrease in volatility (Table 2) with visual evidence for the formation of an oxide on the surface of the indium metal, while the addition of hydrogen to this gas mixture prevented the formation of a protective oxide film.

(c) Indium (III) oxide was reduced to metallic indium in a gas stream of helium/hydrogen; the small weight loss in pure helium (9.8 to 13.3 mg (Table 3)) was attributed to the formation of In_2O gas (the calculated vapour pressure of ~ 40 Pa is comparable with a measured value of 10 Pa at 850°C using a Knudsen effusion cell (40)). These results demonstrate that In_2O is not the dominant vapour species transported during the indium-steam experiments in which the apparent indium vapour pressure varies between 150 to 600 Pa (Table 2).

(d) A negligible weight change was observed when helium/steam was passed over In_2O_3, while heating $In(OH)_3$ in a helium gas stream resulted in simply dehydration to In_2O_3 (Table 3). These studies eliminate the possibility that $InO(OH)$ and $In(OH)_3$ are the major indium vapour species in the indium-steam experiments.

(e) An apparent indium pressure of ~ 25 Pa was measured when a gas mixture of 10 parts carbon monoxide to 1 part carbon dioxide was passed over indium metal (Table 3). This gas composition was used in the absence of steam to maintain the thermodynamic oxygen potential at a level commensurate with the original indium-steam experiments. Vapour transport as In_2O can be inferred (i.e. comparable with a measured value of 10 Pa at 850°C (40)) at a concentration that is significantly lower than the apparent indium vapour pressure in the steam studies.

(f) The transpiration mass spectrometric studies of the indium-steam reaction produced evidence for the presence of InOH and In_2O vapour species (39). However, these observations were made at pressures that were considerably lower than atmospheric conditions (of the order of 10 Pa), and significant changes may have occurred in the stable chemical form of indium.

DISCUSSION

The volatility of indium has been shown to be significantly enhanced in steam, and a series of well-defined transpiration experiments have been undertaken in an attempt to identify the dominant chemical form of the resulting vapour species. Although species such as $In(OH)_2$ could formally account for the observations, it should be noted that the oxidation states of the non-transition elements vary in two-unit increments. Thus In(I) and In(III) are well established, but In(II) is unknown except in compounds with In-In bonds or at very high temperatures in gases (e.g. InO). Furthermore, TlOH is a analogous hydroxide which loses water on heating.

The transpiration studies indicate the vapour transport of an indium hydroxide species represented by the general equation:

$$nIn_{(s)} + nH_2O_{(g)} \rightleftharpoons (InOH)_{n(g)} + \frac{n}{2}H_{2\,(g)}$$

The equilibrium constant (K'_p) is defined by

$$K'_p = \frac{p_{(InOH)_n} \cdot (p_{H_2})^{n/2}}{(p_{H_2O})^n}$$

$$\text{or } \log K'_p = \log[p_{(InOH)_n}] + n \log[\sqrt{p_{H_2}}/p_{H_2O}]$$

If $p_{In(obs)}$ is defined as the effective vapour pressure of indium, and

$$p_{In(obs)} = n\ p_{(InOH)_n}$$

$$\text{then } \log[p_{In(obs)}] = -n \log[\sqrt{p_{H_2}}/p_{H_2O}] + \log(n\ K'_p)$$

assuming the stabilisation of a single species and ideal gas behaviour. A polymeric value (n) of 1.8 can be estimated from the slope of $\log[p_{In(obs)}]$ plotted against $\log[\sqrt{p_{H_2}}/p_{H_2O}]$ in Figure 2. The outlying datum point at $\log[p_{In(obs)}]$ of 2.55 has been discounted because of the greater possibility of forming a protective film of In_2O_3 in this experiment as a consequence of the high steam activity (partial pressure of 2.5 x 10^4 Pa). It should also be noted that this analysis applies only to data obtained from an equilibrium reaction at constant temperature, while the experiments were conducted over

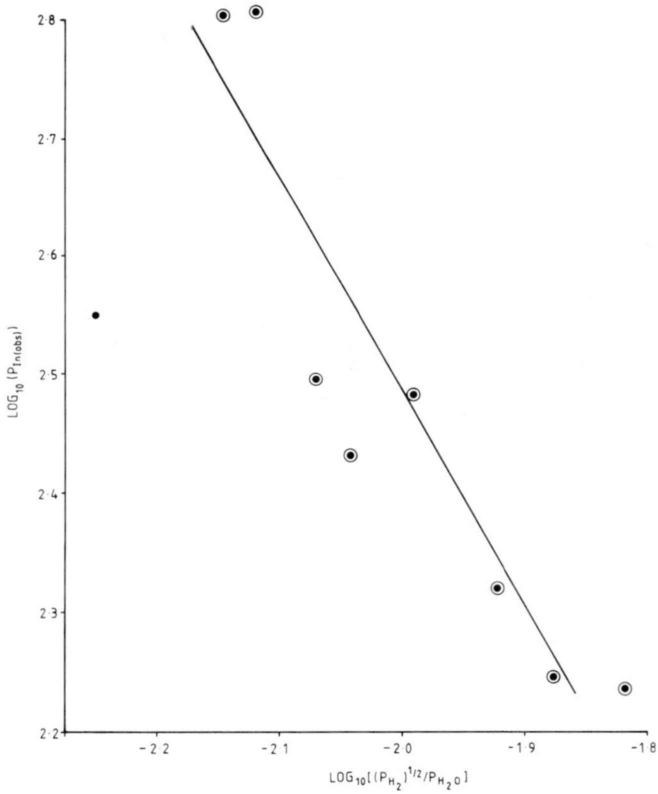

FIGURE 2. Transpiration data for indium metal in helium-hydrogen-steam gas flow: $\log[p_{In(obs)}]$ versus $\log[\sqrt{p_{H_2}}/p_{H_2O}]$ defined on the basis of pressures in pascals

the temperature range 830 to 861°C (Table 2). Furthermore, the uncertainties are difficult to quantify due to a number of experimental problems associated with this system. The derived value of n can only be regarded as a very rough indication of the slope which is sensitive to the data included in the final analysis.

The formation of In_2O cannot account for the observed transport of indium under atmospheric conditions if the vapour pressure measurements of Valderrama-n and Jacob are correct (40). A single well-defined experiment was undertaken with carbon monoxide-carbon dioxide (10:1) to check the validity of these published data; the possible formation of a protective film of In_2O_3 was avoided by maintaining the thermodynamic oxygen potential at an appropriate level in the absence of steam, and an apparent indium vapour pressure of ~ 25 Pa was measured in reasonable agreement with the published In_2O data (40). This value is much lower than the indium vapour pressure in the presence of steam, lending further support to the proposed formation of volatile $(InOH)_n$ species that may be converted to In_2O during the expansion process in the vacuum chamber of the mass spectrometer.

The experimental measurements can be compared with the calculated vapour pressures of $InOH_{(g)}$ and $In_2O_{(g)}$ using the thermodynamic data assessed by Glushko et al (37). In such comparisons it must be remembered that the weight losses and vapour pressures derived from the transpiration studies are integrated values over each experiment, when phases in the sample may be changing due to oxidation or reduction. Data are compared in Table 4 for the hydrogen-steam and other experiments defined in Tables 2 and 3. While the observed and calculated indium pressures agree within the combined uncertainties for the experiments without steam, the calculated indium-bearing pressures in the presence of water vapour are lower than the experimental data by a factor of 3 to 15. These studies imply that the thermodynamic data for $In_2O_{(g)}$ are reasonably correct, and that either $InOH_{(g)}$ is appreciably more stable than the existing data indicate or an additional indium hydroxide species is formed.

Glushko et al (37) used estimated vibration frequencies of 520, 300 and 3600 cm^{-1} in computing the thermodynamic quantities of $InOH_{(g)}$. However, Hauge et al (34) have recently measured ν_1 and ν_2 values of 523 and 422 cm^{-1} respectively; these data can be used in conjunction with flame-spectroscopic studies (35,36) to show that $InOH_{(g)}$ is more stable than Glushko et al suggest by approximately 6.5 kJ mol^{-1} at 850°C, corresponding to a factor of two increase in the vapour pressure. Furthermore, the uncertainty in $\Delta_f H(InOH_{(g)}$, 298.15K) as given by Glushko et al is ± 10 kJ mol^{-1} which leads to a further factor of three in the pressure of $InOH_{(g)}$ at 850° C. Thus, the calculated vapour pressures of $InOH_{(g)}$ could be higher than the values

TABLE 4

COMPARISON OF OBSERVED AND CALCULATED INDIUM VAPOUR PRESSURE

Temperature	Calculated Vapour Pressure (Pa)		Effective Indium Pressure (Pa)	
(°C)	$P_{(InOH)}$	$P_{(In_2O)}$	Calculated	Observed
830	15.7	13.4	43	635
850	22.2	21.8	65	640
852	19.4	19.7	59	175
855	29.6	20.8	71	270
850	24.4	19.8	64	305
859	30.4	26.5	83	355
860	33.6	24.0	82	312
861	39.2	25.3	90	175
857	39.2	25.3	90	210
859	–	26.5	53	50
830	–	27	54	20
861	–	30	60	20

listed in Table 4 by a factor of six, resulting in a variation of 0.6 to 5.3 in the ratio of observed to calculated vapour pressures for the indium-bearing species.

Direct transpiration experiments are the only reliable method of determining the volatility of materials of importance to the analyses of severe reactor accidents. It can be concluded that there are probably dimeric as well as monomeric molecules of indium monohydroxide in the indium vapour transported in steam atmospheres. These indium studies need to be extended to even higher temperatures, and efforts are required to broaden the resulting data base to include the hydroxide species of the actinide elements (U and Pu) and the fission product elements (lanthanides, Mo, Ru and Te).

CONCLUSIONS

Experimental studies have been conducted to characterise the vapour transport of indium in steam using a transpiration apparatus. At ca. 850°C and steam partial pressures between 1×10^4 and 2.4×10^4 Pa, the indium volatility was approximately 600 times that of pure indium metal in an inert atmosphere. While the indium was ultimately stabilised as indium (III) oxide (In_2O_3), an indium (I) hydroxide species (monomeric or polymeric) was inferred as the volatile species:

$$nIn_{(s)} + nH_2O_{(g)} \rightleftharpoons (InOH)_{n(g)} + \frac{n}{2}H_{2\,(g)}$$

Experiments have been conducted over a range of hydrogen to steam ratios to determine the polymeric form of this species; the data imply that the average polymerisation corresponds to a dimer, although monomer and trimer cannot be excluded.

Steam will play a significant role in the vaporisation and transport of fission products and other reactor core materials in a severe accident. However, thermodynamic data for the resulting gaseous oxides, hydroxides and oxyhydroxides are lacking, and their inclusion in mechanistic calculations is not possible at the present time. Greater efforts are required to characterise these species experimentally because of their critical importance in determining the vapour transport of fission products in a hydrogen-steam environment. The thermodynamic properties of these compounds need to be properly quantified before their impact on fission-product behaviour can be adequately assessed. Furthermore, the formation of oxide, hydroxide and oxyhydroxide species from bulk materials within the core (e.g. indium, cadmium, iron, nickel and chromium) needs to be investigated because of the reactive nature of these gas-phase compounds with any radioactive release from the damaged fuel.

ACKNOWLEDGEMENTS

The authors wish to thank A K Brisdon (University of Southampton) for undertaking the transpiration mass spectrometric studies of the indium-steam system. This work was funded by the UK Department of Energy, the Central Electricity Generating Board, and the Commission of the European Communities, Joint Research Centre, Ispra.

REFERENCES

1 Report to the American Physical Society of the Study Group on Radicnuclide Release from Severe Accidents at Nuclear Power Plants, Chairman: R. Wilson, Rev. Mod. Phys., 57 (3), Part II, S1, 1985.

2 Bowsher, B. R., Prog. Nucl. Energy, 20, 199, 1987.

3 Hastie, J. W., High Temperature Vapors Science and Technology, Academic Press, New York, 1975.

4 Brewer, L., Chemistry and Metallurgy of Miscellaneous Materials:Thermodynamics, Editor: L. L. Quill, p.193, McGraw-Hill, New York, 1950.

5 Grossweiner, L. I. and Seifert, R. L., J. Am. Chem. Soc., 74, 2701, 1952.

6 Young, W. A., J. Phys. Chem., 64, 1003, 1960.

7 Dharwadkar, S. R., Tripathi, S. N., Karkhanavala, M. D., Chandrasekharaiah, M. S., Thermodynamics of Nuclear Materials 1974, Volume II, p.455, IAEA-SM-190/32, IAEA Vienna, 1975.

8 Glemser, O. and Wendlandt, H. G., Adv. Inorg. Chem. Radiochem., 5, 215, 1963.

9 Belton, G. R. and Jordan, A. S., J. Phys. Chem., 69, 2065, 1965.

10 Malinauskas, A. P., Gooch, J. W. and Redman, J. D., Nucl. App. Technol., 8, 52, 1970.

11 Stafford, F. E. and Berkowitz, J., J. Chem. Phys., 40, 2963, 1964.

12 Sasamoto, T., Mizushima, K. and Sata, T., Bull. Chem. Soc. Jpn., 52, 2127, 1979.

13 Belton, G. R. and Richardson, F. D., Trans. Faraday Soc., 58, 1562, 1962.

14 Belton, G. R. and Jordan, A. S., J. Phys. Chem., 71, 4114, 1967.

15 Clough, P. N., Dunbar, I. H. and Knights, C. F., CSNI-OECD Workshop on Water-Cooled Reactor Aerosol Code Evaluation and Uncertainty Assessment, Brussels, 9-11 September 1987, Editors: Della Logia, E. and Royen, J., EUR 11351 EN, p.312, Commission of the European Communities, Luxembourg, 1987.

16 Bowsher, B. R., Jenkins, R. A., Nichols, A. L., Rowe, N. A. and Simpson, J. A. H., Silver-Indium-Cadmium Control Rod Behaviour During a Severe Reactor Accident, UK Atomic Energy Authority Report, AEEW - R 1991, 1986.

17 Stubbs, M. F., Schufle, J. A. and Thompson, A. J., J. Am. Chem. Soc., 74, 6201, 1952.

18 Newns, G. R. and Pelmore, J. M., J. Chem. Soc. A, 360, 1968.

19 Chatterji, D. and Vest, R. W., J. Am. Ceram. Soc., 55, 575, 1972.

20 Klinedinst, K. A. and Stevenson, D. A., J. Chem. Thermodyn., 5, 21, 1973.

21 Hochgeschwender, K. and Ingraham, T. R., Can. Metall. Q., 6, 293, 1967.

22 Srivastava, R. D. and Farber, M., Chem. Rev., 78, 627, 1978.

23 Gomez, M., Chatillon, C. and Allibert, M., J. Chem. Thermodyn., 14, 447, 1982.

24 Burns, R. P., De Maria, G., Drowart, J. and Inghram, M. G., J. Chem. Phys., 38, 1035, 1963.

25 Burns, R. P., J. Chem. Phys., 44, 3307, 1966.

26 Demidov, A. V., Gershikov, A. G., Zasorin, E. Z., Spiridonov, V. P. and Ivanov, A. A., Zhur. Strukt. Khim., 24, 9, 1983.

27 Tolmachev, S. M. and Rambidi, N. G., High Temp. Sci., 5, 385, 1973.

28 Hinchcliffe, A. J. and Ogden, J. S., J. Phys. Chem., 77, 2537, 1973.

29 Ovchinnikov, I. V., Serebrennikov, L. V. and Mal'tsev, A. A., Vestn. Mosk. Univ. Khim., 39, 157, 1984.

30 Douglas, M. A., Hauge, R. H. and Margrave, J. L., High Temp. Sci., 16, 35, 1983.

31 Ruscic, B., Goodman, G. L. and Berkowitz, J., J. Electron Spectrosc. Relat. Phenom., 41, 357, 1986.

32 Van Dillen, A. J., Geus, J. W. and de Wit, J. H. W., J. Chem. Thermodyn., 10, 895, 1978.

33 Broch, N. C. and Christensen, A. N., Acta Chem. Scand., 20, 1996, 1966.

34 Hauge, R. H., Kauffman, J. W. and Margrave, J. L., J. Am. Chem. Soc., 102, 6005, 1980.

35 Bulewicz, E. M. and Sugden, T. M., Trans. Faraday Soc., 54, 830, 1958.

36 Kelly, R. and Padley, P. J., Trans. Faraday Soc., 67, 740, 1971.

37 Thermodynamic Properties of Individual Substances, Vol. III, Editors: Glushko, V. P., Gurvich, L. V., Bergman, G. A., Veits, I. V., Medvedev, V. A., Khachkuruzov, G. A. and Yungman, V. S., Nauka, Moscow, 1981.

38 Beattie, I. R., Bowsher, B. R. and Jones, P. J., Vapour Transport of Indium in Steam, UK Atomic Energy Authority Report, AEEW - R 2391, 1988.

39 Brisdon, A. K., Ogden, J. S., Bowsher, B. R. and Dickinson, S., Seminar of the International Centre for Heat and Mass Transfer: Fission Product Transport Processes in Reactor Accidents, Dubrovnik, 22-26 May 1989.

40 Valderrama-n, J. and Jacob, K. T., Thermochim. Acta, 21, 215, 1977.

Radiation and Fission Product Transport in the RCS

D. J. WREN
Atomic Energy of Canada Limited
Whiteshell Nuclear Research Establishment
Pinawa, Manitoba, ROE 1LO, Canada

ABSTRACT

Radiation, as an energy source, has the potential to alter the chemistry in the reactor coolant system (RCS) during a severe reactor accident. This may, particularly in the case of fission product iodine, affect the chemical form released to containment, and the timing of release. This paper shows that the radiation field likely to be present in the RCS is insufficient to alter the chemistry of the anticipated dominant iodine species CsI. This result can be extended to other fission products. Hence, transport through the RCS will not be influenced by chemical changes caused by radiolysis.

INTRODUCTION

The aim of this paper is to review briefly the potential impact of high-energy radiation on the transport of fission products through the reactor coolant system (RCS) under severe-core-damage accident conditions. Radiation is a source of energy. Deposition of this energy in the RCS may cause changes in the chemical speciation of fission products and have an impact on their mobility and volatility. The bulk of the energy from radiation will be deposited as low-grade thermal energy. This thermal energy may lead to localized heating that can affect the thermalhydraulic conditions experienced by fission products in the RCS.

Clearly the complex physics and chemistry associated with radiation under accident conditions, in a system like a reactor coolant loop, is more the subject of an extensive treatise than a paper. This review only attempts to cover some of the more important aspects of the interaction of radiation with fission products in the RCS. The emphasis is on the specific behaviour of CsI. This species is the cornerstone of current reactor safety analyses of iodine behaviour during a severe accident. Not surprisingly, it is also the species for which experimental data are available.

SOURCES OF RADIATION

Before examining the impact of radiation on fission product chemistry in the RCS, it is useful to consider the source, timing and intensity of the radiation field itself. In the event of a severe loss-of-coolant accident, the dominant source of radiation is the radioactive decay of the longer-lived fission-product isotopes. Table 1 gives the core inventory at shutdown of the main volatile and semi-volatile fission product species for a 680 MW(e) CANDU™-600 reactor [1]. Most of the radiation flux is accounted for by beta decay of these fission products. Table 1 gives the total decay energy and the energy of the most prominent beta particles. In addition to a beta particle, most decays are accompanied by one or more gamma rays with a variety of energies.

Part of the core inventory of fission products are alpha emitters. These will be trapped in the oxide matrix of the UO_2 fuel during an accident and will not be released into the RCS except in the event of extremely high fuel temperatures or extensive fuel oxidation and aerosol transport. For this reason, alpha radiation will not contribute significantly to the radiation field in the RCS under most accident conditions, and hence, does not need to be considered further.

TIMING OF RADIATION FIELD IN THE RCS

During an accident, the strength of the radiation field in the RCS is a function of the timing of the passage of the various fission products through the RCS. The core can not be considered as a single homogeneous node. Fuel elements will fail at different times, and local fuel temperatures, which control the release of the fission products, will be very different. As a consequence, there is no single release time for fission products from the core and no simple source for the radiation field.

The largest contributors to the radiation field in the RCS will obviously be those fission products which possess the mobility to enter and transit the RCS during the course of an accident. Table 1 lists the fission products in decreasing order of relative volatility. In general, the noble gas inventory of the core will be released before appreciable quantities of the semi-volatile fission products (Cs,I) and nonvolatile fission products are released. Also, noble gases will not be retained in the RCS, and their contribution to the radiation field will be transitory, with consequent limited impact on subsequently released fission products.

Inspection of Table 1, omitting Kr and Xe, clearly shows that the bulk of the radiation in the RCS in the short term following an accident will be associated with the semi-volatile elements Cs, I, Te and Sr/Ba. The remaining activity, associated with relatively less volatile fission products, will not be transported to the RCS in quantities sufficient to affect the local radiation fields where Cs and I are deposited, because of limited release levels, late release (i.e., after most the the Cs and I have been completely swept through the RCS), or differential deposition temperatures, leading to plateout in different regions of the RCS.

Thus, the radiation field to which CsI will be exposed while transiting the RCS will be a largely self-generated field. There will be low level components from other fission products (transported as aerosols generated in the hotter regions of the degraded core), small amounts of noble gases, and activity transported and deposited during normal operation (cobalt-60). However, the major radiation dose will be due to the short-lived iodine isotopes themselves.

TABLE 1. Total core inventory (CANDU)*

Nuclide	Activity (10^8 Ci)	Half-Life (day)	Decay Energy (MeV)	Most Intense β Energy (MeV)
Kr-85	0.00082	3.95E 03	0.67	0.67
Kr-85m	0.167	1.83E-01	0.31	Isomer Transition
Kr-87	0.323	5.28E-02	3.89	3.8
Kr-88	0.456	1.17E-01	2.8	0.52
Xe-133	1.126	5.28E 00	0.43	0.35
Xe-135	0.055	3.84E-01	1.16	0.92
I-131	0.573	8.05E 00	0.97	0.61
I-132	0.836	9.58E-02	3.56	1.16
I-133	1.205	8.75E-01	1.80	1.27
I-134	1.313	3.66E-02	4.2	2.46
I-135	1.118	2.80E-01	2.8	1.0
Rb-86	0.00017	1.87E 01	1.78	1.78
Cs-134	0.003	7.50E 02	2.01	0.66
Cs-136	0.0086	1.30E 01	2.54	0.34
Cs-137	0.008	1.10E 04	1.18	0.51
Te-127	0.052	3.91E-01	0.69	0.7
Te-127m	0.0033	1.09E 02	0.09	Isomer Transition
Te-129	0.176	4.80E-02	1.48	1.00
Te-129m	0.023	3.70E 01	0.11	Isomer Transition
Te-131m	0.086	1.25E 00	2.46	0.56
Te-132	0.82	3.25E 00	0.50	0.22
Sr-89	0.53	5.21E 01	1.46	1.46
Sr-90	0.006	1.03E 04	0.55	0.55
Sr-91	0.764	4.03E-01	2.67	1.09
Ba-140	1.046	1.28E 01	1.05	0.47
Mo-99	1.07	2.80E 00	1.37	1.19
Tc-99m	0.94	2.50E-01	0.15	Isomer Transition
Ru-103	0.68	3.95E 01	0.74	0.20
Ru-105	0.53	1.85E-01	1.87	1.15
Ru-106	0.052	3.66E 02	0.04	0.04
Rh-105	0.382	1.50E 00	0.57	0.57

*CANDU - 600 reactor, 679 MWe

The actual intensity of the radiation field at any location in the RCS will be very sensitive to the nature of the accident. It will be a complex function of the thermalhydraulics affecting Cs and I release from the core, the transport rate through the RCS and the location of deposition in the RCS. There will be considerable feedback of the Cs, I, Te, Sr and Ba fission products on the thermalhydraulics and hence the field [2]. Table 2 summarizes the decay power available from these fission products as a function of time after reactor shutdown. The heat available is sufficient to revolatilize deposited CsI, if concentrated locally. This will lead to transport of CsI, reducing the local field in a self-limiting fashion.

TABLE 2. Decay power (kw)

Nuclide	Time After Shutdown (h) 1	4	12	24
I-131	328	325	316	302
I-132	1304	528	46	2.1
I-133	1244	1127	865	583
I-134	1484	139	0.2	-
I-135	1673	1228	538	156
Rb-86	0.2	0.2	0.2	0.2
Cs-134	3.6	3.6	3.6	3.6
Cs-136	12.9	12.8	12.6	12.3
Cs-137	5.6	5.6	5.6	5.6
Te-127	19.6	16.0	8.6	3.7
Te-127m	0.2	0.2	0.2	0.2
Te-129	84.3	14.0	0.1	-
Te-129m	1.5	1.5	1.5	1.5
Te-131m	122.5	113.8	94.8	71.5
Te-132	241.1	234.5	218.5	196.6
Sr-89	459	458	456	453
Sr-90	2.0	2.0	2.0	2.0
Sr-91	1126	909	511	217
Ba-140	650	646	633	617

Since the radiation field cannot be unambiguously determined and will never remain a single value throughout an accident, a range of values must be considered. This range is on the order of 10^2 to 10^4 Gy/h (1 Gy = 100 rad). This value can be compared to the field in an operating reactor which is on the order of 10^6 Gy/h.

RADIATION ENERGY DEPOSITION

The radiation field in the RCS will be produced by the beta particles and gamma rays released by the decay of the radioactive fission product isotopes. The primary interactions of these particles and rays with matter is the subject of numerous standard texts (e.g., references 3 and 4). A beta particle is simply a high-energy electron. A gamma ray is a high-energy photon which can interact in several ways, such as Compton scattering, pair production and photoelectron production. In the RCS, for fission decay gammas, Compton scattering will be most important. The result is ultimately a high-energy electron, the same as a beta particle. The high-energy electron, no matter what its source, then interacts by further inelastic and ionizing collisions,

$$M + e \rightarrow M^+ + e + e \,, \qquad (1)$$

which produce a shower of secondary electrons with lower energy.

Ultimately, the energy of a secondary electron approaches the range 10 to 50 eV. Collisions of electrons with this energy with molecules in the gas phase can lead to the breaking of molecular bonds, e.g.,

$$CsI + e \rightarrow Cs + I + e \ . \quad (2)$$

At lower energies, the electrons eventually deposit their energy as heat in the gas and surroundings. Energy will be deposited by radiation in four locations in the RCS:

1) the bulk gas phase molecules (primarily H_2O and H_2);

2) gaseous fission products (e.g. CsI molecules);

3) solid fission products (e.g. CsI - in airborne particles or on a surface); and

4) the RCS pipe wall.

Only energy deposited in the first three locations will be available to alter the system chemistry. Heating of the RCS pipe wall will eventually lead to revaporization of deposited fission products.

CsI CHEMISTRY IN THE RCS

In the RCS during an accident, the dominant gas phase species will be steam and hydrogen (neglecting those unlikely accidents in which an air flow through the RCS is established - such as occurred at Chernobyl). Fission product CsI will be located in the gas phase as either individual molecules (high temperature and/or low concentration) or in condensed-phase aerosol particles. The latter may be pure CsI, but are more likely to be agglomerates, including other fission products or structural/control rod materials. Solid CsI will deposit on RCS pipe walls as either trapped aerosols (location determined by aerosol dynamics) or condensed particles (location determined by local CsI concentrations and temperature gradients). In either case, during the initial phases of a severe accident, there are essentially only two physical states of CsI to consider in the RCS: gas-phase molecules or crystalline condensed particles.

In focussing on CsI, the chemistry of interest is its possible reaction to form alternative iodine species:

$$CsI + H_2O \rightarrow CsOH + HI \quad (3)$$

$$HI + HI \rightarrow H_2 + I_2 \ . \quad (4)$$

Both molecular iodine (I_2) and hydrogen iodide (HI) are substantially more volatile than CsI. As such they will be released more readily from the RCS and behave differently in containment. If a significant fraction of the CsI in the RCS was converted to HI or I_2, the source term to the environment could be affected for particular accident scenarios [5].

The equilibrium gas-phase chemistry of CsI in the RCS is relatively easy to evaluate [6]. When Cs is present in excess over iodine, because of their relative concentrations in the fuel (~ 10:1 Cs:I), and the CsI concentration is appreciable, the dominant gas-phase iodine species is CsI.

The chemistry in the presence of a radiation field is more complex. In this case, the continuous input of radiation energy to the chemical system means that equilibrium calculations are not valid. There are a number of endothermic processes that can be driven by the radiation (indicated by the electron below), e.g.,

$$H_2O + e \rightarrow OH + H + e \quad (5)$$

$$H_2O + e \rightarrow H_2O^+ + e + e \quad (6)$$

$$OH + CsI \rightarrow CsOH + I \quad (7)$$

$$H_2O^+ + H_2O \rightarrow H_3O^+ + OH \quad (8)$$

$$H_2O^+ + CsI \rightarrow H_2O + Cs^+ + I \quad (9)$$

$$CsI + e \rightarrow Cs + I \quad (10)$$

The net result of all of these processes may be considered as an increase in the concentration of I atoms which may then react further to form HI or I_2, or reform CsI via the following reactions:

$$I + I + M \rightarrow I_2 + M \quad (11)$$

$$H + I + M \rightarrow HI + M \quad (12)$$

$$Cs + I + M \rightarrow CsI + M \quad (13)$$

where M is any third-body molecule in the gas.

Chemical kinetics calculations must be performed in order to evaluate the chemistry in a steady-state system with a continuous source of energy. These calculations have been carried out for the chemical system $CsI/H_2/H_2O$ under conditions relevant to the RCS [7].

The chemistry of condensed phase CsI is slightly different. Similar decomposition of CsI is possible via

$$CsI(s) + e \rightarrow Cs^+(s) + I^-(s) + e \quad (14)$$

where the ions Cs^+ and I^- have been separated in the crystal lattice by the impact of a high-energy electron. Subsequent reactions can lead to release of I_2 from the solid to the gas phase.

IMPACT OF RADIATION ON CsI CHEMISTRY IN THE RCS

The impact of radiation on the CsI chemistry in the RCS can be divided into consideration of its impact on condensed phase CsI and airborne molecular CsI.

There have been a large number of studies on the radiolysis of solid CsI and alkali halides in general [8]. Most of these studies focus on the production of colour centres in crystals. These are formed by the dislocation of halide atoms and electrons in the crystal matrix. However, there have been fewer studies of the chemical dissociation of CsI in a radiation field. The latter studies have been stimulated by the need to consider CsI decomposition in the nuclear fuel pellet/clad gap under reactor operating conditions as a contributor to the iodine partial pressure in the gap. The iodine gas in the gap can influence the rate of stress corrosion cracking [9-11].

Cubicciotti and Davies were the first to measure the release of iodine from iodide salts by gamma radiolysis [12]. Their results showed that under gamma radiolysis decomposition of CsI could occur leading to release of iodine. The extent of iodine release was sensitive to impurities in the gas phase, particularly water, which enhanced release. The G value (number of molecules of CsI decomposed per 100 eV of energy absorbed) appeared to peak between 300 and 600 K (data reproducibility poor). However, the G values obtained were small ($G_I = 8 \times 10^{-3}$ to 1.0×10^{-5}). The increase in the G value with water vapour present is likely due to migration of OH^- ions into the solid lattice to take the place of I^- ions displaced by the radiolysis.

Kulikov and Malyshev have also examined the radiolysis of solid CsI [13]. They observed significantly higher G values ($G_{I_2} = 1.5 \times 10^{-2}$).

Aleksandrov et al. [14] have examined the gamma radiolysis of water on the surface of CsI crystals. They found a G value of 5×10^{-3} for decomposition of the water, similar to the G values for release of iodine.

All of the experimental studies have been performed with gamma radiolysis. Since the radiolysis of fission product CsI will occur partially via beta emission in the CsI itself, the G values for iodine release may be slightly higher. The primary source of decay energy will be

$$I^* \rightarrow Xe + \beta + \text{gamma} . \qquad (15)$$

The recoil energy from the decay may displace the daughter Xe atom in the lattice and will create an anion lattice vacancy. This may enhance the migration of other iodine atoms, created by the absorption of the electron energy. The magnitude of this possible effect, on iodine release, is not known. However, the bulk of the radiation energy deposition in the CsI will be effected by secondary electrons which are indistinguishable for beta and gamma radiolysis. Hence, beta recoil effects are expected to be small.

Cubicciotti and Davies showed that the G value for release of iodine from irradiated CsI decreased with increasing temperature, presumably because of more rapid recombination of Cs with I within the crystal. Since the temperature of deposited CsI in the RCS will be high, if only due to the local radiolysis self-heating, this effect will counter other factors, such as the presence of steam, which might tend to increase the G value.

The net impact on the stability of solid CsI may be estimated by considering a range of G values for I_2 release and dose rate. The results are indicated in Table 3. Even a maximum value for G a high dose rate of 10^4 Gy/h and a very long exposure time (24 h) lead to prediction of only 1% decomposition. It thus appears that radiolysis of solid CsI on RCS walls will not lead to appreciable conversion to a more volatile iodide.

It should be noted that aerosol-borne CsI will behave identically to solid CsI. However, aerosol particles that are present in the RCS only in transit will have very short times to accumulate a radiation dose and thus will not decompose much. Also, very small aerosol particles, ca 1 μm, will not be able to absorb much of the energy of iodine atoms decaying within the particle because of their small size. This is not so much of a problem for CsI deposited on pipe walls. There, much of the beta energy will be initially deposited in the pipe walls. The result, however, will be ejection of secondary electrons from the walls back into the CsI.

TABLE 3. Radiolysis of CsI to release I_2

G (molecules/100eV)	Dose Rate (Gy/h)	Fractional CsI Decomposition (%) 1 h	4 h	12 h	24 h
0.015	10^2	0.0004	0.0016	0.0048	0.0097
	10^4	0.04	0.16	0.48	0.97
0.001	10^2	0.00003	0.00011	0.00032	0.00065
	10^4	0.0027	0.011	0.032	0.065

Much less research has been performed on the radiolysis of CsI in the gas phase. Yamawaki et al. [15] have studied the radiolysis of molecular CsI using a low-energy (150-V) electron beam. The experiments were conducted on pure CsI at low pressure (ca 1×10^{-3} Pa) and yielded a cross section for reaction (2) of 2.8×10^{-16} cm^2. This value is difficult to relate to the more practical problem of CsI in the RCS.

Elrick and Powers [16] were the first to examine the behaviour of CsI flowing in high-temperature steam through a stainless steel pipe in a gamma radiation field, conditions similar to those expected in the RCS. They observed significant CsI decomposition, which they originally attributed to the effects of radiolysis. Their experiments were repeated and extended by Wren et al. [17] whose work showed that gamma radiolysis of CsI (at concentrations of approximately 100 μL/L) in steam at 1200 K did not lead to extensive CsI decomposition. Measurable decomposition was observed, but this was attributable to the thermal decomposition of CsI followed by reaction of the CsOH produced with the stainless steel pipe walls [18]. More recently, Elrick and Powers have confirmed this result and shown that thermal reactions can account for the CsI decomposition [19].

The experimental finding that under accident conditions radiolysis does not significantly affect gas-phase CsI chemistry in the RCS is not surprising. The radiation field is too low and the residence time of the CsI is too short. This is most easily demonstrated by considering the radiolysis of the steam in the RCS. Radiolysis of steam will lead to production of radicals (including OH) and ions (Reactions (5) and (6)). The OH radical can react directly with CsI (Reaction (7)). Since there is a large excess of steam over CsI in the RCS, production of the OH radical gives an upper limit for reaction of CsI.

The G value for OH production in steam via gamma radiolysis is 7 [20]. Based on this, the radical production in 100 kPa of steam at 1200 K in a 1 kGy/h field is 2.3×10^{10} molecules·cm^{-3}·s^{-1}. The CsI concentration at 100 μL/L is 6.3×10^{14} molecules·cm^{-3}. Thus the maximum destruction rate of CsI via reaction with radiolytically produced radicals is 0.004 %/s. Since the gas flow rates through the RCS will be high and residence times short, a negligible fraction of the CsI can decompose via this route.

In a similar fashion, the maximum destruction rate of CsI if all of the energy from the radiation field goes directly into breaking the CsI bond (D_0° = 3.56 eV [21]) can be estimated to be 0.015 %/s for a 1-kGy/h field. Even at higher fields, this absolute upper bound indicates that radiation will result in negligible conversion of CsI to more volatile forms of iodine while in the RCS.

In order to evaluate the actual amount of iodine that might be released from CsI in the RCS, it is necessary to consider the reaction kinetics in the system. This has been done by Wren [7] for the $CsI/H_2O/H_2$ system, including the presence of a radiation field. Figures 1 and 2 show the influence of radiation on the reaction kinetics. The addition of a radiation field speeds the approach to steady state for the system but does not significantly alter the fraction of I or I_2 present, for higher CsI concentrations.

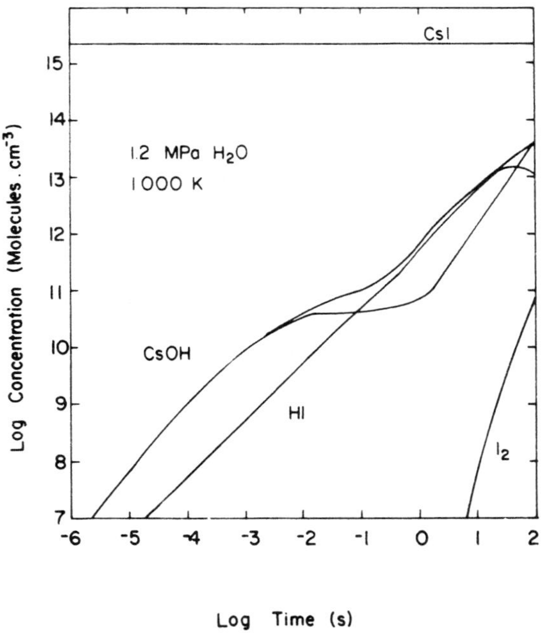

FIGURE 1. Speciation as a function of time for 25 μL/L of CsI introduced into steam at 1.2 MPa and 1000 K

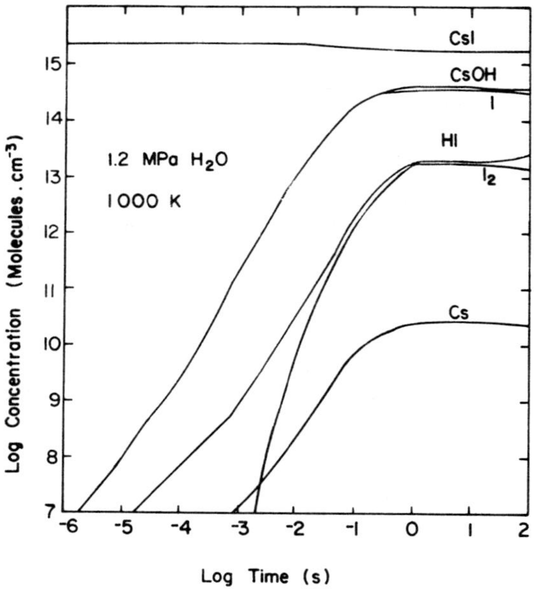

FIGURE 2. Speciation as a function of time for 25 μL/L CsI introduced into steam at 1.2 MPa and 1000 K, with a radiation field of 1 MGy/h

At extremely low CsI concentrations, radiation can lead to decomposition of a greater fraction of the CsI, simply because the ratio of OH radicals produced to the CsI molecules present increases. However, at very low CsI concentrations, the total amount of radioiodine available is low, and hence, the chemical state of that iodine is of limited importance in assessing the overall source term during an accident.

All of the above discussion has been directed toward the behaviour of CsI in the RCS and the influence of radiation on its transport there. This has explicitly excluded the behaviour of CsI in the reactor vessel. One might suspect that radiation could have a greater role to play in the reactor vessel since the fields will be highest there and residence times, due to natural circulation loops, may be longer. However, the temperature in the reactor vessel will also be very high. At high temperatures, water will dissociate on its own. At 1500 K and 100 kPa, the thermal decomposition rate for

$$H_2O \rightarrow OH + H \qquad (16)$$

is 2×10^{-2} s^{-1} compared to the radiolysis rate for Reaction (5) of 1×10^{-5} s^{-1} (for a field of 1 MGy/h). Thus thermal reactions will far outweigh radiolytic reactions in the reactor vessel and, in fact, the very rapid kinetics of thermal reactions will validate the use of equilibrium thermodynamics to predict the fission product chemistry in the reactor vessel.

SUMMARY

Radiation will not significantly alter the chemistry of CsI during its transport through the RCS during a severe reactor accident. This statement is true for all conditions except for very low CsI concentrations. The statement will also be true for the chemistry of all other fission product species. There may be some decomposition of CsI to form HI/I_2 because of thermally driven reactions.

Since radiation will not change the chemical form of iodine in the RCS, it will not alter its transport rate through the RCS or its timing of release to containment because of volatility changes. The thermal energy released by radiation is, however, substantial and this heating will couple to the thermalhydraulics in the RCS altering revaporization rates and aerosol transport. These may affect the timing of fission product release to containment.

REFERENCES

1. S.R. Howieson, H.S. Shapiro, P.J. Allen, J.T. Rogers, T. Mostert and R.W. Van Otterloo, "A Probabilistic Risk Assessment Study of the CANDU-600," Proceedings of Int'l Symposium on Severe Accidents at Nuclear Power Plants, IAEA-FM-296/5, Sorrento, March (1988).

2. B.H. McDonald and D.J. Wren, "Thermalhydraulics and Activity Transport," Proceedings of ICHMT Int. Seminar on Fission Product Transport Processes in Reactor Accidents," Dubrovnik, May (1989).

3. Radiation Chemistry Principles and Applications, ed. Farhataziz and M.A.J. Rodgers, VCH Publishers, NY (1987).

4. A.J. Swallow, "Radiation Chemistry, An Introduction", Longman, London (1973).

5. E.C. Beahm, C.F. Weber, T.S. Kress, W.E. Shockley, and S.R. Daish, "Chemistry and Mass Transport of Iodine in Containment," Proceedings of ACS Symposium on Nuclear Transport of Reactor Severe Accident Chemistry, Toronto, June (1988).

6. F. Garisto, "Thermodynamics of Iodine, Cesium and Tellurium in the Primary Heat-Transport System Under Accident Conditions", Atomic Energy of Canada Limited Report, AECL-7782 (1982).

7. D.J. Wren, "Kinetics of Iodine and Cesium Reactions in the CANDU Reactor Primary Heat Transport System Under Accident Conditions", Atomic Energy of Canada Limited Report, AECL-7781 (1983).

8. V.J. Robinson and M.R. Chandratillake, "Radiation Chemistry of the Alkali Halides", in Radiation Chemistry Principles and Applications, ed. Farhataziz and M.A.J. Rodgers, VCH Publishers, NY (1987).

9. K. Konashi, T. Yato and H. Kaneko, "Radiation Effect on Partial Pressure of Fission Product Iodine", J. Nucl. Mat. 116, 86 (1983).

10. K. Konashi, K. Kamimura and Y. Yokouchi, "Estimation of Irradiation Induced Iodine Pressure in an LWR Fuel Rod", J. Nucl. Mat. 125, 244 (1984).

11. Y.K. Bibilashvili, V.Y. Tonkov, A.S. Sotnikov, I.A. Kulikov, V.V. Novikov and V.I. Solyany, "Out-of-Pile Investigation of Cladding Behaviour for WWER-Type Fuel Elements in Atmosphere Simulating Fission Products (I,Cs)", Res Mechanica 17, 11 (1986).

12. D. Cubicciotti and J.H. Davies, "The Release of Iodine from Iodide Salts by Gamma Radiolysis", Nucl. Sci. and Eng. 60, 314 (1976).

13. I.A. Kulikov and M.L. Malyshev, "Radiation-Chemical Decomposition of CsI", Atomnaya Energiya 55, 316 (1983).

14. A.B. Aleksandrov, A.L. Gusev and N.G. Petrik, "Energy Aspects of the Radiolysis of Water on the Surface of Alkali Halide Crystals", Russ. J. Phys. Chem. 61, 102 (1987).

15. M Yamawaki, T. Yoneoka, H. Kaneko and K. Konashi, "Reactivity of Cesium Iodide Vapor under Irradiation", J. Nucl. Mat. 154, 47 (1988).

16. R.M. Elrick and D.A. Powers, "Effects of Ionizing Radiation on the Transport Chemistry of Cesium Iodide", in Proceedings of OECD Specialist's Workshop on Iodine Chemistry in Reactor Safety, Harwell, September 1985, available as United Kingdom Atomic Energy Authority Report, AERE R-11974 (1986).

17. D.J. Wren, R.K. Rondeau and M.C. Pellow, "The Influence of Radiation on the Stability of CsI in Flowing Steam", Proceedings of ACS Symposium on Nuclear Reactor Severe Accident Chemistry, Toronto, June (1988).

18. R.M. Elrick, R.A. Sallach, A.L. Ouellette and S.C. Douglas, "Reactions Between Some Cesium-Iodine Compounds and the Reactor Materials 304 Stainless Steel, Inconel 600 and Silver, Volume I, Cesium Hydroxide Reactions", NUREG/CR 3197/1 of 3, Sandia National Laboratories Report, SAND83-0395 (1984).

19. R.M. Elrick, R.M. Merrill and A.C. Ouellette, "The Thermal Instability of Cesium Iodide", NUREG/CR-5155, Sandia National Laboratories Report, SAND88-1187, Draft, to be published (1989).

20. R.S. Dixon, "The Dissociation of Water Vapour by Photolytic, Radiolytic and Electron Impact Methods", Radiat. Res. Rev. 2, 237 (1970).

21. K.P. Huber and G. Herzberg, Molecular Spectra and Molecular Structure IV. Constants of Diatomic Molecules, Van Nostrand, New York (1979).

Experimental Studies of Degraded Core Coolability

IVAN CATTON
Mechanical, Aerospace and Nuclear Engineering Department
University of California, Los Angeles
Los Angeles, California 90024-1597, USA

NAIM AFGAN
Boris Kudrec Institute
Belgrade, Yugoslavia

INTRODUCTION

The safety of nuclear reactors has been, and still is, of great concern throughout the world. It is generally assumed that under some situations a particulate bed will form either in the reactor (in-vessel) or on the containment floor (ex-vessel). Some portion of the debris bed may dry out and heat up if insufficient coolant is provided. The high resulting temperature may damage the supporting structure and allow the release of a larger quantity of radionuclides leading to a more severe accident. The question is to determine the heat generation rate, decay heat, which delineates the coolability limits. This paper presents the results of an experimental study that addresses several aspects of the coolability question with emphasis on the effects of pressure and flow from below.

A particulate bed is much harder to cool than an intact core. If the decay heat of the fission products is not too high, the generated heat may be removed by conduction, convection, and boiling processes. Under this situation the temperature of the bed can be maintained around the coolant saturation temperature. The supporting structure will remain intact. However, if the decay heat generates enough vapor, the vapor will prevent liquid from reaching everywhere, and cause dryout. Because a dry bed has a very low thermal conductivity, the temperature in a small dry zone could be very high leading to melting of the core materials. The high temperatures may damage the supporting structure.

Because an LWR is operated at high pressure, knowledge of dryout heat flux at high pressure as well as at atmospheric pressure is necessary if one is to predict coolability. The pressure effect is studied here over a range sufficient to cover much of the range needed to evaluate severe accidents. Flow from below can also significantly impact on the dryout heat flux. The effect of flow is studied in this report to see how much forced flow should be provided to avoid dryout. The pressure drop across the debris bed is also measured to answer another important question, i.e. how much pumping power is required to have sufficient inlet flow. Multi-dimensional geometries similar to those postulated to exist at TMI are studied to address the view that they significantly raise the dryout limit.

In addition, an ex-vessel debris bed is sometimes assumed to form on the concrete. The thermal interaction between a hot debris bed and concrete may be important. The gases from core-concrete interactions may lower the dryout heat flux dramatically and lead to remelting of the core debris. The effect of gas flow from below on dryout heat flux is included in this work.

The importance of dryout phenomena to reactor safety, both for liquid metal fast breeder reactors (LMFBR) and light water reactors (LWR), has led to more than sixty papers being published during the past decade. Reviews of past work have been made by Tsai et al. [1986], Catton and Jakobsson [1987] and Tsai and Catton [1987] and are not repeated here. An interesting comparison of a large number of the contemporary models with data is given by Jakobsson [1983] where it is shown that most of them are not very good.

EXPERIMENTAL APPARATUS AND PROCEDURE

Schematics of the experimental apparatus used to study pressure effects are shown in Fig. 1a and 1b. The apparatus consists of a thick wall glass tube (6.1 cm ID, 7.1 cm OD) with a closed bottom end. The tube contained a debris bed of stainless steel particles and the coolant. The bed was inductively heated with a multi-turn copper coil (radius of curvature 5.8 cm). The coil was powered by a 10 KW, 463 KHz Cycle-Dyne radio frequency generator. A reflux condenser was used to condense the large amount of vapor generated during the runs. The condenser was made of helically coiled copper tubing and was housed above the bed another glass tube. It was fastened to the top of the lower glass tube with a flange and a collar. A nitrogen tank was connected to the apparatus with a gas welder pressure control valve. The line is also connected to a surge tank for pressures greater than 1 atm. The apparatus was connected to a vacuum system to conduct dryout experiments at pressures less than 1 atm.

Four copper-constantan thermocouples connected to an X-Y recorder were buried in the debris bed, both to calibrate the input power and to check the temperatures during the boiling process. Mild steel particles of diameter 0.59-0.79 mm and 404 stainless steel particles of diameter 1.6 mm, 3.2 mm, and 4.8 mm were used to form the porous bed. Freon-113, water, and acetone were used as coolants.

The apparatus for studies of the effects of flow on bed dryout, see Figure 1c, consisted of a 6.9 cm ID Pyrex glass tube to constrain the particulate bed. Subcooled liquid flow up through the porous bed which was supported by a sintered glass filter. Freon-113, methanol, and water were used in this part of the study. The inlet temperature of the coolant was varied by directing the flow through a heat exchanger immersed in a constant temperature reservoir. The bed was inductively heated by a 40 cm long 10 cm ID multi-turn copper coil. The induction coil was connected to a 10 KW, 463 KHz Cycle-Dyne radio frequency generator. Particles of diameter 0.59-0.79 mm, 1.6 mm, 3.2 mm, and 4.8 mm particles were used to form the porous bed. All of the particles were 404 stainless steel, except for the 0.59-0.79 mm particles which were mild steel. The bed height was varied from 8.0 cm to 15.5 cm. The mass flow was through a calibrated flow meter. Constant flow rate was maintained during each run. Two pressure taps were used to measure the pressure difference across the porous bed. The reflux condenser was made of a helically coiled copper tubing and was housed above the bed in the same kind of glass tube that holds the bed. It was fastened to the top of the glass tube with a flange and a collar. Four copper-constantan thermocouples connected to a X-Y recorder were buried in the

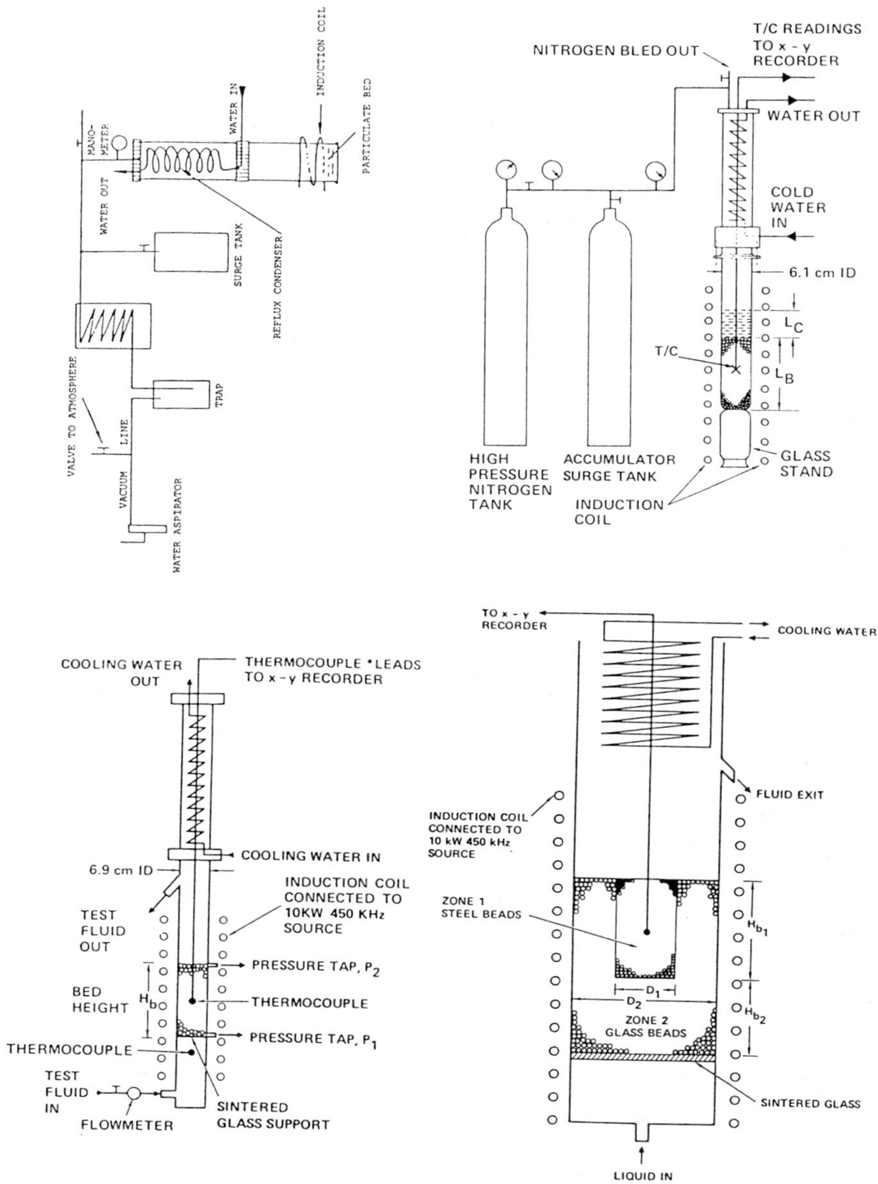

FIGURE 1: (a) Pressure Effects, P > 1 atm.; (b) Pressure Effects, P > 1 atm.; (c) Flow Effects; (d) Multi-Dimensional Effects.

debris bed, both to calibrate the input power and to check the temperatures during the boiling process.

The experimental apparatus for studies of multi-dimensional beds Fig. 1d, was a 15.0 cm or 21.0 cm ID Pyrex glass tube, to constrain the particle bed. Stainless steel particles of diameter 1.6, 3.2, and 4.8 mm, zone 1, were used to form a heated region within the unheated particles, zone 2, which were 6.0 mm glass beds. At the bottom of the steel sphere section there was another layer of glass beads. The stainless steel particles were heated inductively by a multi-turn copper coil which was connected to a 30 kW 450 kHz Westinghouse induction furnace. The bed was supported on a sintered glass plate that had high flow resistance to ensure uniform inlet flow. Freon-113 and methanol were used as coolants so that the experiment could be conducted at relatively low power.

Calibration of the heat generation rate in the particulate bed as a function of generator power setting was made before and after each run. At atmospheric pressure and no through flow, with both the debris bed and the coolant in the tube, the induction furnace was turned on and the temperature history recorded. The input heat flux corresponding to the temperature increase rate is then

$$q = (\varepsilon \rho_{\ell} C_{p\ell} + (1 - \varepsilon)\rho_s C_{ps}) H_b \frac{dT}{dt}$$

The porosity was measured and found to be 0.4 in all cases. The calibration process was repeated for different power settings until a calibration curve was established.

The spatial power distribution within the bed was checked and found to be almost uniform in the axial and radial directions. This was done by placing thermocouples at different locations in the bed. Typically one thermocouple near the middle of the bed could be used for the bed power calibration.

Somerton et al. [1981] suggested the reflood time be used to determine the dryout point. Reflood time is defined as the time it takes the coolant, displaced from the bed by the boiling process, to re-enter when the power is turned off. This time can be measured by watching how the liquid front moves up to a predetermined point at the surface of the bed. Once a dryout spot forms, the reflood time will increase because the local superheat must first be removed by evaporation. If the input power is below the dryout heat flux, the whole bed is at saturation temperature and the reflood time is almost constant. Once the dryout limit has been reached, the reflood time increases linearly as a result of the formation of dried out spots(s). The dryout limit is easily found by plotting the time to rewet.

Commercially available glass, mild steel, and 404 stainless steel particles were used in the present study. Prior to each run the particles were rinsed with acetone and shook vigorously. Then, the particles were dried by heating them on a hot plate to get rid of any dirt or grease sticking to the particles. Dirty particles were found to cause wide variation of the experimental data. The washed and dried particles were poured into the pyrex glass jar to a certain height and weighted to determine the porosity. Several thermocouples connected to an X-Y recorder were then placed into the bed.

The testing procedure was the same for each particle size and different coolant. Coolant was added or pumped into the predetermined bed. The power was calibrated against the induction furnace load meter readings. An arbitrary power setting was chosen to heat the bed for a certain period, e.g. 3 minutes. The system pressure and flow rate were maintained constant. When the power was turned off the reflood time was measured. The reflood time was then plotted as a function of the load meter setting and the dryout point to insure that the power calibration curve had not drifted. This power calibration curve could be used for different system pressure and different bottom flow rates.

RESULTS AND DISCUSSION

Pressure Effects

The debris bed formed following a severe reactor accident is usually under high pressure. The thermophysical properties are functions of system pressure, especially the vapor density. Therefore, the dryout heat flux will vary as a result of different pressures. Jakobsson [1983] measured the effect of up to 10 atmospheres. He adapted the critical heat flux scaling factor proposed by Lienhard and Watenabe [1965] to scale the fluid properties. The scaling factor, q_s, derived by Lienhard and his co-workers is

$$q_s = P_c \frac{(g\sigma)^{\frac{1}{4}}}{(\rho_\ell - \rho_v)} \left[\frac{8MP_c}{3RT_c}\right]^{3/4} \qquad (1)$$

Measured results now include pressures from one-third atmosphere to ten atmospheres. The measured dryout heat fluxes were scaled by the scaling factor, q_s, and are shown in Fig. 2a. It should be noted that within the pressure range of interest the scaling factor, q_s, is almost a constant for each coolant.

The experimental results show that the dryout heat flux increases very rapidly with pressure when the reduced pressure is small. When the reduced pressure, $P_r = P/P_c$, is higher than 10^{-1}, the dryout heat flux becomes almost constant. Because of the high critical pressure of liquid metal, the Liquid Metal Fast Breeder Reactor (LMFBR) operates in a low reduced pressure regime. Dryout heat flux will be very sensitive to the variation of system pressure. Thus, the pressure effect on dryout heat flux will be an important factor in considering coolability of the debris bed formed following a hypothetical accident in a Liquid Metal Fast Breeder Reactor (LMFBR). For Light Water Reactors (LWR), the operating pressure corresponds to a relatively high reduced pressure. As a result, changes in LWR system pressure lead to relatively smaller variation in dryout heat flux. However, the effect is still quite significant.

Even though the experimental error for dryout measurements in a 1-D bed is ± 16%, experimental error alone cannot explain why the scaling factor, q_s, fails to correlate a 3.2 mm bed saturated with acetone. The scaling used assumes thermodynamic similarity among different fluids which is not perfect. Though the lack of perfect thermodynamic similarity leads to some scattering of the scaled test data, it is still a very useful tool and can be used to scale experimental results of one fluid to another at a different pressure.

The scaling fails to deal with dryout of beds of small particles because the mechanisms of dryout in a bed of small particles differs from that of a bed of large particles. In a small particle bed, the flow is limited by friction within the bed whereas in a large particle bed, the flow is limited by counterflowing liquid and vapor processes.

As shown by many models and experiment, the dryout heat flux varies as d_p^2 in the small particle region, whereas it varied as $d_p^{½}$ in the large particle region. Since the heat flux scaling factor works reasonably well for large particle beds and for small particle beds, the strategy is to find a transition diameter, d_p^*, which divides dryout heat flux into large particle and small particle regimes. The transition particle size can then be used to rescale the data to a common fluid. The Ostensen and Lipinski model [1981] and the Dhir and Catton model [1977] were chosen to represent the dryout behavior of large and small particle regimes respectively. Then the transition particle diameter, d_p^*, is found by equating the two models. It is

$$d_p = 184 \frac{(1-\epsilon)}{\epsilon} \left(\frac{\nu^2}{g}\right)^{1/3} \left(\frac{1}{1+(\rho_\ell/\rho_\ell)^{¼}}\right) \tag{2}$$

where g, ϵ, ν, ρ_ℓ, and ρ_v are gravity, debris bed void, viscosity and liquid and vapor density respectively. Some elementary geometry leads to

$$\frac{q_{d1}}{q_{s1}} = \frac{q_{d2}}{q_{s2}} \left(\frac{d_{d2}^*}{d_{p1}^*}\right)^{3/2} \tag{3}$$

where the subscripts 1 and 2 denote two different fluids and q_d is the dryout heat flux. Choosing liquid 1 as the reference liquid, one can define a new scaling factor for small particle sizes

$$q_s^* = q_s(d_{p1}^*/d_p^*)^{3/2}$$

Since water is one of the most used liquids in the nuclear industry, water was chosen to be the reference liquid.

In Fig. 2b, the scaled data of water, Freon-113, and Acetone are shown for small particles (d_p = .59 - .79 mm). As one can see the original scaling method fails to scale the data into a common curve. And the rescaled data were all scaled to the vicinity of a common curve. Thus, the above modifications to heat flux scaling factor can be extended to scale the small particle data.

Flow Effects

When there is flow from below, the physical processes become much more complicated. Some speculations as to the types of flow one must consider are shown in Fig. 3a. To test these speculations, water data is plotted against mass flux from below with the asymptote $\dot{m}_\ell(h_{fg} + c_{p\ell}\Delta T_{sub})$. The difference between the measured value and the asymptote approximately represents the contribution of "top flow", see Fig. 3b. The intersection with the asymptote yields the critical mass flux $\dot{m}_{\ell cr}$, see Fig. 3c.

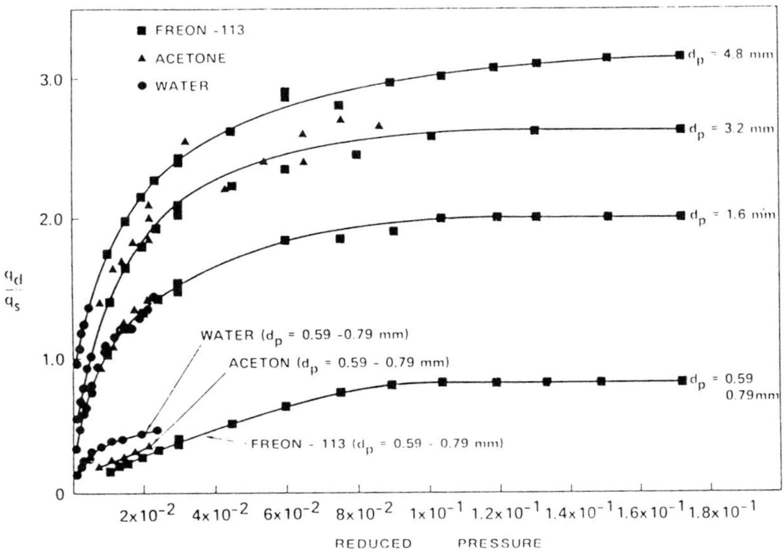

FIGURE 2a: Reduced Dryout Heat Flux vs. Reduced Pressure.

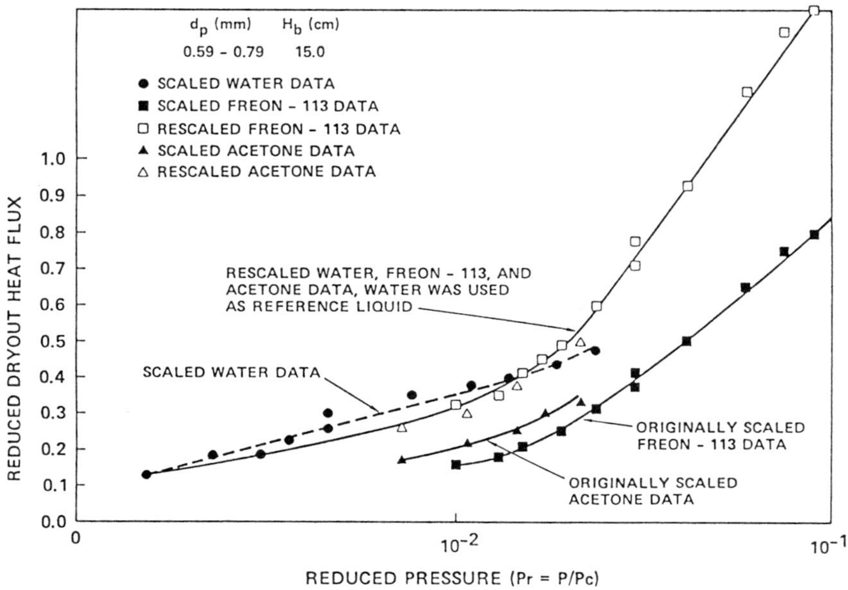

FIGURE 2b: Reduced Dryout Data vs. Reduced Pressure (d_p = .59 - .79 mm).

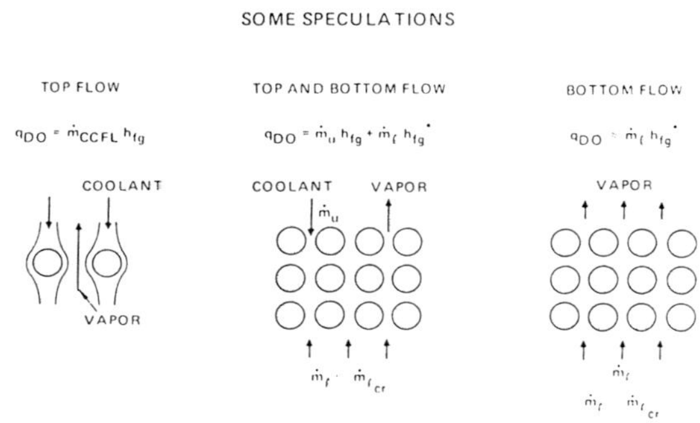

FIGURE 3a: Three Types of Flow in Dryout Process of a Porous Bed.

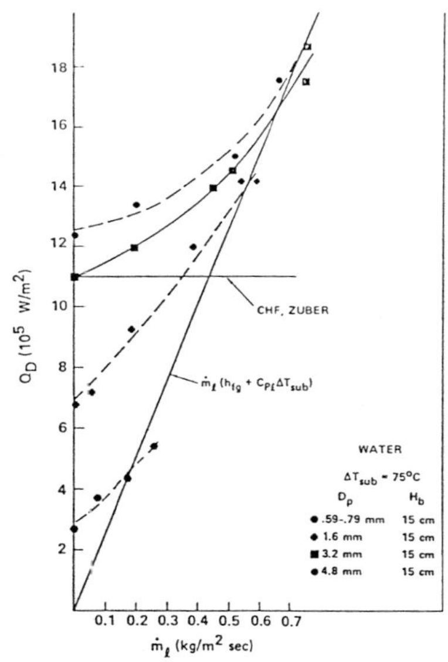

FIGURE 3b: Comparison of Dryout Heat Flux with $\dot{m}_\ell (h_{fg} + C_{p\ell}\Delta T_{sub})$ (Water)

Defining a critical mass flux, $\dot{m}_{\ell cr}$, at which counter flow into the bed becomes insignificant, the dryout heat flux, q_d, can be categorized as follows

1. $q_d = \dot{m}_{CCFL} h_{fg}$, when $\dot{m}_\ell = 0$
2. $q_d = (\dot{m}_{CCF} + \dot{m}_\ell) h_{fg}$, when $\dot{m}_\ell \le \dot{m}_{\ell cr}$
3. $q_d = \dot{m}_\ell h_{fg}$, when $\dot{m}_\ell > \dot{m}_{\ell cr}$

where $\dot{m}_{CCFL}$, $\dot{m}_{CCF}$, and $\dot{m}_\ell$ are shown in Fig. 3a.

The critical mass flux increase as the particle size becomes bigger, and the slopes of the curves decrease, see Fig. 3c. Using the heat flux scaling factor, q_s, the following empirical relationship was found

$$\frac{\dot{m}_{\ell cr} h_{fg}}{q_s} = 5.78\, d_p^{0.514} \left(\frac{P}{P_c}\right)^{0.361} \quad (4)$$

where d_p is in units of mm. The above equation can reproduce the measured critical mass flux data within ± 15%. Again using q_s, the top flow is scaled with

$$(\dot{m}_{\ell cr} - \dot{m}_\ell) h_{fg} / q_s$$

and the contribution of counter current flow to dryout, q_{CCF}, is correlated with

$$\frac{q_{ccf}}{q_s} = 0.448 \left(\frac{(\dot{m}_{\ell cr} - \dot{m}_\ell) h_{fg}}{q_s}\right)^{0.78} d_p^{0.346} P_r^{0.0228} \quad (5)$$

where d_p has units of mm. The dryout heat flux, which is the sum of the two contributions, is correlated with

$$\frac{q_d}{q_s} = \frac{m_\ell h_{fg}}{q_s} + 0.448 \left(\frac{(m_{\ell cr} - m_\ell) h_{fg}}{q_s}\right)^{0.780} d_p^{0.346} P_r^{0.0228}$$

This expression was found to represent the measured data within ±20%, see Fig. 3d. The correlations were derived based on and are limited to the results of the present study. The diameter of the particles used was varied from 0.59 mm to 4.8 mm. The bed height was varied from 8.0 to 15.5 cm. The fluids used were water, acetone, methanol and Freon-113. One should be very careful to properly interpret the results when the correlations are applied beyond the above ranges.

Since forced flow from below requires that a sufficient liquid head exists, the pressure drop across the porous bed is an important parameter in dealing with debris bed coolability. The pressure drop at the dryout point is shown in Figs. 4a and 4b for water and methanol. These figures will allow one to determine whether there is sufficient head to drive the flow through the bed and avoid dryout. This is not, however, the complete picture. If there is room to bypass the debris, then the flow will do little to reduce the dryout potential.

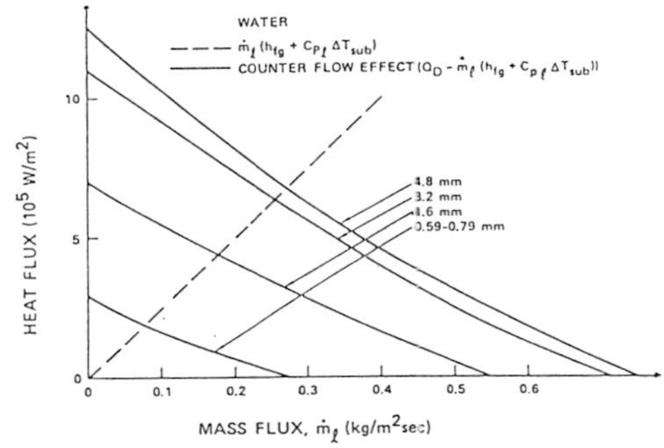

FIGURE 3c: The Contribution to Dryout Heat Flux of Counter Flow (water)

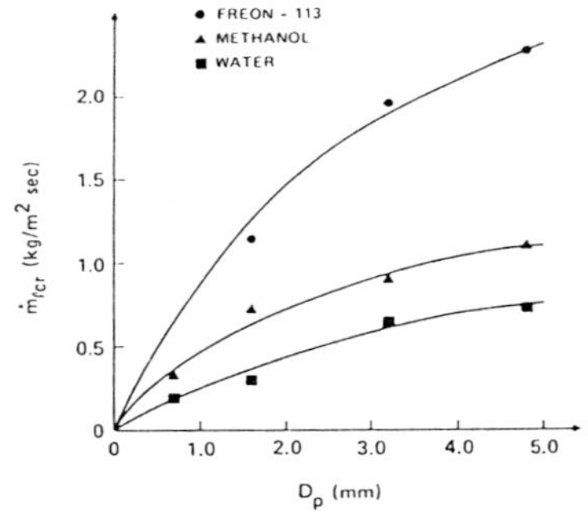

FIGURE 3d: The Critical Mass Flux of Freon-113, Methanol, and Water, Varied with Particle Size.

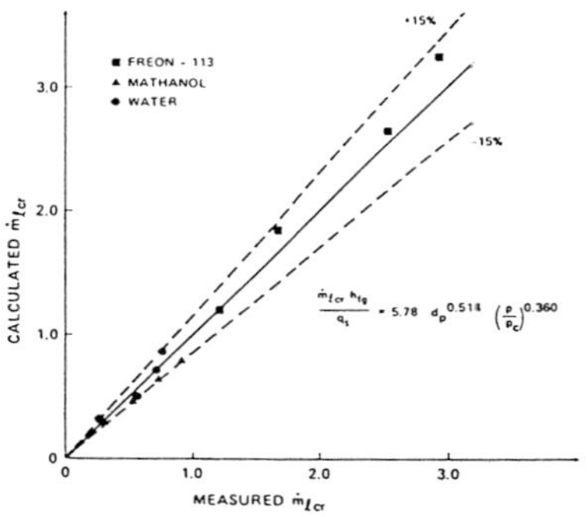

FIGURE 3e: Comparison of Calculated Critical Mass Flux with Measured Data.

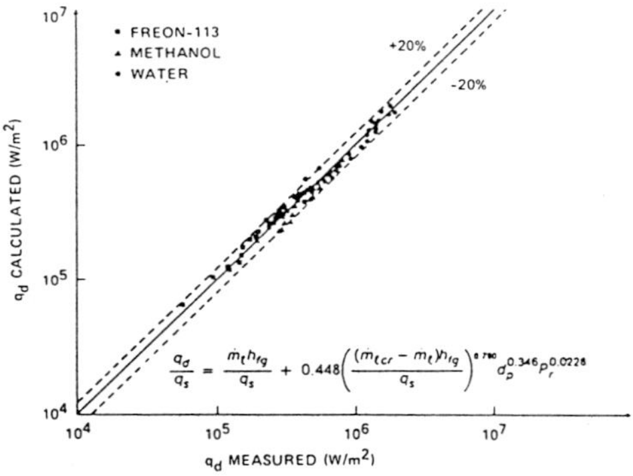

FIGURE 3f: Comparison of Dryout Heat Flux Measurements with Correlation.

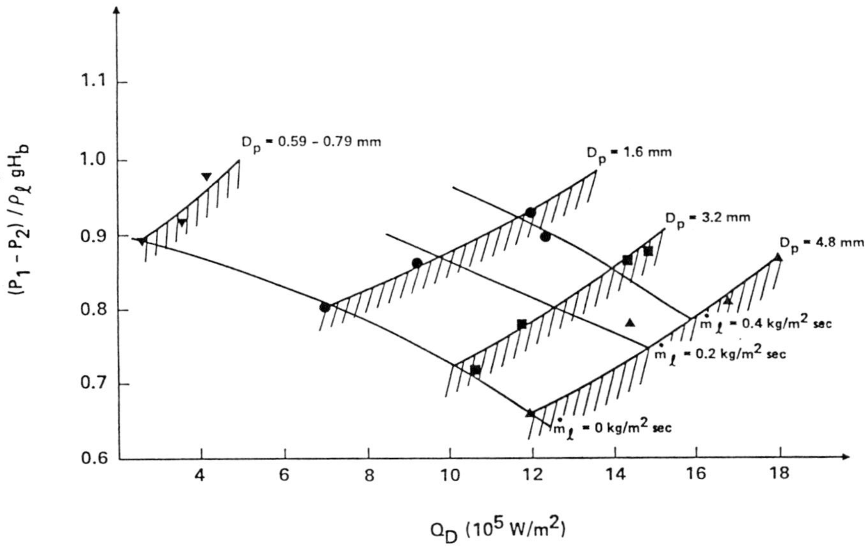

FIGURE 4a: Pressure Drop at Dryout Point (Water).

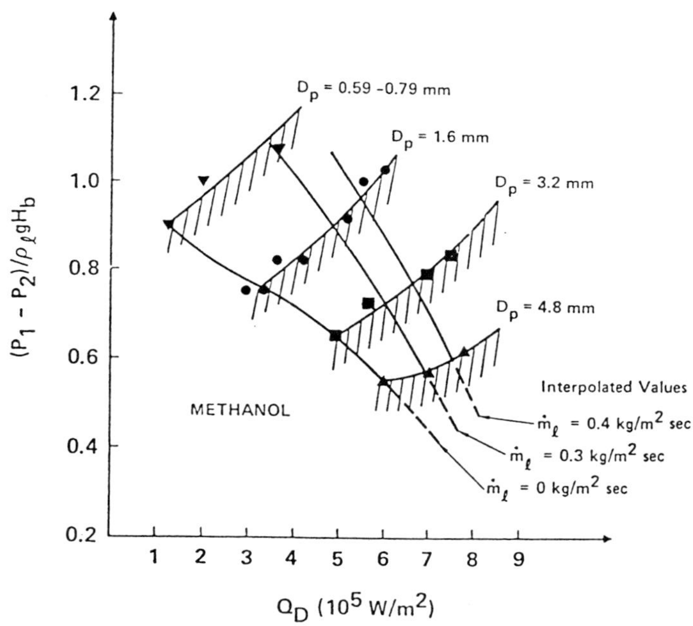

FIGURE 4b: Pressure Drop at Dryout Point (Methanol).

Multi-Dimensional Effects

Studies of multi-dimensional effects were carried out using the apparatus shown in Fig. 1d. The experimental data are shown in Figs. 5(a-d). In Fig. 5a, it can be seen that the dryout heat flux, $Q_d(D_1)$, decreases monotonically to the one-dimensional dryout heat flux, $Q_d(\infty)$, as the heated diameter increases. The rapid decrease indicates that multi-dimensional effects are important only when the heated region is not too large. Furthermore, it can be seen that the dryout heat flux shows no significant change when D_2 is increased if D_1/d_{p1} determines the horizontal flow resistance of the bed and a larger $(D_2 - D_1)/d_{p2}$ doesn't help much in reducing the overall flow resistance. Since the debris bed formed following a nuclear reactor accident has a large D_1/d_{p1} ratio, the resulting debris bed will behave more like a one-dimensional bed. Multi-dimensional effects are not as important to estimates of dryout as they were once thought to be.

Bottom flow was introduced in some of the runs as shown in Fig. 5b. Due to the high permeability in the unheated region, most of the forced flow bypassed the heated zone. As a result, unlike the one-dimensional situation, forced flow from the bottom had little effect on dryout for a multi-dimensional bed.

In the above cases, the heated region consisted of uniform particles. In addition to the experiments with uniform heated beds, some runs with stratified heated beds were carried out. Here, the heated region was divided into layers with the same height. Each layer consists of particles of different sizes. The results are shown in Fig. 5c. The bed consists of 1.6 mm and 3.2 mm diameter particles and results in dryout heat flux higher than that of a bed of 1.6 mm particles and lower than that for a bed of 3.2 mm diameter particles. This is the same as that for a 1-D stratified bed, see Dhir [1979]. However, the existence of the unheated glass particles adds additional complexity. The resistance summation law for one-dimensional beds proposed by Dhir [1979] can not be applied. The sequence of the stratified layers is important. If the small particles were on top of the large ones, the dryout heat flux is lower than that of a bed with large particles on top of the small ones.

Finally, experiments with infinite κ_2 were run, see Fig. 5d. Infinite κ_2 was achieved by removing the glass particles from zone 2. The heated steel bed was immersed directly in the test fluid without the existence of the unheated glass beads. Glass beads under the heated region were kept the same to support the steel particles at the same elevation as previous tests. As shown in Fig. 5d, the removal of the glass beads was found to have very little impact on the dryout process of the heated bed. This result is, however, not surprising. As mentioned earlier, the resistance inside the heated bed is so high that further reducing the resistance in the unheated region will not help much in removing energy from the heated bed.

Effect of Gas Flow From Below on Dryout

The apparatus used for studies of the effect of flow, see Fig. 1c, was modified to allow uniform air flow from below to study its effect on dryout heat flux. An air distributor made of spiral aluminum tubing was used. One end of the tubing was closed and many small holes were drilled to distribute air uniformly. A calibrated air flow meter was used to read the volumetric flow rate. The mass flux of air flow was

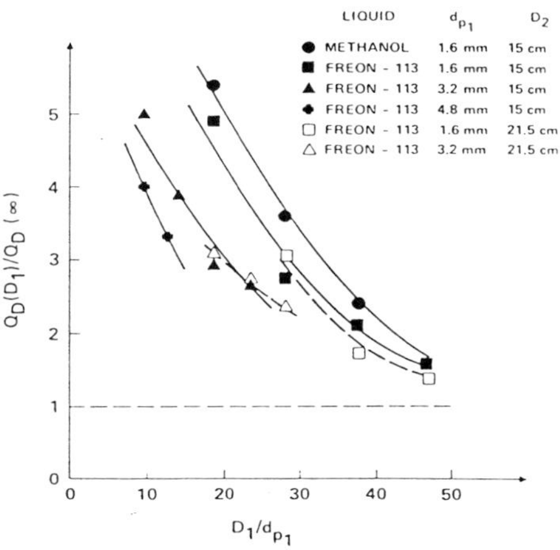

FIGURE 5a: Dryout Heat Flux for Various D_1/d_{p1} with d_{p2} = 6.0 mm, H_1 = 10 cm.

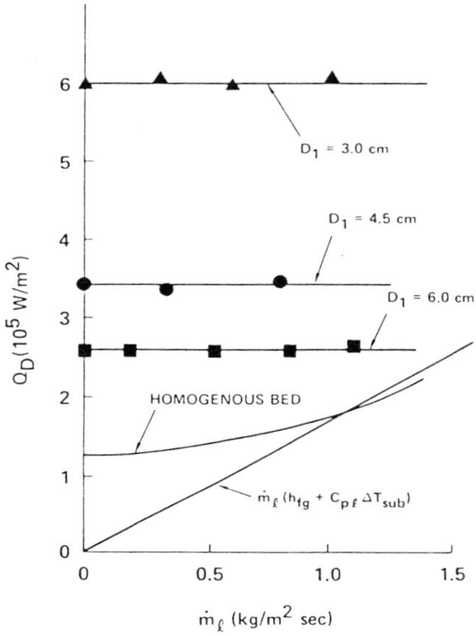

FIGURE 5b: Bottom Flow Effect on Multi-Dimensional Bed Dryout.

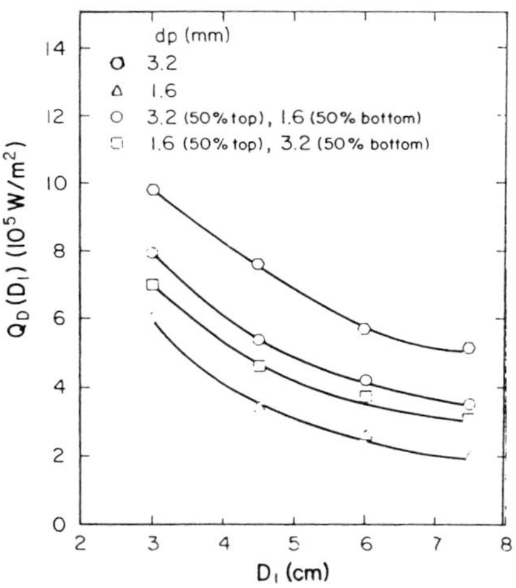

FIGURE 5c: Dryout Data of Stratified Multi-Dimensional Bed.

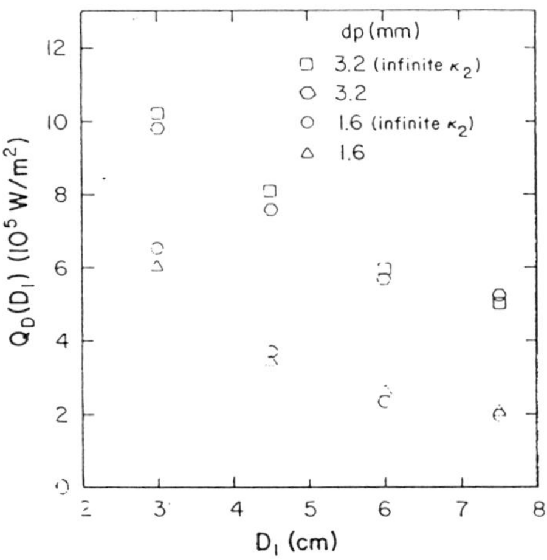

FIGURE 5d: Dryout Heat Flux of Multi-dimensional Bed with infinite κ_2.

varied from 0.0 to 0.211 kg/m^2 sec. (at STP). Since the rewetting time method is not appropriate for determining the dryout point when there is continuous gas flow, thermocouple readings were used to determine the occurrence of dryout.

The experimental data are plotted in Figs. 6a and 6b. In Fig. 6a the results for Freon-113 saturated bed with air flow are shown. The measured results for a water saturated bed are shown in Fig. 6b. The measured data are also correlated to obtain the following empirical formulations:

for Freon-113 $$\frac{q_d(\dot{m}_a)}{q_d(0)} = 1 - 4.401\, d_p^{-1.384} \dot{m}_a^{1.231} \quad (7)$$

for Water $$\frac{q_d(\dot{m}_a)}{q_d(0)} = 1 - 4.707\, d_p^{-1.384} \dot{m}_a^{1.159} \quad (8)$$

where d_p is in units of mm and $\dot{m}_a$ is in units of kg/m^2 sec. It can be seen that the dryout heat flux decreases as a result of incoming bottom air flow. The higher the gas flow is, the lower the dryout heat flux is. The trend shows that if the inlet air flow exceeds some critical value, the dryout heat flux will be equal to zero. The current experimental apparatus did not allow us to reach such a high air flux. One can, however, estimate the critical gas flux, $\dot{m}_{acr}$, by assuming monotonic behavior similar to that observed over the range measured. Setting the LHS of Eqs. 7 and 8 equal to zero yields

for Freon-113 $$\dot{m}_{acr} = 0.300 d_p^{1.248}$$

for Water $$\dot{m}_{acr} = 0.263 d_p^{1.194}$$

For water at atmospheric pressure saturating a bed of 1.6 mm particles, the extrapolated value is 0.46 kg/m^2 sec. For Freon-113 in a similar bed the extrapolated value is 0.54 kg/m^2 sec.

The effect of gas flow from below is more severe for beds of smaller particles than large. The reason appears to be that a small particle bed has lower dryout heat flux, i.e. a smaller amount of vapor flow can keep liquid from reaching some portion of the bed. It is therefore not surprising that a bed of small particles is more sensitive to air flow from below. The present study is not broad enough to fully determine the effect of particle size on dryout. It can be seen from the data that the air flow has a stronger effect on a water saturated bed than a Freon-113 saturated bed. As mentioned earlier, Freon-113 can be used to simulate the behavior of high pressure water. This leads one to conclude that at high pressure the effect of gas flow from below will be weaker.

An ex-vessel debris bed is generally assumed to rest on concrete. The hot debris bed will interact thermally with the concrete. The concrete will decompose and release noncondensable gases (e.g. CO_2). The resulting gases will lower the dryout heat flux of the bed and may lead to remelting of the core debris. When high calcium carbonate concrete is attacked, 44% of the total mass will become gas. If the erosion rate of the concrete ranges from an initial value of 30 mm/min to 10 mm/hr at the final stages, see Schroder et al. [1984], the corresponding gas mass fluxes will range from 0.515 to 0.0029 kg/m^2 sec. respectively. At the upper limit, a debris bed of the particle sizes studied here will

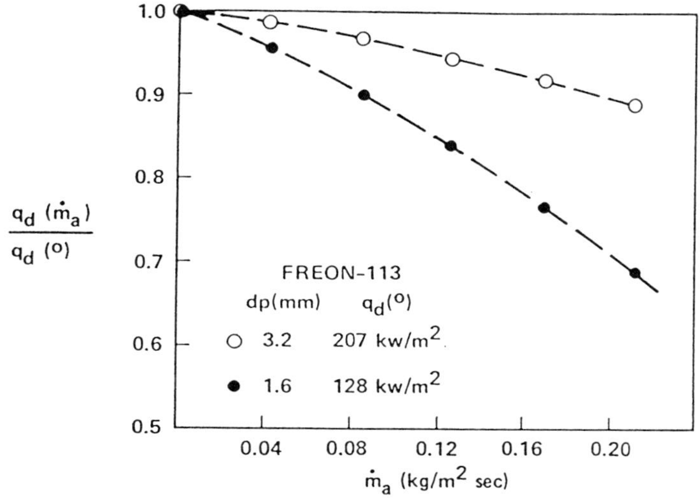

FIGURE 6a: Effect of Air Flow from Below on Dryout Heat Flux (Freon-113).

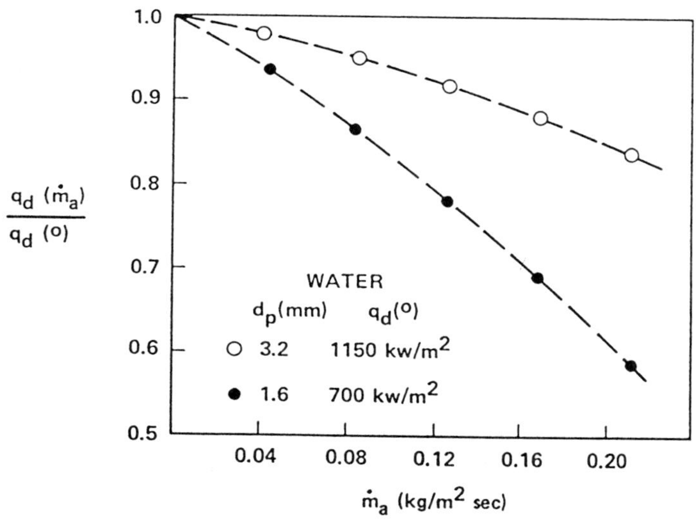

FIGURE 6b: Effect of Air Flow from Below on Dryout Heat Flux (Water).

certainly dryout. The high initial gas release rate is higher than the maximum air mass flux of the present study. A greater reduction in incipient dryout heat flux is expected, i.e. at the high limit of core attack dryout is more likely to occur. Whereas, the current result shows that the low gas release rate in the final stage of core-concrete interaction will have minor effect on bed dryout.

CONCLUSIONS

The following conclusions may be drawn from this study:

1. The different dependence on particle size and the transition of dryout from small particle to large particle regions are taken into account by rescaling all the data to a reference fluid. The modified scaling will scale the results of small particle beds as well as large.

2. The dryout heat flux increases as the inlet mass flux increases. For a sufficiently large inlet flow, dryout heat flux asymptotically approaches $m_\ell(h_{fg} + C_{p\ell}\Delta T_{sub})$, which is the energy required to vaporizes all the inlet flow. Beds of small particles were found to asymptotically approach $\dot{m}_\ell(h_{fg} + C_{p\ell}\Delta T_{sub})$ earlier than beds of large particles.

3. The bottom flow will decrease the counter current flow effect of the top liquid layer and finally dominate the whole bed. For beds of large particles, the bed resistance is smaller and the amount of top counter current flow is larger. Therefore, the critical mass flux, $m_{\ell cr}$, increases as the particle size increases.

4. When there is no flow, the dry zone is near the bottom of the bed. The inlet flow from below will push the position of the dry zone upward and have a tendency to make it smaller. Eventually, it will become a thin layer at the top of the bed.

5. The magnitude of ΔP changes rapidly around the dryout point due to the formation of the dry zone when the power exceeds dryout heat flux.

6. For sufficiently large flow rates, different size particles may have the same dryout heat flux because the asymptotic dryout limit is the total evaporation energy of the inlet flow, but the pressure drop will always be different because it is determined by the bed permeability. Actually, small particles will always have a larger pressure drop.

7. A multi-dimensional bed has a higher dryout heat flux than a "similar" one-dimensional bed because it has more surface area allowing more vapor to escape and more liquid to feed into the bed. The larger the heated bed diameter is, the smaller the dryout heat flux. The asymptotic limit is the one-dimensional bed dryout heat flux. Furthermore, the dryout heat flux decreases very quickly as the heated bed diameter increases. As a result, multi-dimensional effects are not important for a debris bed formed in a reactor.

8. Multi-dimensional dryout heat flux is mainly controlled by the resistance of the heated particles. The surrounding unheated particles do not change the flow pattern much, unless the total radius is close to that of the heated bed. Thus, increasing the flow permeability in the unheated region by removing the glass beads has very little effect on the dryout results. Similarly, bottom flow has little impact on dryout in multi-dimensional beds.

9. The order of the permeabilities in a stratified heated bed will result in different dryout heat fluxes when they are multi-dimensional. This results due to the fact that it leads to different flow paths and, therefore, different total flow resistance.

10. Gas flow from below decreases the dryout heat flux. The decrease in dryout heat flux is monotonic with increasing gas flow rate.

11. There is a critical gas flux that results in zero dryout heat flux. Decreasing particle size reduces the critical mass flux.

12. The effect of gas flow from below on dryout is weaker when the system pressure is high.

ACKNOWLEDGMENTS

The work reported in this paper was supported by the Department of Energy through a Basic Energy Science Division grant and the International Programs Office Yugoslavia Program as well as EPRI. Students contributing were F.P. Tsai and J.O. Jakobsson.

REFERENCES

Catton, I., and Jakobsson, J.O., "The Effect of Pressure on Dryout of a Saturated Bed of Heat-Generating Particles," *J. Heat Trans.*, Vol. 10, pp. 185-195, 1987.

Dhir, V.K., and Catton, I., "Study of Dryout Heat Fluxes in Beds of Inductively Heated Particles," NUREG-0262, June 1977.

Dhir, V.K., "Dryout Heat Fluxes in Debris Beds Containing Particles of Different Size Distributions," *Proc. Int'l Mtg. on Fast Reactor Safety*, Seattle, WA, August 1979.

Jakobsson, J.O., "Pressure Dependence of Dryout Heat Flux in a Volume Heated Porous Bed," M.S. Thesis, University of California, Los Angeles, June 1983.

Lienhard, J.H., and Watanabe, K., "On Correlating Peak and Minimum Boiling Heat Fluxes with Pressure and Heater Configuration," *J. Heat Trans.*, Vol. 87, pp. 94-100, 1985.

Lipinski, R.J., "A One-Dimensional Particle Bed Dryout Model," *Trans. ANS*, Vol. 38, pp. 386-387, June 1981.

Ostensen, R.W., and Lipinski, R.J., "A Particle Bed Dryout Model Based on Flooding," *Nucl. Sci. and Engr.*, Vol. 79, pp. 110-113, Sept. 1981.

Schrock, V.E., Wang, C.-H., Revankar, S., Wei, L.-H., Lee, S.Y., and Squarer, D., "Flooding in Particle Beds and Its Role in Dryout Heat Flux Prediction," Sixth Information Exchange Meeting on Debris Coolability, Los Angeles, November 1984.

Somerton, C., Catton, I., and Thompson, L., "An Experimental Investigation into Dryout in Deep Debris Beds," *ASME Winter Ann. Mtg.*, Washington, DC, 81-WA/HT-17, November 1981.

Squarer, D., and Peoples, J.A., "Dryout in Inductively Heated Beds With and Without Forced Flow," *Trans. ANS*, Vol. 34, pp. 537-538, June 1980.

Squarer, D., Pieczynski, A.T., and Hochreiter, L.E., "Effect of Debris Bed Pressure, Particle Size, and Distribution on Degraded Core Coolability," *NUcl. Sci. and Engr.*, 80, 2-13, Jan. 1982.

Tsai F.P., Jakobsson, J.O. and Catton, I, "Further Studies of Degraded Core Coolability: The Effect of Pressure and Coolant Flow from Below," EPRI NP-4942, Nov. 1986.

Tsai, T.P., Catton, I., "The Effect of Flow From Below on Dryout Heat Flux," *J. Heat Trans.*, Vol. 109, pp. 491-497, 1987.

4. TRANSPORT INTO CONTAINMENT

Direct Containment Heating

MICHAEL L. CORRADINI
Nuclear Engineering and Engineering Physics
University of Wisconsin-Madison
1500 Johnson Drive
Madison, Wisconsin 53706, USA

ABSTRACT

Direct containment heating is a phenomenon identified to occur in a light water reactor during a meltdown accident at high system pressures at the time of vessel breach. Immediate containment failure may result from this process due to overpressure and overtemperature loads. This paper focuses on the "science" of direct heating and the reactor safety criteria and issues that must be addressed to assure the threat to containment is minimized. This discussion attempts to be tutorial in nature with the broader context of melt-coolant-atmosphere interactions being emphasized. From this discussion future research needs may be suggested.

INTRODUCTION

If a complete and prolonged failure of normal and emergency coolant flow were to occur in a light water reactor (LWR) fission product decay heat would cause the fuel to overheat. If no corrective action were taken, the fuel would eventually melt with its heatup exacerbated by metal-water chemical reactions (e.g., $Zr\text{-}H_2O$). Under certain conditions this could lead to slumping of the molten core materials into the lower plenum of the reactor pressure vessel (RPV) followed by failure of the vessel. This would result in the deposition of these materials into the containment (cavity or pedestal and beyond) and interaction of the core materials with the containment atmosphere, water and solid structures (concrete and steel).

During this core melt progression and eventual melt quenching the containment pressure and temperature increases due to physical processes that may challenge its integrity. There have been a number of integrated studies [1-5] which have attempted to systematically identify and quantify these containment pressure/temperature loads and the possible threat of early ($t < 2\text{-}4$ hr) or late ($t > 24$ hr) containment failure. Recently, the US Nuclear Regulatory Commission has issued the Reactor Risk Reference Document [6] (NUREG-1150) containing the most recent integrated study of containment loads and failure modes. This document considers the two LWR reactor types (a pressurized water reactor, PWR, and a boiling water reactor, BWR) and the five reactor containment configurations (PWR large dry, PWR ice condenser, BWR Mark I, II or III) as case studies to estimate the dominant severe accident frequencies, containment loads, source term release paths and the overall risk.

One of the physical processes identified which is estimated to have an impact on the overall risk is direct containment heating (DCH). The process involves the dispersal of some of the fuel melt into the containment atmosphere, at the time of vessel breach. Pressure and temperature loads result as the melt exchanges energy with the atmosphere, available water and surrounding structure, and it is chemically oxidized in flight with possible hydrogen production and combustion.

The focus of this paper is the DCH process in terms of the "science" that may describe it and the reactor safety criteria and issues that must be addressed to assure the threat to containment is minimized. This process involves melt-coolant-atmosphere heat and mass transfer and therefore, should not be viewed in a narrow context. Within a broad context future work may become self-evident. The paper first gives some background on the DCH process and a reference scenario to consider. Next, the important initial conditions and boundary conditions are examined along with the physics one must consider during the process. Finally, reactor safety issues are identified in order to suggest criteria that one must consider to assess the DCH threat to containment.

BACKGROUND

The possibility of a high pressure melt ejection and direct containment heating was first identified in the Zion Probabilistic Safety Study [4] and analyzed by R. Henry. This analysis considered the melt to accumulate in the lower head of the PWR pressure vessel under high-pressure reactor coolant system (RCS) conditions. Failure of the vessel at a location in contact with the melt would then lead to gas pressure-driven ejection of melt from the vessel into the reactor cavity region. The study suggested that the subsequent blowdown of the steam-hydrogen mixture would lead to ejection of a coherent mass of melt from the reactor cavity and eventual deposition of the melt on the water-flooded floor of the containment building. Because of the presence of water in the reactor cavity region and lower compartment of the containment, the analysis assumed good fuel-coolant mixing and the melt ejection resembled a fuel-coolant interaction dispersed in the containment atmosphere. The melt quenched in the presence of water and was calculated to produce a quasi-static pressure rise ($<$ 1 min) in the containment primarily due to steam production, ("steam spike"). Subsequently IDCOR [7] supported this view of the phenomena.

Following this PRA, industry sponsored experiments were begun at Argonne by Spencer et al [8-12]. The purpose of these experiments (Corium-Water Thermal Interactions, CWTI) was to verify under scaled conditions with simulant and prototypic materials the validity of the assumptions of the Zion PRA analysis; i.e. melt sweep out of the cavity region and intimate mixing and quenching by water in the cavity and the lower compartment. The geometric scale was chosen to be 1/30 of the Zion cavity and containment with only the major geometric paths represented. The initial simulant tests used low melting point liquid metals, and results [8] indicated that melt sweepout from the cavity was a primary function of geometry, the gas velocity and material properties. The subsequent CWTI tests (Table 1) used a "corium" melt (UO_2, ZrO_2, Fe, Cr, Ni mixture) with water present over a range of driving pressures and geometric conditions. The results substantiated the simulant test data on sweepout and also suggested that the presence of water changed the process into a dispersed fuel-coolant interaction with good fuel-coolant mixing; i.e., a "steam-spike" was observed with relatively low steam-gas temperatures, with hydrogen produced by metal oxidation.

Subsequent to the start of the CWTI tests the Containment Loads Working Group [13] considered this problem as part of the Zion and Surry standard problems. These early calculations were based upon a conceptual model which postulated, based upon the experimental results, that droplets of molten core material (corium) were ejected from the reactor cavity into the containment atmosphere. These droplets would interact thermally and chemically with the containment atmosphere. From a computational point of view the entire containment building was represented as a single, adiabatic control volume. Thermodynamic equilibrium and rate-dependent calculations were performed (e.g., Ref 14,15) where it was assumed that the metallic constituents of the melt would oxidize by direct reaction with the available oxygen supply in the containment building dome. Interactions between droplets and atmosphere, both chemical and thermal, were computed to be quite efficient because of the large assumed control volume and large possible droplet interaction time. Containment-threatening pressure loadings were predicted when large quantities of melt were assumed in the absence of water. Members of the working group recognized the possibility of the effect of water, but some felt the scale of the ANL CWTI tests were not appropriate to conclusively show its mitigative effect.

At this time Sandia National Laboratory (SNL) began the SPIT and HIPS [16-18] series of experiments which simulated the process of high-pressure melt ejection within the cavity. These experiments provided observations of the jet which enters from an orifice in the experimental melt pressure vessel and the processes of interaction of the blowdown gas with the melt in the 1/20th and 1/10th scale concrete models in the Zion reactor cavity (Table 2). These experiments used iron-alumina thermite as the core melt simulant. The experiments indicated that the blowdown gas very effectively removed the melt from the cavity models and that this process of melt dispersal from the reactor cavity models was accompanied by fragmentation of the melt into droplets, with a spectrum of diameters ranging from aerosol sizes (microns) to several millimeters. This view of high-pressure reactor cavity dispersal phenomena in the Zion cavity was subsequently verified by Tutu, et al [20]. The SNL experimental results also suggested that the high-velocity molten jet from the reactor vessel is fragmented in transit to the base of the cavity under the influence of a high relative velocity at the edge of the jet and perhaps dissolved gas release.

As follow-on to the cavity phenomena tests at Sandia, an enclosed chamber was constructed (SURTSEY) which was a 1/10 geometric scale of the Zion cavity and containment. A number of experiments have been planned with currently four completed [21,22] (Table 3). These tests through preliminary analyses of test results indicated that in open structures in the absence of water the complete dispersal and fragmentation of the melt from the cavity easily occurs and can generate large pressure and temperature peaks over short time scales (1-5 sec). These tests also illustrated that some structure can change the trajectory of the melt dispersal. Subsequent tests are planned with steam driven melts with and without structure present. This is viewed as important since the oxidant is mixed with the melt during dispersal, and possible inerting by nitrogen would not occur. Eventually tests with water present are planned to be conducted.

TABLE 1. Corium-water thermal interaction experiments [12,25]
[1/30 Geometry scale Zion cavity/containment]

CWTI	Corium Mass Injected (kg)	Cavity Sweepout Fraction	Coolant* Mass (kg)	RPV Inj.Press. (MPa)	Cont Gas Comp	ΔT^{+} Atmos (k)	Efficiency** %
1	1.94	0.32	2.65/0	0.49	N_2	+2	0.2
2	2.19	0.01	2.8/0	0.21	N_2	+2	2.5
4	4.06	0.35	5.6/0	0.36	Ar	+25	1.5
5	3.94	0.57	5.6/6.3	5.0	Ar	-4	0
6	3.75	0.32	0/6.26	5.7	Ar	+53	3.0
11	3.52	0.25	0/6.26	5.0	Air	+43	2.4
12	3.75	0.32	4.6/0	5.0	Air	0	0
13	3.1	0.08	0/0	5.0	Air	+185	33
14	3.45	0.1	0/0	5.0	Air	+210	27
15	3.5	0.1	0/0	5.0	Air	+111	24

* First value is water in cavity, and second is water in lower compartment.

\+ Temperature rise in atmosphere ($T_{final} - T_{init.}$)

** Efficiency defined as ΔEatmos/Emelt dispersed where melt is Corium mixture of UO_2, ZrO_2, Fe, Ni, Cr

Table 2. SPIT/HIPS cavity debris dispersal results [15, 19, 30]

Test	Scale	Melt* Mass (kg)	RPV Inj.Press. (MPa)	Cavity Mat'l	Cavity Sweepout Fraction	Water in Cavity
SPIT-18	1:20	10.3	12.6	Alumina	0.58	No
SPIT-19	1:20	10.3	12.6	Concrete	0.95	No
HIPS-1J	1:10	80	10.0	Free Jet Into Gravel Bed		
HIPS-2C	1:10	80	11.7	Concrete	0.99	No
HIPS-3C	1:10	80	4.2	Concrete Slab		No
HIPS-4W	1:10	80	11.7	Concrete	0.95	Yes**
HIPS-5C	1:10	80	6.7	Concrete	0.99	No
HIPS-6W	1:10	80	3.8	Concrete	0.95	Yes**
HIPS-7C	1:10	81.5	5.5	Concrete	0.98	No
HIPS-8C+	1:10	80	3.5	Concrete	0.8	No

* The melt was molten iron-alumina, thermutically generated.

** The cavity was destroyed by pressure generated within it.

\+ This test had an annular gap around melt entry with 20% dispersal from opening.

TABLE 3. SURTSEY DCH test results [21, 22]
[1/10 Geometry scale Zion cavity/containment]+

DCH	Thermite Mass Injected* (kg)	Cavity Sweepout Fraction	RPV Inj.Pres. (MPa)	ΔP Rise (MPa)	Cont Gas Comp	Energy to Atmos (MJ/kg)	Efficiency** (%)
1	20	0.52	2.6	0.095	Air	3.05	82.4
2	80	0.9	6.7	0.225	Air	1.18	31.9
3	80	0.97	6.1	0.2	Air	1000	27.0
4	80	0.94	6.2	0.215	Ar	0.77	20.8

+ This cavity is connected by a chute (DCH-1,2 with extension, 3,4, w/o)

* Thermite mass of iron-alumina (55w/o, 45w/o) @ about 2800K

** Efficiency defined as ΔEatmos/Emelt dispersed where E melt include sensible and chemical energies

At this same time one-dimensional control-volume analyses were developed by a number of researchers [23-28] in support of experimental analysis and severe accident analyses. These analyses have attempted to quantify the direct heating effects based on global mass and energy balances, performed using various sets of parameters that were derived analytically or varied parametrically. Recently, these system analyses have included kinetic rate effects for heat transfer and chemical reactions to determine the effect of various initial and boundary conditions on the results. This work parallels the more detailed developments of Sienicki at ANL and Pilch at Sandia [29,30] in support of the experiments. These analyses have suggested the possible mitigative effects of the containment complex geometries and the presence of water.

It was recognized that the containment dome region, containing the large quantity of oxygen to drive the chemical reactions, is separated from the reactor cavity by intermediate subcompartments which are densely packed with equipment and various other structures. Melt droplets, driven by the steam/hydrogen blowdown gas from the reactor cavity, would likely encounter structure along their trajectories and could be removed from the flow field prior to reaching the containment dome. This trapping of melt is possible but is recognized to be highly geometry dependent.

In addition, the analyses have pointed out that oxidation of the metallic fuel components by water vapor can produce substantial amount of hydrogen [31-33], that may be transported to other regions of containment with possible combustion. Hydrogen generation, would begin in the reactor cavity with reaction rates that are extremely large when calculated assuming, typically, 1-mm diameter zirconium droplets and using accepted rate-limiting mass transfer models. Since no significant quantity of oxygen is available in the reactor cavity, the hydrogen cannot burn until it reaches the oxygen supply available either in the intermediate subcompartments or in the containment dome region. Upon reaching the containment dome the hydrogen may burn, despite large steam concentrations, due to high gas temperature or due to the presence of high-temperature particulate.

Recently, more detailed mechanistic analyses have begun on the DCH process using the three-dimensional compressible, Lagranam-Eulerian, multifluid spray code, KIVA [34,35]. In this computer tool the fuel melt ejection is modelled as a Lagrangian spray jet of droplets which break up and are transported in the Eulerian gas field, undergoing fragmentation, coallesence and interaction with solid surfaces. Preliminary calculations show reasonable agreement with SURTSEY test data (similar to CONTAIN), however many detailed parameters and models are required to be specified and the effect of water droplets cannot be modelled. These calculations are continuing and will be discussed later. A more detailed review of DCH research is given by Ginsberg [35].

DCH SCENARIO

From these experimental and analytical studies some consensus has emerged on a conceptual picture of the DCH phenomenon. This scenario may be classified as involving four stages: (1) forceful explusion of the melt from the RPV, (2) dispersal of a large fraction of this melt mass into the containment atmosphere (i.e., beyond the cavity/pedestal region), (3) rapid oxidation of the metallic components of the melt, (4) transfer of this exothermic chemical energy and melt sensible energy directly to the containment atmosphere. Let us briefly consider this possible scenario.

Molten corium, consisting of oxides and unreacted zirconium and stainless steel, accumulates within the lower plenum of a pressure vessel, with the reactor coolant system at "high" pressure. The containment building may initially contain a mixture of air, steam and hydrogen released from the reactor vessel during core degradation. The vessel fails in the region of the molten material and the corium is forcefully ejected as a molten jet through a breach in the vessel wall into the reactor cavity.

The melt jet begins to fragment as it exits the RPV by atomization processes (i.e., boundary layer stripping and K-H/R-T instabilities) and possibly dissolved hydrogen gas. The unfragmented melt spreads under its own momentum within the cavity region and accumulates. Gases from the RPV (steam and hydrogen), which follow the melt from the vessel, flow at high velocity through the cavity, and fragment the melt carrying some fraction of the melt as droplets into the subcompartments beyond the cavity.

Thermal and chemical energy transfer occurs between the air/stream and droplets within the cavity and the downstream compartments. Hydrogen is produced as the steam reacts with the unreacted molten metallics (e.g., zirconium). As the melt is carried through the containment compartments, a portion of the droplets will be deposited on structures. As the gas flows between the intervening structures to the upper dome region of containment, only a fraction of the molten material which exits the cavity will be carried with it. This material can then interact with the dome atmosphere. A portion of the hydrogen produced in the compartments and the cavity will also be transported to the dome region. Combustion of this hydrogen would occur if conditions permit. The conditions for a deflagration would depend on the propensity of melt droplets to act as local ignition sites, and the local steam/air volume fractions.

This consensus scenario neglects the likely presence of water both in-vessel or ex-vessel in the reactor cavity or the containment lower compartment. The CWTI tests by Spencer [8-12] demonstrated the possible mitigative effects of water. The water coolant (a much lower density liquid) would more easily be dispersed by the gases exiting the RPV and prolong the blowdown. It could mix with the melt. In fact any steam generation from this dispersed melt-water interaction would enhance local mixing and dispersal of the melt. The concern from such an interaction is the added generation of hydrogen locally and transport to oxygen-rich locations in containment. Also the amount of water mixed with melt would alter the pressure/temperature of subsequent discussion; i.e., reduce or increase it depending on the mixed mass proportions.

INITIAL AND BOUNDARY CONDITIONS

Probably the greatest uncertainty in the DCH process involves the initial conditions and boundary conditions at the time of RPV failure:

(a) the melt mass available for ejection as well as its temperature and composition,
(b) reactor coolant system pressure,
(c) reactor pressure vessel (RPV) lower wall failure size/mechanism,
(d) available water mass at the time of RPV breach as well as location and temperature.
(e) geometry of the cavity and downstream containment compartments.

The available melt mass, m_f, available is the key initial condition that must be known to estimate the pressure and temperature loads on containment. The initial temperature, T_f, and composition of the melt determines the thermal and chemical energy per unit mass available from melt quench and metal oxidation in the gas. Given the mass, the total available energy release is

$$E_f = m_f \left[u_f(T_f) - u_o(T_{ref}) + \Delta i_{reac} \right] \quad (1)$$

where u_f is the melt specific internal energy and Δi_{reac} is the heat of reaction for the oxidation of the metallic components, where the reference temperature can be 273K. For the corium melt simulant used in the CWTI tests by Spencer et al., [8-12], E_f/m_f is 3.3MJ/kg for an initial estimated temperature of 3100K. For the iron-alumina thermite simulant used in the DCH SURTSEY tests, by Tarbell et al., [16-18], E_f/m_f is 7.6 MJ/kg for a similar temperature. In theory this temperature limit is due to the boiling of the iron metal phase under test conditions (0.1 MPa). However, the thermite reaction is incomplete (as expected) and heat losses may reduce the temperature (2800K) and this available energy. For a prototypic melt of urania, zirconia and zirconium metal (50% unoxidized) this energy is 3.25 MJ/kg for a temperature of 2800K. One should note from this simple calculation that the iron-alumina simulant has about twice the thermal energy content per unit mass than corium simulant or a particular melt composition. Thus, from a scaling perspective one must be careful to consider the effect of the simulant chosen. If iron-alumina is used the proper scaling would be on a melt volume basis not a mass basis since its density is about one-half of that of corium.

The range of reactor coolant system pressures is important in determining the driving potential for the melt dispersal by the exiting gas. Actually it is the combination of the driving gas pressure and the RPV failure hole size which determines the flowrate of gases out of the RPV which may disperse the melt. In the previous DCH "scenario," failure is assumed to occur locally in the vessel wall at an instrumentation tube weldment and grow due to melt ablation of the hole to diameters of 0.1 to 0.5m [4].

If one considers the conceptual picture of melt ejection advanced in current DCH analysis (e.g., Ref. 16-18) the melt first exits the RPV breach followed by the high pressure gas. In this case the dominant entrainment and dispersal mechanism is thought to be due to hydrodynamic instabilities and boundary layer stripping of the liquid melt as the gas flows over it in the cavity. A dimensionless grouping of variables (product of the Weber and Froude number) sometimes called the Kututaladze number, Ku, is a measure of this dispersal threshold

$$Ku > 3\pi \tag{2}$$

where

$$Ku = \frac{\rho_g V_g^2 L}{\sigma_f} \tag{3}$$

where

$$L = A = \frac{\sigma_f}{(\rho_f - \rho_g)g}, \tag{4}$$

and ρ is the density of the fuel melt, f, and gas, g, v_g is the gas velocity, σ_f is the fuel surface tension and g is gravitational acceleration. One can see the dependency on the initial conditions to determine a dispersal threshold if one rewrites the criterion in terms of the gas mass flowrate out of the RPV.

$$Ku = \frac{\dot{m}_g^2 L}{\rho_g A_H^2 \sigma_f} > 3\pi \tag{5}$$

where A_H is the RPV failure hole area and $\dot{m}_g$ is the gas mass flowrate which is a direct function of the RPV pressure. One notices that the pressure cutoff can be derived by finding the conditions for the gas mass flowrate when Ku falls below 3π for a given RPV failure hole diameter and cavity gas density (containment pressure). A sample calculation for Surry is given as an example in Figure 1 and 2. One should note that this whole dispersal picture is for the cavity only, since the "scenario" has the melt accumulating there as a pool before gas entry and neglects the effect of cavity water as a dispersal/quenching mechanism. These realistic complications will be considered in the following section. Finally, one should consider scaling the cavity dispersal based on this dimensionaless parameter for dispersal as well as possible effects viscosity [20,31],

$$N_\mu = \mu_f / (\rho_f \sigma_f A)^{1/2} \tag{6}$$

and μ_f is the melt viscosity. Typically if the melt is above its liquidus temperature N_μ is quite small and viscosity considerations are negligible. It is when the melt has begun to solidify (slurry formation) that this may drastically change the phenomena.

The availability of water and its location and temperature is quite important. As the CWTI test results suggest, (Table 1) the DCH atmospheric heating efficiency was low (CWTI 1-6, 11, 12) even when water was only in the lower compartment. The dispersal of water and mixing with the melt by the exiting gases or the steam produced from melt quenching would be governed by a similar scaling if the water initially exists as a pool. However, because of its steam production, water dispersal and droplet fragmentation would occur more readily with the water droplets following the gas flow. The scaling of this process would be much more similar to a dispersed FCI and this is considered later.

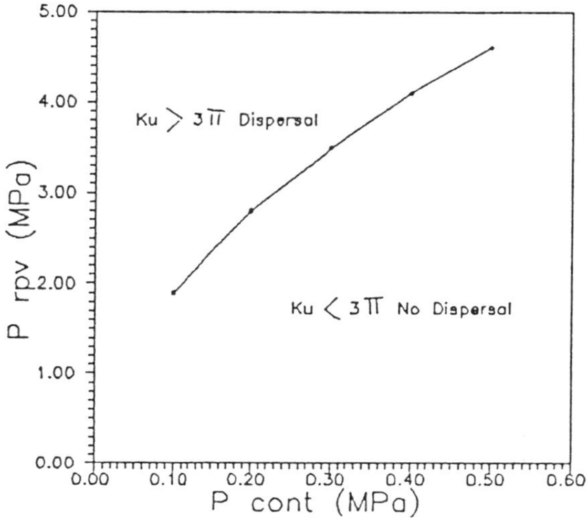

FIGURE 1. Melt dispersal threshold for Surry (D = 0.29M)

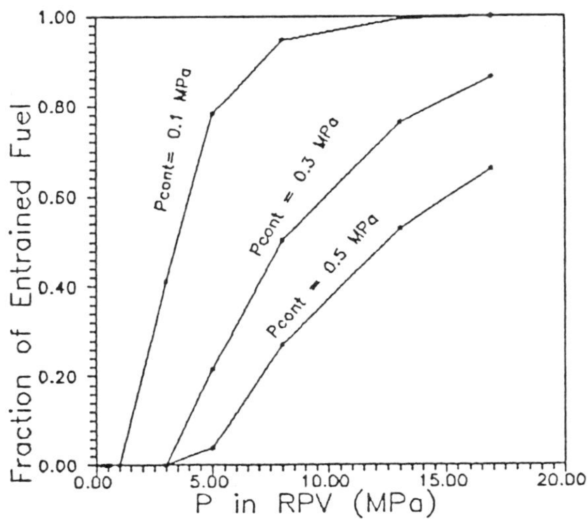

FIGURE 2. Melt dispersal fraction for Surry (D = 0.29M)

The final boundary condition to consider is the geometry of the cavity and the containment compartments. Although this is an important consideration it is difficult to make general statements about its effect. This is because of its multi-dimensional nature and scaling difficulties. IDCOR analyses [7] classified various reactor-containment configurations based on ability to trap melt and reduce its dispersal. However, Tutu and Ginsberg have shown by analyses and by simulant experiments [20,31,34-35] that simple classifications may be misleading and dispersal can occur in quite different geometries (e.g., Zion and Surry cavities). Modelling and scaling must be done on not only geometric scale but also on the relative length scale of the droplets to the structure (which can be distorted) and the time-in-flight for heat and mass transfer within a compartment.

DCH PHYSICS

Let us focus on the DCH scenario by considering each conceptual stage separately and reviewing what is known from past experiments and analyses and what the implications are for understanding the physics and scaling issues. The Zion and Surry reactor-containment configurations are used as example cases [5] for this discussion.

Expulsion of melt from RPV

If the RPV is at high pressure during the in-vessel meltdown, as the melt accumulates and then enters the lower plenum region, the possibility exists for RPV lower plenum wall failure and melt discharge. Consider the scenario previously discussed where the melt accumulates on the lower plenum wall and the RPV failure occurs at the in-core instrumentation tubes. This situation neglects the presence of water and considers the melt to be discharged, driven by the RPV gases (steam and hydrogen), with the gases passing through the hole near melt depletion. Uncertainties center on the time of RPV failure relative to the start of melt accumulation and the number of tube failures which may occur. Even if only one tube failure occurs the melt ablation of this hole could be rapid. The timescale for melt discharge is on the order of a few seconds; about 4 sec for two-thirds of the Surry core for an average hole diameter of 0.3 m for an RPV pressure of 17 MPa. The timescale of gas discharge obviously depends on the initial pressure but is on the order of a little more than half a minute for an isentropic blowdown; e.g., 45 sec for Surry RPV at 17 MPa for a mixture of steam and hydrogen.

This discharge is not completely sequential. Ejection of melt from the vessel would continue until the depth of the molten pool is shallow enough to allow entrainment of gas into the melt, at which point "gas blowthrough" would occur. It is expected that at this point a period of two-phase melt ejection would begin under conditions of two-phase flow at the vessel breach. The two-phase period is believed to be characterized by a process of atomization of the discharging melt at the breach. Pilch summarized correlations to predict the droplet sizes produced during atomization processes [36]. It has not , however, been verified under prototypic DCH accident conditions. Upon exhaustion of the melt supply from the vessel the blowdown of steam and hydrogen from the vessel would continue until the gas supply is exhausted. Figure 3 gives a qualitative picture of this.

One should note, however, that this initially stratified situation (a melt pool with gas above) may not occur due to timing of RPV failure and melt accumulation on the RPV lower plenum wall. For example, a major fraction of the melt may still be in the core region at the time of RPV breach. There would probably be a multi-phase, multi-fluid blowdown throughout the melt explusion (Figure 4). One can estimate a lower bound time for such a blowdown by a homogeneous frozen flow model [37].

$$\dot{m}_{total} = \frac{A}{\upsilon_m} [\gamma X_{go} R_g T_{go}]^{1/2} \left(\frac{2}{\gamma+1}\right)^{1/2} \quad (7)$$

where
$$\upsilon_m = (1-X_{go})\, \upsilon_\rho + X_{go} \upsilon_{go} \eta^{-1/8} \quad (8)$$

for
$$(1-X_{go})\, \upsilon_1 / X_{go} \upsilon_{go} << 1, \; \eta = \left(\frac{2}{\gamma+1}\right)^{\frac{\gamma}{\gamma-1}} \quad (9)$$

where λ is the ratio of specific heat for the gas mixture, Rg is the gas constant for the mixture

X_{go} is the stagnation quality in the RPV
T_{go} is the stagnation temperature
υ is the specific volume of the gas, g_0, and a liquid 1.

For similar Surry conditions to those quoted above the complete discharge time would be about 2.5 minutes; i.e., about 5 times as long. If one now considers the more realistic case of slip between the phases the gases could exit faster than the melt and depressurize before all the melt is discharged. Now this results has some important implications in that the melt would enter (and possibly be dispersed) over a much longer time frame, with perhaps a smaller total mass. This allows for the possibility of more heat transfer to the cold walls and water, which reduce pressure loads in containment. This possibility would be even more important if one considers the likely presence of water in-vessel. This water inventory would be from residual water in the lower plenum and anxiliary water from accumulator discharge. If the RPV failure mode remains unchanged in its presence, water would simply add to the liquid inventory and further slow down the melt ejection during gas depressurization. The second consideration is that the likely presence of water in-vessel will alter the vessel failure mode. This is quite likely in my view and has not been addressed at all in past analyses or experiments.

Melt Dispersal From Cavity

Consider now the second stage of the DCH scenario where the melt accumulates in the cavity from the RPV and may be dispersed from it by gases from RPV blowdown and steam production. Melt droplets produced from the jet breakup and atomization processes would be directed from the vessel into the cavity. These droplets would probably lose their identities and would be absorbed into the bulk of the melt which would be accumulating at this time in the cavity. More detailed mechanistic understanding of these melt transport mechanisms would enable one to track the droplets formed from these mechanisms through the reactor cavity.

Tutu, et al., [20] provided visual observations of the behavior of the melt within the reactor cavity in small-scale experimental involving low-temperature melt simulant materials under conditions which might be prototypic of high-pressure accident conditions. These experiments, indicate that the melt simulant enters the reactor cavity, splashes and spreads in all directions and covers all the surfaces of the cavity prior to "gas blow-through." Gas, which then follows the melt into the cavity, appears to

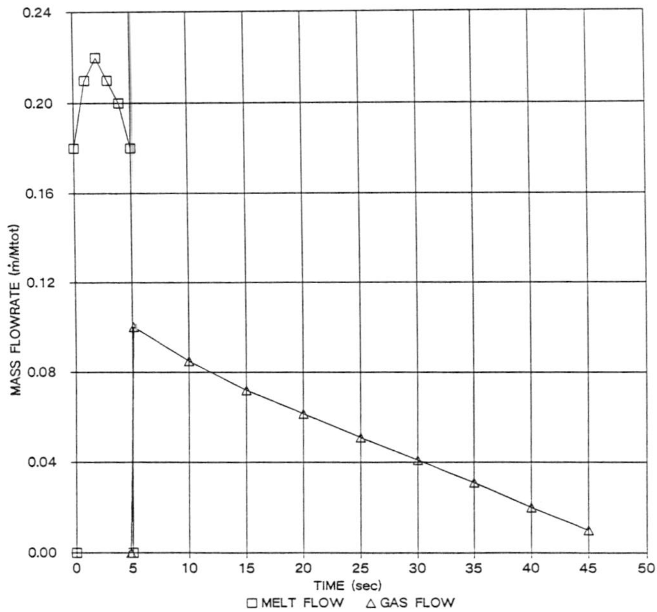

FIGURE 3. RPV melt blowdown (sequential)

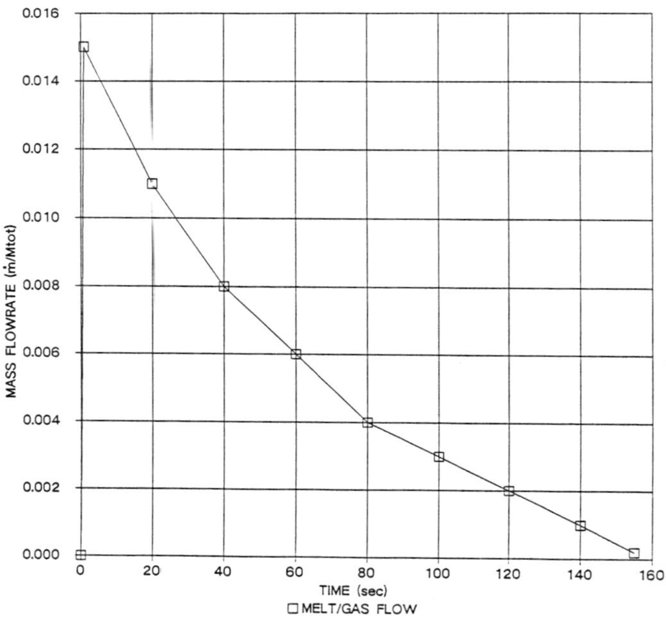

FIGURE 4. RPV melt blowdown (two-phase)

strip the melt from films along the walls, entraining the liquid into the gas stream. The gas carries the droplets out of the cavity exit. The flow patterns within the reactor cavity, deduced from the movies, suggest the existence of a complex, recirculating, three-dimensional flow field within the cavity. The applicability of these observations to prototypic accident conditions, of course, may be influenced by the experimental simulation limitations, particular entrance conditions. This picture of the melt flow pattern presented by these simulant tests cannot be directly verified by experiments with real materials [11,12,21]. Experimental measurements have been limited to pressure and temperature histories as well as qualitiative visual evidence of the chamber at the beginning of melt dispersal. Even x-ray movies of the CWTI tests [9,10] only show qualitatively the general features of the dispersal event. Post-test debris analysis is the only other available means of verification. All of these techniques suggest the phenomena is similar (in the absence of water).

Experiments (simulant and real material) performed to date use scale geometric models of the Zion, and Surry cavities in the US [9,12,20,21,22] and the Sizewell plant in the UK [38]. The experiments related to the US cavities were designed to simulate DCH melt ejection conditions. The experimental conditions of Tutu [20] were chosen to simulate accident conditions where the vessel pressure is initially about 7 MPa and the vessel breach diameter about 0.4m. Experiments related to the Zion cavity geometric scale were performed at SNL, ANL and at BNL. These experiments range from 1/10th to 1/42nd in linear scale. In all but one (DCH-1) of the HIPS and Surstey experiments performed to date, an excess of 90% of the melt was dispersed from the Zion concrete cavities (see Table 1-3). In the ANL CWTI-5 and -6 tests smaller fractions (61% and 30%, respectively) of the melt were dispersed. This was attributed to freezing of the melt on steel cavity surfaces. This result is reasonable since the surface area to volume ratio is one-third as large in the 1/10th scale SURTSEY DCH tests. Also, because the cavity in the ANL tests was made of steel instead of concrete the ability to freeze the melt as a crust was much greater. From this standpoint the empirical experimental results on dispersal sweepout fractions from all the current tests are consistent. In the BNL tests [20] water was used as the melt simulant for the Zion work. The results showed complete sweepout (-100%) of the water. Scaling analyses by Tutu indicate that these results using water overestimate sweepout. The BNL Surry experiment showed nearly complete dispersal using a Wood's metal melt simulant.

The available particle size data from experiments which were designed to simulate high-pressure DCH conditions indicate that the mass-mean particle diameters were in the range of approximately 0.5-1.0 mm [21,22]. Measured particle diameters ranged from aerosol-size (microns) to several millimeters. These data represent distributions resulting from samples of post-test collection of a sampling of debris which were ejected from the cavity during the transient melt ejection process. The droplets were created from the melt by a combination of mechanisms: melt jet breakup by the action of dissolved gases, atomization, entrainment processes from liquid films, and Weber breakup. The droplets were swept from the cavity by the flowing gas and followed trajectories to the nearest wall undergoing collisions and coallescence until solidifying and settling.

It is important to emphasize that the measured debris distributions are the only empirical evidence available of melt droplet sizes. Thus, because detailed mechanisms are not known precisely, one must consider the modelling of these phenomenon on similarity of dimensionless groups between scales.

Because of the apparent consistency of dispersal tests results, and the lack of complete model for melt jetbreakup and dispersal from the cavity, dimensional analysis of the phenomena and scaling to full-scale conditions is important. The importance is not only for the sweepout fraction, but also the melt debris size distribution. Tutu, [20,31] and Pilch [30,36] have performed a scaling analysis based upon a one-dimensional, dispersed flow regime in the reactor cavity. A set of dimensionless parameters have been derived with which the data from small-scale experiments are extrapolated to prototypic conditions. The experimental data and the scaling analysis suggest that the Zion and Surry reactor cavities would retain little melt under conditions of high-pressure melt ejection. Nearly all the melt would be dispersed into the downstream regions of containment.

The main dimensionless parameter is the Kututaladze number, as mentioned previously, although Tutu has identified others of varying importance. The use of this parameter indicates that these experiments all have conditions, in the absence of water, where $Ku \gg 3\pi$; thus, this dispersal threshold is easily satisfied. One must then verify that the rate of entrainment is large enough to cause dispersal during the time of gas blowdown. If one considers Kelvin-Helmholtz instabilities as a major mechanism one finds this rate is much larger than the discharge rate of the gas and melt from the cavity.

The final consideration is the effect of water within the cavity. The only experimental evidence of this effect was from the CWTI tests of, [9-12,-39]. Two qualitiative observations were noted from visual observations at the cavity exit, x-ray imaging of the cavity volume and pressure and temperatures histories. First, when the melt entered the cavity a major fraction of the water pool was ejected ahead of the melt as a slug and was dispersed as droplets in the downstream compartment. The remainder of the water rapidly fragmented into droplets and mixed with the melt droplets dispersed in the cavity. This had a couple of observable effects, the gas phase pressure increased due to the FCI, and the gas temperature remained near the water saturation temperature at the given pressure. This suggests the cavity dispersal produced a great deal of steam from fuel-coolant heat transfer and logically this probably aided fragmentation and dispersal of the melt mixed with water. The second observation was that a mixed melt-water dispersal occurred even at relatively low RPV injection pressures (see Table 1 for CWTI, 1,2 and 4). This is consistent with the observation that steam produced from the dispersed FCI can drive the dispersal and fuel melt-coolant mixing. This also implies with an FCI present and associated steam generation, fuel melt and coolant liquid fragmentation would not necessarily be directly linked to the RPV injection pressure. Thus, a dispersal threshold based on $Ku > 3\pi$ may not be relevant anymore as the towing pressure decreases.

One last point should be emphasized; i.e., continued dispersal in downstream compartments in containment would be determined by a modified Froude number based on the fragmented size of the melt or water droplets,

$$Fr = \frac{\rho_g V_g^{\ 2}}{(\rho_f - \rho_g)\, g\, D_{frag}} \qquad (10)$$

where D_{frag} is the fragmented diameter of the melt (or coolant liquid) generated within the cavity or upon its exit. If $Fr \ll 1$ then deentrainment would occur, and $Fr \gg 1$ would imply continued dispersal and transport. The diameter, particularly, in the presence of water would be determined by Weber breakup in the continuous gas phase within the cavity

$$D_{frag} = \frac{We_{crit}\sigma_f}{\rho_g V_{rel}^2} \quad (11)$$

where V_{rel} is the relative velocity between melt (or coolant liquid) and gas locally within the cavity. These considerations suggest that scaling of the phenomena with water present is quite different than without it, and has not been extensively considered.

Containment heat and mass transfer

Once an estimate has been made of the amount, temperature and composition of melt exiting the RPV and dispersed within the cavity and subsequent containment compartments, one must estimate the pressure and temperature use in the atmosphere. One could classify current analyses in three categories: (1) dimensional, analyses, (2) control volume, 1-dimensional, multi-region analysis, (3) detailed multi-fluid, multi-dimensional analysis (see Table 4). Current detailed DCH modelling was preceded by bounding, simplified, single-cell models. [14,28,35]. The DHEAT, EBAL and CONTAIN codes, were used to perform thermodynamic equilibrium calculations of the DCH containment loading. The DHEAT model, for example, assumes an adiabatic containment, complete chemical reaction of the available iron and zirconium with oxygen and thermal equilibration of the gas atmosphere, reactants and reaction products. Original calculations performed on the time frame of the CLWG did not include the incremental loading attributable to the combustion of hydrogen generated in-vessel prior to the failure of the reactor vessel. Even without this additional loading mechanism, the bounding calculations led to predictions of extremely large pressure loadings, compared with estimates of containment failure pressure, when large quantities of melt were assumed. The results quoted in Table 4 included H_2 chemical reactions in the atmosphere as a point of comparison.

Table 4. DCH Heat-mass transfer analysis models

Model	Ref.	Fluids	Dimension	ΔP Rise (MPa)	Remarks
DHEAT	[14,28]	Thermodynamic Equil.		1.3	Specified Chem. Reactions
EBAL	[35]	Thermodynamic Equil.		1.3	Specified Chem. Reactions
HMC	[26]	Corium Mixture Water Gas Mixture	Cont. Vol;1-D	1.07	Specified Chem. with FCI analysis
CONTAIN	[14,28]	Corium Mixture Water Gas Mixture	Cont. Vol;1-D	1.07	Specified Chem. Reaction with Aerosol Transport
DIRHEAT	[25]	Corium Mixture Water Gas Mixture	Cont. Vol;1-D (5 Volume Max.)	1.05	Specified Chem. Reaction
KIVA	[40,42]	Corium Mixture Gas Mixture	Fin. Diff; 3-D		Specified Chem. with La Grangian tracking

* This is the CLWG Standard Problem 2 for Surry [5] with 100% melt release.

Subsequent to these thermodynamic calculations one finds that the large majority of the analysis has been done with these control volume, one-dimensional models (e.g., CONTAIN) [14,28]. Only recently has there been more detailed analysis using the spray code, KIVA [40,42]. In all of these analyses the focus is to determine the pressure and temperature rise as a function of time and location within the containment for a given set of initial conditions. Recently a large parametric study was completed by Williams et al [28] using CONTAIN. The objective of this summary is not to review all these analyses. Rather the focus is on the type of modelling and how it compares to present experiments and possible DCH loading in containment. In this way the major effects can be demonstrated and illustrate possible scaling considerations.

The current modelling of melt droplet/atmosphere/water heat and mass transfer is best described by a conceptual picture (Figures 5 and 6). As Figure 5 indicates the flow regime assumed in all the analysis is a dispersed droplet regime. The continuous phase is the gas mixture and the melt and water droplets are carried along by the gas flow. In all the control volume modelling to date the basic assumption is that the droplets are of a small enough size that they move at about the same velocity as the gas phase. In some of the computer modelling allowance is made to input a velocity difference parametrically [14,25,28]. KIVA tracks the droplet field as a set of Lagrangian material volumes ("parcels") with their own momentum equation.

The ability for the droplets to settle out of the atmosphere is accounted for in different ways. DIRHEAT considered a simple gravitational settling model similar to that used in aerosol mechanics without collisions or coalescence. CONTAIN has a user input trapping fraction which can be made to behave like gravitational settling or parametrically allow for much more droplet deentrainment and trapping due to structures. KIVA considers the complete physics of droplet collisions, coalescence and settling with 3-D wall interations. Obviously, this complex physics approach requires sophisticated models which are only in the early stages of development, with an extremely sparse data base.

The final point is the size of droplets in the gas mixture. The approaches taken are based on the sophistication of the momentum balance for each tool. For control volume formulations the droplet sizes are input as constant values based on the initial breakup in the cavity due to hydrodynamic instabilities; i.e., Weber breakup. For KIVA the original model did not have a breakup model but one was included in the initial work by Marx [40] based also on Weber breakup.

Given the size and relative velocity of the droplets in the gas field one can then use these values to estimate the heat and mass transfer from the melt to the surrounding gas and coolant droplets (Figure 6). The figure indicates that simple empirical correlations are used to model the heat flux from the fuel drop to the surrounding gas and from the gas to coolant droplets. This heat flux is due to convective heat transfer as well as radiative heat transfer. The mass transport of oxygen (or steam) to the metallic melt surface for oxidation is again described by convective diffusion transport through the gas mixture and liquid or solid diffusion within the particle. Chemical reactions considered are prescribed based on

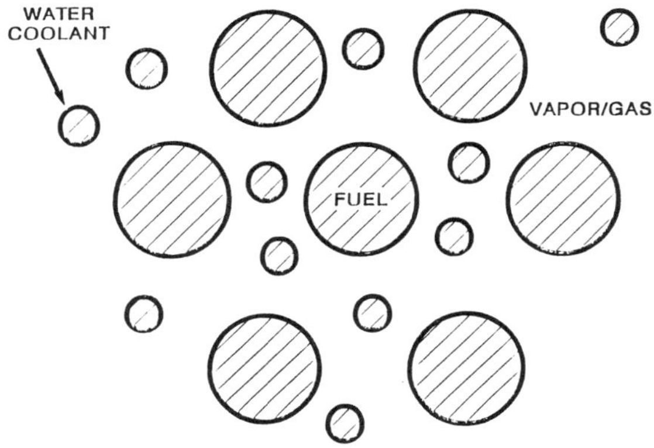

$$D_i = \frac{\sigma_i We_{crit}}{\rho_g V^2_{rel_f}} \; ; \; We_{crit} \doteq 12$$

FIGURE 5. Conceptual picture for high pressure dispersal

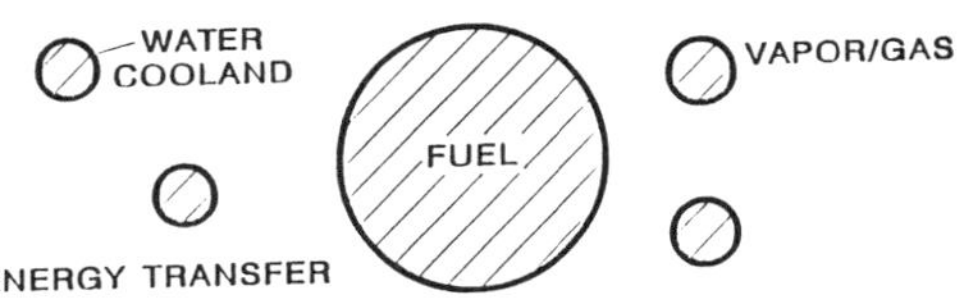

- RATE OF ENERGY TRANSFER

  $\dot{q}'' = Nu\,(k_v/D)\Delta T$ $\quad Nu = f\,(Re, Pr)$

- RATE OF REACTION IS CONTROLLED BY DIFFUSION

  $\dot{m}''_{oxid} = N_m \rho_m (Di/D)\,\Delta Y$ $\quad N_m = f(Re, Sc)$

  D_i - DIFFUSION OF OXIDANT THRU OTHER GASES

- REACTIONS CONSIDERED IN SERIES:

$$Zr + O_2 \rightarrow ZrO_2 \qquad Zr + 2H_2O \rightarrow 2H_2 + ZrO_2$$

$$2Cr + \tfrac{3}{2}O_2 \rightarrow Cr_2O_3 \qquad 2Cr + 3H_2O \rightarrow 3H_2 + Cr_2O_3$$

$$2Fe + O_2 \rightarrow 2FeO \qquad Fe + H_2O \rightarrow FeO + H_2$$

$$3UO_2 + O_2 \rightarrow U_3O_8$$

FIGURE 6. Conceptual picture of direct heating model

the most chemically reactive species proceeding first until the oxidant or metal is consumed, then the next most reactive species begins. Although this heat and mass transfer modelling is simple it is considered to be sufficient since the predicted characteristic sizes of the melt and water particles are small (0.001 to 1mm), and thus the kinetic rates of transfer are large.

Probably the major uncertainty in regard to heat and mass transfer is not the local kinetic rates, but the bulk mixing of the fuel melt droplets with the gas oxidizer (and water) and the heat and mass loss to the surrounding walls. The analysis of mixing requires subdivision of the containment into compartments and perhaps smaller computational volumes (perhaps multi-dimensional) while the effect of containment structures is inherently complex and multidimensional if precision is required. Let us use sample calculations for SURRY and analysis of SURTSEY tests to illustrate the importance of these various effects.

Consider the pressure history for SURTSEY test DCH-1. Figure 7 shows the experimental data from DCH-1 for the early times up to the peak pressure and as the heat is transferred to the cold walls. Although about 20 kg of iron-alumina melt was injected from the simulated pressure vessel a much smaller mass actually was dispersed in the atmosphere (~ 10 kg). The mass decrement was found primarily as a frozen mass on the ceiling of the SURTSEY chamber. This was unexpected and attributed to the trajectory of melt directly out of the cavity, because of the "chute extension," causing immediate impact and freezing on the chamber walls. This effect is a good qualitative illustration of mass loss due to trapping on structure; however, the quantitative amount may be misleading because the surface was a large steel surface (versus concrete) and the surface to volume ratio is larger than full scale by an order of magnitude. Using the model DIRHEAT one can observe the relative effect of this initial dispersed mass difference in peak pressure relative to adiabatic conditions. It has a much larger effect on the predicted peak pressure when compared to gravitational settling or the diabatic wall effects. The absence of chemical reactions between melt and atmosphere has a large effect and this is used here to illustrate a lower bound given the mass dispersed.

These same qualitative effects can be demonstrated in a simulation for SURRY (Figure 8). In this case the containment was modelled as three control volumes (cavity, lower compartment and upper dome) with a sequential blowdown of the melt and gases. One notes that the mitigative effect of diabatic boundaries and gravitional settling are of second order because of the smaller surface area to volume ratio. The initial mass dispersed still has the largest contribution to peak pressure variability.

Melt-water interaction and hydrogen generation

During the high pressure meltdown and a DCH scenario the steam and hydrogen would eject the melt from the RPV. This steam inventory along with steam generated by the FCI by the dispersed melt-water droplet interaction can chemically react with the melt droplets to produce hydrogen. This is an adverse consequence of any metal-water interaction due to possible condensation. The benefit of enhancing melt-water mixing by the presence of water is that the melt will be quenched and the melt sensible and chemical heat will be absorbed by the latent heat of steam production. The net result of steam production causing quenching and hydrogen generation with possible combustion may increase or decrease the containment pressure load depending on the relative masses of melt and water dispersed in the

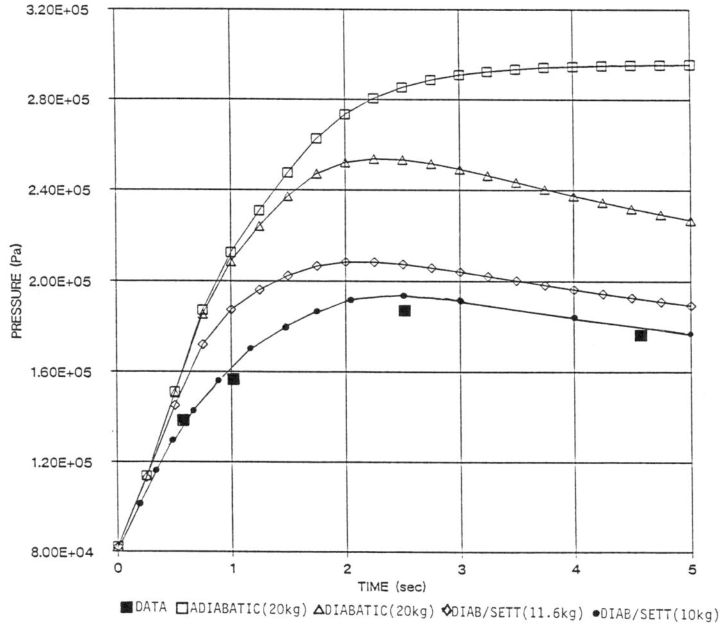

FIGURE 7. DCH-1 analysis

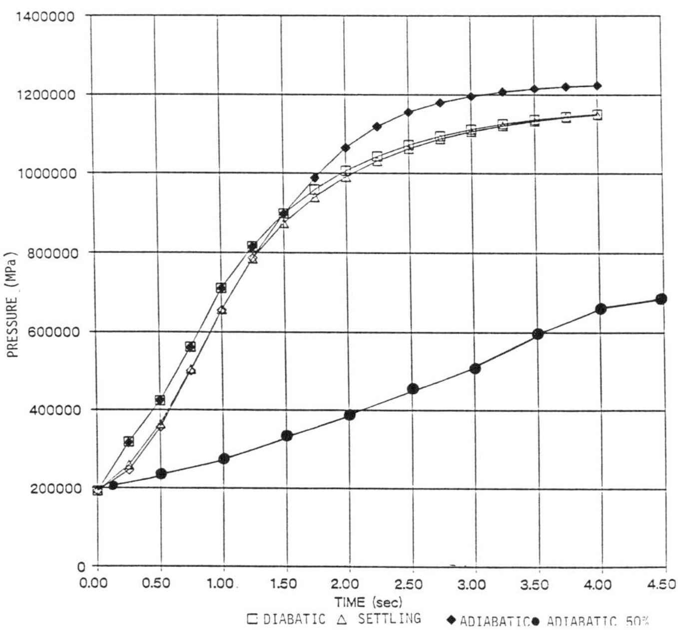

FIGURE 8. Surry 3-volume

atmosphere.

This is demonstrated in Figures 9 and 10, where the amount of water dispersed and mixed with the melt was varied parametrically. These calculations used the DIRHEAT control-volume model for the SURRY containment (3-volumes) for all the melt dispersal. As the figures indicate the addition of water initially causes the pressure to increase. This is because the amount of water is too small to quench the melt completely. The water vaporized becomes the working fluid and causes the total pressure to increase. As the amount of water is increased further the melt is quenched and the water droplets are not totally vaporized; thus the total pressure decreases. This maximum occurs at about equal volumes of melt and water. The behavior for hydrogen generation shows the quenching effect of the water regardless of the water mass. Now this observed behavior should not be surprising and is not unique. With the addition of water in the droplet form, assumed to be mixed with the melt droplet the DCH process becomes an FCI dispersed in the gas atmosphere. One of the fundamental observations from the thermodynamic analysis of the FCI is that the coolant has its maximum effect as a working fluid at equal liquid volumes. It is thus imperative to assure ar ample water supply if water is to be used as a mitigative agent during the accident.

Now these observations are based on simple control volume analyses where the droplet diameters are small enough ($D << 1$ mm) that local transport and kinetic rate effects do not limit the heat and mass transfer. These transport models (see Figure 6) assume a dispersed droplet flow regime in the reactor cavity, and that the steam supply in the local is effectively unlimited. The reaction rates are governed by either steam diffusion through a steam-hydrogen layer external to the droplets or by diffusion through the molten liquid. If the melt droplet begin to solidify droplet side diffusion would control. The relative velocity of steam and droplet is either specified by assumption or through solution of a momentum equation for a droplet in a specified gas flow field. Because of these expected small droplet sizes due to hydrodynamic droplet fragmentation, these results are consistent for all the models considered.

These simplified models such as those described above may overestimate the extent of metal-steam reaction. Several potentially-important effects are not considered. Mass transport rates may be affected by the presence of neighboring droplets. If the local concentration of droplets in the cavity is large, then local steam depletion could occur, thereby either turning off or reducing the rate of reaction. Finally, the presence of diluent oxides should have some effect on the reaction rates. These effects could be accounted for in more detailed models of the interaction scenario, if this becomes the controlling uncertainty.

REACTOR SAFETY ISSUES

For a reactor safety application the DCH phenomenon must be understood insofar as the threat to containment integrity is minimized. Based on the work of NUREG-1150 [6] this implies that the DCH process, if it occurs, does not cause early containment failure. Probabilistically, this means that the likelihood of early containment failure would be low (e.g., $p_f << 0.1$) given the uncertainties in the process. From this perspective, one ought to first research and understand better the portions of the DCH process with greatest uncertainty. From the past discussion one may conclude that the issues of greatest uncertainty involve;

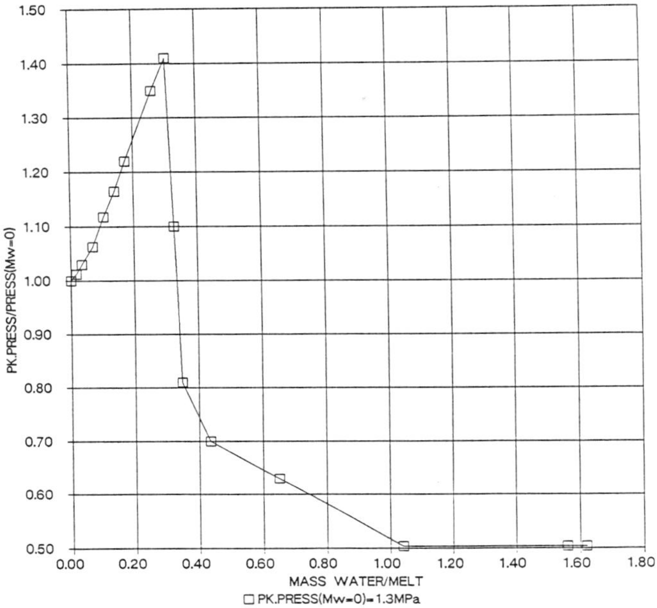

FIGURE 9. Water effect on DCH (Surry-Adiab.)

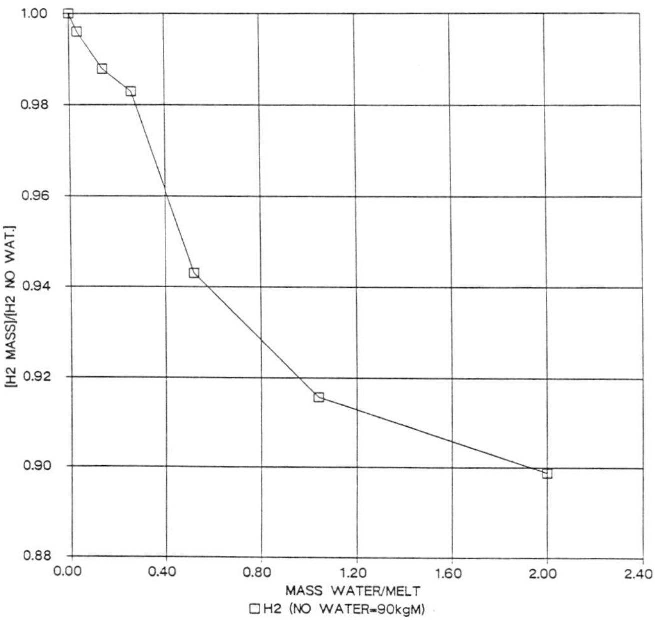

FIGURE 10. Water effect on H2 prod. (Surry)

1) the available melt mass, composition and temperature,
2) the RPV lower plenum wall failure mechanism
3) the reactor coolant system pressure at RPV failure
4) the effect of water on the DCH process
5) the geometry effects on dispersal and pressurization.

The first two are initial conditions which would affect the loading of containment and the final three, while also initial and boundary conditions, are parameters one may be able to control by design modification and/or operator actions. Let us conclude by discussing each briefly.

As the parametric results indicate the melt mass and its initial temperature/composition are the most important determinants of containment loading from DCH. Also because of the lack of complete understanding of in-vessel meltdown processes these are still quite uncertain. Most past research work on in-vessel melt progression has focused on in-core processes. This is quite reasonable if one is considering accident management and methods to arrest core degradation. However, it seems that the in-vessel melt progression in the lower plenum of the RPV is more pertinent to the DCH process and many other ex-vessel loading conditions (e.g., MCCI). In this area, there is a relative scarcity of data and understanding. Melt pouring through water and interacting with structure are the two most relevant phenomena to consider that would affect melt mass and its thermodynamic state upon exiting the RPV lower plenum. In conjunction with this is the failure mechanicm of the RPV lower plenum wall. In the ZPSS it was proposed that melt accumulation on the RPV wall would lead to a thermally induced failure of the welds on the in-core instrumentation tube penetration. However, at TMI about 20-30 tons of molten material poured into the lower plenum (at relatively high pressure) and no thermal failure occurred. The contrast of the ZPSS analysis and the observations at TMI should be resolved by some study of the possible processes in the lower plenum and their effect on the RPV vessel wall. By focusing on this phase of the accident it appears one would obtain a clearer picture of the melt state at RPV failure as well as the lower plenum wall-failure mechanism.

The RPV pressure at the time of vessel breach can be a large determinant in the amount of melt that may be dispersed into the containment. Past and current work on the subject focuses on the possibility that the reactor coolant system may fail in some other location and depressurize due to natural convection effects. This would reduce the potential for melt dispersal. In conjunction with this, research has begun into specifying emergency operating procedures (and possible design modifications) which would instruct the operator to depressurize the RPV and preclude melt dispersal. These investigations seem to be a possible way to reduce DCH uncertainty by knowing the RPV pressure better. However, one omission in trying to determine a DCH pressure cutoff is the effect of water. If water is in the lower plenum of the RPV or in the cavity region, the FCI that would result as the melt is ejected would promote dispersal (and melt quenching). The results of the CWTI tests are a good indication of this (Table 1). In the first tests the simulated RPV pressure was only a few bars (2-5), yet the sweepout fraction was large. Thus any investigation of a DCH pressure cutoff should consider the role of water.

In addition, the blowdown history of the RPV during melt ejection ought to be focused on more closely. In the sample calculations previously discussed, it was noted a multi-phase blowdown would be at least three to five times longer than the sequential melt-gas blowdown currently modelled and tested. More thought ought to be given to the likely blowdown history from the RPV, to determine if a multi-phase blowdown is more likely. This may reduce the driving pressure at the end of the blowdown for dispersal and it would also lengthen the time for heat transfer to cold walls to reduce the pressure load.

The effect of water has been alluded to throughout this paper, with the focus being on the role of water to quench the melt and reduce the pressure load, even considering hydrogen generation. To summarize research must focus on verifying the mitigative effect of water on DCH under various circumstances, and develop management strategies to assure that it is available during the accident in the proper proportions to maximize mitigation; e.g., as Figures 9 and 10 indicate the relative mass of water mixed with melt must be above some threshold to assure its quenching effect. Because of the lack of test data on the mixing of melt and water in the gas phase more efforts should be focused in this area. Because scale effects are so important, one might consider developing a test matrix at two different geometric scales to observe the mixing effect and verify the scaling of the phenomena.

The final area of uncertainty is the effect of geometry on melt trapping or dispersal, both within the cavity or in the lower compartments of containment. Of all the areas of uncertainty identified this has received the most investigation both experimentally and analytically. It is also the most complicated effect to reduce uncertainty, because of the complex structures within containment and their multi-dimensional effect on melt dispersal and DCH phenoma. Past discussions have indicated that work has focused on melt dispersal from the cavity [29,35,41] as well as more detailed analysis of melt droplet transport within containment [14,28,-40,42]. The work involving KIVA [40,42] is probably the most complicated analysis possible and can involve 3-D computations. The level of sophistication here approaches that in internal-combustion engine design. One might wonder about the necessity of this detail given the uncertainty of many of the other parameters. The general effects of structure on the DCH process need to be considered but too narrow of a focus may not significantly reduce uncertainties.

The main observation from this paper is that a well balanced approach on major uncertainties must be the focus for reactor safety issues and future research. It is hoped this discussion can help in developing such an approach.

REFERENCES

[1] USNRC, Reactor Safety Study - An Assessment of Accident Risks in US Commercial Nuclear Power Plants, WASH-1400, NUREG-75/014 October 1975.

[2] W.B. Murfin et al, Report of the Zion (Indian Point Study: Volume I, NUREG/CR-1410, SAND 80-0617, Sandia National Laboratories August 1980.

[3] USNRC, Preliminary Assessment of Core Melt Accidents at the Zion and Indian Point Nuclear Power Plants and Strategies for Mitigating Their Effects, NUREG-0850, VI November 1981.

[4] Zion Probabilistic Safety Study, Commonwealth Edison Corp 1981.

[5] USNRC, Containment Loads Working Group Report, NUREG-1079 December 1985.

[6] USNRC, Reactor Risk Reference Document, Draft Report NUREG-1150 February 1987.

[7] R.E. Henry et al., Technical Documentation Issue Resolution, 85-2, IDOOR July 1985.

[8] B. Spencer et al., "Sweepout Threshholds in Reactor Cavity Interactions," ANL/LWR/SAF 82-1, Argonne National Laboratory April 1982.

[9] J.J. Sienicki, B. Spencer, "Debris Dispersal in Reactor Material Experiments on CWTI in an Ex-Vessel Geometry," Proc of NHTC, Niagra Falls, NY 1984.

[10] B. Spencer et al., "Corium/Water Dispersal Phenomena in Ex-Vessel Cavity Interactions," Proc Intl Mtg LWR Severe Acc, Cambridge MA August 1983.

[11] B. Spencer et al., "Results of Scoping Tests on CWTI in Ex-Vessel Geometry," AIChE Symposium Vol of NHTC, V79, No 225, Seattle WA 1983.

[12] B. Spencer et al., "Hydrodynamics and Heat Transfer Aspects of Corium/Water Interactions," ANL/LWR/SAF 85-1, Argonne National Laboratory 1986.

[13] Nuclear Regulatory commission, Containment Loads Working Group Report, NUREG-1079 1985.

[14] K.D. Bergeron et al., "Contain Calculations of Containment Loading of Large Dry PWR's," Nucl Engr Des, V90, p153 1985.

[15] M. Pilch et al., "Preliminary Calculations of Direct Heating of a Containment Atmosphere Airborne Core Debris," Sandia National Laboratories, NUREG-CR-4555, SAND 85-2439 July 1986.

[16] W.W. Tarbell et al., "The Role of Ex-Vessel Interactions in Determining the Severe Accident Source Term," Proc Intl Mtg LWR Severe Acc, Cambridge MA August 1983.

[17] W.W. Tarbell et al., "High Pressure Melt Streaming Program Plan," SAND82-2477, NUREG/CR-3025, Sandia National Lab August 1984.

[18] W.W. Tarbell et al., "Melt Expulsion and Direct Heating in Realistic Plant Geometries," Proc Intl ANS/ENS Topical Mtg on Thermal Reactor Safety, San Diego, CA February 1986.

[19] W.W. Tarbell et al., Pressurized Melt Ejection Into Scaled Reactor Cavities, SAND 86-0153, NUREG/CR-4512, Sandia National Laboratories October 1986.

[20] N.K. Tutu et al., Debris Dispersal from Reactor Cavities During High Pressure Melt Ejection Accident Scenarios," BNL-NUREG-52147, Brookhaven Nat'l Lab July 1986.

[21] W.W. Tarbell et al., Results from the DCH-1 Experiment, SAND 86-2483, NUREG/CR-4871, Sandia National Laboratories June 1987.

[22] W.W. Tarbell, et al., DCH-2: Results from the Second Experiment in the SURTSEY Test Facility, SAND 87-0976, NUREG/CR-4917, Sandia Nat'l Lab January 1988.

[23] K.D. Bergeron et al., "Development and Application of the Interim Direct Heating Model for the CONTAIN Computer Code," Proceedings 14th WRSI Mtg, Gaithersburg MD October 1986.

[24] M.L. Corradini et al., "A Preliminary Analysis of the ANL Corium-Water Experiments," UWRSR-19, University of Wisconsin June 1984.

[25] M.L. Corradini, et al., "A Model for Direct Heating in a LWR Containment," UWRSR-35, University of Wisconsin, February 1986.

[26] M.L. Corradini et al., "Surry S2D Severe Accident Containment Loads Calculations using HMC," UWRSR-34, University of Wisconsin January 1986.

[27] T. Ginsberg et al., "A Direct Heating Containment Vessel Interaction Code (DHCIVIC) and Prediction of the SNL SURTSEY Tests," Trans Amer Nucl Soc, V53, Washington D.C. November 1986.

[28] D.C. Williams et al., Containment Loads Due to Direct Containment Heating and Associated Hydrogen Behavior, NUREG/CR-4896, SAND 87-0633, Sandia National Lab May 1987.

[29] J.J. Sienicki, B. Spencer, "A Multifluid, Multiphase Flow and Heat Transfer Model for the Prediction of Sweepout from a Reactor Cavity," Proc Mtg on Multiphase Flow and Energy Conversion, Miami 1987.

[30] M. Pilch et al., The Influence of Selected Containment Structures on Debris Dispersal and Transport Following High Pressure Melt Ejection from a Reactor Vessel, SAND 87-0940, NUREG/CR-4914, Sandia National Laboratories September 1988.

[31] Tutu, N.K., Safety Research Programs Sponsored Office of Nuclear Regulatory Research, Quarter Progress Report July 1-September 30, 1985, NUREG/CR-2331, BNL-NUREG-51454, Brookhaven National Laboratory, Upton NY March 1986.

[32] Baker, L., "Droplet Heat Transfer and Chemical Reactions During Direct Containment Heating Proceedings of the International ANS/ENS Topical Meeting on Thermal Reactor Safety, San Diego, CA February 1986.

[33] Williams, D.C., et al., "Impact of Chemical Phenomena in Direct Containment Heating," Proc at American Chemical Society Severe Accident Chemistry Symposium, Anaheim, CA September 1986.

[34] T. Ginsberg, "DHCVIM-A Direct Heating Containment Vessel Interactions Module," AICHE Symposium Series, V83, No 257, p 347, August, 1987.

[35] T. Ginsberg, N.K. Tutu, "Progress in Understanding of Direct Containment Heating Phenomena in Pressurized Water Reactors," Proc. of NUPTHO-3, Seoul, Korea, November 1988.

[36] M. Pilch, W.W. Tarbell, "High Pressure Ejection of Melt from a Reactor Pressure Vessel--The Discharge Phase, "NUREG-4384, SAND 85-0012, September, 1985.

[37] R. Henry, H.K. Faushe, "Two-Phase Critical Flow through Short Pipes, Nozzles and Orifices," Journal of Heat Transfer, V90, p 183, May, 1971.

[38] MacBeth, R.V. and R. Trenberth, "Experimental Modeling of Core Debris Dispersion from the Vault Under a PWR Pressure Vessel,: AEEW-R 1888, UKAEA-Winfrith December 1987.

[39] B.W. Spencer et al., Results of EPRI-ANL DCH Investigations and Model Development," Proc. of ANS/ENS Conf Termal Reactor Safety, Avignon, FR, October, 1988.

[40] K.D. Marx A Model for the Transport and Chemical Reaction of Molten Debris in DCH Experiments, NUREG/CR-5120, SAND 88-8213, May, 1988.

[41] R.T. Nichols, W.W. Tarbell, "Low Pressure Debris Dispersal from Scaled Reactor Cavities," Proc. of THD ANS Winter Annual Mtg, Washington, D.C., November, 1988.

[42] W.W. Tarbell, et al., "Direct Containment Heating and Aerosol Generation During High Pressure Melt Ejection Experiments," Proc of Thd ANS Winter Annual Mtg, Washington, D.C., November, 1988.

Ex-vessel Melt/Water Interaction

H. E. UNGER
University of Bochum
FRG

M. BÜRGER
IKE University Stuttgart
FRG

E. F. HICKEN
GRS Garching
FRG

1. INTRODUCTION

Ex-Vessel melt-water interactions can take place within cavities outside of the reactor pressure vessel (RPV) in both cases, boiling water reactors (BWR) and pressurized water reactors (PWR).

These events are summarized under the phenomenology of "Cavity Phenomena". This means, it is dealt with a PWR-cavity geometry, which consists of a cylindrical region beneath the vessel, connected to an annular region around the RPV and other compartments within the lower part of the containment.

There are several possibilities for deliberately or unavoidably at the time of the vessel failure partly or totally water-filled cavities, which will be discussed later.

After core meltdown, and vessel failure, there also exist many possibilities (scenarios!) for an interaction of molten or solid (resolidified) core material with water.

In case of a melt-water-interaction within the cavity (water filled cavity at the time of a RPV-failure or entrance of water into a "dry" cavity - containing molten material - later in the course of the accident), there are a number of events to be considered:

1) <u>Quenching.</u> Most likely, molten or solid core debris will simply be quenched (like in case of the TMI-2 accident). In this case, the speed of the quenching has to be assessed, e.g. with respect to the steam production rate within the compartment (a large amount of cool water may act as a heat sink and mitigate or prevent direct containment heating/DCH).

2) <u>Material Transport and Oxidation.</u> In case of rapid steam production, radioactive material may be resuspended, dispersed and relocated. By means of heating and steam production, chemical processes, like oxydation may be enhanced.

3) Steam Explosion. In case of ex-vessel steam explosions, the cavity may be exposed to very highly transient or quasi-static pressure loads. The possible "scenario" is manyfold, the situation is complex and not easy to be assessed in detail. However, there are many limiting or mitigating effects, which make a large explosion, endangering the integrity of the containment, nearly impossible.

It is to be expected, that only part of the water reacts with the melt, this effect will be dealt with in this paper. Starting from the region, where an interaction takes place, pressure pulses penetrate the compartment. There are several effects, mainly

- acceleration of water slugs, which may exert a high load on the concrete structure (water hammer effect) and/or prevent rapid venting of the cavity in case of pressurization due to rapid steam production;

- acceleration of solid debris (generation of missiles) during highly transient expansion (explosion), which must be included into containment integrity considerations;

- impact of shock waves on structures.

This paper deals predominantly with the last item.

2. GENERAL CONSIDERATIONS

The safety assessment of the interaction between melt or debris and water outside the reactor pressure vessel is important for several reasons:

1. The steam production and the resulting pressure increase can threaten the integrity of structures or even the containment. A simple estimate shows, that the production of 50 - 100 tons of steam which may result if a completely molten core is cooled down, is by itself no threat to a PWR or BWR containment. It is, however, plant specific if surrounding walls are threatened due to a pressure build-up.

2. The slug resulting from a fast melt/water interaction can threaten the integrity of containment structures. This behavior is very plant- and sequence-specific.

3. The steam produced can eject melt or debris into the containment atmosphere (this process is often called "direct containment heating").

4. Debris or melt can impinge side walls or pipes as a result of a fast melt/water interaction.

5. Chemical reaction of hot metals with water, producing hydrogen, can take place.

6. Due to the fragmentation of melt or debris, fission products may be released and relocated.

While the designs of the reactor pressure vessels and the coolant systems have become relatively similar for PWR's, respectively BWR's, containments show a variety of designs.

With regard to a fast melt/water interaction to be investigated, the designs may be grouped as follows

1) <u>Initially dry cavities</u>
   In some designs, within minutes, water from a compartment outside the cavity below the RPV may flood the melt or mixture of melt in concrete.
   For some other designs, flooding will occur several hours after the contact between melt and concrete as a result of a melt-through of side walls. In this case, in general, the mass flow will be small. PWR's of KWU-type have a dry cavity, initially.

2) <u>Cavities with initially</u>
   - a few m^3 of water
   - several hundred m^3 of water
   - a few (2 - 4) thousands of m^3 of water (e.g. condensation pools of BWR's).

The water could be saturated.

The assessment of a melt/water interaction will be further complicated, if "obstacles" influence the melt or debris escaping the RPV. This is the case for containment designs with a floor between RPV and water pool or for BWR's, which have a thick steel plate for radiation and control rod ejection purposes below the RPV.

For completeness it should be added that for ruptures, for instance in the lower vessel head at system pressure above the accumulator pressure setpoint, the inventory of the accumulators (about 200 m^3) is expelled onto the melt within several ten seconds as soon as the pressure starts to drop below this setpoint. Additional variables are the melt temperature and the mass flow.

Of further importance is the melt temperature. In contrary to an in-vessel core slump, melt or debris being expelled out of the reactor vessel have been in contact with water in the lower plenum with exception for lower head pipe breaks. In addition for BWR's, the melt or debris may have molten lower plenum structures. For BWR's, control rod material will lower the melt temperature. Therefore, melt being expelled from the RPV is a mixture of fuel, steel (more for BWR's (up to several tons) than for PWR's), zirconium (more for BWR's than for PWR's) and control rod material. It is therefore assessed that the melt temperature usually is considerably lower than the temperature of a molten core.

The mass flow from the RPV into the cavity below is evidently mainly dependent on the break size. The gross failure of the lower head, possible for low pressure scenarios but also not ruled out for high pressures, seems to have a low probability of occurence. Therefore, the failure of a penetration of the lower head or the lower portion of the RPV seems to be a major failure mode. Usually the diameter of a penetration is about 5 - 7 cm. This results initially in a mass flow of 1 -2 tons/s at high system pressure and several 10 - 100 kg/s at low pressure. This value decreases with pressure, but increases with time due to an expected ablation of the steel vessel wall.

3. ESTIMATION OF SHOCK WAVE LOADS ON CONTAINMENT WALLS DUE TO FCI

3.1 The Thermal Detonation Event.

In order to estimate the structural load in the cavity, the course of severe FCI's has to be investigated. Such highly energetic FCI's are very unlikely. It is assumed, that their probability to occur is lower than 10^{-1} to 10^{-2} in case of a severe core meltdown accident [1], because the necessary conditions for such a large interaction, like matching of triggering time, premixing conditons, melt temperature, ambient pressure level etc., must coincide.

Nevertheless, such cases have been investigated at the IKE [2]. In order to assess the structural loads and the possibility for a containment damage even for such extremely rare cases, the results of these investigations, performed with the steady state and transient thermal detonation models STADEMO and IDEMO [2, 3], are presented and discussed here (Table 1 gives an overview on the sequence of events involved and the models used; see also Fig. 1).

For the further investigations, an example of an "ideal premixture" of coarsely fragmented melt, distributed uniformly in a sufficiently large volume of coolant is assumed.

The most important parameters of this mixture and the melt (corium) are given in Table 2.

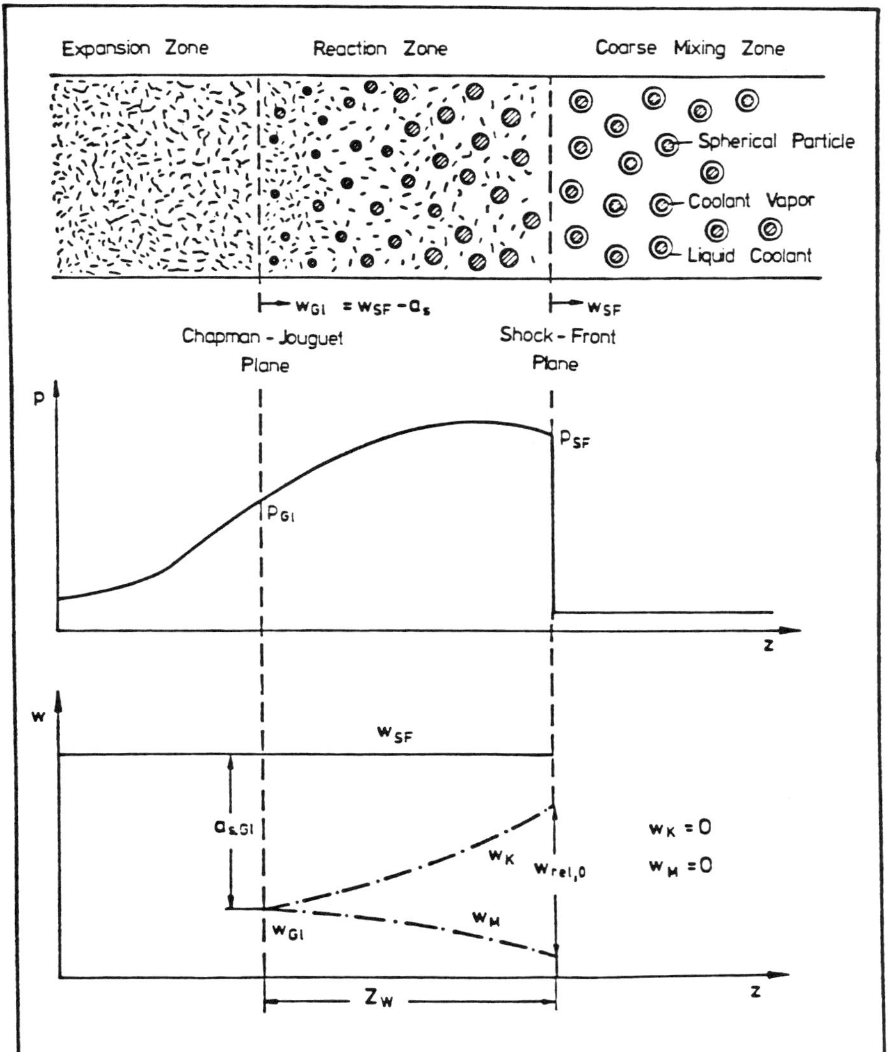

FIGURE 1: Qualitative description of a thermal detonation

TABLE 1: Sequence of Events and Models

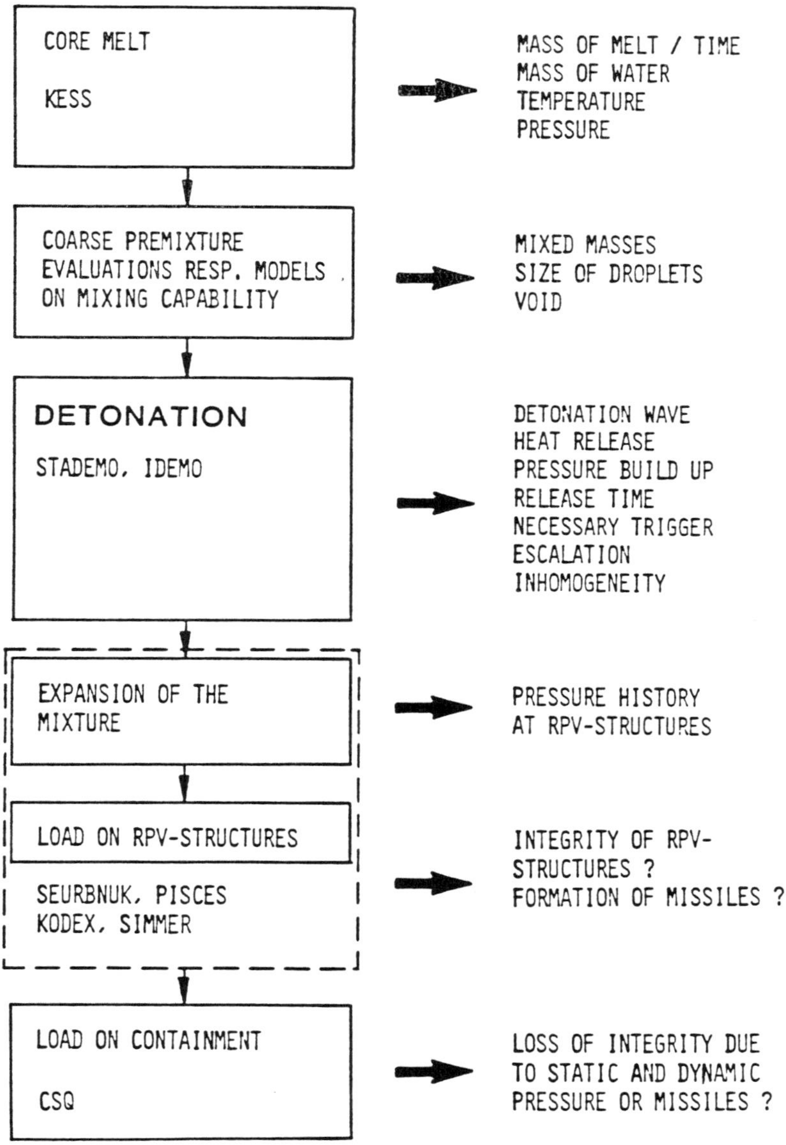

TABLE 2: Data of Premixed Corium-Water Array (from [2])

A **Mixture Data**

Mass ratio melt/coolant:	f_{mk}	$= 5$
Melt drop radius:	0,5 cm	
Steam volume fraction in the water:	γ_{st}	$= 0.3$
Temperature of the melt:	T_m	$= 3000$ K
Temperature of the coolant:	T_c	$= 370$ K
Ambient pressure:	P_U	$= 0.1$ MPa

B **Corium Data**

Density:	ρ_m	$= 8610$ kg/m^3
Specific heat:	c_p	$= 0.54$ kJ/kg K
Melting temperature:	T_{fm}	$= 2673$ K
Latent heat of fusion:	H_f	$= 200$ kJ/kg

If the extension of this premixed zone is large and a trigger pulse is high enough and given at a suitable position (e.g. center of the array), a detonation wave can be triggered. For the example chosen the maximum energy release under subcritical conditions at the end of the wave (i.e. under corresponding CJ-conditions) is given, according to [2] in Table 3.

TABLE 3: Example of Heat Transfer from the Melt into the Coolant, Caused by a Thermal Detonation Wave.

Shock front pressure:	61 MPa
Maximum pressure:	63.5 MPa
Shock front velocity (wave velocity):	539 m/s
Wave length up to CJ-plane:	4.8 cm
Corresponding time span:	89 μs
Degree of fragmentation:	13.5 %
End pressure (CJ-Plane) = pressure in the highly pressurized mixture:	10.3 MPa
Corresponding temperature of the water (saturation temperature):	585.7 K
Thermal energy transfered into the water/steam region:	2.11×10^5 J/kg melt, respectively 1.56×10^6 J per kg of fragmented melt

It can be seen, that only a small portion of the enthalpy of the melt is discharged within the relevant detonation time (see also Fig. 2); but that the shock front pressure can still be very high, depending on the energy of the trigger (Fig. 3 shows a few additional examples for the shock front pressure development in the reaction zone, calculated by the transient detonation model).

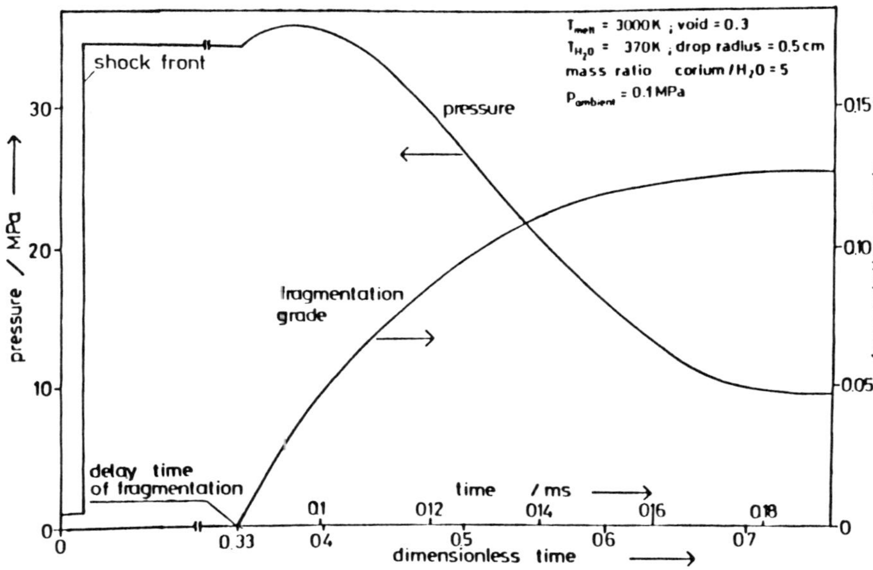

FIGURE 2: Theoretical results of steady state detonation model (STADEMO) for a corium-water case: pressure and fragmentation development in detonation wave

Cases with supercritical pressure levels at the end of the wave seem to be possible (see e.g. V 10* = VJ from Table 4.2 of [2]). The chance for an escalation in such cases depends on the triggering energy, resp. - pressure. In relation to accident scenarios, this means, that it has to be checked, if the triggering pressure, resp. - energy, is sufficiently high enough to initiate a very strong shock wave [3].

In general, the pressure pulse, generated by the extended and escalated triggering shock wave is high, but of short duration. Therefore it is not so much a major threat to the containment or compartment structure than the effect of the expansion of the highly pressurized reaction volume, left behind by the shock wave (see Chapter 3.3). This will be discussed next.

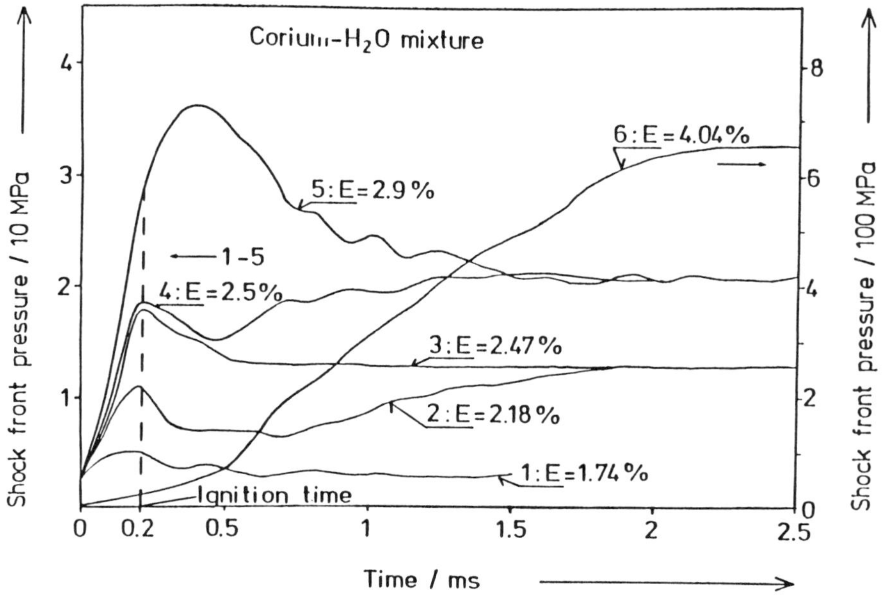

FIGURE 3: Histories of the shock front pressure calculated by the transient model for different values of E (expressed in percent of the initially fragment melt mass)

3.2 Expansion of the Highly Pressurized Mixture, Left Behind by the Shock Wave

Once the shock wave has passed through the premixed melt-coolant-region, then, caused by fragmentation and extremly rapid heat transfer, resulting in extensive steam production, a highly pressurized system is left behind.

Due to the pressure drop between this region and the surrounding area, a rapid expansion takes place.

For the case of an adiabatic expansion of a mixture as given in Tables 2 and 3 to an environmental end pressure of 0.1 MPa, the mechanical energy release calculated by Müller [4] amounts to 5.1×10^4 J per kg melt in a 13,5 % fine fragmented mixture; this corresponds to $3.8 \cdot 10^5$ J/kg of fine fragmented melt.

For comparison: Calculations performed by Benz [5] based on the Hicks-Menzies-model with isothermal fuel-coolant mixture adiabatic expansion showed 7.3×10^5 J per kg of corium, if expanded to 5 MPa (not accounting for some minor differences in the corium data).

With the assumption of a volume of 4 m^3 of the premixed region - which seems to be extremely conservative - an "engineering judgement", based on data given above can be performed with respect to the efficiency of the reaction and the magnitude of pressure loads to be expected. From the data given by Table 2, it follows that the melt volume is 1.16 m^3 and therefore the mass of melt appr. 10 t. Using the data from Table 3, this means with the degree of fine fragmentation of 13.5 % - that 1.24 t of melt discharged 2.09×10^9 J thermal energy to the coolant. The mechanical work portion, in case of an expansion to 0.1 MPa is therefore - based on the entire mass of fuel - $5.1 \cdot 10^4$ J/kg $\times 10 \cdot 10^3$ kg $\approx 5.1 \cdot 10^8$ J. This corresponds to a degree of conversion of thermal in mechanical energy of appr. 24 % for the expansion event behind the initial shock wave. It can be assumed, that the remaining energy, which still is stored in those parts of the melt, which are only coarsly fragmented, will be transfered to the coolant much slower. If the estimation of the degree of efficiency is based on the heat, which is discharged from the entire amount of fuel into the coolant, then it is 24 % $\times$ 0.135≈3.2 %. This value seems to be in the order of experimental results. It has to be noted however, that the configuration assumed here is based on very conservative data. This means, it can be considered as an estimate of an upper bound for cases to be discussed.

Next, pressure loads on structures are assessed.

3.3 Pressure Load on Walls

According to the previous chapters, two different pressure load cases have to be considered:

1) Direct impact of the detonation (shock) wave, which caused fine fragmentation and more intensive melt-coolant-mixture.

This wave is generated and increased to full strength within the reaction zone geometry (see Fig. 4). It migrates afterwards into the surrounding water area, the non-flooded area of the cavity and hits the surrounding structure (RPV, walls, etc.). Although the pressure peak is very high, the total impact on the solid concrete wall structure is assumed to be of limited effect, because the time duration is short (e.g. pressure drop from 60 MPa to 10 MPa within ≈ 0.1 ms. There is also a strong decreasing effect of the surrounding environment to be considered (without consideration of weakening effects, estimations show that the momentum $\int p dt$ of the wave would amount to appr. 2500 Pa s; exponential pressure decrease at the end of the wave assumed).

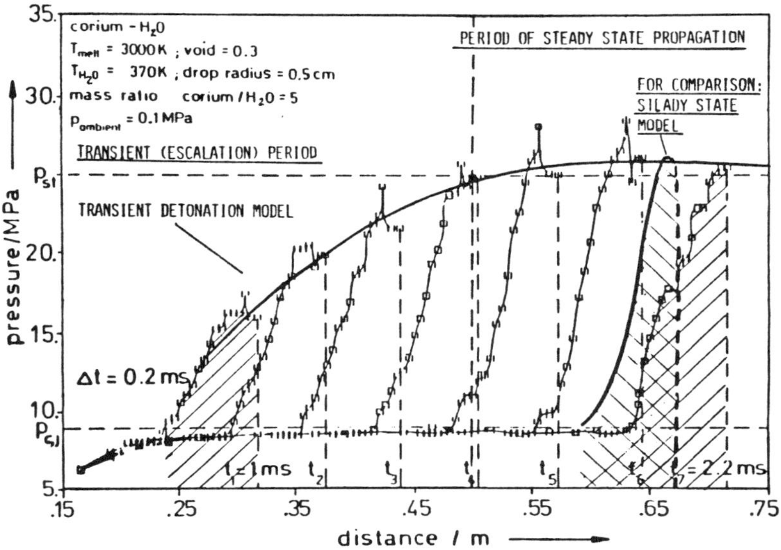

P_{ST} = SHOCK FRONT PRESSURE FROM STEADY STATE MODEL

$P_{C,J}$ = CHAPMAN, JOUGUET (END) PRESSURE FROM STEADY STATE MODEL

FIGURE 4: Escalation into steady state propagation calculated with the transient detonation model (IDEMO)

2) The "left-behind", highly pressurized volume tends to rapid expansion and pressurization of the surrounding region. This event is estimated as follows:

It is assumed, that the mechanical energy produced by expansion of the highly pressurized region is predominantly transfered to form a new shock wave, which is extending through the steam- resp. gas-volume. The main pressure load of the surrounding walls is expected to result from this secondary expansion shock wave (the effect of water present around the reaction zone is not yet considered here).

Table 4 shows mechanical energy values calculated for different volumes of the initial reaction zones, if these zones expand after premixture and partial fine-fragmentation (caused by means of a detonation wave) into a volume of - as an example - 235 m^3 (cylindrically assumed model compartment, e.g. around, resp. below the RPV). This volume is rather arbitrary, as expansion into the other compartments of the containment is to be expected. However, the first compartment is subject to the impact of the detonation primarily; therefore this consideration may give an estimation of the magnitude of effects to be expected.

As seen from Table 4, the volume V_{tot} of the premixed reaction zone is parameterized. Because of the limited size of the compartment to 235 m^3, the expansion is terminated before reaching the end pressure of 0.1 MPa (as previously considered) in all cases investigated here. The fraction of the additionally released energy in case of an expansion to 0.1 MPa would however be limited, as can be seen from Table 4 by means of a comparison of $W_{kg}[kJ/kg_m]$ = 35.....51, with the specific expansion work of 53.4 kJ/kg_m in case of an expansion to 0.1 MPa. The specific expansion work is decreasing with increasing size of the reacttion zone (due to higher end pressures), but the total expansion work is increasing due to the higher amount of heat transferred to the coolant.

TABLE 4: Mechanical Energy Released During Expansion of Highly Pressure Loaded Reaction Zones in a Compartment of 235 m^3 (According to [4]).

Total volume of the reaction zone (in m^3)	1	2	3	4
Volume of the melt in the reaction zone [m^3]	.28	.58	.86	1.12
Mass of melt in kg (density of melt = 8610 kg/m^3)	2475	4927	7378	9643
Specific work released W_{kg} [kJ/kg_{melt}]	51	43	38	35
Total mechanical energy released during expansion to 235 m^3 (W_{tot} [MJ])	126	211	281	344

The maximum value of $W_{tot} = 344 \times 10^6$ J for $V_{tot} = 4.0\ m^3$ is 32 % lower than that estimated before (appr. 5×10^8 J) for the expansion to 0.1 MPa end-pressure.

Now, an impact due to expansion in planar geometry at a distance of 9 m from the reaction center (assumed volume: 4 m^3) is considered. This seems to be a relatively large distance for a compartment geometry; however this example case has been selected in order to assess for some dissipative effects in a rather coarse and global manner. The wave, generated in the cavity is expected to diverge and rapidly decrease with increasing distance from the reaction zone, but may be reflected several times by internal walls within the compartment also.

According to Table 4, because of the limited degree of fragmentation etc., the amount of 9.8 tons of corium has to be compared to 0.5 tons, fully reacting in case of Benz [5]. From this comparison (see Fig. 5), a peak pressure of appr. 2.5 MPa results for the shock front.

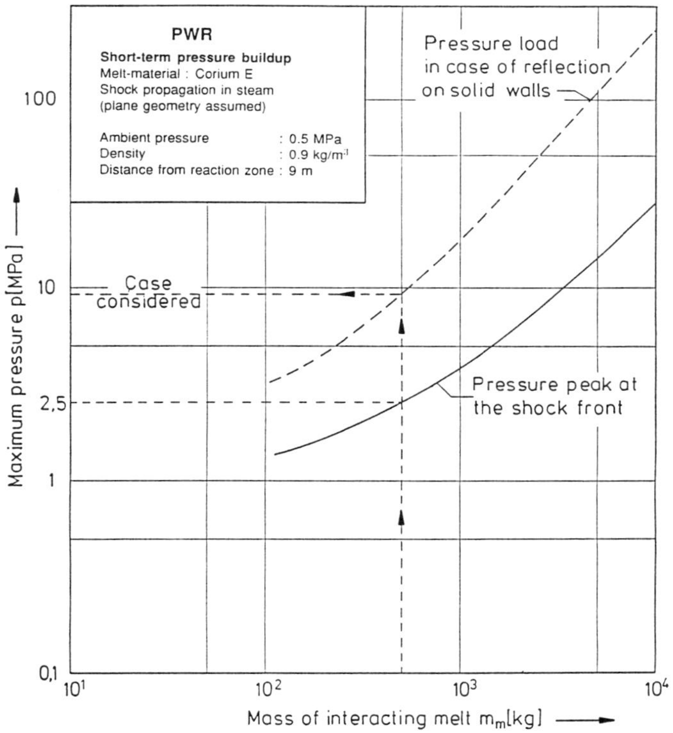

FIGURE 5: Example of a short-time plane wave maximum pressure peak load on a solid wall at appr. 9 m distance (effects of internal structure neglected); from [5]

Pressure peak estimations based on a "TNT-equivalent consideration" and a relation used by Anderson and Armstrong [6] and Cole [7] resulted in pressure levels of the same order of magnitude, namely 8.3, resp. 6 MPa.

The momentum load of the shock wave is even more important.

According to Fig. 6 the half width of the pressure pulse is much longer as in case of the inital shock wave, namely 2.6 ms.

The momentum load of the shock wave is (exponential decay with decay constant λ assumed):

$$I = \frac{p_o}{2\lambda} \quad \text{, using } t_h = \ln 2/\lambda = 2.6 \text{ ms :}$$

$$\rightarrow \lambda = 265 \ \frac{1}{s} \ \text{. With } p_o = 2{,}5 \text{ MPa follows } I = 4.7 \times 10^3 \text{ Pa s.}$$

In case of a reflection at the wall, an extension of the pressure load on the wall by a factor of four (!) results according to [5] (see Fig. 5). In this case, the pressure load amounts to 10 MPa; the corresponding momentum load is almost eight times as high as in case of the detonation wave, discussed in Ch. 3.1 (effects of dissipation and distance not considered). Estimations show, that this value approaches the critical pressure respectively momentum for the compartment walls.

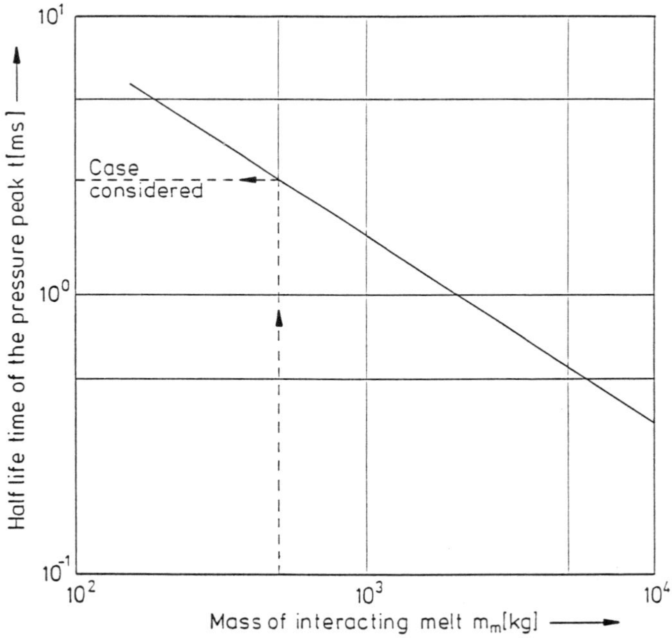

FIGURE 6: Half-life-time of the highly transient pressure load on a solid wall at approx. 9 m distance; from [5] (influence of structure and additional inhomogeneities not considered; data as in Fig. 5)

In general - if at all - much smaller reaction volumes with lower mechanical energy released can be expected. In such cases, the detonation waves generated do not reach their full strenght before they leave the reaction zone. In addition, since the degree of fragmentation is lower in case of smaller waves, it can be expected that the average degree of fragmentation in the reaction zone decreases with decreasing volume.

Since reaction zones of 4 m^3 are very unlikely to exist homogeneously premixed and in addition are unlikely to be triggered at the right moment, the estimations performed here seem to be extremly conservative. Therefore it can be concluded, that formerly undamaged and unloaded (!) reactor compartments are expected to withstand the impact of shock wave caused by ex-vessel fuel-coolant interactions.

Nevertheless, more detailed investigations, especially of secondary effects, like water hammer, or missile formation etc., are necessary.

References

[1] Unger, H., Bisanz, R., Bürger, M., Schwalbe, W., The Role of Steam Vapor Explosions During Core Meltdown of LWR's, *International Meeting on Thermal Nuclear Reactor Safety, Chicago*, August/September 1982.

[2] Bürger, M., Carachalios, C., Cho, S.H., von Berg, E., Kim, D.S. and H.E. Unger, Theoretische und experimentelle Untersuchungen zur Eingrenzung von Dampfexplosionen im Rahmen von Sicherheitsbetrachtungen von Leichtwasserreaktoren, Abschlußbericht zum Forschungsvorhaben BMFT 1500 639 0, Institut für Kernenergetik und Energiesysteme (IKE) der Universität Stuttgart., Report-Nr.: IKE 2 TF-77 (1985).

[3] Karachalios, K., *Ein transientes Modell zur thermischen Detonation*, Dissertation, Universität Stuttgart, April 1986, IKE-Rep.: IKE 2-75.

[4] Müller, K., and Fröhlich, G., Modelle zur Bestimmung der freisetzbaren mechanischen Arbeit bei Wasserdampfexplosionen, Institut für Kernenergetik und Energiesysteme, Universität Stuttgart, IKE 2-72, October 1985.

[5] Benz, R., Schwalbe, W., and Unger, H., Ingenieurmäßige Abschätzung der Energiefreisetzung und des Druckaufbaus bei Explosionen in Reaktorgeometrie, IKE-Report BMFT-RS 206, December 1976.

[6] Anderson, R.P., and Armstrong, D.R., Comparison Between Vapor Explosion Models and Recent Experimental Results, *AICHE Symposium Series 70* (1974) pp. 31 - 47.

[7] Cole, R.H., *Underwater Explosions*, Princeton, New Jersey, 1948.

Resuspension of Fission Product Aerosols from the Boiling Sump

W. SCHÖCK and M. WAGNER-AMBS
Kernforschungszentrum Karlsruhe
Laboratorium für Aerosolphysik und Filtertechnik
Postfach 3640
D-7500 Karlsruhe 1, FRG

1. DEFINITION OF THE PROBLEM

The contribution of aerosols to the source term of core melt accidents in LWRs has been thoroughly investigated by code calculations. Experimental investigations of separate effects and large scale code validation programs as DEMONA and LACE improved the confidence in the codes.

Up to date, these activities were mainly concerned with transport and behavior of aerosols released early in the accident from the core during meltdown and melt through phases. It was found that retention mechanisms inside the reactor building are very effective. For certain accident sequences only very small fractions of the initially released aerosol escape to the environment up to the time of containment failure.

Earlier calculations stopped at the time of containment failure and only rough estimates of additional long term contributions to the overall source term wer made. Now, since source terms have decreased more and more, even very weak additions become noticeable in certain scenarios, and a better assessment of the behavior of aerosols from long term sources is necessary.

In this paper the contribution of resuspended aerosol to the overall source term of low consequence accident sequences will be discussed. For low probability scenarios with large source terms the weak resuspension aerosol source does not contribute much to the overall source term.

Two areas will be covered. First, an assessment is made of the relevance to the source term of aerosol particles resuspended from the sump, particles which were depleted from the airborne state and subsequently collected in the sump water. Second, first results of the REST experimental program will be presented.

2. RELEVANCE OF RESUSPENDED AEROSOLS TO THE OVERALL SOURCE TERM

The resuspension of particles from the sump is a potential long term source of aerosols. All particles depleted from the airborne state are collected in the sump sooner or later. In order that resuspension contribute significantly to the overall aerosol balance, the sump needs to be boiling. This occurs only when sump water ingression on the molten core takes place and when simultaneously long term cooling of the sump water by accident management measures is unavailable. The decay heat sources distributed in the sump are too weak to cause noticeable effects.

The relevant parameters which determine the time dependent rate of resuspension are: surface area of the molten core in the cavity, total mass of sump depending on the scenario, the depth of the pool, decay heat production time function, the thermodynamic state of the atmosphere over the pool (temperature, pressure, saturation), and the amount of incondensible gases produced by core concrete interaction which bubble through the pool.

The available experimental literature data have been reviewed [1]. They all suffer from being not directly applicable to the problem of resuspension in an LWR accident because the experimental conditions were much different from those of a core melt accident. Even the data which are close to our problem show very large differences among different authors, data range over two orders of magnitude. Recently an investigation of droplet entrainment was made under realistic accident conditions [2], in this experiment resuspension from solutions was measured.

Nevertheless, based on the literature data from [1], some estimates of resuspension rates could be made for a 1300 MW PWR. For the long term sump boil off rate a value of 5 kg/s was kept constant throughout the calculation. Assuming a resuspension fraction of 10^{-4}, and taking into account the concentration of materials in the sump, one arrives at a possible range of resuspension rates between 0.5 and 20 mg/s.

As an example of the results Fig.1 shows the mass leak rate from the reactor building as a function of time. The scenario was an FK5 of the German Risk Study which is a late overpressure failure of the containment with no accident filters available. The containment failure occurred at 124 hours at 9 bar overpressure. The initial sump volume of 2000 m^3 was allowed to evaporate completely.

The narrow shaded peaks between 0 and 30 hours and between 124 and 133 hours are leakages of the initially released aerosol from the melting core, all the rest is aerosol from resuspension. It has to be kept in mind that the mass leak rate shown in the figure is the product of volumetric leak rate and concentration. So the mass leak rate of 10^{-4} g/s of resuspended aerosol is to a large part due to the failed containment. The much larger inventory of aerosols at the beginning of the sequence leads to low mass leak rates only because the containment volumetric leak rates are very low before failure. If one would take into account accident filters the mass leak rate of the aerosol after containment failure would be reduced by the retention factor of the accident filters, e.g. 10^3, and the mass leak rate would be in the range of 10^{-7} g/s, still contributing the major part to the source term.

From this kind of assessments it was concluded that, in some scenarios, the resuspension of aerosols from the sump could be either a dominant contributor to the source term or an impact to accident filter systems. Both cases need more accurate information on the release rates and particle size distributions of the resuspended aerosol than is available from the literature.

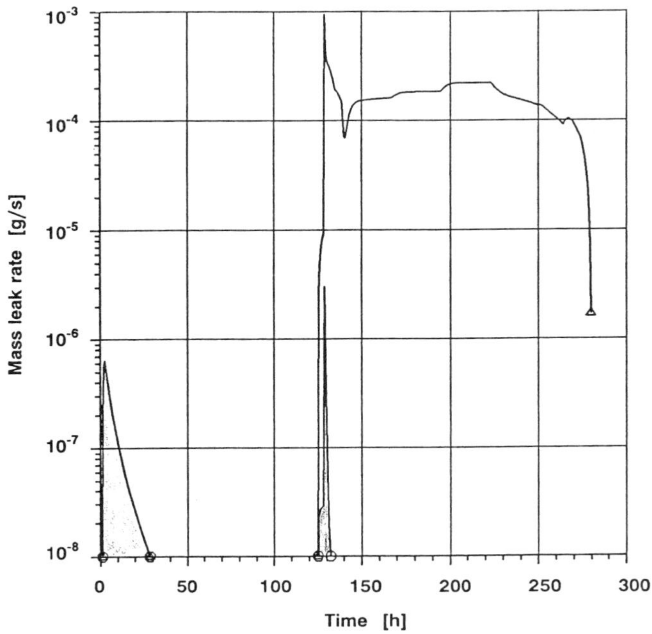

FIGURE 1. Mass leak rates from the containment, with resuspended aerosol

3. EXPERIMENTAL INVESTIGATION OF RESUSPENSION PHENOMENA

3.1 The REST Program

Since 1987, the physics of resuspension is being investigated in the REST (resuspension source term) program conducted at KfK. The program is conducted in the framework of the Shared Cost Action Reactor Safety Programme of Euratom, Ispra. The main aim is to develop a model and computer code, which is generally applicable and comprises all relevant effects, some of which have been investigated separately and independently in the past. To support the model development, separate effects experiments are being conducted in iteration with the model development.

From the phenomenology of resuspension it is kown [3] that bubble bursting at the boiling surface is the mechanical process which ejects droplets into the gas space above the surface. These droplets are carried away with the evaporating steam. Two different fractions of droplets are generated: small film droplets in the size range of a few μm and large jet droplets which are > 100 μm in diameter.

The ejected droplets contain material dissolved and suspended in the boiling liquid. While the concentration of dissolved material in the droplets is equal to that in the bulk solution, the content of suspended material may be much enriched. Because of this principal difference the experimental part of the REST program is structured in three phases, experiments with solutions, experiments with suspensions, and mixed experiments. To date the first phase has been completed.

3.2 The Experimental Procedure

Fig.2 shows a schematic diagram of the experimental setup. The containment vessel has a volume of 3.7 m^3 and is thermostated by two bath type thermostats. The only function of this vessel is to provide the ambient conditions of the experiment and a large enough free volume above the sump unit.

The sump unit is mounted on one of the flanges. It consists of a cylindrical cup with a volume of one liter. The filling tube is simultaneously used for support. Boiling is achieved by electrical heating. The heater consists of a titanium foil of 1 mm^2 cross section, the maximum heating power is 1 kW (18 V and 55 A). The heating is done with a.c. power in order to minimize electrolytic effects

An area specific power density of 250 kW/m^2 is typical for the sump heating during the late phases of a core melt accident in a German PWR of the KWU 1300 MW class. The sump unit is capable of reaching this level, although during most of the experiments lower power is used.

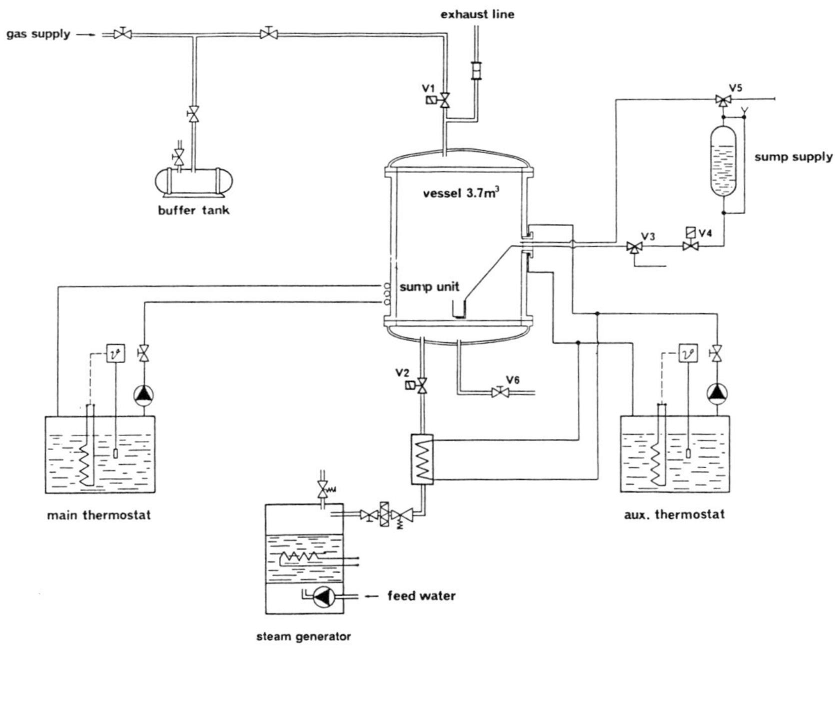

FIGURE 2. Schematic diagram of the REST facility

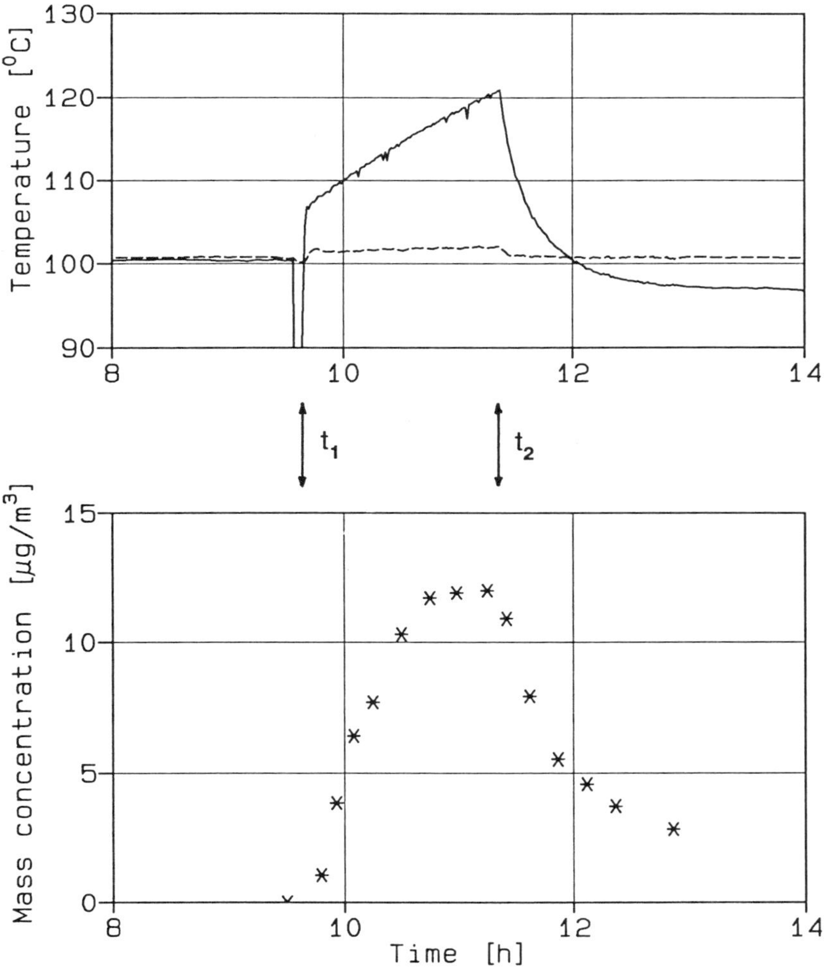

FIGURE 3. Example of experimental results of a resuspension test

During experiments the sump level has to be kept constant. The linear boiloff rate at 1 kW power is .125 mm/s. Therefore, it is necessary to regulate the liquid level in the sump. This is done by automatic feed controlled by closely spaced maximum and minimum level sensors at the sump unit.

Aerosol measurement is done by filter sampling. This method, although somewhat tedious, turned out to be the most reliable of all the methods tested. During the experiments photometric on-line monitoring is used for control purposes. The aerosol mass on the filter samples is determined by different analyses, the particle size distribution is obtained from scanning electron micrographs.

The experimental procedure is illustrated with the aid of Fig.3 which shows the raw data of one of the tests. In the upper part of the figure the measured temperatures of the sump (solid line) and of the atmosphere (broken line) are shown. The times t_1 when boiling is started and t_2 when boiling stops can be identified from the sump temperature data. The difference between sump and atmospheric temperatures also gives the degree of saturation of the atmosphere, 100% saturation was not reached in this particular experiment. In the lower part of the figure the measured aerosol mass concentration values are shown. In all experiments they were in the range of a few $\mu g/m^3$ or even less, the total masses on the filter samples always were in the range of nanograms.

3.3 The Evaluation Scheme

Let m_s be the mass of dissolved and/or suspended material in the sump and M_s the mass of the sump itself. Above the boiling surface, dm_s/dt is the release rate of resupended material and dM_s/dt the steam boiloff rate. Then the resuspension fraction R_s is defined as

$$R_s = \frac{(dm_s/dt)/m_s}{(dM_s/dt)/M_s} \tag{1}$$

It is likewise undesirable and impossible to measure R_s directly. R_s is the released fraction immediately above the boiling surface, it contains particles with a very broad size spectrum. The large jet droplets will not reach the containment and are thus irrelevant for use in a containment aerosol behavior calculation. They can also not be sampled and measured quantitatively. Instead we define a resuspension fraction R for a somewhat aged aerosol and evaluate it from the mass concentration, measured some distance away from the boiling surface, according to the following scheme.

Assuming a constant aerosol source rate S and constant aerosol depletion mechanisms for long enough time intervals, which means that no coagulation takes place because of the low mass concentration, the time dependence of airborne mass during an experiment has the very simple form

$$dm/dt = S - \alpha \cdot m$$

where α summarizes all depletion mechanisms. The general analytic solution is

$$m = S/\alpha + (m_0 - S/\alpha) \cdot \exp(-\alpha(t-t_0))$$

Two cases lead to simple forms

Case 1, at the start of sump boiling ($t_0=t_1$, $m_0=0$, $S=S_0$), gives

$$m = S_0/\alpha \cdot (1 - \exp(-\alpha(t-t_1))) \tag{2}$$

Case 2, at the end of sump boiling ($t_0=t_2$, $m_0=S_0/\alpha$, $S=0$), gives

$$m = S_0/\alpha \cdot \exp(-\alpha(t-t_2)) \tag{3}$$

Both are identical exponential functions and from a two parameter fit of the experimentally measured masses m the values of S_0 and α can be obtained.

S_0 is analogous to the release rate dm_s/dt as defined in eq(1). Using the known concentration in the sump $X_s = m_s/M_s$ and the boil-off rate dM_s/dt one can calculate the experimental resuspension fraction R from

$$R = \frac{S_0}{X_s \cdot (dM_s/dt)} \tag{4}$$

3.4 First Results

The experiments with solutions have been completed, a total of 43 test were done with different heating power, sump concentration, temperature and saturation of the atmosphere. For each test the values of S_0 were evaluated from the up and down slopes of the measured mass concentration values (eqs(2) and (3)). As an example Fig. 4 shows the resuspended fractions vs the power input calculated from the ascending slopes of the tests done at 100^0C and in dry atmosphere. It is seen that resuspension is higher for low boiling power. The values range approximately between $5 \cdot 10^{-4}$ when using 250 W power input and $2 \cdot 10^{-5}$ when using 1000 W. This already shows that the value of 10^{-4} assumed for the calculation presented in Fig.1 may be exceeded in some cases. On the other hand, measurements under condensing conditions give lower resuspension fractions. Also the resuspension from suspensions is still to be investigated before a quantitative statement can be made.

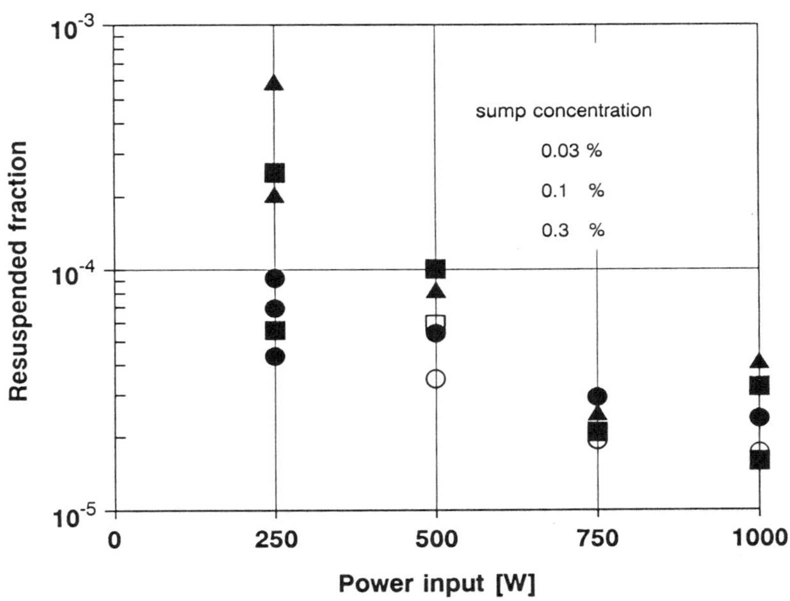

FIGURE 4. Measured resuspended fractions at 100^0C in dry atmosphere

The evaluation of particle size distributions from scanning electron micrographs (Fig.5) showed a few large particles in the range of >>1 μm diameter, obviously jet drops, while most particles are <0.5 μm in diameter.

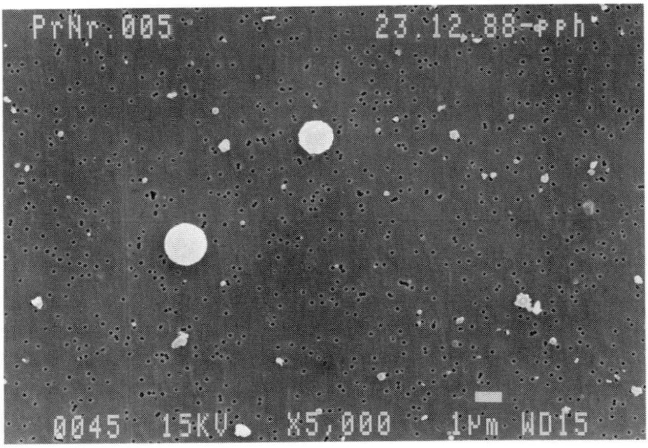

FIGURE 5. Scanning electron micrograph of resuspended aerosol

4. CONCLUSIONS

The contribution of resuspended aerosols as a weak long term source to the overall source term has been assessed in code calculations and is under experimental investigation. First experimental results show that resuspension fractions are in the expected range but a better understanding of their dependence on the influence parameters is obtained.

It may be concluded that for certain scenarios long term aerosol sources would contribute significantly to the overall source term as long as countermeasures such as filtered venting are disregarded. With filtered venting the source terms are reduced to values which are comparable to earlier source terms, then the load for the vent filters is caused by the long term aerosol source. Research efforts are continuing to improve the understanding of resuspension phenomena and the data base for application in source term assessment.

REFERENCES

[1]R.Borkowski, H.Bunz, W.Schöck, Resuspension of Fission Products during Severe Accidents in Light-Water Reactors, KfK 3987, EUR 10391EN, 1986.

[2]F.Richter, R.Rippl, G.Pröbster, O.Fernholz, Experimental analysis of the behavior of Iodine in the event of hypothetical accidents, Part 1: Droplet carry-over, KWU Report R917/86/009, 1986

[3]M.Tomaides, K.T.Whitby, in *Fine Particles*, ed. B.Y.H.Liu, Academic Press, New York, 1975.

Failure Modes of the Primary System of PWR

H. SCHULZ, P. GRUNER, and E. HICKEN
Gesellschaft für Reaktorsicherheit (GRS) mbH
Köln, FRG

INTRODUCTION

For the analysis of the structural response of primary system components for core melt-down sequences different questions have to be answered, for example

- Most probable sequence of loading.
- Probable sequence of failure. This is mainly of importance if post-failure behaviour results in different paths of further accidents sequences.
- Failure loads or times connected to a high confidence of low probability of failure.

Because there are large uncertainties in the determination of loads as well as in the structural response in the non-linear regime a combined effort between the various disciplines involved is necessary to get the best answers to the questions cited above. To identify the weak points of the components to be studied with respect to different loads it is helpful to perform a sensitivity type of analysis. The sensitivity analysis should point out where changes of failure modes take place or safety margins decrease rapidly. For example the structures of the primary circuit made of ferritic steels are more sensitive to the parameter temperature than the structures made of austenitic steels.

PRINCIPLE DESIGN OF PWR SYSTEMS

The principle design of the primary circuit of pressurized water reactors show a large degree of similarities but also some important differences with respect to the questions cited above.

The similarities are:

- The reactor pressure vessel (RPV) is the fixed point of the system. The vessel is fabricated using a low alloy ferritic steel and an internal austenitic cladding.
- The steam generator (SG) rests on a flexible support system which allows for thermal movement. The material of the shell and tube support plate is similar to the material used for the RPV. For the steam generator tubes a high alloy material is used (INCONEL or INCOLOY)

which is widely used in the industry also for high temperature applications.

- The main cooling pump (MCP) is placed in the cold leg of the system.
- The system pressure is controlled by a pressurizer which is connected to the hot leg of the main coolant system.
- The main components are placed within a short distance and connected by large diameter piping with only a few or none extra supports.
- All branch connections or small bore piping connected to the main components are made of austenitic steels.

The differences are:

- Material of the main coolant piping.
  For some designs the material is equivalent to the material used for the main components, so that the main welds are between ferritic materials on both sides. The bi-metallic welds between ferritic and austenitic steels are then in the branch connections.
- In other designs austenitic materials are used. The bi-metallic welds are between the components and the piping. The welds to the branch connections consist of austenitic materials on both sides.
- Reactor pressure vessel support.
  In some designs the reactor pressure vessel is supported on separate locks between the nozzles. For a large number of pressure vessels the vessel is supported on the vessel nozzles.
- Penetrations in the pressure vessel lower head.
  In one design there are no penetrations of the reactor pressure vessel lower head. In the design of the other vendors of PWRs the reactor pressure vessel lower head has numerous penetrations for the incore instrumentation.

The differences in design details and materials are important for the questions regarding failure mode, time to failure and failure location. The calculations performed in our studies are directed to the Siemens-KWU design and are not transferable in their results to other designs.

GOAL OF THE INVESTIGATIONS

Two main accident conditions are to be distinguished.

- core melt-down under high pressure conditions (HPC) which may arise e.g. in case of intact primary side and failure of system functions on the secondary side,
- core melt-down under low pressure conditions (LPC), caused e.g. either by leakages in the primary circuit or by purposely enforced conditions, such as accident management measures.

With respect to early failure of the containment the first condition (HPC) is much more severe than the second one (LPC).

It is therefore important to investigate the structural behaviour of those sections of the circuit which might have a reasonably high potential for leading into the low pressure regime and to predict the time span between initiation of further effective counter measures and failu-

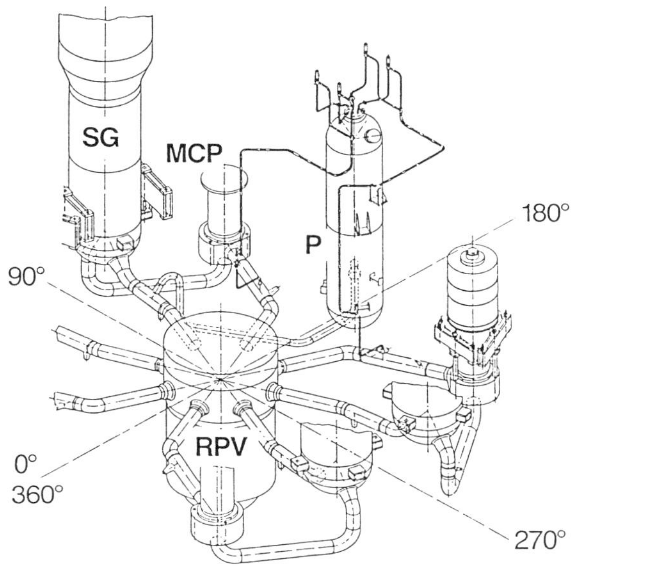

FIGURE 1.

VIEW OF PRIMARY COOLANT LOOP

re of the vessel under LPC. The main components of a German 4 - loop coolant circuit are depicted in Figure 1: the reactor pressure vessel (RPV), the steam generators (SG), the main coolant pump (MCP) and the pressurizer (P). Pressurizer and hot leg (between RPV and SG) are connected by the surgeline piping. The following system structures were considered in the analyses:

- The upper head of the reactor pressure vessel including flange connections,
- the steam generator tubes,
- the surgeline and its connection to the hot leg,
- the pressure vessel lower head.

The failure potential of the main coolant pump seals up to times when failure of the bottom head is to be expected was rated to be low. In this time domain the temperature level in the cold leg is considerably lower than that of the hot leg. Despite of high temperature input failure of the RPV nozzle was not seen to precede loss of integrity of the lower head because of the thick walls of the nozzle and its partial embedment into the vessel shell.

More recent analyses of the surgeline and of the lower head will be described in the following sections. Investigations of the other components listed above were dealt with before and will here be summarized in short only.

THERMAL-HYDRAULIC BOUNDARY CONDITIONS

Assumptions about the thermal-hydraulic conditions during core degradation, core slump and heat-up of the bottom head of the RPV are necessarily required for the assessment of RPV failure, the location as well as the failure mode.

The sequences to be considered can be grouped as follows:

- high pressure core melt (HPC)
  with pressure about the operating pressure
- low pressure core melt (LPC)
  with pressure about 1-2 MPa due to an intentional depressurization of the coolant system
- core melt at containment pressure due to a medium or large break in the primary circuit.

Superheated steam and hydrogen at a later stage of core degradation are produced inside the RPV. During high pressure sequences they are relieved through valves on the pressurizer. Therefore, a cyclic or continuous flow of hot gas is flowing through a hot leg and the surge line to the steam generator and thus heating up the walls. Natural convection flows are lengthening the core degradation process by about 30-45 minutes. The prediction of core degradation, core exit temperature and the heat up of internal structures, the hot leg and the surge line are relatively uncertain due to missing experimental data. Some more research is under way in this area.

Steam generator tubes can be heated up by hot gas, if pump seals fail or a small break exists in a cold leg. The heat-up by convection flows during high pressure sequences has to be studied further; it is expected that these flows are of minor importance.

It is planned to install the capability in German PWR's to intentionally depressurize the coolant system to about 1-2 MPa. As a major core slump cannot be excluded, the RPV pressure will increase because - for some time - the steam produced will exceed the mass flow through valves on the pressurizer. Several models exist for the energy transfer from the melt and/or debris into the residual water in the lower plenum. In Figure 2 the enveloping pressure in the RPV is shown as a function of time.

The core material slumped into the lower plenum cannot be cooled down totally to saturation temperature. Therefore, the inner surface of the bottom head of the RPV is exposed to a specific temperature; in Figure 3 the temperature history used for the assessment of the RPV integrity is shown. Because the core slump process and the evaporation of the residual water cannot be predicted with sufficient accuracy it has to be assumed that a major part of core material is equally distributed in the plenum.

For core melt scenarios at containment pressure the lower head of the RPV will be molten through.

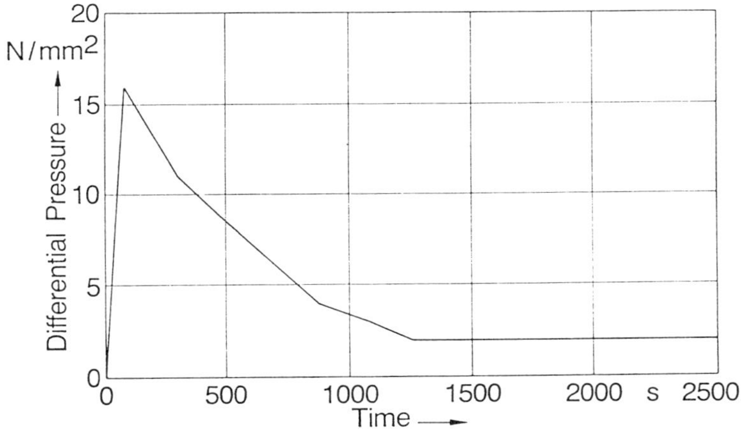

FIGURE 2. Internal pressure enveloping results of various models of core slump into residual water.

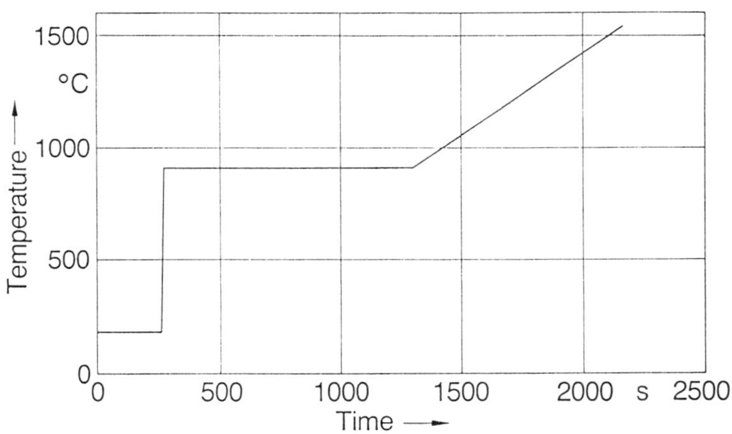

FIGURE 3. Temperature versus time at inner surface of lower head.

UPPER HEAD AND FLANGE CONNECTIONS

Investigations of the upper head of the vessel including the flange were performed using both finite element techniques and simplified methods for strength calculations and heat propagation as well. Loading conditions were 17.5 MPa internal pressure and time depending temperature profiles based on a temperature at the head inside of about 760 °C.

Compressive strains of below some 0.7 p.c. were confined within less than 20 p.c. of wall thickness near the inner surface. There was no yielding in tension. Distortion of flange and bolt connections does not bear any potential for leakage at times below the time when the molten core slumps into the residual water in the lower head.

A potential for early depressurization of the primary coolant system can therefore not be seen in this area.

STEAM GENERATOR TUBES

Within the German risk studies steam generator tubes were investigated under loading conditions which may result from core melt-down [AZODI 1987]. As worst case conditions (main steamline break) a pressure level of some 16 MPa and temperatures around 600 °C were assumed. At this temperature the yield point and the ultimate strength for INCOLOY 800 tube material is still very high. The studies yielded a failure pressure of some 35 MPa for undefected tubes. In case of wall degradation due to corrosion, even at a reduction of wall thickness by some 50 p.c., rupture of the tubes under core melt-down conditions may be excluded.

ESTIMATION OF THE TIME TO FAILURE OF THE SURGELINE AND THE CONNECTION TO THE HOT LEG

The nozzle connecting the ferritic main coolant piping (external diameter: 900 mm, wall thickness without cladding: 45 mm) and the austenitic surgeline (external diameter: 425 mm, wall thickness: 38.5 mm) is manufactered by ferritic steel. It is protected from being in contact with flowing fluid by a thermal sleeve which covers also the bi-metallic weld. Therefore highest temperatures arise just before the nozzle entrance in the surrounding ferritic material of the hot leg and in the austenitic section just behind the weld.

Enhancement of mechanical loading due to geometric effects at the penetration was considered leading to equivalent stresses between some 150 MPa and about 180 MPa in the main coolant piping at an internal pressure of 16.3 MPa. At the same pressure level the equivalent stress in the austenitic piping section amounts to some 95 MPa.

The time to rupture for austenitic material is larger by at least a factor of 50 than for ferritic material, under otherwise equal conditions. From this consideration and the loading level it should be expected that failure in the main piping precedes failure in the surgeline. Yet, results from thermo-hydraulic analyses show the tendency that the austenitic section reaches higher temperatures in shorter times than the main coolant piping. Hence, times to failure had to be determined in a more detailed manner.

Based on the Larson-Miller method [LARSON 1952] which will be described in the next section and on the concept of linear damage accumulation times to failure were determined.

The time ranges for which failure must be assumed overlap to a large extend. Hence, a decision as to the sequence of failure of both pipe sections cannot be made without narrowing further the thermodynamic uncertainties.

These results will be discussed further following the presentation of analyses for the lower head.

ANALYSIS OF THE REACTOR PRESSURE VESSEL LOWER HEAD

The investigations to be described in this section are primarily related to the low pressure path. However, it will be also possible to draw conclusions from the results concerning the time to failure of the lower head under high pressure conditions.

θ (C)	λ W/(m·K)	c W·s/(kg·K)	ϱ (kg/m³)	E·10⁻⁴ (N/mm²)	$E_T \cdot 10^{-3}$ (N/mm²)	$R_{p0,2}$ (N/mm²)	ν	$\alpha \cdot 10^6$ (1/K)	
100	43.8	467	7825	19.2	3.8	480	0.30	11.0	
300	43.1	522	7765	19.0	4.4	440	0.30	12.8	
500	36.2	572	7690	17.4	2.7	380	0.31	13.9	
700	30.0	650	7625	9.5	1.5	150	0.32	13.4	
900	27.0	700	7550	1.0	1.5	50	0.33	12.8	
1100	28.5	695	7425	1.0	1.4	30	0.33	12.6	Extra- polated Values
1300	31.0	688	7275	1.0	1.2	5	0.33	12.2	
Sources: EPRI, MPA, Stahleisen mbH, VdTÜV									

FIGURE 4. Material properties of steel 22 NiMoCr 37.

There is a strong dependence of thermal conductivity with temperature as may be seen from Figure 4 which contains material properties of the lower head steel. The temperature dependence of the thermal properties determines spatial variations of the through-wall heat conduction.

The results of heat propagation computations as presented in Figure 5 are based on the conservative assumption that the outer surface of the bottom head does not transmit energy. The graphs show that computations of profiles in the high temperature regime must incorporate the dependence on temperature of the material properties thermal conductivity, specific heat, and mass density in order to obtain realistic data concerning the rate of temperature rise. The solid curves are based on temperature dependent material properties as given in Figure 4. Compared to profiles which are based on constant properties, in this case on 500 °C, deviations range up to some 100 °C. The difference is mainly due to a decrease in thermal diffusivity with increasing temperature.

In view of the strong dependence on temperature of the strength properties of the lower shell steel and with respect to reliable estimations for the time to failure it is very important to analyse the heat propagation process properly. Variation of yield point and ultimate stress with temperature are presented in Figure 6. The data obtained by EPRI and KWU were used to obtain an average flow stress which was employed in simplified considerations. These earlier measurements did not contain values at temperatures above some 650 °C. Values for the limiting flow stresses RFL1 and RFL2 had therefore to be obtained by extrapolation.

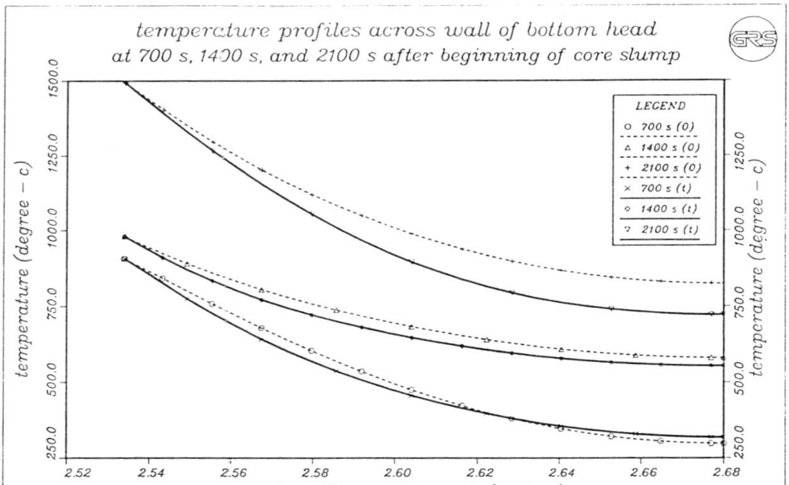

FIGURE 5. radial position within wall (meters)
(0) : constant material properties (referred to 500 deg-c)
(t) : temperature dependent material properties

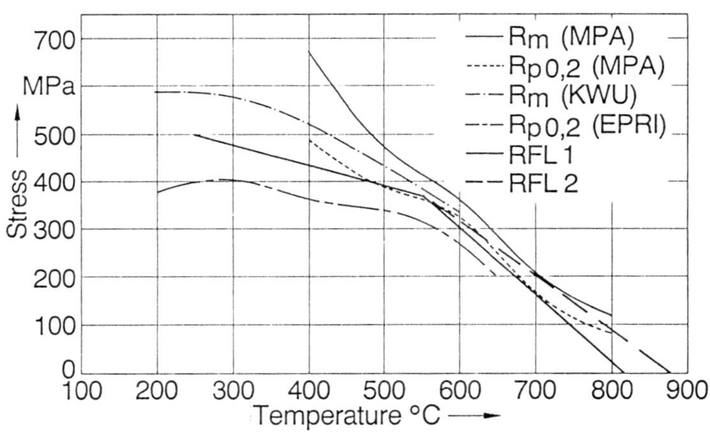

FIGURE 6. Yield point ($R_{p0,2}$), ultimate strength (R_m) and extrapolated average strengths (RFL 1, RFL 2) of steel 22 NiMoCr 37 versus temperature according to several sources.

The simplified analyses were based on two limiting stresses to cover the uncertainty of strength of the material. More recent measurements by MPA include temperatures up to some 800 °C. The data by MPA were utilized in a more advanced analysis with finite elements.

The structural behaviour is analysed using a rather simple modified elastic-plastic method and a finite element calculation considering creep effects, as well. Estimates of the time to failure are also determined employing as a third procedure the Larson-Miller concept.

Following core slump, extended areas of the spherically shaped lower head are assumed to be covered by molten core material which transmits heat directly into the steel structure. The structure does not contain any geometric irregularities or penetrations. As a result of heat conduction possible differences of temperature at the covered inside surface will be equalized with increasing depth through the bottom head wall, leaving a more or less homogeneous circumferential temperature distribution in colder outer zones of the shell which predominantly control progressive deformation. Thus, in the most heavily heated sections of the shell spherically symmetric conditions concerning geometry and loading by heat and pressure may be assumed without any greater loss of accuracy in the determination of the time to failure of the bottom head.

Simplified Method

This approximate method used to determine the behaviour of the bottom head was described in detail before [GRUNER 1988]. Only a short summary of the procedure will be given here.

The basic items to be provided are:

- the temperature distribution through the bottom head $\theta(r,t)$ as a function of time,
- the temperature dependence of the yield point and the ultimate stress of the bottom head material 22 NiMoCr 37 [MAILE 1988]. From these data and knowing $\theta(r,t)$, an average flow stress can be derived as a function of the radial coordinate r (RFL (r), Figure 6),
- average temperature independent data for young's modulus, Poisson's ratio, and the coefficient of thermal expansion.

The goal of the computation is to determine maximum pressure levels which can be carried by the lower head at successive times after core slump has taken place. The so-called tolerable pressures are conservatively defined in such a manner that the equivalent stress does not exceed the flow stress at any position within the shell. This requirement will be satisfied by modifications of linear-elastically determined stress distributions and by equilibrium iterations.

Estimates of the time to failure are obtained by comparison of the tolerable pressures with the pressure transient resulting from core melt slump into residual water. Respective data are presented later in connection with other results.

Analysis Including Creep

The simplified method presented before does not consider any creep effects and describes the plastic behaviour of the structure and associated stress redistributions only approximately. Hence, comparison with results of a more sophisticated approach is required. For that purpose a finite element analysis was performed on a segment of the bottom head.

Measured data of the temperature dependent strength characteristics-modulus of elasticity (E), tangential modulus (E_T), Poisson's ratio (ν) and thermal expansion coefficient (α) of 22 NiMoCr 37 up to temperatures of approx. 800 °C were used in the analysis. Above 900 °C, the lack of experimental information was compensated by extrapolated values.

The employed numerical creep model was adapted to measured creep strains in the relevant ranges of variability of stress and temperature.

Approach According to Larson-Miller

Plotting rupture times as taken from uniaxial creep tests on a logarithmic scale versus reciprocal values of temperature yields for a number of metallic materials nearly straight lines. Family parameter is the mechanical stress. Experience shows that within a certain range of scattering the straight lines intersect in a common point (Figure 7). These empirical results were first used by Larson-Miller (LARSON 1952) to extrapolate rupture times for temperature and stress conditions not realized in the experiments.

Using creep test results for material SA 508 which is comparable to the bottom head steel 22 NiMoCr 37 the constant C was found to be nearly equal to 21. A best-fit analyses yielded the dependence on stress of P_{LM} according to

$$P_{LM}(\sigma) = 5.049 \cdot 10^3 \cdot (6.078 - \log(\sigma/\sigma_o)) \text{ (K)},$$

where

σ_o : reference stress (1N/mm^2).

Based upon the magnitude of equivalent stresses ($\sigma_v \approx 20\text{N/mm}^2$) as obtained in the finite element analyses at times above 1200 s when creep-tension strains begin to form the Larson-Miller parameter was found to be

$$P_{LM} = 24190 \text{ (K)}.$$

To account for the case of continuously increasing temperatures in the determination of the time to failure the concept of linear accumulation of damage is applied (Figure 8).

Results

The tolerable pressures which have been obtained by the simplified elastic-plastic method and by the finite element analysis are in good agreement as can be seen from Figure 9. The first column in this table refers to times following the beginning of core slump. The second column lists internal pressure values from Figure 2 at these times. Columns 3 and 4

Creep-Rupture Test:

□ σ = const., T = const. → $t_B(\sigma,T)$

Rupture Time t_B Versus Temperature T approximately given by:

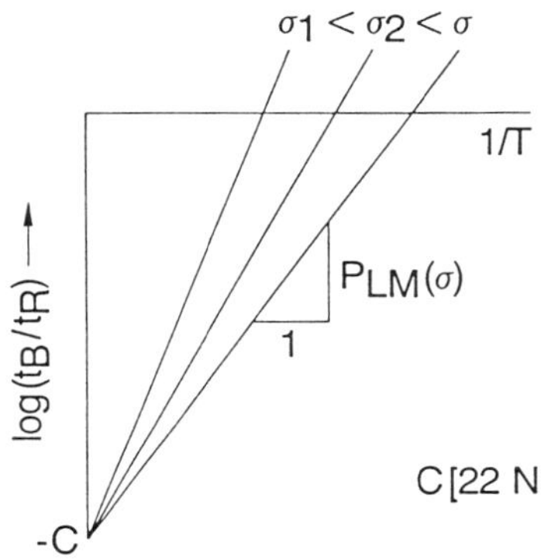

$C[22\ NiMoCr\ 37] \approx 21$

$t_R = 1\ h$

□ Slopes Larson-Miller Parameter within Wide Limits Independent of Temperature

⇒ Time to Rupture: $t_B/t_R = 10^{\left[\frac{P_{LM}}{\overline{\overline{T}}} - C\right]}$

FIGURE 7. Creep rupture time (Larson-Miller).

Linear Damage Accumulation assumed:

$$\sum_{1}^{N} \frac{\Delta t_i}{t_B(\sigma, \overline{\overline{T}}_i)} = 1 \rightarrow N \rightarrow t_B^* = \sum_{1}^{N} \Delta t_i$$

□ Evaluation for RPV-Lower Head Yields under Given Temperature- and Pressure-Conditions the Time to Rupture

$$t_B^* \approx 2300\ s$$

FIGURE 8. Procedure to account for damage accumulation due to changing temperature.

t (s)	p N/mm2	Rp0,2 N/mm2	σv N/mm2	FE-Analysis p* N/mm2	Simplified Analysis p* N/mm2
400	9.8	435	415	10.9	10.5
500	8.4	435	435	8.4	8.1
600	7.2	445	420	8.6	8.3
700	5.9	445	380	9.4	9.0
800	4.8	430	325	10.5	10.5
900	3.8	415	275	11.4	11.6

FIGURE 9. Comparison of tolerable pressures obtained from different analysis methods.

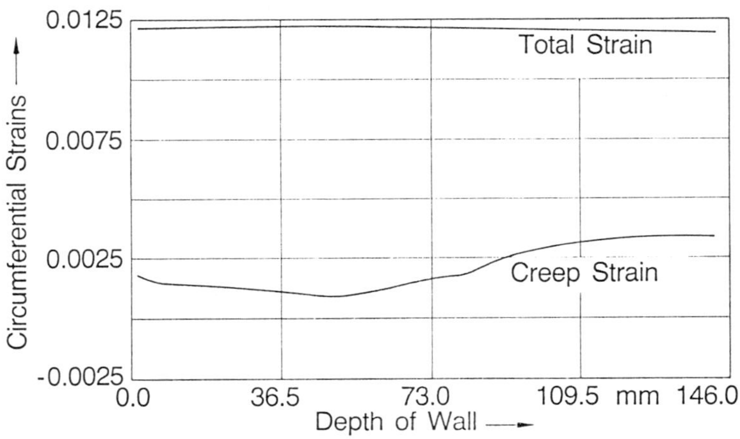

FIGURE 10. Total and creep strains at 2260 s after beginning of core slump.

present respective data for the yield point and for the equivalent stress, respectively. The reduction of the tolerable pressures in the temporal range between some 450 s and 750 s is a consequence of the high temperature gradients which build-up initially. Later, an equalization of the temperature gradients develops, so that higher pressure loads can be carried again until finally the derease of the material strength with temperature and growing creep strains reduce the tolerable pressures.

Except for the temperature dependence of the flow stress the simplified method is based on otherwise temperature-independent material parameters. Parametric studies have shown that best results are obtained if these material parameters are based on a mean wall temperature which is not too high.

If the tolerable pressures drop below the internal pressure failure of the lower head must be assumed. Depending on the core slump model used and on the assumed slope of the flow stress (RFL1, RFL2 in Figure 6) times to failure for the lower head were obtained ranging from 1800 s to 2200 s.

Conclusions concerning core melt-down under high pressure conditions can also be drawn from Figure 9. At 400 s after beginning of core slump the allowable pressure has already dropped below the system pressure which is 16.3 MPa. The analyses show that about 5 min following beginning of core slump the integrity of the lower head is challenged under high pressure conditions. With respect to a time scale starting at the beginning of core overheating failure of the lower head must be assumed between 40 min and 1 h, depending on the thermodynamic assumptions made. This range may be compared with the times to failure for the surgeline and the main piping which was described earlier. Due to the overlap of the time ranges it is not possible as of now to decide on the sequence of failure, i.e., whether the confinement of internal pressure is first lost in the piping or at the lower head.

The finite element analysis of a section of the bottom head shell showed rather moderate creep strains within the investigated time span of up to some 2260 s. Maximum values reach 0.3 p.c.in compression in zones near to the inner surface. Tensile creep in the outer half of the wall begins to develop after roughly 1800 s and spreads over the entire thickness at about 2200 s. At 2260 s again some 0.3 p.c. creep strain is reached, as is to be seen from Figure 10. The total strain amounts to some 1.2 p.c. at that time. As reasons for this low level creep are seen:

- the confinement of compressive creep by outer shell zones
- the low stress levels in hot regions due to low yield point
- the rather short time span

Initially the plastic zone propagates from the inside surface into more centrally located shell areas. A maximum of 0.9 p.c. in compression is reached here.

As a result of thermal stresses, the outer fibers of the shell show minor plastic strains (approx. 0.02 p.c.). However, these do not continue to grow any further due to the successive reductions of the temperature gradient in time. The plastic compression zone also comes to a halt. This, too, is mainly due to the reduction of the temperature gradients and the decrease of the internal pressure.

Up to some 2000 s the deformation process is mainly controlled by the temperature dependence of the yield point. Subsequently, an added factor is the increase in creep which starts in outer areas of the shell and, eventually, in the entire shell. These two factors lead to a rapid loss in load carrying capability between 2000 s and 2300 s, as is evidenced by the displacements in Figure 11.

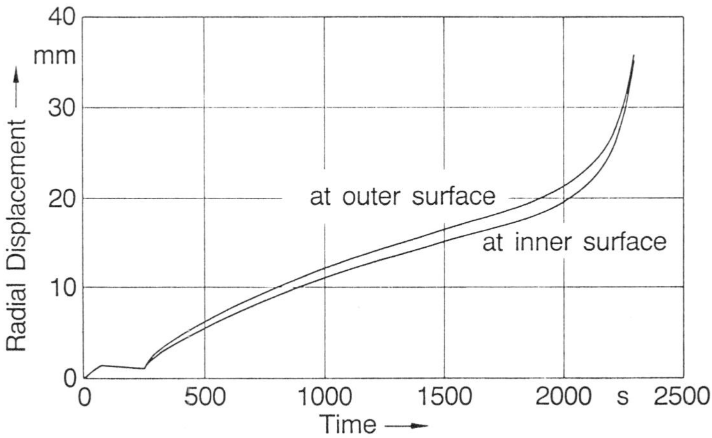

FIGURE 11. Radial displacement of lower head.

As a consequence of the low influence of creep, the results obtained by the simplified method are confirmed by the detailed investigation.

Concerning failure, all three methods yielded equivalent results within narrow bounds. With respect to the time of beginning core slump the following times to failure were obtained:

- 1800 s to 2200 s with the simplified method
- 2300 s to 2400 s with the finite element analysis
- 2150 s to 2450 s with the Larson-Miller method

The methods applied in the analysis of the bottom head were in parts used to predict the time to failure of a section of the primary circuit piping which in a large scale test was subjected to loads by a temperature transient and by internal pressure (MAILE 1988). Agreement between prediction and test result was achieved.

CONCLUSION

The failure modes of the primary circuit system and the respective times to failure were investigated for core melt-down under high pressure (HPC) and low pressure (LPC) conditions.

Special interest focussed on the questions:
Firstly, is there a potential for failure modes leading to depressurization of the primary circuit system prior to failure of the vessel and secondly what is the time to failure of the lower head under LPC.

Core melt down sequences are connected to various temperature-pressure-time-histograms. To determine this influence on the various structures of the primary system the investigations were based on simplified and finite element methods using limit load considerations as well as fracture mechanics and creep rupture analyses.

Particular interest was directed towards the behaviour of steam generator tubes, to surgeline and main coolant piping, to the upper head and flange connections of the reactor pressure vessel, and to the lower head.

The comparison of the respective times to failure yields to the following conclusions:

For HPC:

- A potential for depressurization prior to core slump which might be caused either by rupture of steam generator tubes or by leakage at the upper flange connection or by failure of the upper head assembly can not be seen.
- The variety of physical processes involved in simulations of heat transport phenomena through the primary circuit during core degradation leads to large uncertainties concerning the magnitudes and rates of energy transfer to structures. On the other hand, rupture times of these structures are very sensitive to temperature changes in the high temperature regime. The parametric studies performed to determine the time to failure of the ferritic main coolant piping and the austenitic surgeline exhibit therefore rather broad variations. Although there is some potential that failure of the main piping near the branching of the surgeline might precede failure of the surgeline and perhaps also precede failure of the lower head, a definite decision can not be made on basis of the available data. Hence, the present state of investigations does not yield the basis of a reasonably high probability for the assumption that high pressure scenarios could lead to core melting under low pressure without any additional measures.

For LPC:

Assuming core slump under low pressure conditions - e.g., enforced by accident management measures - the behaviour of the lower head using analyses methods of varying degree of sophistication leads to the following conclusions:

- The load carrying behaviour of the reactor pressure vessel bottom head subject to temperature loads as they occur in the course of core melt-down accidents can be calculated with a sufficient degree of accuracy.
- Creep processes will only gain a greater influence on the deformations when the entire bottom head wall has been heated up to high temperatures around approximately 750 °C to 800 °C.
- The simplified method supplies sufficiently accurate and conservative results. When using this method, however, care should be taken to ensure that the material parameters which are considered to be independent of temperature are related to a mean wall temperature which is not too high.
- Times to failure obtained from three analysis methods agree satisfactorily within narrow bounds of uncertainty.
- The predictive capabilities have been verified by large scale experiments.

ACKNOWLEGDEMENT

This work was supported by the German Bundesminister für Forschung und Technologie (BMFT).

REFERENCES

AZODI 1987: Azodi, D. and Schulz, H., Gesellschaft für Reaktorsicherheit, Köln, FR Germany, R. Arenz, Kraftwerk Union A.G. Erlangen, FR Germany, "On the Integrity of Steam Generator Tubes and Plugging Assessment", Transact. 9-th SMiRT, Lausanne, 17-21 August, 1987

GRUNER 1988: Gruner, P. and Schulz,H. "Load Carrying Behaviour of the Primary System of PWR for Loads beyond the Design Limits, Part 1: Analytical Determination of the Load Carrying Capacitiy". 14-th MPA-Seminar 1988, Universität Stuttgart

LARSON 1952: Larson, F.R. and Miller, J. "A Time-Temperature Relationship for Rupture and Creep Stresses", Transactions of ASME 74 (1952) No. 5

MAILE 1988: Maile, K., Obst, V., Klenk, A., and Julisch, P. MPA-Stuttgart, Forschungsvorhaben, "Versuche zum Versagen einer Hauptkühlmittelleitung infolge Kriechbruch unter hohem Systemdruck" (Final Report), February 1988

BWR Reactor Vessel Bottom Head Failure Modes

STEPHEN A. HODGE
Boiling Water Reactor Severe Accident Technology (BWRSAT) Program
Oak Ridge National Laboratory
P.O. Box 2008, MS 8057
Oak Ridge, Tennessee 37831-8057, USA

INTRODUCTION

Boiling water reactors (BWRs) incorporate many unique structural features that make their expected response under severe accident conditions very different from that predicted in the case of pressurized water reactor accident sequences [1]. The effect of the BWR procedural and structural differences upon the progression of a severe accident sequence during the period preceding movement of core debris into the reactor vessel lower plenum has been discussed previously [2]. It is the purpose of this paper to briefly address the events occurring after debris relocation past the core plate and to describe the subsequent expected modes of bottom head pressure boundary failure. As an example, the calculated timing of events for the unmitigated short-term station blackout severe accident sequence at the Peach Bottom Atomic Power Station is also presented.

BWR LOWER PLENUM STRUCTURES

The internal structure of the BWR reactor vessel is illustrated in the cutaway drawing of Figure 1. For a large 1050 MW(e) BWR such as Peach Bottom, the distance from the low point (vessel zero) of the bottom head to the bottom of the core plate is 5.207 m. Depending on the water temperature, the mass of water in the bottom head varies between 72,000 and 90,000 kg, which is more than enough to quench an entire mass of molten core and associated structural material.

Forced circulation flow during power operation is downward through the jet pumps into the water volume surrounding the control rod guide tubes in the lower plenum, then through orifices in the upper portion of the guide tubes and upward into the fuel bundles of the core. The structure of a single control rod guide tube is shown in Figure 2, where the control blade is shown withdrawn, as in normal power operation. The guide tubes are 3.912 m in length, have an outer diameter of 27.94 cm, and are located on a 30.48 cm pitch. It follows that two-thirds of the lower plenum volume immediately beneath the core is blocked by the control rod guide tube cluster. Since the free cross-sectional area between

"The submitted manuscript has been authored by a contractor of the U.S. Government under contract No. DE-AC05-84OR21400. Accordingly, the U.S. Government retains a nonexclusive, royalty-free license to publish or reproduce the published form of this contribution, or allow others to do so, for U.S. Government purposes."

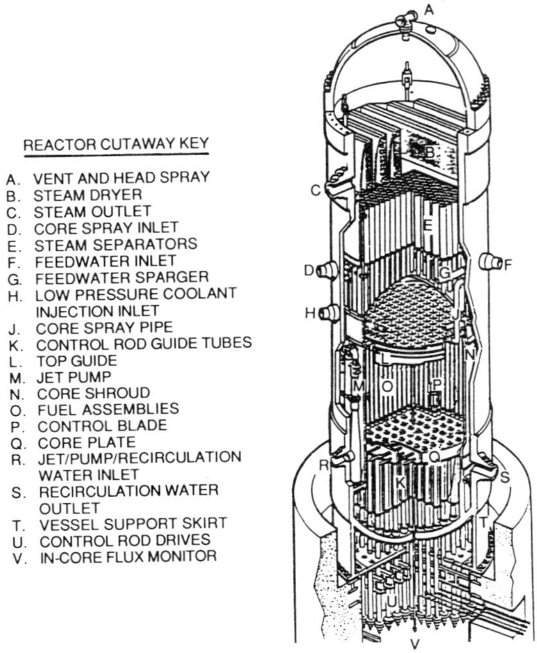

FIGURE 1. Internal structural assembly of the BWR reactor vessel. The control rod guide tubes are located immediately beneath the core.

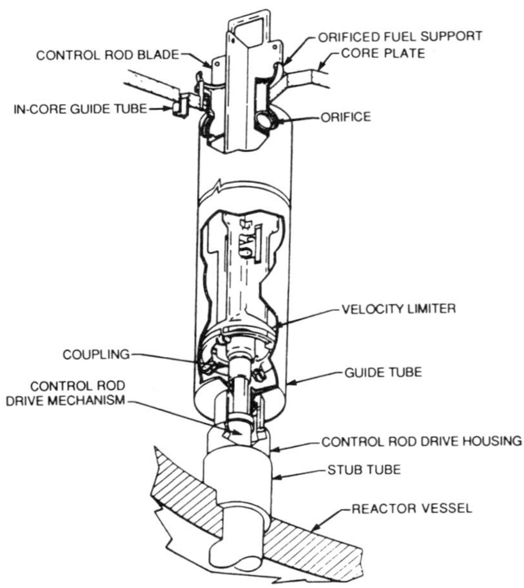

FIGURE 2. The control rod guide tubes hold the control blades when they are withdrawn from the core into the lower plenum.

any four control rod guide tubes is only 0.0316 m^2, it is obvious that the underwater forest of guide tubes in the BWR lower plenum would limit the size of downward moving debris masses and thereby promote quenching of the relocating material.

DEBRIS BED FORMATION IN THE BWR LOWER PLENUM

After structural deformation and downward relocation of molten control blade, channel box, and candling clad material (in that order) onto the dry core plate [2], local creep rupture failures of the core plate would introduce relocating material into the lower plenum water and begin the accumulation of quenched debris in the reactor vessel bottom head [3,4]. Relocation of the metal structure of the core is expected to leave the fuel pellet stacks standing until weakening, by overtemperature, of the ZrO_2 sheaths surrounding the fuel pellets and similar loss of strength by the previously molten material that tends to weld the fuel pellets together. It should be noted, given the progressive relocation methodology outlined above, that the majority of the debris entering the lower plenum is expected to be in the solid state when it enters the water.

As the relocated core material accumulates in the BWR reactor vessel bottom head, it is expected that the composition of the quenched debris bed would vary with height. Lowermost in the bed would be the mostly metallic debris (control blades, canisters, candled clad and dissolved fuel) that had either accumulated on the core plate before local core plate failure or had subsequently relocated downward above the core plate failure locations before fuel pellet stack collapse. Higher, within the middle region of the bed, would be the collapsed fuel and ZrO_2 from the central region of the core. The initial local core plate structural failures would cause temporary bursts of steaming as the relocated metallic debris was quenched; however, with the collapse of the central core fuel pellet stacks, a constant heat source (the decay heat associated with the pellets) would be introduced to the lower plenum reservoir, initiating a rapid continuous boiloff of the lower plenum water.

After bottom head dryout, the debris bed temperature would increase, causing thermal attack and failure of the control rod guide tube structure in the lower plenum, which the debris would completely surround to a depth of two or three meters. Since the control rod drive mechanism assemblies and the control rod guide tubes support the core, the remaining standing outer regions of the core would be expected to collapse into the vessel lower plenum when these support columns fail. Thus, the uppermost portion of the completed bottom head debris bed should be composed of the collapsed metallic and fuel material from the relatively undamaged outer regions of the core. The stainless steel of the control rod guide tubes and mechanism assemblies would be subsumed into the surrounding debris as it becomes molten.

BWR BOTTOM HEAD PENETRATIONS

There are more than 200 reactor vessel bottom head penetrations in a BWR reactor vessel of the size employed at Peach Bottom, where there are 185 control rod drive mechanism assembly penetrations, 55 instrument guide tube penetrations, and a 5.08 cm drain line penetration near the low point in the bottom head. The general arrangement of the in-core instrument housings and the stub tubes for the control rod drive mechanism assemblies is indicated in Figure 3. It seems certain that the initial pressure boundary failure after bottom head debris bed dryout would occur through the vessel penetrations and not by melt-through of the 21.43 cm-thick bottom head itself. This point will be discussed in the next two sections.

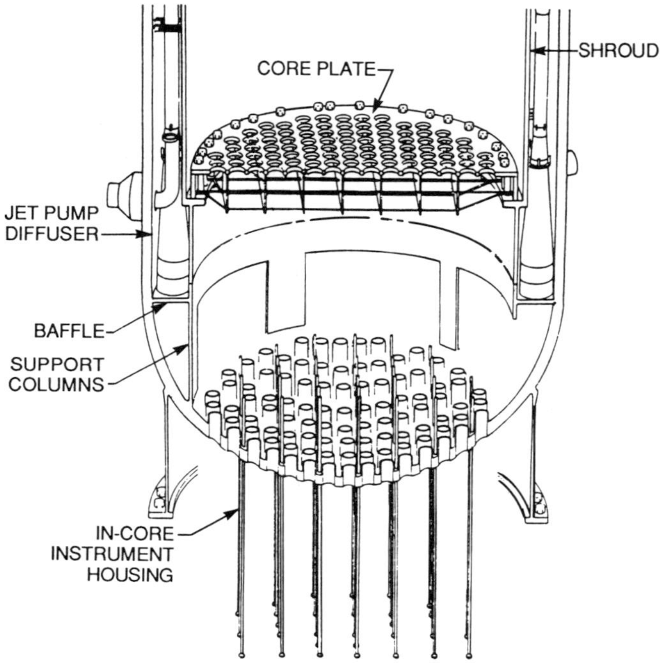

FIGURE 3. The BWR reactor vessel bottom head accommodates 241 penetrations and therefore is thicker than the remainder of the reactor vessel pressure boundary.

The bottom head of a BWR reactor vessel is clad with Inconel (thickness 0.3175 cm) while the control rod drive mechanism assembly and instrument guide tube penetrations are stainless steel. Cross sections of the control rod drive mechanism assembly and instrument tube penetrations and their weldments are illustrated in Figure 4. It should be noted that each in-core instrument tube is held in place by an Inconel-stainless steel weld located at the inner surface of the bottom head wall, whereas the control rod drive mechanism assemblies are held in place by similar welds at the upper ends of the Inconel stub tubes. These latter welds would be located about 10 cm within the bottom head debris bed expected to be formed during an unmitigated BWR severe accident.

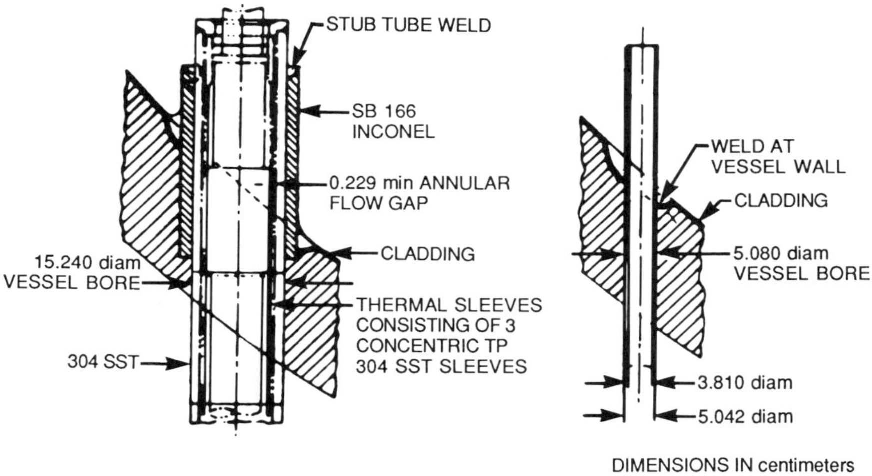

FIGURE 4. The BWR control rod drive mechanism assemblies are held in place by stainless steel-to-Inconel welds at the upper ends of the stub tubes whereas the in-core instrument tubes are supported by stainless steel-to-Inconel welds at the vessel wall.

FAILURE AT THE BOTTOM HEAD PENETRATIONS

Since the lower portion of the debris bed would be composed almost entirely of metallic materials while the UO_2 fuel pellets would constitute more than half of the central portion of the bed, the central portion would heat up much more rapidly after bottom head dryout than would the lower portion, and heat transfer within the debris bed would be toward the wall. As the temperature of the bed increased, materials in the central portion would begin to melt, migrate within the bed, freeze, and subsequently melt again. Eventually, temperatures near the walls would be sufficient to induce penetration failure and thereby open a path for gas blowdown and passage of molten material from the vessel. (In general, it is expected that most of the bottom head debris bed would still be solid at the time of penetration failure and initial vessel blowdown, so that relatively little of the debris would be expelled during the initial vessel blowdown.)

Since the stainless steel-to-Inconel welds supporting the control rod drive mechanism assemblies are located above the vessel wall, at the top of the stub tubes and within the adjacent portion of the debris bed, it is expected that these welds would reach failure temperatures first. The failure mechanism would be creep-rupture and would occur at lower temperatures if the reactor vessel remained pressurized at the time of failure. J. T. Han provides guidance on the time required at various temperatures and pressures for this failure mechanism for Inconel-to-stainless steel welds [5].

Although reactor vessel bottom head pressure boundary failure should occur first at the upper stub tube welds, this failure is less important to debris relocation from the vessel than the subsequent instrument tube failures. This is because BWRs are required to have a structure beneath the vessel bottom head that would limit the downward movement of any control rod mechanism assembly to about 3 cm in the event of failure of its stub tube weld. (The concern is to guard against the expulsion of a control blade from the core during power operation.) Since the vessel bottom head is 21.43 cm thick, this limited downward movement could not open a wide path through the vessel wall even if the control rod drive mechanism assembly were melted within the debris bed. This is not true for the instrument tubes, for which there is no provision to limit their downward movement.

Temperatures at the inner surface of the reactor vessel wall would eventually become sufficiently high to cause failure of the welds that hold the instrument tubes in place. However, it is probable that a different mode of failure for the instrument tubes would occur first. This predicted initial failure of the in-core instrument housing guide tubes for the source, intermediate, and power range detectors (55 penetrations in all) involves melting of the portions of these guide tubes within the central portion of the bottom head debris bed; then, when the downward relocation and freezing of molten metals has progressed to the point that molten metals are standing in the central portion of the bed, these metals could spill into the failed instrument tubes and pour through the vessel wall.

Would movement of molten metals through an instrument tube result in tube failure outside the vessel wall? Although it is known that small amounts of metallic debris did exit the vessel by this means at Three Mile Island (TMI), tube failure did not occur in this accident. This feature of the accident sequence has been extensively analyzed [6–9]. With regard to consideration of the applicability of the TMI results to the case of a BWR undergoing a severe accident, it should be recognized that the BWR instrument tube internal diameter is more than twice as large (3.810 vs 1.560 cm) while the BWR tube wall thickness is only slightly larger (0.616 vs 0.554 cm). In addition, the TMI reactor vessel bottom head was always filled with water, whereas for the BWR, instrument tube penetration failure is only predicted to occur after bottom head dryout, when the portion of the instrument tubes immediately beneath the vessel would be dry as well.

L. J. Ott at Oak Ridge National Laboratory has recently applied the approach of A. W. Cronenberg for TMI [6] to the BWR severe accident situation, substituting the appropriate BWR structural dimensions for the TMI values. This work provides the following observations:

1. The penetration distance for refreezing of molten debris within the BWR instrument guide tube walls is at least twice that of TMI; that is, the melt would be expected to travel more than twice as far ex-vessel as did the TMI melt.

2. The estimated peak temperature of the BWR instrument tube wall ex-vessel is significantly higher than for the TMI case. This establishes the need for a more precise calculation of the BWR instrument tube response than can be provided by Cronenberg's steady-state constant heat source approach.

In recognition of the need, L. J. Ott has developed a detailed transient model of the melt, vessel wall, instrument tube wall, and structures interacting (radiation heat transfer) with the instrument tubes beneath the reactor vessel. A metallic pour with no superheat and no heat generation at a pour rate estimated by the BWRSAR code [4] was used to drive this BWR instrument guide tube failure analysis model. The metallic pour is estimated to freeze and thereby plug the instrument tube at a distance from the reactor vessel about twice that sustained at TMI. Although the tube wall is not predicted to melt, the BWR instrument tube is predicted to sustain temperatures above 1478 K for a period of minutes in the simulation. Creep-rupture considerations ensure that the tube wall could not mechanically survive these temperatures for long. With an estimated weight of 90 kg for ex-vessel guide tube, internals, and debris plug, stress in the wall area for a depressurized reactor vessel would be slightly more than 1×10^6 N/m^2 which for 304 stainless steel at temperatures above 1478 K would produce rupture on the order of tens of seconds. Thus, instrument tube failure after bottom head dryout for an unmitigated BWR severe accident seems assured.

Downward relocation of molten material from the central portion of the bottom head debris bed through the instrument tube locations is expected to cause ablation of the lower portion of the debris bed as well as ablation of the vessel wall itself. This ablation is modeled in the BWRSAR code [4] based upon experimental observations at Sandia National Laboratories [10,11]. See also the work with regard to bottom head penetration failure in Ref. 12.

CREEP-RUPTURE FAILURE OF THE BOTTOM HEAD ITSELF

After bottom head dryout, heat transfer from the central portion of the bottom head debris bed would increase the temperature of the reactor vessel bottom head wall, eventually to the point of failure by creep-rupture. However, about 95% of the wall stress under normal operating conditions is due to the internal vessel pressure and the BWR Owners Group Emergency Procedures Guidelines [13] direct the control room operators to manually depressurize the reactor vessel during a severe accident sequence long before the onset of debris relocation into the lower plenum. The wall stress after bottom head dryout with the reactor vessel depressurized and taking into account the weight of debris resting on the bottom head and the weight of the bottom head itself is approximately 1×10^6 N/m^2. At this low stress level, creep rupture failure would occur only at temperatures approaching the melting temperature of the ASME SA-508 Class 2 carbon steel wall, and the vessel instrument tube penetrations are predicted to fail long before this. Thus, most of the metallic debris would have left the vessel by means of the penetration failures before failure of the bottom head itself.

It should be recalled that one of the dominant BWR severe accident sequences is Long-Term Station Blackout, for which the reactor vessel could not be depressurized [2]. For this accident sequence, the tensile stress in the bottom head wall would be approximately 26×10^6 N/m^2 so that creep-rupture failure would be expected to occur about four hours after the wall temperature reached 1225 K. Nevertheless, BWRSAR code calculations again predict that penetration failure would occur within a few minutes after bottom head dryout when the maximum wall temperature is about 800 K. Therefore, it is expected that most of the metallic debris would have left the reactor vessel by means of the instrument tube penetration failures for this case as well.

SHORT-TERM STATION BLACKOUT

The recent ASEP results assign 30% of the total risk of Peach Bottom core melt to the short-term station blackout accident sequence [14]. The estimated timing of events for this relatively fast-moving accident sequence as recently calculated with the BWRSAR code at Oak Ridge National Laboratory [3] is provided in Table 1. Events prior to central fuel column collapse have been discussed in a previous paper [2].

TABLE 1. Timing of events for Peach Bottom short-term station blackout

Event	Time after scram, s
Swollen level below top of core	2 412
ADS system actuation	4 800
Core plate dryout	4 854
Debris relocation begins	7 944
First local core plate failure	7 962
Central fuel column collapse	13 368
Lower plenum dryout	15 294
Bottom head penetration failure	15 300
Failure of the reactor vessel bottom head wall	28 020

As indicated, the BWRSAR code predicts a delay of about 3½ hours between the time of initial bottom head penetration failure and the time that the bottom head (beneath the vessel skirt) is melted through. The calculation takes into account the effect of ablation of the vessel wall as molten materials pour through the instrument tube penetrations.

SUMMARY

This paper has briefly described the approach taken by the BWRSAT Program at Oak Ridge National Laboratory towards understanding the probable sequence of events for an unmitigated BWR severe accident. There are many associated uncertainties, and experimental verification of the approach is certainly desirable.

For an unmitigated BWR severe accident involving the progressive relocation of material from the core region into the lower plenum of the reactor vessel, the control rod guide tube structure and the large amount of water in the lower plenum would be expected to provide for distribution and quenching of the relccating debris. Since the earliest relocation of materials from the core region would consist of metals from the control blades, channel boxes, and cladding, the lower portion of the bottom head debris bed should be metals-rich. The subsequent collapse of fuel pellet stacks into the lower plenum would provide an underwater decay heat source and provide for continuous boiloff of the surrounding water. After bottom head dryout, the debris bed temperature would begin to increase.

The cluster of control rod guide tubes in the lower plenum would be heated by the surrounding debris bed and would be weakened at high temperatures to the point of failure. Loss of control rod guide tube strength would cause collapse of the remaining standing outer regions of the core that are supported by the guide tubes. This collapse would form the upper portion of the bottom head debris bed while the stainless steel mass of the control rod guide tubes would be subsumed into the surrounding debris bed as they melt. Thus, there is expected to be a large amount of stainless steel included in BWR bottom head debris.

As the bottom head debris reaches high temperature, failure of the bottom head pressure boundary would occur at some point. Penetration failures can occur by weakening of the stub tube welds supporting the control rod drive mechanism assemblies or by failure of the instrument tube welds at the reactor vessel wall. However, failure of a stub tube weld would only cause a small downward motion of the associated control rod drive mechanism assembly, and therefore, although gas blowdown would be initiated by such a failure, gross release of debris from the vessel would not.

For the instrument tube, although there is nothing to prevent its complete detachment from the vessel given weld failure at the vessel wall, it seems probable that an earlier failure would be by opening of the tube in the middle (hottest) point of the bottom head debris bed with subsequent spillover of molten material into the tube with passage through the vessel wall, causing heatup and creep-rupture of the tube just outside the wall. Instrument tube failures in this manner would provide pathways for release of molten debris from the vessel.

The individual components of the debris bed would be expected to leave the vessel in the order in which they reach their melting points and transform to the liquid state. Solid metallic material surrounding the lower portion of the original instrument guide tube locations would be ablated into the molten material flowing from the reactor vessel via these pathways.

Gross failure of the portion of the reactor vessel bottom head underneath the vessel support skirt would be expected to occur long after the penetration failures discussed above. The reactor vessel bottom head wall is thick, and there is relatively little wall stress after the vessel is depressurized. BWR severe accident sequence calculations with the BWRSAR code predict failure of the bottom head wall only after the majority of the metallic debris has left the vessel.

REFERENCES

1. Greene, S. R., Realistic Simulation of Severe Accidents in BWRs — Computer Modeling Requirements, NUREG/CR-2940, ORNL/TM-8517, April 1984.

2. Hodge, S. A., Thermal Hydraulic Processes in the Reactor Coolant System of a BWR Under Severe Accident Conditions, *Proc. ICHMT International Seminar on Fission Product Transport Processes in Reactor Accidents, Dubrovnik, Yugoslavia, May, 1989,* Forthcoming.

3. Hodge, S. A., Hyman, C. R., and Ott, L. J., Primary Containment Response Calculations for Unmitigated Short-Term Station Blackout at Peach Bottom, NUREG/CR-5317, ORNL/TM-11057, Forthcoming.

4. Hodge, S. A., and Ott, L. J., Boiling Water Reactor Severe Accident Response (BWRSAR) Code Description and Assessment, NUREG/CR-5318, ORNL/TM 11058, Forthcoming.

5. Uncertainty Papers on Severe Accident Source Terms, Chapter 2, p 2-23, NUREG-1265, May 1987.

6. Cronenberg, A. W., Behling, S. R., and Broughton, J. M., Assessment of Damage Potential to the TMI-2 Lower Head Due to Thermal Attack by Core Debris, EGG-TMI-7222, June 1986.

7. Reactor Safety Research Semiannual Report July—December 1986 Volume 36, NUREG/CR-4805 (2 of 2), SAND86-2752 (2 of 2), Section 1.2, November 1987.

8. Reactor Safety Research Semiannual Report January—June 1987 Volume 37, NUREG/CR-5039 (1 of 2), SAND87-2411 (1 of 2), Section 1.2, January 1988.

9. Cronenberg, A. W. and Tolman, E. L., Thermal Interaction of Core Melt Debris with the TMI-2 Baffle, Core-Former, and Lower Head Structures, EGG-TMI-7811, September 1987.

10. Powers, D. A., Erosion of Steel Structures by High-Temperature Melts, *Nuclear Science and Engineering*, 88, 357-366, June 1984.

11. Pilch, M., and Tarbell, W. W., High Pressure Ejection of Melt From a Reactor Pressure Vessel — The Discharge Phase, NUREG/CR-4383, SAND85-0012, September 1985.

12. Kim, S. H., Kim, D. H., Koh, B. R., Pessanha, J., SiAhmed, El-k, Podowski, M. Z., Lahey, R. T., The Development of APRIL.MOD2 — A Computer Code for Core Meltdown Accident Analysis of Boiling Water Nuclear Reactors, NUREG/CR-5157, ORNL/Sub/81-9089/3, July 1988.

13. The General Electric Company and the BWR Owners Group, EMERGENCY PROCEDURES GUIDELINES, BWR 1 through 6, Revision 3, December 1982.

14. Kolaczkowski, A. M., Cramond, W. R., Daniel, S. L., Maloney, K. J., Sype, P. T., and Wheeler, T. A., Analysis of Core Damage Frequency: Peach Bottom, Unit 2 Internal Events, NUREG/CR-4550 Rev. 1, Vol. 4, SAND 86-2084, Forthcoming.

5. CONTAINMENT TRANSPORT PHENOMENA

LECTURE

Containment Transport Behavior: Status and Needs

T. S. KRESS, E. C. BEAHM, and A. L. WRIGHT
Oak Ridge National Laboratory
P.O. Box 2008, MS 6008
Oak Ridge, Tennessee 37831-6008, USA

INTRODUCTION

The containment vessel of a nuclear reactor is the last barrier against the release of radioactive material in the event of a severe accident. For Light Water Reactors (LWRs), there are more than a half-dozen different types of containments. The risk significant differences among them are of geometry, size (volume), strength, and construction details. They vary in volume from the smallest (the Boiling Water Reactor Mark-I) of about 4500 m^3 to the largest (the FRG BIBLIS type) of about 72000 m^3. The U.S. large-dry Pressurized Water Reactors (PWRs) have a volume of about 51000 m^3. Some containments are free standing steel structures (e.g. the BIBLIS type), while others derive their strength as concrete structures lined on the inside with steel for leak tightness. Some have special features designed to condense steam and/or to remove heat as a means of reducing the pressure loads produced by an accident. These features include: fan coolers in large-dry PWRs; ice beds in ice-condenser PWRs; suppression pools in BWRs; and sprays in PWRs and in the BWR Mark-III. Some of the reactor plants in Europe are equipped with the capability to vent through a filter system. Mark-I and Mark-II BWRs have their containments inerted with nitrogen and are housed within a substantial reactor building that is often considered to be a "secondary" containment for severe accident source term analyses.

In the U.S., the original basis for selecting the size, strength, and pressure suppression capability of a containment was the "containment design basis accident" which consisted of the blowdown of steam from a large break (double ended rupture) in the largest coolant line in the reactor coolant system (RCS). In conjunction with such a break, the applicable containment feature to suppress the pressure rise is assumed to be operable [1]. Consequently in the U.S., the design pressure capability varies from about 12 psig for an ice condenser to about 50 psig for a large dry containment. In PRA analyses, the failure pressures are considered to be several times greater than the design pressures. Thus, BMI 2104 [2] assumed failure pressures of 60 psig and 150 psig respectively for an ice condenser containment and a large-dry containment.

"The submitted manuscript has been authored by a contractor of the U.S. Government under contract No. DE-AC05-84OR21400. Accordingly, the U.S. Government retains a nonexclusive, royalty-free license to publish or reproduce the published form of this contribution, or allow others to do so, for U.S. Government purposes."

In hindsight, the "design basis" concept for containments has led us to the predicament in which we now find ourselves. An intact containment is highly effective in reducing the consequences of an accident. However, in PRAs, containments are calculated to fail under the simultaneous effects of multiple loads imposed by risk dominant severe accidents. If, during the time of original design, we had had the knowledge and analytical capability to estimate the maximum severe accident loads and had designed the containments accordingly, then we could confidently, and with small uncertainty, declare the severe accident source terms to be inconsequential and we could focus our attention and resources more appropriately on the prevention of such accidents and on the more efficient operation of the plants.

As we were not blessed with such foresight, we have had to resort to the detailed calculation of source terms that has involved the full range of phenomena from core heat-up to atmospheric transport. Parenthetically, even if containments were designed for severe accident loads, we would probably have had to do this anyway to provide the assurance that containment would not fail. Therefore the important phenomena that must be considered with respect to the source term aspects of containment behavior can be characterized as follows:

Containment Loads/Threats:

- Initial blowdown of steam from the RCS.
- Continued generation of steam during the meltdown and as material falls into the bottom head.
- Hydrogen generation from steam/Zr reactions during the heat-up and meltdown processes and during the melt ejection from the bottom head after vessel failure.
- High pressure melt ejection and subsequent direct heating of the containment atmosphere as well as additional hydrogen generation.
- Molten core material interaction with water in the reactor cavity to produce a steam spike and perhaps more hydrogen.
- Generation of heat and noncondensable, combustible gases due to molten-core concrete thermal interactions.
- Hot melt contacting containment boundaries.

Containment Failure Mode:

- Structural overpressure failure.
- Over-temperature failure of penetration seals (BWR MK-I and MK-II).
- Meltthrough of BWR MK-I shell on contact with molten core materials.
- Melt penetration of concrete basemat.

Containment Fission Product Transport Phenomena:

- Sources of fission products/aerosols:
  * direct from RCS during heat-up and core melt
  * revaporization from RCS surfaces
  * release during high pressure melt ejection events
  * fuel/coolant interaction induced release
  * core-concrete interactions
  * Iodine/Tellurium re-volatilization from water pools
  * possible resuspension as a result of hydrogen explosions or blowdown following containment failure

- Aerosol behavior:
  * agglomeration (Brownian, turbulent, gravitational)
  * gravitational settling
  * diffusiophoresis
  * water absorption by hygroscopic aerosols
  * water condensation onto aerosols
  * efficiency, capacity, and survivability of vent/filter systems
  * removal effectiveness of sprays, ice condensers, suppression pools
  * effects of water overlying MCCI

- Vapor Iodine species behavior:
  * water hydrolysis/radiolysis kinetics
  * re-release from vaporizing/flashing pools
  * diffusiophoretic and turbulent mass transport to aerosols
  * diffusiophoresis to containment surfaces
  * interaction with wet aerosols
  * effects of hydrogen burns

The question before us now is, which of these important aspects of containment transport behavior still need additional research?

RESEARCH STATUS AND NEEDS:

The focus of research on containment transport behavior has been on providing the data and capability to calculate the various severe accident loads, the containment response, and the detailed behavior of fission products as aerosols (more on loads and behavior, and much less on response).

Many experts believe that the ability to calculate the containment fission product transport behavior is the best developed and least uncertain aspect of source term determination. We agree with this assessment. However, We do not agree with the declaration that this is sufficient reason to suspend research on the transport mechanisms. The reason for this statement is the belief that in the near future we will have developed the capability to more precisely determine the containment failure times and modes rather than having to treat these parametrically. The calculated accident consequences could then prove to be very sensitive to the behavior of the airborne fission products (as aerosols) prior to the failure. We would like to see a sensitivity analysis done with respect to aerosol uncertainties that assumes precise knowledge of the sources and of the containment failure. We believe that such a sensitivity analysis will reveal that there is a window of containment failure times in which the uncertainties associated with our ability to calculate aerosol behavior will lead to significant uncertainties in the estimated risk.

This brings us to a point that has been made before but which bears repeating. The objective measure of the status of research on any particular phenomena with respect to the ability to estimate source terms is the contribution to the overall uncertainty that results because of deficiencies in our knowledge about and how we treat that phenomena. Therefore, any discussion of the research status and needs, as this lecture purports to be, becomes merely subjective opinion unless it is based on a detailed uncertainty analysis.

The most recent and inclusive uncertainty analysis on source term estimation is that of NUREG-1150 [3] sponsored by the U.S. Nuclear

Regulatory Commission. What does NUREG-1150 say about the contribution of containment related phenomena to the overall uncertainties in source terms?

In general, we see in NUREG-1150 that the time of containment failure is the most important determinant of risk. Firstly, then this tells us indirectly that the most important area of knowledge is the quantification of the extent and timing of the various loads that threaten to fail containment. In particular, these are: hydrogen burning and/or detonation, direct containment heating (DCH), steam and gas sources, and (for BWR Mark-I) potential for the core melt to attack and melt through the containment shell.

Secondly, NUREG-1150 tells us that the quantification of radioactive and aerosol sources are important. These include: direct release combined with deposition and revaporization from RCS, MCCI releases, and (somewhat surprisingly) late sources of iodine.

Following these, a number of items have roughly equal but lower rank: For Mark-Is and Mark-IIs, the ability of the reactor building to act as a secondary containment is important. This, then, involves the questions of survivability and the interplay between aerosol behavior and hydrogen burning; For Mark-IIIs, the suppression pool decontamination factor (DF) proved to be an important uncertainty.

For all plants, the agglomeration and deposition behavior of aerosols proved to be of importance but not to be a risk driver.

An updated version of NUREG-1150 is being developed and may have been released by the time of this seminar. At the time of writing of this lecture, we were not privy to the findings of this second version. However, for the purposes of this lecture we believe it is appropriate to speculate that the ordering of important elements will remain as: timing of containment failure/threatening loads/radioactivity and aerosol sources/aerosol transport behavior.

Granted that these are the important elements of containment behavior under severe accidents, then it would be of interest to review the status of research in these particular areas.

The USNRC has recently developed a new Severe Accident Research Plan that has not yet been released. However, the draft report issued for comment divided the severe accident research program into two categories, near-term issues and longer-term issues:

The near-term issues include

* direct containment heating (DCH) and depressurization of the reactor coolant system,
* containment shell meltthrough (BWR Mark-I issue), and
* hydrogen transport and combustion.

Included within the longer-term program plans are

* in-vessel hydrogen generation,
* fuel coolant interactions,
* molten core/concrete interactions, and
* containment fission-product re-evolution chemistry.

Given our belief that the timing and mode of containment failure is the most important risk determinant factor, it is gratifying to see this degree of focus on containment issues.

Direct Containment Heating:

This issue is being reviewed in another session of this seminar. Therefore, it will not be discussed here other than to put forth the opinion that we believe in the future it will occupy a less prominent position because of increased ability to specify the quantities of core melt that could be involved, as additional experimental data are brought forth on the loads produced, and as accident management strategies are developed for venting the RCS. We would also like to note that we do not think it wise to rely on uncertain calculations of natural convection induced RCS failure to provide depressurization and that there is a need to evaluate dedicated relief systems and procedures.

Fuel/Coolant Interactions:

When molten core material falls into water, its internal energy can be rapidly transferred to the water generating steam which can pressurize the system boundaries and result in a steam explosion. In the process of intimate mixing of the molten core materials with the water and steam, any metallic phases present can react to produce hydrogen. In addition, the interaction of the hot metallic particles with the steam can oxidize associated fission product Ru to greatly increase its volatility and possibly enhance its release into containment. Although considerable past research has addressed this particular phenomenon, the focus has generally been on the potential for alpha mode failure (generation of a missile that penetrates containment) as identified in WASH-1400 [4]. The fundamental details of the interaction processes, therefore, are still not well known.

With the recognition that FCI is also a source of hydrogen and fission product release as well as of steam and pressure and that it establishes the initial conditions for subsequent molten core concrete interactions, there has been renewed interest in it. A number of recent attempts have been made to develop improved mechanistic models of FCI [5-8]. These models generally treat the process as proceeding through several stages: a coarse pre-mixing stage of relatively large fragments surrounded by a semi-stable steam film; a triggering event (such as melt impact on the floor); collapse of the voids around the particles accompanied by further fragmentation and propagation of the resultant steam pressure wave; and expansion of the high pressure vapor against any inertial constraints of the surroundings and the mixture.

With the present focus of these models being on hydrogen production, steam production, and Ru release, there is need for more experiments to evaluate the mechanistic details of the models – in particular, the events associated with the void collapse and subsequent fine fragmentation of the debris.

Molten Core Coolant Interactions (MCCI):

When molten core materials fall into the cavity beneath the reactor vessel and come into contact with the concrete basemat, a vigorous interaction can occur. The hot core materials can heat the concrete to its melting temperature and incorporate the molten concrete components into the melt while releasing gases (CO_2 and H_2O) from the concrete. The gasses sparge up through the melt and chemically react with metallic phases to produce energy that keeps the system hot and tends to maintain the interaction for a considerable time. The sparging gases are converted by the chemical reactions into combustible CO and H_2 that are released into containment along with vaporized fission product and aerosol species sparged by the gases from the melt. Most recent risk analyses have identified MCCI as a dominant factor in creating containment loads as well as being a sustained source for fission product release.

Ongoing experimental programs at Sandia National Laboratory [9], at KfK in the Federal Republic of Germany [10], and in the international cooperative ACE program [11] led by EPRI, are providing realistic data using prototypic melt materials on different concretes.

The separate effects work at Brookhaven National Laboratories, as summarized by Greene et al. [12], has included experimental and analytical progress on interlayer heat and mass transfer and liquid-liquid boiling processes. These studies have shown that the dominant internal melt conditions are in the regimes of bubbly and churn-turbulent flow and that the void fraction data are in excellent agreement with the drift flux model. In addition, these studies have evaluated the importance of entrainment between liquid layers and have developed new correlations for the rate of entrainment and the rates of heat transfer.

We believe the understanding and ability to calculate this aspect of containment behavior is now becoming very good with perhaps two possible exceptions: 1. the behavior during the initial transient as the melt first leaves the reactor vessel and spreads out across the concrete basemat floor in the presence of existing water, and 2. the effects of an overlying layer of water on the interaction.

Both of the above areas have been under investigation by Greene et al. [13] at BNL using simulants in small scale tests. This work has observed that the melt spreading behavior can be categorized into at least five separate regimes as characterized by the final melt average thickness (a measure of how efficiently it has spread). The five categories depend on the water depth and the melt superheat. As might be suspected, small depths of water and high melt superheats lead to rapid and extensive spreading whereas deep pools with low melt superheat can essentially freeze the melt without extensive spreading. Greene et al. have developed an empirical correlation of these results relating a dimensionless thickness (after spread) to a dimensionless spreading parameter that includes the melt mass, density, water depth, the latent and sensible heat required to vaporize water, and the latent and sensible heat required to solidify the melt.

Greene et al. have also used simulants to evaluate the bubble driven boiling heat transfer behavior of a volatile liquid overlying a hot liquid metallic melt. They have observed that without gas sparging of the system, the top surface boiling heat transfer is well described by the Berenson film boiling of liquids on top of flat plates. However, when gas bubbles are sparged through the system, the film boiling heat transfer rates are enhanced linearly with increasing superficial gas velocity to as much a 5–6 times the Berensons film boiling model at a superficial gas velocity of less than 3 cm/s.

The SWISS series of tests [14] at SNL have also investigated the effects of overlying water on steel melts undergoing sustained inductive heating on limestone/common sand concrete. In these tests, the rate of concrete erosion and the rate of combustible gas generation was unaffected by the presence of the overlying water and the top surface heat transfer corresponded closely to that of nucleate boiling on a flat plate. However, aerosol emissions were attenuated significantly [factors of 10 to 30] on passing through the 50 cm deep water pool overlying the melt.

We believe these efforts represent a fine beginning for this aspect of MCCI but that there are still needs to validate the correlations using more prototypic melts and to demonstrate the behavior in full integral tests. There are plans being formulated to conduct such integral tests in the NRC MCCI programs at Sandia and as part of the EPRI MCCI ACE program at Argonne National Laboratory.

Part of the motivation for Greene's melt spreading work has been to evaluate a potential early containment failure mode of BWR Mark-I containments in which the melt is hypothesized to spread laterally over the drywell floor, directly contact the relatively close-in steel containment shell, and essentially develop a melt-through failure. The thermal response of the shell to such direct contact has been evaluated by Kress and Cleveland [15], Weingardt and Bergeron [16], Greene [17], Corridini [18], and others. There seems to be general agreement among these that the question of whether or not the shell reaches melting (or a lower failure temperature) depends critically on four parameters: 1. the melt superheat, 2. the melt depth, 3. the top surface boiling heat transfer rates, and 4. the heat transfer rates between the melt and any crust that is formed along the shell inclined surface.

The previously cited work by Greene directly addresses the first three of these. Greene [13] has also investigated the fourth parameter above using simulants to evaluate the bubble stirred heat transfer between the melt and a vertical wall. He found the heat transfer rates to be high with heat transfer coefficients ranging from 2000 to 5000 W/M^2K depending on the gas superficial velocity. The data were well described by the Konsetov [19] and Blottner [20] correlations for bubbling heat transfer to vertical walls. However, Theophanus [21] has observed in his tests that the heat transfer rates can be significantly reduced when the wall is inclined from the vertical (the BWR MARK-I shell is inclined at a 45° angle). In addition, experiments conducted by Henry of FAI [22] in which thermite melts were injected into a 2-D concrete channel containing water indicated very high heat transfer rates (higher than critical heat flux) during the early transient conditions as the melt flowed toward the wall.

There is need, therefore, to continue such tests to determine the mechanisms responsible for the observed high heat transfer rates in Henry's experiments and to better establish the heat transfer coefficients to inclined walls under bubble stirred conditions using more realistic melt materials. The LINER-QT and the WITCH-LINER test [23] at SNL are providing vertical wall heat transfer data with metallic melts undergoing MCCI. However, inclined walls and ceramic melts apparently are not in the plans.

Hydrogen Combustion:

The production of hydrogen as a result of steam chemically reacting with Zircaloy and other metal phases in-vessel and ex-vessel (FCI, DCH, and MCCI) represents a major potential containment loading threat. The BWR MARK-I and II containments are inerted to eliminate this threat and the ice-condenser and BWR MARK-III containments are equipped with hydrogen igniters to assure controlled burning of any hydrogen rather than allow the build up to potentially detonation levels.

The issues associated with hydrogen combustion have been extensively investigated [24-26] and there have been a number of excellent recent reviews of the state of the art [27-30].

It is not practical in this paper to provide a comprehensive review of the extensive past and ongoing work related to hydrogen combustion in containments. In general, the research has dealt with the major aspects of hydrogen burning: deflagrations, diffusion flames, transitions from deflagration to detonation (DDT), and detonations. Data and correlations have been developed for:

* flammability and detonation limits in air and steam-air mixtures,
* flame speeds
* effects of turbulence, obstacles, and geometry,
* multi-volume effects,
* completeness of the burn, and
* the associated thermal hydraulics (pre-mixing and heat transfer/ thermodynamics during burns).

Recent USNRC sponsored work at SNL [31] has shown that: 1. Detonation limits are broader than previously thought. Mixtures of H_2-air-water as lean as 13% H_2 were detonated. 2. The detonability is increased with increasing temperature but is decreased with increasing steam and non-condensible gas content.

Data are missing at the conditions of high H_2-air-steam mixtures that are predicted in some small-break LOCA and TMLB' sequences in which high-temperature H_2/steam mixtures can be injected int pre-existing mixtures.

Although it is believed that auto-ignition of a mixture to cause detonation will not occur in reactor accidents, it is possible for deflagration flames to accelerate due to passage through restricted openings or around obstacles. The acceleration can advance to the point of detonation. Such transitions from deflagration to detonation need an experimental data base. At present, no reliable model exists to predict the conditions for possible DDT. It is believed that the research being conducted at Sandia [32,33], in the Federal Republic of Germany [34], and in Canada [35,36] will help resolve the issues associated with DDT.

Thermal Hydraulics:

Containment thermal-hydraulic modelling related to the atmosphere and surface responses to heat and mass additions make use of an extensive literature and data base on forced and natural convection heat and mass transfer, thermodynamics, convection, and condensation behavior. Significant comparisons of the coded models have been made with measured temperatures, pressures, flows, concentrations, etc. in large vessels that have included Demona and the Heiss Dampf Reaktor (HDR) in FRG and the Marviken facility in Sweden. These comparisons have been so remarkably favorable that additional work in this area does not appear to be necessary.

There are, of course, significant couplings between the thermal hydraulics and the aerosol behavior in severe accidents which will be discussed in the presentation following this one.

In addition to a need to assure that such couplings are made in the integral codes, there is a need for the thermal hydraulic codes to better predict the super-saturation levels and to provide an internal prediction of the changing turbulence levels in containments as driven by the natural convection forces. Such information is needed for the aerosol behavior models.

Aerosol Behavior:

The current state of the art of the containment behavior codes that deal with aerosol transport is being reviewed in another session of this symposium. In this review, we would just like to observe that a significant data base now exists on aerosol behavior under severe accident conditions. This data base includes such experimental programs as: SAUNA, NAUA, NSPP, TRAP-MELT Validation, Vapor/Aerosol Interactions, DEMONA, MARVIKEN, LACE, DCH (SURTSEY), the French aerosol program, the Japanese aerosol program, and extensive filter system efficiency and capacity work.

In spite of this extensive data base, there are still some research needs with respect to aerosol behavior in containments. These needs include:

- behavior of hygroscopic aerosols in steam environments*,
- turbulent agglomeration efficiencies,
- resuspension via hydrogen burns*,
- shape factor measurement as functions of particle size, and
- chemical interactions with vapor iodine species*.

The EPRI led ACE program (Phase B) is conducting a set of experiments at Whiteshell Nuclear Research Establishment in Canada, at Oak Ridge National Laboratory, and at Hanford Engineering Development Laboratory to investigate the items above that are marked with an asterisk.

In addition, the proposed French PHEBUS-PF program slated to begin after 1990 will conduct completely integral experiments that use real fuel subjected to LOCA conditions in the PHEBUS reactor. The system will include simulated RCS pathways and a simulated containment vessel. The release and transport of fission products and aerosols will be followed throughout the entire pathway.

Containment Chemistry:

Our last topic addresses the issue of fission product iodine behavior in containment. Fission product iodine released from the RCS is generally believed to be primarily as CsI which is a relatively non-volatile aerosol at containment conditions. The natural aerosol attenuation processes in containment will deposit such aerosols onto containment surfaces (primarily horizontal surfaces) where they generally contact water pools. The question is, will the iodine be permanently sequestered in the water or can aqueous chemistry processes re-partition the iodine back into the gas space in a more volatile form (e.g. I_2 or CH_3I).

When CsI contacts water it will immediately disassociate into a cesium ion and an iodide ion. The iodide ion can enter into a number of hydrolysis reactions that can be represented by the overall system,

$$3I_2 + 3H_2O \rightleftarrows IO_3^- + 5I^- + 6H^+$$

This overall reaction actually represents the end-effect of a sizable number of intermediate reactions that go in both forward and reverse directions. Therefore, the CsI that first enters the water as Cs^+ and I^- can be converted to volatile I_2. This conversion is enhanced by radiolysis effects that produce additional species that tend to drive the above reaction to the left.

Organic species, if present, can also convert the I^- to volatile organic iodides.

There have been a number of code systems recently developed to describe this rich aqueous chemistry. These include the TRENDS code at ORNL, the LIRIC code at WNRE, the IMPAIR code of FRG and EIR, the French IODE code, and the British general chemistry code, FAXIMILE.

There is need to include these models as separate modules within the containment thermal hydraulic and aerosol codes and to provide integral experimental validation. The ACE Phase-B and the planned PHEBUS-PF programs should be useful in this respect.

Summary Comments:

The research needs identified in this review related to containment transport behavior are summarized in Table 1.

At this point, we believe one final observation is in order. Recall that the various serious threats to containment integrity include:

1. Hydrogen produced from steam/Zircaloy reactions in-vessel.

2. Hydrogen produced on high pressure melt ejection when the steam continues to react with metallic Zircaloy.

3. Fuel coolant interactions (mostly possible because of high content of unreacted metallic phases in the melt) with the concomitant hydrogen generation from steam reaction with the metal phases.

4. Molten core concrete interactions that release fission products and combustible gases and which are sustained mostly because of steam and carbon dioxide reactions with unreacted metallic phases (zirconium) still present in the melt.

5. Melt-through of BWR MARK-I shells when metallic melts directly contact the shell.

6. Potential FCI events when metallic melts enter into the suppression pool of BWR MARK-IIs (not discussed).

The observation that we think needs to be made here is that the one thing all of the above have in common is that they are primarily possible because of the use of Zircaloy in the core of our reactors. It

TABLE 1. Summary of Containment Behavior Research Needs

Phenomenological Area	Research needs
FCI	* Separate effects experiments focused on the premixing and void collapse fragmentation and propagation behavior * Integral experiments to measure the hydrogen and aerosol production
MCCI	* Incorporation of new heat transfer results into the code models * Effects of water overlying the melt * Integral validation experiments with prototypic melts
MCCI + Vessel Melt-through	* The thermal hydraulic aspects of the initial pour and the spreading behavior using prototypic melts on concrete with water present * Additional tests on the bubble stirred sidewards heat transfer coefficient to inclined walls
Hydrogen burns	* High-Temp/High Steam flammability and detonation limits * Transition from deflagration to detonation due to acceleration brought about by obstacles and restricted openings between chambers
Thermal hydraulics	* Assured integral coupling with aerosol and chemistry models * Internal prediction of turbulence levels
Aerosol behavior	* Hygroscopic aerosol behavior in steam environments * Turbulent agglomeration efficiencies * Effects of hydrogen burns * Interaction with vapor species * Shape factors as functions of size

is not the fission products that are driving the risk, it is the Zircaloy! If we did not have Zircaloy in our cores we could declare victory over severe accidents and closeout the source term research. It is time, then, that research be initiated to develop new fuel as a replacement for what we are now using in LWRs. Replacing the Zircaloy with a chemically nonreactive material (perhaps a cermet) would be the one single most effective action to increase safety and resolve the severe accident issues.

REFERENCES

1. Code of Federal Regulations 10CFR100/ Reg. Guides 1.3 and 1.4.

2. J. A. Gieseke et al., "Radionuclide Release Under Specific LWR Accident Conditions," BMI-2104, Vols. IV and V, July (1984).

3. USNRC, "Reactor Risk Reference Document," Draft Report NUREG-1150, February (1987).

4. USNRC, "Reactor Safety Study – An Assessment of Accident Risks in U.S. Commercial Nuclear Power Plants," WASH-1400, NUREG-75/014, October (1975).

5. M. F. Young, "IFCI: An Integrated Code for Calculation of All Phases of Fuel-Coolant Interactions," NUREG/CR-5084, SAND87-1048 R3, September (1987).

6. M. L. Corradini et al., "TEXAS-II", ANS Proceedings, February (1986). Nucl. Science & Engr., Vol 101, January (1989).

7. D. F. Fletcher and A. Thyagaraja, "A Mathematical Model of Premixing," Proc. 25th National Heat Transfer Conference, Houston, Texas, 24-27th July, 1988, HTC-3, 184-190, ANS, (1988).

8. T. G. Theofanous et al., Nucl. Science & Engr. Vol. 97, December (1987). ANS Proceedings, August (1988).

9. D. A. Powers and F. E. Arellano, "Large Scale, Transient Tests of the Interaction of Molten Steel with Concrete," SAND81-1753, NUREG/CR-2282, Sandia National Laboratories, Albuquerque, NM, January (1982).

10. H. Alsmeyer, M. Riemann, and J. P. Hosemann, "Preliminary Results of the KfK MCCI Experimental BETA Facility," 11th NRC Water Reactor Safety Information Meeting, Gaithersburg, Md., October (1984).

11. D. H. Thompson and J. K. Fink, "Advanced Containment Experiments System Design and Operations Description, ACE-TR-C3, Argonne National Laboratory, October (1988).

12. G. A. Greene, C. Finfrock, and S. B. Burson, "Phenomenological Studies on Molten Core-Concrete Interactions," Nucl. Engr. & Design 108, pp. 167-177 (1988).

13. G. A. Greene, C. Finfrock, J. Klages, C. E. Schwarz, and S. B. Burson, "Experimental Studies on Melt Spreading, Bubbling Heat Transfer, and Coolant Layer Boiling," 16th NRC Water Reactor Safety Information Meeting, Gaithersburg, Md., October (1988).

14. R. E. Blose, J. E. Gronager, A. J. Suo-Anttila, and J. E. Brockmann, "SWISS: Sustained Heated Metallic Melt/Concrete Interactions With Overlying Water Pools," NUREG/CR-4727, SAND85-1546 R3, R4, R7, Sandia National Labortories, Albuquerque, NM, July (1987).

15. T. S. Kress and J. C. Cleveland, "Analysis of the Thermal Response of a BWR Mark-I Containment Shell to Direct Contact by Molten Core Materials," 16th NRC Water Reactor Safety Information Meeting, Gaithersburg, Md., October (1988).

16. J. J. Weingardt and K. D. Bergeron, "TAC2D Studies of Mark I Containment Drywell Shell Melt-Through," NUREG/CR-5126, SAND 881407, Sandia National Laboratories, Albuquerque, NM, August (1988).

17. G. A. Greene, "Review of the Mark I BWR Steel Containment Shell Melt-Through Analyses," draft, Brookhaven National Laboratories, Upton NY, February (1988).

18. M. L. Corradini, "BWR MARK I Containment Loading After Reactor Vessel Breach", Presented at the BWR MARK I Workshop, Baltimore, Md. February 24-26 (1988).

19. V. V. Konsetov, "Heat Transfer During Bubbling of Gas Through Liquid, " Int. J. Heat Mass Transfer, 9, pp. 1103-1108, September (1966).

20. F. G. Blottner, "Hydrodynamics and Heat Transfer Characteristics of Liquid Pools with Bubble Agitation," NUREG/CR-0944, November (1979).

21. T. G. Theophanus, Presentation at the NRC 16th Water Reactor Safety Research Information Meeting, Gaithersburg, Md. October (1988).

22. R. E. Henry, R. J. Hammersley, B. Malinovic, R. Sehgal, and E. Fuller, "Experiments Relating to Drywell Shell-Core Debris Interactions," Presented at the BWR MARK I Workshop, Baltimore, Md. February 24-26 (1988).

23. D. A. Powers, J. E. Brockmann, and E. R. Copus, "Experiments to Validate Models of MARK I Liner Melting by Core Debris," Sandia National Laboratories, Albuquerque, NM, Personal Communication.

24. M. Berman, "A Critical Review of Recent Large-Scale Experiments on Hydrogen-Air Detonations," J. Nuclear Science and Engr., V93, No. 1 (1986).

25. M. P. Sherman et al., "The Behavior of Hydrogen During Accidents in LWRs," SAND 80-1495, NUREG/CR-1561, Sandia National Laboratories, Albuquerque, NM, August (1980).

26. A. L. Camp et al., "Light Water Reactor Hydrogen Manual," SAND 82-1137, NUREG/CR-2726, Sandia National Laboratories, Albuquerque, NM, August (1983).

27. M. W. Jankowski et al., "Hydrogen in Water-Cooled Nuclear Power Reactors," IAEA and CEC Report, 1987.

28. National Research Council, "Technical Aspects of Hydrogen Control and Combustion in Severe Light-Water Reactor Accidents," National Academy Press, Washington, D.C., (1986).

29. M. Berman, "Hydrogen Behavior and Nuclear Reactor Safety," Hydrogen Combustion Technology, Chap. 2, CRC Press, (1985).

30. P. Worthington, "Hydrogen Combustion," USNRC Report Uncertainty Papers on Severe Accident Source Terms, NUREG-1265, May (1987).

31. M. Berman, "A Critical Review of Recent Large-Scale Experiments on Hydrogen-Air Detonations," Proceedings of 23rd ASME/AICHE/ANS National Heat Transfer Conference (Denver, CO), American Nuclear Society, p. 316, August (1985).

32. M. P. Sherman et al., "The Effect of Transverse Venting on Flame Acceleration and Transition to Detonation in a Large Channel," Dynamics of Explosion-Progress in Astronautics and Aeronomics, 106, pp. 66-84, (1985).

33. M. P. Sherman, "Hydrogen Flame Acceleration and Detonation," 12th Water Reactor Safety Research Information Meeting, Gaithersburg, MD, October 22-26 (1984).

34. M. Schildknecht et al., "Versuche zur Freistrahlsundung von Wasserstoff-Luft-Semischen im Hinblick auf den Ubergang Deflagration-Detonation," Abschlussbericht fur Interatom, International Atomreaktorbau GmbH, Bergisch Gladbach, BleV-R-65.769-1, March (1984).

35. I. O. Moen and P. A. Thibault, OECD/IGUS Working Group Meeting on Unconfined Explosions and Blast Effects, Defense Research Establishment, Suffield (DRES), Ralston, Alberta, Canada, August 20-21 (1984).

36. I. O. Moen et al., "Transition to Detonation in a Large Fuel-Air Cloud," Comb. and Flame, May (1984).

The Interface between Containment Thermal Hydraulics and Fission Product Transport

I. H. DUNBAR
SRD/AEA Technology
Culcheth, Warrington, WA3 4NE, UK

1. INTRODUCTION

Containment thermal hydraulics concerns itself with the flows of heat and matter into and from the containment atmosphere, and with the circulation of the gases within the containment. Containment fission product transport studies the behaviour of fission product vapours and aerosols within the containment atmosphere. Special attention is paid to the competition between deposition onto surfaces within the containment building and leakage from the containment to the external environment. The two subjects have traditionally been seen as separate ones. The role of thermal hydraulics has been seen as assessing the threat to the containment structure, while that of fission product transport has been seen as calculating the radiological source term to the environment. Even without a coupling between the two sets of phenomena this dichotomy is a false one; the reason for wanting to predict containment failure is that its timing and mode is an important determinant of the timing and magnitude of the release. In addition there are in fact important couplings between the two sets of phenomena [1].

To begin with, thermal hydraulics is a major source of input for the fission product transport calculation. For example rates of heat and mass transfer to the wall are needed if one wishes to model aerosol removal by the processes known as thermophoresis and diffusiophoresis. Fission product transport specialists cannot produce a source term without supporting calculations on a whole range of accident phenomenology, of which thermal hydraulics is a part. A good example of a failure to match fission product transport modelling with supporting thermal hydraulics work can be found in the field of turbulent agglomeration modelling. Aerosol codes have for many years contained a model for turbulent agglomeration, in which the rate of the process depends on the value of the turbulent energy density dissipation rate, but for a long time even the crudest estimates of the magnitudes of this quantity were unavailable. Only recently have flow calculations for containment buildings, including turbulence models, begun to appear.

If the flow of information were only from the thermal hydraulics to the fission product calculation, then the two sets of calculations could be performed separately. There is however also question of feedback from the fission product behaviour to the thermal hydraulics. The fission product is not simply a passive contaminant. Their effects may not be important when calculating the pressure for the purposes of estimating the threat to the containment, but if more accurate thermal-hydraulic calculations are needed, then some form of coupling should be implemented. As we shall see, the effect of the thermal hydraulics on the aerosol is itself an area which does need this enhanced accuracy.

Three instances of ways in which the presence of fission products can influence the thermal hydraulics are the following:

1. DECAY HEAT - the fission products, both gaseous and aerosol, are a potentially important source of heat within the containment;

2. THERMAL RADIATION - the transfer of heat by thermal radiation between the atmosphere and structures and from one structure to another will depend on the concentration and size of the aerosol in the atmosphere;

3. VAPOUR CONDENSATION IN THE BULK ATMOSPHERE - the rate and even the extent of the condensation of a vapour within the bulk atmosphere, and the particle size distribution of the condensate so formed, depends on the nature of the aerosol already present in the atmosphere.

Taking proper account of the release of the decay heat in the containment atmosphere needs a simultaneous calculation of heat balances and the whereabouts of the fission products, but once such a coupled calculational framework has been set up, the effect is in principle simple to model. The subject of the interaction of aerosols with electromagnetic radiation is a complicated one, but once suitable models of the emissivity of aerosols are obtained, it is comparatively simple to include them within a calculational framework dealing with both aerosols and heat balance. The condensation of vapour onto aerosols induces a particularly tight coupling between aerosol behaviour and thermal hydraulics, and it involves problems both of detailed physical modelling and of the incorporation of the models within a computer code. Of all the coupling phenomena, it is the condensation of vapours in the bulk atmosphere, and more specifically the behaviour of steam in the containments of LWRs, which has attracted the most attention in recent years.

For these reasons the rest of this paper concentrates on the modelling of condensation. (Note that unless explicitly stated otherwise, "condensation" is used here to include negative condensation, i.e. the evaporation of liquid aerosols.) The following section looks at the potential importance of condensation on the source term, and then section 3 reviews the basic physics of the process. How this physics has been embodied in computer codes is the subject of section 4, and section 5 consists of concluding remarks.

2. THE IMPORTANCE OF STEAM CONDENSATION ONTO AEROSOLS

The potential importance of steam condensation to LWR severe accident source terms arises from the following facts. Even if the containment building eventually fails, it can still reduce the release to the environment if fission products airborne within the containment have time to deposit onto surfaces, thereby becoming much less mobile. With the exception of the noble gases, and some chemical forms of iodine, airborne fission products under containment conditions will be in the form of an aerosol, that is small liquid or solid particles suspended in the gas phase. Although there are a number of mechanisms which can cause aerosol particles to deposit onto surfaces, the most efficient of these under containment conditions is sedimentation under gravity. Larger particles sediment more rapidly than small ones; for the radii of greatest interest in source term calculations the rate is proportional to the square of the radius. Therefore growth of particles due to the condensation of steam (or some other vapour) onto the particles can be an important mechanism for enhancing the gravitational sedimentation of the particles. Evidence that this effect is a real one came from the Three Mile Island accident, and from experimental studies of the phenomenon of steam condensation.

2.1 The Three Mile Island Experience

In the aftermath of the accident at Three Mile Island it was noted that the fission product source term was much lower than would have been predicted using the methodology in use at the time. It was argued that the reason for this was the wet environment in the containment atmosphere [2]. It was claimed that iodine and caesium fission products would exist in the containment in form of water-soluble aerosols (caesium iodide for the iodine and caesium hydroxide for the rest of the caesium). The presence of the soluble materials would promote the condensation onto the aerosol particles, thereby making them grow and deposit more rapidly from the atmosphere by gravitational sedimentation. In effect, the fission products would be washed out of the atmosphere by the condensation they themselves helped to promote.

2.2 Experimental Studies

The possibility of being able to reduce the predicted source terms for severe accidents in LWRs, apparently borne out by the experience at Three Mile Island, motivated a large amount of experimental and modelling work on the the process of steam condensation onto aerosols. The development of models is described in detail in section 4 below. Among the first experimental studies of the phenomenon was the NAUA series of experiments [3], carried out at the Kernforschungszentrum Karlsruhe in the Federal Republic of Germany. These tests created supersaturated conditions by adiabatic expansion, thereby forcing condensation on the aerosols. They validated the Mason equation (see section 3 below) for condensation rates. This was the equation used in the NAUA code, and subsequently in all other aerosol codes with condensation models.

The NAUA experiments looked at how an aerosol would react to a given supersaturation. A separate series of experiments looked at the behaviour of aerosols in constant-volume vessels containing steam-air atmospheres. This situation is closer to that in a containment building. The heat loss is via the walls, and the steam can condense either on the walls or on the particles. The Nuclear Safety Pilot Plant (NSPP), a 38 cubic metre vessel at Oak Ridge National Laboratory in the United States of America, was used between 1980 and 1986 for an extensive series of tests on the behaviour of various insoluble aerosols in dry and saturated conditions [4]. A 640 cubic metre model containment at the Battelle-Frankfurt Institute in the Federal Republic of Germany was used between 1983 and 1986 for the DEMONA series of experiments [5], involving again insoluble aerosols in dry and saturated atmospheres. Finally the LWR Aerosol Containment Experiments (LACE) programme [6] was carried out in the 850 cubic metre Containment Systems Test Facility (CSTF) at the Hanford Engineering Development Laboratory (HEDL) in the United States of America, between 1985 and 1987. These experiments, which attempted to simulate different accident sequences, used mixtures of soluble and insoluble aerosols in saturated steam atmospheres.

The results of these experiments showed that significant enhancements of aerosol deposition rates could indeed be achieved by having saturated atmospheres. Figure 1 shows the aerosol mass concentrations as functions of time in the experiments DEMONA B2, B3 and B4. In B2 the atmosphere was dry, while B3 and B4 had the same aerosol source but a saturated steam atmosphere. The figure shows the accelerated fall-out in saturated conditions. In B2 it took approximately 23 hours for the aerosol mass concentration to be reduced by a factor of one thousand, whereas in B3 and B4 the time to effect the same reduction is only 5 hours. In the LACE programme there were no tests with dry conditions, so the same sort of direct comparison is not possible. However in Figure 2 we see the comparison between the fall-out actually observed in the experiment and two calculations by Jokiniemi [7], one assuming no condensation and the other using a model for the condensation of steam onto a hygroscopic aerosol. The assumption of dry conditions leads to a prediction of fall-out much slower than that observed, whereas the model for condensation gives good agreement. The time for the aerosol mass concentration to fall by a factor of one thousand in the experiment was approximately 5 hours. The curve for the dry calculation stops before this reduction is achieved, but a simple extrapolation gives 23 hours for the timescale. The acceleration of fall-out, and indeed the actual values of the times are almost the same as those in DEMONA.

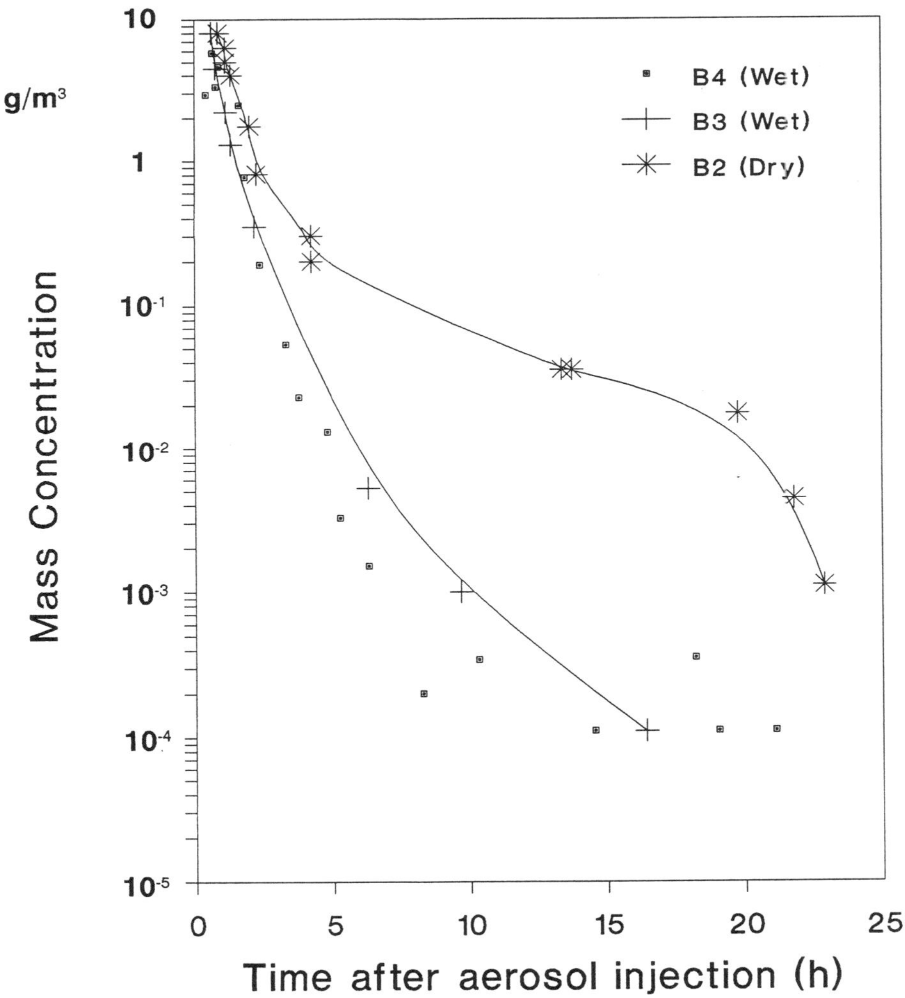

FIGURE 1. Comparison of wet and dry DEMONA experiments.

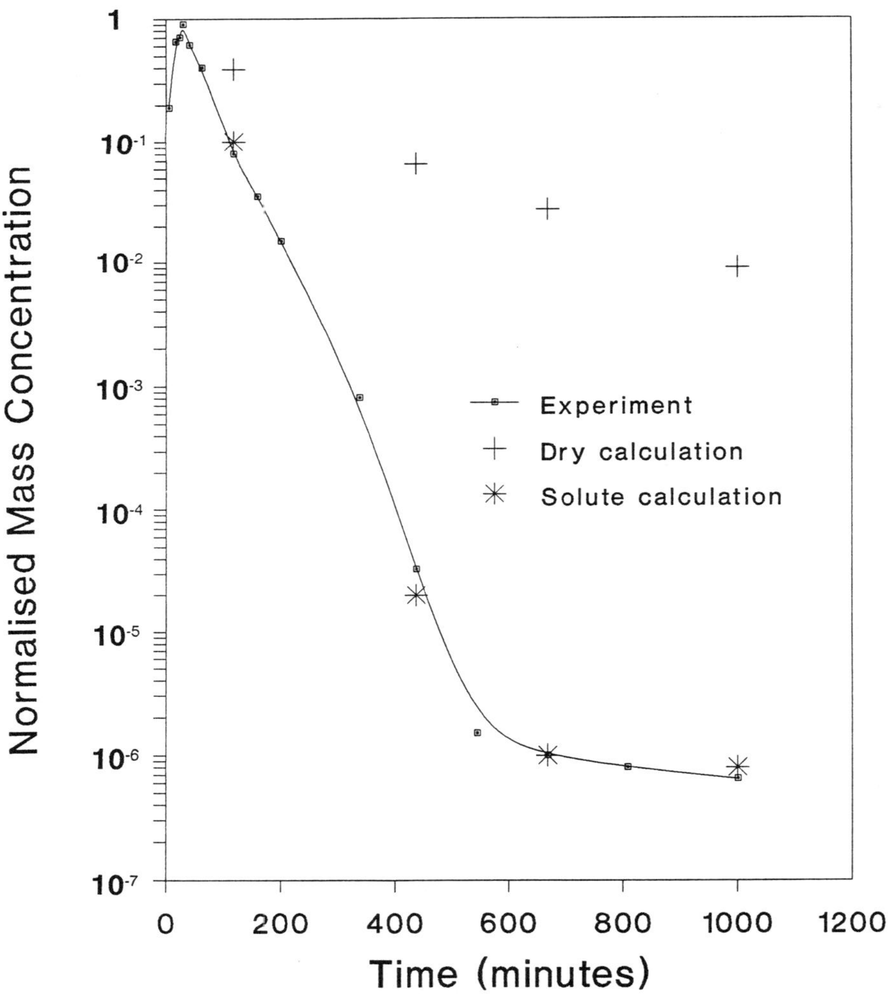

FIGURE 2. Calculations for LA-2, with and without steam condensation - from Reference [7].

These experiments show that less aerosol remains airborne at later times under saturated conditions than when the atmosphere is dry. These observations have provided a renewed motivation to develop computer codes which can accurately model steam condensation. But it should be noted that these experiments were designed so that the atmosphere would be saturated. To be able to claim reduced source terms on the grounds of steam condensation on the aerosol one still has to show that under accident conditions the steam in the containment atmosphere would indeed be saturated.

3. PHYSICAL PRINCIPLES

The central assumption in the modelling of steam condensation onto aerosols is that any condensation that can occur in the bulk atmosphere will go onto the aerosol particles already present in the atmosphere. Once this assumption has been made, models for the process have to address the following three questions:

1. what is the saturation ratio in the containment atmosphere?

2. for a given a saturation ratio, what is the condensation rate on a particle of given size and composition?

3. what does a given amount of condensation onto each particle do to the aerosol size distribution as a whole?

These questions are now addressed in turn, in reverse order.

3.1 Changes in the Aerosol Size Distribution

An aerosol can be characterised by a size distribution. This expresses either the number concentration or the mass concentration as a function of some measure of the particle size, such as its mass or its radius. Ignoring for the moment the effect of condensation, we find that the size distributions of aerosols in containments change with time under the influence of three processes:

- agglomeration, that is the process whereby particles collide and stick to one another, thus converting two smaller particles into one larger particle;
- deposition, that is the transport of particles to one of the surfaces within the containment, to which the particles become attached;
- aerosol sources, that is the injection of new aerosols into the containment atmosphere.

This time-dependence is given by an integro-differential equation, the form of which can be found, for example, in reference [8], along with a derivation of the additional term needed to account for condensation. An aerosol modelling code is essentially a means of solving this equation numerically.

The equation contains, for each of the processes listed above, a rate coefficient, which depends upon the detailed behaviour of individual particles. It is important to keep in mind when discussing aerosol modelling this distinction between modelling the rates of the individual processes, and then, given these rates, devising methods for solving the equation for the size distribution.

The growth of particles due to the condensation on them of some vapour brings a new term into the equation for the time-dependence of the size distribution. This term causes the size distribution as a whole to move up the size axis; mathematically it can be described as an advective term. It is quite different in form from the agglomeration term, and this fact causes difficulties for the numerical solution of the equation. Most of the numerical solution techniques involve approximating the size axis by some form of discrete grid. The agglomeration term is solved most efficiently using a fixed grid, whereas the condensation term is better dealt with using a grid which moves with the particle distribution [9]. Some of the compromises made in aerosol codes are discussed in section 4 below.

3.2 The Condensation Rate

Like the other terms in the equation, the condensation term contains a rate coefficient, namely the mass rate of condensation on a single particle. This rate, as a function of the saturation ratio, is given by the Mason equation [10]. This equation is derived by setting up the steady state diffusion equation for steam diffusing into the particle surface and the steady state heat conduction equation for the removal of heat from the particle. The latter is needed because the condensation on the particle releases latent heat. Unless this heat is removed from the particle, the temperature of the particle would rise and condensation would be suppressed. The actual rate of condensation is determined by a balance between the mass flux into the particle and (latent) heat flux out away from it. Given the two equations, one can eliminate the particle temperature between them and obtain the condensation rate as a function of bulk atmosphere properties and of the size of the particles. A simple form of the derivation, along the lines described above, can be found in reference [8].

If the water condensed onto aerosol particles had the same thermodynamic properties as a plane surface of pure water, then the condensation rate would simply be proportional to (S-1), where S is the saturation ratio. In other words, the rate would be proportional to the difference between the actual steam partial pressure in the atmosphere and the saturation partial pressure of steam at the atmosphere temperature. There are however factors which modify the saturation partial pressure of steam above the surface of an aerosol particle. The first is the so-called Kelvin effect. The water surface on a spherical particle is curved, and this raises the saturation partial pressure. The particles are therefore more ready to evaporate (and conversely, less ready to receive condensation) than a plane surface of water. Smaller particles have more highly curved surfaces, and so are even more likely to evaporate than the larger particles. This means that the Kelvin effect tends to destabilise aerosol size distributions. Consider, for example, a distribution of particles consisting of pure water surrounding an insoluble core. Let us further suppose that the ambient steam is in equilibrium with the water on particles of one particular radius. Particles with radii larger than this value will have saturation partial pressures smaller than the ambient partial pressure and will tend to grow by condensation. Conversely the smaller particles will shrink due to the evaporation of their water, until there is nothing left but the dry insoluble core. The net result is that the distribution splits in half, with the smaller particles getting smaller and the larger particles getting larger.

The thermodynamics of water on aerosol particles is also altered if the aerosol contains substances which are soluble in water (for example caesium hydroxide or caesium iodide). Solutes reduce the saturation partial pressure of the water. Steam is therefore more likely to condense on particles containing solutes than onto pure water. A list of the soluble substances which could form aerosols following a severe accident in an LWR can be found in reference [11]. To model the effect of a given soluble species on aerosol behaviour, one needs the following items of solution thermodynamics data as functions of temperature and solution concentration.

Vapour pressure of steam in equilibrium with the solution: this can be presented in the form of a phase diagram, or in terms of the water activity of the solutions. Measurements of the water activities for caesium hydroxide systems have been reported recently [12].

Heats of solution: the release of heat during condensation is an important factor in determining the amount of condensation. We ought therefore to be able to estimate how the presence of solutes affects amount of heat released.

Partial molar volumes: within an aerosol calculation one has always to derive from the masses of substances in a particle the radius of the particle, so strictly speaking we need information on partial molar volumes. The assumption that volumes are additive may however be good enough for the purposes of aerosol modelling.

Once we have a method of calculating condensation rates, we can ask the question of how sensitive these rates are to small changes in the thermodynamic conditions. By "small" here we mean small compared with the typical level of accuracy to be expected from thermal hydraulics calculations. The relevant thermodynamic variables are the steam partial pressure and the temperature. The latter determines the value of the equilibrium partial pressure for a given particle. Two factors are relevant to the question of sensitivity:

- the Mason equation predicts that small deviations of the steam partial pressure from the equilibrium value can give rise to large condensation or evaporation rates;

- the equilibrium partial pressure is sensitive to small changes in temperature.

These two facts combine to produce what might be called the sensitivity problem for the steam condensation on aerosols. Fission product transport calculations including steam condensation onto aerosols ask the thermal hydraulics calculations for a much greater accuracy than is needed to assess the threat to the containment. The level of accuracy needed may in fact be impossible to achieve. In particular we have spoken of <u>the</u> steam partial pressure in the containment as if this were a unique number. For a large volume like the containment building there are in fact likely to be significant spatial variations in this quantity. One of the tasks of the new generation of coupled aerosol and thermal hydraulics codes will be to test whether the sensitivity problem is an important restriction in practice on our ability to predict accurately the behaviour of aerosols in a saturated steam environment.

3.3 The Saturation Ratio

Having seen how to calculate the rate of steam condensation as a function of the saturation ratio, we now consider how this quantity is itself calculated. Firstly, the saturation ratio depends on the following boundary conditions:

- heat and mass inputs to the containment atmosphere, for example from the primary circuit, from hot pools of water in the containment sumps, and from the decay of radioactive substances in the atmosphere (heat only);

- fluxes of heat and steam to walls of the containment, the walls typically being colder than the containment atmosphere;

- leaks of gas from the containment.

In addition the saturation ratio depends on the overall rate of bulk condensation. This sets up a feedback loop, which causes the tight coupling between thermal hydraulics and aerosol behaviour. Bulk condensation reduces its own driving force in two ways. The removal of steam from the atmosphere moves its partial pressure down towards the equilibrium value. The release of latent heat into the atmosphere raises its temperature, and this moves equilibrium pressure up towards actual partial pressure. Both of these effects drive the system towards equilibrium, and therefore act to suppress further condensation. (A similar negative feedback, with the directions reverse occurs for evaporation.) In fact a simple calculation [11, 13] shows that the release of latent heat is the main determinant of how much steam is condensed during this relaxation process. This is a clear argument for the need to couple the equations for both heat balance and mass balance to the aerosol equation.

A second conclusion is that bulk condensation can proceed only if there is a removal of latent heat from the atmosphere to the ultimate heat sink, namely the containment walls. Otherwise condensation will switch itself off. It has been shown by Clement [14,15] that the amount of bulk condensation in a saturated atmosphere being cooled through the walls is governed by the relative efficiencies of the transport of heat and steam to the walls. Put simply, if the mass transfer of steam is more efficient then the condensation will occur on the wall surfaces, the heat being removed as the latent heat released at the walls. Conversely, if heat removal is more efficient, then the condensation will occur in the bulk atmosphere, and the latent heat thus released will be transported to the wall as sensible heat. This effect was seen clearly by Ricchena in an analysis of the LACE experiment LA-4 [16]. He found that in order to predict the saturation conditions observed in the experiment, one needed to model thermal radiation from the gas to the wall. This radiation formed only small part of total heat flux to the walls, the greater part of which is carried as latent heat by the steam diffusing to and condensing on the walls. The radiation was however the largest fraction of the "dry" heat flux, and so was crucial in determining the extent of bulk condensation.

It is important to distinguish between the two different heat transport effects discussed above. The Mason equation condensation rate for the individual particles depends, among other things, on the rate at which the latent heat can be transferred from the particle to the bulk atmosphere. The way the saturation ratio evolves with time, and therefore the total extent of condensation on the aerosol as a whole, depend on the rate at which the latent heat can be transferred from the atmosphere to the wall. One should also note that the strong negative feedback effects associated with bulk condensation may mitigate the sensitivity problem by damping out errors in the calculation of the saturation. The overall rate of bulk condensation may be governed by what goes on at the walls and not by the transient details of the internal state of the atmosphere.

4. THE MODELLING OF STEAM CONDENSATION ONTO AEROSOLS

Having looked at the basic physics underlying the process of bulk condensation and its effects on the aerosol, we now move on to a discussion of how these phenomena have actually been modelled in the computer codes which analyse severe accident phenomenology in LWR containments. These codes can be divided into those that model thermal hydraulics only, those that model aerosols only, and those in which the two aspects are coupled. Established codes of these three types are described below, followed by a discussion of some more recent developments.

4.1 Thermal-Hydraulic, Aerosol and Coupled Codes

In traditional containment thermal hydraulic codes, for example the MACE module of the MARCH code [17], a distinction is made between the condensation on the walls and the condensation in the bulk. Surfaces are not assumed to be in equilibrium with the bulk atmosphere. Instead there is a kinetic treatment of transport processes to surfaces, based on heat and mass transfer coefficients. By contrast, the water aerosol suspended in the bulk atmosphere is assumed to be in equilibrium with the steam. Within each time-step, once heat and mass transfers to and from sinks and sources have been dealt with, the steam remaining in the atmosphere is equilibrated with the airborne water. If the steam is initially supersaturated then some of it condenses. If the steam is initially subsaturated then any airborne water present evaporates. If there is no airborne water then the atmosphere is allowed to remain subsaturated. These stand-alone thermal hydraulics codes usually have what amounts to a simple aerosol model for the airborne water. The water is assumed to fall out with fixed rate, that is, as if it had a single fixed drop size.

The modelling of steam condensation in aerosol codes has been reviewed in detail in reference [8]. The first aerosol code to model the effect was NAUA [18]. It includes the condensation growth term in the aerosol size distribution equation, and the Mason equation with the curvature effect but not the solute effect. The total bulk condensation rate as a function of time is required as an input (for example, from a thermal hydraulic code). It is converted by the code at each time step into an effective saturation ratio, which is then used in Mason equation to produce condensation rates for individual particles. These rates can be integrated over the aerosol size distribution to recalculate a total condensation rate, which should be consistent with input value.

The term for condensation growth has been included in a generalisation of the AEROSIM code [19], originally written to study sodium fire aerosols. As with NAUA, AEROSIM takes as input a total condensation rate. This is however translated directly into a rate per particle without going via a saturation ratio. The Mason equation is used only to give the distribution of the total condensation over the particle sizes. The price paid for this simplification is that the code cannot then model the solute effect, and can treat the Kelvin effect in only the most rudimentary manner.

AEROSOLS [20], like AEROSIM and NAUA, is a stand-alone aerosol code which includes the condensation growth term and Mason equation rates. It was however designed specifically to look at the behaviour soluble aerosols, particularly when the steam in the atmosphere is slightly below its saturation pressure. It takes as input from a thermal hydraulic code the saturation ratio.

In contrast to the codes described above, CONTAIN 1.0 [21] is an integrated thermal hydraulic and fission product transport containment code. It models steam condensation onto aerosols, and the consequent coupling between thermal hydraulics and aerosol behaviour in the following way. Each time-step the thermal hydraulics module does an equilibration calculation, similar to that described above for stand-alone thermal hydraulic codes. It then passes the amount available for condensation (or needed for evaporation) on to the module dealing with heat and mass transfer to the wall. The steam mass remaining after condensation on the wall is finally passed on to the aerosol module where it is converted into a saturation ratio, for use in the Mason equation, and the resulting changes to the aerosol size distribution are computed. The solute and curvature effects are not treated. The net result of all the steam transfers is a new effective steam source for the equilibration calculation in the next time-step. For a more detailed description of this procedure, see reference [8].

4.2 Recent Developments

The discussion so far describes the situation in 1985, as reviewed in reference [8]. A number of further developments were described at a workshop held in Brussels in 1987, the proceedings of which include the references [5, 9, 11, 15, 16, 22 and 23]. There were two independent extensions of NAUA to include solute effects: SW-NAUA [22] and NAUA HYGROS [7,23]. The latter code performs a steam mass balance calculation, but not a heat balance calculation. It relies instead on the atmosphere temperature being supplied as input, either from a thermal hydraulics code, or, in the case of an experiment, from the actual measurements. It can therefore be thought of as "half coupled" to the thermal hydraulics.

Also described at the Brussels workshop was new modelling intended for the CONTAIN code [9]. The problem of numerical diffusion had been found to be particularly serious when fixed grid methods were used to model condensation growth. A moving grid method was developed to eliminate this problem. The method involves following the growth of particles corresponding to the grid points across the time-step. The growth equations for the different grid points are decoupled from each other by assuming a constant background steam concentration during the time-step. The integral which relates the time-step length to the change in the position of the section boundary is singular at equilibrium, so an equilibrium calculation is needed to ensure that the integration stays on one side of the singularity. Once the condensation calculation is completed, the new distribution is mapped back onto a fixed grid so that the agglomeration and deposition calculation can be performed for the time-step. It is planned to implement this method in a new version of CONTAIN, which will also include the Kelvin and solute effects.

A different approach is being adopted in the development of the code ITHACA, based on the work reported in reference [13]. The intended scope of this code is more restricted than that of CONTAIN. It will model the containment atmosphere thermal hydraulics up to but not including the wall. Conduction in the wall is not modelled, and the user will have to supply the code with a wall temperature as a boundary condition.This means that ITHACA will not be a fully predictive model of containment phenomena. It is instead intended to be a method for looking in detail at physical and numerical issues associated with the coupling between thermal hydraulics and aerosol behaviour. ITHACA can be seen as the next stage in the development of aerosol codes beyond SW-NAUA and NAUA HYGROS, performing the heat balance calculation as well as that for the mass balance. The specification for ITHACA includes the requirement that it should solve the aerosol and thermodynamic equations strictly simultaneously, rather than sequentially within the time-step. This enforces the choice of a fixed grid, without which the agglomeration and condensation terms have to be treated sequentially. Care is being taken to devise methods to reduce numerical diffusion to an acceptable level within the context of a fixed grid approach.

5. CONCLUSIONS

An accurate calculation of containment fission product transport requires input concerning the thermal hydraulics of the containment atmosphere. There are also ways in which the fission products have an influence back on the thermal hydraulics (the so-called "back-coupling" effects). This means that the thermal hydraulic and fission product transport equations have to be solved in a coupled manner (for example, within the same computer code), at least when one is modelling phenomena like steam condensation on aerosols which require highly accurate thermal hydraulic information.

Two of the back-coupling effects: the release of decay heat and the effect on thermal radiation of an aerosol, are comparatively straightforward to include within coupled containment codes once adequate physical models are available. Treating steam condensation onto aerosols required greater modifications to previously existing methods. Over last few years there has been a great deal of work on steam condensation, both in performing experiments and developing models. This has been motivated by the potentially large reduction in source terms one can claim if steam condensation on aerosols can be proven, because steam condensation onto particles causes them to grow and they are then removed more rapidly from the atmosphere by gravitational sedimentation. This effect which has been clearly seen in experiments.

The condensation rate onto particles is very sensitive to the saturation ratio. There is a fear that it could be too sensitive, and that it may not be possible to calculate thermal hydraulics accurately enough to predict steam condensation onto particles reliably. Condensation has however a strong negative feedback on the saturation ratio, and this may have the effect of reducing the sensitivity. Computer codes of increasing sophistication have been developed in recent years; the current generation of codes should be able to investigate the problem of sensitivity, and also make meaningful comparisons with the body of experimental evidence now available.

6. REFERENCES

1. Dunbar, I. H. Some Aspects of the Interaction Between Thermal Hydraulics and Fission Product Transport in Reactor Containments under Severe Accident Conditions. *Proceedings of the ANS/ENS International Topical Meeting*

on Probabilistic Safety Methods and Applications, San Francisco, 24 February-1 March 1985, paper 125.

2. Levenson, M. and Rahn, F. Realistic Estimates of the Consequences of Nuclear Accidents. *Nuclear Technology*, 53, 99-110 (1981).

3. Schoeck, W., Bunz, H. and Koyro M. Messung von Wasserdampfkondensation an Aerosolen unter LWR-unfalltypischen Bedingungen. KfK 3153 (1981).

4. Adams, R. E., Tobias, M. L. and Petrykowski, J. C. Aerosol Behaviour in a Steam-Air Environment. *Proceedings of the CSNI Specialist Meeting on Nuclear Aerosols in Reactor Safety, Karlsruhe*, 4-6 September 1984, pp 435-447, KfK 3800, CSNI 95.

5. Haschke D. and Schoeck W. Results of the DEMONA Aerosol Removal Demonstration Experiments. *Proceedings of the the OECD/CEC Workshop on Water-Cooled Reactor Aerosol Code Evaluation and Uncertainty Assessment, Brussels*, 9-11 September 1987, pp 88-99, EUR 11351 EN.

6. Rahn, F., Sher, R. and Vogel, R. C. Summary of the LWR Aerosol Containment Experiments (LACE) Programme. *Proceedings of the IAEA/NEA International Symposium on Severe Accidents in Nuclear Power Plants, Sorrento*, 21-25 March 1988, pp 185-200.

7. Jokiniemi, J. The Effect of Selected Binary and Mixed Solutions on Steam Condensation and Aerosol Behaviour in Containment. *Proceedings of the Symposium on Nuclear Reactor Severe Accident Chemistry, toronto,* 6-10 June, 1988 (to be published).

8. Dunbar I. H. Steam Condensation Modelling in Aerosol Codes.
EUR 10584 EN, 1986

9. Gelbard, F. Aerosol Growth in a Nuclear Reactor Containment Environment. *Proceedings of the the OECD/CEC Workshop on Water-Cooled Reactor Aerosol Code Evaluation and Uncertainty Assessment, Brussels*, 9-11 September 1987, pp 256-266, EUR 11351 EN.

10. Mason, B.J., '*The Physics of Clouds*' Oxford University Press, 1971.

11. Clough P. N., Dunbar I. H. and Knights C. F. The Chemical Composition of Containment Aerosols and its Effect on Steam Condensation. *Proceedings of the the OECD/CEC Workshop on Water-Cooled Reactor Aerosol Code Evaluation and Uncertainty Assessment, Brussels*, 9-11 September 1987, pp 312-328, EUR 11351 EN.

12. Johansson, L.-G. and Kazikowski, S. An Experimental Study of Equilibrium Water Vapour Pressure in the CsOH-Water System. *Proceedings of the Symposium on Nuclear*

Reactor Severe Accident Chemistry, 6-10 June, 1988, Toronto (to be published).

13. Dunbar, I. H., Preparatory studies for modelling steam condensation on soluble aerosols. EUR 11807 EN, 1988

14. Clement, C.F., Aerosol Formation from Heat and Mass Transfer in Vapour-Gas Mixtures' *Proc. Roy. Soc. Lond.* A389, 307-339, 1985.

15. Clement C. F. The calculation of Aerosol Growth Rates in Conjunction with Heat and Mass Transfer. *Proceedings of the the OECD/CEC Workshop on Water-Cooled Reactor Aerosol Code Evaluation and Uncertainty Assessment, Brussels*, 9-11 September 1987, pp 156-169, EUR 11351 EN.

16. Ricchena R. CEC Comparative Exercise on the Performances of the Computer Codes Simulating the LA-4 Experiment. *Proceedings of the the OECD/CEC Workshop on Water-Cooled Reactor Aerosol Code Evaluation and Uncertainty Assessment, Brussels*, 9-11 September 1987, pp 21-36, EUR 11351 EN.

17. Wooton, R. O., Cybulskis, P. and Quayle, S. F., MARCH 2 (Meltdown Accident Response Characteristics) code description and user's manual. NUREG/CR-3988, BMI-2115 (1984).

18. Bunz, H., Koyro, M., NAUA Mod 5 und NAUA mod 5-M : Zwei Computerprogramme zur Berechnung des Aerosolverhaltens im Containmentsystem eines LWR nach einem Kernschmelzunfall. KfK 4278 (1987).

19. Ramsdale, S. A., AEROSIM-M user manual. SRD R 401 (1987).

20. Gauvain, J., AEROSOLS B2 - Version 1. Code for computing aerosol behaviour. Presentation report. Technical report SASC/87/50 (1987).

21. Bergeron, K. D., Clauser, M. J., Harrison, B. D., Murata, K. K., Rexroth, P. E., Schelling, F. J., Sciacca, F. W., Senglaub, M.E., Shire, P. R., Trebilcock, W. and Williams, D. C. User's Manual for CONTAIN 1.0, a Computer Code for Severe Nuclear Reactor Accident Containment Analysis. NUREG/CR - 4085, SAND84-1204, 1985.

22. Drozd, A. and Baron, J. Hygroscopic Aerosol Growth at Near Saturation Conditions. *Proceedings of the the OECD/CEC Workshop on Water-Cooled Reactor Aerosol Code Evaluation and Uncertainty Assessment, Brussels*, 9-11 September 1987, pp 291-303, EUR 11351 EN.

23. Jokiniemi, J. and Sher, R. Modelling the Growth of Water Soluble Particles in the NAUA Code. *Proceedings of the the OECD/CEC Workshop on Water-Cooled Reactor Aerosol Code Evaluation and Uncertainty Assessment, Brussels*, 9-11 September 1987, pp 280-290, EUR 11351 EN.

Validation of Core Debris/Concrete Interactions and Source Term Models

D. A. POWERS, D. R. BRADLEY, J. E. BROCKMANN, and E. R. COPUS
Sandia National Laboratories
P.O. Box 5800
Albuquerque, New Mexico 87185, USA

G. A. GREENE
Brookhaven National Laboratory
Upton, New York, USA

S. B. BURSON
U.S. Nuclear Regulatory Commission
1717 H Street, NW
Washington, D.C. 20555, USA

1. INTRODUCTION

Severe nuclear reactor accidents--accidents involving the melting of the reactor core--dominate the residual risk associated with the use of nuclear power [1]. The uninterrupted progression of a severe reactor accident is expected to lead to the expulsion of core debris into the reactor containment. Many safety-significant phenomena may be hypothesized to occur when core debris is expelled from the reactor coolant system [2]. The exact nature of these events depends on whether or not the coolant system is pressurized at the time of melt expulsion and whether or not expulsion is into water. Regardless of what transient events are associated with the initial expulsion of core debris from the reactor coolant system, a protracted period of core debris interactions with the structural concrete of the reactor is expected in most analyses of severe reactor accidents.

The physical consequences of core debris interactions with concrete have been routinely considered in severe reactor accidents for a long time. The first comprehensive account of core debris/concrete interactions appeared in connection with the Reactor Safety Study [1]. It was assumed in this analysis that the entire core inventory was instantly deposited in the reactor cavity as a constant temperature, homogeneous mass. Heat was thought to be generated in the core debris strictly as a result of radioactive decay, and this heat was arbitrarily partitioned into the concrete and into the containment atmosphere. Gases (H_2O and CO_2) liberated from the concrete were assumed to pass around or though the core debris without chemical reaction and to contribute to the pressurization of containment. If containment pressurization or some other event did not rupture the reactor containment, the analyses showed erosion of the concrete would, eventually, result in a loss of containment integrity.

Experiments [3] to examine high temperature melt interactions with concrete did not validate the model used in the Reactor Safety Study. These experiments showed the melts did not remain at constant temperature. Of more importance, gases liberated from the concrete reacted with metals to form combustible gases (H_2 and CO).

This work was supported by the U.S. Nuclear Regulatory Commission and performed at Sandia National Laboratories which is operated for the U.S. Department of Energy under Contract Number DE-AC04-76DP00789.

Combustible gases generated during the interactions not only augmented the deflagration threat to containment integrity, they also implied that there was an additional, chemical, heat source augmenting the decay heat source and driving the attack on concrete.

The experimental results showed that a substantially revised model of core debris/concrete interactions was required for severe accident analyses. This need was amplified by advances in the understanding of containment response to severe accident phenomena with regard to both the magnitude of threats to containment integrity and the behavior of radionuclides suspended in the containment atmosphere and available for release should containment integrity be lost [4]. Application of this improved understanding to accident analyses required more detailed input concerning the source terms of heat, gas and aerosols generated by the interactions of core debris with concrete.

The need for improved models of core debris interactions with concrete has been met by several computer codes [5,6,7]. In this paper, the status of validation of one of these codes, CORCON [7], will be discussed. The phenomenon that the CORCON code and the other core debris/concrete interactions codes attempt to model is the thermal attack on concrete by an internally heated, chemically reacting, molten mixture of metals (steel and unoxidized zirconium cladding) and oxides (UO_2 and ZrO_2, predominantly). Two key features of the models that must be validated if confidence is to be placed in the model predictions are the (1) heat transfer from core debris, and (2) heat generation especially that caused by chemical reactions. A third key feature, the release of radionuclides during core debris/concrete interactions, is not explicitly addressed in this paper.

2. HEAT TRANSFER FROM CORE DEBRIS

Heat transfer from core debris during its interactions with concrete can be broadly categorized as (1) axial heat transfer into the concrete, (2) radial heat transfer into the concrete or other materials, and (3) upward heat transfer into the containment atmosphere or overlying water pool. Because the heat sources are not uniformly distributed between the immiscible phases of core debris (metallic and oxidic), a fourth area, heat transfer between phases, must be considered in addressing these broad areas.

2.1 Axial Heat Transfer

The magnitude of the heat flux to the concrete, especially in the axial direction, has been emphasized throughout the history of the study of core debris interactions with concrete. The intense sparging of melts by gases from the concrete has provoked many analogies to boiling. The first, widely used, model of the axial heat transfer from core debris to concrete was an analogy to film boiling [8,9]. Though the existence of a gas film limiting heat transfer could not be established experimentally [10], data on features of melt interactions with concrete pertinent to reactor accident analyses, such as concrete ablation and melt temperatures, were for many years too crude to disprove this model. Most accident analyses done prior to 1983 entailed descriptions of core debris interactions with concrete based on the gas film model of heat transfer. One of the most important results obtained in the BETA tests of melt/concrete interactions [11] conducted in West Germany was the data base necessary to conclusively demonstrate that the gas film model underpredicted heat transfer to concrete.

Replacements for the gas film model have been based largely on analogies to nucleate boiling. Perhaps the most sophisticated of these models is one devised by Bradley [12]. This model, which is being incorporated into the CORCON code, considers the serially related resistances to heat transfer inherent in melt convection to a boundary layer, formation, growth and detachment of gas bubbles, and conduction across films of molten concrete. The model successfully predicts the concrete erosion observed in tests with molten steel interacting with concrete. Predictions of this model are compared to experimental results from two such tests in Figure 1.

2.2 Radial Heat Transfer

Radial heat transfer from molten core debris has not received the attention devoted to axial heat transfer. It is widely accepted that the radial heat transfer may be different than axial heat transfer. It is also recognized that radial heat transfer, because it can involve attack on critical internal structures of the reactor containment, may be at least as safety-significant as axial heat transfer. Further, radial heat transfer may be to materials other than concrete. Some data have been obtained on radial heat transfer from gas-sparged steel melts and simulant melts to structural steel and materials that do not evolve gas when under thermal attack [13,14].

2.3 Upward Heat Transfer

When there is no water overlying the core debris, the upward heat transfer from core debris to the confined spaces of typical reactor cavities is thought to be dominated by radiation with a modest enhancement by convection [7]. Complexities in the analysis of this heat transfer arise in estimating how aerosols produced by the core debris interactions might affect radiation. Crude models have shown that the opacity of the gas phase caused by aerosols is not a negligible effect. Neither definitive models nor experimental data have appeared in this area.

The presence of water over core debris poses more substantive heat transfer questions. Certainly, there is the possibility that the overlying water could lead to a steam explosion [15]. Unpublished arguments have been made suggesting water would quench the core debris and terminate the attack on concrete. Neither steam explosions nor complete quenching has been observed in core debris/concrete experiments. Reported tests of simultaneous core debris/concrete/coolant interactions [16,17] have shown that overlying water quenches a porous, surface crust of debris, attenuates the aerosol emissions and otherwise little affects the attack on concrete. It appears, however, that the heat flux to the water pool cannot be predicted accurately by simple boiling. Current thinking suggests that the non-condensible gas generation produced by core debris attack on concrete augments the film-boiling heat flux to water (barbotage

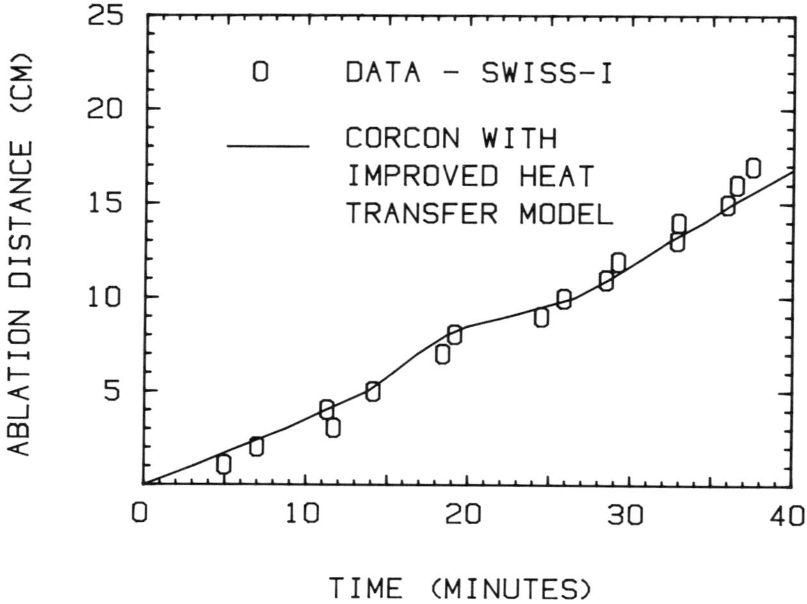

FIGURE 1a. Comparison of CORCON predictions to the concrete erosion observed in the SWISS-I test [16].

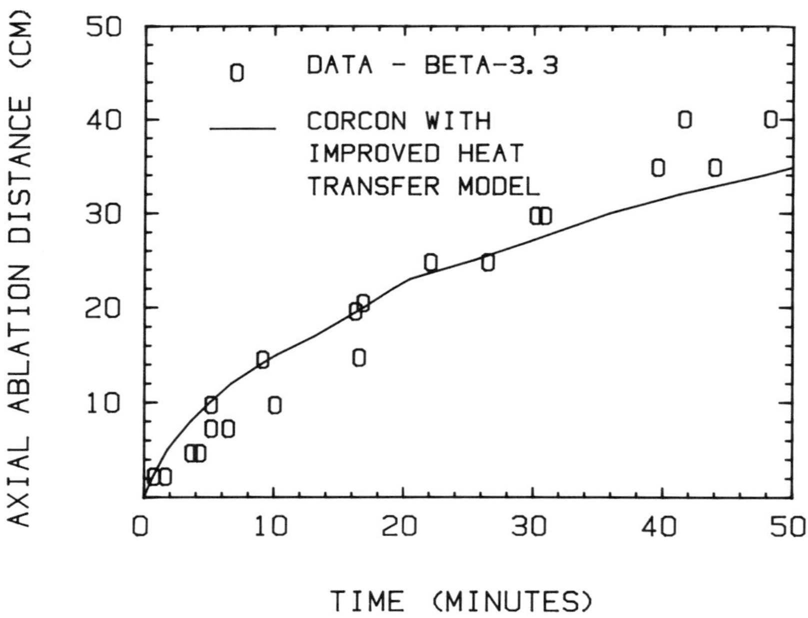

FIGURE 1b. Comparison of CORCON predictions to the axial concrete erosion observed in the BETA 3.3 test [11].

effect). Experimental validation of this idea has been undertaken using simulants [18] and a barbotage enhancement of film boiling is being incorporated into the CORCON code.

3. HEAT GENERATION IN CORE DEBRIS

The augmentation of decay heating of core debris by exoergic reactions especially with metallic constituents of core debris is a critical aspect of reactor accident analyses. A remarkable evolution in the understanding of the composition of core debris that might be expelled from a reactor and engage in attack on concrete has occurred in the last 10 years. It had been thought that any zirconium cladding on fuel would be completely oxidized to ZrO_2 during core degradation. Some models of core degradation even provided non-mechanistic enforcement of this belief [19]. Advances in the understanding of core meltdown have now shown that not all of the zirconium will oxidize within the reactor coolant system [20]. Especially in the cases of accidents at boiling water reactors, a substantial amount of unoxidized zirconium may be present in core debris attacking concrete. This highly reactive zirconium can have a pandemic effect on the interactions of core debris outside the reactor vessel. It certainly augments the potential chemical heat source as a result of reactions such as:

$$[Zr]_{metal} + 2H_2O \rightarrow ZrO_2 + 2H_2 \qquad (1)$$

The reactions of zirconium can more completely reduce to combustible forms the gases evolved from the concrete. The unoxidized zirconium may even affect the relative densities of the oxidic and metallic phases of core debris. The presence of zirconium metal challenges the assumption in the CORCON code that chemical interactions between the phases of core debris can be neglected.

Only recently have tests been done with zirconium metal in melts attacking concrete [21,22]. Results of one of these tests, the SURC-4 test, are shown in Figure 2. In this test, a sustained interaction of 200 kg of molten stainless steel with a siliceous concrete (55 w/o SiO_2) was established. Then, 20 kg of zirconium metal were added. Admission of the zirconium promptly initiated dramatic rises in the melt temperature, the rate of gas production, the rate of aerosol evolution, and the rate of concrete erosion.

As shown in Figure 2, the results of the SURC-4 test could not be well predicted with the CORCON code using either the gas film model or Bradley's improved model of heat transfer to the concrete. The sudden temperature rise and concomitant phenomena can be predicted only if there is an additional source of heat. This additional heat source has been identified as the reaction of zirconium with silicon dioxide in droplets of ablated concrete rising through the metal melt:

$$[Zr]_{metal} + SiO_2 \rightarrow ZrO_2 + [Si]_{metal} \qquad (2)$$

The silicon metal product of reaction also reacts exoergically with gases from the concrete:

$$[Si]_{metal} + 2H_2O \rightarrow SiO_2 + 2H_2 \qquad (3)$$

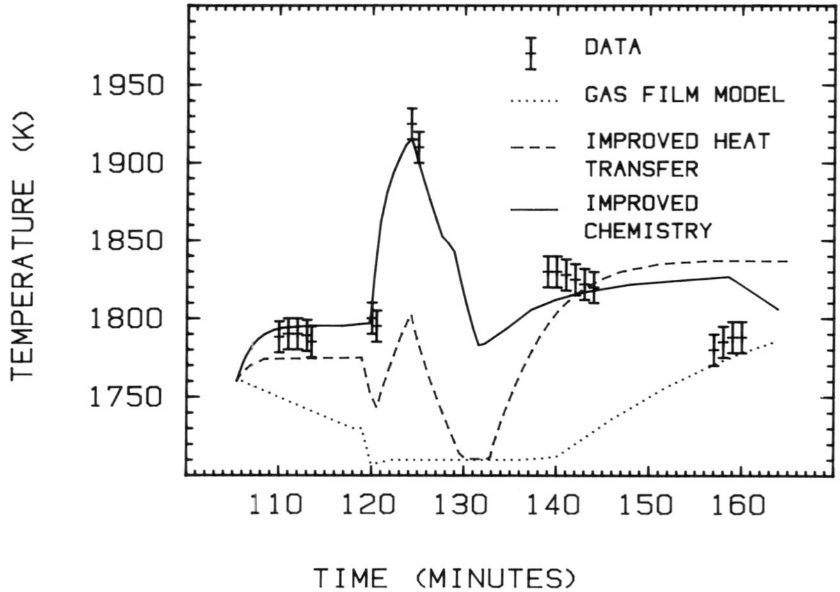

FIGURE 2a. Melt temperature data from test SURC-4 and predictions obtained with the CORCON code. Zr added at 119 minutes.

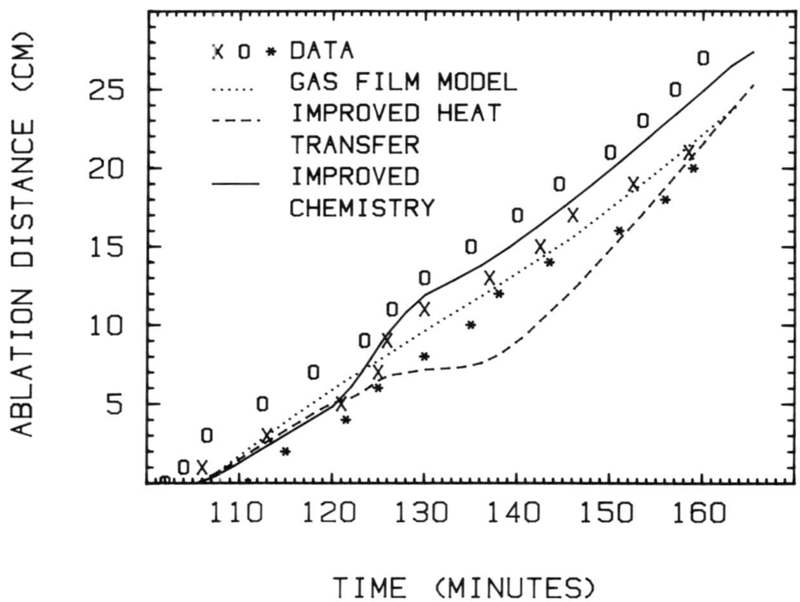

FIGURE 2b. Concrete ablation data from test SURC-4 and predictions obtained with the CORCON code. Zr added at 119 minutes.

Recognition in CORCON of the additional oxidant fluxes provided by silicon dioxide and available to react with zirconium yields predictions of the sudden temperature rise in excellent agreement with data from test SURC-4 as shown in Figure 2. There is, however, an indication in this comparison of data and predictions that details of the model predictions of heat production and reaction rates need to be further refined to achieve complete agreement with the data. These refinements may be quite worthwhile since the additional reactive flux provided by silicon dioxide from ablated concrete can significantly alter conclusions concerning source terms to the containment during core debris interactions with siliceous concretes.

4. EXTENSION TO OXIDE MELTS

Models of core debris interactions with concrete, including the CORCON code, have been developed and are being validated based on data obtained in tests of molten metals interacting with concrete. A crucial test of these models is to see if they can be extended without modification to predict the interactions of oxide melts with concrete. Only recently have data become available to make this test of the model's predictive capabilities [23].

Test SURC-1 involved the sustained interaction of 206 kg of a molten UO_2-ZrO_2-Zr mixture with a calcareous concrete. Results obtained in this test are shown in Figure 3. Also shown in this figure are the predictions obtained with CORCON modified with the improved heat transfer model and an allowance for condensed-phase chemical reactions as discussed in connection with the results of the SURC-4 test. The qualitative agreement between the test data and the code predictions is quite good. As in the case of the comparison to the results of the SURC-4 test, there is an indication from this comparison that descriptions of the heat generation and reaction rates when metallic zirconium is present need to be further refined to improve the quantitative agreement with the experimental data.

5. SUMMARY

The current state of validation of models of core debris interactions with concrete involves examination of heat transfer processes and the chemical generation of heat. Axial heat transfer models developed for the CORCON code yield predictions in good agreement with data from experiments. Radial heat transfer and heat transfer to water overlying core debris attacking concrete merit further examinations. Core debris containing the very reactive material zirconium has been found to interact with concrete both thermally and chemically. Condensed-phase chemical reactions between the oxide and metallic constituents of melts interacting with concrete augment the total heat generation and the rate of heat generation significantly. An allowance for these condensed-phase reactions needs to be included in the models of core debris/concrete interactions. Once these improvements are made, it appears that the models based on data for metallic melt interactions with concrete will also be applicable to the prediction of oxide melt interactions with concrete.

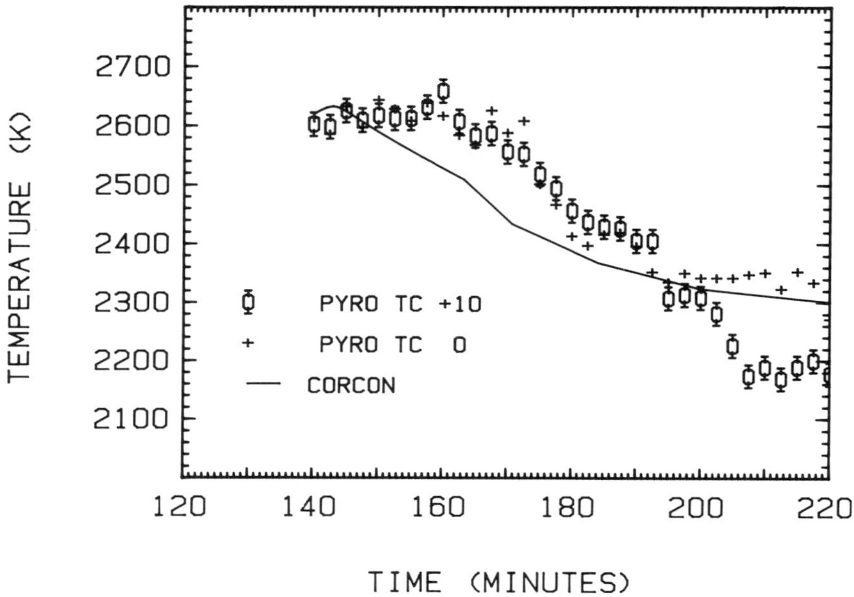

FIGURE 3a. Melt temperatures observed in test SURC-1 and predictions of these temperatures obtained with the improved CORCON code.

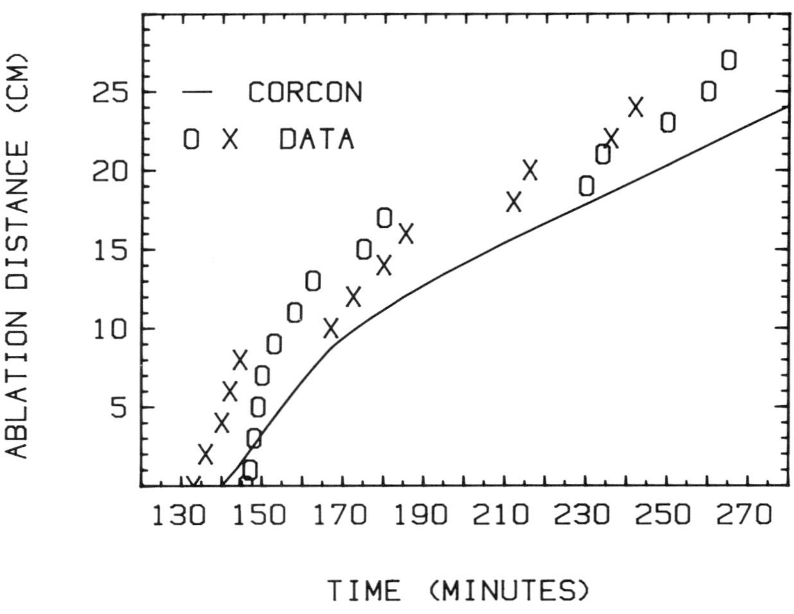

FIGURE 3b. Concrete erosion observed in test SURC-1 and predictions of the erosion obtained with the improved CORCON code.

6. REFERENCES

1. U. S. Nuclear Regulatory Commission, Reactor Safety Study-An Assessment of Accident Risks in U. S. Commercial Nuclear Power Plants, WASH-1400, NUREG-75/014, October 1975.

2. Speis, T. P., Sheron, B. W., Eltawila F., and Theophanous, T. G., The Examination of Containment Systems Performance in Individual Plant Examinations. *International Topical Meeting on Probability Reliability and Safety Assessment*, Pittsburgh, PA, April 2-7, 1989.

3. Powers, D. A., Dahlgren, D. A., Muir, J. F., and Murfin, W. B., Exploratory Study of Molten Core Material/Concrete Interactions, SAND77-2042, Sandia National Laboratories, Albuquerque, NM, February 1978.

4. Bergeron, K. D., Clauser, M. J., Harrison, B.D., Murata, K. K., Rexroth, P. E., Schelling, F. J., Sciacca, F. W., Senglaub, M.E., Shire, P. R., Trebilcock, and Williams, D. C., User's Manual for CONTAIN 1.0, A Computer Code for Severe Nuclear Reactor Accident Containment Analysis, NUREG/CR-4085, SAND84-1204, Sandia National Laboratories, Albuquerque, NM, May 1985.

5. Alsmeyer, H., Barleon, L., Koster, J., Michael, I., Miller, V. and Reimann, M., A Model Describing the Interaction of a Core Melt with Concrete, KfK-2395, English translation NUREG/TR-0039, U. S. Nuclear Regulatory Commission, October 1977.

6. Henry, R. E., A Model for Core-Concrete Interactions, Paper TS12.10, *Proc. Int'l. Mtg. on LWR Severe Accident Evaluation*, Volume II, Cambridge, MA, 1983.

7. Cole, R. K., Kelly, D. P., and Ellis, M. A., CORCON Mod2: A Computer Program for Analysis of Molten Core Concrete Interactions, NUREG/CR-3920, SAND84-1246, Sandia National Laboratories, Albuquerque, NM, August 1984.

8. Dhir, V. K., Castle, J. N., and Catton, I., Role of Taylor Instability on Sublimation of a Horizontal Slab of Dry Ice, Journal of Heat Transfer 99 (1977) 411.

9. Blottner, F. G., Hydrodynamics and Heat Transfer Characteristics of Liquid Pools with Bubble Agitation, NUREG/CR-0944, SAND79-1132, Sandia National Laboratories, Albuquerque, NM, November 1979.

10. Powers, D. A., and Arellano, F. E., Direct Observation of Melt Behavior During High Temperature Melt/Concrete Interactions, NUREG/CR-2283, SAND81-1754, Sandia National Laboratories, Albuquerque, NM, January 1982.

11. Alsmeyer, H., Hagen, S., Oehmann, M., Skokan, A., Schock, W., and Ruhmann, H., Beta Experimental Results on Melt/Concrete Interaction: Silicate Concrete Behavior, and Beta Experimental Results on Melt/Concrete Interaction: Limestone Concrete Behavior, Proc. CSNI Specialists' Mtg. on Core Debris-Concrete Interactions, EPRI NP-5054-SR, Electric Power Research Institute, Palo Alto, CA, February 1987.

12. Bradley, D. R., Modeling of Heat Transfer Between Core Debris and Concrete, *Proc. 25th National Heat Transfer Conference*, Houston, Texas, July 24-27, 1988.

13. D. R. Bradley, E. R. Copus, J. E. Brockmann, and D. A. Powers, Experiments to Assess the Thermal Response of the Mark I Drywell Shell, *Proc. LWR Safety Information Meeting*, Gaithersburg, MD, October 1988.

14. Greene, G. A., Heat Transfer from a Liquid Pool in the Bubbly Flow Regime to a Vertical Boundary, forthcoming, *1989 National Heat Transfer Conference*, Philadelphia, PA, August 1989.

15. Marshall, B. W., Berman, M., and Krein, M. S., Recent Intermediate Scale Experiments on Molten Fuel Coolant Interactions in an Open Geometry, *Proc. ANS/ENS Topical Meeting on Thermal Reactor Safety*, San Diego, CA, February 1986.

16. Blose, R. E., Gronager, J. E., Suo-Anttila, A. J., and Brockmann, J. E., SWISS: Sustained Heated Metallic Melt/Concrete Interactions with Overlying Water Pools, NUREG/CR-4727, SAND85-1546, Sandia National Laboratories, Albuquerque, NM, July 1987.

17. Tarbell, W. W., Bradley, D. R., Blose, R. E., Ross, J. W., and Gilbert, D. W., Sustained Concrete Attack by Low Temperature, Fragmented Core Debris, NUREG/CR-3024, SAND82-2476, Sandia National Laboratories, Albuquerque, NM, July 1987.

18. Greene, G. A. and Irvine, T. F., Jr., Film Boiling of R-11 on Liquid Metal Surfaces, *8th International Heat Transfer Conference*, 86-IHTC-305, San Francisco, CA, August 1986.

19. Rivard, J. B., Interim Technical Assessment of the MARCH Code, NUREG/CR-2285, SAND81-1672, Sandia National Laboratories, Albuquerque, NM, November, 1981.

20. Leonard, M. T., Cybulskis, P., Lee, K. W., Kelly, R. F., Jordan, H., Schumacher, P. M., and Curtis, L. A.. Supplemental Radionuclide Release Calculations for Selected Severe Accident Scenarios, NUREG/CR-5062, BMI-2160, Battelle Columbus Division, Columbus, Ohio, February 1988.

21. Copus, E. R., Blose, R. E., Brockmann, J. E., Gomez, R. D. and Lucero, D. A., Experimental Results of Core Concrete Interactions Using Molten Steel with Zirconium, NUREG/CR-4794, SAND86-2638, Sandia National Laboratories, Albuquerque, NM (unpublished).

22. Copus, E. R., Blose, R. E., Brockmann, J. E., Gomez, R. D., and Lucero, D. A., Core-Concrete Interactions Using Molten Steel with Zirconium on a Basaltic Basemat: The SURC-4 Experiment, NUREG/CR-4994, SAND87-2008, Sandia National Laboratories, Albuquerque, NM (unpublished).

23. Copus, E. R. (unpublished results).

Core Debris/Concrete Interaction During a Severe Reactor Accident

HANS ALSMEYER
Kernforschungszentrum Karlsruhe
Institut für Reaktorbauelemente, PRS
Postfach 3640, D-7500 Karlsruhe 1, FRG

1. INTRODUCTION

The ex-vessel phenomenology of a severe Light-Water Reactor accident after release of the core melt from the reactor pressure vessel is very much affected by the melt concrete interaction. Substantial erosion of the concrete foundation does occur if early coolability of the core debris in the reactor cavity cannot be achieved. While some recent concepts for accident mitigation propose flooding of the cavity to gain debris coolability, they include the possibility of serious steam explosions or, at the minimum, rapid melt/water interactions which might endanger the containment integrity early in the accident with severe consequences. If, however, the initial presence of water is excluded, the long term basemat erosion would result in mainly four major consequences for the risk evaluation:

(i) The long term pressure increase of the containment atmosphere
(ii) The accumulation of burnable gases in the containment
(iii) Release of aerosols and fission products from the melt
(iv) The melting or erosion of important internal static structures and of the basemat

For source term evaluation, the most important quantity is the time of containment failure or depressurization relative to the time and amount of aerosol production. As substantial aerosol and fission product deposition processes are taking place in the containment, late depressurization would result in a small source term to the environment. Therefore, special attention has been given in the melt/concrete interaction studies to the release and composition of gases. It is the subsequent task of the containment codes to consider the gas and energy fluxes into the containment and to calculate the pressure history, taking into account the plant specific containment volume and the heat absorbing structures. Furthermore, during melt concrete attack, composition of the concrete and of the melt and its initial temperature as well as the geometrical configuration influence the gas production rates, concrete erosion, and aerosol release. Consequently, plant and accident specific analyses must be carried out to determine the consequences of individual accident sequences.

2. FUNDAMENTAL PHENOMENOLOGICAL ISSUES

The high temperature core material when released into the reactor cavity is composed of an oxidic and a metallic melt. The interaction with the structural concrete is driven by the sensible and latent heat from the initial debris, by the decay heat power from the fission products which are partly in the metal phase but mainly in the oxide phase, and by the chemical reaction enthalpies which are releasing a considerable amount of energy mainly from the zirkonium oxidation in the early interaction process. For concrete with high siliceous aggregates, the dominant gaseous and liquid species are steam and silica, while for high limestone concrete, carbon dioxide and CaO resulting from the high temperature lime burning are dominant. The erosion processes in siliceous concrete are somewhat faster because of the lower decomposition enthalpie of this concrete, while for limestone concrete the volume fluxes of gases are higher with the consequence of earlier pressurization of the containment.

The gases, and possibly some of the oxides from the concrete, will undergo chemical reactions with the high temperature metallic components of the melt. The oxidation of metallic zirconium from the cladding and core structure is of special importance because of the high energy input associated with this chemical reaction which may contribute substantial more energy in the early phase of melt/concrete interaction than the decay heat. Present modelling in most of the relevant computer codes assumes reaction of Zr with the gases only, while a recent experiment of SNL /1/ indicates additional reactions with liquid or solid oxides, possibly with SiO_2, leading to a faster Zr oxidation. The resulting gaseous reaction products from metal oxidation are hydrogen and carbon monoxide, which under certain conditions may accumulate in the containment with the possibility of deflagration. The gases may transport condensation aerosols from the high temperature melt, which form due to associated evaporation processes of the various species in the melt and possibly some aerosols generated by mechanical processes such as bubble bursting. The aerosol generation rate is especially sensitive to the temperature of the melt and its chemical constitution.

The liquid products of concrete decomposition, mainly SiO_2 and CaO, are soluble in the oxidic fuel melt. This leads to an increase of the melt mass as the erosion of the concrete propagates and to a substantial change of the oxidic properties: Densities, viscosities and freezing temperature are in the long term dominated by the concrete material. The admixture of this material lowers the density of the decay heat source in the oxidic melt, while however the integral decay heat remains uneffected and follows the slowly decreasing decay heat level.

From the phenomenological point of view the interaction of the melt with the concrete may be visualized in terms of two rather distinct regimes:

1) The high temperature regime is characterized by a liquid pool with insignificant freezing processes in the melt. Heat transfer to the melting concrete is high and the melt is highly agitated by the gases evolving from the decomposing concrete. Duration of this regime is short but depends on the content of Zr metal.

2) The low temperature regime is characterized by a balance of decay heat produced in the melt and the heat fluxes at the boundaries leading to a low, however remarkable continuous erosion of the basement over many days. Although freezing of the metal phase is expected, erosion of the basemat would persist, as the "melting temperature" of the concrete is higher than the temperature of the solidified steel. In most cases, the oxidic phase would remain liquid at some hundred Kelvin above the "melting" temperature of the concrete. Low pool temperature and low gas release rates limit the aerosol production from the melt.

3. EXPERIMENTAL INVESTIGATIONS

Historically, the experimental melt/concrete investigations were separated into the thermohydraulic behavior of erosion processes on the one side and the aerosol release on the other side. Indeed, coupling between these processes is rather one-sided in that the aerosol release is controlled by the thermodynamic and thermochemistry in the melt.

In the Federal Republic of Germany intense investigations mainly on the thermodynamic behavior of melt concrete interaction have been carried out. Similar experiments with more emphasis on fission product behavior have been conducted in the US and are still underway in the SNL-SURC /1/ and ANL - ACE /2/ test series. The large scale BETA experiments /3/ were conducted at Nuclear Research Center Karlsruhe, FRG, to validate the WECHSL computer code /4/ with various models which had been developed on the base of a series of small scale phenomenological experiments, and aimed mainly at the conditions expected in a severe accident of a German standard PWR. The BETA facility which was especially designed for this purpose, was built as a large melt facility with sustained induction heating of a simulated core melt in a concrete crucible. The experiments concentrated on the behavior of siliceous concrete with three experiments in limestone crucibles. As the simulated melt did not include simulants for fission products, the aerosol release in the test is representative for the concrete and inactive melt aerosols only.

The most important experimental results for siliceous concrete may be summarized as follows:

- For melts with high power input and small influence of crusts at the melt-concrete interface, propagation of the melt is predominantly downward (Fig. 1, test 1.8).

- Downward heat transfer is considerably higher than predicted by earlier modelling. The high heat transfer results in fast cooling of the melt, and in steady melt temperatures close to the solidification temperature of the metal, even for power input to the melt being an order of magnitude higher than according to decay heat level.

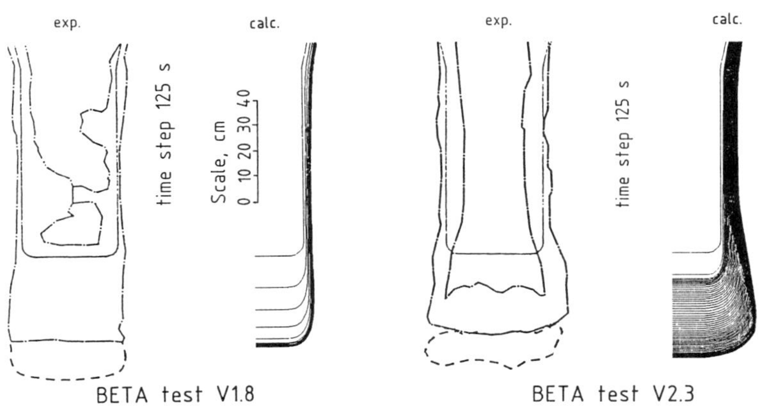

Figure 1: Experimental and calculated cavity formation in the BETA experiments. High temperature test V1.8 and low temperature test V2.3

- Dispersion of the steel into the oxidic melt may occur for high gas fluxes, mainly dependent on density difference of the melt phases and on viscosity of the oxide. For low temperature melts, the segregated state is the stable configuration.

- Experiments with low power input show the role of solidification processes. Sideward and downward propagation are found to be more balanced (Fig. 1, test 2.3). As crusts are always permeable for the gases from the concrete, the gases continue to agitate the melt, thus establishing an effective heat transfer mode to the melting concrete which keeps the long term temperatures of the melt phases low, that is, close to the solidification temperature of the metal.

- The aerosol release is low, with the exception of an initial aerosol peak during and shortly after pouring of the melt.

- Material investigation shows the fast oxidation of chromium, followed by the iron oxidation. The admixture of molten concrete to the oxide reduces its freezing temperature to some 1200°C, as is expected for the reactor material after some interaction time.

- Gas analysis gives the release rates of H_2, H_2O, CO, and CO_2 with a small amount of CH_4. H_2 production continues also after crust formation.

For concrete with high limestone content, an increased release of carbondioxide and carbonmonoxide is observed resulting form the decarburization of the $CaCO_3$. The same chemical process seems to be responsible for an increased aerosol production from pure limestone concrete. Additionally, spalling is observed to be an important process during long term limestone concrete erosion, when melt front propagation is low.

Up to now, the BETA experiments did not investigate the special influence of metallic Zr behavior. Future tests will include this topic.

4. PRINCIPLES OF THE WECHSL CODE AND ITS APPLICATION

Following the objectives of the BETA experiments the WECHSL computer code was validated on the base of the experimental results /4/. This code is a general model describing the thermal and chemical interactions between the molten or freezing core material and structurel concrete. Aerosol release is not considered in this code and may be estimated e.g. from the VANESA code /5/ with the WECHSL results as input. In its modelling structure, WECHSL is similar to other existing melt/concrete codes, e.g. the CORCON code /6/.

WECHSL has been developed specifically for melt/concrete interaction in two-dimensional cylinder geometry (Figure 2). It assumes the existence of an underlying metal layer covered by an oxide layer of fuel, metal oxides, and/or the concrete decomposition products. Mixing of the metal and oxide layer is presently not modeled in the code, as this is considered a transient process, which is possibly important in the early interaction phase only. The gases released from the decomposing concrete may oxidize the constituents of the metal layer. Energy can be internally produced by decay heat in the metal and oxide layer and by chemical reactions. Energy is lost to the melting concrete and to the upper environment by a variety of heat tranfer mechanisms. Moreover,

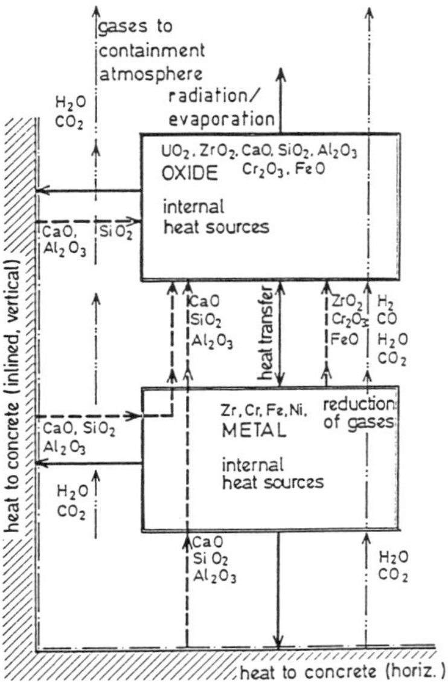

Figure 2: Structure of the WECHSL code

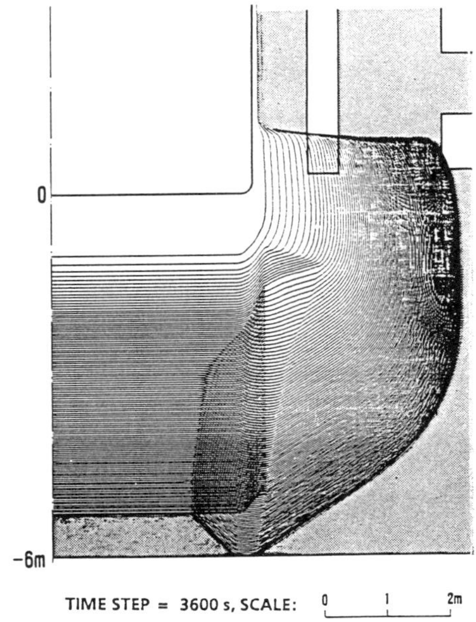

Figure 3: Basemat erosion during 5 days as calculated by WECHSL

energy can be exchanged between the molten layers.

One of the key phenomena during melt concrete interaction is the heat transfer from the melt to the decomposing concrete. The large volume flux of the released gases causes phenomena which are described in analogy to boiling processes on heated surfaces. Depending on the gas flux of the decomposing concrete and the inclination of the surface, heat transfer is controlled by a gas film, nucleate boiling, or transition type of heat transfer.

Crust formation starts at the melt boundary layer with thin skins floating with the movements of the melt. With increasing thickness, the crust stabilizes while the interior of the layer remains liquid and is continously agitated by the gases passing through the porous crust. Thus, heat transfer is by gas driven convection in the pool, and by conduction through the crust. Solidification takes place, however, predominantly in the metal phase while the oxide phase, by the admixture of the concrete decomposition products has a freezing temperature which drops below the melting temperature of the concrete. Therefore, the oxide melt remains liquid over considerable periods of time.

In the course of the WECHSL code validation, improvements were introduced concerning mainly heat transfer and solidification phenomena, oxidation behavior, void behavior and details of cavity formation late in the interaction phase. For the BETA siliceous concrete experiments, the WECHSL code gives good agreement with the important experimental results

/4/ as is shown by the experimental and calculated cavities in Figure 1. After successful completion of this large scale experimental validation, the WECHSL code is now applied to LWR accident analysis.

The accident situation to be analyzed here is the PWR sequence in a German standard reactor. The accident may be initiated e.g., by a large break in the primary loop and subsequent failure of the emergency core cooling system or by a high pressure sequence after controlled depressurization of the primary circuit.

The development of the basemat erosion as calculated by WECHSL is given in Fig. 3. The cavity shapes are plotted every hour from start of the interaction until final penetration of the 6 m basemat after 5 days.

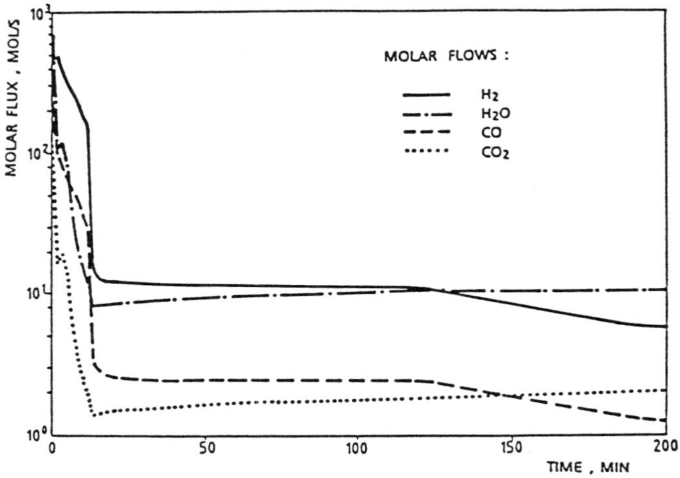

Figure 4: Gas release rates from basemat erosion

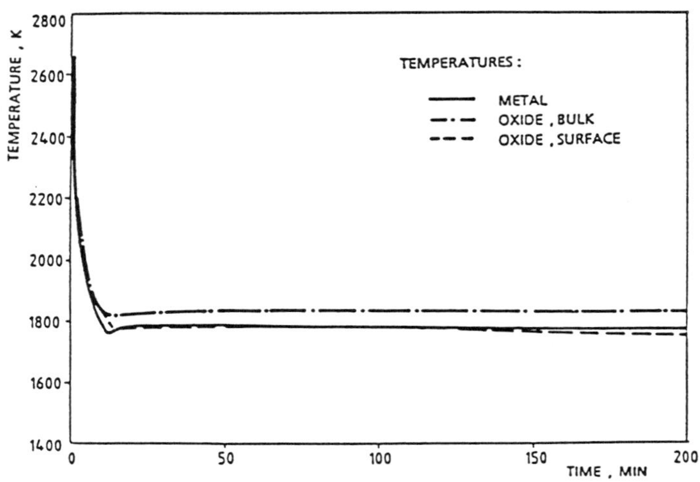

Figure 5: Melt temperatures during basemat erosion

In the initial phase of the melt/concrete interaction, the prevailing heat source in the melt is the zirconium oxidation in the metal, which is controlled by high gas release from the decomposing concrete, and whose chemical energy release is considerably higher than the nuclear decay heat. In this early phase, the high temperature melt attacks the basemat predominantly in the downward direction (Fig. 3), and roughly 80cm of concrete are eroded in the first 1000s, whereas the radial erosion remains limited. The very high erosion velocity results in the release of large gas fluxes from the melt surface into the containment. The gas composition, as given in Fig. 4, shows hydrogen as the dominant gas species, resulting from the metal oxidation. The actual H_2 release rate is considerably higher than has been calculated in earlier models and causes a fast hydrogen enrichment in the containment. Because of the high heat flux to the concrete associated with the gas agitation of the melt, the heat losses from the melt overcome the heat generation in the melt. The melt cools down in the first 1000s to a temperature close to the freezing temperature of the metal phase (Fig. 5).

With the onset of solidification of the metal phase at the concrete boundaries, a different heat transfer mode controlled by the heat resistance of the frozen metal crust reduces the penetration rate, and a balance of decay heat in the melt and melting processes at the concrete interface is established, with a long term temperature of the oxidic and metallic melt phases close to 1800K (Fig. 5). The ratio between vertical and radial erosion is much more balanced (Fig. 3), and the reduced concrete erosion results in a lower gas release rate (Fig. 4). However, the production of hydrogen from the ongoing metal oxidation is still important and will increase the concentration of burnable gases in the containment. Long term H_2 release is calculated, in accordance with experiments, also after solidification of the metal due to oxidation of Zr, Cr, or Fe, respectively. After 2 hours, Zr is calculated to be burnt out and the less exothermal Cr oxidation becomes dominant. Some 8 hours after start of the MCCI, the sideward penetration of the melt has eroded the inner shield structure that it could fail. However, the failure mode of this cylinder is unknown, especially with respect to the role of the sump water separated by the concrete cylinder. Massive flooding of the melt by sump water might be delayed or even prevented by interfacial oxidic crusts, which form at the cylinder wall by the intense cooling of the evaporating sump water.

If flooding by the sump water is anticipated, additional steam production from the melt surface does occur with the consequence of a faster pressure increase in the containment leading to a containment failure after 4 to 5 days. The sump water evaporation would reduce the flamability of the burnables and finally prevent hydrogen burns in the containment. However, the sump water is unable to stop further penetration of the melt into the basemat, as cooling of the core melt by the water atop is limited by the formation of an oxidic surface crust, and substantial fragmentation of the melt is unlikely to occur. Therefore, the upward heat flux density is not much higher than it was before by radiation only and concrete erosion by the deep melt pool will continue. The very small influence of flooding on downward propagation was demonstrated in the SWISS experiments /7/ and confirmed by model experiments at KfK /8/.

The calculations in Fig. 3 postulate the existence of water on top of the melt after 8 hours. The long term erosion is characterized by a solidified however gas permeable hot metal slug moving slowly downward into the melting concrete as was observed experimentally /9/. The highly viscous however liquid glassy oxidic melt which is continuously agitated by the concrete gases, is penetrating into the radial and outer axial direction. The typical erosion rate is 2 to 4 cm/h, where the oxidic melt with the higher content of fission products is moving faster and in by-passing the metal slug penetrates the basemat after 5 days. The overall radial erosion in the basemat at this time is 6.5 m radius.

4. HEAT CONDUCTION ESTIMATES ON LONG TERM BASEMAT BEHAVIOR

To check the results on basemat penetration and to gain further insight of the long term behavior, a complementary calculation has been made using a modified version of the HEATING 5 transient heat conduction code /10/. This estimate neglects the details of melt behavior, but is exact in solving the global mass and energy equations. As heat transfer in the HEATING code is by conduction only and independent of the geometrical orientation, the results give a certain approximation only. Nevertheless, they are especially useful in estimating the penetration over very long times where the WECHSL code can not be applied. Calculations take into account the effect of cooling water on top of the melt.

For a semi-infinit reactor basemat, the maximum downward erosion is 19 m after 1050 days. Thereafter, solidification of the melt at the lower boundaries begins. These calculations show that for existing reactors with a confined reactor cavity penetration of the basemat has to be expected, if coolability of the core material cannot be established early after basemat contact.

As reactor foundations are often situated in the groundwater, an additional calculation with HEATING was carried out, with an isothermal boundary condition of 20°C at the lower basemat surface at 7 m, maintained by groundwater contact (Fig. 6). The downward propagation of the melt stops, when it approaches the lower boundary with the heat flux density in the order of 50 kW/m^2 only. The remaining thickness of the concrete layer through which heat of the melt is transferred by conduction to the water, is typical 5 cm with 1300°C at the inner and

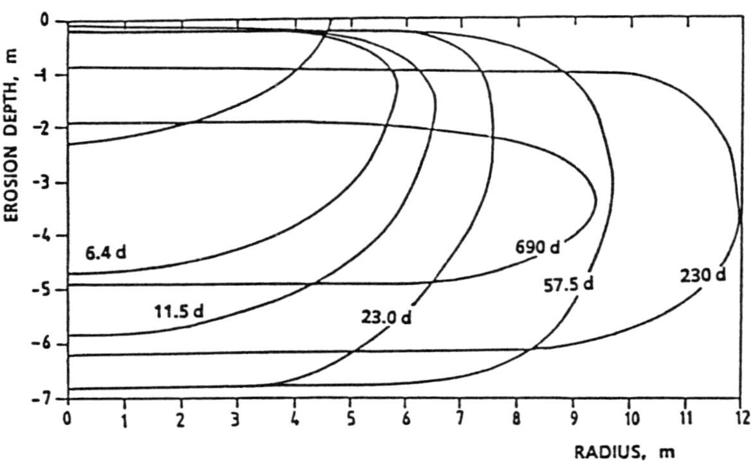

Figure 6: Erosion of a basemat with groundwater contact at the lower side. Maximum radial erosion at 230 days. HEATING calculation.

20°C at the outer side. As this layer is under considerable thermal and mechanical stresses, it can not be assumed to prevent direct contact of the glassy and encrusted melt with the groundwater, and dissolution of radionuclides in the groundwater must be considered /11/.

In spite of stabilization of the melt front in the downward direction, the heat transfer to the water is insufficient to remove all the decay heat, and radial basemat erosion will continue, driven by the energy of the central pool. The maximum radial erosion is some 12 m at 230 days, and thereafter the cavity volume reduces because of starting solidification.

5. CONCLUSION AND PERSPECTIVE

Under source term aspects, the most important result of the preceeding calculations is the consequence of gas release into the containment. The slow pressure increase would fail the large PWR containment under consideration not before 4 or 5 days, which may be prevented by controlled venting. The increasing hydrogen concentration may need hydrogen ignition or recombination to prevent deliberate deflagration and pressure spikes exceeding the failure pressure of the containment.

From the temperature history of the core melt during the basemat erosion, release of important fission products into the containment may occure during the first 30 minutes of interaction. In the long term when the oxidic melt has cooled down to less than 1500°C, a small aerosol source of 0,1g/s into the containment may persist. These aerosols are mainly concrete aerosols with potentially some permille of active fission products. Filtered venting of the containment would retain these aerosols to an acceptable level.

Basemat penetration must be expected as was confirmed by independent heat conduction analysis. Cooling at the lower side of the basemat, either by engineered heat sinks or by existing ground water would not necessarily stop the sideward progression of the melt into the basemat. This might threat the statics of the internal containment structure. Concerning the potential ground water source term by leached fission products, the slow transport with the ground water flow and the retardation by physio-chemical processes in the soil would act to reduce the consequences which, however, must be analyzed plant specific.

So far, code development and experimental investigations were oriented to PWR accident situations. Some additional questions need further consideration. As there is an uncertainty in the failure mode of the biological shield and the possibility of sump water ingression, experimental clarification of this process would be valuable. For a melt with a high content of metallic zirconium, the chemistry of Zr oxidation and the consequences on temperature behavior, gas and aerosol release needs further consideration. Additionally, the presence of boroncarbide from control blades in BWR's may influence the material behavior and solidification processes in the melt and must be quantified in the course of future analysis.

Finally, it should be pointed out, that large uncertainties exist in the initial conditions of the melt slumping from the reactor pressure vessel into the reactor cavity. Initial temperature of the core material and its composition especially in Zr-metal and fission product content may considerably influence the aerosol release mainly in the early interaction period. For early containment failure, the environmental source term is strongly influenced by these processes. If containment integrity is maintained over several days, plate out of the early aerosols would reduce the source term to the minor aerosol rate, which may persist during the low temperature concrete erosion over a long period of time.

REFERENCES

1. Bradley, D.R., Powers, D.A., Recent Advances in CORCON/VANESA Model Development, 16th Water Reactor Safety Information Meeting, Gaithersburg 1988, Paper 20-11

2. Rubio, A., Vogel, R.C., Meeting Evolving Reactor Safety Requirements, 16th Water Reactor Safety Information Meeting, Gaithersburg 1988, Paper 6-1

3. Alsmeyer, H., BETA-Experiments in Verification of the WECHSL-Code: Experimental Results on the Melt Concrete Interaction, Nuclear Engineering and Design 103 (1987), pp 115 - 125

4. Reimann, M., Verification of the WECHSL-Code on Melt-Concrete Interaction and Application to the Core Melt Accident. Nuclear Engineering and Design 103 (1987) pp 127 - 137.

5. Powers, D.A., Brockmann, J.E. and Schiver, A.W., VANESA: A Mechanistic Model of Radionuclide Release and Aerosol Generation during Core Debris Interactions with Concrete, NUREG/CR-4308 (1986)

6. Cole, R.K., Kelly, D.P. and Ellis, M.A., CORCON-MOD2: A Computer Program for Analysis of Molten Core-Concrete Interactions, NUREG/CR-3920 (1984).

7. Blose, R.E. et al., SWISS: Sustained Heated Metallic Melt/Concrete Interactions with Overlying Water Pools. NUREG/CR-4727 (1977)

8. Alsmeyer, H. et al., Modellentwicklung zur analytischen Beschreibung von Kernschmelzunfällen, in Sicherheitsorientierte LWR-Forschung, Jahresbericht 1987, KfK 4450 (1988), pp 113 - 128

9. Copus, F.R., Bradley, D.R., Interaction of Hot Solid Debris with Concrete, NUREG/CR-4558 (1986)

10. Turner, W.D., Elrod, D.C., Siman-Tor, I.I., Heating 5 - An IBM 360 Heat Conduction Program, ORNL/CSD/TM-15

11. Bayer, A., Tromm, W., Abschätzung der Radionuklidverteilung und der Strahlenexposition bei Auslaugung einer erstarrten Kernmaterial-Beton-Schmelze durch Grundwasser, KfK 4512 (1989)

Heat Transfer in Reactor Cavity during Core-Concrete Interaction

B. ADROGUER and G. CENERINO
Protection and Nuclear Safety Institute
French Atomic Energy Commission
France

ABSTRACT

In the unlikely event of a severe accident in a nuclear power plant, the core may melt through the vessel and slump into the concrete reactor cavity. The hot mixture of the core material called "corium" interacts thermally with the concrete basemat. The WECHSL code, developped at K.f.K. Karlsruhe in Germany is used at the Protection and Nuclear Safety Institute (I.P.S.N.) of CEA to compute this molten corium concrete interaction (MCCI). Some uncertainties remain in the partition of heat from the corium between the basemat and the upper surrounding structures in the cavity where the thermal conditions are not computed.

The CALTHER code, under development to perform a more mechanistic evaluation of the upward heat flux has been linked to WECHSL-MOD2 code. This new version enables the modelling of the feedback effects from the conditions in the cavity to the MCCI and the computation of the fraction of upward flux directly added to the cavity atmosphere.

The present status is given in this paper. Preliminary calculations of the reactor case for silicate and limestone common sand (L.C.S) concretes are presented. Significant effects are found on concrete erosion, gases release and temperature of the upper part of corium, particularly for L.C.S concrete.

1. INTRODUCTION

The interaction between molten core debris and concrete is an important aspect of the overall severe accident because of its consequences :

- A release of radioactive and non radioactive aerosols.
- A generation of concrete decomposition gases, combustible and non combustible and an energy transfer to the containment which contributes to the risk of overpressurization.
- A containment failure through penetration of the basemat.

This paper deals with our ability to provide predictions in the two last areas of interest.

In its analytical effort, the CEA uses and contributes to the development of the WECHSL computer code /1/ for MCCI studies. WECHSL is one constituent of the ESCADRE code package /2/ used in France to study severe accidents. The last version MOD2 of WECHSL includes the long term period of the MCCI. WECHSL was validated on BETA experiments, mainly with concrete of silicate type /3/.

Currently some uncertainties still remain in the treatment of the heat and mass transfer and on the melt configuration. The main feedback from the cavity conditions to the MCCI which is the influence of the temperature and emissivity of the structures of the cavity on the temperature of the top surface of the corium is not mechanistically modelled. The upper boundary condition is only given by the user via a surrounding temperature versus time. This value depends on the heat exchange between the top surface of the corium and the cavity. This dependence influences in turn the partition of energy between the basemat and the structures in the cavity which can be damaged by ablation.

Preliminary computations show that a change in the upper boundary condition can cause substantial changes in the fraction of heat that is used to ablate the basemat concrete and in the gas generation. Similar conclusions were obtained by UKAEA with reactor calculations performed with CORCON MOD2 code /4/.

These conclusions demonstrate the need for an integrated code including WECHSL with a cavity model describing the thermal behavior and ablation of the walls, the thermal and mass transfer exchanges, and the cooling of its atmosphere by natural convection via the venting shaft.

For these reasons, the CEA decided to develop a cavity code named CALTHER. A first version of CALTHER presently under development is linked to the German WECHSL code.

2. DESCRIPTION OF THE CALTHER CODE

2.1 General description and modelling

The reactor cavity is modelled by different components.

- The top surface of the corium which is the interface between the WECHSL and CALTHER codes.
- The concrete walls of the cavity which is divided in different annular surfaces each one assumed at a uniform temperature.
- The remaining reactor vessel can be modelled with different steel slabs.
- The gas mixture of absorbing gases H_2O, CO_2 and CO coming from the concrete decomposition.

At the present stage of development the aerosol effects on the radiative heat exchanges are not taken into account.

The figure 1 gives a simple example of a cavity configuration used in CALTHER.

At the beginning of a time step all the surface temperatures are known (the surface temperature of the corium is given by WECHSL).

A gas absorption model calculates the mean transmittance of the gas mixture from the characteristics of each absorption band. Then a radiative heat transfer model determines the net heat flux on each surface as well as on the gas mixture presently supposed to be at an intermediate temperature between the different wall components of the cavity (assumption of no cooling by natural convection through the cavity).

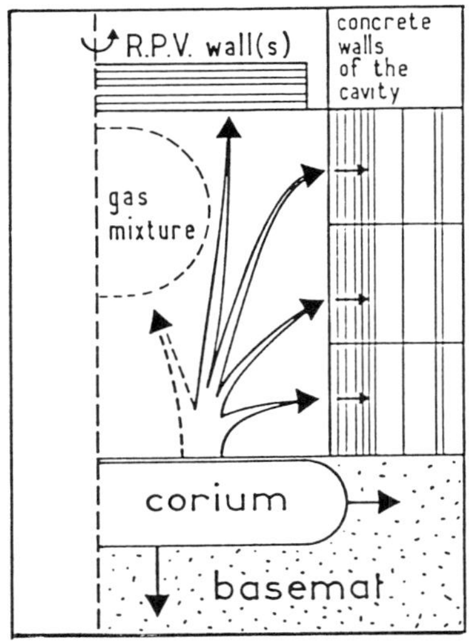

FIGURE 1 - Cavity configuration with 6 components : corium, 3 concrete walls, 1 RPV wall, the gas mixture

The net surface fluxes previously computed by the radiative heat transfer model are used as internal boundary conditions by the thermal conduction model to calculate the thermal behavior and the ablation of each wall component. This model is linked to a mechanistic model of gas release for concrete walls.

In spite of the possible ablation of the walls, the view factors and the mean beam lengths are kept constant during the calculation.

2.2 The radiative heat transfer model of the cavity

This model provides :

- The net radiative heat flux on each component of the cavity enclosure as function of wall and gas temperatures.
- The energy deposited in the gas mixture which may absorb and emit radiation.

The geometrical configuration of the cavity (fig. 1) is described by view factors and geometric mean beam lengths between the different surfaces of the enclosure. All the surfaces are treated as gray diffuse emittors and reflectors. The gas mixture is assumed isothermal.

The general steady-state equations for determining radiative exchanges in a gray diffuse enclosure of N surfaces are based on the net-radiation enclosure model /5/ used with the concept of the total interchange area S_iS_j between the different wall surfaces and gas mixture /6/. The surfaces S_iS_j, which characterize the cavity geometry and reflectivity represent the part of the surface A_i whose radiative emission is absorbed after reflections at all surfaces by the surface j :

$$Q_{i \to j} = S_iS_j \, E_i \qquad E_i = \sigma \, T_i^4$$

These S_iS_j quantities are the solution of the following relations solved by the Gaussian elimination.

$$\sum_{m=1}^{N} \left[\frac{\delta mk}{\xi_m . A_m} - \frac{(1-\xi_m)}{\xi_m . A_m} F_{km} \, \bar{T}_{km} \right] S_jS_m = \xi_j \, F_{kj} \, \bar{T}_{kj}$$

k = 1, N   j = 1, N

ξ_m : emissivity of surface m
F_{km} : view factor between the two surfaces k and m
$\bar{T}_{km}$: mean gas transmittance between surfaces k and m
Am : surface area

The S_iS_j surfaces being a weak function of the temperatures of the enclosure, its calculation is only performed with a lower frequency given as input data.

The total interchange area S_iS_g between a surface and the gas mixture results from the conservation equation of the energy emitted by this surface :

$$A_i \, \xi_i = \sum_{j=1}^{N} S_iS_j + S_iS_g$$

The net radiative flux on a surface S_i and on the gas mixture are :

$$Q_i = \sum_{j=1}^{N} (S_iS_j \, E_i - S_jS_i \, E_j) + S_iS_g \, E_i - S_gS_i \, E_g$$

$$Q_g = - \sum_{j=1}^{N} Q_j \text{ (energy balance in the cavity enclosure)}$$

2.3 <u>The gas absorption model</u>

This model provides for each absorbing gas :

- The characteristics of each absorption band.
- The transmittance T_{ij} (B) between the surfaces i and j for each band B.

A mean transmittance $\bar{T}_{ij}$ is then calculated by summing the contribution of the (2B+1) spectral intervals delimited by the B bands of the gas mixture (B=5 for H_2O alone). This sum is weighted by the

Planck black body radiation distribution. Outside an absorption band, T_{ij} (B) is equal to 1. The transmittance T_{ij} (B) can be evaluated at the temperature T_i of the surface emitting the radiation through the gas or at the mean temperature (choice in the input data).

The transmittance of each band is obtained from the statistical Goody-Mayer model assuming within a band a random distribution of spectral lines. In the conditions of MCCI, the line profile obeys the Lorentz profile (collision broadening lines) and Tij (B) is a simple expression of the characteristics of the absorption bands (band width, line width-to-spacing ratio, band intensity). These band properties which depend of the gas nature, partial and total pressure and gas temperature, are given by the Thomson correlations.

The only gas component calculated in the present version is H_2O. The non-condensable gases CO_2 and CO will be introduced later. A user option enables calculations without an absorbing gas mixture.

2.4 Thermal conduction model

The one dimensional heat conduction equation is solved implicitly in each wall by a finite difference approximation of the following equation :

$$C r^{\alpha} \frac{\partial T}{\partial t} = - \frac{\partial \emptyset}{\partial r} + q\, r^{\alpha}$$

$$\emptyset = -r^{\alpha}\, k(T)\, \nabla T(r)$$

T(r)	:	temperature at position r ..... K
C	:	volumetric heat capacity ...... J/m^3
q	:	volumetric heat source .... J/m^3
α	:	geometrical factor (plane : α=0 ; cylindric : α=1)
k(T)	:	wall conductivity .... W/(mK)

The changes in T in the radial direction being much greater than the changes in the axial direction the 1D approximation is sufficient.

Boundary conditions

The flux $\emptyset s$ calculated by the heat exchange model is imposed at the inner surface and a temperature is given in the input data for the outer surface. The inner surface may move with time due to ablation. In this case, the velocity of the melt front is given by :

$$\dot{X} = \frac{1}{L} \{\emptyset s + r^{\alpha}\, k(T)\, \nabla T\} \qquad \text{L : specific latent heat of melting}$$

and the temperature of the face mesh is limited to the melting temperature. This mesh is eliminated when it is completely molten. The melt front can also move back in case of freezing.
The multilayer wall slabs use a node-size distribution given in the data.

2.5 Gas release model

Only the most important aspects are computed : the energy sinks due to concrete decompositions, the location of the three decomposition fronts and the mass flow rates of free water, bound water and CO_2. This model is strongly linked with the thermal conduction model. Continuity equations describe the composition distributions. Momentum equations are not considered so the pressure distribution and the gas flows inside the walls are not calculated.

When a concrete wall is heated the released conditions are :

- Free water at the local saturation temperature inside the wall. This value is given in the data (approximately 400 K for MCCI conditions).
- Bound water between 530 K and 770 K due to hydroxides decomposition.
- CO_2 between 900 K and 1300 K due to calcium carbonates decomposition.

The free water release mechanism is similar to the mechanism of ablation. Bound water and CO_2 are released with Arrhenius kinetic laws which enable with the concrete decomposition heats, the calculation of the corresponding energy sink.

Validation of the thermal and gas release models

A series of realistic scale experiments have been conducted in the CEA to study the thermal response and gas release of concrete cylindrical samples (1 m length and 1.13 m diameter). The slabs were heated on one side and quite well instrumented to measure the temperature on different axes and at different levels. The CO_2 kinetic correlation developed with these results have been incorporated into the code. Figures 2 and 3 illustrate a comparison between data and the calculation of a test performed with a L.C.S concrete sample. The comparisons show that this simple model is sufficient to calculate the thermal cavity behavior during MCCI conditions.

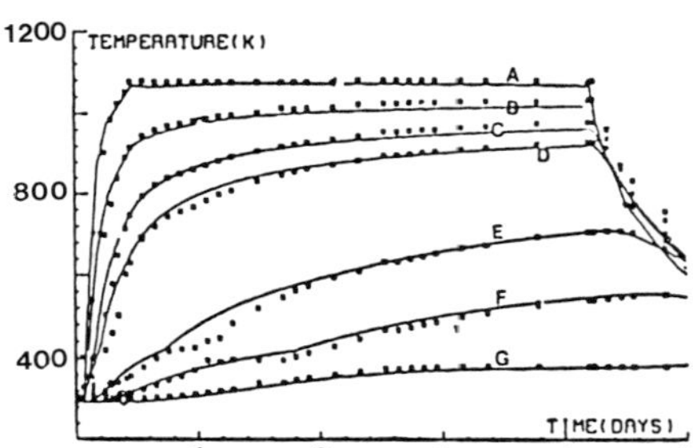

FIGURE 2 - Calther-experiment comparisons. Temperature evolution at different elevations.

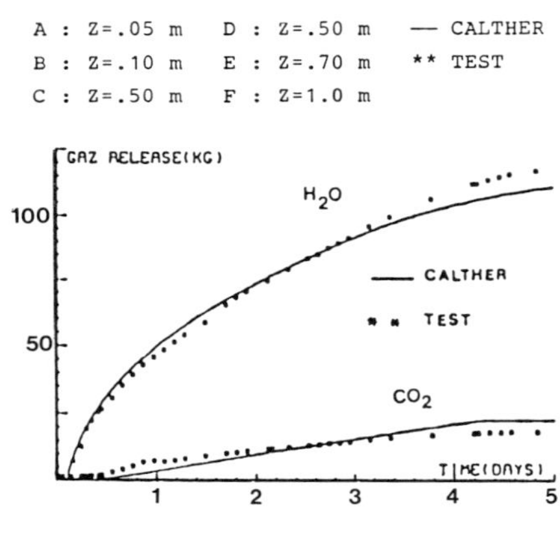

FIGURE 3 - Calther-experiment comparisons. Total mass of H_2O and CO_2 released.

2.6 Linking between WECHSL MOD2 and CALTHER

The two codes are linked at three levels :

- The top surface heat flux from the corium is now calculated by the CALTHER module from all the surrounding surface temperatures.
- The decomposition products from the cavity (molten concrete or molten steel) are added to the corium during the MCCI calculation.
- The gases released from the cavity walls are added to the gas release calculated by WECHSL.

CALTHER is used as a user's option model. After the short-term of the MCCI, the slow variation of the wall thermal response enables cavity calculations with greater time steps than those of WECHSL.

3. PRELIMINARY COMPUTATIONS OF THE REACTOR CASE

WECHSL calculations of MCCI have been performed with and without the CALTHER module. The cavity is a concrete cylinder (4 m high, 6 m inner diameter) divided in 3 annular surfaces and the vessel is represented by a 0.6 m thick plane steel slab (fig. 1). A constant ambient temperature of 773 K is imposed in boundary condition for the standard WECHSL calculations (value used to represent the cavity until now).

The runs are performed with a silicate and a L.C.S basemat concrete (Table 1). The gas mixture is assumed transparent. The corium composition is typical of a 900 MWe PWR with an initial temperature of 2673 K (79.6 t UO_2 ; 17.2 t ZrO_2 ; 34.9 t Fe ; 5.4 t Zr ; 6.2 t Cr ; 4.3 t Ni). The decay heat of 22.4 MW at the beginning of the MCCI is mainly located in the oxide phase (88 %).

TABLE 1 - Concrete composition in % mass

	Ca CO_3	Ca$(OH)_2$	SiO_2	Free H_2O	Al_2 O_3, ...
Silicate concrete	1.5	10.3	81.8	4.2	2.2
L.C.S concrete	56.	2.	25.	5.7	11.3

3.1 Short-term phase (t < 1 h)

The effect of the CALTHER module is not significant for two reasons : first, the vertical heat transfer from the corium to the basemat is so high (a few hundreds MW) than the upward heat loss is always less than 10 % of the total heat loss. Second, when this vertical heat transfer decreases, the walls of the cavity are still "cold" and at a comparable temperature-level with the one in the standard WECHSL runs.

3.2 Long-term phase

The cavity effect increases gradually with its wall temperatures leading to a significant decrease of the upper heat flux loss (fig. 4). This very important effect of CALTHER leads to higher temperatures in the upper oxide layer of the corium especially for the surface temperature (fig. 5). In consequence the upper crust of the oxide layer never appears although always predicted by WECHSL in standard MCCI calculations.

Silicate concrete cases.

With CALTHER the upward heat loss decreases by a factor 2 from 15 % to 8 % of the total heat loss. After 4 days the mean cavity and top surface temperatures (1450 K and 1500 K respectively) are much greater than without the cavity (770 K and 950 K). The low value of the upward flux explains the weak effect on the global concrete erosion. A higher temperature in the oxide phase leads to a lightly greater radial erosion (fig. 6).

Basemat penetration times for a 4.7 m thickness are similar in the two cases (3.8 days with and without CALTHER for oxide phase penetration and 4.6 days instead of 4.4 days for metal phase penetration). After three days of MCCI, the axial erosion by the metal phase is slightly delayed because the dilution effect becomes more important than the one of the upward heat flux reduction. The total gas release (fig. 8) is increased with CALTHER due to more radial erosion and to the addition of the gases released (12 t of H_2O) by the walls of the cavity. The addition of oxides from the ablation of the walls (25 t) increases the level of the corium.

LCS cases.

The important upward heat loss predicted by the standard WECHSL calculation is strongly reduced by the cavity model from 35 % to 10 % of the total heat loss after 4 days (fig. 4). This effect is more important than with silicate concrete because the top surface temperature of the corium calculated by the standard WECHSL code is significantly greater (fig. 5). Upper temperatures after 4 days are also quite different (1550 K and 1600 K for the mean cavity and corium surface temperatures instead of 770 K and 1100 K).

Radial and axial erosion are significantly increased (fig. 7). Basemat penetration times for a 4.7 m thickness are lower : 4 days instead of 5.7 days for oxide phase and 7 days instead of 9 days for metal phase. The total gas release is also strongly increased (fig. 9). The cavity model adds 20 t of oxides into the corium, 13 t of H_2O and 15 t of CO_2 into the containment.

3.3 Absorbing gas mixture effects

Sensitivity calculations performed with an absorbing gas mixture (50 % of H_2O) show significant effects on the upward heat flux and on the energy distribution between the cavity structures and the gases. These energy transfers are mainly governed by the gas mixture temperature.

For natural convection through the cavity with a "cold mixture" imposed at 1300 K, the upward flux is two times greater than with the transparent mixture and 85 % is directly evacuated to the containment by the gases. Without cooling by convection the energy is almost entirely absorbed by the structures like in the transparent mixture case.

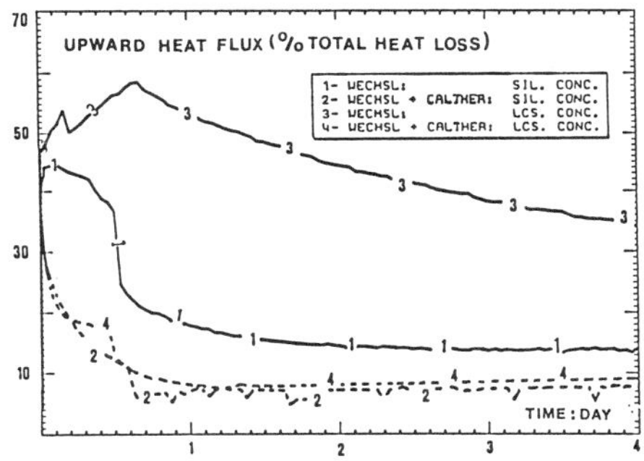

FIGURE 4 - Upward heat flux.

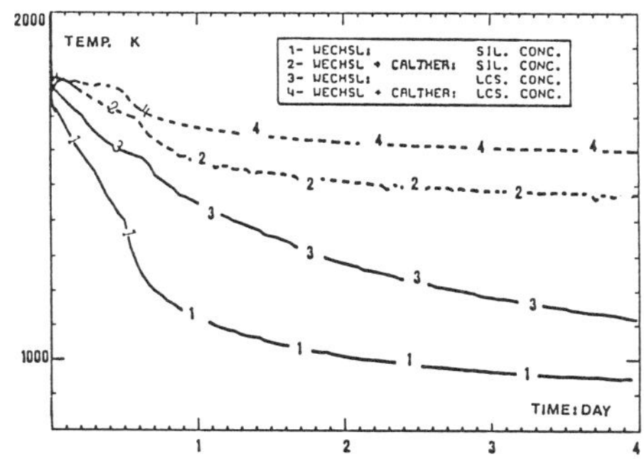

FIGURE 5 - Corium surface temperature.

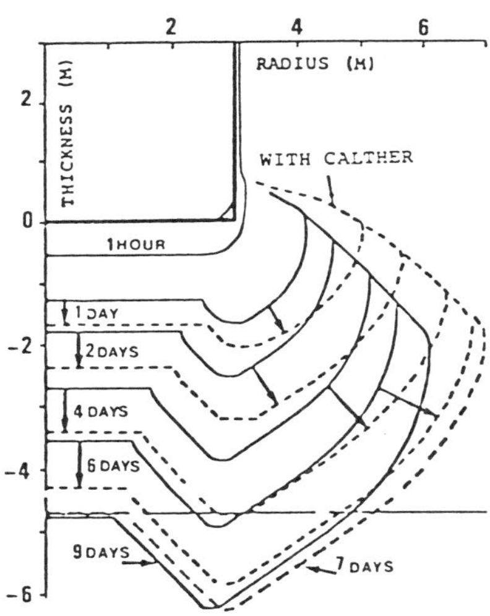

FIGURE 7 - LCS basemat erosion with and without CALTHER.

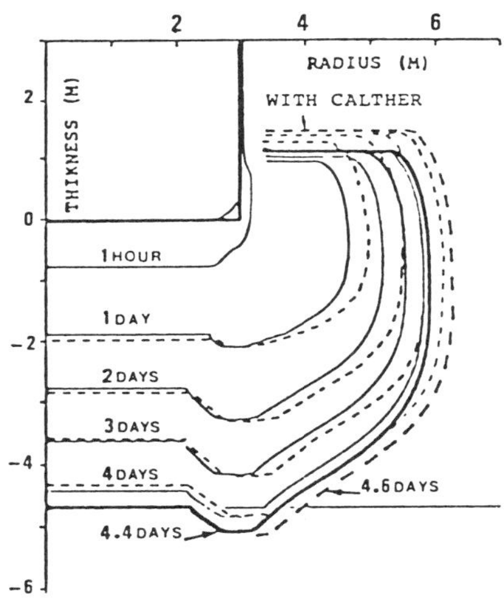

FIGURE 6 - Silicate basemat erosion with and without CALTHER.

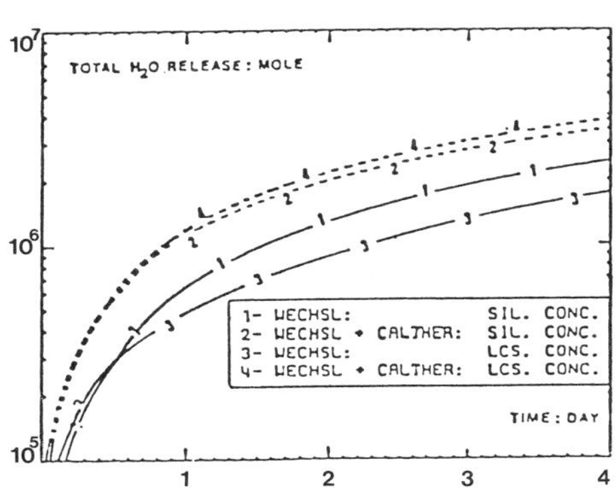

FIGURE 8 - Total H_2O release.

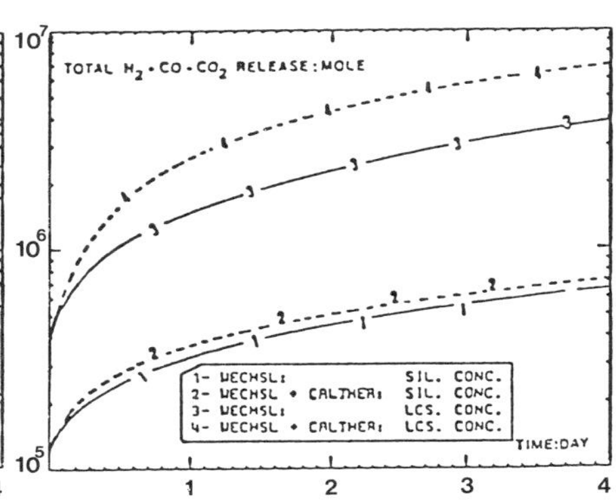

FIGURE 9 - Total H_2 + CO + CO_2 release.

Realistic evaluations of the fraction of upward flux added to the atmosphere require a more complete gas mixture model to calculate the mixture temperature from the energy and mass balance equations in addition to the modelling of the absorption by CO and CO_2 gases.

4. CONCLUSION

During the development of CALTHER the thermal and gas release models have been validated. A first version has been linked to the last version MOD2 of WECHSL code.

Preliminary WECHSL reactor calculations performed with and without CALTHER show significant effects on the concrete erosion and consequently on the gas release particularly with LCS concrete basemats which are completely penetrated 30 % earlier. Damages to the cavity structures stay limited with a maximum radial ablation of 0.4 m in the concrete walls and no further significant melting of the RPV. Significant effects are also found on the temperature level and on the crust formation of the upper part of the corium. These effects could be of particular relevance for the calculation of aerosols formation and fission products release during a MCCI.

The effects on the containment pressurization and on the accumulation of burnable gases will be examined subsequently with the ESCADRE code.

These results must be considered with caution especially for the long-term phase of MCCI and for the LCS concrete because the WECHSL code is not yet completely validated.

REFERENCES

1. M. REIMANN, W.B. MURFIN, The WECHSL code. KfK 2890. Kernforschungszentrum Karlsruhe. 1981.

2. J. DUFRESNE, G. CENERINO, J.M. DUMAS, J. FERMANDJIAN, J. GAUVAIN, J.P. L'HERITEAU, G. LHIAUBET, Presentation du système ESCADRE. Symposium on severe accidents. Sorrento 21-25 March 1988.

3. H. ALSMEYER : BETA Experiments in verification of the WECHSL code. Nuclear Engineering and Design 103 (1987). 115-125.

4. P.M. KEEPING, B.D. TURLAND, N.J. BREALEY, and R.L.D. YOUNG, CSNI specialist's meeting on core debris concrete interactions. PALO ALTO. 09-86.

5. SIEGEL R. and HOWELL J.R., Thermal radiation heat transfer. 2nd ed. Mc Graw Hill. New York. 1981.

6. HOTTEL H.C. and SAROFIM A.F., Radiative transfer. Mc Graw Hill. New York. 1967.

Theoretical and Experimental Investigations on Large-Scale, Long-Term Thermohydraulics Relevant to Severe Accident Conditions in Containment

L. WOLF and L. VALENCIA
Kernforschungszentrum Karlsruhe GmbH
Projekt HDR
FRG

T. KANZLEITER
Battelle-Institut eV
Am Romerhof 35
D-6000 Frankfurt/Main 90, FRG

INTRODUCTION

In the course of a severe accident, the containment of a nuclear power plant constitutes the last barrier in the concept of defense - in-depth approach to protect the public from the consequences. Because of this outmost importance, emphasis has shifted from previous DBA-related containment research towards severe accident containment research. In this context the thermohydraulic phenomena play a major role in determining the source term as a potential thread to the public.

The containment building is designed to contain the energy and material released in a complete, double-ended break of the largest pipe of the reactor coolant system and to withstand the impact of internally generated missiles.

In terms of thermal-hydraulic issues, the containment can be generally characterized by a multiply connected domain subdivided into complexly shaped 3-D subcompartments with different surfaces and structural internals, complex, 3-D recirculating flows in large rooms, two-phase, multi-component, multi-field flow, combined mass, momentum, energy and species transfer, widely differing transport processes, detailed flow patterns in vents, jets, along walls with associated momentum and energy exchanges.

Figs. 1 and 2 schematically illustrate the DBA-related short-term flow effects in vents and between subcompartments and the long-term transfer processes in the containment. Whereas Fig. 1 reveals the momentum fluxes as driving forces during the short-term period, Fig. 2 reveals the importance of a variety of couplings between mass, energy and momentum transfer processes. In addition, energy, mass and momentum transfer to structural surfaces and sump pool interfaces become very important and dominant. For the heat transfer the structural walls and subcompartmental internal metallic structures play a major role. The major physical dependencies of the heat transfer indicate rather complex interactions.

For hypothetical accident scenarios the interactions are more complicated through the existence of a variety of gases and particles. Thus, for these scenarios, the complexity and interactions dramatically increase.

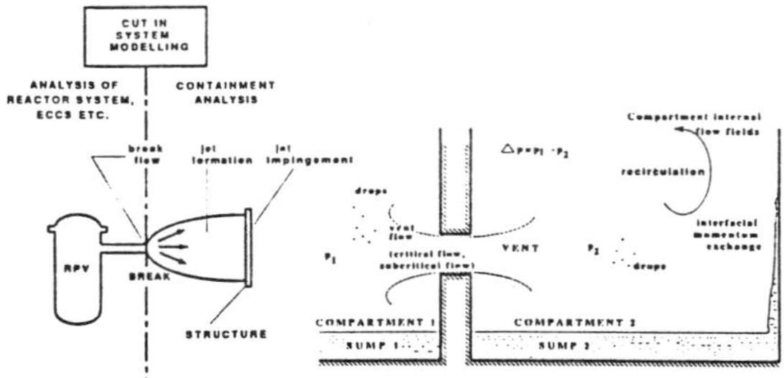

Fig. 1: Fluiddynamic Aspects

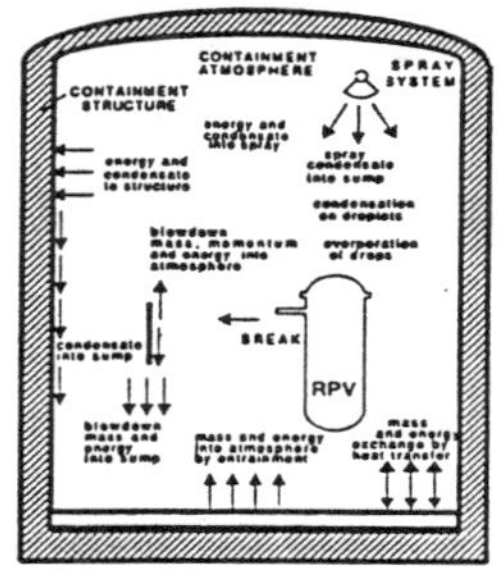

Fig. 2: Mass, Momentum and Energy Transfer Processes

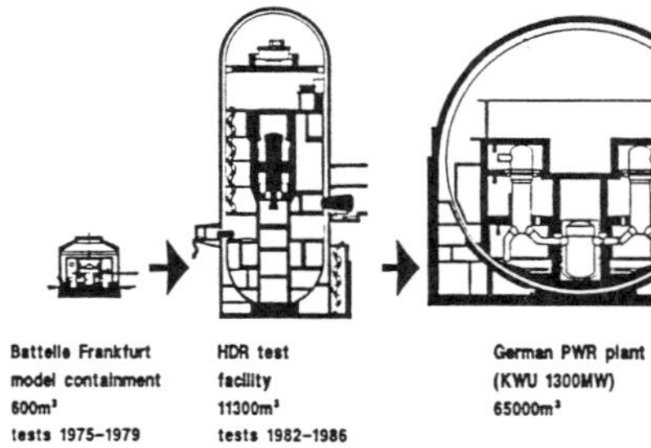

Fig. 3: Containment Test Facilities

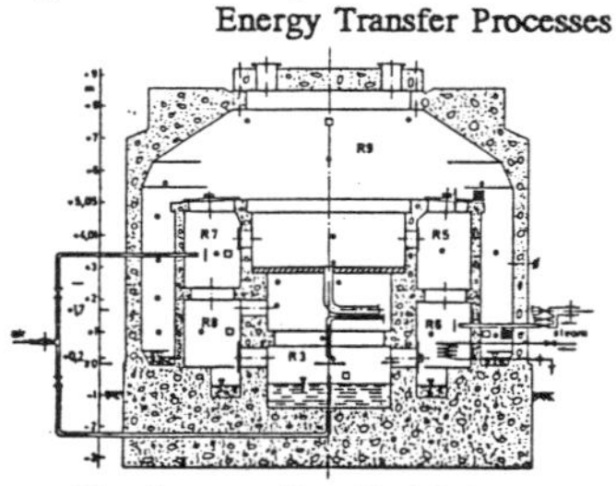

Fig. 4: Battelle Model Containment

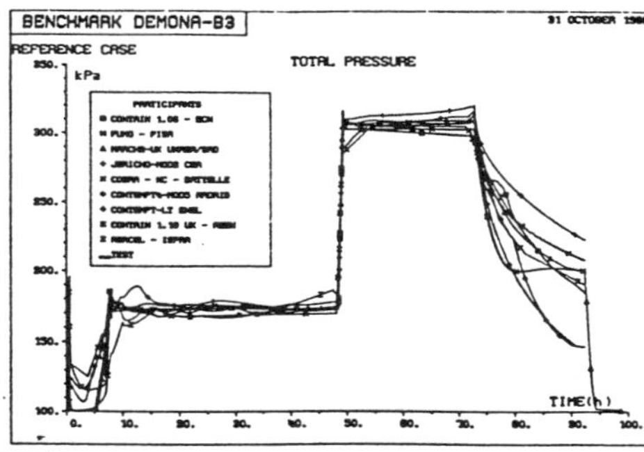

Fig. 5: CEC T-H Benchmark

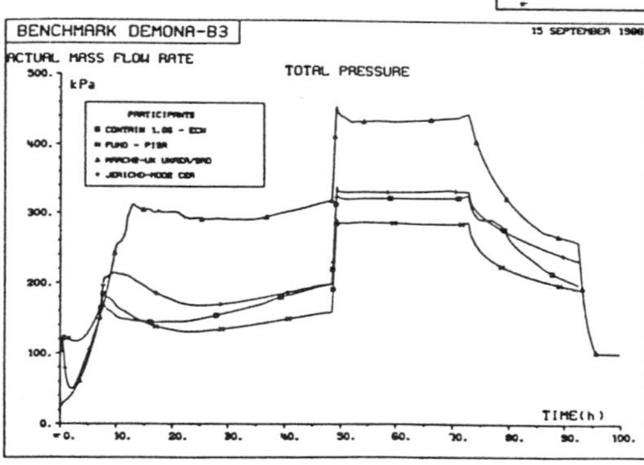

Fig. 6: CEC T-H Benchmark

It is obvious, that in order to comply with guidelines and rules, integral and separate effect experiments are needed under realistic conditions to cope with the complexities cited above and schematically illustrated in Figs. 1 and 2. Because these experiments can only be performed in facilities smaller than real size, computer codes have to be developed, qualified and validated in order to generate reliable results for the design, licensing and best-estimate quantification for the real containments.

The present review focuses on containment research and experiments performed for Pressurized Water Reactors (PWR) in the Federal Republic of Germany (FRG), because of its major contributions to past and present international code validation efforts. For this purpose, results of the experimental test series performed at the Battelle Model Containment (BMC) and the Heißdampfreaktor (HDR) as well as th accompanying computational results are being assessed and the rational behind them developed. The two German facilities are shown to scale compared with a German PWR plant of 1300 MWe design in Fig. 3. The arrows indicate the direction for the transfer of knowledge from one facility size to the other with increasing complexities.

BMC-FACILITY AND RESULTS OF LONG-TERM EXPERIMENTS

The Battelle test facility (see Fig. 4) was especially designed and built with prestressed concrete in order to serve the purpose of conducting scaled containment experiments. By opening or closing flow openings different flow networks can be examined.

The DEMONA test series /1/ was performed in the BMC-facility to demonstrate aerosol retention capability in the containment and produced at the same time very interesting long-term thermal-hydraulic data. According to the German Risk Study Phase A the most probable type of core meltdown accident in the PWR in the Federal Republic of Germany with its large dry containment involves late overpressure failure of the containment after a time period of about five days /2/.

As aerosol behavior is coupled with the thermodynamic and thermohydraulic conditions of the containment atmosphere, thermodynamic/thermohydraulic measurements and codes were also included in the DEMONA investigation. One particular objective of the DEMONA test program was to check the applicability of the single, well-mixed volume concept which was the underlying modeling feature of both the NAUA-code as well as the COCMEL-code which provided the thermal-hydraulic input to NAUA. For the first eight of the nine DEMONA experiments, some of the original internal partition walls of the Battelle Model Containment were removed, and the remaining compartment were connected by maximum vent flow areas to approximate a quasi-single compartment geometry in order to remove uncertain complexities possibly introduced by a multi-compartment geometry upon thermohydraulic and aerosol processes. All DEMONA-experiments were generally performed in the following sequence of four phases:

Phase 1: Expulsion of air by steam injection at 1 bar
Phase 2: Two days of heat-up of the structure by saturated steam atmosphere (1.7 bar, 115 °C)
Phase 3: Aerosol injection together with hot air (1 hour, 3 bar, concentration $10/m^3$)
Phase 4: Aerosol depletion under thermal steady-state conditions by controlled continuous steam injection over a time period of 24 to 30 hours

Besides of the domestic code application efforts with NAUA, COCMEL and later FIPLOC the CEC organized and performed two separate benchmark exercises on an international scale on behalf of DEMONA Test B3 with respect to the aerosol behavior /3/ and the thermalhydraulic issues /4/, respectively. The most interesting feature of the latter is the fact that for uniquely defined single-compartment conditions, both typical severe accident analysis as well as conventional containment analysis codes were applied. The list of participating codes included: CONTAIN 1.10, CONTAIN 1.06, CONTEMPT-LT, JERICHO-MOD2, COBRA-NC and AERCEL, MARCH 2-UK, which spans the spectrum (with the exception of MAAP) of presently available codes for long term severe accident containment safety analyses both with respect to modeling and numerical features.

Figs. 5 through 10 present a cross-section of the major results of the code comparisons with the data. As it was recognized by the CEC-working group that phase 1 of the B3-experiment could not be properly handled under the strict single-compartment assumption, the initial thermohydraulic values were specified at the beginning of phase 2. Because all of the DEMONA-experiments were performed under tight pressure control the participants were asked to replicate this procedure and Figs. 5, 7, 9 and 10 result from this option which was exercised by all institutions in the first place. During the exercise another alternative approach was suggested for a second computation using the measured steam mass flow rate as code input. This additional option was used by only four participants and Figs. 6 and 8 present some of these results.

As can be seen from Fig. 6, all of the codes met the pressure control acceptably well although, as the details show, different approaches (e.g. iterative procedures) were used. Code predictions differ rather markedly during phase 5 when total containment pressure decreased. On the other hand, Fig. 7 indicates large differences in the total pressure for the steam supply input case which can be primarily attributed to the computed differences at the end of phase l. Partial steam pressures are shown in Figs. 7 and 8 for both input cases considered. The spread in predictions in Fig. 7 is about ± 20 % around the experimental value and obviously depends upon a number of assumptions exercised with the individual codes with respect to leakage, sump water withdrawl etc.. The partial steam pressure resulting from the steam supply input is depicted in Fig. 8 and shows rather large discrepancies, again, mostly attributable to the false end of phase 1 predictions. Rather excellent agreement has been achieved for the atmospheric temperature, Fig. 9, whereas the predicted steam condensation rates on the structure substantially differ, Fig. 10, especially during phases 1 and 2 of the experiment. Overall, the CEC T-H Benchmark Exercise achieved its major objectives by evaluating the present status of codes used in this areas as well as in identifying still open issues for the thermal-hydraulic analysis alone (volumentric condensation during supersaturation, heat transfer). For this purpose, experiment B3, posed a number of stringent constraints upon the code application (pressure control, sump removal, supersaturation) which were only partially offset by the open post-test character.

A much more sophisticated approach to analyse the thermohydraulics of B3 is possible by the application of a multi-compartment code such as demonstrated by the FIPLOC-nodalization, flow pattern and mass flow through vents /5/ presented in Figs. ll and 12, which allow a much more detailed interpretation of the experimental data in terms of flow patterns and recirculation flows during the various phases.

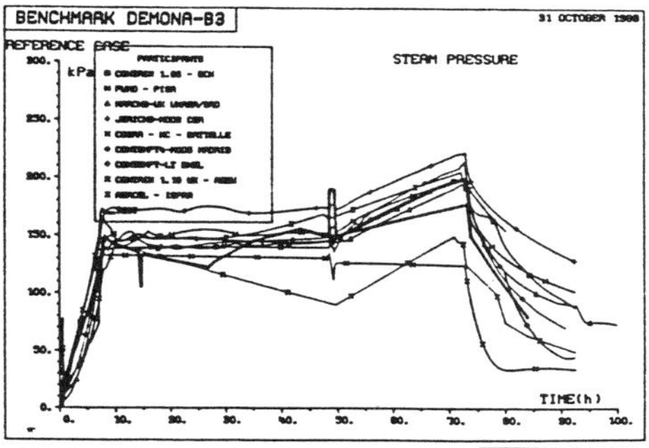

Fig. 7: CEC T-H Benchmark

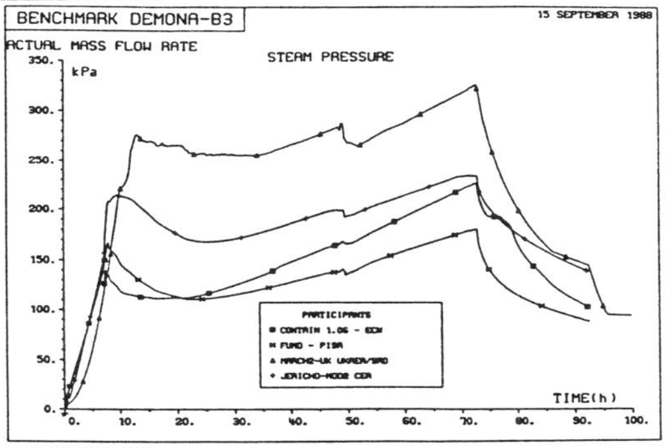

Fig. 8: CEC T-H Benchmark

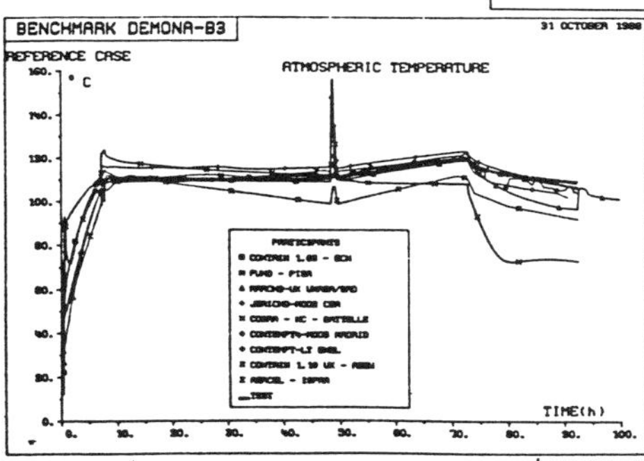

Fig. 9: CEC T-H Benchmark

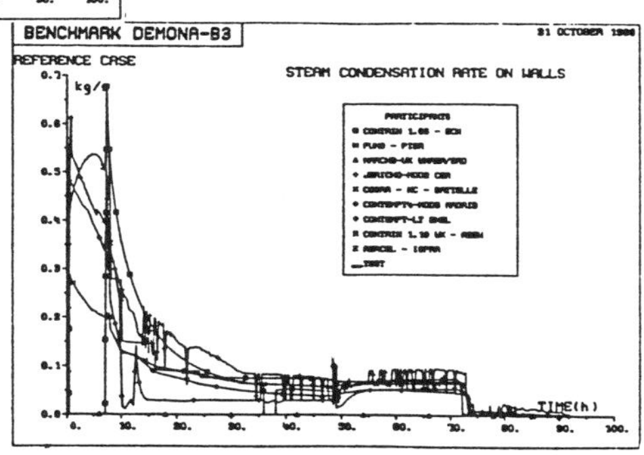

Fig. 10: CEC T-H Benchmark

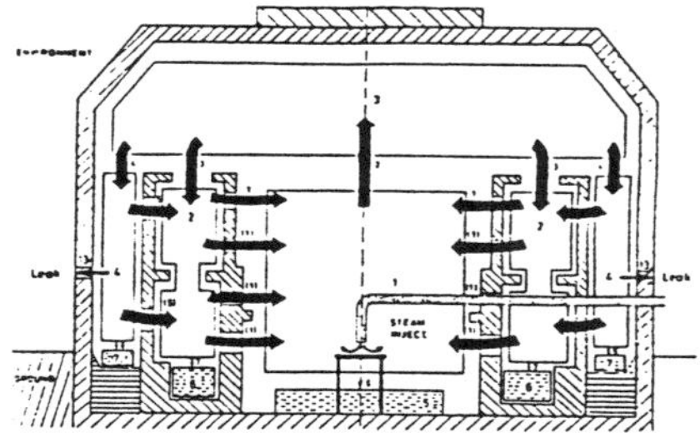

Fig. 11: B3: FIPLOC Nodalisation and Flow Pattern During Depletion Phase

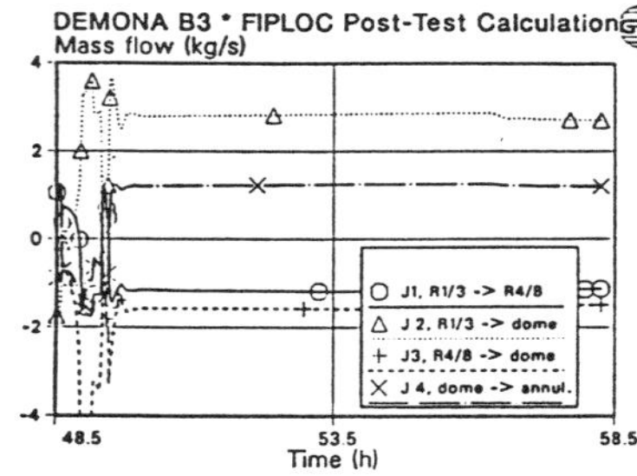

Fig. 12: B3 FIPLOC- Mass Flows During Injection and Depletion Phases

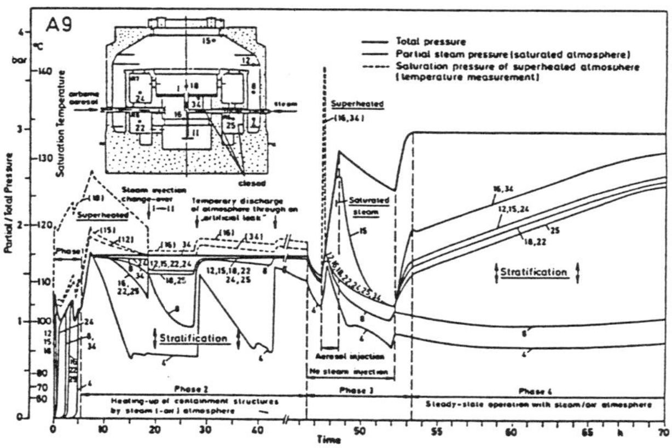

Fig. 13: DEMONA A9: Temperatures and Steam Distributions

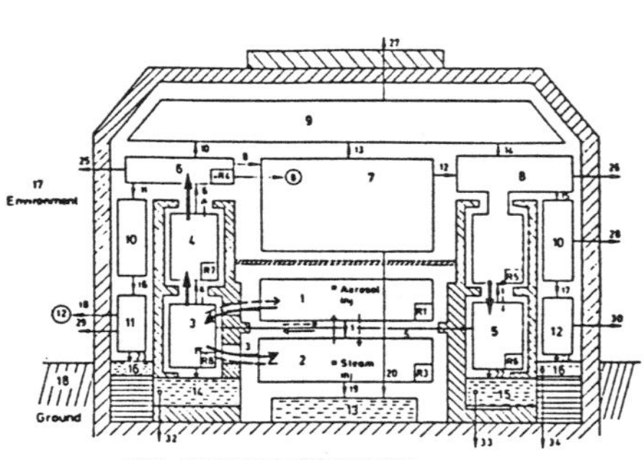

Fig. 14: A9: FIPLOC Nodalization and Flow Patterns

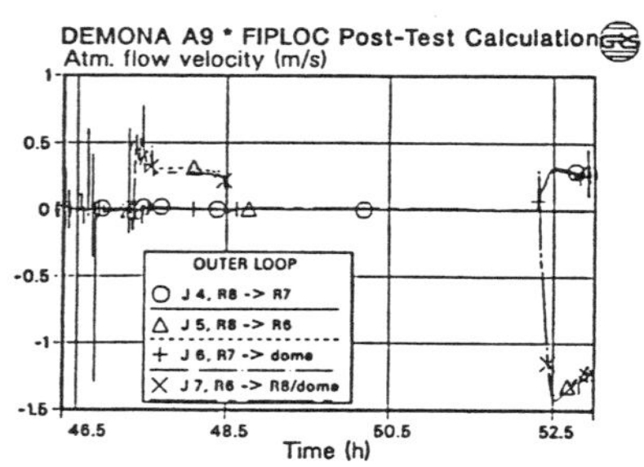

Fig. 15: A9: Atmospheric Velocities in Outer Loop During Injection, Depletion and Steam Injection Phases

For the last DEMONA experiment, A9, the simplifying initial- und boundary conditions exercised for the previous tests were given up in order to study the impact of multi-compartment geometry upon thermohydraulic and aerosol behavior and the resultant complexities of global and local coupling phenomena. From an analytical point of view it was envisioned that it would generate a first data base for the verifications of more advanced codes such as FIPLOC, NAUA-M, CONTAIN etc..

Steam and aerosol mixture ($Fe_2 0_3$, $Sn 0_2$) were injected into the first containment zone (50 m^3). Between the injection region and the large dome region (467 m^3), two intermediate zones (82 m^3 and 41 m^3) were introduced. A second steam injection position was installed. For test A9, the proven, general 4-phase procedure was varied with respect to the following aspects:

Phase 2: Steam-injection switch-over from upper to lower position, temporary discharge at bottom of outer ring comparments

Phase 3: Temporary stop of steam supply resulting in a superheated atmosphere during aerosol injection and the first subsequent 3.8 hours of aerosol depletion; restart of steam supply resulting in subsequent supersaturation and intensive fog formation

Fig. 13 summarizes the major experimental findings with respect to the thermodynamic quantities over all four experimental phases of test A9 at different positions in the Battelle Model Containment. The vertical axis gives the total pressure and local partial steam pressure or the steam-air composition of the atmosphere. Steam injection through the upper injection port during Phase 2 leads to:

- superheating in the upper central zone and upper dome region
- substantial air enrichment in the outer annular space and less pronounced in the neighboring subcompartments; overall a clearly stratified containment atmosphere developed

Switching from the upper to the lower port somewhat reduces the stratification pattern. In order to appreciate the importance of test A9 in the context and in comparison to all of the other less complex DEMONA-experiments, the interested reader is referred to Refs. /6, 7/. It is quite obvious that this situation cannot be handled by the single-compartment methodology, rather multi-compartment codes such as FIPLOC, COBRA-MAEROS or CONTAIN must be applied. A typical FIPLOC nodalization as well as computed flow directions and flow velocities through vents /5/ are depicted in Figs. 14 and 15. From Fig. 14 it becomes apparent that flow directions reverse from aerosol injection to aerosol depletion phase. The complexity of flow patterns during the A9 test in various phases suggest that for multicompartment geometries more advanced codes capable to analyze the coupling of thermohydraulics and aerosol physics in multi-compartment geometries should be applied on a case by case basis because no generic generalization is a priori possible.

FIPLOC VERIFICATION TEST PROGRAM

With the open questions raised by the complex results of the DEMONA-A9 experiments, it was decided to perform two additional test series solely for the purpose of revealing more details upon the thermohydraulic behavior in multi-compartment geometry under changing mass and energy input conditions without aerosol injection.

The FIPLOC Verification Test Program /8, 9/ consisted of two test groups, F1 and F2. F1 was a series of 8, F2 was a series of 20 consecutive experimental phases. Both test groups were preceded by a heat-up period of two days. The particular objectives of the two test series (Fig. 16) are:

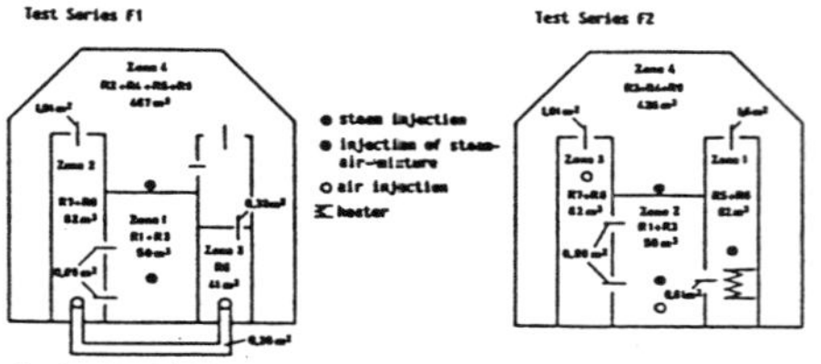

Geometry: Branched chain (like DEMONA Test A9)

Objectives: Atmosphere removal effects following variations of steam-air mixture injections

Loop with slight asymmetry

Natural convection loop behaviour following variations of steam injection site and heat supply

Fig. 16: FIPLOC-Verification Experiments F1 and F2

Table 1: Test Series F1

Test phase No.	Duration h	Fluid	Fluid Injection Mass flow kg/s	Ratio air/steam	Location
1 (heat-up)	46.4	steam	<0.5	0	upper site (R2)
2.1	6.8	steam	0.11	0	
2.2	2.5	steam + air	0.024 / 0.036	1.5	
3.1	4.5	steam	0.05	0	
3.2	0.4	steam + air	0.03 / 0.08	2.7	lower site (R3)
4.1	1.8	steam	0.05	0	
4.2	0.4	steam + air	0.02 / 0.07	3.5	
5	8.4	steam	0.05-0.08	0	
6	12.0	steam	0.05-0.07	0	upper site (R2)

containment pressure 2.5 bar during phase 2 through 6

Fig. 17: Test Matrix F1

Table 2: Test Series F2

Test phase No.	Duration h	Containm. pressure bar	Fluid injection Fluid	Mass flow kg/s	Location	Heat release in R6 kW	Resulting convection loop cm/s *
1 (heat up)	48.0	1 - 3	steam	$\leq$0.5	upper site (R2)	-	-
2.1	4.5	3	steam + air	0.03-0.[illegible]4 / 0.03-0.[illegible]6	R3 / R7	-	<8 up
2.2	3.6	3	steam	0.07	R3	12	20 down
2.3	2.4	3	steam	0.09	R6	-	21 up
2.4	2.0	3	steam	0.09	R3	-	18 down
3	7.2	3 - 1.8	-	0	-	9 (1.3h)	<10 up
4.1	1.8	1.8	steam	0.04	R3	-	22 down
4.2	2.2	1.8	steam	0.04	R3	16	25 up
4.3	1.8	1.8	steam	0.05	R6	-	<17 up
4.4	1.8	1.8	steam	0.05	R3	-	10 down
5	5.2	1.8 - 3	steam		R6, R3	-	22 down
6.1	2.1	3.0	steam	0.11	R3	-	30 up
6.2	2.5	3.0	steam	0.11	R3	25	30 up
6.3	2.0	3.0	steam	0.12	R6	-	22 up
6.4	2.0	3.0	steam	0.12	R3	-	12 down
7	5.8	3 - 1.8	-	0	-	8 (1.2h)	-
8.1	2.5	1.8	air	0.05	R3	-	-
8.2	2.7	1.8	steam	0.02	R3	-	<25 down
8.3	2.5	1.8	steam	0.03	R3	10	20 down
8.4	2.0	1.8	steam	0.04	R6	-	12 up
8.5	2.0	1.8	steam	0.04	R3	-	27 down

* measured in vent opening U 59 A (1,6 m2)

Fig. 18: Test Matrix F2

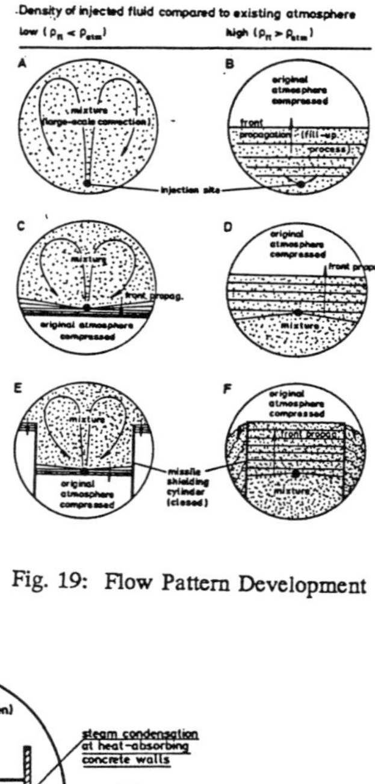

Fig. 19: Flow Pattern Development

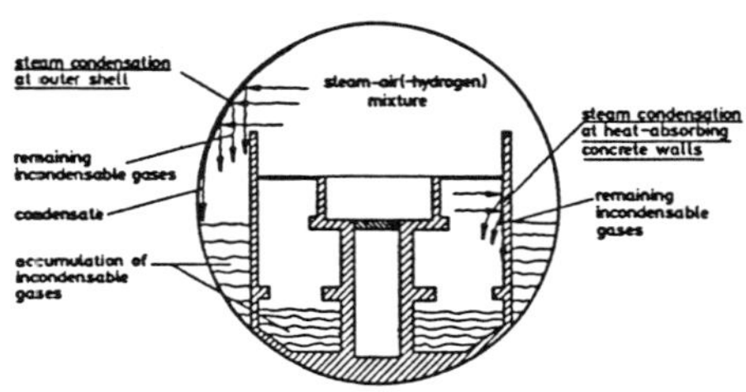

Fig. 20: Condensation Enhanced Stratification

Fl: Atmospheric removal effects with a chain-type multi-compartment geometry following variations of steam-air mixture injections (test matrix in Fig. 17)

F2: Natural convection phenomena in a loop-type multi-compartment geometry affected by variations of steam injections and heat supply into the various compartments at different times (test matrix in Fig. 18).

Fl+F2: Atmosphic composition in the annulus between containment shell and missile shielding cylinder

The chain-type arrangement of compartnents in test series F1 covers different vertical levels (cf. Fig. 16). The test conditions of the individual experiments of test series F1 differ mainly in the composition of the injected fluid. Depending on the density difference between the injected steam-air mixture and the existing steam-air atmosphere in the containment, as well as on the levels of compartments and vent openings, the following differences in the behavior were observed (compare Fig. 19):

- A steam-air mixture of relatively high density (i.e. with high air concentration) which is injected into a low-level compartment "fills up" the injection compartment and the adjacent low-level compartments almost completely, whereas the lighter atmosphere that originally was there moves upwards.
- Mixtures of low density that are injected at the same location show the tendency to immediately escape directly upward into higher-level compartments and to very slowly entrain the original atmosphere from the injection compartment.

The experimental observations on atmospheric removal effects made in test series F1 and similar containment experiments are qualitatively summarized in Fig. 19.

It is evident that analytical simulation of the observed effects by multinode codes needs an adequately fine nodalisation, i.e. further division of the individual containment compartments into several computational zones. Further more, very stable numerical solution procedures are necessary to handle the very small pressure differentials of the buoyancy effects involved.

Because in PWR containments various geometric configurations exist where convective flow loops may develop (formed by the two steam generator compartments connecting to the dome region) the loop-type test geometry used for test group F2 and shown in Fig. 16 should approximate this situation by virtue of the central injection zone (No. 2 of Fig. 16) at a low level, a large dome compartment (zone 4) and - in between - two identical 'steam generator' compartments (zone 1 and 3; each 5 m high). Convective loop flow which results from atmospheric density differentials between the right-hand and left-hand branches of this geometry, can be simulated in the experiments either by heat release (via a steam-powered heat exchanger) or by steam and air injection at different locations (cf. Fig. 16). Size and location of the connecting openings of the test loop geometry are chosen somewhat asymmetric to impose a preferential flow direction in case of steam release into the central zone No. 2.

The comprehensive F2 test matrix summarized in Fig. 18 shows four similar sequences, each consisting of four comparable experiments (test phase 2.1, etc.: 4.1, etc.; 6.1, etc. and 8.2 etc.; variation of steam and heat release) at different levels of containment pressure and steam-air atmosphere composition.

Comparison of the experimentally determined convective flow results obtained from these 4 x 4 sequences reveals:

- Dependent upon the actual test conditions under consideration convective loop flow pattern may develop in either direction or stagnation prevails.
- During stagnant phases or at low flow velocities significant temperature or steam concentration gradients develop along the flow path
- Generally, high flow velocities result in homogeneous atmospheres in respective containment zones
- The loop flow velocity and its direction are primarily affected by the actual flow and energy released into the respective containment zone but are also determined by the structural temperature (stored energy) which resulted from the preceeding sequence.
- The observed delay between changed experimental conditions and the resulting changes in flow patterns ranged between approximately 1 and 100 minutes.

During a sequence of four experiments (Nos. 4.1 throuqh 4.4) with variation of steam release and heater operation, various convective flow patterns are measured by the installed turbine flowmeters as shown in Fig. 21. During Sequence 4.1 (period Δ t_1), steam is injected into the central compartment at the bottom of the loop geometry. As the injected steam has a lower density than the existing steam-air mixture, buoyancy-enhanced natural convection is induced which - due to the asymmetric arrangement of the vent openings - ascends on the left-hand branch of the loop geometry and descends on its right-hand branch. As a result, temperature differences between the structures of the left-hand and right-hand branches develop, and the direction of convective flow stabilizes. This becomes evident in Sequence 4.2 (period Δ t_2) when dry heat is introduced into the right-hand branch with the intention to reverse the direction of flow. First, the additional dry-heat source merely affects a significant decrease in flow velocity. Simultaneously the atmosphere temperature differences within the loop region increase from about 3 to 6 K. This affects the temperatures of the structures, and after a delay of 1.5 hours the convective flow loop develops in counter-clockwise direction.

During Sequence 4.3 (Δ t_3), steam is injected into the right-hand branch to maintain the actual flow direction. It is very interesting to see that transition to Sequence 4.4 (Δ t_4) with central steam injection results in a second flow reversal. Comparison with the previous - delayed - flow reversal indicates the strong influence of the actual asymmetric arrangement of openings upon the buoyancy enhanced convection.

Only containment zones which participate in the large-scale convection loops as discussed above have an approximately homogeneous steam/air, gas distribution. By contrast, stagnant zones either separated by partition walls or affected by stratification phenomena do not exhibit any mixing effects. As a result, inhomogeneous atmospheric compositions may develop in these zones because of the following effects:

- Injection and 'zone fill-up' processes discussed above and schematically shown in Fig. 19.
- Air and gas distribution due to steam condensation effects as indicated in Fig. 20, where the steam condensation at structures naturally leads to continuous steam depletion and air enrichment in the adjacent atmosphere resulting in its increasing density and thereby to its downward movement finally resulting in an accumulation of air and other non-condensable gases in lower, especially separated containment regions.

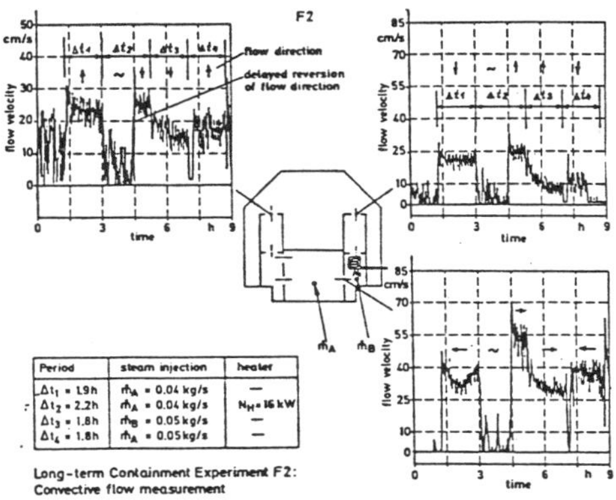

Fig. 21: F2-Test: Measured Changes of Flow Velocities at Different Positions During Different Phases

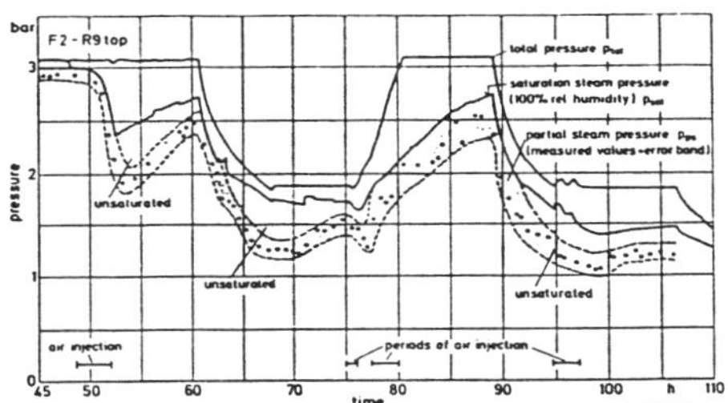

Fig. 22: F-2-Test: Local Atmospheric Conditions at Top (left) and Bottom (right) of Subcompartment R9

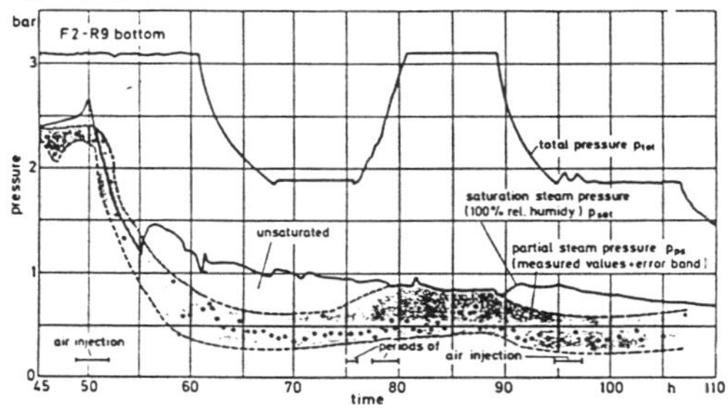

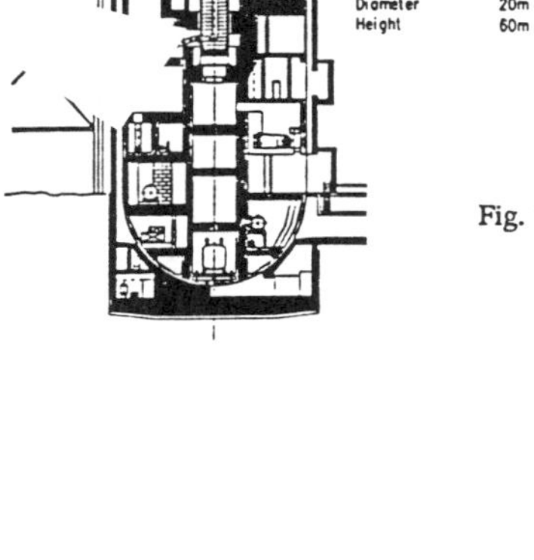

Fig. 23: HDR-Facility

In both test series F1 and F2 distinct air enrichment effects - caused by the two processes mentioned above - were observed in the annulus compartment of the model containment between outer cylinder wall and missile shielding cylinder. This annulus is open to the dome at its top and closed at the bottom, thus forming a 'dead-end' compartment. The resulting differences in steam-air composition at different elevations (stratification) during test series F2 are depicted in Fig. 22. This finding is very important for assessing the hydrogen behaviour in a containment atmosphere. Due to the gas accumulation effect described in Fig. 20, hydrogen and air can locally form combustible mixtures in spatially separated containment zones and thus may endanger the containment integrity due to the potential of local DDT.

MODEL CALCULATIONS

To check the capabilities of the available multi-node long-term containment codes, 'blind' post-test model calculations (i.e. calculations after performance of the test, using the actual boundary conditions) of the series F2 using the FIPLOC and CONTAIN codes were performed by GRS Munich and AEE Winfrith, respectively. The FIPLOC calculations, for example, predicted the typical natural convection flow velocity values correctly and also yielded, the correct predictions for the loop flow directions of most of the experiments. On the other hand, air accumulation in the annulus was significantly underestimated. The FIPLOC code computations have been documented in /10/.

In total the calculational results indicate general applicability of these codes to those types of problems, but additional analytical work will be necessary to further qualify the codes and to extend reliably the experience of the code users.

In summary, the test series F1 and F2 constitute a very comprehensive unique experimental data base for a variety of different issues related to thermal-hydraulic long-term containment phenomena. This data base certainly warrants the attention of computer code developers and users concerned about severe accident modeling in containments. The F1 and F2 experimental data can surely serve as a rather stringent test for multi-compartment code capability verification.

This potential has been recognized by the CEC, Brussels, in the meantime and consequently another international T-H Benchmarck Exercise on parts of the F_2-experiment has been initiated with distinct milestones /11/.

HDR-FACILITY AND RESULTS OF LONG-TERM EXPERIMENTS

The HDR containment experiments are part of a large-scale experimental research program at the HDR (Heißdampfreaktor) in Karlstein, near Frankfurt. Containment experiments were performed during Phase I of the HDR Safety Program (1980-1984) as well as in the Phase II running from 1985 through May 1988. A new series of containment experiments related to hydrogen distribution/deflagration will be performed in Phase III (1988-1991), including long-term thermal-hydraulic issues.

The HDR containment (11,300 m^3 volume) consists of a steel shell 60 m in height and 20 m in diameter (see Fig. 23), which is subdivided into 62 compartments by concrete partition walls. These compartments are interconnected by about two hundred vent flow openings. The HDR containment test instrumentation consist presently of a total of 700 thermal-hydraulic and structural measuring channels for both short-term and long-term measurements. It also includes special measuring systems for heat transfer, steam/air composition and vent mass flow and velocities through the major vents out of the break subcompartment and between major containment zones, e.g. staircases and dome region.

From the vent measurement system, die parameters of passing liquid droplets can be inferred during the short-term period. Thus, the instrumentation presently applied in the HDR-facility is a major step forward.

Two kinds of test series have been performed consisting of two vapor test series (COND, CON) and one water test series (CONW). In the first vapor test series COND consisting of the experiments V42-V45 the experimental parameter variations were the liquid level in the reactor pressure vessel (RPV) which determines the duration of the vapor inflow into the containment and the area and configuration of overflow openings in the break compartment (cf. Fig. 24).

All of the aforementioned HDR-experiments were planned, performed and analytically assessed with the common Design Basis Accident short-term containment behaviour in mind. Fortunately though, experimental data for the most typical experiments (Figs. 24, 25) were also recorded for up to 20 hours, thereby allowing a direct comparison between DBA and non-DBA (small breaks) containment behaviour.

A critical review of present and future requirements of containment research and computer code verification revealed that in addition to the conventional LOCA-scenarios, small break LOCAs into the containment with long blowdown periods are needed. Thus, it was decided ad hoc by the Project HDR to perform two such experiments in the aftermath of other test groups where the large pressurized HDR-RPV was available for blowdown. It was decided that both experiments were of scoping character, thus only part of the usual containment instrumentation was applied with emphasis on temperature measurements over the height of the HDR containment. The first experiment of this kind was T30.4 performed after the so-called RPV-I (Reactor Pressure Vessel Internal Test Series) experiments, simulating a 0.1-F break situation in terms of the HDR-200 mm pipe diameter used previously in other RPV-I test series. The second experiment, (0.025-F-break) termed T31.6 was performed in the aftermath of cyclic thermal shock experiments. It was more carefully instrumented and partly co-sponsored by EPRI, USA, for the purpose of testing turbine windmill-type velocity meters under 'realistic' containment conditions.

Finally, experiment T31.5 (Figs. 24, 25) designed for the International Standard Problem No. 23 (DBA) was performed for 20 hours by additional steam and H_2-mixture injections to examine the hydrogen distribution behavior over long times and as a testing ground for measurement techniques, operational procedures and first large-scale code validation efforts /12/ in view of the forthcoming five HDR-H_2-distribution experiments.

Figs. 26 and 27 compare the pressure behaviors as result of the different blowdowns with the various break sizes into the HDR-containment and indicate substantial differences with respect to peak pressures and temporal characteristics.

Figs. 28 throug 31 compare the temperature for left and right flow paths at the same axial elevations for large (Figs. 28, 29) and small (Figs. 30, 31) LOCAS, respectively. As becomes apparent from Figs. 28 and 29, no temperature difference develops between both flow paths over the 20 hours, for the large break blowdowns, whereas Figs. 30 and 31 indicate more or less pronounced temperature differences over the complete time period dependent upon the break size. The existence of this temperature difference as driving force for natural convection flows is more clearly depicted in Figs. 32 and 33 for experiment T31.6 along the heights of the left and right flow paths for two different instants in time.

Test Group	Initial Conditions in RPV			
	Test No	Pressure bar	Temp. °C	Liq. Level m
CONW	V21.1	110	313	full
COND	V43	110	316	6
RDBE	T30.4	110	308	full
CON	T31.6	110	308	full
CON	T31.5	110	318	7.8

Fig. 24: HDR Long-Term Containment Experiments

Test Group	Break Nozzle		Spray System		Variation of Test Parameters
	Length mm	Diameter mm	On min	Off min	
CONW	6091	453	–	–	1–F–Break, Water
COND	6091	453	10	70	1–F–Break, Steam
RDBE	5285	63.25	–	–	0.1–F–Break
CON	5295	25	–	–	0.025–F–Break
CON	9584	453	–	–	1–F–Break, Steam +Steam Injection +H_2/He–Injection

Fig. 25: Major Characteristics of Long-Term Experiments

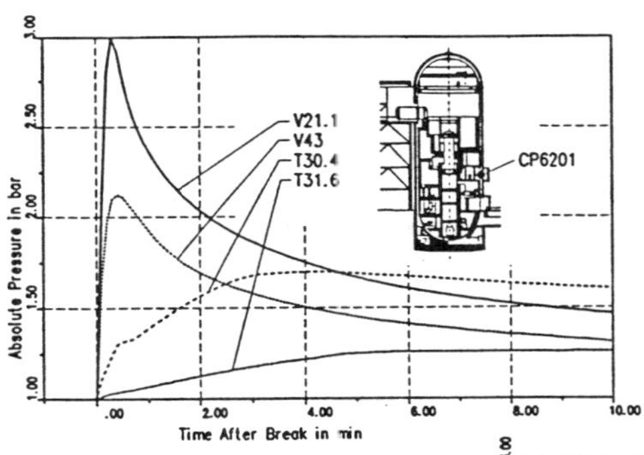

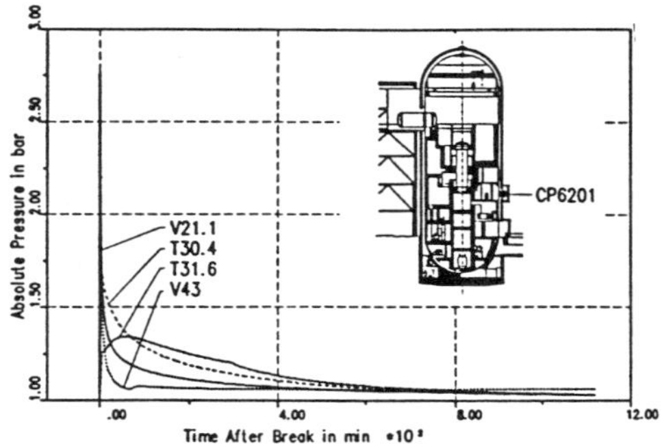

Fig. 26, 27: Containment Pressure Behaviors for 10min and 20h for large and small LOCAs

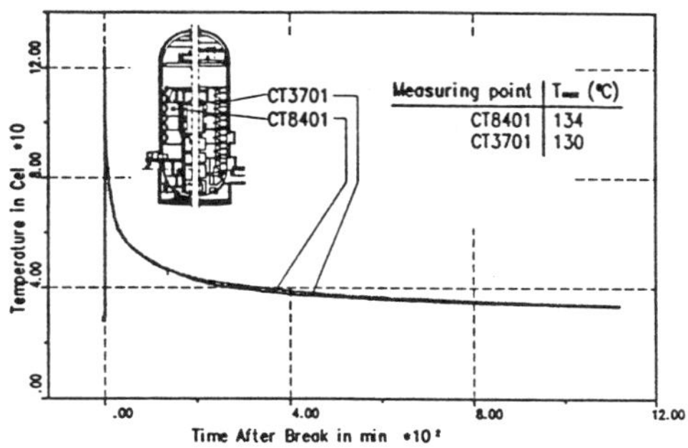

Fig. 28: Test V21.1: Transient Temperatures in Left and Right Flow Paths for Large LOCA

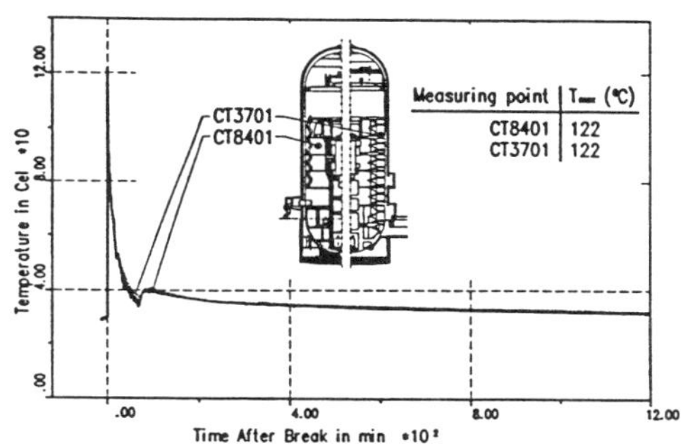

Fig. 29: Test V43: Transient Temperatures in Left and Right Flow Paths for Large LOCA

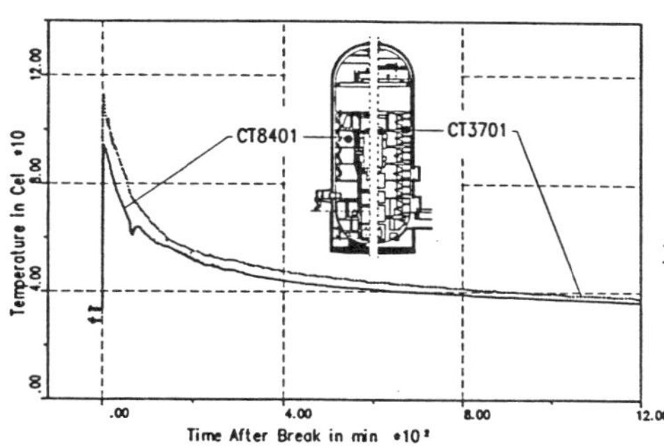

Fig. 30: Test T30.4: Transient Temperatures in Left and Right Flow Paths for 10% LOCA

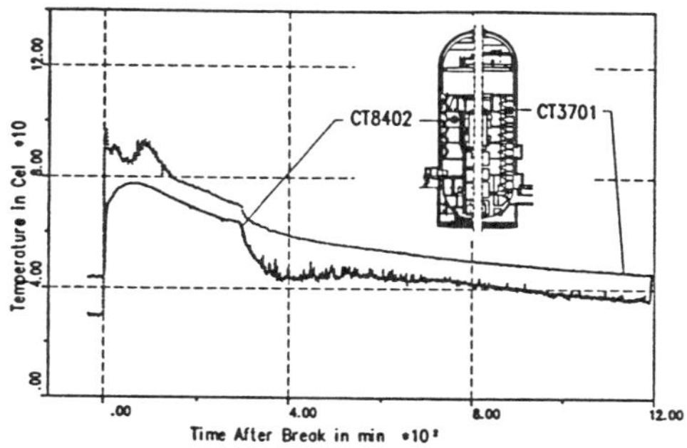

Fig. 31: Test T31.6: Transient Temperatures in Left and Right Flow Paths for 2.5% LOCA

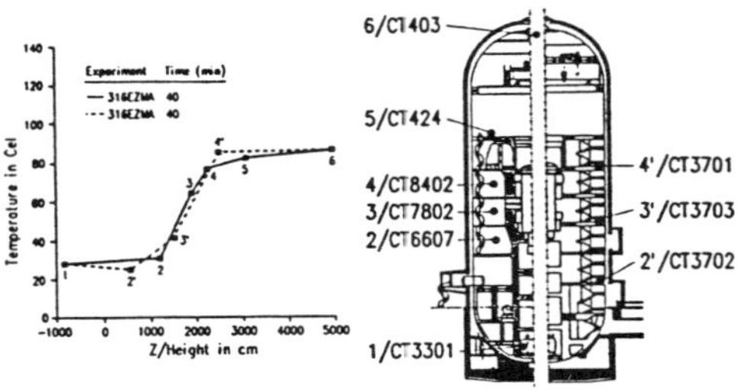

Fig. 32: Test T31.6: Axial Tempertures Profiles for t=40min

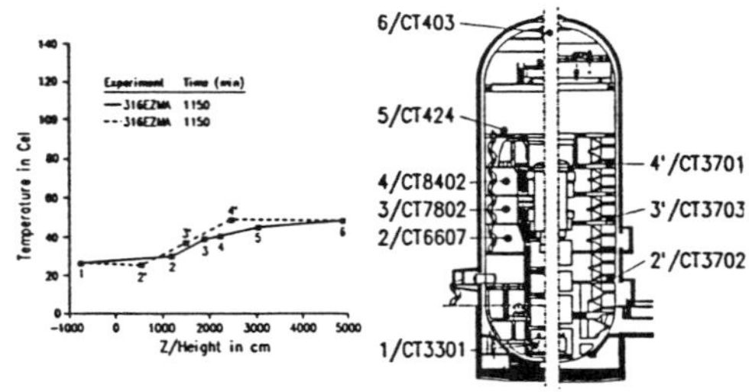

Fig. 33: Test T31.6: Axial Temperatures Profiles for t=1150min

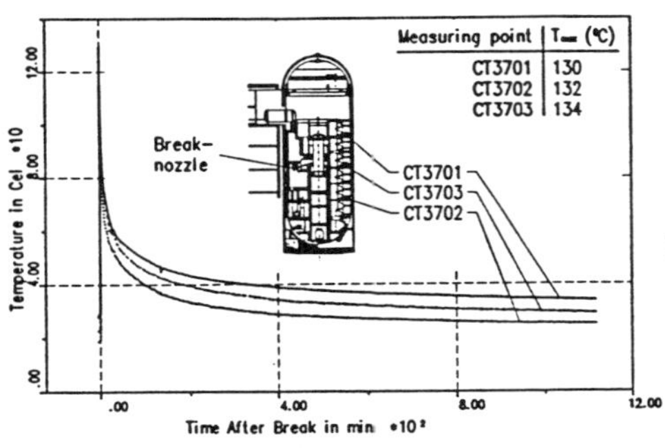

Fig. 34: Test V21.1: Transient Temperatures at Different Axial Positions for Large LOCA

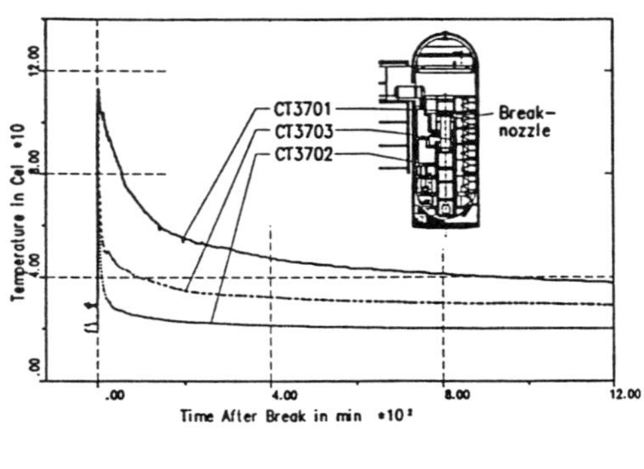

Fig. 35: Test T30.4: Transient Temperatures at Different Axial Positions for 10% LOCA

The non-existence of transverse temperature gradients does not preclude axial temperature gradients also for large break blowdowns as Fig. 34 demonstrates for V21.1. However, the axial temperature gradients are much more pronounced for smaller sized breaks as Fig. 35 reveals for T30.4.

Stratifications also develop in the large dome region for both large and small break LOCAS as demonstrated by Figs. 36 and 37. They are more pronounced and long lasting for smaller-sized breaks.

MODEL CALCULATION

Thus far, few efforts have been undertaken to simulate the effects of small break blowdowns into the HDR-containment. Figs. 38 and 39 present two different approaches taken for CONTAIN-simulations of experiment T31.6. The 8-node model shown in Fig. 38 was set up for demonstration purposes of CONTAIN 1.05 whereas the 40-node model (including the annular gap between steel shell and outer concrete) was recently developed by Battelle-Frankfurt to perform a best-estimate CONTAIN 1.10 computation. A similar but somewhat different model was selected by GRS for CONTAIN 1.10, results of which are also included in some of the following figures.

As Fig. 40 clearly shows, the few node computation does not follow the experimental data over longer times. It needs a much higher spatial resolution to match more closely the experiment over the total of observation times as Figs. 40 and 41 demonstrate. Accounting for even more details (annular gap) as in the Battelle model results in even improved agreements shown in Figs. 42 and 43. As presented in Fig. 44, the CONTAIN 1.10 - simulation with the Battelle-model meets the data quite well even in the lower containment region which remains unaffected by the thermal plume, although the initial temperature peak is overpredicted by a factor of 2. In order to improve and to follow the plume-affected and not affected regions finer nodalizations are needed such as used in a COBRA-NC prediction for T30.4 and shown in Fig. 45 at an elevation of 25 m. This figure shows the importance of accurate modeling of mixing effects.

From the velocity data obtained during T31.6, a period of complete stagnation was deduced which after hours was resolved by the onset of natural convection currents. Thus, extremely sensible processes take place within the containment involving energy storage and release in the structure as well as heat transfer across the steel shell and convective currents in the annulus between steel shell and secondary concrete structure surely will develop. In conclusion then, in contrast to the tacitly made simplifying assumptions concerning one or two node models for long-term analysis, the HDR data suggest that a much more refined approach is needed to account for a variety of phenomena usually not modelled by standard, only DBA-oriented containment codes.

Therefore, it is not surprising that a trend in containment analysis is developing to account for 30-50 nodes, such as for instance by the FIPLOC-, CONTAIN and COBRA-NC-codes or even for thousands of nodes by the HMS-code employing a distributed parameter model with finite-difference nodalization and accounting for turbulence models.

Additional insights in the predictive qualities of present state-of-the-art containment analysis codes are expected from the international PHDR blind post-test exercise on behalf of the hydrogen distribution experiments E11.2 and E11.4, results of which should become available by mid-1990.

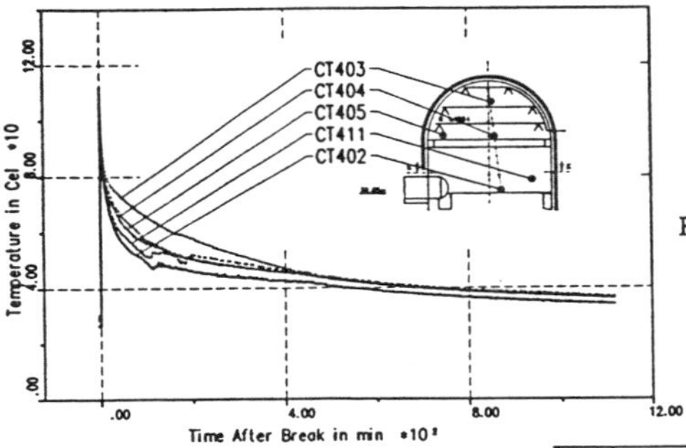

Fig. 36: Test V21.1: Transient Temperatures at Different Done Positions for Large LOCA

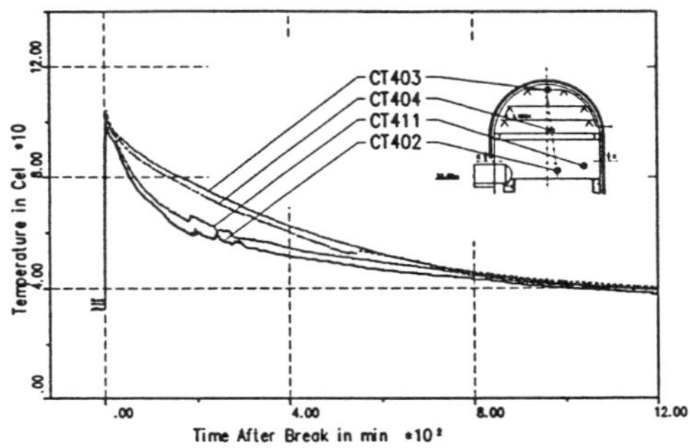

Fig. 37: Test T30.4: Transient Temperatures at Different Dome Positions for 10% LOCA

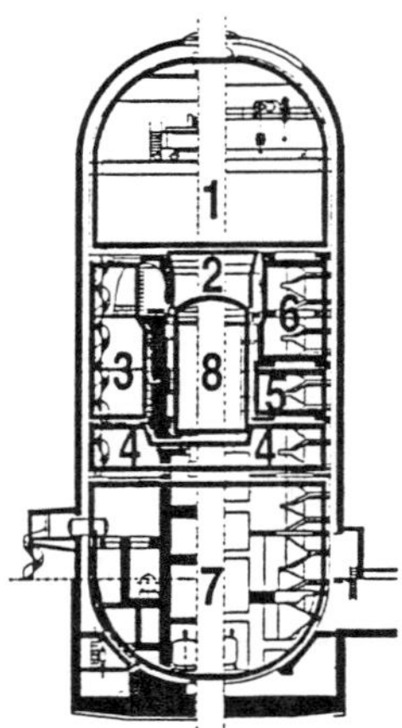

Fig. 38: 8-Zone Nodalization for DEMO-CONTAIN 1.05

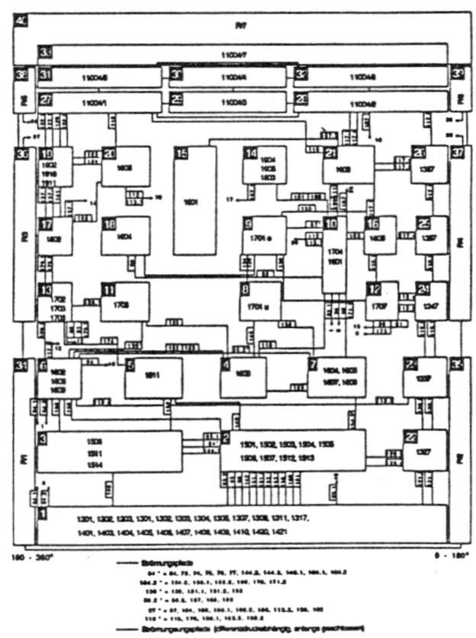

Fig. 39: 40-Zone Nodalization for CONTAIN 1.10

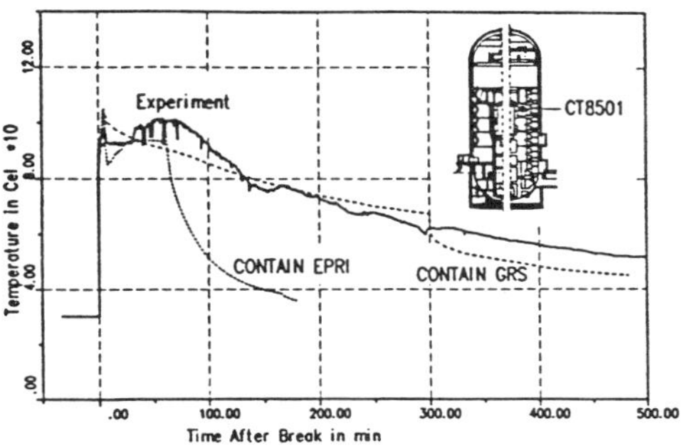

Fig. 40: Comparison Between Data and Few Node DEMO- and Multiple Node CONTAIN Computations

Fig. 41: T31.6: Axial Measured and Computed Temperature Profiles at Late Times

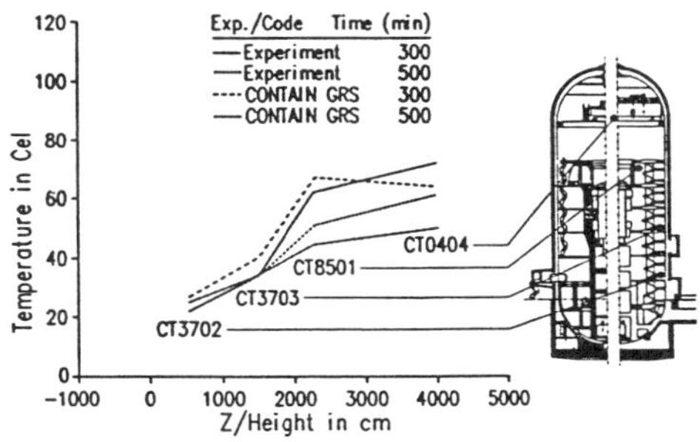

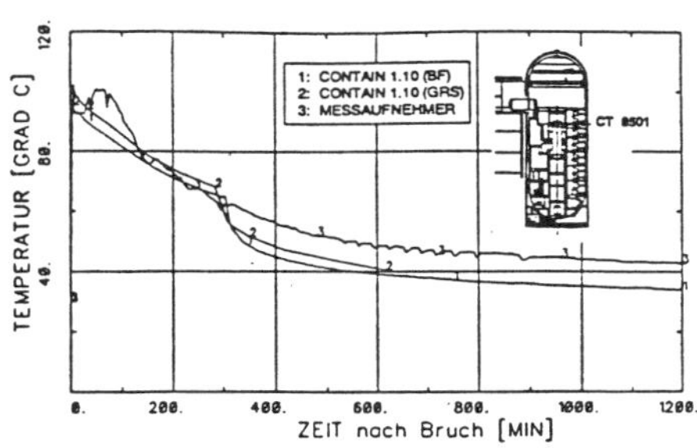

Fig. 42: T31.6: Comparisons Between Temperature Data und Two Differently Nodalized CONTAIN 1.10 Computations

Fig. 43: T31.6: Comparison Between Measured Pressure and Computations by Two Differently Nodalized CONTAIN 1.10 Models

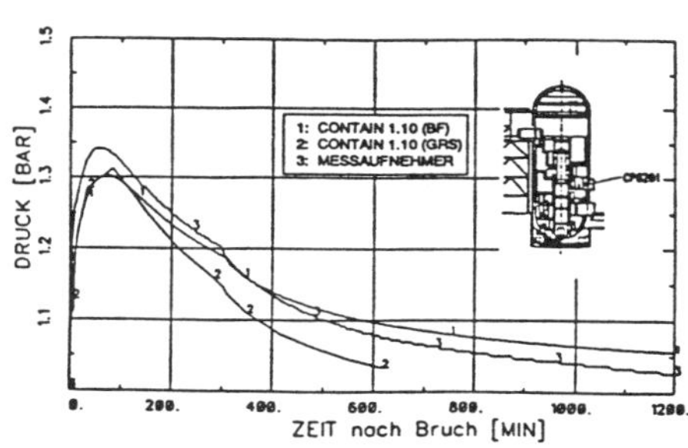

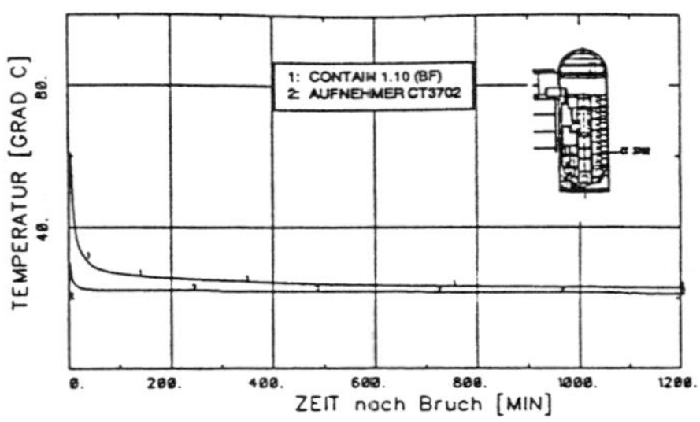

Fig. 44: T32.6: Comparison Between Temperature Data and 40-Zone CONTAIN 1.10 Computation in Lower HDR-Region (+5.5m)

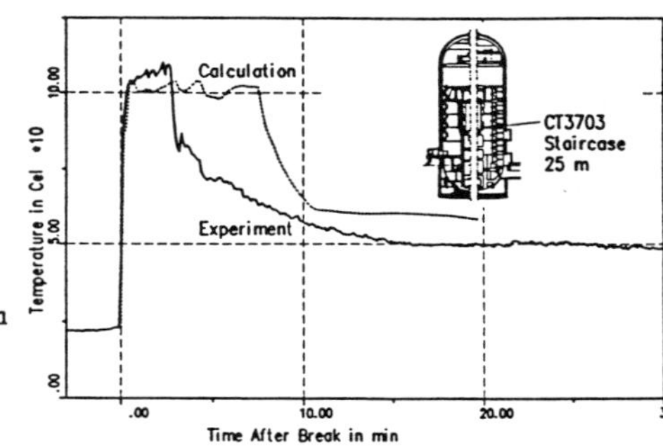

Fig. 45: T30.4: Comparison Between Temperature Data and 82-Zone COBRA-NC Computation at 25 m

REFERENCES

1. Kanzleiter, T.: Technical-Scale Integral Experiments in the Battelle Containment Concerning the Behavior of Aerosol Fission Products Under Core Meltdown Conditions (DEMONA), (in German), Final Report, Battelle Report, BIeV-R-65.523-01, Sept. 1987

2. GRS: German Risk Study Nuclear Power Plants, vol. 5, Examination of Core Melt Accidents, (in German), TÜV Rheinland, 1979

3. Schöck, W.: Post-Test Calculations of Behavior in DEMONA Experiment B3 With Various Computer Codes Used in CEC Member States, EUR 11374 EN, 1988

4. Gauvin, J.: Post-Test Calculation of Thermal-Hydraulic Behavior in DEMONA B3 With Various Computer Codes Used in CEC Member States (Draft Final Report), Nov. 1988, to be published

5. Weber, G.: FIPLOC-Multicompartment Analysis of the Spatial Aerosol Distribution Phenomena Observed in the DEMONA-Experiments, in: Water Cooled Reactor Aerosol Code Evaluation and Uncertainty Assessment, EUR 11351 EN, 1988, pp. 356-365

6. Kanzleiter, T.: Results of the Multi-Compartment Experiment A9, (in German), Battelle-Report BIeV-R-65.523-30-9, 1986

7. Wolf, L.: Kanzleiter, T.: Multi-Compartment DEMONA-Experiment A9 and Conclusions for Coupled Thermohydraulic/Aerosol Codes, in: Water Cooled Reactor Aerosol Code Evaluation and Uncertainty Assessment, EUR 11351 EN, pp. 329-345

8. Kanzleiter, T.: FIPLOC-Verification Experiments Concerning Thermal Hydraulic Long-Term Phenomena Within a Multicompartment Containment, (in German), Battelle-Report BIeV-R-66.614-01, 1988

9. Kanzleiter, T.; Wolf, L.: Thermal-Hydraulic, Natural-Convection, Multi-Compartment Experiments in the Battelle Model Containment, in: Water-Cooled Reactor Aerosol Code Evaluation and Uncertainty Assessment, EUR 11351 EN, 1988, pp. 346-353

10. Schwarz, S.: FIPLOC Verification Test F2 - Blind Post-Test Calculation Compared to the Experiment, GRS-A-1490, Oct. 1988

11. Kanzleiter, T.; Wolf, L.: CEC-Thermal-Hydraulic Benchmark Exercise on FIPLOC-Verification Experiment F2 in BMC, Battelle-Frankfurt, July 6, 1989

12. Wolf, L.; Valencia, L.: Experimental Results of the Preliminary HDR-Hydrogen Distribution Test T31.5, 4th Intl. Topical Mtg. on Nuclear Reactor Thermal-Hydraulics, Karlsruhe, FRG, Oct. 10-13, 1989

The Influence of the Sensible Heat Removal on the Condensation onto Aerosols inside a Light Water Reactor Containment

H. M. VAN RIJ and V. VONKA
Netherlands Energy Research Foundation (ECN)
P.O. Box 1
1755 ZG Petten
The Netherlands

In the event of a severe nuclear accident in a light water reactor the release of radioactive materials into the biosphere may occur. The radioactive release into the biosphere, also called the source term, is formed by the radioactive aerosols which leave the containment building during venting, leakage or failure. Therefore it is important to know how and in which quantities aerosols are removed from the containment atmosphere during the period that the containment is intact, in order to determine the source term following a containment failure. The DEMONA [1] and LACE [2] experiments were performed to determine this removal rate. As these experiments have demonstrated an aerosol on which water has been condensed will behave quite differently from a dry aerosol. This immediately indicates that the thermodynamic conditions, particularly the saturation ratio, are important factors for the aerosol behaviour.

The purpose of this paper is to show that the transport of sensible heat forms one of the governing parameters for the rate of aerosol deposition. Furthermore it is shown that the heat and mass transfer analogy, used in some computer codes, has its limitations.

The experimental results of DEMONA are used to underline the large difference in aerosol deposition between "dry" and "wet" cases. Then some ECN analyses are included in order to describe the heat balance of the "wet" case. The importance of the sensible heat transfer on the deposition of the aerosols is demonstrated by these analyses. Thereafter the heat and mass transfer analogy used in computational codes will be discussed.

Figure 1 shows a comparison between the results of the DEMONA experiment B2 versus experiments B3 and B4 as presented in the paper by Kanzleiter et. al. [1]. It shows the decrease of mass concentration of the aerosols as a function of time. The "wet" experiments B3 and B4, in which steam was injected, resulted in a much faster deposition of aerosols than the "dry" experiment B2. The exponential decay constant averaged over a 6 hour period starting at a mass concentration level of 1 g/m^3 is for the B2 experiment .38 h^{-1} and for the experiments B3 and B4 1.22 h^{-1}.

In figure 2 the dominant aerosol removing mechanisms, i.e. gravitational settling, diffusiophoresis and thermophoresis, are depicted over this 6 hour period. Using the relations of [3] together with the thermodynamic conditions of DEMONA B3 [4] and B4 it is derived that 2% of the exponential decay constant is due to thermophoresis and 18% is due to diffusiophoresis. The lacking 80% is contributed to gravitational settling. Since experiment B2 was "dry" and the temperatures were not elevated the ex-

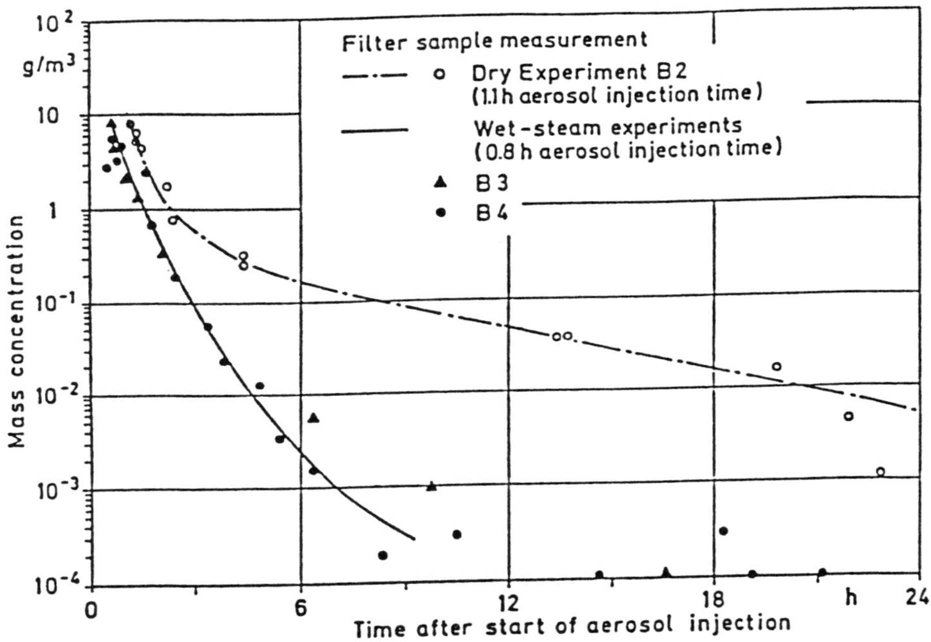

FIGURE 1. Dry experiment B2, wet-steam experiments B3 and B4: measured aerosol mass concentration; from [1] fig. 10.

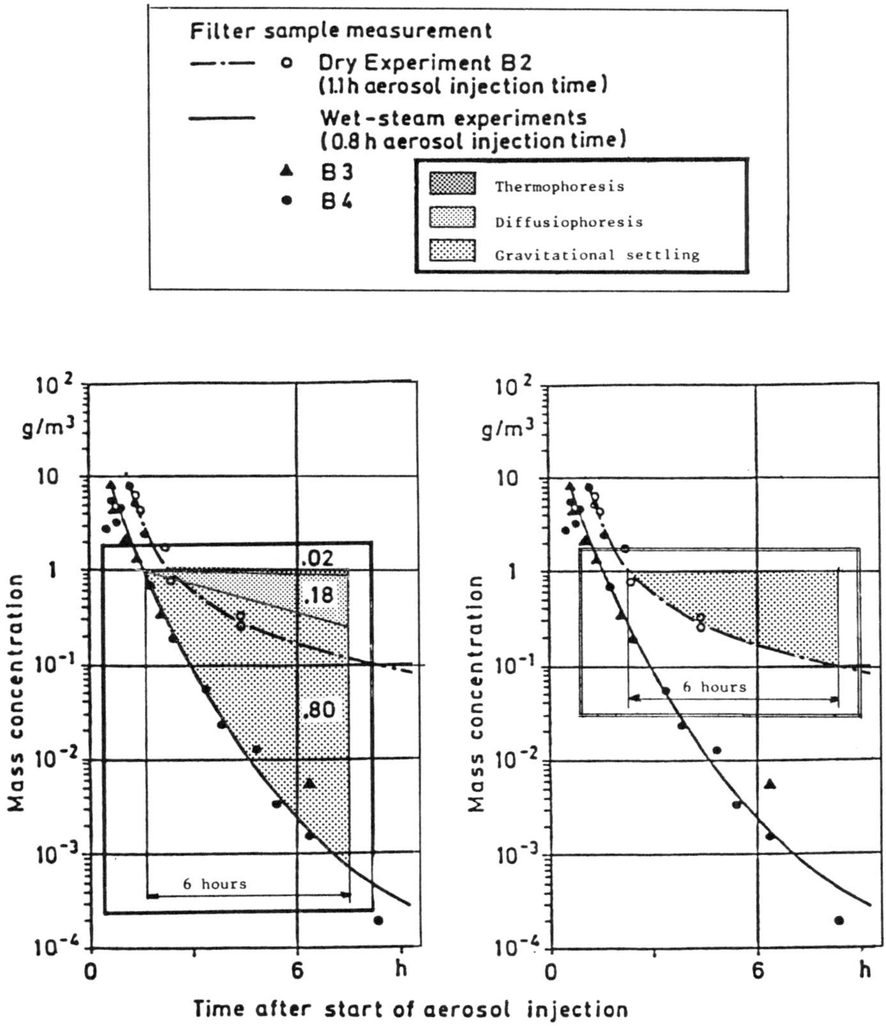

FIGURE 2. Aerosol depletion according to removal mechanism in dry experiment B2 and wet-steam experiments B3 and B4.

ponential decay constant is assumed to be caused by the gravitational settling only. The amount of aerosols removed by the three removal mechanisms are accentuated with various patterns in this figure. The "wet" sedimentation is by far the dominant aerosol deposition process in the B3 and B4 experiments. This is due to steam condensing onto the aerosols, thus increasing their mass and their settling velocity.

Hence it is important to know what parameters determine the condensation of steam onto the aerosols. In order to do that consider the heat balance in the DEMONA vessel. In figure 3 the balance of latent heat and sensible heat is shown schematically for experiment B3. The balance is based on the CONTAIN code [5] calculations [4], which were performed in the frame of the CEC benchmark on DEMONA B3 [6]. The numbers are percentages of the total heat source due to the injected superheated steam. Steam is injected continuously, transporting 97 units of latent heat into the experimental vessel. The 3 additional units of heat, transported into the vessel, represent the sensible heat above saturation. In the vessel steam condenses predominantly on the walls, but a small amount at the rate of 4.0 g/s condenses onto the aerosols. This latter condensation transfers 5 units of latent heat into sensible heat. Due to condensation onto the walls, at a rate of about 70 g/s, 85 units of latent heat, and, at a much smaller rate due to convection and conduction, 8 units of sensible heat are being removed from the vessel. The total of 93 units of heat leave the vessel as sensible heat by means of conduction through the walls and the pool. The additional 7 units of latent heat are due to leakage from the vessel and due to the steam mass increase caused by the change of gas composition, i.e. air is replaced by steam.

In figure 4 a similar balance is shown for a "hypothetical" case B3′, in which the sensible heat removal of 8 units is decreased to 3 units. In this case the sensible heat produced by steam condensation onto the aerosols is not removed, and consequently the thermodynamic conditions of the atmosphere change in such a way, that condensation onto the aerosols stops. This demonstrates that the dominant aerosol deposition process, - the "wet" sedimentation -, not only depends on the condition whether or not steam is involved, but also on the amount of sensible heat removed from the vessel. If this amount drops from 8 to 3 units in case B3′ then the condensation onto the aerosols is inhibited, hence aerosol growth stops and consequently "wet" sedimentation is impossible.

Figure 5 shows "hypothetical" case B3", in which an improved sensible heat removal from the atmosphere to the walls causes an increased bulk condensation rate. As is shown in these examples the total heat flux through the walls is assumed to be constant.

Sensible heat removal, although relatively small, is one of the dominant factors for bulk condensation. These examples show that the ratio of sensible and latent heat removal determines the occurrence and magnitude of bulk condensation. Bulk condensation in its turn determines the magnitude of aerosol sedimentation.

The second part of this paper is focussed on the way computer codes deal with this heat and mass transfer problem. During condensation at the walls, the latent heat removal is much larger than the sensible heat removal. The total heat removal can be very well predicted, even with a rough model for the sensible heat removal. Hence, an accurate model for the sensible heat removal in computer codes is, one might think, not needed. However, as already is demonstrated, it is essential for the aerosol deposition rate.

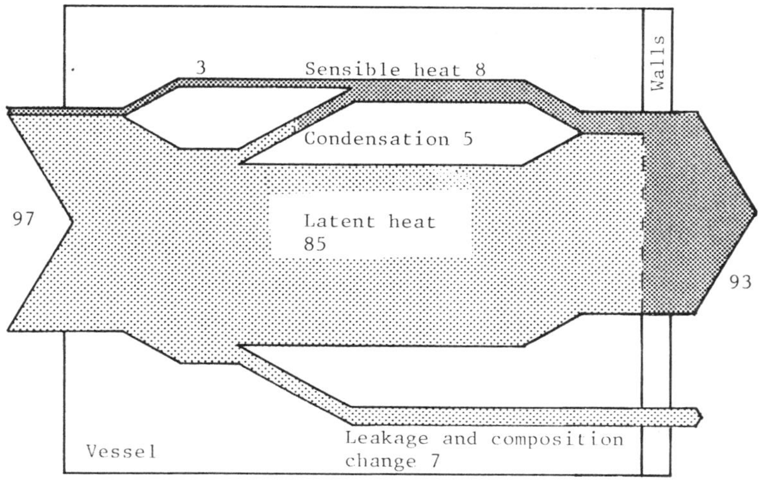

FIGURE 3. Schematic heat fluxes DEMONA B3.

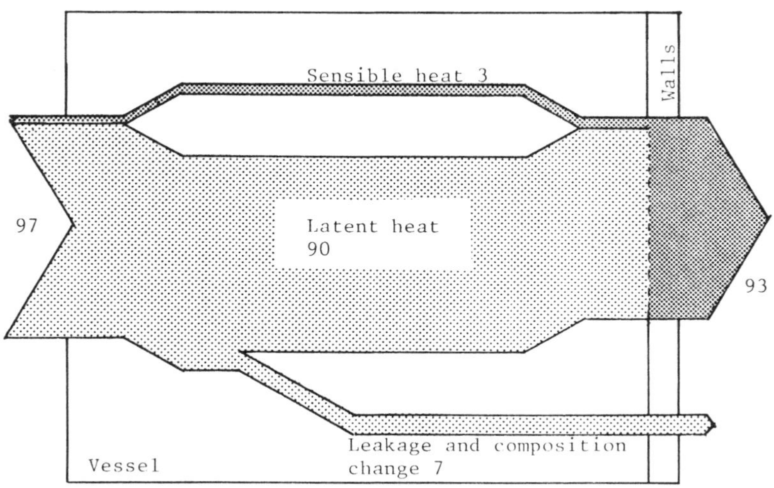

FIGURE 4. Schematic heat fluxes DEMONA B3'.

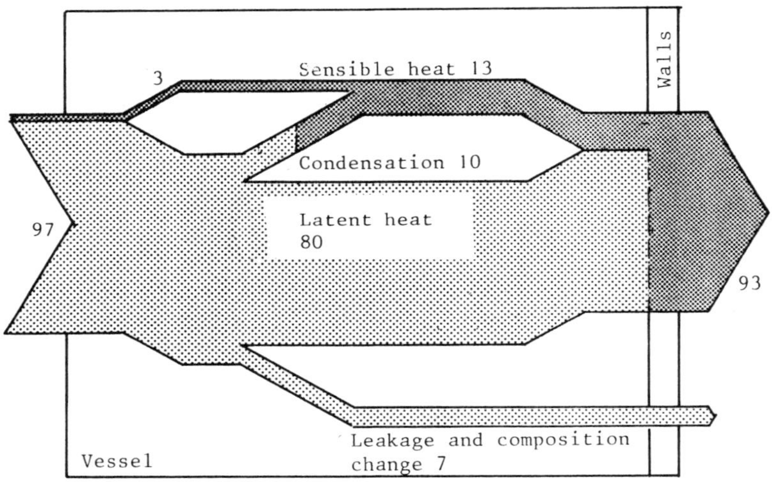

FIGURE 5. Schematic heat fluxes DEMONA B3".

In codes like CONTAIN the latent heat removal is modelled based upon the analogy between heat transfer and mass transfer [7]. The analogy is based on a number of assumptions, which imposes limitations on the use of the heat and mass transfer analogy. Five limitations of the analogy are postulated [8]. Firstly, the analogy is based on the fact that the boundary conditions are comparable. Physically speaking this implies that the boundary layer thicknesses for heat and mass transfer are equal. In thermodynamic terms this implies that the Lewis number is 1.0 . However, in reality this number is 0.7 . Secondly, in the analogy the assumption is made that, the mass transfer to the wall is small. Unfortunately, this is not observed in the analyses of the DEMONA experiments. Thirdly, the analogy assumes a stagnant layer, which is not the case in large containment buildings in which turbulent mass flow conditions are likely to occur. Although the temperature gradients in these buildings are relatively small, the natural circulation leads to turbulent mass flows due to the large size of the containment buildings. Fourthly, the assumption is made that the physical properties remain constant. Fifthly, the analogy is based upon zero velocity of the gas at the interface at the wall, which is not true in case the gas interfaces with an intensive condensate flow at the wall surface. Using the analogy outside its limitations there exist a possibility that the computed latent and sensible heat flux has a ratio which is not correct. Small changes in these heat fluxes may seriously effect the aerosol behaviour, even when the total heat flux is computed correctly.

This paper has demonstrated that:

- besides the presence of steam also a sufficient rate of sensible heat removal is essential for a fast aerosol deposition mechanism,
- while the violation of assumptions, on which the heat and mass transfer analogy is based, is not critical for the prediction of the total heat and mass transfer, it may seriously influence the sensible heat balance and hence the aerosol removal rate.

Due to what is presented in this paper, further evaluation seems necessary. Currently ECN is in the process of assessing a research proposal. This research project intends to investigate this problem in further detail.

REFERENCES

1. Kanzleiter, T.F., et. al., DEMONA experiments investigating the aerosol behaviour within a containment, IAEA conference, Columbus OH, 1985.

2. LACE, LWR Aerosol Containment Experiments, EPRI Palo Alto CA 94303.

3. Ernstige reactorongevallen opnieuw bezien, de bronterm, bijlagen, ECN-PB-87-13/suppl., Petten, The Netherlands, pp 101,104, December 87.

4. Van Meulenbroek, B.M., Analyses of the thermal hydraulics in DEMONA-B3 with CONTAIN 1.06, ECN memo 1.055.00-GR27 (84-1), 1988.

5. Bergeron, K.D., et. al., User's manual for CONTAIN 1.0, a computer code for severe nuclear reactor accident containment analysis, SAND84-1204, March 1985.

6. Gauvain, J., Post-test calculation of thermal-hydraulic behaviour in DEMONA experiment B3 with various computer codes used in CEC member states, Commission of the European Communities and nuclear technology, to be published.

7. Bird, R.B., et. al., Transport phenomena, Wiley, New York, Chap. 21, Nov. 1966.

8. Bird, R.B., et. al., Transport phenomena, Wiley, New York, pp 465, Nov. 1966.

TRENDS: A Code for Modeling Iodine Behavior in Containment during Severe Accidents

C. F. WEBER
Computing and Telecommunications Division
Oak Ridge National Laboratory
P.O. Box 2008
Oak Ridge, Tennessee 37831, USA

T. S. KRESS
Engineering Technology Division
Oak Ridge National Laboratory
P.O. Box 2008
Oak Ridge, Tennessee 37831, USA

E. C. BEAHM and W. E. SHOCKLEY
Chemical Technology Division
Oak Ridge National Laboratory
P.O. Box 2008
Oak Ridge, Tennessee 37831, USA

S. R. DAISH
Oak Ridge National Laboratory
P.O. Box 2008
Oak Ridge, Tennessee 37831, USA
and British attaché from
AERE Harwell, UK

1. INTRODUCTION

The chemistry and transport of iodine in light-water reactor (LWR) containment are best viewed as a dynamic system in which iodine species enter from the reactor coolant system and interact with a complex gas mixture, with transport to and from surfaces and aqueous solutions. Iodine would enter containment as some combination of cesium iodide (CsI), hydrogen iodide (HI), elemental iodine atoms (I_2), silver iodide (AgI), and perhaps some organic iodide. The containment atmosphere would consist of steam, air, hydrogen, some hydrocarbons, aerosol material, other fission products in various chemical forms, and the iodine species. The containment atmosphere is in contact with water pools, water films, surfaces, and, in some cases, with sprays.

The ultimate aim of a description of iodine behavior in severe LWR accidents is a time-dependent accounting of iodine species released into containment and to the environment. Factors involved in the behavior of iodine can be conveniently divided into four general categories: (1) initial release into containment, (2) interaction of iodine species in containment not directly involving water pools, (3) interaction of iodine species in, or with, water pools, and (4) interaction with special systems such as ice condensers or gas treatment systems.

A review of the state of knowledge of iodine chemistry important to severe accidents was given in Chapter 5 of NUREG-0772 [1]. This review identified deficiencies in information and gave the starting point for the work in this program. While indicating the potential importance of radiation chemistry to the chemical form of iodine, this review stated "textbooks hardly acknowledge this subject, and specific information in the literature is sparse." Additional factors that were cited included uncertainties in the extent of organic iodine formation and the importance of pH in determining the chemical form of iodine in water pools.

To fill the large gaps in knowledge and to provide a means for assaying the iodine source term, this program has proceeded along two paths:

Research was sponsored by the Office of Nuclear Regulatory Research, U.S. Nuclear Regulatory Commission, under Interagency Agreement DOE 1886-8045-4B with the U.S. Department of Energy under contract DE-AC05-84OR21400 with the Martin Marietta Energy Systems, Inc.

1. Experimental studies of the chemical behavior of iodine under containment conditions.

2. Development of TRENDS (Transport and Retention of Nuclides in Dominant Sequences), a computer code for modeling the behavior of iodine in containment and its release from containment. The main body of this report consists of a description of TRENDS.

These two parts to the program are complementary in that models within TRENDS use data that were produced in the experimental program; therefore, these models are supported by experimental evidence that was obtained under conditions expected in severe accidents.

2. DESCRIPTION OF EXPERIMENTAL STUDY

The two main processes that control the fraction of iodine in water pools that will be in a volatile form are hydrolysis and radiolysis.

A comprehensive series of tests on hydrolysis and radiolysis were carried out to evaluate (1) release of volatile iodine when water pools or puddles evaporate to dryness and (2) the effect of impurity silver (from control rod material) on controlling iodine volatility.

Details of the experimental results have been reported previously [2–4]. The following conclusions were obtained from this work:

1. Each accident sequence has unique conditions that will determine the pH values of water pools that form in containment. It is possible for the pH to vary between approximately 3 and 9.

2. Low pH and ionizing radiation are favorable conditions for formation of I_2 by radiolytic oxidation processes. Molecular iodine will partition to the gas phase more quickly if gas is bubbled through the solution [e.g., boiling-water reactor (BWR) suppression pool].

3. Organic iodides, dominated by methyl iodide, will form when I_2, organic material, and radiation are present in an aqueous environment. The conversion is most efficient when organic gas mixtures are bubbled beneath the solution surface. They quickly partition to the gaseous phase.

4. Lightly buffered solutions can be acidified by the nitric acid that is formed by radiolysis of air, resulting in accelerated production of I_2 as the irradiation dose to a containment pool increases.

5. In order to reduce the extent of iodine release to the environment, it is important to maintain a high pH within a well-buffered system in water pools in containment.

6. Secondary sources of iodine release are possible from the decomposition of I_3^-, reduction of IO_3^- or IO_4^-, and desorption of I_2 or organic iodides from containment surfaces.

7. Iodine volatility can be greatly reduced by the presence of silver powders or colloids in a solution, even when a pool evaporates to dryness. Other materials (e.g., copper) may have similar effects and could be deliberately placed in a containment sump.

3. DESCRIPTION OF TRENDS

3.1 Basis for Code Design

The TRENDS code for calculation of iodine behavior and release in containment is a system of FORTRAN routines whose objective is the quantitative determination of fission product transport and retention characteristics during severe reactor accidents. It represents a best estimate of nuclide behavior under the hypothetical conditions of various severe accident sequences. Every attempt has been made to account for all significant physical processes. Since many physical processes are modeled in greater detail than in previous studies, the results should yield considerably more insight into the mechanisms and consequences of fission product redistribution.

The TRENDS code is designed so that a combination of input data and internally calculated parameters provide the physical and chemical environment for a given severe accident sequence. Information concerning the system thermal hydraulic response and physical behavior of aerosols is obtained from other codes in the United States Nuclear Regulatory Commission's source term code package. The environment in water pools, in terms of pH and radiation dose rates, is calculated within TRENDS. Thus, TRENDS is able to calculate iodine behavior under changing conditions during a severe accident.

3.2 Models and Equations

Chemical forms. Five iodine species are explicitly considered, as shown in the first column of Table 1. The phases in which each form can exist are also listed. The release of iodine from either the reactor coolant system (RCS) or fuel rubble can occur in any of the forms listed. Subsequent interconversions between different species are modeled as follows:

$$I_2(g) \leftrightarrow CH_3I\ (g)\ , \quad (1a)$$

$$I^-(aq) \leftrightarrow CH_3I\ (aq)\ , \quad (1b)$$

$$I^-(aq) \leftrightarrow I_2\ (aq)\ , \quad (1c)$$

$$I_2(aq) \leftrightarrow AgI\ (s)\ , \quad (1d)$$

$$I^-(aq) \rightarrow AgI\ (s)\ , \text{ and} \quad (1e)$$

$$CH_3I(aq) \rightarrow AgI\ (s)\ . \quad (1f)$$

In addition to the five iodine species that are tracked throughout the accident sequence, aqueous iodine with oxidation numbers of -1 to +5 are considered in the modeling of hydrolysis and radiolysis.

The iodine model follows 20 inventory locations within each containment volume.

Gas-liquid partitioning. Iodine entering water as iodide (i.e., CsI or HI) should remain dissolved as long as it is in ionic form. However, the species I_2 and CH_3I can evaporate from water into the containment gas space and from there be convected to other control volumes or even to the outside environment.

TABLE 1. Iodine Chemical Forms in Containment Following an Accident

Species	Phases considered
I_2	Gas, aerosol, liquid, surfaces (paint, steel, concrete), deposited aerosol
CH_3I	Gas, liquid, surface (paint)
CsI or I^-	Gas, aerosol, liquid, surfaces, deposited aerosol, liquid (as I^-)
HI	Gas, aerosol, surface (steel), deposited aerosol
AgI	Precipitate

Within the TRENDS code, solubilities are measured in terms of partition coefficients (Henry's Law constants), defined as the equilibrium ratio

$$P = \frac{C_\ell^*}{C_g^*} \quad . \tag{2}$$

For I_2 and CH_3I, partition coefficients have been measured and are used as functions of temperature only [5].*

Because Eq. (2) models only equilibrium, and volatile gases may take time to diffuse to the surface before evaporation, it is necessary to consider the transient delaying behavior. Liss and Slater [6] studied the evaporation of various gases from ocean water and have applied the following model:

$$\frac{1}{A}\frac{dN_g}{dt} = h_e\ (P\ C_g - C_\ell) \quad . \tag{3}$$

From these data, values of h_e for I_2 and CH_3I can be determined as

$$h_e(I_2) = 1.84 \times 10^{-3},\ h_e(CH_3I) = 3.0 \times 10^{-3} \quad . \tag{4}$$

TRENDS uses Eq. (3) to describe evaporation processes and currently uses the values from Eq. (4) for coefficients.

Mass transport models. The modeling of the mass transport of vapor species of iodine in containment is an integral part of the TRENDS code. The mass transport processes that are being modeled include:

1. Turbulent convection between:

   - the bulk atmosphere and pools of water,
   - the bulk atmosphere and containment surfaces,

*Data Calculated from FACT, a copyrighted product of Thermfact Ltd/Ltee., 447 Berwick Ave., Mount-Royal, Quebec, Canada, H3R 1Z8.

– the bulk atmosphere and aerosol surfaces, and
– containment volumes.

2. Diffusiophoresis to:

– containment surfaces and
– aerosol surfaces.

Organic iodide formation. The equation for organic iodide formation by gas or gas/surface reactions can be expressed as

$$\frac{dc}{dt} = \lambda(C_g^* - C_g) \ . \qquad (5)$$

The value of λ was taken [7] as 5.1×10^{-3}, and C^* was related to the gaseous I_2 concentration, C_{I_2}, by the relation

$$C^* = 1.89 \times 10^{-2} \ C_{I_2}^{0.8} \ mg/m^3 \ , \qquad (6)$$

based on tests involving irradiated fuel.

TRENDS also includes a model for the radiation-induced conversion of iodide ion, I^- (aqueous), into methyl iodide, CH_3I.

Silver iodide formation. If sufficient quantities of silver are available, then I_2, I^-, or CH_3I could react with it to form AgI. The reactions are essentially irreversible, and the resulting AgI forms a solid precipitate that immobilizes the iodine completely. As in the initial description of radiolysis of I^- to form more volatile species, the TRENDS code uses empirical, rather than detailed reaction, models.

The conversion of I^- and CH_3I into silver iodide is catalyzed by water radiolysis products; hence, no reaction occurs without radiation. The rate of formation of silver iodide by this mechanism is modeled by

$$\frac{d[AgI]}{dt} = \alpha\beta D^{\beta-1} \ \dot{D}[I^-] \ Ro \ . \qquad (7)$$

In practice, TRENDS uses Eq. (7) if there is more silver than iodine in the water pool and if the total dose exceeds 0.8 mrad.

The reaction of silver and I_2 to form AgI is not affected by radiation and is modeled by the first-order irreversible reaction

$$\frac{d[I_2]}{dt} = k \ [I_2] \ . \qquad (8)$$

The rate constant taken from the IMPAIR code is $k = 8.36 \times 10^{-6} \ s^{-1}$. This is applied in TRENDS whenever the silver molarity exceeds ten times the iodine molarity.

4. APPLICATION TO SEVERE ACCIDENT ANALYSIS

Operation of TRENDS was tested by running calculations of TMLB' (station blackout) sequences. In these calculations, in order to enable evaluation of all models, it was assumed that iodine entered containment as either CsI or HI. Figure 1 gives an example of the type of output that was obtained. This shows iodine inventories that were obtained through- out a TMLB' sequence when it was assumed that the entire core inventory of iodine entered containment as HI. In this sequence, there was time for rate controlled processes to become effective before containment failure. Figure 1 shows that (I_2 + CH_3I) aqueous, I_2 deposited on surfaces, and AgI have increased to their maximum values by the time of containment failure (~75 min into the sequence). The overall effect of the rate processes is to decrease the airborne iodine by aerosol settling, deposition onto surfaces, and conversion into AgI or (I_2 + CH_3I) aqueous. This results in lowering the overall release of iodine. Clearly, the extent of iodine release and the distribution of iodine in containment will depend on the specific sequence and on the knowledge or assumption of the amount and the chemical form(s) of iodine that enter containment.

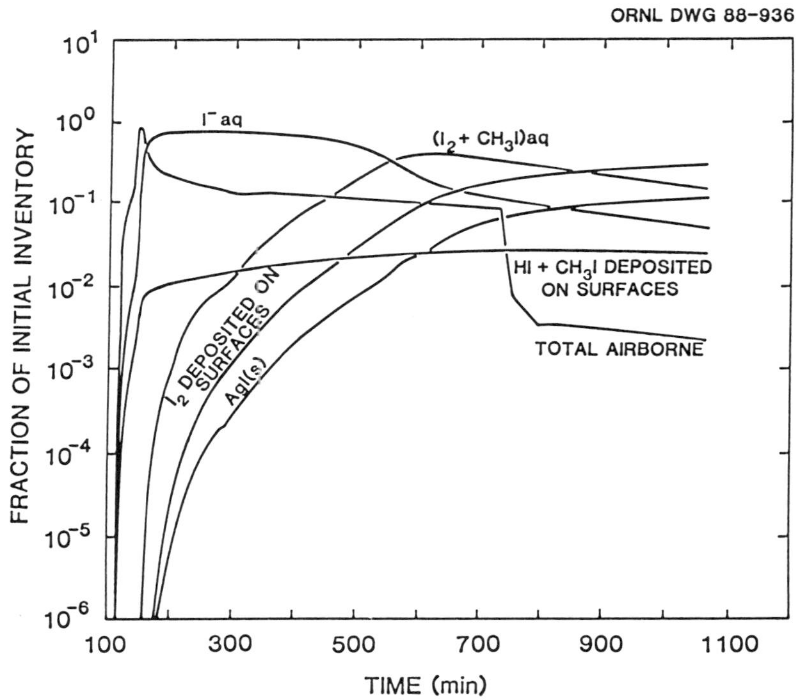

FIGURE 1. Surry TMLB'ε pH (HI) case containment.

5. CONTINUING DEVELOPMENT OF TRENDS

The accident sequence mentioned in Sect. 4 that was calculated with TRENDS was run with iodine volatility data that were experimentally measured for the specific sequences. This approach is valid and it permitted testing of the models in TRENDS, but it severely limits the application of TRENDS for use by other investigators. To overcome this, we are in the process of making TRENDS portable so that it may be used by others. The three additions necessary for portability are (1) internal calculation of pH, (2) calculation of radiation dose rates, and (3) internal calculation of volatility effects – hydrolysis and radiolysis.

5.1 Calculation of pH

Experimental tests have shown that solution pH is a major factor in determining the extent of I_2 and organic iodide in solution.[2] In a severe accident, the materials that will determine pH will be chemical additives, fission products, core-concrete aerosols, boric acid, carbon dioxide, and nitric acid produced by radiation. All of these factors are being included in a computer routine for use in TRENDS. This routine uses a variation of the well-known SOLGASMIX program to calculate chemical equilibrium in water pools. The initial calculations with borates and phosphates show good agreement between calculated and experimental values.

5.2 Calculation of Hydrolysis/Radiolysis

In pure water, I_2 hydrolyzes to form I^- and IO_3^-, and can be described by the overall reaction

$$3I_2 + 3H_2O \rightleftharpoons 5I^- + IO_3^- + 6H^+ \ . \quad (9)$$

However, this form is inadequate to fit all data, or to describe the detailed processes that occur in reactor accidents, where radiation and chemical impurities will be present.

If all oxidation states are included in the reaction sequence, the simplest pattern of reaction steps would involve the following premises:

1. only oxidation state changes of ±1 occur, and

2. no species can be oxidized by another species with a lower oxidation state.

The overall forward reaction of iodine hydrolysis [cf. Eq. (9)] results in a "spreading out" of the occupied oxidation states, beginning at zero and moving to the extremes -1 and +5. Thus, the reaction substeps should also reflect this pattern. The resulting reactions for the forward process are those involving like species, as modeled by the following irreversible reaction steps:

$$I_2 + H_2O \rightarrow I^- + HIO + H^+ \ , \quad (10a)$$

$$2HIO \rightarrow 1/2I_2 + IO + H_2O \ , \quad (10b)$$

$$2IO + H_2O \rightarrow HOI + HIO_2 \ , \quad (10c)$$

$$2HIO_2 \rightarrow IO + IO_2 + H_2O \ , \text{ and} \quad (10d)$$

$$2IO_2 + H_2O \rightarrow HIO_2 + HIO_3 \ . \quad (10e)$$

The substeps for the overall reverse reaction (also known as the Dushmann reaction) also utilize the two assumptions stated above. This basic process involves oxidation of I^- and reduction of HIO_3 and other intermediates to form I_2. Because I^- is the strongest possible reducing agent, and because it is generally present in greater abundance than other iodine species, the individual reaction steps use only I^- in the stepwise reduction of species in each oxidation state from HIO_3 to I_2:

$$H^+ + I^- + HIO_3 \rightarrow 1/2I_2 + IO_2 + H_2O \ , \tag{10f}$$

$$H^+ + I^- + IO_2 \rightarrow 1/2I_2 + HIO_2 \ , \tag{10g}$$

$$H^+ + I^- + HIO_2 \rightarrow 1/2I_2 + IO + H_2O \ , \tag{10h}$$

$$H^+ + I^- + IO \rightarrow 1/2I_2 + HIO \ , \text{ and} \tag{10i}$$

$$H^+ + I^- + HIO \rightarrow I_2 + H_2O \ . \tag{10j}$$

Recent experiments in this program have yielded over 130 data points at temperatures of 298 K, 323 K, and 363 K and pH values 2.8 to 9.0. Optimal choices for rate constants and activation energies have been obtained by fitting the solutions for Eqs. (10a)–(10j) to these data using a complicated optimization procedure. Results are shown below in Table 2.

TABLE 2. Calculated Rate Constants

Reaction	Rate constants at 298 K	
1	1.06×10^8	min^{-1} $(gmol/L)^{-1}$ hydrolysis
2	3.47×10^6	
3	8.93×10^4	
4	1.04×10^5	
5	1.10×10^5	
6	3.27×10^5	min^{-1} $(gmol/L)^{-2}$ "Dushmann reaction"
7	1.06×10^6	
8	7.39×10^5	
9	5.36×10^5	
10	3.97×10^{11}	

5.3 Calculation of Radiation Dose Rates and Radiolysis Effects

The reaction of iodine species with water radiolysis products makes it necessary to calculate water pool radiation dose rates to evaluate the formation of volatile I_2 and organic iodides. The problem is to calculate radiation dose to a water pool due to suspended or dissolved fission products. In any such pool, some fraction of the radiation energy will be absorbed by the pool, while some will escape and be absorbed elsewhere. For a given source energy, it has been observed that there is a linear relationship between fraction absorbed and surface area to volume ratio. Thus, the energy fraction absorbed can be expressed as

$$\text{energy fraction absorbed} = a + b\ (S/V)\ . \qquad (11)$$

Values for the constants have been calculated as

$a = 1.062 - 0.051\ \sqrt{E}$ and $b = 0.0355 - 0.0216\ \sqrt{E}$.

Therefore, calculation of the dose rate is a matter of summing the contributions from each radionuclide to obtain the total energy absorption by the water pool.

From a practical standpoint, radiation effects are the most important factors in determining iodine volatility in a severe accident. The chemical reactions resulting from radiolysis may convert nonvolatile iodine forms, such as I^-, into volatile chemical forms (at low pH), or, conversely, volatile I_2 could be converted into nonvolatile forms (at high pH).

The interaction of radiation with aqueous solutions leads principally to the dissociation of water molecules into primary radiolysis products, such as the radicals, H·, OH·, and the hydrated electron, e_{aq}^-. Secondary radiolysis products, including molecular species such as H_2O_2 and O_2, then form by reaction between the reactive radical species. The radicals establish standing concentrations of about 10^{-7} $\underline{M}$ due to their short lifetime and can interact with other solutes by oxidizing or reducing them (e.g., $OH\cdot + I^- \rightarrow I\cdot + OH^-$). The molecular species are more durable and can reach millimolar concentrations, thus exceeding the concentration of the iodine species. The radiation can also interact directly with organic material to produce reactive fragments such as R·, or these species might be formed by interaction with water radiolysis products (e.g., $R\text{-}H + H\cdot \rightarrow R\cdot + H_2$).

Experimental data on radiation effects obtained in this program are being used to determine optimized rate constants using the same techniques that were described in Sect. 5.2 for hydrolysis. Initially, this will be done by considering the hydrolysis rate expressions and one additional reaction, the oxidation of iodine by hydroxyl radical. Subsequently, other reactions will be added to this basic set, but only as needed, in order to avoid a proliferation of unnecessary and uncertain reactions.

The resulting model will provide a description of aqueous iodine volatility behavior using optimized parameters that are based on experimental values measured under conditions expected during severe accidents.

6. SUMMARY

TRENDS has been designed as a detailed mechanistic code to calculate iodine behavior in containment. However, it is easy to use so that it may be inserted into codes such as MELCOR or CONTAIN. The work remaining on TRENDS before it can be released for general use includes (1) insertion of a routine to calculate pH, (2) insertion of a hydrolysis rate model, (3) optimization of rate constants for an iodine radiolysis rate model, and (4) some improvements in modeling of mass transport of gaseous iodine.

NOMENCLATURE

Symbol	Quantity	SI Unit
(Aq)	aqueous	
(s)	solid	
g	gas	
C	concentration	
C^*	equilibrium concentration	
$\dot{D}$	dose rate	
D	cumulative dose	
E	source energy	MeV
Ro	silver to iodine mass ratio	
S/V	surface to volume ratio	m^{-1}
α, β, λ, a, b, k	constants	

7. REFERENCES

1. Elrick, R. M., Bell, J. T., Sallach, R. A., Toth, L. M., Campbell, D. O., and Malinauskas, A. P., Technical Bases for Estimating Fission Product Behavior During LWR Accidents, Chapter 5, NUREG-0772, June 1981.

2. Wisbey, S. J., Beahm, E. C., Shockley, W. E., and Wang, Y.-M., Iodine Behavior in Containment Under LWR Accident Conditions, *Proc. Symp. on Chemical Phenomena Associated with Radioactivity Releases During Severe Nuclear Plant Accidents*, Session 6, p. 29, 1986.

3. Beahm, E. C., Shockley, W. E., Weber, C. F., Wisbey, S. J., and Wang, Y.-M., Chemistry and Transport of Iodine in Containment, NUREG/CR-4697, ORNL/TM-10135, October 1986.

4. Beahm, E. C., Wang, Y.-M., Wisbey, S. J., and Shockley, W. E., Organic Iodide Formation During Severe Accidents in Light Water Nuclear Reactors, *Nucl. Technol.*, vol. 78, p. 34, 1987.

5. Hasty, R. A., Partition Coefficient of Methyl Iodide Between Vapor and Water, *Can. J. Chem.*, vol. 46, p. 163, 1968.

6. Liss, P. S. and Slater, P. G., Flux of Gases Across the Air-Sea Interface, *Nature*, vol. 247, p. 181, 1974.

7. Beahm, E. C., Shockley, W. E., and Culberson, O. L., Organic Iodide Formation Following Nuclear Reactor Accidents, NUREG/CR-4327, ORNL/TM-9627, March 1985.

Effect of Radiation on Iodine Transport and Chemistry

K. ISHIGURE and H. SHIRAISHI
Department of Nuclear Engineering
University of Tokyo
Japan

1. INTRODUCTION

Radioactive iodine is one of the most important fission products in relation to their release to environment in nuclear reactor accidents. There has been vigorous discussion on chemical forms of radioactive iodine released from failed nuclear fuels in accidents, but it is not possible so far to know it with confidence. One of the largest possibility is laid on cesium iodide which is thermodynamically the most stable species. A recent study(1), however, gave no evidence for the existence of cesium iodide within irradiated fuels, suggesting the predominant presence of iodine atom and molecular iodine. The iodine species released from fuels, whatever they may be, will get into aqueous phase in containments according to accident scenario. Cesium iodide and hydrogen iodide are very soluble in water and dissociate almost completely in the aqueous phase, giving iodide ions. Molecular iodine is less soluble than cesium iodide and hydrogen iodide, but is readily hydrolysed to iodide and hypoiodous acid in the aqueous phase. It is well known that chemical behavior of iodine is very complicated in the aqueous phase since iodine has a variety of valence states. There have been some thermodynamical and kinetic approaches to understanding of chemical behavior of iodine in relation to nuclear reactor accidents. On the other hand, when iodine species are subject to radiation in the aqueous phase, the radiolysis has a possibility to give a large effect on the chemical form of the iodine species. This radiolytic effect is curretly one of the great concerns in the source term problems of nuclear reactor accidents.[2] Relatively a little work has been carried out regarding the radiolysis effect on iodine species under the condition of nuclear reactor accidents. In this paper a brief review is given on the radiolysis effect on chemical forms of iodine under the condition of nuclear reactor accidents, especially in containments, with the main emphasis on the recent works in our labolatory.[3],[4],[5]

2. MODEL EXPERIMENTS

One of the specific features of the radiolytic effects on chemical behavior of iodine under reactor accidents is that we have to deal with very dilute iodine solutions. This makes the situations more complicated with many factors affecting the chemical behavior of iodine, and leads to some difficulties in the experimental procedures. Model experiments of radiolysis were carried out under the simulating conditions of reactor accidents.

It should be pointed out that the post-irradiation effects have been very often observed in these model experiments, leading to significant errors in the experimental results. Therefore, a care has to be taken to eliminate this effect. In our experiments a special loop system was used for gamma irradiation at the dose rate of 3.8×10^3 Gy/hr which enabled us to make chemical analysis of the iodine species about 30 seconds after the irradiation of the samples. The post-irradiation effect was attributed to the slow reactions of the iodine species with hydrogen peroxide accumulated in the systems by the radiolysis of water.

3. RADIOLYSIS OF IODINE SOLUTIONS

Cesium iodide or hydrogen iodide gives aqueous iodide solutions by its dissolution in water. Iodide ion itself is not volatile, but is oxidized to volatile species in the presence of radiation. This radiolytic oxidation process was found to be affected many factors, which are described below in more detail.

a. Concentration of iodide ion

The initial concentrations of cesium iodide were varried from 1×10^{-4} mol dm^{-3} to 1×10^{-6} mol dm^{-3} in the gamma irradiation experiments of the deaerated solutions at neutral pH. An approximate steady state was almost always attained for the distribution of iodine species after some duration of the irradiation, indicating that a subtle oxidation-reduction balance was made rather rapidly in the kinetic scheme of the competing reactions.

In the 10^{-4} mol dm^{-3} solutions, I_2 is the major product in the steady state, while IO_3^- is the most predominant in the 10^{-6} mol dm^{-3} solutions as shown in Fig.1. Thus the oxidation proceeds more progressively with the solutions of the lower CsI concentration.

b. pH

It was found that the pH of the system is the most important factor to control the oxidation-reduction balance in the radiolysis of the dilute iodine solutions. When the pH was changed from 7 to 5 in the irradiation of the deaerated 10^{-4} mol dm^{-3} CsI solutions the level of I_2 was increased in the steady state by a factor of 5 to 6, whereas almost no net oxidation took place in the irradiation at pH 9. The redox balance shifts to the reducing side in the kinetic reaction scheme with increasing pH.

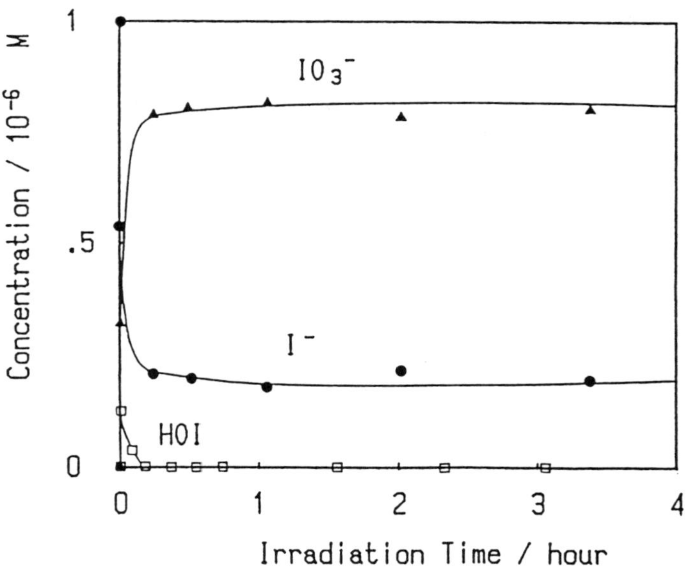

Fig. 1 Variation of iodine product distribution in neutral CsI solution. The initial CsI concentration is 10^{-6} M.

c. Dissolved oxygen

Air oxidation of I^- is known to be a slow reaction at ambient temperature except in strongly acidic solutions. The radiolytic effect, however, was found to be significantly affected by the presence of oxygen. No radiolytic oxidation of iodide was observed with the neutral 10^{-5} and 10^{-6} mol dm^{-3} CsI solutions saturated with simulated air (Ar gas containing 20% O_2). On the other hand, the level of I_2 as an oxidation product is decreased by a factor of three in the radiolysis of the neutral 10^{-4} mol dm^{-3} CsI solution saturated with the simulated air in comparison with the corresponding deaeratd solutions. This result indicates that the presence of oxygen makes the reducing power stronger in the redox balance of the radiolysis and appears to be paradoxial at a first glance.

d. Temperature

The radiolysis experiments were carried out at 80° C with deaerated CsI solutions to investigate the effect of temperature. In Fig.2 the distribution of the iodine species is shown for the 10^{-4} mol dm^{-3} CsI solutions irradiated at 80° C. Iodine molecule is a major product at 80° C as is the case at 25° C, but its level declines fairly rapidly at 80° C after an initial maximum yield. A similar situation was also observed with the 10^{-5} mol dm^{-3} CsI solution.

4. RADIOLYSIS OF IODINE SOLUTIONS

As already mentioned previously, the possibility for molecular iodine to be released as an initial species from failed fuels is not completely eliminated in nuclear reactor accidents. Therefore, it seems to be worth while to carry out comparative radiolysis experiments with I_2 solutions. In Fig.3 is shown the

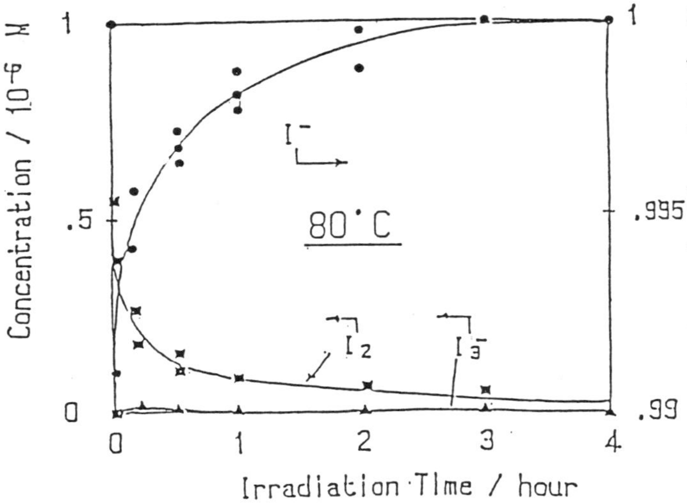

Fig. 2 The distribution of iodine species in the deaerated 10^{-4} mol dm^{-3} CsI solution irradiated at 80° C at pH 7. Dose rate 0.38 Mrad/hr

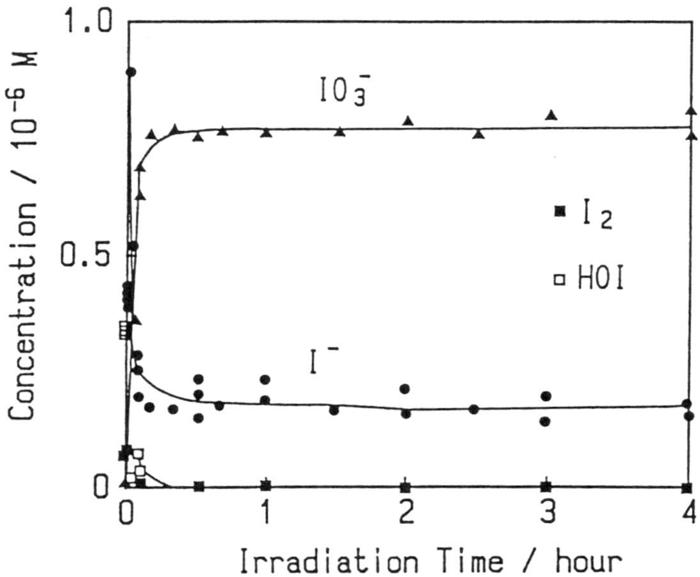

Fig. 3 Variation of iodine product distribution in initially deaerated I_2 solution. Initial $[I_2]$ + [HOI] : 5 × 10^{-7} mol dm^{-3}, 25° C.

distribution of the iodine species in the irradiation of the deaerated 5 × 10^{-7} mol dm $^{-3}$ I_2 solution. The measurd initial pH was ~ 5.8, corresponding to the hydrosis equilibrium reaction of I_2. In the initial 1 to 2 minutes I_2 and HOI, a hydrolysis product, are reduced almost completely to I^- and the oxidation reactions follow after this initial period. The final steady state distribution of the iodine species is very similar to that observed for the near neutral 1 × 10^{-6} mol dm^{-3} CsI solution.

In more concentrated I_2 solutions the steady state or near steady state distribution of the iodine species is also nearly the same as that observed in the comparative experiment with the corresponding CsI and final pH solution.

5. REACTION MECHANISM

As described in the previous section the radiolytic behavior of iodine is very complicated. Its quantitative understanding is not possible at the moment, but the qualitative explanation is applicable to some aspects of the experimental results.

The radiolysis of water is expressed by the following equation.

$$H_2O \rightsquigarrow e_{aq}^-, H, OH, H_2O_2, H_2, H^+$$

The active species produced react with each other, and the reactions involved are tabulated with their rate constants in Table I. When iodine species are present in the system, the active species react also with the iodine species. The reactions presumed to be involved are listed in Table II. It was experimentally confirmed that the radiolytic products other than iodine species, that is, H_2O_2, O_2 and H_2, accumulate in the systems with increasing dose. The accumulation of these products gives a large influence on the oxidation-reduction balance and, conse-

Table I. Elementary Reactions of Water Radiolysis Products

* e_{aq}^- REACTIONS		
F1	$H_2O + H_2O + e_{aq}^- + e_{aq}^- = H_2 + OH^- + OH^-$	1.6 E06
F2	$H_2O + e_{aq}^- + H = H_2 + OH^-$	4.5 E08
F3	$e_{aq}^- + OH = OH^-$	3.0 E10
F4	$H_2O + e_{aq}^- + O_2^- = HO_2^- + OH^-$	2.3 E08
F5	$e_{aq}^- + HO_2 = HO_2^-$	2.0 E10
F6	$e_{aq}^- + H_2O = H + OH^-$	1.6 E01
F7	$e_{aq}^- + H^+ = H$	2.3 E10
F8	$e_{aq}^- + H_2O_2 = OH + OH^-$	1.2 E07
F9	$H_2O + e_{aq}^- + HO_2^- = OH + OH^- + OH^-$	6.3 E07
F10	$e_{aq}^- + O_2 = O_2^-$	1.9 E10
* H REACTIONS		
F11	$H + H = H_2$	6.0 E09
F12	$H + OH = H_2O$	2.0 E10
F13	$H + HO_2 = H_2O_2$	2.0 E10
F14	$H + O_2^- = HO_2^-$	2.0 E10
F15	$OH^- + H = e_{aq}^- + H_2O$	2.0 E07
F16	$H + H_2O_2 = OH + H_2O$	4.5 E07
F17	$H + O_2 = HO_2$	1.9 E10
* OH REACTIONS		
F18	$OH + OH = H_2O_2$	5.2 E09
F19	$OH + HO_2 = O_2 + H_2O$	1.2 E10
F20	$OH + O_2^- = O_2 + OH^-$	1.0 E10
F21	$OH + H_2 = H + H_2O$	3.5 E07
F22	$OH + H_2O_2 = HO_2 + H_2O$	3.3 E07
F23	$OH + HO_2^- = OH^- + HO_2$	5.0 E09
* O_2 , HO_2 REACTIONS		
F24	$HO_2 + O_2^- = O_2 + HO_2^-$	8.5 E07
F25	$HO_2 + HO_2 = H_2O_2 + O_2$	7.6 E05
* EQUILIBRIA		
F26	$H^+ + O_2^- = HO_2$	5.0 E10
F27	$HO_2 = H^+ + O_2^-$	8.0 E05
F28	$H^+ + OH^- = H_2O$	1.4 E11
F29	$H_2O = H^+ + OH^-$	2.6 E-5
F30	$OH^- + H_2O_2 = HO_2^- + H_2O$	1.8 E08
F31	$H_2O + HO_2^- = OH^- + H_2O_2$	1.8 E04
F32	$H^+ + HO_2^- = H_2O_2$	2.0 E10
F33	$H_2O_2 = H^+ + HO_2^-$	3.6 E-2
F335	$H_2O_2 = OH + OH$	8.5 E-9

quently, the distribution of the iodine species in the kinetic steady state. For instance, H_2O_2 converts e_{aq}^-, the most powerful reducing species to OH, the strongest oxidising species through reaction F8. On the other hand, H_2O_2 reduces HOI and OI^- to I^- through reactions F63 and F64. These reducing reactions of H_2O_2 are considered to play an important role in determining the distribution of iodine species, and for instance, the pH effect described above is attributed mainly to these reactions. The dissociation of HOI(F73) proceeds more progressively at high pH, and this results in a significant increase in the reducing power in the redox balance at high pH because of the high rate constant of reaction F64.

The suppression of the oxidation by oxygen seems to be very paradoxial as mentioned earlier, but may be explained by preponderance of reaction F10 in the presence of O_2 at the expense of reaction F8. The product of reaction F10, O_2^-, rapidly reduces I_2, I_2^- and I_3^- through reactions, F53, F44 and F57, respectively, while reaction F8 converts, e_{aq}^-, the powerful reducing species, to a strong oxidising species.

The above explanation is only qualitative, and a computer simulation is essential to establish the quantitative understanding of the phenomena. Such a simulation is already under way in our laboratory, and the result will be presented elsewhere.

In order to make the computer simulation more reliable, the input data for the simulation such as kinetic data base have to be more refined and more reliable. The data base for the radolysis of water is based on the abundant experimental results. However, the kinetic data of the iodine system are less abundant. For instance, the data in Table II are generally the measured values based on the experimental results, but significant part of them are the evaluated values with the absence of the experimental ones. Further experimental works are required in this field, especially kinetic processes including iodine species of higher oxidation state.

Table II. Reactions of Iodine Species

• OXIDATION TO I, I_2		
F34	$I^- + OH = I + OH^-$	1.5 E10
• EQUILIBRIA		
F35	$I + I^- = I_2^-$	1.1 E10
F36	$I_2^- = I + I^-$	1.0 E05
• REDUCTION OF I, I_2		
F37	$I + e\bar{a}q = I^-$	2.4 E10
F38	$I + H = H^+ + I^-$	2.7 E10
F39	$I + O_2^- = I^- + O_2$	1.0 E10
F40	$I + HO_2 = I^- + H^+ + O_2$	1.0 E09
F41	$I + H_2O_2 = HO_2 + H^+ + I^-$	3.0 E03
F42	$I_2^- + e\bar{a}q = I^- + I^-$	1.3 E10
F43	$I_2^- + H = I^- + I^- + H^+$	1.8 E07
F44	$I_2^- + O_2^- = I^- + I^- + O_2$	7.5 E08
• PRODUCTION OF FREE IODINE		
F45	$I + I = I_2$	1.0 E10
F46	$I_2^- + I = I_3^-$	8.4 E09
F47	$I_2^- + I_2^- = I_3^- + I^-$	2.5 E09
F48	$I_2^- + HO_2 = I_2 + HO_2^-$	4.0 E09
F49	$I_2^- + OH = I_2 + OH^-$	3.8 E10
F50	$I + OH = HOI$	1.6 E10
• REDUCTION OF FREE IODONE		
F51	$I_2 + e\bar{a}q = I_2^-$	5.1 E10
F52	$I_2 + H = I_2^- + H^+$	3.5 E10
F53	$I_2 + O_2^- = I_2^- + O_2$	5.0 E09
F54	$I_2 + HO_2 = I_2^- + H^+ + O_2$	5.0 E07
F55	$I_3^- + e\bar{a}q = I^- + I_2^-$	2.0 E10
F56	$I_3^- + H = I_2^- + I^- + H^+$	8.0 E09
F57	$I_3^- + O_2^- = I_2^- + I^- + O_2$	2.5 E08
F58	$HOI + e\bar{a}q = I + OH^-$	2.0 E10
F59	$IO^- + e\bar{a}q + H_2O = I + OH^- + OH^-$	2.9 E08
F60	$HOI + H = I + H_2O$	5.0 E09
F61	$HOI + O_2^- = I + OH^- + O_2$	1.0 E06
F62	$HOI + HO_2 = I + O_2 + H_2O$	1.0 E05
F63	$HOI + H_2O_2 = I^- + H_2O + O_2 + H^+$	2.0 E02
F64	$IO^- + H_2O_2 = I^- + H_2O + O_2$	1.4 E07
• OTHER REACTIONS		
F65	$I_2 + OH = HOI + I$	1.1 E10
F66	$H_2O_2 + I^- = HOI + OH^-$	1.1 E-2
• EQUILIBRIA		
F67	$I^- + I_2 = I_3^-$	6.2 E09
F68	$I_3^- = I^- + I_2$	8.5 E06
F69	$I_2 + H_2O = HOI + I^- + H^+$	5.4 E-2
F70	$HOI + I^- + H^+ = I_2 + H_2O$	k69*MH20/4.6 E-13
F71	$I_2 + OH^- = HOI + I^-$	6.0 E09
F72	$HOI + I^- = I_2 + OH^-$	k71*KW/4.6 E-13
F73	$HOI = IO^- + H^+$	1.2 E-1
F74	$IO^- + H^+ = HOI$	k73/2.4 E-11
F75	$OH^- + HOI = IO^- + H_2O$	1.0 E08
F76	$IO^- + H_2O = OH^- + HOI$	k75*KW/MH20/2.4 E-11

6. REFERENCE

(1) C. A. Alerander and J. S. Ogden
Proc. of ACS Symposium on Chemical Phonomena Associated with Radioactivity Release During Severe Nuclear Plant Accidents, NUREG/CP-0078, 2-21(1987)

(2) R. R. Bowsher, Prog. in Nucl. Energy, 20, 199(1987)

(3) H. Shiraishi and K. Ishigure, Ref. 4, p.6-119

(4) K. Ishigure, H. Shiraishi, and H. Okuda
Radiat. Phys. Chem., 32, 593(1988)

(5) H. Shiraishi, T. Kamiya, Y. Ohmae and K. Ishigure
Proc. of ACS Symposium on Nuclear Reactor Severe Accident Chemistry, Tront, 1988(in press)

Influence of Minor Materials on Iodine Behavior

JAN-OLOV LILJENZIN and OLIVER LINDQVIST
Kjemisk Institutt
Postboks 1033
Blindern
N-0315 Oslo 3, Norway

INTRODUCTION

Besides the materials normally considered in ex-vessel phenomena during hypothetical severe nuclear reactor accidents, several other important materials are also present in a real nuclear power station. In a typical Swedish BWR containment there are large amounts of aluminium, copper, zinc and various synthetic rubbers around and below the vessel. The atmosphere in the containments of these reactors is inerted during operation and consists of nitrogen and steam. When a core melt situation arises with molten corium exiting through the bottom of the pressure vessel, much of these secondary materials are subjected to high temperatures by direct contact with the melt or by thermal radiation from the melt during its downwards flow into the pedestal. This will pyrolyse the organic materials and may even melt some of the aluminium and zinc present near the melt passage.

Such processes and their effect on the source term have seldom been considered up til now. The purpose of our paper is therefore to point out the possible effects these materials may have on the chemical environment in the containment and the possible effects on iodine chemistry and not so much a quantitative evaluation. This is largely due to the lack of models for these effects in most codes for severe accident modelling known to us and ecpecially in MAAP.

PYROLYSIS OF INSULATION MATERIALS

The amounts of organic polymer materials present in a typical modern Swedish PWR containment has been determined by ABB ATOM AB on behalf of the RAMA and NKA/AKTI-150 projects [1]. Some of their results are given in Table 1. It is known from experiments [2] that the synthetic rubbers used, mainly as electrical insulation, begin to release considerable amounts of condensable and non-condensable gases and acidic substances already at 400 °C in a nitrogen/steam atmosphere. The experimental results are recalculated to correspond to the amounts of insulation material used in Swedish BWR containments, see Table 2.

TABLE 1. Inventory in a typical BWR containment of some materials important in severe core melt accidents. All amounts are in metric tons and areas in square meters, and are accurate within about 10% [1].

	Pedestal		Drywell		Wetwell	
Material	Amount	Area	Amount	Area	Amount	Area
Rubber insulation	2.8	-	1.6	-	0.0	-
Copper wires	1.0	-	0.5	-	0.0	-
Aluminium	0.7	40	8.3	4 500	0.0	-
Zinc plating	0.6	>1200	17.2	>35 000	0.0	

TABLE 2. Amount, volume and acid equivalents of pyrolysis products produced in a nitrogen/steam atmosphere from typical rubber insulation corresponding to the amounts used in Swedish BWR containments [2].

Temperature (°C)	Non-condensable gas (Nm^3)	Condensable gas (kg)	Acidic compounds (moles of H^+)
400	68	270	1 600
600	120	150	13 000
800	350	250	12 000
1 000	480	70	11 000

The gases are evolved in such amounts that they can influence the pressures and streaming pattern in the containment significantly. Thus an estimate of their transport and ultimate fate in the containment based on the output from a normal MAAP calculation will be very approximate and only of a qualitative nature.

The acidic substances, mainly hydrochloric acid, sulfur dioxide and carbon dioxide, will change the acidity of condensates and water pools and thereby influence the iodine chemistry, especially in the radiation fields present.

The non-condensable gases comprise various amounts of several unsaturated organic compounds like etene, acetylene and propene, depending on the pyrolysis temperature, see Figure 1. These substances are prone to add hydrogen iodide, if present, to their double or triple carbon-carbon bonds forming organic iodides. From Figure 1 it is evident that a high pyrolysis temperature, as expected, yields more of low molecular weight products than a low temperature.

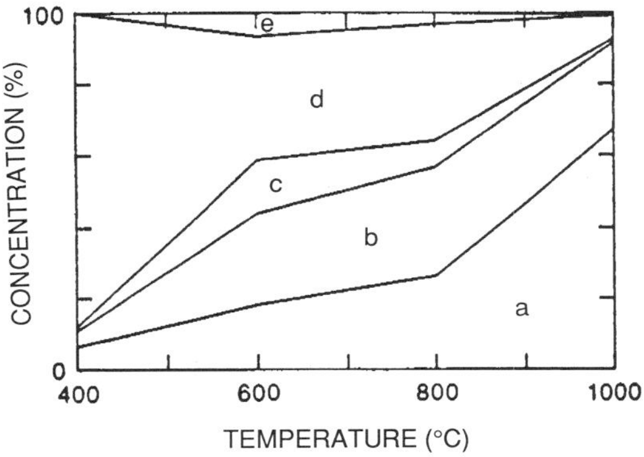

FIGURE 1. Composition of non-condensable gas produced by pyrolysis in a nitrogen/steam atmosphere of a typical insulation rubber used in Swedish BWR containments. Data are given as percent by volume at room temperature measured by gas chromatography [2].
a= sum of hydrogen and methane, b= sum of ethene and acetylene
c= ethane d= propene e= propane

Based on the results from two MAAP calculations [3, 4] we have tried to illustrate the pH-values expected in the pedestal, drywell and wetwell compartments of a Swedish BWR containment, see Figure 2, during two accident sequences, TB and AB.

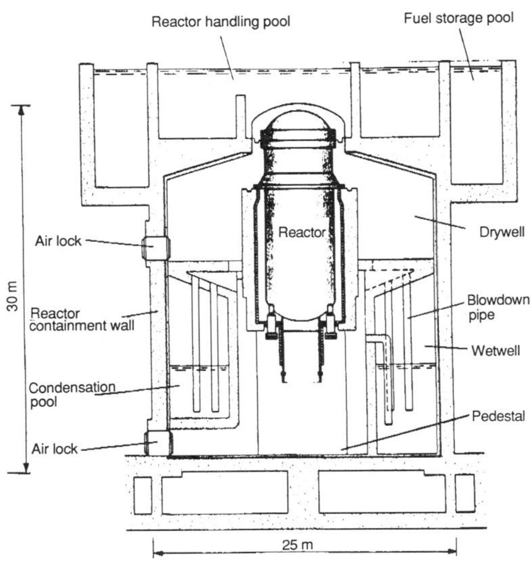

FIGURE 2. A schematic drawing of a modern Swedish BWR containment showing the interconnections and names used in this paper for the various compartments.

The pyrolysis gases produced from the organic material present below the vessel are assumed to correspond to a pyrolysis temperature of 800 °C. Furthermore, we assumed that the amount pyrolysed increased from zero to half of the maximum amount possible as linear function of the amount of core melt leaving the vessel. To compute the distribution of pyrolysis gas in the different compartments we assumed that the transport of the gases produced would be similar to the movement of nitrogen between the compartments. An acid-base balance was then made in each compartment for a series of times. All effects from corrosion were neglected when making these balances. The excess/deficit of acid, in hydrogen ion equivalents, was then divided by the volume of liquid water in respective compartment to obtain a nominal hydrogen concentration in each compartment as a function of time. Values for time periods without any liquid water present were arbitrarily set to 10^{-7}. The results are shown in Figures 3 and 4 as function of time into the accident.

As can be seen from Figures 3 and 4, the water in all compartments becomes acid at one time or the other. The final state is always acidic. This will favour the evolution of hydrogen iodide from the aqueous solutions, and subsequently, of elementary iodine by radiolysis in the gas phase. However, as we neglected corrosion phenomena, especially on zinc plated surfaces, the final state may in reality be less acidic that indicated in these figures.

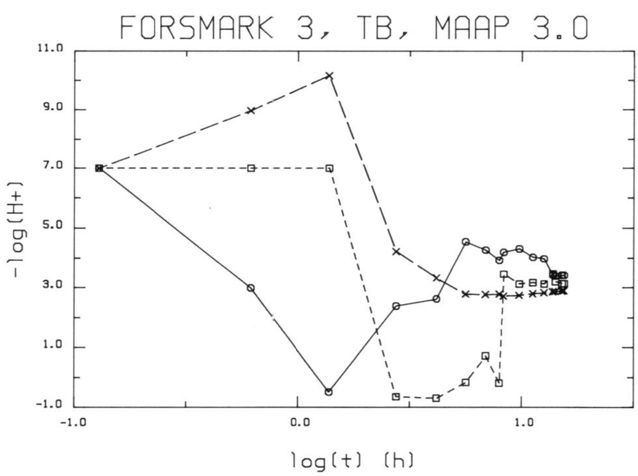

FIGURE 3. Approximate calculation of the acidity in the three compartments in a Swedish BWR containment as a function of time during a TB Sequence based on thermo-hydraulics modelled by MAAP 3.0 [2]. The logarithm of the free hydrogen ion concentrations was arbitrarily set to 7.0 at any time when either no liquid water was present or no acidic/alkaline substances were present in a compartment.

○ = Pedestal, □ = Drywell, × = Wetwell

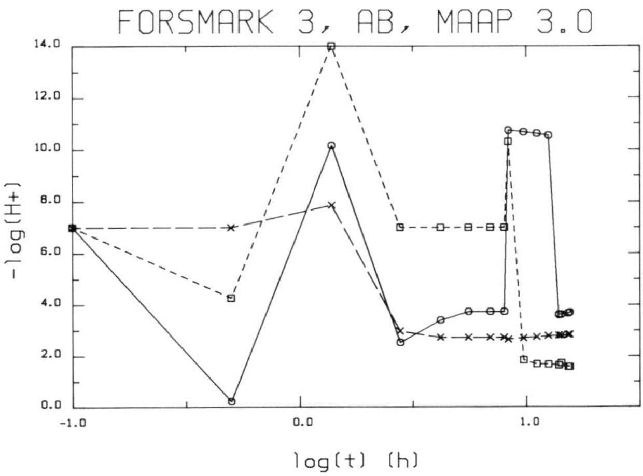

FIGURE 4. Approximate calculation of the acidity in the three compartments in a Swedish BWR containment as a function of time during a AB sequence based on thermo-hydraulics modelled by MAAP arbitralily set to 7.0 at any time when either no liquid water was present or no acidic/alkaline substances were present in a compartment.

○ = Pedestal, □ = Drywell, × = Wetwell

CORROSIN OF ALUMINIUM AND ZINC

Aluminium and zinc are present i Swedish BWR containments in considerable quantities, mostly as thin plates and surface coating, respectively. Table 1 gives some typical data.

A high local pH, caused by an initial deposition of an aerosol containing hygroscopic cesium hydroxide, will lead to rapid corrosion with hydrogen evolution. The corrosion will stop, or decrease strongly, once the alkali metal hydroxide, mainly cesium hydroxide, has been used up in the process. Alkaline corrosion at high pH is important and rapid for both aluminium and zinc. The alkaline corrosion will bind some cesium in the corrosion products.

When the acidic gases from the pyrolysis of insulating materials spread in the containment the situation will be reversed. Most wet surfaces will then soon reach a low pH. At low pH, corrosion of zinc is normally more rapid than corrosion of aluminium. Also acid corrosion will produce hydrogen and some of the acidic substances will become neutralized in the process. At the same time some cesium may be leached from earlier formed corrosion products.

The large areas covered with aluminium and zinc and the large amounts of these materials give them the potential to be the major source of hydrogen in the containment. Catastrophic corrosion of these metals could produce 2 500 kg of hydrogen, should it occur. However, there seems to be several factors limiting the amount of corroded material. The major limitation will be a limited supply of the alkaline and acidic substances necessary for large scale corrosion as these are consumed by the corrosion process. The situation would be much worse if there was any source of catalyst, e.g. mercury or mercury salts in the containment. Mercury and its salts acts as catalysts for corrosion of aluminium by water even at neutral pH.

Because of the incomplete information available from the MAAP calculations on the deposition of fission products onto various surfaces in the compartments, it is presently not possible to make any quantitative estimation of the amount of corroded aluminium and zinc. We can only point out the importance of corrosion as it will produce hydrogen, consume hydroxide at high pH and consume acid at low pH. The very large surfaces covered with these metals indicates that the effects from corrosion will be seen fairly rapidly. The net effect of metal corrosion will be a reduction in the span of pH-values and an extra production of hydrogen.

COPPER

When electric wiring below the vessel comes into direct contact with the exiting very hot core melt it is likely that some copper will vaporize. The copper vapor will condense onto cooler surfaces. The effects from the presence of finely divided copper can be manyfold.

Copper condensed on zinc plated surfaces will give galvanic corrosion at low pH. This will accellerate corrosion of zinc. It is not clear at present if copper also will have any large effect on the corrosion rate of aluminium.

It is known from experiments (e.g. ref.2) that metallic copper acts as a good getter for elementary iodine. Hence, the vaporized and condensed copper will react with and bind any elementary iodine present in the same compartment as the copper. This action is similar to that of silver from PWR control rods. Even if we assume that of the copper present immediately below the vessel is only vaporized and only 2% available as iodine getter, this corresponds to about 20 kg copper which could bind up to 40 kg of iodine as CuI. However, there is only about 21 kg of iodine normally assumed present in the fuel. This makes us believe that no elementary iodine could exist as such for any length of time in the pedestal atmosphere after the core melt has penetrated the vessel bottom. Hence, we believe that any radiolytic production of elementary iodine in the final acidic water of the pedestal is of no concern.

SUMMARY AND CONCLUSIONS

In this paper, we have discussed three possibly important effects on iodine chemistry from the presence and reactions of secondary materials.

1. Acidic compounds from the pyrolysis of organic polymers sooner or later cause low pH-values in all water pools

inside the containment. Low pH-values will lead to formation of hydroiodic acid and secondary radiolytic production of elementary iodine .
2. Hydrogen iodide is known to add to unsaturated organic compounds such as those found in the pyrolysis gases forming organic iodides of various kinds.
3. Elementary iodine is, at least in the pedestal, probably captured by finely divided copper from vaporized electric wiring.

Corrosion of aluminium and zinc will produce hydrogen, but will at the same time consume alkaline and acidic substances thereby reducing the possible swing in pH-values. Catastrophic corrosion could possibly make aluminium and zinc the major source of hydrogen in the containment.

We have mainly concentrated on the effects of secondary materials present in a BWR containment, but similar questions must exist also for PWR containments even if they could be of smaller importance because of a larger containment and a smaller amount of materials immediately below the pressure vessel.

As seen from our discussion, the secondary materials can have large effects on the iodine source term by changing the iodine speciation by binding cesium in corrosion products and they can even have some important effects on the thermal-hydraulic processes during a severe core melt accident. Hence, we believe that there is a need to include such materials and phenomena in future nuclear accident analysis codes.

REFERENCES

1. Kelen, T., RAMA III - Kompletterande materialinventering inom inneslutningen för bedömning av pH och redoxpotential vid svåra haverier, ABB ATOM AB, Report RM 88-1099, rev. 1, 1988.

2. Fridemo, L., Liljenzin, J.O., Produkter från kabelpyrolys, Department of Nuclear Chemistry, Chalmers University of Technology, Göteborg, Report KKR 841108 Göteborg 1988.

3. Enerholm, A., FRISK - FORSMARK 3 - MAAP-beräkning för totalt elbortfall, designsekvens (Fall 1), Swedish State Power Board, Report PK-93/87, 1988.

4. Enerholm, A., FRISK - FORSMARK 3 - MAAP-beräkning för stort ångledningsbrott och totalt elbortfall (Fall 3), Swedish State Power Board, Report PK-93/87, 1988.

Large-Scale Experiments on Hydrogen Distribution

T. KANZLEITER and G. LANGER
Battelle-Institut eV
Am Romerhof 35
D-6000 Frankfurt/Main 90, FRG

L. VALENCIA
Kernforschungszentrum Karlsruhe GmbH
Projekt HDR
Karlsruhe, FRG

INTRODUCTION

The generation of hydrogen by radiolysis of water or metal-water reactions during an accident within an LWR and the resulting possibility of hydrogen explosions are an old concern in reactor safety research. This was substantiated by the accident at Three Mile Island during which the reactor core was degraded to a large extent, some hundred kg hydrogen were produced, and burnt in the reactor containment building resulting in a short-term pressure increase of 1.9 bar [1].

As a consequence of this event, hydrogen research was intensified and some additional research programmes were started in the USA and in Western Europe, covering the following subjects:

(1) Hydrogen sources, particularly in case of severe reactor accidents (degraded core accidents, core meltdown accidents)
(2) Hydrogen distribution, in particular within a compartmentalised containment
(3) Hydrogen combustion (deflagration, possibility of detonations)
(4) Hydrogen mitigation procedures

The present paper is focussed on experimental investigations for item (2), hydrogen distribution within the containment.

HYDROGEN GENERATION AND RELEASE INTO THE CONTAINMENT

Design Basis Accident (DBA). Due to correct operation of the Emergency Core Cooling System, ECCS, hydrogen generation is small during a DBA, and mainly determined by the long-term effects of water radiolysis in the core or in the containment sump, and metal(zinc, aluminium)/aqueous spray reactions in the containment building. It takes typically 30 days to produce (maximum) 140 kg H_2 [2], corresponding to an average H_2 concentration of approximately 2.5 vol.% in a 65,000 m^3 PWR containment. As the lower flammability limit of hydrogen amounts to ~ 4 vol.%, the hydrogen mass generated in a DBA can become safety relevant only for smaller containment types (BWR and ice-condenser containments) or in case of an extremely inhomogeneous distribution (with high local H_2 concentration values) in a PWR.

Degraded Core Accident (DCA). In DCAs ECCS fails to function, the Zircaloy metal cladding of the fuel rods reaches extremely high temperatures and reacts with the present hot steam to produce hydrogen in very large rates. In the TMI accident, for example, the amount of hydrogen generated is estimated to range from 150 to 600 kg, corresponding to an oxidation from 15 to 58 % of the Zircaloy cladding in the short term [1]. Dependent on the actual DCA scenario the hydrogen generated in the reactor pressure vessel, RPV, is released in the containment building - where air is available to form an inflammable H_2-air mixture - either immediately (in case of a large pipe break) or delayed (small leak, or activated pressuriser relief valve).

Core Meltdown Accident, CMA. This type of accident is assumed to involve two phases of intense hydrogen release in the short term and an additional third low-rate phase in the long term. During the first short-term phase, the core degradation and melting phase, similar amounts of H_2 (e.g. 660 kg in a 1300 MW PRS [3] will be generated by Zircaloy-steam reaction in the core region as during a DCA.

In the further course of a CMA the core melt is presumed to penetrate the RPV bottom and to attack the concrete basemat below the reactor cavity. During this so-called core-concrete interaction period, water vapour and carbon dioxide are released from the attacked concrete, pass up through the melt, oxidise metallic constituents of the melt, and are converted to hydrogen and carbon monoxide, both combustible gases which are directly released into the containment atmosphere. The German BETA core-concrete interaction experiments revealed that concrete decomposition initially proceeds very fast and within half an hour enough gas is produced to oxidise another 40% of the zirconium mass and to generate another 540 kg hydrogen [3] in this time period.

Finally, in the long term, the residual zirconium and a portion of the iron and chromium from the core-melt and the molten reinforcing bars will be oxidised and yield a continuous small H_2 release rate, roughly another 100 kg H_2 within 24 hours [3].

HYDROGEN DISTRIBUTION

As described in the previous chapter, generation of 600 kg hydrogen is typical for a DCA of a 1300-MW LWR, while for a CMA even 1,200 kg must be expected. The release of 600 kg H_2 into a large dry PWR containment of 65,000 m^3 will result in an average hydrogen concentration between 7 vol.% H_2 (at 2 bar/90 ^oC containment atmosphere conditions) and 11 vol.% H_2 (at 1.2 bar/ 40 oC). These hydrogen concentration values exceed the lower flammability limits of ~ 4 vol.% H_2 for upward flame propagation or ~ 9 vol.% H_2 for downward flame propagation. In case of an inhomogeneous hydrogen distribution even a local concentration value of 15 vol.% H_2 may be reached or exceeded, above which a transition from a relatively smooth hydrogen burn mode ("deflagration") to a very violent mode ("detonation") must be expected, whose extremely high pressure peaks severely attack the integrity of the containment structures.

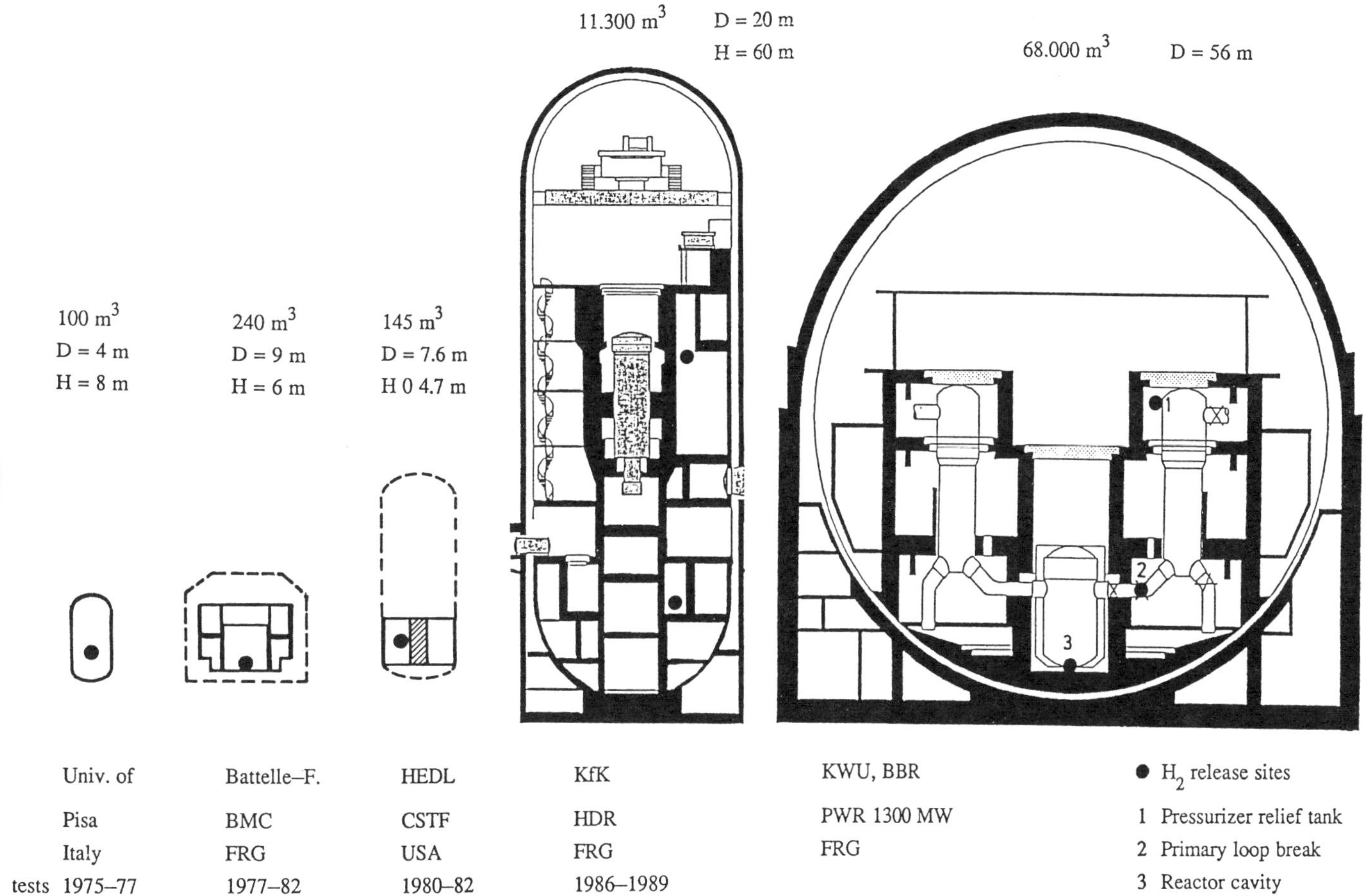

Fig. 1: Hydrogen Distribution Test Facilities. Comparison in Size to a German PWR

These data illustrate why investigations on hydrogen distribution are necessary to either make sure that high local H_2 concentrations cannot occur in a given accident situation, or as a basis for the design of hydrogen mitigation procedures.

Hydrogen distribution investigations cover both code development and experiments. Typical codes used for calculation of hydrogen transport and mixing processes are MARCH, RELAP 4, HECTR, RALOC or WAVCO. Some of the individual models of these codes are empirical and based on the results of separate-effect tests. For code validation integral experiments in large-scale test facilities are used, which will be described in the following.

Hydrogen Distribution Experiments (cf. Fig. 1)

University of Pisa, Italy. Basic H_2 mixing effects following DBA typical hydrogen release rates were investigated in a 100 m^3 steel vessel without partitions. Atmosphere conditions: air + hydrogen, 1 bar, 16-85 °C, isothermal or with a positive temperature gradient ($T_{bottom} > T_{top}$), [4].

Battelle-Institut Frankfurt, FR Germany. Basic experiments on hydrogen transport and mixing effects in various multi-compartment geometries (2 to 6 compartments) were performed in a 240-m^3 part of the Battelle model containment. The H_2 release rates were similar to DBA conditions or higher by a factor of ten. Atmosphere conditions: air + hydrogen, 1 bar, 14-53 °C, rel. humidity up to 100 %, isothermal or with a negative temperature gradient ($T_{top} > T_{bottom}$), [5,6]. It should be noticed that these and the University of Pisa experiments were initiated before the accident at TMI happened in 1979.

HEDL, Hanford, USA. H_2 and He mixing and distribution tests were conducted in a special 1/3-scale model ice-condenser containment geometry (150 m^3, one compartment) using DCA-typical H_2 release rates. Atmosphere conditions: air + hydrogen/helium, 1 bar, 20-94 °C, quasi-isothermal, rel. humidity up to 100 %, [7].

KfK-PHDR, Karlstein, FR Germany. As a part of the HDR Safety Research Programme two groups of large-scale H_2 transport and distribution experiments were and are being performed in the 11,300 m^3 compartmentalised containment building of the decommissioned HDR reactor under thermal-hydraulic conditions relevant for severe accidents. For safety reasons a He-H_2 mixture is used in the experiments as a simulant for hydrogen, the release rates corresponding to DCA or CMA conditions. The atmosphere conditions are: steam + air + He-H_2, ~ 1.5 bar, 30-100 °C, local steam and temperature distribution effected by natural convection.

The aim of the HDR hydrogen distribution experiments is to cover all H_2 transport and mixing phenomena relevant for severe accident conditions, while the former tests in the smaller facilities could only partly meet these requirements.

Hydrogen Distribution Effects

Based on the finding of the above H_2 and some steam distribution experiments, this chapter gives an overview on the relevant phenomena effecting the distribution of hydrogen after its release into the containment volume.

Steam-air distribution of the atmosphere before hydrogen release. In the most accident scenarios steam will be released into the containment prior to hydrogen, yielding a homogeneous or inhomogeneous steam-air and temperature distribution in the containment atmosphere, dependent on the respective accident conditions. Figs. 2 and 3 depict some basic distribution effects, derived from the results of steam-air experiments performed in the Battelle model containment [9]. For steam release into an air atmosphere the lefthand part of Fig. 2 (density of the released fluid lower than density of the existing atmosphere) may be relevant. A completely homogeneous steam-air atmosphere (case A) can be obtained only if the steam source is located close to the containment bottom and steam transport by natural convection is not restrained by internals. Case C and E yield an inhomogeneous atmosphere with stable stratification, density of the atmosphere at the top of the containment volume (steam-air mixture) being lower than at the bottom (cold air).

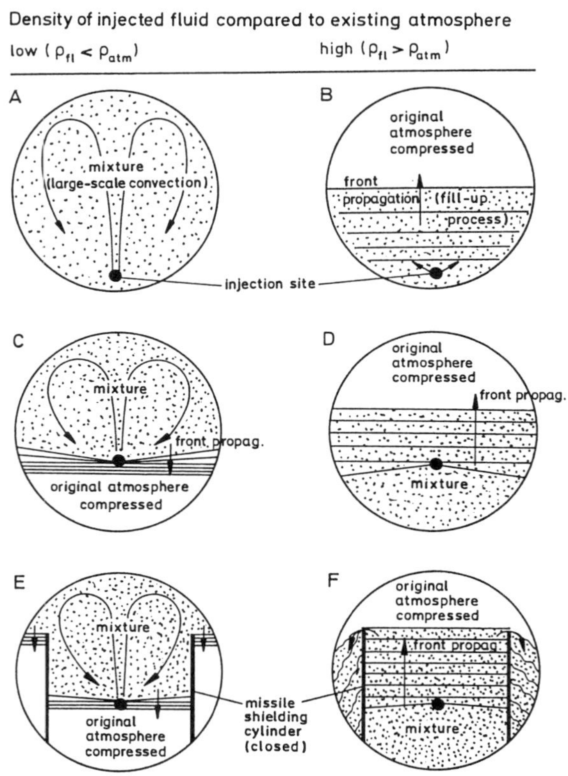

Fig. 2:
Formation of a homogeneous or inhomogeneous containment atmosphere distribution due to injection of fluid (hydrogen, steam, air)

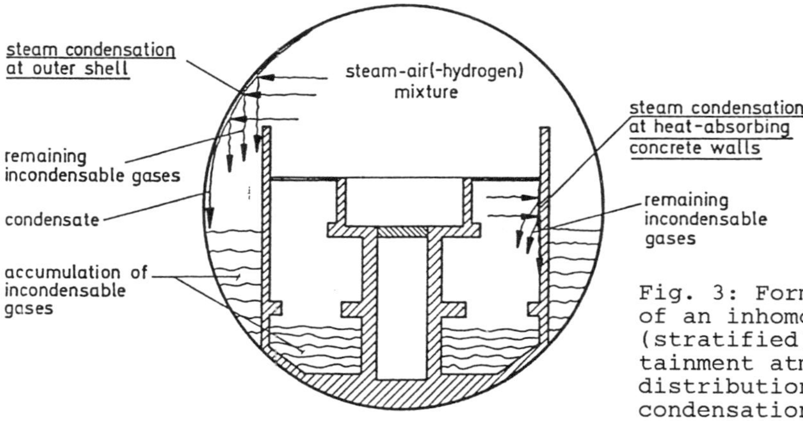

Fig. 3: Formation of an inhomogeneous (stratified) containment atmosphere distribution due to condensation effects

Most of the internal concrete structures are located in the lower part of the containment building where they act as heat sinks, lowering the temperature of the adjacent atmosphere and its steam content by condensation, and thus supporting the formation of a stable atmosphere stratification, see Fig. 3.

The described phenomena finally result in a certain steam-air distribution of the atmosphere (either homogeneous or stratified or a combination of homogenised and stratified regions) which forms the starting condition for the subsequent hydrogen release and distribution period.

Mixing and transport of hydrogen by density controlled natural convection. Released into a containment volume with an initially homogeneous atmosphere distribution and no internals, hydrogen mixes up with the atmosphere above the injection site, while the region below remains almost unchanged, as depicted in Figs. 2A and C. Partitions inside the containment will restrain the transport and mixing process (cf. Fig. 2E; compare Figs. 2A and 4A) or not (as in case of a "loop geometry" as in Fig. 4C). Fig. 4E depicts a more complicated case: In a first period the released hydrogen mixes up with the dome atmosphere, in a later period buoyancy effects in the lower vertical ducts can stimulate a loop flow with subsequent mixing also in that region.

Stagnant stratification. An existing stable steam-air atmosphere stratification (vertical density gradient) prevents a large-scale mixing of the released hydrogen. Fig. 4B depicts a situation with a cold lower compartment and a hot humid atmosphere (lower density) in the upper region. In this case the hydrogen injected at the bottom is trapped in the lower compartment as long as the density of the cold H_2-air mixture formed remains smaller than the atmosphere density in the upper region. If more than one opening exists in the partition between the upper and the lower region (cf. Fig. 4D) the situation is basicly the same but less stable; small local differences in the vertical density gradient can result in a partial mixing and formation of a transition zone.

Fig. 4:
Fluid transport and mixing processes following hydrogen release into a containment of initially homogeneous or stratified atmosphere distribution

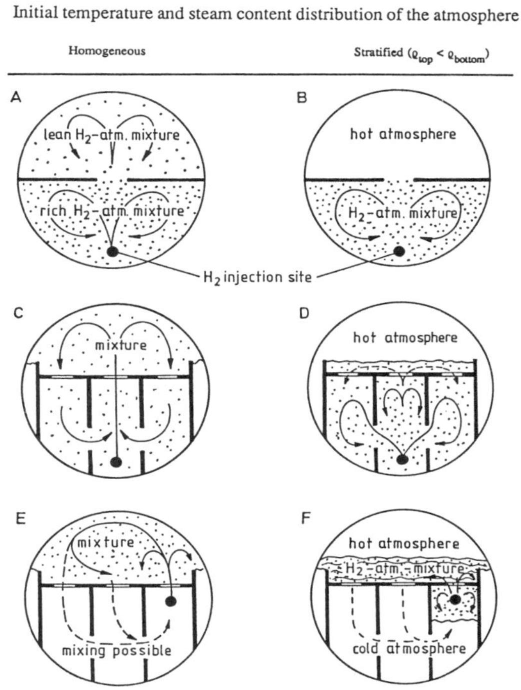

Other hydrogen distribution effects. Compared to the already mentioned transport effects, diffusion is negligible. H_2 release jets can play only a local role for mixing because they are affected by obstacles and internal walls. More important is the phenomenon of hydrogen and air accumulation in lower containment regions (if no remixing exists) and in the annulus between missile shielding cylinder and containment shell, as described in Fig. 3.

Samples of Experimental Data

Battelle-Frankfurt tests. Figs 5 and 6 depict two samples of results from the H_2 distribution tests in the Battelle model containment (hydrogen-air 6-compartment tests). Under isothermal conditions the hydrogen released is homogeneously distributed and there are no remarkable concentration differences at any time, see Fig. 5 and Fig. 4C. For those tests with an inverse thermal stratification established (temperature in the upper regions higher than in the lower ones) but all other test conditions the same, H_2 concentrations are significantly higher in the lower compartments, cf. Fig. 6. The reasons for this are the differences in atmosphere density: As long as the density of the hot hydrogen-lean atmosphere in the upper region is higher than that of the cold hydrogen-rich atmosphere below, no convection takes place. With H_2 release continuing the densities become almost the same and natural convection starts to mix both regions (at t = 8,000 s in Fig. 6).

Fig. 5:
Battelle Frankfurt tests:
Homogeneous hydrogen distribution under isothermal conditions

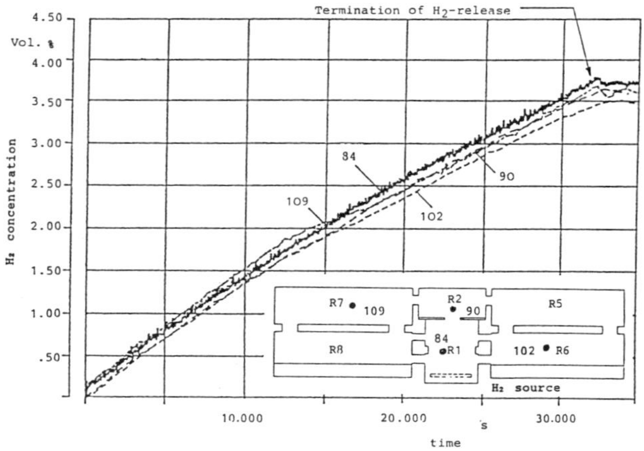

Fig. 6:
Battelle Frankfurt tests:
Inhomogeneous hydrogen distribution in case of inverse thermal stratification

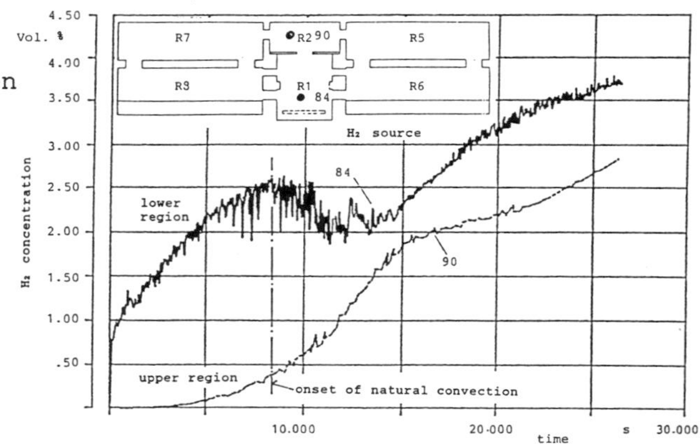

HDR tests. Figs 7 and 8 show measured temperature and hydrogen-concentration profiles from the first H_2 distribution test performed in the HDR containment. The steam and hydrogen release site was located at a level of +23 m, at a polar angle of about 90° relative to the depicted cross sections. From t = 0 to 1 min a large amount, between t = 20 and 35 min a smaller amount of steam was released there, followed by a H_2 release period from t = 35 to 47 min. The measured temperature profiles show maxima at that level, the gradients below being negative, above close to zero. This behaviour is comparable to Fig. 2C. Due to natural convection by buoyancy effects in the vertical ducts of the internal structures a transition zone develops between the cold downstairs region and the hot upper part. Temperature at the lowest measuring point is elevated by warm sump water, cf. Fig. 7. According to the given atmosphere density distribution, the H_2 released at level +23 m from t = 35 min ascends in the left-hand spiral stair case into the dome (cf. Fig. 4E) and initiates a downward backflow of initially H_2-lean air through the right-hand staircase into the compartmentalised section of the containment building, see 40-min curve of the lower plot in Fig. 8.

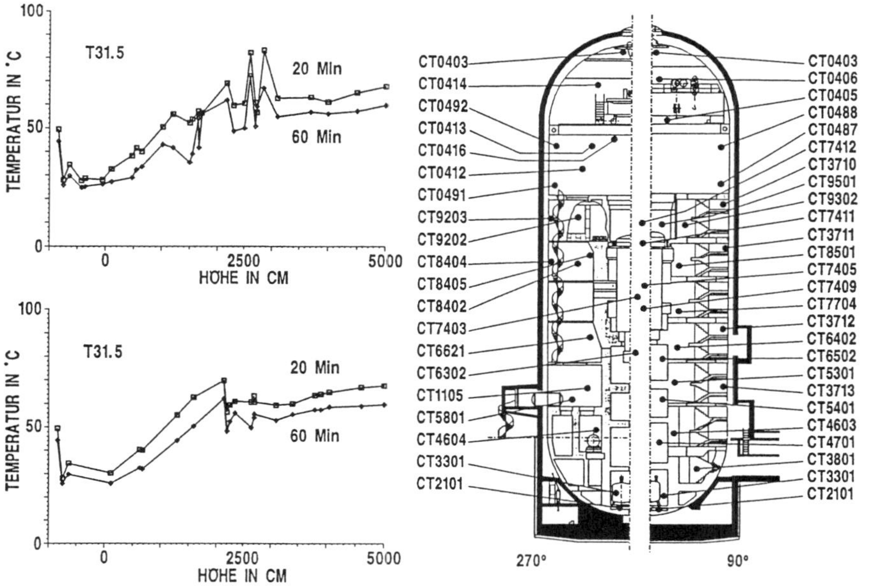

Fig. 7: HDR hydrogen distribution test T31.5:
Vertical temperature profiles in the righthand staircase (upper plot) and at the lefthand spiral stairs (lower plot)

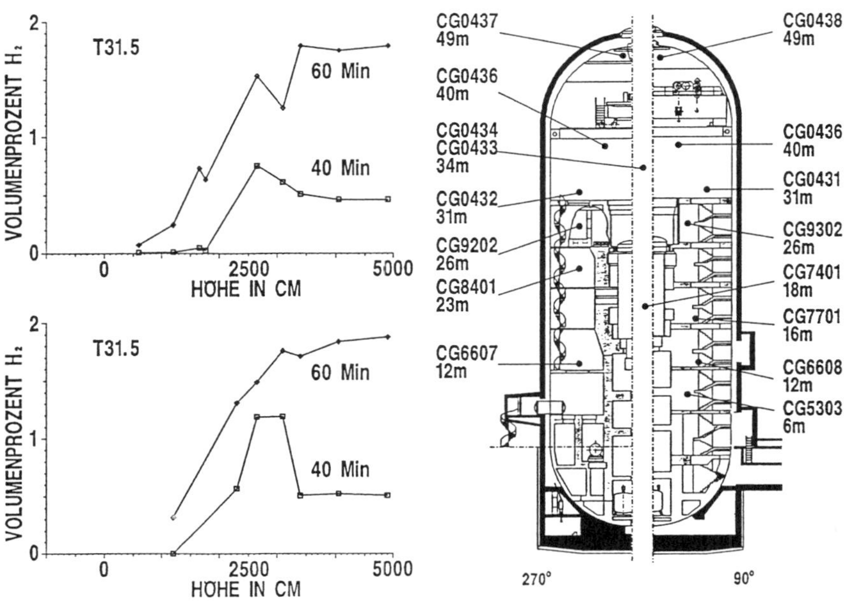

Fig. 8: HDR hydrogen distribution test T31.5:
Vertical hydrogen concentration profiles in the righthand staircase (upper plot) and at the lefthand spiral stairs (lower plot)

RESUME, FURTHER INVESTIGATIONS

Together with the older experimental hydrogen distribution investigations (e.g. Battelle-Frankfurt and HEDL), which formed a good basis to understand the principal hydrogen mixing and transport effects and to develop adequate codes, the most recent HDR hydrogen distribution experiments (test group E11, performed in 1989) form a unique data base for code validation and further improvement, as well as for the further development of the code users' qualification.

Future experiments in the same test facilities; Battelle model containment and HDR, are directed to investigate hydrogen deflagration phenomena in multi-compartment containment geometries (under way) and the efficacy of hydrogen mitigation methods (igniters, catalysts) in a large scale.

REFERENCES

1. Berman, M., Cummings, J.C., Hydrogen Behavior in Light-Water Reactors, Nuclear Safety, vol. 25, No. 1, pp. 53-74, 1984

2. Jahn, H. (ed.), Hydrogen Generation, Distribution and Explosion Potential Associated with Accidents in LWR. Commission of the European Communities Rpt. XII/107/1985, Sept. 1985

3. Rohde, J., Schwinges, B., Wasserstoff-Freisetzung und -Verteilung während schwerer Unfälle in LWR-Anlagen, 12th Status Report Meeting of the HDR Safety Programme, PHDR/KfK, Karlsruhe, FR Germany, Dec. 8, 1988

4. Fineschi, F. et al, Mixing of Hydrogen in a Closed Containment, Instituto di Impianti Nucleari di Pisa, Rept. RP 296(77), 1977

5. Langer, G. et al, Experimentelle Untersuchung der Wasserstoffverteilung im Containment eines LWR nach einem Kühlmittelverlust-Störfall, Battelle Frankfurt Rept. BF-R-63.363-3, May 1979

6. Langer, G., Experimentelle Untersuchung der Wasserstoffverteilung in einem Modell-Containment (Grundlagenversuche Phase II, Battelle-Frankfurt Rept. BIeV-R-64.036-3, July 1982

7. Bloom, G.R., Muhlestein, L.D., Postma, A.K., Claybrook, S.W., Hydrogen Mixing and Distribution in Containment Atmosphere, EPRI Rept. NP-2669, March 1983

8. Wolf, L., Valencia, L., Results of the Preliminary Hydrogen Distribution Experiment at HDR and Future Experiments for Phase III, 16th Water Reactor Safety Information Meeting, Gaithersburg, MD, USA, Oct. 24-28, 1988

9. Kanzleiter, T., DEMONA Experiments, Final Report, Battelle Frankfurt Rept. BIeV-R65.523-01, September 1987

Heat and Mass Transfer from Bubbles Rising through Pools

F. ABBEY and S. A. RAMSDALE
UKAEA Safety and Reliability Directorate
Culcheth, Warrington, WA3 4NE, UK

1. INTRODUCTION

Two computer codes have been developed at SRD to model the decontamination of bubbles rising through water. This phenomenon is relevant in many nuclear accident scenaria, as well as certain processes in the chemical industry. BUSGAS considers a spherical cap shaped bubble containing a relatively insoluble gas such as hydrogen and contaminated by a soluble gas such as I_2, and calculates the mass transfer of the soluble species out of the bubble as it rises through the pool. The important feature is the build up of a quasi stagnant wake behind the spherical cap bubble, which limits the removal of the soluble contaminant. The second code, BUSCA, considers aerosol removal from a bubble containing a mixture of gases including steam. The bubble thermal hydraulics are modelled as well as the aerosol behaviour, since aerosol removal mechanisms such as diffusiophoresis and thermophoresis are highly dependent upon steam condensation/evaporation and the temperature gradient at the bubble surface.

An informal collaboration on bubble scrubbing codes and experiments has been set up between SRD and the Paul Scherrer Institute (Switzerland), ENEA (Italy), GRS (Germany), and CEA Cadarache (France). This has led to a code comparison exercise between the UK code BUSCA and the corresponding Source Term Code Package code, SPARC, based on the planned test matrix for the SPARTA experiments at Casaccia. BUSCA has recently been adopted by PSI for Source Term applications, and has undergone further development there.

Both BUSCA and BUSGAS are described in the paper, along with the results of the code comparison exercise.

2. THE BUSGAS CODE

BUSGAS was developed in order to calculate the decontamination of large bubbles containing a soluble gas pollutant as they rise through water. Bubbles of over 5 cm^3 in volume take on a spherical cap geometry[1] as opposed to the spherical or ellipsoidal shape of smaller bubbles. Although spherical cap bubbles may not be numerous in a typical bubble swarm, they may nevertheless carry a significant fraction of the discharged gas, because of their large volume.

A special feature of spherical cap bubbles is that they carry behind them a relatively stagnant wake. This leads to the build up of contaminant behind the bubble which inhibits diffusion through the flat base. Consequently, although spherical cap bubbles have a larger surface-to-volume ratio than spheres of the same volume, they exhibit lower decontamination factors than would be expected for a spherical bubble, and the simple exponential equation applicable in the spherical case is incorrect.

2.1 The Boundary Value Problem

Diffusion of a trace contaminant from a predominantly insoluble gas bubble is somewhat more complicated than the single component case, since the concentration of the contaminant within the bubble varies with time. This means that liquid-side mass transport is not necessarily the rate-limiting step. Consider first the upper, curved surface (radius of curvature R) of a spherical cap bubble, which is rising with speed U. As the bubble rises through the water, a circulation pattern is set up inside the bubble, and two boundary layers are formed, one on the gas-side of the bubble:pool interface and one on the liquid-side. Let C(t) be the bulk concentration of contaminant in the the bubble , and let c_G and c_L be the concentrations in the gas- and liquid-side boundary layers respectively. Then we must solve two, coupled diffusion equations

$$u_s\frac{\partial c_L}{\partial s} + u_n\frac{\partial c_L}{\partial n} = D_L \frac{\partial^2 c_L}{\partial n^2}, \qquad 0 \leqq n < \infty$$

$$u_s\frac{\partial cG}{\partial s} + u_n\frac{\partial c_G}{\partial n} = D_G \frac{\partial^2 c_G}{\partial n^2}, \qquad -\infty < n \leqq 0 \qquad (1)$$

subject to the boundary conditions

(i)	$c_L \rightarrow 0$	as $n \rightarrow \infty$	
(ii)	$c_G \rightarrow C(t)$	as $n \rightarrow -\infty$	
(iii)	$c_G = H\, c_L$	on $n = 0$	(Henry's Law)
(iv)	$D_G \frac{\partial c_G}{\partial n} = D_L \frac{\partial c_L}{\partial n}$	on $n = 0$	(continuity)

where u_s and u_n are the velocity components normal to and parallel to the curved surface, $u_s=\frac{3}{2}\, U \sin\theta$, $u_n=-3\, U\, \frac{n}{\theta}$, D_L and D_G are the liquid- and gas-side diffusivities, and H is the inverse of the partition coefficient of the contaminant. The spherical cap bubble rise speed is derived in[1] and is given by $U = \frac{2}{3}\sqrt{(gR)}$. The solution of this problem is described in detail in[2]. Suffice to say that the total flux of contaminant through the curved surface of the bubble can be shown to be

$$F_c(t) = \frac{4\pi^{\frac{1}{2}}R^{7/4}\, D_L^{\frac{1}{2}}\, g^{1/4}\, C(t)}{H + (D_L/D_G)^{\frac{1}{2}}}\, f(\theta_B) \qquad (2)$$

where $f(\theta_B) = \{\frac{2}{3} + \frac{1}{3}\cos\theta_B\}^{\frac{1}{2}}(1 - \cos\theta_B)$ and θ_B is the bubble semi-angle of a spherical cap bubble, typically 55^0.

In order to calculate the bulk concentration as a function of time, we also need to know the flux of contaminant through the bubble's flat base. If we assume that the bubble wake is essentially stagnant, then c_L satisfies the diffusion equation for time-dependent concentration in a semi-infinite region

$$\frac{\partial c_L}{\partial t} = D_L \frac{\partial^2 c_L}{\partial z^2} \qquad (3)$$

subject to the boundary conditions

(i) $c_L(z,t) = H^{-1}C(t)$ on $z=0$ (Henry's Law)
(ii) $c_L(z,t) \to 0$ as $z \to \infty$
(iii) $c_L(z,t) = 0$ at $t=0$

where z is measured vertically downwards from the base of the bubble. Note that the first condition assumes that there is no laminar boundary layer on the gas-side of the flat base, so that the bulk concentration extends down to the interface with the wake.

Clough et al show that the Laplace transform (variable, p) of the total flux across the bubble base is then given by

$$\mathcal{F}_B(p) = \pi\, R^2 \sin^2\theta_B \frac{p^{\frac{1}{2}} D_L^{\frac{1}{2}} \mathcal{C}(p)}{H} \quad . \qquad (4)$$

However, this expression gives rise to a physically unrealistic maximum in the decontamination factor at $H = (D_L/D_G)^{\frac{1}{2}}$. This maximum is removed if we replace H by $H + (D_L/D_G)^{\frac{1}{2}}$ in the denominator of equation (4), which is equivalent to shifting the well-mixed condition a distance of the order $(D_G t)^{\frac{1}{2}}$ into the bubble. The final result is then

$$\mathcal{F}_B(p) = \pi\, R^2 \sin^2\theta_B \frac{p^{\frac{1}{2}} D_L^{\frac{1}{2}} \mathcal{C}(p)}{H + (D_L/D_G)^{\frac{1}{2}}} \quad . \qquad (5)$$

This correction is important for small values of H only, ie. high partition coefficients.

2.2 The Decontamination Factor

The concentration of contaminant in the bubble, C(t), is governed by the conservation equation

$$\frac{d}{dt}\{CV\} = -F_C - F_B \qquad (6)$$

where V is the spherical cap bubble volume $V = \frac{2}{3}\pi\beta\, R^3$ and

$\beta = 1 - \frac{3}{2}\cos\theta_B + \frac{1}{2}\cos^3\theta_B$. BUSGAS makes the approximation that the bubble volume remains constant throughout the bubble rise time. Equation (6) can be solved using Laplace transforms to give C(t) in terms of complimentary error functions. However, provided that the characteristic time-scale for diffusion through the curved surface, τ, is small compared to the bubble rise time, t_R, the asymptotic form of the complimentary error function[3] for large argument can be used to give a decontamination factor of

$$DF = \frac{C(0)}{C(t_R)} = 2\pi^{\frac{1}{2}} \frac{(t/\tau)^{3/2}}{\lambda}, \qquad 0 < \lambda < 2 , \tag{7}$$

where

$$\tau = \frac{\pi^{\frac{1}{2}}\beta\, R^{5/4}\, (H + (D_L/D_G)^{\frac{1}{2}})}{6\, D_L^{\frac{1}{2}}\, g^{1/4}\, f(\theta_B)} , \tag{8}$$

and

$$\lambda = \frac{3}{2} \frac{\sin^2\theta_B D_L^{\frac{1}{2}} \tau^{\frac{1}{2}}}{\beta R\, (H + (D_L/D_G)^{\frac{1}{2}})}$$

The special case $\lambda = 0$ corresponds to a spherical bubble ($\theta_B = 180^0$). The decontamination factor in this case is well-known and given by the exponential form

$$DF = \frac{C(0)}{C(t)} = e^{-t/\tau} \tag{9}$$

3. THE BUSCA CODE

BUSCA[4] (BUbble SCrubbing Algorithm) is a mechanistic code which calculates the removal of aerosol from a bubble as it rises through water. It also models the bubble thermal hydraulics, since these have a strong influence on aerosol behaviour, especially in condensing or evaporating conditions. The bubble may contain a mixture of non-condensible gases and steam. Like BUSGAS, BUSCA assumes a spherical cap bubble geometry, but is being modified at the Paul Sherrer Institute to include ellipsoidal and spherical geometries as well as other features.

3.1 Bubble Thermal Hydraulics

In the situations of interest, the bubbles are likely to be hotter than the pool into which they are released. In addition, they may contain steam as well as non-condensible gases. This means that steam will be condensing rapidly when the bubbles first enter the pool, but thereafter some evaporation into the bubble can be expected as the hydrostatic pressure falls and the bubble expands during the bubble's rise. Changes in bubble size affect the aerosol removal processes, since they alter the surface to volume ratio for deposition. The bubble volume is related to the hydrostatic pressure by the Perfect Gas Law. The depth of the bubble relative to the pool surface is tracked by integrating Batchelor's spherical cap bubble rise speed (see Section 2.1).

In order to calculate the steam condensation/evaporation rate, BUSCA uses Sherwood, Pigford and Wilke's[5] expression for the flux of water vapour across a phase change interface of area A

$$\frac{dm_S}{dt} = -\, k_m^{*}\, A\, \ell n \left(\frac{1-X_{Si}}{1-X_S}\right) \tag{10}$$

where

$k_m^* = \frac{M_s p_B}{R_G T_B} k_m$

M_s = molecular mass of steam
p_B = total bubble pressure
R_G = gas constant
T_B = bubble temperature
k_m = mass transfer coefficient
X_s = bulk mole fraction of steam in bubble
X_{si} = mole fraction of steam at the interface

BUSCA assumes that the bubble interface is in equilibrium with the surrounding pool, so that

$$X_{si} = p_{sat}(T_p) / p_B \;. \qquad (11)$$

Note that equation (10) implies that steam condensation and evaporation are driven by the pool temperature, T_p, which determines the equilibrium mole fraction of steam at the interface.

Clift et al[6] give correlations for the mass transfer coefficient k_m for spherical cap bubbles. The upper, curved surface of the bubble and its flat base are treated separately, to give the following overall Sherwood Number

$$Sh = \frac{k_m d_e}{D_G} = A_e \left\{\frac{1.1284}{A_c} + \frac{0.357}{A_B}\right\} Pe^{\frac{1}{2}} \qquad (12)$$

where Pe is the bubble Peclet Number, $Pe = Ud_e/D_G$, and

d_e = volume equivalent sphere diameter
A_e = volume equivalent sphere surface area
A_c = area of upper, curved surface
A_B = area of flat base
D_G = molecular diffusivity of gas
U = bubble rise speed

By analogy, the heat transfer coefficient from the spherical cap bubble, k_h, is given by the Nusselt Number correlation

$$Nu = \frac{k_h d_e}{K_G} = A_e \left\{\frac{1.1284}{A_c} + \frac{0.357}{A_B}\right\} Pe'^{\frac{1}{2}} \qquad (13)$$

where Pe' is the heat transfer Peclet Number, $Pe' = Ud_e/a$, a is the thermal diffusivity of gas and K_G the thermal conductivity.

The total bubble enthalpy, H, changes as a result of both conduction of sensible heat into the pool and steam condensation/evaporation at the interface :

$$\frac{dH}{dt} = - k_h A (T_B - T_P) + h_s \frac{dm_s}{dt} \qquad (14)$$

where h_s is the specific enthalpy of the outgoing/incoming steam respectively. This is calculated at the bubble temperature under condensing conditions and at the pool temperature under evaporating conditions. Note that BUSCA assumes that the pool temperature remains constant during the time taken for the bubble to rise to the surface.

The rate of change of bubble temperature is derived by differentiating the expression for total bubble enthalpy

$$H(m_S, m_N, T_B) = m_S\, h_S(T_B) + m_N\, h_N(T_B)$$

where m_N and h_N are the mass and specific enthalpy respectively of the non-condensible gases in the bubble. This gives

$$\frac{dT_B}{dt} = \frac{1}{\langle mc_p \rangle} \left\{ \frac{dH}{dt} - h_S(T_B)\frac{dm_S}{dt} \right\} \tag{15}$$

where $\langle mc_p \rangle$ is the total heat capacity of the bubble,

$$\langle mc_p \rangle = m_S c_{pS} + m_N c_{pN} \; .$$

Given the initial bubble size and properties, BUSCA integrates equations (10), (14) and (15) numerically with respect to time in order to calculate the bubble temperature and steam condensation or evaporation rate during its rise to the pool surface. These equations are coupled to the aerosol removal equations described below.

3.2 Aerosol Removal Equations

BUSCA calculates the change in the aerosol mass distribution, $m(r,t)$, using the equation

$$\frac{dm(r,t)}{dt} = - \lambda(r,t)\, m(r,t) \tag{16}$$

where $\lambda(r,t)$ is the removal rate for particles of radius r at time t corresponding to the combined effects of diffusiophoresis, thermophoresis, gravitational settling, inertial impaction and diffusive deposition. Clearly, over the upper, curved surface of the bubble, gravity will be acting away from the deposition surface, and any evaporation will also reduce deposition. This is taken into account by adding all five deposition velocities at right angles to the bubble surface, and setting the total deposition rate to zero at points where the net velocity towards the surface is less than or equal to zero. The deposition velocities corresponding to the five processes are described below.

<u>Diffusiophoresis</u>. Steam condensation or evaporation at the bubble surface causes the diffusion of steam and non- condensible gas down their concentration gradients, as well as the bulk flow of steam towards or away from the surface. The net effect of the bulk and diffusive flows is often referred to as diffusiophoresis[7], and gives rise to the following particle velocity:

$$v_{cond} = \frac{X_S}{X_S + \sum_i X_i (M_i/M_S)^{\frac{1}{2}}} \; \frac{-dm_S/dt}{m_S} \; \frac{V}{A} \tag{17}$$

where M_i and X_i are the molecular mass and mole fraction respectively of the i'th non-condensible gas in the bubble, and V and A are the volume and surface area of the bubble. The condensation rate dm_S/dt is defined by equation (10). It is important to note that v_{cond} is negative under evaporating conditions and then acts against deposition.

Thermophoresis. Thermophoresis is the movement of particles down a temperature gradient. Brock[8] derives the following expression for the particle velocity due to this effect:

$$v_{th}(r) = \frac{3\eta}{2\rho_G T_B} \frac{k_h(T_B - T_P)}{K_G} C(\ell/r) \, Br(\ell/r) \tag{18}$$

where $C(\ell/r)$ is the Cunningham correction factor for molecular effects and $Br(\ell/r)$ is the so-called Brock factor

$$C(\ell/r) = 1 + 1.246\ell/r + 0.42\ell/r \exp(-0.87r/\ell)$$

$$Br(\ell/r) = \frac{(K_G/K_P + 2.48\ell/r)}{(1 + 3\ell/r)\{1 + 2(K_G/K_P + 2.48\,\ell/r)\}} \tag{19}$$

ℓ = molecular mean free path in the bubble gas
ρ_G = bubble gas density
K_p = thermal conductivity of the particle

Thermophoresis will be most important during bubble release, when the temperature difference between the bubble and the pool is greatest.

Gravitational settling. The terminal velocity of particles falling freely under gravity is given by Stokes' Law[1]

$$v_g(r) = \frac{4}{3} \pi \rho_p g r^3 B(r) \tag{20}$$

where ρ_p is the effective material density of the particles and $B(r)$ is the particle mobility function defined by

$$B(r) = \frac{1}{6\pi\chi\eta r} C(\ell/r).$$

χ is the mobility shape factor, which is used to correct for non-spherical particles.

Inertial impaction. Fuchs[9] describes a model for the inertial impaction of particles in a spherical bubble due to centrifugal forces set up by the circulation of the gas. The centrifugal force is equated to the Stokes' Law drag to give an impaction speed of

$$v_{imp} = \frac{9}{4} U^2 v_g(r) \frac{\sin^2\theta}{Rg}. \tag{21}$$

This form of deposition is assumed to apply over the curved surface of the bubble only, not the base.

Diffusive deposition. Fuchs'[9] describes an approximate model for diffusive deposition in a spherical bubble. This is based on the distance diffused by a particle during the contact time of a fluid element with the pool as it circulates the bubble. BUSCA uses an adaptation of Fuchs' model for a spherical cap geometry, and assumes diffusive deposition takes place between θ = 10 and 50 . The resultant deposition velocity is

$$v_d(r) = 1.24 \left\{\frac{D_p U}{R^3}\right\} \frac{V}{A_d} \tag{22}$$

where A_d is the surface area between $\theta = 10^0$ and 50^0, D_p is the particle diffusivity

$$D_p = kT_B\ B(r)\ ,$$

and k is Boltzmann's constant. Diffusive deposition at the bubble base is neglected, because of uncertainties in the flow rates in this area.

3.3 BUSCA Versus SPARC Code Comparison Exercise

An informal collaboration on bubble scrubbing codes and experiments has been set up between SRD and the Paul Scherrer Institute (Switzerland), ENEA (Italy), Gesellschaft fuer Reactorsicherheit (Germany) and CEA Cadarache (France). As part of this collaboration, a code comparison exercise was carried out, based on the proposed test matrix for the SPARTA[10] experiments which are to take place at ENEA's Casaccia laboratories. This enabled the predictions of ENEA's version of the Source Term Code Package code SPARC[11] to be compared with those of BUSCA. Since these two codes were developed completely independently and use a very different approach to the bubble thermal-hydraulics, the comparison of their predictions sheds some light on the degree of uncertainty surrounding bubble scrubbing phenomena.

The SPARTA test matrix used for the code comparison consists of twelve small pool tests (SP1-SP12) and seven large pool tests (LP1-LP7). The inlet gas temperature for all the tests was specified as 170^oC, and either caesium iodide, caesium hydroxide or manganese oxide aerosol was specified. A mixture of steam, non-condensible gas and aerosol is injected at the base of the pool, using either an X-quencher device or a horizontal vent discharger. Table 1 shows the thermal hydraulic conditions for each test, along with the predictions of the two codes.

Comparisons between SPARC and BUSCA are hampered by the different input requirements of the two codes. SPARC calculates the initial bubble size from the gas injection characteristics using engineering correlations, whereas BUSCA requires initial bubble size to be supplied as input. The correlation specified in [10] was used to calculate this input, for consistency with SPARC. The SPARC code also uses inlet flow characteristics to estimate the removal of particles due to impaction during injection.

The effect of impaction at the entrance is shown vividly in Tests SP11 and SP12 where SPARC's predicted DF increases from 52 to 96 when the flow rate increases from 45 to 90 g/s. SPARC predicts higher DF's than BUSCA for all the quencher discharge tests. The version of BUSCA used for the code comparison study did not include aerosol impaction at the entrance, but it is planned to include this effect in future versions.

Apart from the discrepancy in some of the tests due to the effect of impaction, the agreement between the codes is good, particularly in relation to the uncertainties in the experiments themselves. This is encouraging, since the two codes were developed independently and approach the problem from different angles, BUSCA mechanistically and SPARC using mainly engineering correlations.

TABLE 1. Code Comparison Results

Test	Pool Depth (m)	Pool Temp. (K)	Inlet Flow Rates(g/s): Non-Cond.	Steam	$X_s(0)$	DF Predictions[1]: SPARC-ENEA	BUSCA
SP1 (HV[2])	1.8	293	45	0	0	1.5	2.34
SP2 (HV)	1.8	372	45	0	0	1.1	1.11
SP3 (HV)	1.8	293	27	230	0.932	76.3	145
SP4 (HV)	1.8	372	27	230	0.932	8.6	5.21
SP5 (HV)	1.8	293	45	0	0	1.6	2.36
LP1 (Q[3])	5.7	293	1248	0	0	85	5.00
LP2 (Q)	5.7	293	58	800	0.957	3270	2803
LP3 (Q)	5.7	293	728	200	0.305	273	8.89
LP4 (Q)	5.7	293	82	20	0.281	157	8.36
LP5 (HV)	1.8	293	1248	0	0	1.1	2.24
LP6 (HV)	1.8	293	82	20	0.281	6.1	3.39
LP7 (HV)	1.8	293	728	200	0.305	4.9	3.57
SP6 (HV)	1.8	293	45	45	0.615	11.3	8.44
SP7 (HV)	1.8	372	45	45	0.615	1.1	1.28
SP8 (HV)	1.8	293	45	0	0	1.6	2.23
SP9 (HV)	1.8	293	45	45	0.615	12.3	7.75
SP10 (Q)	1.8	293	45	0	0	79	2.36
SP11 (Q)	1.8	372	45	0	0	52	1.11
SP12 (Q)	1.8	372	90	0	0	96	1.11

have been compared with the SPARC code's, using the proposed test matrix for the SPARTA experiments. Agreement is good, with the exception of those tests where impaction at the entrance is important. This effect will be included in future versions of BUSCA. It is hoped to verify BUSCA against experimental data from the ACE programme as well as SPARTA and other relevant experiments when data becomes available.

---

[1]Aerosol properties are :

ρ_p=4.5 g/cm^3, χ=1, r_{50}=0.41µm, σ_g=1.75 (SP1-SP4, SP6,SP7,SP11,SP12)
ρ_p=3.7 g/cm^3, χ=1, r_{50}=0.46µm, σ_g=1.75 (SP5, SP10)
ρ_p=4.42 g/cm^3, χ=1.85, r_{50}=0.56µm, σ_g=1.75 (LP1-LP7)
ρ_p=4.8 g/cm^3, χ=1.5, r_{50}=0.47µm, σ_g=1.75 (SP8-SP9)

[2]HV denotes horizontal vent discharge
Small Pool tests - 1 hole, 15 cm diameter
Large Pool tests - 1 hole, 70 cm diameter

[3]Q denotes quencher discharge device
Small Pool tests - 432 holes, 0.2 cm diameter
Large Pool tests - 1498 holes, 1 cm diameter

4. SUMMARY

Theoretical models have been developed for the decontamination of bubbles rising through pools. In the case of removal of soluble gases, analytical solutions are available for both spherical and spherical cap bubbles of fixed volume. These could be incorporated into the BUSCA code, so that it could handle soluble gas DFs, aerosol DFs and thermal hydraulic effects combined.

The equations governing the removal of aerosol particles from spherical cap bubbles have been coupled to the bubble thermal hydraulics in the BUSCA code, which then integrates them numerically. BUSCA's DF predictions

5. REFERENCES

1. Batchelor, G.K., *An Introduction to Fluid Mechanics*, Cambridge University Press, 1970.

2. Clough, P.N., Brighton, P.W.M. and Ramsdale, S.A., Decontamination Factors for Gas Mixtures in Spherical Cap Bubbles, SRD R 311, July 1984.

3. Abramowitz, M. and Stegun, I.A., *Handbook of Mathematical Functions*, Dover, 1965.

4. Clough, P.N., Ramsdale, S.A. and Smith, P.N., Aerosol Decontamination Factors in Pools Overlying Molten Core-Concrete - Code Modelling, *Proc. OECD (NEA) CSNI Specialist Meeting on Core Debris/Concrete Interactions*, Palo Alto, September 1986.

5. Sherwood, T.K., Pigford, R.L. and Wilke, C.R., *Mass Transfer* McGraw-Hill, 1975.

6. Clift, R., Grace, J.R. and Weber, M.E., *Bubbles, Drops and Particles*, Academic Press, 1978.

7. Dunbar, I.H., Steam Condensation Modelling in Aerosol Codes, CEC Rept. EUR 10584, 1986.

8. Brock, J.R. *J.Colloid.Sci.*, 17, 1962.

9. Fuchs, N.A., *The Mechanics of Aerosols*, Pergamon Press, 1964.

10. Passalacqua, R. SPARC Code Calculations for the SPARTA Experiments, ENEA Rept. TERM/ISP NEEC 1TS4B 88016, 1988.

11. Owczarski, P.C., Schreck, R.I. and Winegardner, W.K., SPARC5: A Code for Calculating Aerosol Particle Capture in Suppression Pools, NUREG/CR-4285, PNL-5495, May 1985.

6. THE ACCIDENTS AT THREE MILE ISLAND AND CHERNOBYL

LECTURE

Radioactive Releases Due to the Chernobyl Accident

S. N. BEGICHEV, A. A. BOROVOJ, E. V. BURLAKOV, A. J. GAGARINSKY, V. F. DEMIN, and A. A. KHRULEV
I. V. Kurchatov Institute of Atomic Energy
Moscow, USSR

I. L. KHODAKOVSKY
V. I. Vernadsky Institute of Geochemistry and Analytical Chemistry
USSR Academy of Sciences
Moscow, USSR

ABSTRACT

General data on radioactive releases of the damaged Chernobyl NPP unit (ChNPP--4) which were presented by the soviet delegation to the post accident IAEA meeting in August 1986 are given. The method of its obtaining is described. These data are analysed and revised on the basis of investigation results obtained in the USSR and abroad after August 1986 on the issues of the accident course and consequences. Fission and neutron activation products inventory data in the ChNPP—4 core before the accident are presented including isotopes of U and transuranic elements. Dynamics, physical and chemical processes of the radioactive releases formation are considered including urgent mitigation measures. Major scientific on-side and off-side investigations planned and performed are described.

1. INTRODUCTION

The accident at Unit 4 of the Chernobyl NPP (ChNPP—4), occured on April 26 1986, its causes and course, measures on mitigating the accident and eliminating its consequences were described in detail in the Soviet report prepared for the IAEA Experts' Meeting in August 1986 [1].

The accident happened prior to shut-down of the unit for planned maintenance during operations mode tests on one of the turbogenerators. As a result of rough errors and violations of regulation rules on the part of the plant's staff the reactor was taken to an unsteady, unregulated state. Suddenly its power sharply rose. This has resulted in destruction of the reactor and releases of a portion of radioactive subatances accumulated in the reactor core to the atmosphere. Further development of the accident could make the reactor a long-acting source of radioactive releases. A fire started in several places at the ChNPP—4 and a graphite burning in the core complicated saverely the situation.

Urgent measures undertaken made it possible to shut the damaged reactor as a source of radioactive releases on the 10-th day after the initial explosion. The achievement of this result was a very difficult task because of considerable destructions.

A long-term isolation is performed by the special „shelter" which was built to the end of 1986. Its construction and purpose are described in [2].

In the paper [1] detailed data are given on radioactive releases resulting from the accident at the ChNPP—4. They were obtained up to the middle of July 1986 in the course of extremely complicated works on mitigating the development of the accident and eliminating its consequences.

Obtaining accurate and complete data on radioactive releases was impeded due to the impossibility of direct measurements on the damaged unit, the complex character of physico-chemical processes of formation of releases determined by both the internal processes of the accident course and external effects (depositing materials on the core), complicated and varying weather conditions in the ChNPP region during the period of the source action, the complex radiation situation inside the ChNPP—4, on the site and in the close-in zone.

Data on releases were calculated with the use of atmospheric dispersion models. To carry out calculations the results of the following were used as an initial information:

— systematical studies of radionuclide composition of aerosol samples taken over the damaged unit for 26.04.86;

— aero-gamma survey of the ChNPP region;

— analysis of fallout samples;

— systematical measurements of meteorological conditions in regions of contaminated air mass transport.

It should be noted that at that time sufficiently complete data were not available on radioactive environment contamination outside the USSR territory.

As a result of these calculations the data were obtained on the total amount of radioactive releases, its radionuclide composition and on the time dynamics of releases during 10 days of the source action. These data were presented in the mentioned report [1] together with a considerable body of initial data on the environment contamination and weather conditions in the close-in and distant zones of the ChNPP.

The total release was estimated to be equal to 4% (rounded) of the initial inventory in the core. The accuracy of results was estimated by the authors as ±50%. In accordance with this, the range of releases uncertainty was 2—6%.

Radioactive noble gases were not taken into account in this sum. It was assumed that their release amounted to 100% of their inventory.

After August 1986 many expert's analysis, research works were performed in the USSR and abroad on all aspects of the accident including the issues of radioactive releases. A lot of materials were published. Among them are:

— the report of INSAG experts to IAEA General Conference [3],

— Materials of international conferences [4, 5],

— the WHO report [6],

— the report of a number of USA organizations [7] (NUREG—1250),

— the UNSCEAR report to the UN General Assembly 1988 [8],

— the report of the USSR State Committee on Hydrometeorology and Control of Natural Environment [9].

The main directions of researches on radioactive release issues are:

— the detailed study of radioactive contamination of the environment,

— modelling atmospheric transport and environment migration of radioactive substances on the local, regional and global scales.

Two main objectives of these researches could be pointed out:

— obtaining data for making decision on measures of eliminating the accident consequences and on ensuring radiation safety of the population,

— obtaining data to analyze processes of formation of radioactive releases.

The last objective corresponds to the task of the present seminar.

The main objectives of this paper is to reanalizing the Chernobyl radioactive release data of the first soviet report [1] on the basis of new data obtained after July 1986.

The subsequent material are presented as fallows. At first the generalized data on the Chernobyl radioactive releases from the report [1] are described. Then data concerning different aspects of the Chernobyl radioactive releases and obtained in the USSR and abroad both before and after July 1986 (1986—1988 yrs) are given. On this basis the data of the report [1] are analyzid from the view-point of considering the total amount, the radionuclide composition, the time dynamics, the formation mechanism etc. of radioactive release. These aspects is often called the „accident source term".

2. RADIOACTIVE RELEASES DATA FROM THE FIRST SOVIET REPORT

General data of the report [1] on the Chernobyl radioactive releases are given in Tables 2.1. and 2.2. In the first Table the data are presented on daily releases from the damaged reactor. The second one contains the radionuclide composition of releases as well as the radionuclide inventories in the ChNPP—4 core before the accident.

Here it should be mentioned that in the report [1] there were some misprints, inaccuracies and vaguenesses. The report was translated into English and other languages and distributed before the IAEA Experts' meeting began. Therefore, unfortunately, the Soviet experts had no opportunity to make necessary corrections in the report's text. Some of them were reported verbally and after the meeting they were informed of in written answers to enquiries of specialists from different countries. A number of such corrections in the first report [1] will be indicated below in appropriate places. Here in Table 2.2 there is one correction: the value 0.3 MCi for the total release of ^{99}Mo was placed instead of misprint 3.0 MCi in [1].

TABLE 2.1. Daily release, q, of radioactive substances to the atmosphere from the damaged unit (excluding radioactive inert gases)*

Date	Days after the accident	q, MCi**
26.04	0	12
27.04	1	4.0
28.04	2	3.4
29.04	3	2.6
30.04	4	2.0
01.05	5	2.0
02.05	6	4.0
03.05	7	5.0
04.05	8	7.0
05.05	9	8.0
06.05	10	0.1
09.05	14	0.01
23.05	28	20.10^{-6}

* Release evaluation error is ±50%

**The values of q were calculated on 6 May 1986 taking into account radioactive decay.

Destruction of the reactor due to the accident made it an intensive source of radioactive releases into the atmosphere. They lasted about ten days and were ceased due to urgent mitigation measures. The process of radioactive releases can be divided into several stages.

In the first stage (at the beginning of the accident) there was a release of dispersed fuel from the damaged reactor. It was caused by processes of an explosive type. The radioisotopic composition at this stage corresponded roughly to that of the irradiated fuel, but enriched by volatile isotopes of iodine, tellurium, caesium and inert gases. The second stage went on from 26 April to 2 May 1986. The release rate considerably falled and continued to decrease. During this period the composition of the radionuclides being released was again similar to that in the fuel. During this stage finely dispersed fuel was being carried out of the reactor by a flow of hot air, graphite combustion products and graphite dust which was formed due to the initial explosion. The third stage was marked by a sharp increase in the rate of radioactive material release. In the initial phase of this stage, the release was composed mainly of volotile components, especially iodine. But then the radionuclide composition again became similar to that of the irradiated fule. Features of this stage were determined by gradual heating fuel in the core due to the reactor after-heat. To the end of this stage the temperature exceeded 2000 °C (as it was estimated). This stage finished on 5 May.

In the last, fourth stage a rapid drop in releases ahd occured by about hundreds during one day (from 5 to 6 May) followed by further its gradual decrease to a very low level in the end of May. This was a consequence of the special measures taken on the accident mitigation.

From 30 April to 10 May with the use of helicopters about 5000 t of various materials were deposited on the core, among them:

TABLE 2.2. Radionuclide composition of the accident releases and of the ChNPP—4 core inventory*

Radionuclide		Half—life		Release activity, MCi 26.04.86	Release activity, MCi total** release	Core inventory, MCi 06.05.86**	Core inventory, MCi 26.04.86	Percentage of inventory released
85	Kr	10.8	years	—	0.9	0.9	0.9	100
133	Xi	5.27	days	5	45	45	170	—"—
131	I	8.05	days	4,5	7,3	37	86	20
132	Te	3.25	days	4	1.3	8.7	73	15
134	Cs	2.05	years	0.15	0.5	5.0	5.0	10
137	Cs	30.1	years	0.3	1.0	7.7	7.7	13
99	Mo	2.8	days	0.45	0.3	13	150	2.3
95	Zr	65.5	days	0.45	3.8	120	130	3.2
103	Ru	39.5	days	0.6	3.2	110	130	2.9
106	Ru	368	days	0.2	1.6	55	56	2.9
140	Ba	12.8	days	0.5	4.3	77	130	5.6
141	Ce	32.5	days	0.4	2.8	120	150	2.3
144	Ce	284	days	0.45	2.4	86	88	2.8
89	Sr	53	days	0.25	2.2	55	63	4.0
90	Sr	27.9	years	0.015	0.22	5.5	5.5	4.0
239	Np	2.35	days	2.7	1.2	38	720	3.2
238	Pu	86.3	years	$0.1.10^{-3}$	$0.8x10^{-3}$	$2.7.10^{-2}$	$2.7.10^{-2}$	3.0
239	Pu	$2.4.10^{4}$	years	$0.1.10^{-3}$	$0.7.10^{-3}$	$2.3.10^{-2}$	$2.3.10^{-2}$	3.0
240	Pu	$6.6.10^{3}$	years	$0.2.10^{-3}$	1.10^{-3}	$3.3.10^{-2}$	$3.3.10^{-2}$	3.0
241	Pu	13.2	years	0.02	0.14	4.7	4.7	3.0
242	Pu	$3.8.10^{5}$	years	$0.3.10^{-6}$	2.10^{-6}	$6.7.10^{-5}$	$6.7.10^{/5}$	3.0
242	Cm	164	days	$0.3.10^{-2}$	$2.1.10^{-2}$	0.7	0.7	3.0

* Data on the activity of the main radionuclides measured in the radiometric analysis
** Reduced from 26.04.86 to 06.05.86 taking into account radioactive decay.

boron carbide	40 t
dolomite	800 t
lead	2400 t
sand and clay	1760 t

From the composition of materials deposited it is not difficult to see primary purposes of this measure:

— to stop graphite burning,

— to insert additional neutron absorbers for excluding the possibility of the chain fission reaction,

— to ensure an additional heat extraction from a heated-up fuel due to melting of materials and endothermic chemical reactions,

— to reduce and finally to stop radioactive releases from the damaged reacotr.

Besides, the core was purged with nitrogen.

3. ChNPP—4 UNIT AND ITS CORE BEFORE THE ACCIDENT

The ChNPP—4 was put into operation in December 1983. By 26 April 1986 it worked for 865 days. The capacity load factor amounted to 0.82. Other data on the unit and its core are given in Table 3.1 and 3.2. They were taken both from the report [1] and from results of later investigations.

Most of the fuel assemblies (≅75%) were first load bundles with burn-up of 11—17 MWd/kg. From the more precise calculation data the average burn-up over the core was 10.9 MWd/kg. A slightly less value — 10.3 — was presented in [1]. Besides additional more precise calculations of radionuclides inventory in the ChNPP—4 fuel were carried out. Their results somewhat differ from those presented in Table 2.2. But for the most radionuclides these

TABLE 3.1. The ChNPP–4 UNIT CHARACTERISTICS ON 26.04.86

Characteristics	Units	Value
Type of reactor		RBMK–1000
Power		
– thermal	MW	3200
– electric	MW	1000
Time in operation	days	865
Core dimensions		
– height	m	7
– diameter	m	11.8
Graphite mass in the core and reflectors	t	2000
Initial enrichment	%	2.0
Burn-up		
– maximum (before discharge)*	MWd/kg	20
– average**	"	10.9
Number of fuel assemblies	pc.	1659
Fuel content (U) in an assembly	kg	114.7

* On 26.04.86 it was not reached
**On 26.04.86

TABLE 3.2. Calculated data for the ChNPP–4 reactor core on 26.04.86

Group of FA	Number of FA	Burn-up		Content, kg/t (U)*	
		MWd/FA	MWd*/kg(U)	^{235}U	$^{239-242}Pu$
I	721	1700–1500	13.7	8.4	3.8
2	392	1500–1300	12.3	9.4	3.5
3	154	1300–1100	10.5	11	3.2
4	101	1100–900	8.8	12	2.9
5	35	900–700	7.0	13	2.5
6	43	700–500	5.4	15	2.0
7	41	500–300	3.5	16	1.5
8	172	300–0	1.2	19	0.5
Total	1659		10.9**	9.7**	3.56**

* Average over FA group
**Average over the core (separately for ^{239}Pu, ^{240}Pu, ^{241}Pu and ^{242}Pu 2.35, 0.91, 0.24 and 0.055 kg/t respectively)
FA – fuel assembly

differences are not essential. It should be noted here that corrected results for ^{239}Np and ^{242}Cm: 1340 MCi and 0.4 MCi, respectively (instead of 720 MCi and 0.7 MCi in Table 2.2). Corrected data for plutonium are given in Table 3.2.

Radionuclieds ^{137}Cs and ^{134}Cs should be considered in more detail. Firstly, caesium attracts special attention due to its relatively great radioecological and radiobiological significance. Secondly, there are attempts by some specialists to use measured activity ratios of these radionuclides as an index of fuel burn-up in application to the concrete situation at the ChNPP–4.

Fig. 3.1 shows the dependence of ^{134}Cs and ^{137}Cs activity ratio R on a burn-up, obtained by calculations at the I. V. Kurchatov Institute of Atomic Energy. Here, for comparison, is given one of published calculated curves obtained by authors from FRG [11]. The last one lies slightly lower.

Using the measured values of R for radioactive fallouts in FRG and their calculated curve the authors [11] obtained the average burn-up in the ChNPP–4 fuel, being equal to 12.85 ± 0.15 MWd/kg.

It could be recognized as being rather close to the quoted corrected value 10.9: one should take into account the roughness of estimate [11] due to the non-homogeneity of burn-up in the ChNPP–4 reactor core and possible inaccuracy of the model calculation [11] of the R value in application to the RBMK–1000.

From the corrected data the content of ^{134}Cs and ^{137}Cs in the CNPP–4 reactor core was 3.8 and 7.2 MCi (2.9 and 83.5 kg), respectively, the average of R equals to 0.524. This differs slightly from data of the report [1].

By the moment of the accident about 680 kg of plutionium and trans-plutonium elements (among them 450 kg — ^{239}Pu) was accumulated in the ChNPP–4 fuel. Their total α–activity ($\cong$0.5MCi) was mainly determined by ^{242}Cm.

The ChNPP–4 unit worked at the full power up to 01.00 p.m. April 25. Then the staff began to reduce the reactor power in connection with a preparation to the planned maintance and test of the turbogenerator. Fig. 3.2 demonstrates the time dependence of power before the beginning of the accident. This time interval of approximately 24 hours duration didn't considerably affect the content of radionuclides in the reactor core which are interesting from the point of view of their radiological significance.

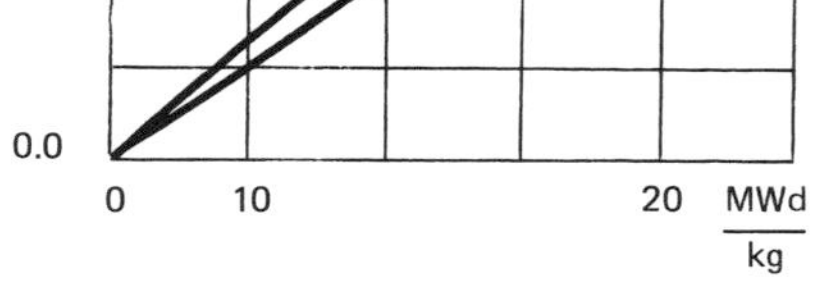

Fig. 3.1. Ratio R of radioactivities of ^{134}Cs and ^{137}Cs as a function of the burnup

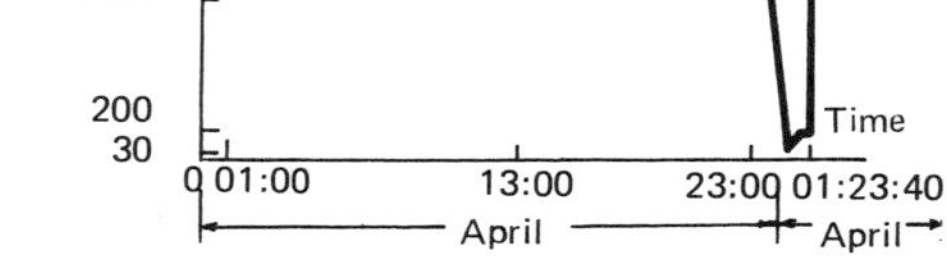

Fig. 3.2. The reactor power during the day before the accident

4. RADIOACTIVE CONTAMINATION OF ENVIRONMENT

4.1. General

As a result of the accident considerable territories of the BSSR, UkSSR, RSFS regions were subjected to radioactive contamination. Some amount of radioactive substances was transported outside the USSR territory. Contaminated air masses and radioactive fallouts were fixed in the West European countries as well as in many places of the Northern hemisphere [8].

The close-in and remote zones of radioactive contaminations were formed mainly from April 26 to May 8 1986. But the complete formation of the radioactive „trace" and individual „spots" continued during the whole May. Features of their formation were determined by:

– the dynamics and height of radioactive releases from the damaged reactor;

– meteorological conditions nearby the ChNPP and in remote regions in the direction of transport of air masses (the direction and strength of wind, precipitations etc.).

The following features [1, 9, 10] should be noted:

– radioactive releases of extremely high and variable intensity (hundreds of thousands of curie per hour) took place during 10 days;

– the high rise of radioactive plume above the damaged reactor for the first 2 days (according to aircraft measurements within the USSR, the plume height on 27 April exceeded 1200 m in the north-west direction at the 30-km distance from the ChNPP site), in subsequent days the plume height did not exceed 200-400 m;

– varying meteorological conditions during the period of action of the source with various directions of wind at different altitudes.

The simplified picture of the radioactive contamination of the European part of the USSR and West European countries is given in Fig. 4.1 [8]. One can see 3 plumes of contaminated air masses that spread over Europe: the north, west and south. Inital times of arrival of a contaminated air mass are indicated by figures 1–8. They mean numbers of days after the beginning of the accident: 1–26.04.86, 2–27.04.86, etc. The north plume has been

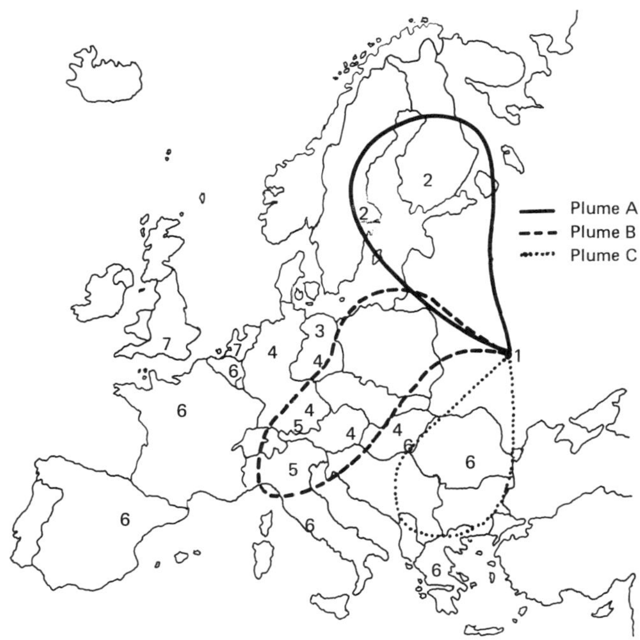

Fig. 4.1. Descriptive plume behaviour and reported initial arrival times of detectable activity in air. Plumes A, B and C correspond to air mass movements originating from Chernobyl on 26 April, 27–28 April, and 29–30 April, respectively. The number 1 to 8 indicate initial arrival times: 1 (26 April), 2 (27 April), 3 (28 April), 4 (29 April), 5 (30 April), 6 (1 May), 7 (2 May), 8 (3 May). UNSCEAR 1988 [8].

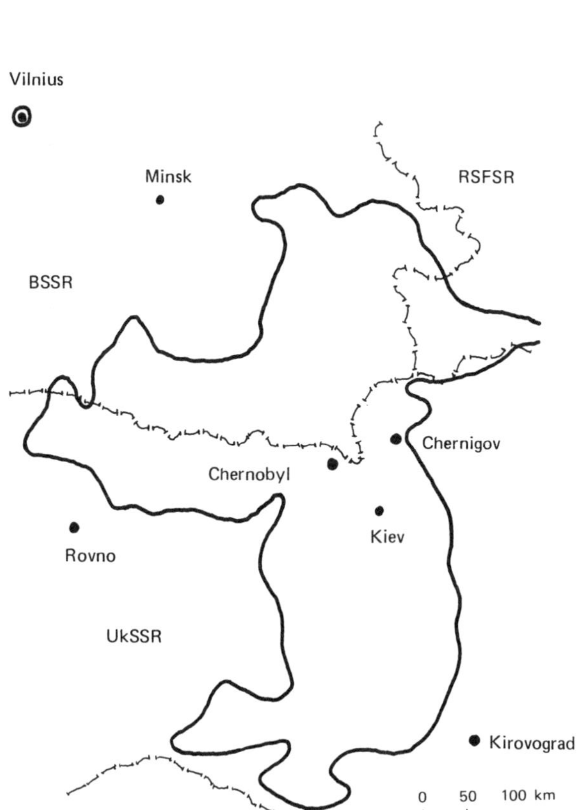

Fig. 4.2. Dose rate 0.05 mR/h isolevel on the USSR territory

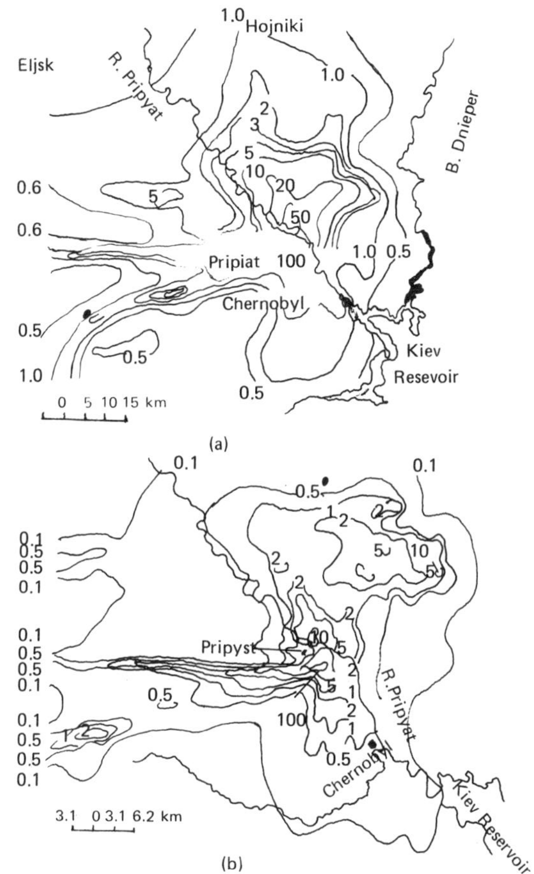

Fig. 4.3. Gamma-field distribution, mR/h, on 29 May 1986 (a) and 1 May 1987 (b)

formed mainly by the release due to the initial explosion to heights >1200 m.

During 1986 and 1987 the national and international organizations have made a great deal of work on investigation of the level and character of the radioactive contamination of natural media in the north hemisphere countries as well as of its possible radiobiological consequences.

The results of this investigations are used for various generalizing assessments and forecasts. One of the most complete works in this respect was carried out by UNSCEAR. Its results were published in the report for 1988 [8]. They are used in the present paper.

Especially great deal of work on investigating consequences of radioactive releases from the ChNPP–4 was done on the USSR territory by organizations of the USSR State Committee on Hydrometeorology, Ministry of Health, the USSR Ministry of Defence, the USSR State Committee on the Utilization of Atomic Energy, the USSR Ministry of Atomic Energy, the USSR Academy of Sciences, etc. [1, 9]. Monitoring of radioactive contamination of natural media on the regular basis is in progress up to now and will continue even for a long time, especially in zones of moderate and serious radioactive contamination.

This monitoring includes ground sampling from a contaminated territory with their subsequent gamma-beta-spectrometric and radiochemical analysis in laboratory conditions. By the spring 1989 beginning from May 1986 over 100000 samples had been analyzed [10]. The environmental radioactive contamination monitoring was added with research on radioecological, radiobiological and other problems.

The results of natural media radioactive contamination monitoring were used to draw maps of the USSR territory radioactive contamination. The first map of a close-in trace was prepared by 2 May 1986. Later on, radioactive contamination maps for the close-in and remote zones around the ChNPP were prepared including maps for individual radionuclides.

Some general picture on radioactive contamination of the USSR territory and its change in time can be got from Figures 4.2–4.5 and Table 4.1. On the gamma-field distribution map on 10.06.86 three radioactive traces are seen clearly: the south, west and north-north-east.

For the first weeks after the beginning of the accident the gamma-field maps in the close-in zone were used to evaluate roughly the total fallouts activity. In this case the following ratio was used between exposure dose rate above the earth surface and radioactive fallouts density: 1mR/h ↔ 100-200 Ci/km^2.

After 8 May 1986 the change in radioactive contamination was mainly determined by radioactive decay, wind transport, wash-out by rain and high waters, diffusion in a soil, etc.

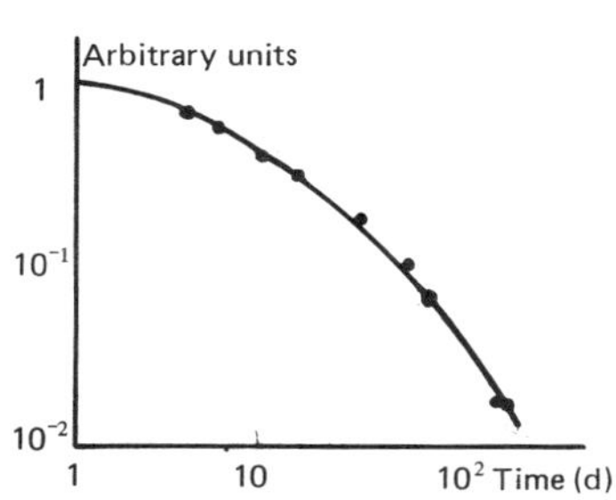

Fig. 4.4. Change in gamma dose rate from radioactive materials in the close-in zone as a function of time

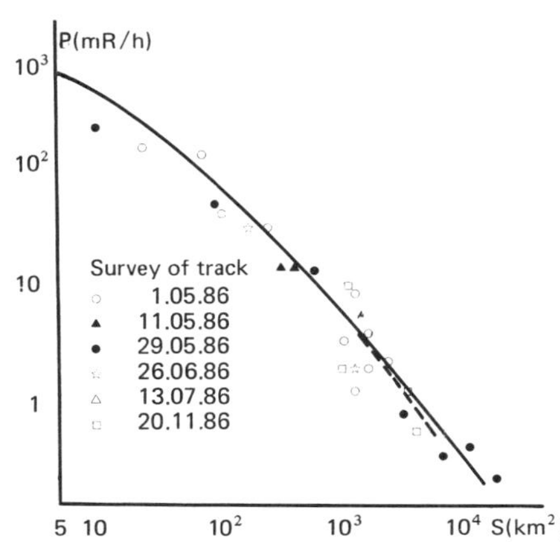

Fig. 4.5. Correlation between the area of radioactive deposits in the close-in zone and the dose rate isolevel (data reduced to 1986–05–29): full line, average measurements, broken line, calculated.

All measures have been taken where it was necessary and possible to prevent againt the horizontal transport of activity [1]. As it was shown from investigations during 1986—88, the horizontal migration of radionuclides was insignificant: radioactive contamination isolines changed mainly due to radioactive decay and the vertical migration in a soil [10].

In a year after the accident the areas limited by dose rate isolines $P = P_o = const$ were reduced by about two orders of magnitude.

TABLE 4.1. Areas (km^2)* of contaminated territories limited by isolines of dose rates $P = P_o = const$

P_o, mR/h	29.05.86	01.05.87
20	500	8
10	1000	20
5	2000	70
I	10000	500

*in rounded figures

For the first two weeks after the accident ^{131}I and some other relatively short-living radionuclides were of great importance.

By the middle of 1987 after tdecay of short-living radionuclides the most significant were caesium-134, 137, strontium and plutonium isotopes.

4.2. The detailed picture

Due to the above-mentioned features the picture of the environment contamination both in terms of activity level and radionuclides composition was and is very complicated. In many places of the close-in zone the radionuclide composition of fallouts approaches to its composition in the fuel. At distances exceeding 15—30 km the significant fractioning of radionuclides is observed. For instance, on the northern trace the considerable enrichment of fallouts by caesium-137 up to 10 and over times was found.

On the close-in trace the density of plutonium contamination is 0.1—1.0 Ci/km^2. Near the ChNPP site the density exceeding 10 Ci/km^2 is observed in some places. With the increase of distance from the ChNPP site the level of plutonium contamination reduces rapidly.

It should be noted that plutonium, strontinum and a number of other long-living and non-volatile radionuclides are present mainly in the fuel particles. This must be taken into account when evaluating radiobiological consequences.

These conclusions on the radionuclide composition of fallouts were obtained on the basis of radiometric and radiochemical studies of aerosol samples from air and soil samples, carried out until May 30, 1986. In paper [1] a considerable body of these research data is given. They kept fully their significance and agree with results of later investigations. Only corrections for a relative content of ^{134}Cs and ^{137}Cs in the ChNPP fuel should be introduced into tables 4.2 and 4.10 and renormalization to 100% should be made in two columns of Table 4.10 in the report [1].

„Hot particles"

„Hot particles" are one of the main forms of releases from the damaged reactor. Their fallouts have taken place mainly in the close-in zone. Some small amount of them was found in fallouts on other country territories. A number of works was published, dedicated to the analysis of the found „hot" particles, see, for example, [12—14], as well as [15]. In the USSR a significant attention is paid to the study of „hot particles". Samples of these particles were taken from the damaged unit rooms and close-in zones. The objective, of research in the USSR and abroad are the estimate and forecast of their radiobiological hazard and obtaining the information to describe processes in the damaged reactor during the accident:

Based on the results of studying particles extracted from samples of ground and specimens from the damaged units one can find two forms of fuel releases, differing in particle sizes and their radionuclide composition.

The first form contents relatively large (with diameter of tens of microns) fragments of a fuel matrix. The radionuclide composition is close somewhat to their composition in the fuel. In them there were identified radioisotopes of barium, cerium, zirconium, niobium etc. Depletion of caesium and ruthenium isotopes is observed in the majority of particles. Distribution of particle sizes is submitted to the logarithmic-normal law. The diameter median value decreases with the increase of a distance from the unit. Among large particles are present Halso so-called ,,ruthenium particles''. Their activity for the first weeks after the accident was determined mainly by ^{103}Ru, Later on-by ^{106}Ru being in the metal matrix (ruthenium, molibdenium with a touch of some of (ther metals). A fraction of ,,ruthenium particles'' in close fallouts does not exceed 10%.

The second form of the fuel release is presented by a finer fraction of particles with diameter of several microns. The size distribution of particles is also submitted to the approximately logarithmic-normal law. The radionuclide composition varies strongly from one particle to another. Data are available on the significant change in the initial fuel structure. The stoichiometric composition of uranium matrix in some particles is close to UO_3. In the particles were found iron, zirconium, silicate compounds. In γ-radiation spectra ^{60}Co is identified regularly. All this makes it possible to assume that the second form was developed while a fuel portion melting occured in the core.

In this group of particles ,,ruthenium'' particles were found as well, including ,,pure'' rutenium one's.

From the results of studying ground samples in close-in fallouts one san conclude that the fractions of released activity in the form of ,,hot'' particles of the both forms are values of the same order.

According to data available, the ,,hot particles'' found in fallouts beyond the USSR territory, belong to the second group (with diameter not exceeding several microns) [12–14].

An activity of individual ,,hot'' particles of the both forms lies over a wide range of values from tens to thousands of Bq (reducing to 26 August 1986). Single particles had activity up to 20000 Bq.

Caesium fallouts

Radioactive caesium (mainly, ^{137}Cs) deserves specific attention. Key problems of ensuring the radiation safety of population in a long-term plan have to do with it. It has a very complicated picture of fallouts:

— spot character;

— relatively large area.

The picture caesium fallouts (^{137}Cs) on the BSSR, RSFSR and UkSSR territories is shown in fig. 4.6 [10]. The trace of radioactive contamination with ^{137}Cs is spread to the north and partially to the west far beyond the 30-km zone. In this case, two large areas could be distinguished: one occupies the central and noth-west parts of the 30-km zone, partially beyond its boundary; the other one is located far to the north-north east from this zone, occupying parts of the Gomel, Mogilev regiones of the BSSR and Bryanskaya region of the RSFSR. In every of these areas there is its ovn complicated inherent structure. It is shown in fig. 4.6, b.c. There are numbers of separate spots, especially in the second area.

There are individual ,,spots'' of a moderate or weak contamination far beyond the main areas as well. On the map one can see such ,,spots'' to the nort from Orel-city, to the east from Belaya Tserkov-city, to the south from Vinnitsa-city, etc.

The total territory with contamination level (^{137}Cs) greater or equal to 15 Ci/km^2 amounts to about 10^4 km^2 (2.10^3 km^2 in the RSFSR; $1.5.10^3$ km^2 — in the Uk SSR, including $0.5.10^3$ km^2 in the evacuation zone; 7.10^3 km^2 — in the BSSR including 3.10^3 km^2 in the evacutaion zone).

For a contamination level within a range between 5 and 15Ci/km^2 the total area amounts to $2.1.10^4$ km^2.

It should be noted that densities of contamination with ^{137}Cs 15 and 40 Ci/km^2

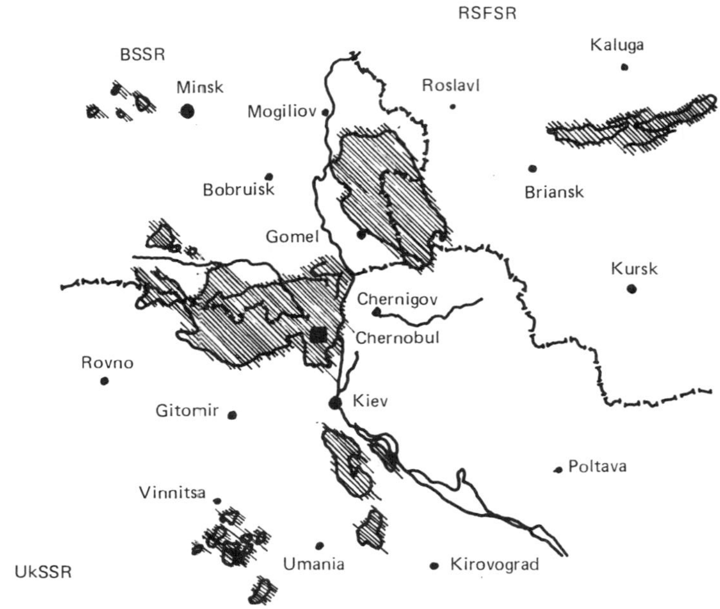

Fig. 4.6. ^{137}Cs fallout distribution on the territory of the USSR (a) and the detailed picture of it in two regions (b and c)

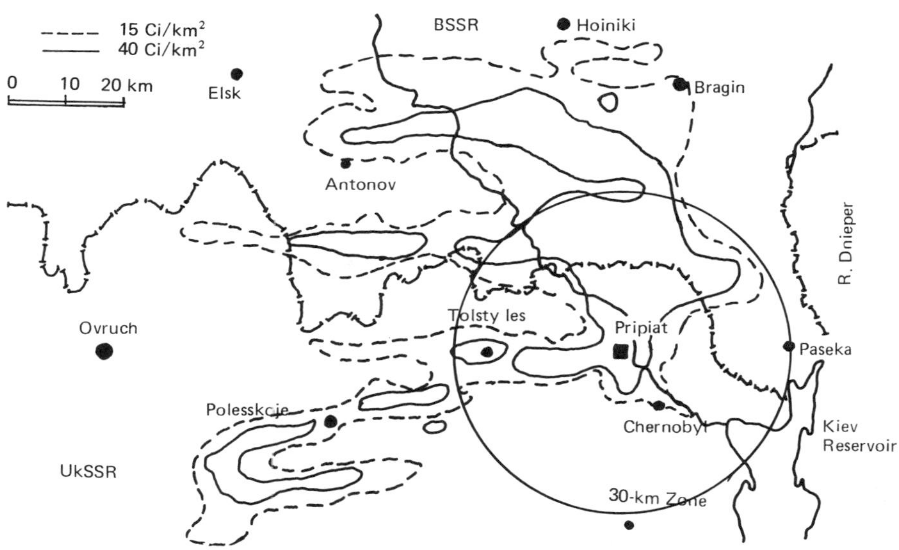

Fig. 4.6, b

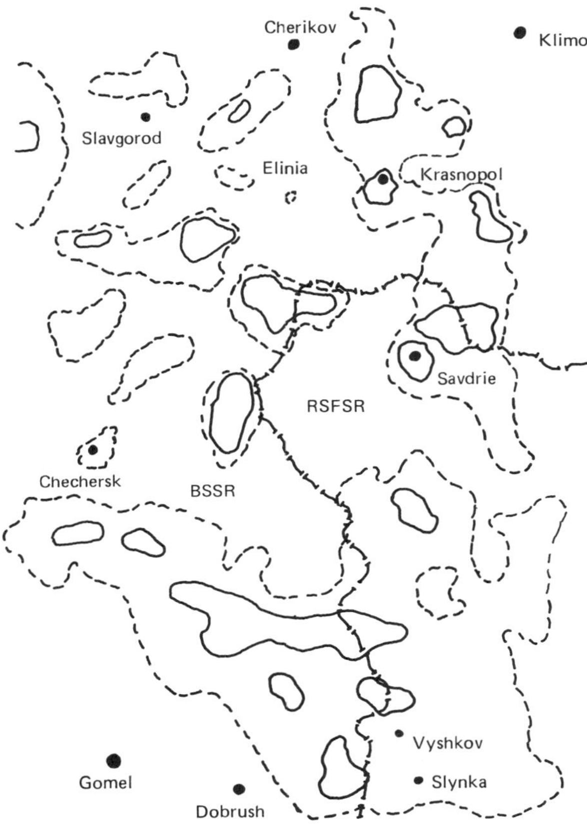

Fig. 4.6,c

are control levels in decision-making on ensuring the radiation safety of the population. On May 1986 the upper limit for ^{137}Cs contamination was established as 7 (and later on 15) Ci/km^2 *).

In regions with contamination density σ (^{137}Cs) over 40 Ci/km^2 the use of lands for agricultural production was prohibited. On lands with less contamination the agricultural production is permitted with some limitations and special rules.

According to full and revised data the total amount of ^{137}Cs fallenout on the USSR territory was eastimated by the USSR State Committee on Hydrometeorology specialists as 0.8 MCi [9].

Caesium fallouts of a relatively weak level took place in a number of European countries as well. The highest levels were fixed in Sweden: 2 Ci/km^2.

In maximum fallout places in Austria, Switzerland and FRG the σ value amounts to 1–1.5 Ci/km^2.

In the majority of parts of radioactive traces the contamination levels are much more lower than these values[8]. For comparison it should be noted that the contamination level (^{137}Cs) by globar fallouts as a result of nuclear tests comes up to $\cong$0.1 Ci/km^2.

The UNSCEAR Experts, based on the analysis of data on falouts in the northern hemisphere, have estimated the total release of ^{137}Cs from the ChNPP–4 damaged unit as

*3.0 and 0.1 Ci/km^2 for ^{90}Sr and $^{239,240}Pu$ respectively

approximately 1.9 MCi. Among them 42% they attributed to fallouts on the USSR territory; 37% — in Europe (without the USSR), 6% — in the Ocean and the rest part — in other regions of the Northern hemispehere.

Radioactive iodine

^{131}I is relatively short-living radionuclide. It disappeared rapidly — for some weeks — from contaminate areas. About this radionuclide there are only that data which was obtained till July 1986. News one could be only the results of its analysis. Here only the general result on ^{131}I from UNSCEAR report [8] obtained on the basis of radioactive fallout investigations in the Northern hemisphere is given. Total release of ^{131}I was estimated to be equal approximately 8.9 MCi (reduced to 06.05.89). It is rather close to the data of the report [1], if considerable uncertainties of both estimations would be taken into account.

5. PHYSICAL—CHEMICAL PROCESSES IN THE DAMAGED REACTOR

As it has already been mentioned, the direct measurements inside the unit after the beginning of the accident couldn't been performed. Physical-chemical processes of formation of radioactive releases were considered using only measurement data obtained outside the reactor. The last ones can be analyzed on the basis of available scientific knowledges on behavior of fission products, structural and other materials, as well as their compounds over a wide range of changes in temperature and pressure. They include also the results of a number of special experimental studies.

The results of one of such analyses are presented in the report by Ponomarev-Stepnoy N. N. and Khrulev A. A. at this seminar. We also note the detailed analysis [15] of Soviet report data [1] carried by D. Powers, T. Kress and M. Yankowski at the end of 1986. Their results were partially used here. Attention should be paid to proceedings of the symposium [16], especially, papers [17, 18].

Such analysis is extremely difficult due to specific features of the Chernobyl accident course and its radioactive releases. The last are characterized by:

- substantial duration (10 days);
- multistage;
- fast changes in time;
- the presence of relatively large amount of non-volatile fission products.

The release as a result of the initial explosion, the acceleration in the release five to nine days after the start of the accident and its sudden sharp decrease — almost by two orders of magnitude — between 10-th and 11-th days should be emphasized especially.

Such features of radioactive releases have not been considered before in any analysis of accidentes at NPP with light-water nuclear reactors.

Under these conditions the physical-chemical processes of formation of the radioactive releases during the accident at the ChNPP—4 could be considered only qualitatively.

At the initial stage of the accident the release formation can mainly be explained by physical processes:

- rapid heat-up and fragmentation of fuel;
- partial melting of fuel;
- interaction between molten fuel, water and structural materials;
- additional fragmentation of fuel particles in the process of release.

A substantial experimental material is available on the fuel behavior in such processes (see in [15, 16]). It could be used to analyze the inital stage of the accident and its radioactive release.

All the above-mentioned physical processes to either degree are responsible for a formation and release of „hot" particles, mainly, of the first fraction (with diameter d is equal to several tens of micrometers). In [1] the average energy deposition in fuel at the initial stage was estimated to be equal to 300 kal/g. Sizes of particles of this fraction seem to point to the fact that in some part of the fuel there was a considerably higher energy deposition.

The physical processes considered are also responsible for the release of some part of noble gases and volatile fission products.

In the irradiated fuel (UO_2) there are small metallic inclusions (d ≅5–50 μm) consisting, mainly, of ruthenium and molybdenum [15]. One can assume that ruthenium particles from the first fraction were formed by these inclusions when dispersal of the fuel occured.

Subsequent stages of the accident are characterized by more complicated processes of radioactive release formation.

The physical processes

– graphite burning;
– fuel heat-up;
– melting of structural materials and, possibly, fuel;
– evaporation of gases and volatile fission products from the fuel;
– sorption-desorption;
– the carrying-out of aerosols by a hot air flow;
– the filtration through material deposited on the core, etc.

are insufficient for an explanation of the radioactive releases on these stages.

For instance, in [15] with using the data on relative rates of various volatility class radionuclide release and evaporation models for radionuclide escaping from fule strongly inconsistent fuel temperature values were obtained: 1200–1400K, 1800–2300K and 2500–3000 K respectively from data on volatile, semi-volatile and non-volatile elements.

Chemical processes in fuel

It is evident that various chemical transformations played essential role in radioactive releases formation. The fuel, products of fission and neutron activation, graphite, structural materials, air oxygen and nitrogen as well as various substances contained in materials deposited on the core could take part in these transformations. Among them the following two principal chemical reactions should be noted:

– oxidation of fuel and fission products;
– fuel carburization.

The oxidation of fuel and fission products in the air flow passing through the damaged reactor core can explain, particularly, the ruthenium release.

In the form of oxides (RuO_3 and RuO_4) it converts to a gas phase at relatively low temperatures – within the range of 1300–1800 K. While passing through various parts of the damaged reactor core these gaseous compounds could come into a reduction zone and be reduced interacting with CO to metallic ruthenium with a formation of spherical ruthenium particle of micron sizes. The last ones were carried out from the damaged reactor by the air flow together with other aerosol particles.

This complex multiphase physical-chemical process can pretend to explain the presence of purely ruthenium „hot particles'' in radioactive fallouts.

Generally speaking, the role of oxidation process is limited in the presence of a large amount of graphite. But the space and time non-uniformity of physical-chemical processes, including the graphite burning, could create separate zones in the core with a strongly differentiated oxidation potential – from negative to positive values. As a result, for some elements an increased release would be possible to explain by the oxidation process, for some others – by the reduction process. In particular, the relatively extensive release of radioactive barium could not be explained by the action of the oxidation process only.

The essential role can be played by the process of fuel carburization, which in a simple way is presented by the following reaction:

$$UO_2 + 4C \rightleftarrows UC_2 + 2CO$$

The fuel carburization takes place when the fuel temperature exceeds 2040 K, and it is very sensitive to changes in temperature, and to changes in CO concentrations, as well. These properties of fuel carburization could give an explanation of the process of radioactive releases formation at the last two stages of the course of the accident [15].

Interaction with materials deposited on the core

Possible physical-chemical transformations of the deposited materials and their interactions with fission products are shown in fig. 5.1. It should be noted that here a small part of possible chemical reactions in such a multicomponent system is shown.

Te and J evaporated from fuel would react with a molten lead and produce the compounds PbTe and PbJ_2 melting at 1197 and 1353 K, respectively. The generation of such compounds may have slowed the release of these elements.

The boron carbide oxidation leads to the formation of boron oxide. The last one at an interaction with ceasium compounds may produce the salt $CsBO_2$, promoting in retention of this element. However, the interaction between boron oxide and CsJ may result in the iodine freeing.

Various components containing ceasium, strontium or barium, formed in the damaged reactor core, could interact with sand and clay, producing various silicate compounds. It promotes to retention of these elements. One may think that sand and clay were heated-up in their some part up to a sufficiently high temperature (over 900 K), in order the role of the mentioned chemical interaction could be essential. It should be noted that in lower rooms adjoining the reactor core, molten sand leakages were found. The silicate compounds were found in „hot particles" of the second form.

The most likely effect of deposited materials consisted in retention of aerosols evolved from the damaged reactor core due to filtration process through the relatively thick bed formed by these materials above the core.

5. CONCLUSIONS

In 1986—1988 a large complex of post-accident activities was carried out. Some of them directly or indirectly related to the radioactive releases from the damaged unit. Radioactive contamination maps on the USSR territory were prepared. Continuous monitoring has been organized. Of special interest are the detailed maps of caesium deposition.

A great deal of research has been performed. All this enabled us to better understand all aspects of radioactive releases from the ChNPP—4 accident.

The data of the report [1] related to both the radioactive releases and the radionuclide inventory of the ChNPP—4 core have been analized and improved. Some misprints

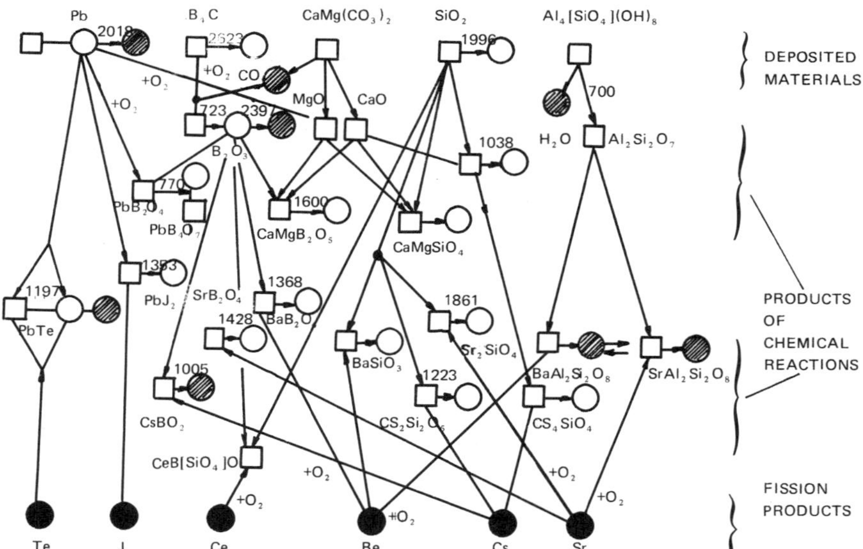

Fig. 5.1 Some possible chemical reactions between fission products and the deposited materials; □ — solid, ○ — liquid, ● — gas, 2018—transition temperature

and inaccuracies were eliminated. Data on certain nuclides have been slightly corrected: ^{131}I, $^{134,137}Cs$, transuranic elements etc.

UNSCEAR estimates of ^{137}Cs and ^{131}I releases are higher than in [1]. However these differences are within the limits of both estimates errors.

In a whole the results and conclusions of the report [1] remain valid. Understanding of the physical and chemical processes of the radioactive releases formation during the Chernobyl accident has been improved. However this understanding still remains qualitative. A more concrete description of these processes would be achieved after completion of in-reactor research. The last started in 1986 and continue now in accordance with the long--term program. One topic of these reseaches is connected with the fuel and other materials in the damaged unit including their physical and chemical conditions. Very important information can be obtained from analysis of samples taken out of the reactor rooms.

The issues of chemical processes in a damaged reactor are of great importance not only for reactors of RBMK type.

In the and of 1987 the USA Academy of Sciences in cooperation with other organizations organized a workshop on chemical processes and products in severe nuclear reactor accidents. Its goal was to review the current state of knowledge in this area and to elaborate recommendations on future reseach programs.

Conclusions and proposals of this workshop are contained in [19]. Its general conclusion is as follows: modern knowledge in this area having been considerably improved during the last decade nevertheless requires some further development.

The in-reactor studies are a part of a comprehensive long-term research program on all aspects of the Chernobyl accident.

This program includes also investigations of the radioactive contamination of the environment and its consequences:

– physical and chemical states,
– transport in the atmosphere and hydrosphere,
– migration in soil,
– caesium spots,
– „hot" particles,
– migration through ecological and food chains,
– radioecological effects,
– radiobiological effects,
– modelling and development of computer codes.

All these topics may be covered by international bilateral or multilateral cooperation. Proposals along these lines are being developed. Some of them have already been included in the USSR–USA bilateral cooperative program.

REFERENCES

1. USSR State Committee on the Utilization of Atomic Energy, The Accident at the Chernobyl NPP and its Consequences (Information compiled for the IAEA Post-Accident Review Meeting, Vienna, 25–29 August 1986), IAEA, Vienna, August 1986, Part I and II.
2. Asmolov, V. G., Borovoj, A. A., Demin, V. F. et al., the Accident at the Chernobyl NPP' one year after, Proc. Int. Conf. on Nuclear Power Performance and Safety, Vienna, 28 September–2 October 1987, IAEA, Vienna, 1988, v. 3, pp. 103–147.
3. Summary Report on the Post-Accident Review Meeting on the Chernobyl Accident, INSAG, IAEA, Vienna, 1986, Safety series, No. 75–INSAG–1.
4. Nuclear Power Performance and Safety, Proc. Int. Conf., Vienna, 28 September – 2 October 1987, IAEA, Vienna, 1988, v. 3.
5. Medical Aspects of the Chernobyl Accident, Proc. Conf., Kiev, 11–13 May 1988, Kiev, Sdorovia, 1988 (in Russian).
6. Beninson, D. And Lindell, B., Chernobyl Reactor Accident, WHO Report ICP/CEH–129 (PB86–195740), NTIS, May 1986.
7. The Accident at the Chernobyl NPP, report of US DOE, EPRI, EPA, FEMA, INPO, NRC, NUREG–1250, January 1987.
8. Sources, Effects and Risks of Ionizing Radiation, UNSCEAR 1988 Report to the General Assembly, UN, New York, 1988.

9. Radioactive contamination of the environment in the Chernobyl emergency zone, USSR State Committee for Hydrometeorology and Control of Natural Environment, Moscow, 1987.
10. Izrael, Yu. A., Chernobyl: Prognostic on the Future, newspaper „Pravda", Moscow, N 79, 20 March 1989, p. 4 (in Russian).
11. Kichner, G., and Noack, C. C., Core History and Nuclied Inventory of the Chernobyl Core at the Time of the Accident, Nuclear Safety, v. 29, N 1, pp. 1–5, 1988.
12. Devell, L., Tovedal, H., Bergström, U., Appelgran, A., Chyssler, J. and Andersson, L., Initial Observations of Fallout from the Reactor Accident at Chernobyl, Nature, v. 321, 15 May 1986, pp. 192, 193.
13. Raunemea, T., Lehtinen, S., Saari, H. and Kulmala, M., 2–10 m Sized Hot Particles in Chernobyl Fallout to Finland, J. Aerosol Sci., v. 18, N 16, pp. 693–696, 1987.
14. Hofmann, W., Crawford–Brown, D. J. and Martonen, T. B. The Radiological Singnificance of Beta Emitting Hot Particles Released from the Chernobyl Nuclear Power Plant, Radiation Protection Dosimetry, v. 22, N 3, pp. 149–157, 1988.
15. Powers, D. A., Kress, T. S. and Jankowski, M. W., The Chernobyl Source Term, Nuclear Safety, v. 28, N 1, pp. 10–28, 1987.
16. Proc. of the Symp. On Chemical Phenomena Associated with Radioactivity Releases During Severe Nuclear Plant Accidents, Toronto, Canada, June 1988, US NRC, Washinton, D. C., 1988.
17. Garisto, F., Thermodynamic Behavior of Fission Products at High Temperatures: Ruthenium and Tellurium, in 16.
18. Cox, D. S., Iglesias F. C., Hunt, C. E. L., Keller, N. A., Barrand, R. D., Mitchell, J. R., O'Connor, R. F., Oxidation of UO_2 in Air and Steam with Revelance to Fission Product Releases, in 16.
19. Chemical Processes and Products in Severe Nuclear Reactor Accidents, Report of a Workshop, National Academy Press, Washington, D. C., 1988.

Effect of the Annealing Temperature on Escape of Metal Fission Products from Different Media (Features of Experimental Data Analysis)

N. N. PONOMAREV-STEPNOI and A. A. KHRULEV
I. V. Kurchatov Institute of Atomic Energy
123182, Moscow, USSR

Abstract

The methodical issues concerning analysis of the experimental data on fission product (FP) release from irradiated fuel are considered to derive constants required for predicting FP behaviour in accident situations. The features of the analysis are compared in terms of the diffusion and activation models of FP behaviour as well as of the one-group FP release model used for the analysis of accident situations.

In the analysis of FP release from the fuel in accident situations the authors use the activation FP release model where the release rate is meant as the superposition of FP releases from two groups of defects with the activation energies

$$\varepsilon > K_oT \quad \text{and} \quad \varepsilon < K_oT$$

the fraction of the products in each group being determined by the spectral characteristic "m" The relation between the basic constants of the diffusion one-group and activation models of FP behaviour is established.

On example of the amalysis of the experimental data on Cs release in heating irradiated fuel it is shown that use of the traditional models for determination of the radionuclide release in the accident may lead to inaccurate conclusions on predicting the evolution of the radionuclide release and quantitative estimates of the fission product balance in the damaged reactor.

The release evolution in a severe NPP accident is analyzed in terms of the activation model of FP behaviour in the fuel.

A two-stage character of the release is substantiated. For determination of temperature fields in the fuel and in the reactor allowance is necessary for transfer of FP, sources of the radioactive decay energy.

Specific features of an express method for evaluating quantitative characteristics, radionuclide composition, and release dynamics basing on the analysis of characteristic nuclide fractionation coefficient are considered. Use of the model in predicting the dynamics of the radionuclide release at the Chernobyl-4 accident realized at the accident initial stage is discussed.

The results are presented and the explanation is given of some features typical of an accidental release radionuclide composition which cause the presence of cesium spots and in fall-outs, "hot" particle etc.

Proposals are made on organizing the system of the initial express evaluation of the accident parameters, and also on the sampling methodology .

Introduction

Safe operation is the most important requirement in development and construction of the nuclear power plant. The main safety aspects under nominal and accidental operation conditions of the plant are:

- prediction of the dynamics, quantity and radionuclide composition of fission products (FP) contained in the release;
- prediction of behavior of FP remaining inside the damaged reactor.

The Chernobyl events and earlier TMI accident have made the problem even more acute, because they showed that occurence of severe accidents is possible.

Decisive in the safety analysis is the problem of FP behavior in the fuel, fuel elements and other media containing fuel and FP. It appears that the most important feature characteristic of the fuel state under accidental conditions is the temperature instability. Therefore the physical model of FP behavior, which is used, must describe in the best possible way the FP release under non-stationary temperature conditions. A number of models describing FP behavior in the fuel are available. The concentration diffusion model is the most used one; in the analysis of experiments modelling accidental conditions the approach is used where the release is presented as the first-order reaction /1,2/.

At the I.V.Kurchatov Institute of Atomic Energy the phenomenological activation model, representing FP behavior in the fuel, is used /3,4/ (Table I).

1. ACTIVATION MODEL OF FP RELEASE /3,4/

The activation model of FP release has been developed on the basis of the analysis of numerous experiments on investigation of the temperature dependence of FP release from various materials (Table 2). Release of FP from the gaseous phase under the experiment conditions was studied; the typical character of volatile FP release is given in Fig. 1. Theve ave two region chavacterising two rate of release (a), release of FP from irradirtion fuel at the temperature T_{irr} below is small, and rise at the move high temperature; the same at the cycle annealing. Also it was shown experimentally that the main features of release are the same for all radionuclides investigated.

In the activation model all FP in the fuel (fiel or other materials containing FP) are supposed to be in the defects of the substance, which may be divided into two groups.

The first group of defects containing FP is characterized by the activation energy $\mathcal{E} < K_0 T_{irr}$, while the second one has $\mathcal{E} > K_0 T_{irr}$, where K_0 is the Boltzman constant, T_{irr} is the working temperature of the fuel (Fig. 2).

The FP quantity in each group is determined at the temperature T_{irr} by the so-called spectral characteristic "m" which is the fraction of FP release from the fuel after escaping from the defects at the activation energy $\mathcal{E} < K_0 T_{irr}$, at heating at the temperature T_{irr}. The FP fraction remaining in the fuel is (1-m).

Each FP group is characterized by its own constant release - α_q, α_s and $\alpha_s << \alpha_q$. The lifetime of FP of each group is equal to $\tau_q = 1/\alpha_q$, $\tau_s = 1/\alpha_s$. The experimental studies show that the lifetimes differ by 10^2-10^4 times depending on the temperature.

The main relations of the activation model are given in Table 3 and Fig. 1. Other important processes leading to FP release are effects occuring in the zone of fission fragment track (of the knock-out type) as well as the processes proceeding in the temperature gradient field, such as heat stress, may move FP together with the defects. Additional effects can be taken into account by introducing appropriate additional components in the equation of FP balance in the fuel.

Let us compare the two main models used in treatment of the experimental data obtained in modelling the accidents:

- FP release from the fuel is determined by the first order reaction $R = -kA$
- FP release occurs by the activation model

$$R = -\sum_{q,s} \alpha_i A_i$$

The difference between these two approaches is illustrated in Fig. 3.

In carrying out the experiments on studying release of FP (particularly non-volatile) from the fuel the annealing of the irradiated fuel element (fuel) is used as the main investigation method, when the residual FP content is periodically measured by the gamma-spectrometric methods A_j. Since in the annealing experiments the analysis of FP with long half-lives is usually used, the FP content in the fuel during the experiment changes mostly due to FP release in annealing. Thus the value of the residual content of the radionuclide under investigation is meausred, the value $1-F = A_j/A_c$ (τ=0) is calculated (F is the relative release). The release constants are determined from the analysis of the residual content value. The basic relations for determination of the FP release parameters in different models are presented in Table 4. Table 5 lists the relations determining the interaction of the parameters of different models. It follows from the relations presented that the release velocity α_q is essentially higher, particularly at relatively low temperatures, than K.

Fig. 4 presents the experimental data on the release constants Cs and J, obtained in the USA by the accident condition investigation program /2/ and the data recalculated by relation (1, Table 5) using the domestic experimental data on the spectral characteristic "m" value obtained in experiments on annealing of the irradiated dioxide. The results of α_q determination obtained from the same experiments are also shown there.

It follows from comparison of the results that use of the one-group model for evaluation of FP release from the fuel in fast processes will result in underestimation of the FP output and in some other dynamics of radionuclide release from the damaged reactor than in the activation(two group) model application. In terms of the activation model the radionuclide release will be essentially higher and, hence, the temperature of fuel overheating in the damaged reactor as well as the probability of the under-reactor structure melting will be lower.

Thus, the literature data on the constants of the diffusion and one-group models, characterizing FP release from the fuel in accidental situations require correction.

2. BYHAVIOR OF FP IN THE FUEL DURING THE ACCIDENTS

Fig. 5 presents the scheme of fuel temperature variations and release dynamics during the accident, which is based on the activation model of FP behavior. The release dynamics has some main stages.

First stage: the release is connected with increase in the fuel temperature due to the self-sustaining reaction (SSR) (or to mechanical release of FP together with the fuel in explosion). The quantity of the thermoinitiated release is characherized by the value of the spectral characteristic $m(T_I)_{max} - m(T_{irr})$.

Second stage: the release dynamics and quantity are determined by variations of the fuel temperature due to change energy of FP accumulated in the fuel during the reactor operetion. In the beginning of the second stage FP release from the fuel decreases rapidly because of fuel cooling due to heat transfer to the surrounding structures; when the temperature stabilives over the reactor volume and possibly increases in the core, secondary FP release can occur if the fuel temperature exceeds that of the first stage. The value of the second release in characterized by the value of the spectral characteristic difference $F \approx \Delta m \approx m(T_{II}) - m(T_I)$

After the maximum temperature of the second stage is reached FP release from the fuel decreases rapidly and its value is determined by the constant release value

$$R \sim \alpha_S(T) \cdot A(\tau)$$

The calculation analysis of FP behavior and temperarure in the fuel in the core and reactor materials reveals that the following foatures important for the accident analysis must be taken into account:

- redistribution of FP and, hence, transfer of FP radioactive decay energy over the fuel volume; FP release from the zone with high temperature and condensate in the zone with lower temperature the process enables equalization of the fuel temperature and limitation of its maximum value (Fig. 6), appearance of local microzones with higher concentrations of some radionuclides ("quantization"); the process is more efficient in fuels having a low thermal conductivity and may proceed both under the accidental and normal operation conditions of the fuel; in the samples taken in the places of fall-out of dispersed fuel released from the reactor as a result of an accident, microparticles with very high concentrations of some radionuclides may be observed (Fig. 7) particularly in the zones of low activity;

- similar redistribution of FP, energy transfer and temperature equalization may take place over the core and the entire reactor; FP will concentrate in the zones with the best cooling; the maximum core temperatures will be determined by the temperature at which an intense release of radionuclides, the least volatile at high temperature such as Ce, Zr, Nb, Ru takes place;

- FP release is affected by chemical processes: interactions of fuel with oxygen, graphite etc., resulting in FP release from the fuel, for example, use of UO_2 leads to carbidization and oxidation up to the highest oxide; as a result of these reactions FP releases may contain non-volatile radionuclides such as Ce, Zr, Ru (together with volatile ones) at relatively low temperatures.

3. METHODS FOR INITIAL (EXPRESS) ESTIMATION OF THE MAIN ACCIDENT PARAMETERS

The following methods can be used:

- analysis of dose fields of radioactive fall-outs;
- analysis of dose fields near and inside the reactor unit;
- analysis of the radionuclide composition of fall-outs.

The restriction of the first method is that knowledge is necessary of detailed distribution of activtty over the contaminated territory in the local and global scales. The work can only be done on completion of the accidental release.

The restriction of the second method is gamma-ray self-absorption in materials containing FP and FP transfer process leading to separation of FP from the fuel. Therefore it is difficult in practice to establish the quantitative relation between the gamma-field and the fuel.

In the third method the fractionation coefficient is analyzed

$$A_K^i = \left(A_i / A_r \right)_{exp} \cdot \left(A_i / A_R \right)^{-1}_{cal}$$

where $A_{i\ exp}$, $A_{i\ cal}$ are the measured values activity of the radionuclide activity "i" in the sample and the mean calculation value in the reactor at the time of accident onset; $A_{r\ exp}$, $A_{r\ cal}$ are the same values for the reference (standard) radionuclide.

The analysis of the fractionation coefficient may give useful information on the first stage of accident development, basing on which possible dynamics of the release at the second stage can be predicted (Table 6).

As the reference radionuclide one most stable at high temperatures is chosen, which is produced only in fission, e.g. Ce.

For the analysis samples of high activity fall-outs in the direction of the main cloud trace must be used. Low-activity samples are not representative: their activity may be determined by one (of several) microparticles with the may of radionuclide accumulation characteristic of this very particle.

Tables 7, 8 present information on a method for using the fracitonation coefficient for the analysis of the fuel state during the first stsge of accident; temperature conditions and quantitative release estimation. In the analysis the data on the temperature dependence of the spectral characteristic "m" obtained in annealing of the irradiated fuel (under laboratory conditions) are used (e.g. Fig. 8). Some additionsl information on the above parameters can be obtained in the quantitative analysis of the content of activation radionuclides, e.g. Zr-95, as well as in measuring the residual content of Kr-85 in fall-out samples (e.g. in annealing of samples containing some non-volatile radionuclides (Ce, Zr etc) whose presence is indicative of the presence of fuel in the samples.

4. USE OF THE TWO-STAGE RELEASE MODEL IN THE ANALYSIS OF THE CHERNOBYL NPP ACCIDENT

The model proposed was used in the analysis of the parameters of the Chernobyl accident release.

The initial information: in samples, taken in 3-4 days after the beginning of the accident at a distance of about 5 km from the damaged reactor, almost-all radionuclides characteristic of the reacotr (Cs, I, Zr, Mo, Ce, Np, Ba, Ru etc) were identified by the cloud trace, the relations being $A_K^{Cs}=3-4$; $A_K^{I}=7-8$; $A_K^{Zr}\approx 1.5$ and the rest $A_K^{I}\approx 1$ /5/.

The analysis of air sampled revealed reduction of the release from the damaged reactor unit; sampled are enriched by volatile radionuclides /5/. The residual content of Kr-85 in the fall-out sample was about 0.75.

The results of the preliminary analysis are listed in Tables 9, 10. Fig. 9 shows the results of estimation of the expected dynamics of volatile radionuclide release, obtained in the same days assuming that the core heating-up rate is (5-10) degree/hr. (At high heating-up rates increase in the release would have begin earlier, before 30.04). The estimates showed the possibility of an additional release of radionuclides. The consequent observations and analysis confirmed the assumptions made (Fig. 9). After May 3 in addition to the release of volatile radionuclides (Cs, I, Te) that of high-melting radionuclides (Ce, Ru, Zr). was observed (Table 10) /5/, which could be accounted for by chemical processes of graphite-oxygen interaction. However a rapid decrease in the amoune of radioactive releases after the assumed attainment of the maximum temperature as well as a high fractionation coefficient of I are indicative of an insignificant contribution of chemical processes to the total release. It appears that the oxygen interacted mostly with the stacking graphite; the carbidization process was restricted because of a relatively low temperature and limited contacr of uranium dioxide with graphite.

Conclusions

1. A two stage model of FP release in severe (hypothetic) accident based on the activation model of fission product behavior in fuel is proposed.

2. The method for express analysis of the accident parameters based on the analysis of the radionuclide composition of fall-outs is proposed.

3. The model has been used for analyzing the release dynamics during the accident at the Chernobyl NPP-4; it permitted the main release features to be explained and an additional activity release from the damaged reactor unit to be predicted.

4. It is shown that in analyzing the FP balance and the fuel and core temperature fields the transfer of the FP souces of radioactive decay energy should be taken into consideration.

Autors express gratitude to Vedenov A.A. and t. Ivanov O.P. for theis help in the analises of the results.

References

1. C.D.Andriesse. Interpretation of Fission Producr Release from overheated Fuel. IAEA Columbus USA 1985.

2. Source Term Evaluation for Accident Conditions. Proceeding of an International Symposium on Source Term Evaluation for Accident Conditions IAEA, Vienna, 1986.

3. N.N.Ponomarev-Stepnoy and A.A.Khroulev. Release of Products from High Temperature Fuel Materials. III Int. Conf. on Thermoionic Electrical Power Generation. Julich, 1972.

4. A.O.Gol'tsev and A.A.Khroulev. Calculation investigation of the influence of various factors on HTGR primary circuit activity-VANT ser. AVF-T N 2, 1986.

5. The Chernobyl NPP accident and its consequences Parts 1,2. Information presented at the IAEA Expert Meeting (August 25-29, 1986, Vienna GK IAE, USSR, 1986).

TABLE I

FP BALANCE IN FUEL

Characteristic	Model		
	diffusion	one group	activation
Leakage rate R	$-\mathcal{D}' \frac{\partial C}{\partial r}\Big\vert_{r=a}$	$-KA$	$-\sum_{\delta, M} \alpha_i A_i$
Mass balance equation	$\nabla(\mathcal{D}' \nabla C) + B - \lambda C \pm \dots$	$B - KA - \lambda A \pm \dots$	$A = A_\delta + A_M$ $mB - \alpha_\delta A_\delta - \lambda A_\delta \pm \dots$ $(1-m)B - \alpha_M A_M - \lambda A_M \pm \dots$

$\mathcal{D}' = \mathcal{D}/a^2$

a is the characteristic size
D is the diffusion coefficient
C is the FP concentration

TABLE 2

MAIN RADIONUCLIDES, FUEL MATERIAL, TEST CONDITIONS INVESTIGATED

Materials investigated	
Fuel	UO_2, UC, UC_2, UN, UP, UO_2+UC, UO_2 + C, UC_2 + C etc.
Structural	Graphite, stainless steel, tungsten, molybdenum alloys, sand (soil)
Fuel elements	Rod, spherical
Coated particles	Oxide and carbide cores with multilayer PyC-SiC-PyC and PyC-PyC coatings
Irradiation conditions	
burnup (fluence)	10^{13} n/cm^2-20% FIMA
temperature	20°C up to 2100÷2200°C
Annealing conditions	
medium	noble gas (He, Ar), air
temperature	20°C up to melting temperatures
Radionuclides investigated	Xe, Kr, I, Cs, Sr, Ce, Zr, Ru, Co.

TABLE 3

THE ACTIVATION MODEL OF GASEOUS FP LEAKAGE

Characteristic	Activation model
Relative leakage	
Solution for spherical source	
<u>Annealing of fuel under irradiation</u>	
<u>Stationary conditions</u> short-lived radionuclides	$m = k m_0 (\lambda_0/\lambda) \exp[-\frac{\varepsilon_a}{K_0}(\frac{1}{T}-\frac{1}{T_0})]$ for Xe $K=1$, Kr $K=2$
long-lived (stable) radionuclides	$m + \frac{\alpha'_M}{\lambda+\alpha'_M}(1-m)$
<u>Nonstationary conditions</u> short-lived radionuclides	$m f(\alpha'_\delta \tau) + (1-m) f(\alpha'_M \tau)$ $f(\alpha'_i \tau) = 1 - \frac{1-\exp(-\alpha'_i\tau)}{1-\exp(-\lambda\tau)} \cdot \frac{\lambda \exp(-\lambda\tau)}{\lambda+\alpha'_i}$
long-lived (stable) radionuclides	$m + (1-m)[1 - \frac{1-\exp(-\alpha'_M \tau)}{\alpha'_M \tau}]$
<u>Annealing of irradiated fuel</u>	
Nonstationary conditions, T	$m(T)\cdot f(\alpha_\delta \tau) + (1-m)\cdot f(\alpha_M \tau)$ $f(\alpha_i \tau) = 1 - \exp(-\alpha_i \tau)$ at $\tau > 1/\alpha_\delta$ $m(T) + [1-m(T)][1-\exp(-\alpha_M \tau)]$
Effect of irradiation intensity	$\alpha'_i = \alpha_i + \sigma\phi$

TABLE 4

MAIN PARAMETERS OF FP BEHAVIOR MODELS

Model	Main relations
Diffusion $D'=D/a^2$	$[1-A(\tau)]^2 \cdot A_0^{-2} \cdot \tau^{-1}$
One group, K	$-\dfrac{\ln[A(\tau)\cdot A_0^{-1}]}{\tau}$
Activation (two-group), m	$1-A(\tau_1)\cdot A_0^{-1}$ at $\tau_1 \sim \alpha_\delta^{-1}$
α_δ	$-\tau^{-1}\ln[A(\tau_1)\cdot(mA_0)^{-1}]$
α_M	$-(\tau_2-\tau_1)^{-1}\ln[A(\tau_2)\cdot A_0^{-1}(1-m)^{-1}]$ at $\tau_2 \gg \dfrac{1}{\alpha_\delta}$

TABLE 5

INTERACTION BETWEEN THE PARAMETERS OF THE ACTIVATION AND OTHER MODELS OF FP BEHAVIOR

Model	Main relations
Diffusion	$F \simeq m \sim \sqrt{D'\tau}$ $\varepsilon_D \simeq 2\varepsilon_a$ at $\dfrac{1}{\alpha_\delta} < \tau < \dfrac{1}{\alpha_M}$
One group	$m = \dfrac{\exp(-\alpha_M\tau)-\exp(-k\tau)}{\exp(-\alpha_M\tau)-\exp(-\alpha_\delta\tau)} \simeq \dfrac{1-\exp(-k\tau)}{1-\exp(-\alpha_\delta\tau)} \simeq \dfrac{k}{\alpha_\delta} < 1$ at $\tau < \alpha_M^{-1}$ at $\alpha_\delta\tau < 1$ (1)

TABLE 6

Fraction coefficient

$$A_K^i = (A_i/A_p)_{exp} \cdot (A_i/A_p)_{coe}^{-1}$$

stable to the fall-out zone features

stable to calculation of radionuclide activity in the reactor

Requirements to the sample

high activity in the direction of the main cloud trace from the zone of highest contamination

Reference radionuclide

the least movable at high temperatures

formed only in fission, e.g. Ce.

TABLE 7

FRACTIONATION COEFFICIENT AND FUEL STATE

Radionuclides in sample		Characteristics of fuel release			
		carrier	temperature	release mechanism	amount and place of fall-out
1 ± 0.2	holds for all	fuel particle	$T \sim T_{oper}$	mechanical processes	anyone near NPP
	does not hold for all	free radionuclides (gas, phase, aerosols)	$T_{eff} > T_{melt}$	SCR, fuel dissociation	largest part of fuel and FP far away from NPP
	holds for refractory non-volatile				
$A_k^i < I$	for I, Cs Xe, Kr	fuel particles	$T > T_{oper}$ (see Table 8)	explosion with heating of fuel	partly fuel and FP release (see Table 8)
$A_k > I$	for I, Cs	fuel particles + free radionuclides			
if $A_k^I > A_k^{Cs}$		(gas, phase, aerosols)	T< 1800°C		
if $A_k^I \sim A_k^{Cs}$			T> 1800°C		

TABLE 8

PARAMETERS CHARACTERISTIC OF A DAMAGED REACTOR

Fuel temperature

$$T_2 = \frac{T_{melt} \cdot \mathcal{E}_a}{\mathcal{E}_a - K_o T_{melt} \ln(1 - A_k^{J,Cs})}$$

Fuel release

$$\frac{\Delta G}{G_o} \leq K \left\{ (A^{J,Cs} - 1) \exp\left[\left(\frac{1}{T_2} - \frac{1}{T_{melt}} \right) \frac{\mathcal{E}_n}{K_o} \right] \right\}^{-1}$$

at T $\gtrsim$ 1800°C

activation energy $\mathcal{E}_a^{UO_2}$ = (2.3 2.9).10^{-19}

$$\Delta G / G_o = n K / A_n^{Zr} - 1$$

n - is the fraction of activation Z_r in the reactor (assumed to be ~ 5%)

k - is the fraction of radionuclide escaped out of the reactor building

TABLE 9

Radionuclide composition of samples

A_K^{Cs}	$= 3-4$	within 3-4 days after accident
A_K^{J}	$= 7-8$	distance $\sim$ 5 km
A_K^{Zr}	$= 1,5$	
A_K^{Kr}	$= 0,75$	
A_K^{i}	$= I$	

Conclusions

Explosion with dispersion of fuel (or its part)

Effective temperature of remaining fuel	up to 1800°C
Temperature of released fuel	$\sim$ 2000°C
Fraction of released	insignificant
fuel	3-6%
additional radionuclide	
release should be expected	

TABLE 10

RELATIVE CONTENT OF RADIONUCLIDES IN THE ATMOSPHERE ABOVE THE CHERNOBYL-4 . FRACTIONATION COEFFICIENT VALUE

Radionuclide	29.04	2.05	3.05	4.05	5.06	9.05	11.05	13.05	24.05
Zr-95	2	1,3	0.5	1.1	1.5	0.77	0.85	1.4	1.8
Ru-103	1.02	0.6	6	2	1.2	5	4	2.1	5.5
Ru-106	0.96	3.3	5	-	1.5	-	-	-	-
I-131	4.2	2	50	4	2.5	30	25	15	23
Te-132	10	4	70	-	6.3	-	-	-	-
Cs-134	5.5	2	31	2.1	-	-	5		2
Cs-137	3.5	3	43	3.3	2.6	-	13	12	3.3
Ba-140	1.4	1.4	3.3	3.3	1.5	-	-	-	-
Ce-141	0.92		0.8	I	I	0.8	0.91	0.92	-
Ce-144	1		I	I	I	I	I	I	I

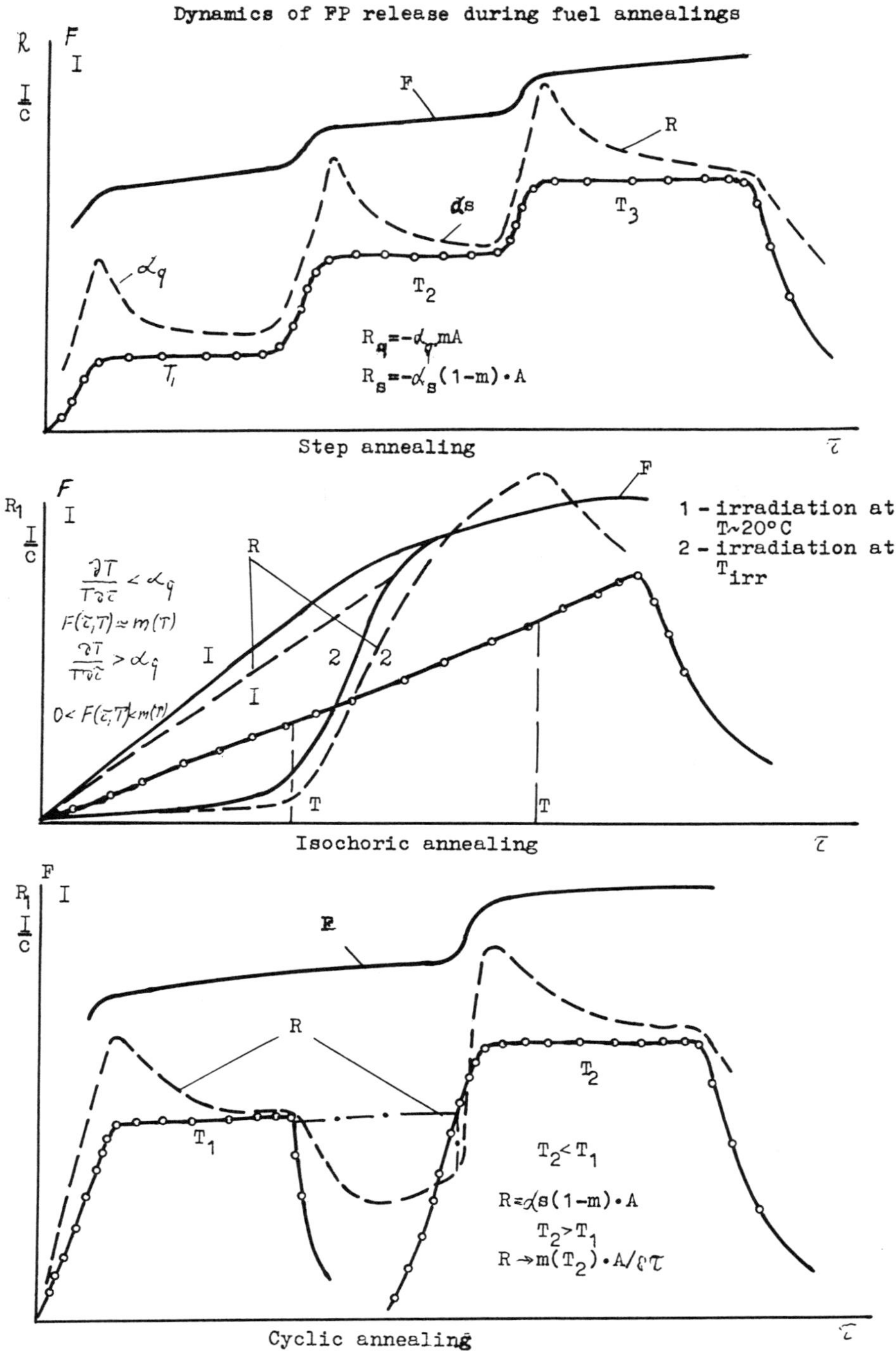

Fig.1. Dynamics of FP leakage during fuel annealing

Distribution of FP over defects during irradiation

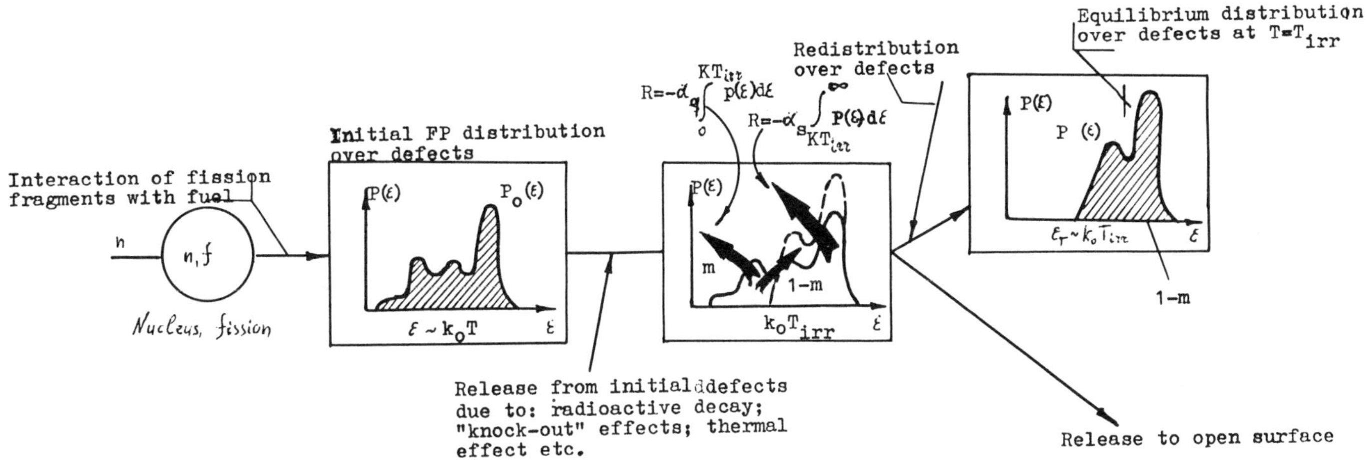

Fig.2. Distribution of FP over defects during irradiation

Isothermal annealing, T=const

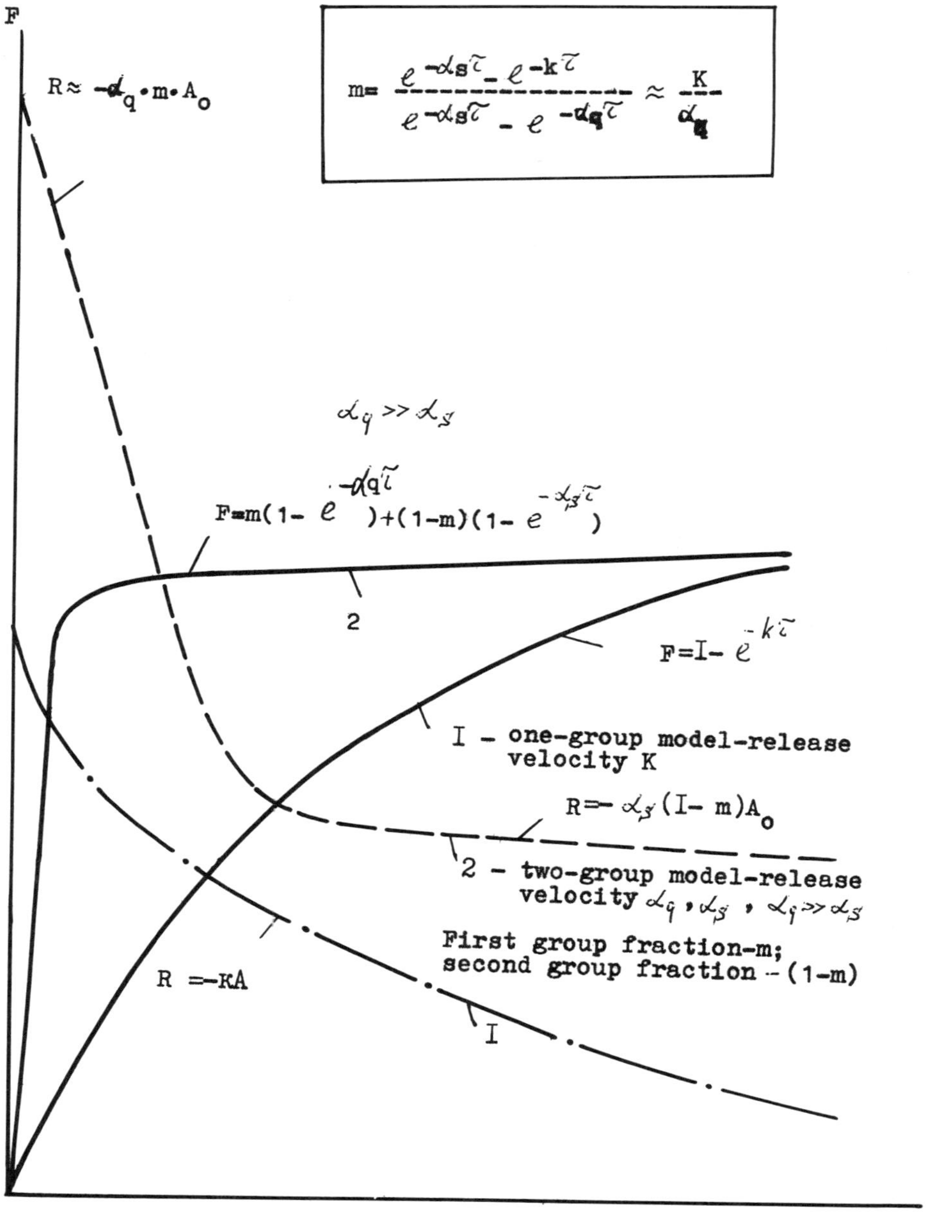

Fig.3. Isothermal annealing, T = const

C$, Te, I release constants abtained using different methods

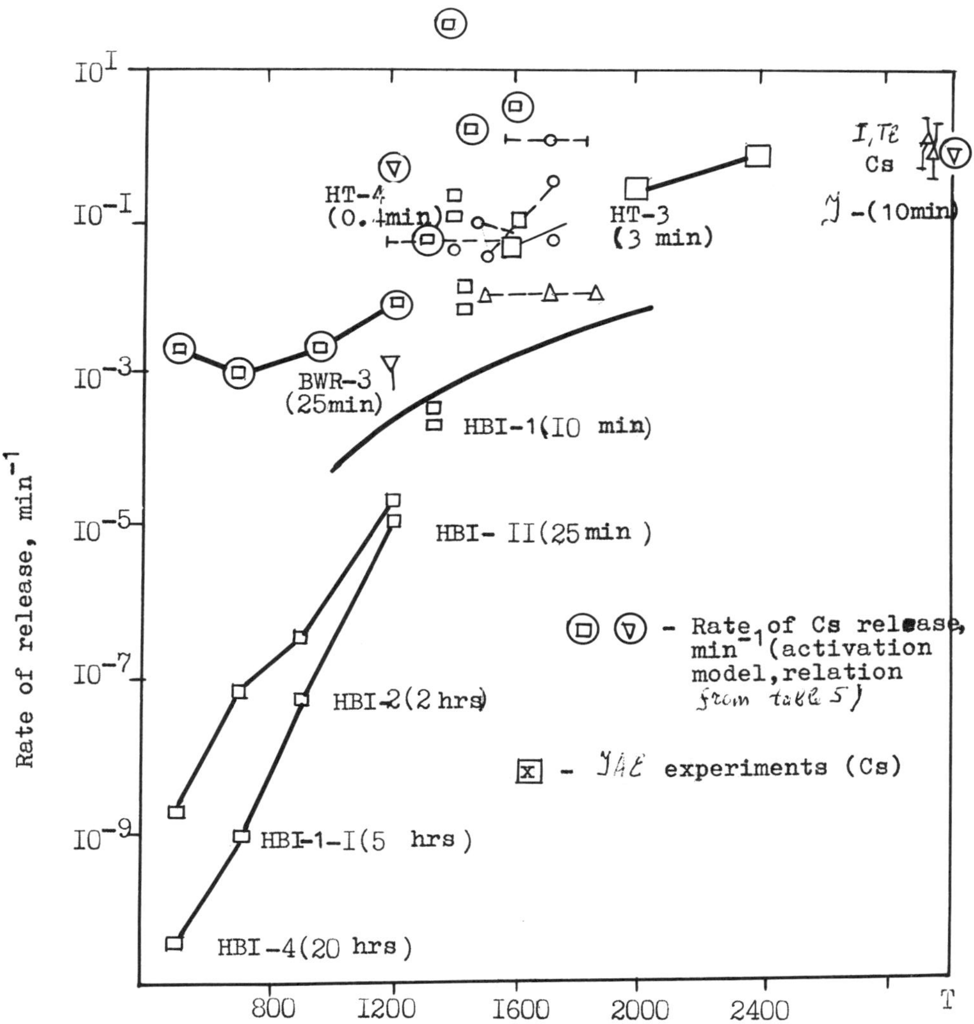

Fig.4. Cs, Te, J leakage constants obtained using different methods

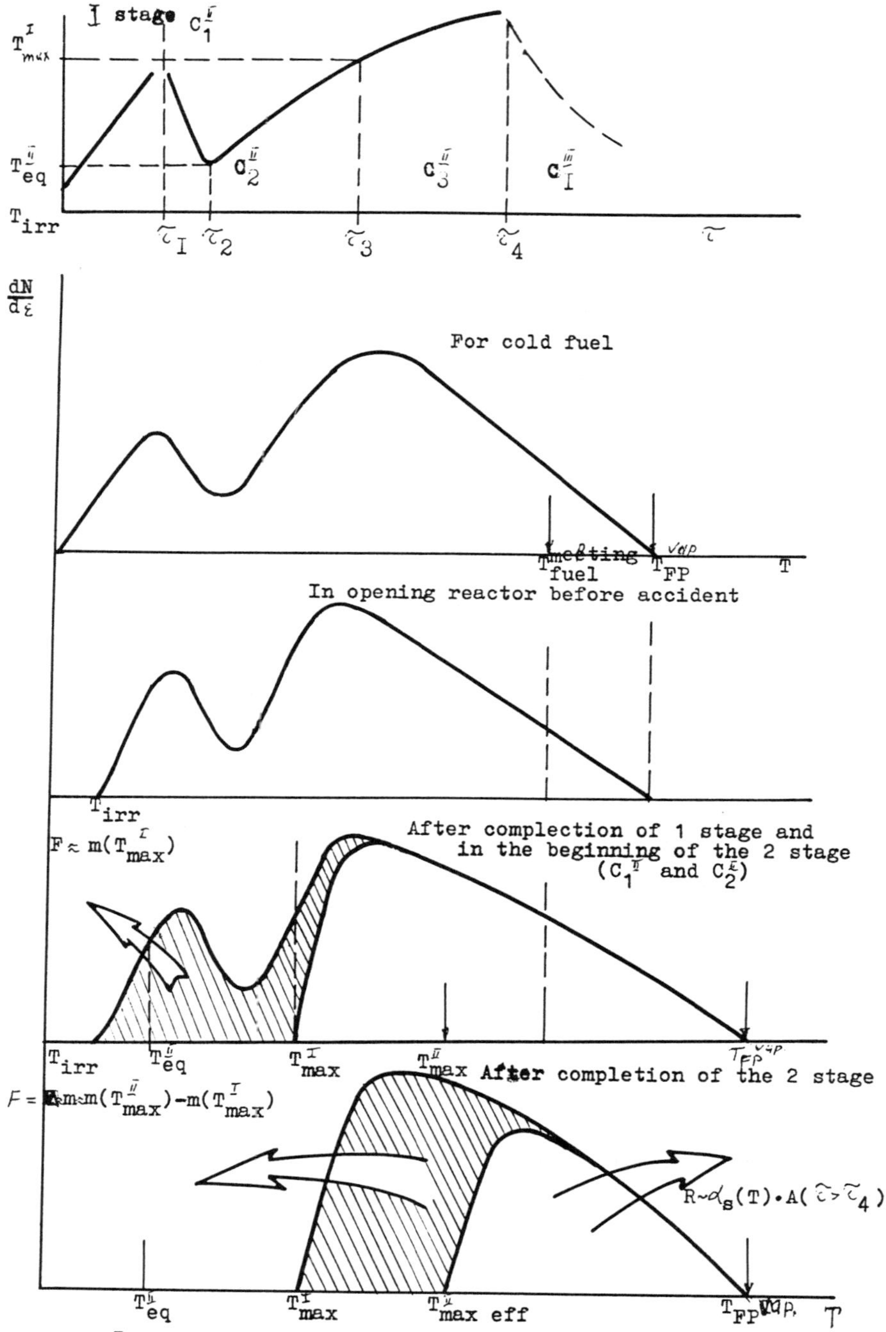

Fig 5. FP distribution and temperatures at different accident stages

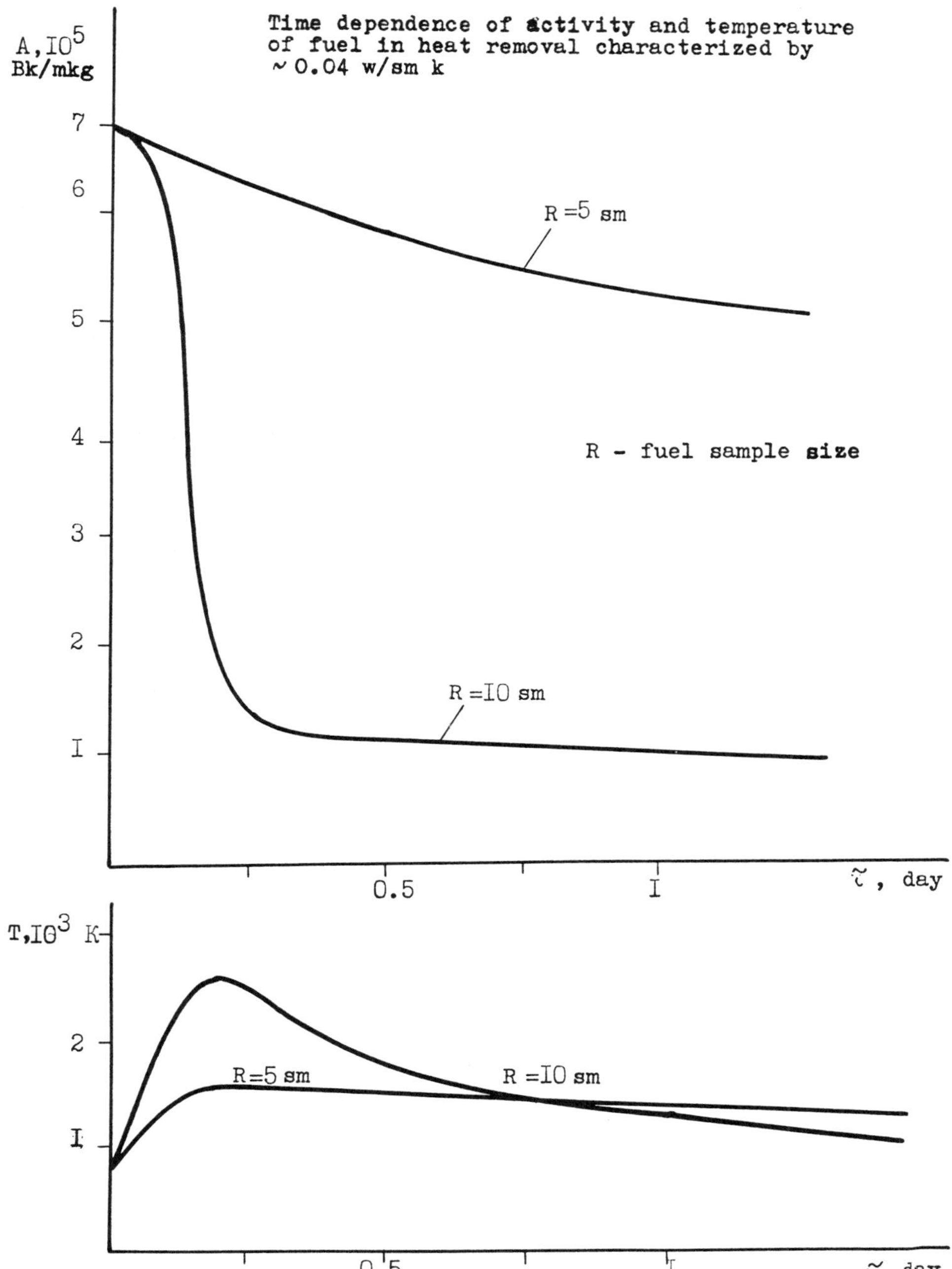

Fig.6. Time dependence of activity and temperature of fuel in heat removal characterized by 0,004 W/cm °K

Radionuclide composition of fractions observed in sample flotation from the Chernobyl NPP site

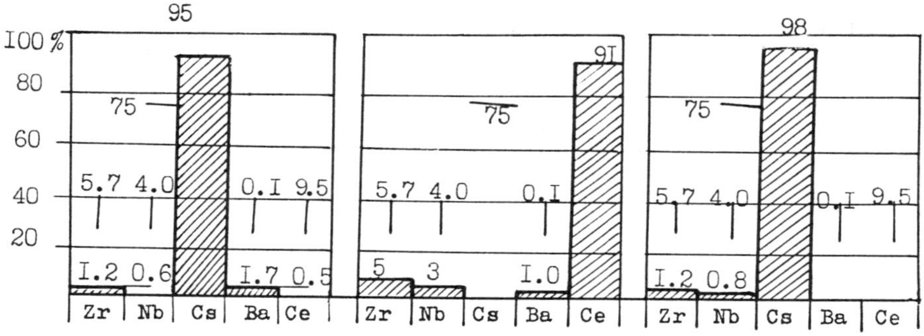

a) density of particles > 4 g/sm³   в) density of particles ~ 3÷4 g/sm³   c) density of particles 2÷3 g/sm³

5.7 / 75 – calculated content

Fig.7. Radionuclide composition of fractions observed in sample flotation from the Chernobyl NPP site

Temperature dependence of ^{85}Kr and ^{137}Cz releases

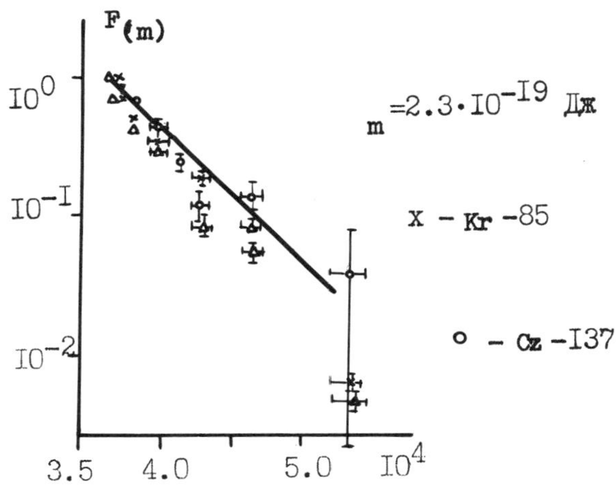

Fig.8. Temperature dependence of ^{85}Kr and ^{137}Cs releases

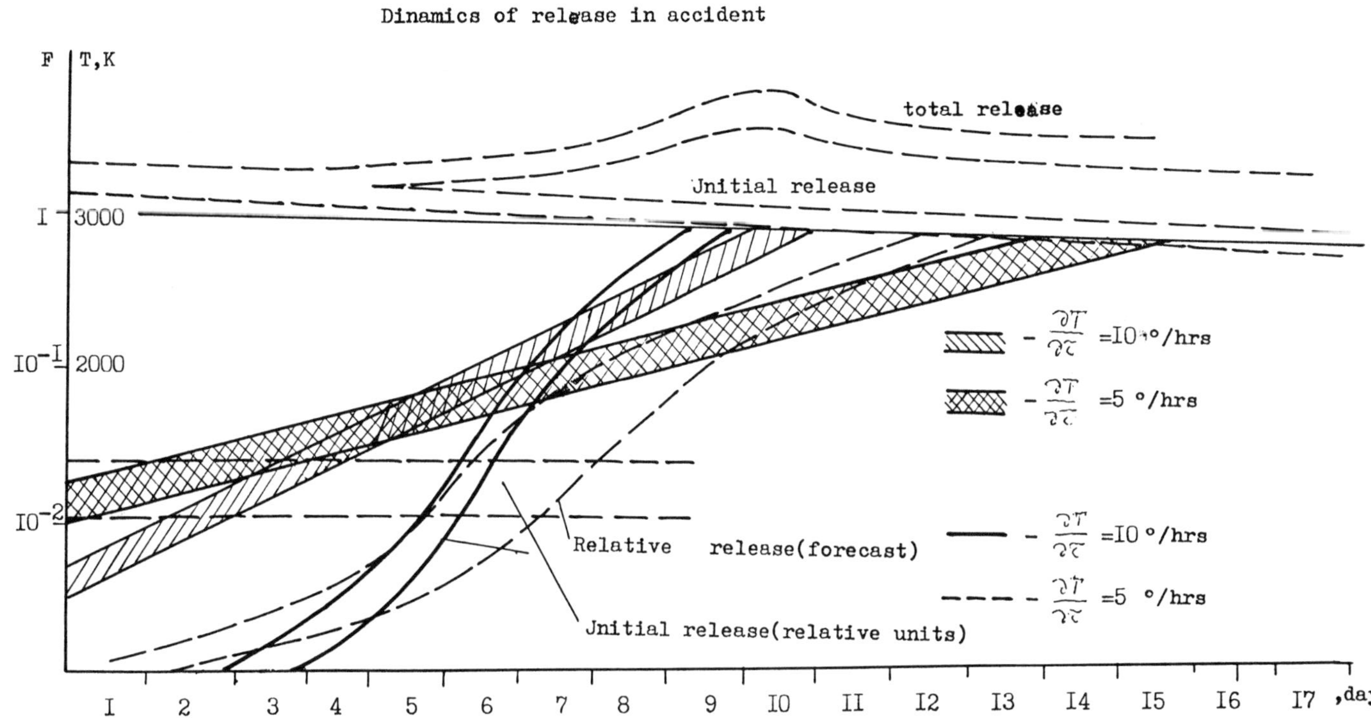

Fig.9. Dynamics of release in accident.

FIGURE CAPTIONS

Fig.1. Dynamics of FP leakage during fuel annealing

I. Step annealing

II. Isochoric annealing

1) irradiation at T ~ 20°C

2) irradiation at T_{irr}

III. Cyclic annealing

Fig.2. Distribution of FP over defects during irradiation

1. Nucleus, fission
2. Interaction of fission fragments with fuel
3. Initial FP distribution over defects
4. Release from initial defects due to: radioactive decay; "knock-out" effect; thermal effect etc.
5. Redistribution over defects
6. Equilibrium distribution over defects at $T = T_{oper}$.
7. Release to open surface; leakage R_i.

Fig.3. Isothermal annealing, T = const.

FP leakage during annealing. Two models:

1. One-group model-leakage velocity k
2. Two-group model-leakage velocity $\alpha_\delta, \alpha_M, \alpha_\delta \gg \alpha_M$

First group fraction m; second group fraction (1-m).

Fig.4. Cs, Te, J leakage constants obtained using different methods.

1. Rate of Cs leakage, min^{-1}

2. Rate of Cs leakage, min^{-1} (activation model, relation (Table 5) is used for calculation.

3. IAE experiments (cesium)

4. NVI-4 (20 hrs)

5. NVI-1 (5 hrs)

6. NVI-2 (2 hrs)

7. NVI-11 (25 min)

8. NT-1 (10 min)

9. BWR-3 (25 min)

10. NI-2 (7 min)

11. NI-4 (0,4 min)

12. NT-3 (3 min)

13. J-10 min.

Fig.5. FP distribution and temperatures at different accident stages

1. Variation of fuel element temperature during the accident

2. T_{oper}

3. T_2^{II}

4. T_I^{max}

5. First stage

6. Second stage

7. Third stage

8. For cold fuel

9. In operating reactor before accident

10. After completion of first stage and in the beginning of the second stage.

$(C_1^{II}$ and $C_2^{II})$

11. After completion of the second stage

12. FP distribution at different stages of accident

Fig.6. Time dependence of activity and temperature of fuel in heat removal characterized by 0.004 W/cm °K.

1. Activity
2. Temperature
3. R-fuel sample size.

Fig.7. Radionuclide composition of fractions observed in sample flotation from the Chernobyl NPP site

a) density of particles > 4g/cm^3

b) density of particles ~ 3-4 g/cm^3

c) density of particles 2-3 g/cm^3

57 ↓ → 75 - calculated content.

Fig.8. Temperature dependence of ^{85}Kr and ^{137}Cs releases

Fig.9. Dynamic of release in accident

1. Total release outside reactor
2. Initial release
3. Relative J leakage (forecast)
4. Initial release (relative units).

Filtering Cooling Model for the Pile-Up in the ChAPP Accident Block

V. P. MASLOV, V. P. MYASNIKOV and V. G. DANILOV
USSR Academy of Sciences
Moscow, USSR

INTRODUCTION

How the heat regime in the active zone of the accident block in the Chernobyl atomic power plant (ChAPP) will evolve was the main question from the very beginning of the accident. The answer to this question was to determine the strategy for suppressing the accident's after-effects. An unfavourable sharpening of the situation, e.g. burning-through of the reactor pit bottom, of the concrete ceilings under the reactor or of the block foundation by incandescent fragments of heat releasing elements (HREl) with subsequent radioactive contamination of underground waters (the so-called "China syndrome"), all this might have led to a sharp aggravation of the already severe radiation scene. The conceivable threat of such events alone already required the urgent building of a concrete plate with forced cooling in the ground under the foundation of the accident block.

Not less important was the problem of preserving the ChAPP accident block and choosing a cooling system for the "sarcophagus" which would guarantee the safety of the block's subsequent operation.

The analysis of different physical models describing the cooling of the active zone in the accident ChAPP block required a large amount of detailed estimations and calculations, demanded the physical modelling of cooling processes and special experiments in the accident block, after which the scheme of convective filtering cooling of the ChAPP accident block was accepted as the base model.

From the physical point of view this model describes the gas filtration process through self-heating porous medium in the gravity field. The equations of this model are classical and were used in different modifications in many papers dealing with the filtration theory. A new type of boundary conditions for these equations which appeared when certain specific conditions of the pile-up cooling in the reactor pit were considered is the principal distinction in our case.

THEORY

In dimensionless form the stationary equations of filtering cooling can be written as follows:

$$\frac{1}{\varepsilon}\nabla P = -\vec{e}_z P^{1/\gamma} e^{-S} - \frac{\mu}{k} U, \tag{1}$$

$$\langle \nabla, P^{1/\gamma} e^{-S} U \rangle = 0, \tag{2}$$

$$P \langle U, \nabla S \rangle = Q = M^2 q \tag{3}$$

where P , S , U , μ are pressure, entropy, velocity and gas viscosity, k is penetrability. The boundary conditions are the following: the pressure is equal to atmospheric on the upper surface and at the bottom of the pile-up, the sides are impenetrable for gas, entropy at the bottom is equal to zero.

We introduce the following notations:

$$\varepsilon = gH/RT_0 \quad , \quad M^2 = \frac{H\mu_0}{V c_p T_0 \rho_0^2 k_0 g} \int_V Q\,dv ,$$

$$q = Q \Big/ \int_V Q\,dv$$

where V is the volume of the pile-up; Q is the heat release distribution; H , k_0 , T_0 , ρ_0 , μ_0 are characteristic thickness and penetrability of the pile-up, temperature, density and dynamical viscosity of gas; R is the gas constant, g is free fall acceleration, $\vec{e}_z$ is the vertical unit vector, c_p is heat capacity of the gas (under constant pressure).

This model is based on the assumption that the main role in the cooling process of the pile-up in the reactor pit is played by heat transfer by means of the filtering flow due to the action of buoyancy forces caused by heating (in the pile-up) of free air coming through the reactor bottom. This point of view is supported by certain constructive distinctions of the reactor pit; in particular, the pit has 1676 technological openings in its bottom, which connect it with the subreactor room, known as the "cross", which is connected, in its turn, with the room under the level of the reactor pit bottom.

However, we must be rather careful in applying Darcy's law in order to describe the process of gas filtering in the physical model when one is dealing with a natural object. It was possible that the pile-up in the reactor pit of the accident block, formed by fragments of the reactor's constructive elements and by large amounts of sand, clay, dolomite, and lead thrown into the pit from April 27 to May 2, 1986, might have "fitulae", i.e. channels of sufficiently large diameter with small hydrodynamic drag. In this case, even though the gas flow through such channels might have been large enough to explain the air flow velocities (of the order of several meters per second) measured in some of the subreactor quarters (those where radioactivity conditions made such measurements possible), the application of Darcy's model would have been incorrect.

The authors proposed experiments with smoke cartridges thrown to the bottom of the reactor. If there had been any "fistulae", concentrated smoke ejections would have appeared over the pile-up. The absence of such ejections in the experiments thus showed that it is possible to use Darcy's law for describing the filtering processes of free air through the pile-up in the reactor of the accident block.

Measurements in the physical model gave results which were confirmed by direct measurements of the temperature level in the gas just over the pile-up, effected by measuring buoys thrown there, the results having been recalculated by using rules taking into account the difference in scale between model and nature. The rules of scale change from model to nature coincide with theoretical similarity laws obtained in the case when the gas filtering process is described by Darcy's law.

At the outset the goal of mathematical modelling was to study the influence of different spacial singularities in the distribution of heat sources caused by HRE1 fragments inside the pile-up, and first of all in those configurations, whose physical modelling was difficult. The mathematical model chosen was intended, first of all, to give an adequate description of the physical model of the destroyed reactor. But if a mathematical model adequately describes the laws of the physical phenomena under consideration, then it usually explains all the facts related in one way or another with the process under investigation, even those facts whose description was not planned at the beginning of the investigation. Moreover, the results obtained on the basis of such a model and later confirmed by experiment can be unexpected both for the authors and for the specialists directly observing the phenomena, and explain some facts which seem obscure before the analysis of the mathematical model. Such results cannot be obtained on the base of simple estimates carried out on an intuitive level.

This was precisely the situation encountered by the authors when they analysed the model of convective filtering cooling of the ChAPP accident block's active zone. The model possessed a number of properties whose analysis gave some practically useful recomendations for fighting the accident's after-effects. In particular, it thwarted attempts to hermetically close all openings in the "sarcophagus" in order to protect against radiation. The violation of established air cooling conditions in the active zone as a result of such actions would have led to unfavourable changes in the heat regime of the "sarcophagus" and to unpredictable complications, including ejections of radioactive matters.

Let us briefly consider some of the model's most interesting properties. From the physical point of view, the convective cooling model convincingly explains, why HREl fragments did not burn through the bottom of the reactor pit. Indeed, the gas flow which arises in such a system under the action of buoyancy forces leads to anisotropy of the total heat flow and to the related anisotropy in spacial temperature distribution induced by a lumped source.

It is well known from everyday experience how effective the convective mechanism of heat transfer is. It is difficult to imagine that even a very powerful heater can warm up water running in a stream above the heater. At any rate, hot sewage from factories destroy the ice cover of a river only below the place where the hot liquid pours out. Similarly, the "chimney effect" is well known: though coal in the furnace can be red hot, the fire grate between the wind box and the furnace is not heated to a very great extent. In large boilers the fire grate easily withholds the fuel load and is not heated up to those temperatures for which its structural resistance yields.

However if the above mentioned anisotropy property of the heat transfer process in a convective cooling model is rather evident, its following property turned out to be rather unexpected and could not have been predicted on the basis of intuitive considerations.

At first, the result obtained puzzled the authors. It turned out that the stationary solution of the problem exists only for certain values of the parameter M, a natural similarity parameter for the problem.

$$M^2 = \frac{H\mu_0}{V c_p T_0 \rho_0^2 k_0 g} \int_V Q \, dv$$

In order to visualize our considerations we take the distribution of active elements mass and penetrability which are homogeneous in section. In this case there is no dependence on coordinates in horizontal section and system (1)-(3) can be reduced to the following system of ordinary differential equations:

$$\frac{1}{\varepsilon}\frac{dP}{dz} = e^{-S} P^{1/\gamma} - \frac{U}{k},$$

$$\frac{d}{dz}\left(P^{1/\gamma} e^{-S} U\right) = 0, \qquad (4)$$

$$PU\frac{dS}{dz} = M^2 q$$

with the boundary conditions:

$$P|_{z=0} = 1, \; S|_{z=0} = 0, \; P|_{z=1} = e^{-\varepsilon}, \; \varepsilon > 0 \qquad (5)$$

Obviously, the second equation of our system can be integrated easily:

$$U = U_0 P^{-1/\gamma} e^{S}, \; U_0 = const \tag{6}$$

The constant U_0 means the gas discharge entering the pile-up and is not known beforehand. We assume that q, k are constants, and without loss of generality we set $q = 1$, $k = 1$.

First we consider the most simple case $\gamma = 1$.

THEOREM 1. The solution of problem (4)-(5) in parametric form is the following for $\gamma = 1$:

$$P(\xi) = \left[\left(1 + \frac{\lambda U_0}{2+\lambda}\right)\xi^{-\lambda} - \frac{\lambda U_0}{2+\lambda}\xi^{2}\right]^{1/2}, \; \lambda = \frac{2\varepsilon U_0}{M^2},$$

$$S(\xi) = \ln \xi,$$

$$U(\xi) = U_0 \xi P^{-1/\gamma}(\xi), \; \xi = \frac{M^2}{U_0} z + 1$$

It exists under the condition $M^2 < M_*^2 = (1 - e^{-2\varepsilon})/\varepsilon$

For $M \to M_* - 0$ the relations hold:

$$P \to \left[1 - (1 - e^{-2\varepsilon}) z\right]^{1/2},$$

$$S \to \infty, \; z > 0$$

$$U \to M_*^2 z \left[1 - (1 - e^{-2\varepsilon}) z\right]^{-1/2}.$$

PROOF. We divide the first equation in system (4) by the third equation taking into account (6):

$$\frac{dP^{2/\gamma}}{d\xi} = -\frac{2\varepsilon U_0}{\gamma M^2}\left(\frac{P^{2/\gamma}}{\xi} + U_0 \xi\right), \; \xi = e^{S} \tag{7}$$

From (5) we have that $P(\xi)$ is defined uniquely by the condition:

$$P(\xi)\Big|_{\xi = 1} = 1 \tag{8}$$

As the result we obtain:

$$P(\xi)=\left[\left(1+\frac{\lambda U_0}{2+\lambda}\right)\xi^{-\lambda}-\frac{\lambda U_0}{2+\lambda}\xi^2\right]^{\gamma/2}, \quad \lambda=\frac{2\varepsilon U_0}{\gamma M^2} \tag{9}$$

and the equation for $\xi(z)$ has the form:

$$\left[\left(1+\frac{\lambda U_0}{2+\lambda}\right)\xi^{-\lambda}-\frac{\lambda U_0}{2+\lambda}\xi^2\right]^{(\gamma-1)/2}\frac{d\xi}{dz}=\frac{M^2}{U_0} \tag{10}$$

The last condition in (5) yields the connection between the problem parameters:

$$\xi_*=e^{S(1)}, \quad e^{-2\varepsilon/\gamma}=\left(1+\frac{\lambda U_0}{2+\lambda}\right)\xi_*^{-\lambda}-\frac{\lambda U_0}{2+\lambda}\xi_*^2 \tag{11}$$

Now by integrating equation (10) on the interval $[1,\xi_*]$ we obtain:

$$\frac{M^2}{U_0}=\int_1^{\xi_*}\left[\left(1+\frac{\lambda U_0}{2+\lambda}\right)\xi^{-\lambda}-\frac{\lambda U_0}{2+\lambda}\xi^2\right]^{(\gamma-1)/2}d\xi \tag{12}$$

By expressing ξ_* from (11) and substituting it into (12), we get the equation which connects M^2 and U_0.

The case $\gamma=1$ is the most simple for investigation. From (12) it follows immediately:

$$\begin{aligned} \xi(z)&=1+\frac{M^2}{U_0}z, \\ \xi_*&=1+\frac{M^2}{U_0}, \\ S(z)&=\ln\left(1+\frac{M^2}{U_0}z\right) \end{aligned} \tag{13}$$

and after easy manipulations equation (11) can be represented in the form:

$$\frac{M^2}{2\varepsilon}=\frac{2+\lambda}{\lambda^2}\,\frac{1-e^{-2\varepsilon}((\lambda+2\varepsilon)/\lambda)^{\lambda}}{((\lambda+2\varepsilon)/\lambda)^{2+\lambda}-1}=f(\lambda) \tag{14}$$

It is easy to see that $f'(\lambda)<0$ and

$$f(0)=(1-e^{-2\varepsilon})/2\varepsilon^2 \tag{15}$$

Thus, since $\lambda \geq 0$, equation (14) has non-negative solutions if the following condition holds:

$$M^2 \leq M_*^2 = (1-e^{-2\varepsilon})/\varepsilon \tag{16}$$

For small ε (16) yields immediately that $M_* = \sqrt{2}$.

We denote by $\lambda(M^2)$ a root of equation (14). From (14),(16) we have that $\lambda(M^2) \to 0$ for $M \to M_* - 0$. Hence, $U_0 \to 0$ for $M \to M_* - 0$, and therefore, $S(z) \sim \ln(M_*^2 z/U_0)$ for $z > 0$. Similarly, we obtain from(9):

$$P(z) \xrightarrow[U_0 \to 0]{} [1 - M_*^2 \varepsilon z]^{1/2} = [1-(1-e^{-2\varepsilon})z]^{1/2}$$

Now we calculate the limit value of the velocity:

$$U(z) = U_0\left(1+\frac{M^2}{U_0}z\right)P^{-1}(z) \xrightarrow[U_0 \to 0]{} M_*^2 z[1-M_*^2 \varepsilon z]^{-1/2}$$

The theorem is proved.

The problem of obtaining the critical value M_* can be formulated similarly for any arbitrary γ. As the result we have the existence condition for the stationary regime, and the critical value of the parameter M is defined by the relation:

$$M^2 \leq M_*^2 = \frac{\gamma}{\varepsilon}(1-e^{-2\varepsilon/\gamma})\left[\int_0^1 [1-(1-e^{-2\varepsilon/\gamma})\ell^2]^{(\gamma-1)/2} d\ell\right]^2 \tag{17}$$

Obviously, formula (17) yields $M_* = \sqrt{2}$ for $\varepsilon \to 0$ so that for small ε the critical value M_* depends weakly on adiabatic exponent.

Formulas (9) and the following relations:

$$S = \ln \xi,$$

$$z = \frac{U_0}{M^2}\int_1^{\xi}\left[\left(1+\frac{\lambda U_0}{2+\lambda}\right)\xi^{-\lambda} - \frac{\lambda U_0}{2+\lambda}\xi^2\right]^{(\gamma-1)/2} d\xi$$

represent the exact solution of our initial problem (4) in the parametric form for $q(z)=1$. These solutions exist for $M<M_*$.

The parameter ε is always small for real parameters of atmosphere and equations (4) can be simplified by means of the following change:

$$P=e^{-\varepsilon z-\varepsilon \Pi} \tag{18}$$

By substituting (18) into (4) and passing to the limit for $\varepsilon \to 0$, we obtain:

$$\frac{d\Pi}{dz}=e^{-S}-1+\frac{U}{k}$$
$$\frac{d}{dz}\left(e^{-S}U\right)=0 \tag{19}$$
$$U\frac{dS}{dz}=M^2 q$$

According to (5) the boundary conditions for system (19) have the form:

$$\Pi(0)=0,\ \Pi(1)=0,\ S(0)=0 \tag{20}$$

The following statement holds:

THEOREM 2. The solution of problem (19)-(20) exists for

$$M^2<M_*^2=\left(\int_0^1\frac{dz'}{k(z')}\int_0^{z'}q(z'')dz''\right)^{-1}$$

and has the form:

$$S=\ln\left(1+\frac{M^2}{U_0}\int_0^z q(z'')dz''\right),$$

$$U=U_0+M^2\int_0^z q(z')dz',$$

$$\Pi=U_0\int_0^z\frac{dz'}{k(z')}+M^2\int_0^z\frac{dz'}{k(z')}\int_0^{z'}q(z'')dz''-\frac{M^2}{U_0}\int_0^z\frac{\int_0^{z'}q(z'')dz''}{1+\frac{M^2}{U_0}\int_0^{z'}q(z'')dz''}dz'$$

PROOF. Actually, (19) yields:

$$U = U_0 e^S, \quad e^S = 1 + \frac{M^2}{U_0}\int_0^z q(z')dz', \quad U_0 = const$$

It is easy to obtain the distribution $\Pi(z)$ in the explicit form. By taking into account the fact, that we have $\Pi(0)=0$ according to (20), we obtain:

$$\Pi(z) = U_0\int_0^z \frac{dz'}{k(z')} + M^2\int_0^z \frac{dz'}{k(z')}\int_0^{z'} q(z'')dz'' - \frac{M^2}{U_0}\int_0^z \frac{\int_0^{z'} q(z'')dz''}{1+\frac{M^2}{U_0}\int_0^{z'} q(z'')dz''}dz' \tag{21}$$

Conditions (20) yield that we must have $\Pi(1)=0$, which is an equation defining the constant U_0. Due to (21) we have:

$$U_0\int_0^1 \frac{dz}{k(z)} + M^2\int_0^1 \frac{dz'}{k(z')}\int_0^{z'} q(z'')dz'' = \frac{M^2}{U_0}\int_0^1 \frac{\int_0^{z} q(z')dz'}{1+\frac{M^2}{U_0}\int_0^{z} q(z')dz'}dz \tag{22}$$

Now we investigate the solvability conditions for this equations. For this purpose we tranform its right-hand side. By denoting $\int_0^z q(z')dz' = R(z)$, we obtain:

$$\frac{M^2}{U_0}\int_0^1 \frac{\int_0^z q(z')dz'}{1+\frac{M^2}{U_0}\int_0^z q(z')dz'}dz = \int_0^1 \frac{R(z)}{\frac{U_0}{M^2}+R(z)}dz$$

Evidently, the expression in the right-hand side of the last equality does not exceed unity and is equal to its maximum value (equal to unity) for $U_0=0$ for all the values of parameters.And (22) yields the following expression which defines the critical constant M_* :

$$M_*^2 = \left(\int_0^1 \frac{dz'}{k(z')}\int_0^{z'} q(z'')dz''\right)^{-1} \tag{23}$$

So, the solution of problem (19), (20) exists under the condition $M_* \geqslant M$.

Now we investigate the behaviour of the solution in the limit as $M \to M_* - 0$. Similarly as above the equalities:

$$e^S = 1 + \frac{M^2}{U_0} \int_0^z q(z')dz',$$

$$U = U_0 + M^2 \int_0^z q(z')dz',$$

yield that for $M \to M_* - 0$ we have:

$$e^S \to \infty, \ z > 0$$

$$U \to M_*^2 \int_0^z q(z')dz'$$

$$\Pi \to M_*^2 \int_0^z \frac{dz'}{k(z')} \int_0^{z'} q(z'')dz''$$

The theorem is proved.

Equation (22) allows to investigate the dependence of the critical value of the parameter M_* on the distribution character of heat release $q(z)$ and penetrability $k(z)$.

We consider

$$k(z) = e^{-z}, \ q(z) = e^{-z}/(e-1) \tag{24}$$

Heat release is concentrated, in the main, in a domain near to the entrance of the pile-up, and the pile-up penetrability is minimal in its upper part. By substituting (24) into (23), we obtain:

$$M_*^2 = (e-1)/e^2 < 1 \tag{25}$$

In general case such distributions $q(z)$ and $k(z)$ exist that M_* is very small. For example, let

$$q(z) = \delta(z - z_0), \ k(z) = e^{-nz}$$

In this case:

$$M_*^2 = \frac{n(1-z_0)}{e^n - e^{nz_0}} \tag{26}$$

and the value of M_* tends to zero for large n. Similarly in the case:

$$q(z)=\delta(z-z_0),\ k(z)=e^{-n(1-z)}$$

we obtain:

$$M_*^2=\frac{n\,e^{nz_0}(1-z_0)}{e^n-e^{nz_0}} \tag{27}$$

The comparison of (26) and (27) shows that if we change the penetrability distribution of the system by its mirror image the critical value M_* varies essentially. The loosening of the upper part of penetrable layer increases the stability margin of the stationary regime considerably. This shows that in real situation dynamic regimes of changes of the pile-up penetrability can lead to such formations when the existence condition for the stationary regime does not hold. Yet some external effects which lead to decreasing of the pile-up penetrability in its upper part can start the stationary regime.

The following theorem given here without the proof allows to estimate the value M_* in general three-dimensional situation.

We remind that the system of filtration equation has the form:

$$\frac{1}{\varepsilon}\nabla P=-\vec{e}_z P^{1/\gamma}e^{-S}-\frac{M}{k}U$$

$$\langle\nabla, Pe^{-S}U\rangle=0 \tag{28}$$

$$\langle PU,\nabla S\rangle=Q$$

The boundary conditions for system (28) have the form: (see Fig.1)

$$\left.\frac{\partial P}{\partial n}\right|_{\Gamma}=0,\ \left.P\right|_{\Omega}=P_a^+,\ \left.P\right|_{\Omega_1}=P_a^-,\ \left.S\right|_{\Omega}=0 \tag{29}$$

where $P_a^{\pm}$ is the atmospheric pressure (note that $P_a^+>P_a^-$).

We assume that the surface Ω_1 (see Fig.1) is described by the following equation:

$$\Omega_1=\{z=h\}$$

THEOREM 3. Let the solution of problem (28), (29) exist such that the condition $\langle U,\vec{e}_z\rangle\geqslant 0$ holds. Then we have the following inequality:

$$2M\int_0^h\int_0^{z_0}\int_{z=z_0}Q\,d\Omega\,dz\,dz_0\leq\frac{1}{\varepsilon}\left(\min_V k\right)\left((P_a^+)^2\int_{z=0}d\Omega-(P_a^-)^2\int_{z=h}d\Omega\right) \tag{30}$$

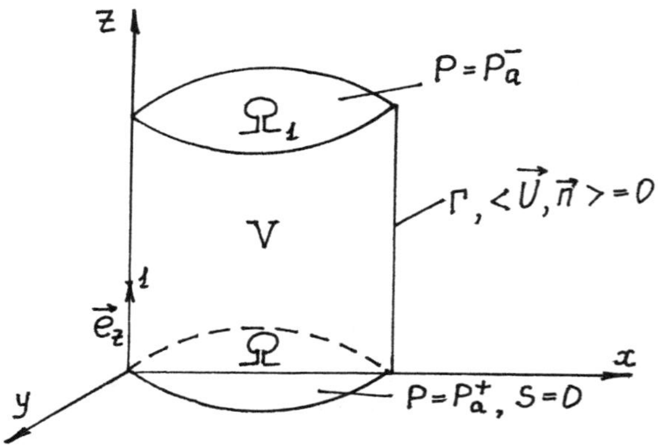

FIGURE 1.

We consider the particular case $\mu = 1$, $h = 1$, $P_a^- = e^{-\varepsilon}$, $P_a^+ = 1$, $\int_{z=z_0} d\Omega = 1$, $Q = M^2$, $k = 1$. We calculate the values which are contained in (30), then inequality (30) has the form:

$$M^2 \leq (1 - e^{-2\varepsilon})/\varepsilon \qquad \text{i.e.} \quad M_*^2 = (1 - e^{-2\varepsilon})/\varepsilon$$

CONCLUSIONS

For $M > M_*$ the solution does not exist. The value M_* was calculated exactly in the particular case of a homogeneous pile-up with uniform distribution of heat sources and turned out to be equal to $\sqrt{2}$. When we took into account possible variants of penetrability changes with respect to the thickness of the pile-up, assumed homogeneous along the horizontal direction, we obtained values $M_* > \sqrt{2}$, when the penetrability was minimal at the bottom of the pile-up, and $M_* < \sqrt{2}$, when the penetrability was minimal in its upper part.

The estimates of M_* in the general case, which coincide in case of the one-dimensional problem with values obtained from exact solutions, show without doubt that the fact obtained is connected with a certain physical effect.

Careful analysis showed that for $M > M_*$ "dry boiling" arises, a new effect probably not described in the literature earlier.

First of all, the analysis of how exact solutions behave as $M \to M_*$ gave the following results: the gas entropy in the pile-up tends to infinity, the pressure and the velocity of gas filtering through the pile-up are finite, the total gas transfer rate through a horizontal section of the pile-up is equal to zero.

It became clear that when the pile-up is formed, an unstationary development of the process if $M > M_*$ will lead to a phenomenon like an explosion. Thus the critical value $M = M_*$ is a criterion of the regime's change from heat equilibrium to heat explosion. Here the analogue may be : the explosion of a heated tin can or the boiling of the milk it contains. Unexpected is the fact that if the heating is sufficiently intensive, then the tin can will explode even when it is not closed hermetically.

The process taking place in the pile-up when the heat release is large was investigated by using a model in which the mechanical properties of the heap are taken into account, namely, the model of elastic granular continuous medium. Actually, the fact that the pile-up consisted of disconnected elements suggested that the porous body formed in the pile-up cannot admit any extending stresses. The investigation of the appropriate equations (describing the mathematical model of elastic granular medium) led to the discovery of moving discontinuities, like bubbles in a liquid, in such a system under dynamical conditions there. Although detailed modelling of the cooling process in the supercritical domain $M > M_*$ (involving phase transitions) was not carried out, the results obtained, as it happens when a sufficiently adequate mathematical model of some process is used, allowed us to explain some important facts whose investigation was not intended at first and which were unexpected for the physicists directly occupied in suppressing the after-effects of the accident.

Subsequent analysis showed that the filtration cooling model of the pile-up completely explains all the phenomena taking place in the reactor after the accident.

First of all, right after the explosion, intensive burning of graphite in the reactor pit could be observed; this would have been impossible if the pit bottom was isolated from the atmosphere. The existence of air flows in the subreactor quarters of the damaged block was also confirmed by subsequent direct measurements.

From the point of view of the results obtained the general picture of how the accident block behaved can be represented as follows. Just after the explosion gas filtering through the active zone of the reactor began due to the gas being heated both by the active mass of damaged and destroyed HREl and by the burning of graphite. The sand, clay, dolomite and lead which were th own into the active zone decreased the temperature there and at the same time decreased the penetrability of the pile-up which was thus formed. The decrease of penetrability in the pile-up and the decrease of air flow through the active zone, led to a reoccurence of heating, but now under small penetrability. In this case the existence condition for stationary regime was violated, so that the number M exceeded the critical value, $M > M_*$. This situation led to the development of"dry boiling" regime and to an increase of radioactive ejections during May 3-5, 1986. "Dry boilinh" in the pile-up created conditions for its loosening, and, possibly, for additional ejections of the active fuel mass from the reactor. After the May 3-5 ejections the configuration of heat sources and the penetrability in the pile-up became such that the natural filtering ventilation of the pile-up became sufficient for the realization a stationary cooling regime. The transition to the stationary filtering regime led to a decrease by several orders of magnitude of the intensity of radioactive ejections.

The behaviour of the pile-up was influenced by melting and evaporation processes of materials volatile at the temperature level in the active zone of the reactor at the beginning of the accident. The ejection of vapours of these materials by filtering flow intensified the active zone's cooling. However, the melting process and the mobility of melts could have stimulated the processes of "dry boiling", decreasing the penetrability of the pile-up.

The stationary regime of filtering cooling changed, of course, as the "sarcophagus" was constructed, and its subsequent evolution depends on the character of the changes taking place in the pile-up under the action of various physical-chemical factors and accompanying mechanical processes. The existence of the critical value M_* and the character of the dependence of M on the diagnostic parameters of the problem show that the regime of the "sarcophagus" should be carefully watched. An increase in temperature of the gas coming out from the pile-up may be related, for example, to a penetrability decrease in the pile-up due to growth of dust content in porous space or to its changes as a consequence of creeping in the pile. When the temperature increases, the pile-up should be loosened artificially, otherwise, as the results of modelling show, nature can take care of it itself.

The mathematical models considered are a direct generalization of classical models of gas filtration involving non-isothermal conditions and elastic granular properties of the porous medium skeleton. The only essentially new element in setting the problem was the condition that the system opens into atmosphere and, as a consequence, that the rate of gas flow through the porous system caused by its self-heating is self-consistent.

The new results obtained here were unexpected, since similar problems were never considered in technical applications. There was no technical and laboratory experience on which intuitive construction of plausible scenarios for the behaviour of such systems could be based. Therefore, specialists directly concerned with suppressing the after-effects of the accident in the fourth ChAPP block encountered certain physical phenomena which were new to them and their work was considerably complicated by these circumstances.

Of course, the specific situation which arose after the ChAPP accident will not necessary repeat itself in eventual accidents of the same type, although the authors certainly hope that such accidents will never occur again. For example, the appearance of "fistulae" would have resulted in an essentially different course of events. However, the tragic experience of the ChAPP accident must be taken into account, and the necessary theoretical and experimental preparations for fighting similar accidents in the future must be carried to the full extent, as an important component of the general security programme for power saturated technical systems.

REFERENCE

Maslov, V.P., Myasnikov, V.P., and Danilov, V.G., Mathematical modelling of ChAPP accident block, "Nauka", Moscow, 1987, pp. 1-141.

Method and Results of Aerosol Particle Size Definitions in 30-KM Zone of Chernobyl Nuclear Power Plant in 1986–1987

V. I. SKITOVICH, A. K. BUDYKA, B. I. OGORODNIKOV, and I. V. PETRYANOV
Karpov Institute of Physical Chemistry
Moscow, USSR

ABSTRACT

During the period from May, 1986 to November, 1987 the aerosol samples were collected around the Chernobyl nuclear power plant to define radioactive aerosol particle sizes by filter pack technique. The parameters of lognormal distribution (activity median aerodynamic diameter, AMAD, and geometric standard deviation) were defined for various radionuclides in aerosol form. The submicron aerosol particle were only shown to be produced by the damaged block-4 during the initial period after the accident. AMAD was increased up to 4-6 μm at the following stages when the secondary updraft of aerosol particles from contaminated surfaces became the dominant source of radioactive aerosols.

INTRODUCTION

The information about the sizes of radioactive aerosol particles resulting from a nuclear reactor accident is of primary importance for solving some problems of the transport of the radioactive aerosol particles and their land fallouts. The particle size distribution is the basis for the choice of individual and collective protective respiratory systems and the calculations of the deposition of inhaled radioactive aerosols in human respiratory tract. It should be noted that the effect of radionuclide fractionation may result from the differences in the mechanisms of the formation of radioactive aerosols. Hence a method is needed for simultaneous definition of both nuclide composition and aerosol particle size. This problem could be solved by collecting of aerosol particles with their subsequent particle size analysis. The traditional impactor method of aerosol size analysis is not effective in this case because it is too time consuming. Moveore, the information about particle size distribution is somewhat indefinite once the particle size does not exceed 1-2 μm.

METHOD

The filter pack technique was used for the determination of aerosol particle size spectra. The basic principles of this method were described in ref. [1]. Aerosol particles pass through the set of filters FP consisting of ultrafine fibers with the high velocity (from 0.5 to 2.0 m/s). If the velocity of aerosol particle is big, the main mechanism of aerosol filtration is the inertion capture and the efficiency of filters is the function of particle sizes. The filter efficiency as a function of size, density and velocity was investigated for the filter pack technique [2]. The results of the investigation of the efficiency for filter pack, consisting of three layers are presented in Fig.1.

In the case of lognormally distributed radionuclide size spectrum

$$f(r) = (2\pi)^{-1/2}(r\ln\sigma)^{-1} \exp(-(\ln r-\ln r_0)^2/2\ln^2\sigma) \qquad (1)$$

with r_0 being the median radius, σ - standard geometric deviation and r - the radius of particle, the number of particles collected of i-th layer is

$$N_i = \int_0^\infty \varepsilon_i(r)f(r)dr \qquad (2)$$

Here $\varepsilon_i(r) = \varepsilon_i'(r)\prod_{j=1}^{i-1}(1-\varepsilon_j'(r))$, where $\varepsilon_i'(r)$ is the efficiency of the i-th layer of filter pack.

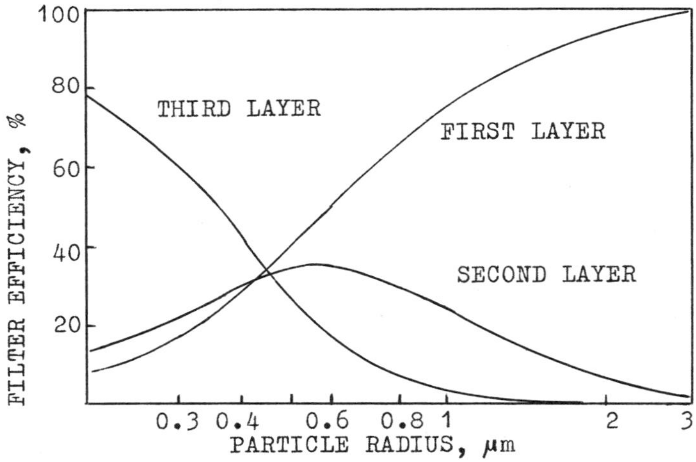

FIGURE 1. The efficiency of fiber filter layers in filter pack vs particle size. First layer - ФПА-70-0.1; second layer ФПА -70-0.17; third layer - ФПА-15-2.0.
The flow rate is 0.5 m/s, the density of particles is 1 g/cm^3

Three filters are required for a correct determination of the distribution parameters. On calculating the integrals (2) for all possible magnitudes r_o and $б$ and measuring the activity of aerosol particles on i-th layer N_i' the comparison between N_i and N_i' gives the parameters of the lognormal distribution. The set of nomograms were calculated for various compositions of fiber filters.

The size range measured is to be a function of both linear velocity of filtration and thickness of the filter layers. If the flow rate is about 1,5 m/s the particle diameter within the interval 0.1 - 2 μm might be calculated. Lowering the flow rate down to 0.5 m/s and using extremely fine filters at the first stage (the aerodynamic resistivity of this layer being about 1 Pascal at the flow rate of 1 cm/s) allow one to extend the upper limit up to 5-6 μm. The basic advantage of this method is the aerosol sampling from a large volume of pollutant air. On the other hand the optimum parameters of sampling (among them are the flow rate and the structure composition of the filter pack could be determinated to by a more precise method. To choose the optimum parameters of filter pack the premilinary results on the aerosol particle size are required. As a rule the magnitudes of N_i' are measured by the gamma-spectrometer. The absolute activity of the sample is determinated by the alpha-activity method with the correction coefficients [3].

EXPERIMENTS AND RESULTS

The sampling of aerosols was performed inside the Chernobyl nuclear power plant, over the damaged block-4, on the industrial site near the station and in some points of the 30-km zone. For the sampling both portable and stationary aspiration installations were used. Among them were ПВ-20М("Shmel") (the fan capacity is 1.2 m^3/h, the filter area is 2 or 3 cm^2), 12- ЦС-34(10 m^3/h, 0.3 m^2), ШЛА-2(7 m^3 /h, 20 cm^2). These aspiration installations were chosen to provide the optimum flow rate through the filter pack and the total volume of filtrating air. The operation time depended on the concentration of radionuclides in aerosol form, that had been found premilinary by the analytical filters AFA(USSR). For correct determination of the particle size distribution the minimum air volume passing through the filter pack was to be twice as large as that for the determination of aerosol concentration. After the sampling of aerosols the filter pack was separated into three layers and each of them was studied separately by a gamma-spectrometer. For the control of validity of the results some samples were remeasured in 2-3 months. Then the nomograms were used for determination of the parameters of lognormal particle size distribution: activity median aerodynamic diameter AMAD and geometric standard deviation. As a rule the density of a radioactive aerosol particle was not known, so the concept of aerodynamic size was used in treating our experimental results. AMAD was defined as the diameter of a particle having the same deposition velocity as the particle with the density 1 g/cm^3. This parameter is commonly used for calculation of inhaled radioactive aerosol particles deposition in human respiratory tracts. Once the density of aerosol particle is known their geometric size is readily defined.

The measurements of the radioactive aerosol particle size distributions were performed twice: July-September, 1986 (14 samples seven of which- on the industrial site, five- inside the station building and two - over the damaged block-4) and September-November,1987 (11 samples from various points inside the 30-km zone after the completion of "Sarcofag" over the damaged block). The experimental data(AMAD of radioactive aerosols, carrying gamma- and alpha-nuclides) are presented in Tables 1 and 2. The geometric standard deviation varied within the range 1.2-2.6 for all experiments. Table 1 clearly shows that radioactive isotopes were concentrated on particles whose diameter exceeded 2 μm for all 7 samples from the industrial site. In some cases the AMAD exceeded 4 μm (e.g. sample 4). As is known [4] the radioactive aerosols-the products of nuclear explosions consist of particles with the diameter of 0.4-0.8 μm. The radioactive aerosol particles produced in Chernobyl accident were considerably larger. This fact was obviously connected with the mechanism of their production. We suppose the main contribution to air pollution was yielded by secondary updraft of aerosol particles from contaminated surfaces, civil engineering works and traffic. It is interesting to note that all radionuclides except for Cs-137 and Ru (in samples 1 and 7) were collected by the particles of practically equal sizes. Nevertheless the nature of activity distribution on the fibrous layers indicated that there were two fractions of aerosol particles. The presence in the atmosphere of both coarse and fine particles led to the increase of activity on the first and third layers of the filter pack. Such distribution of activity on the layers led to considerable errors in attempting to treat the results numerically. The parameters of both coarse and fine fractions may be evaluated if the coarse particle size spectrum is similar to that of the unimodal coarse aerosols. The final results on the activity of fine and coarse fractions are also summarized in Table 1. The specific character of the particle size distributions for Ru-103 and Ru-106 can be accounted for by the presence of volatile compound(RuO_4), which is deposited from the air on the most numerous particles with the diameters of 0.2 μm (this value commonly corresponds to the maximum in the particle surfaces distribution). Depending on meteorogical situation and dust loading the percentage of fine particle fraction of Ru was varied within 0÷100. The ruthenium fine disperse fraction was observed in sample 6 only, collected under the conditions of sharp decrease in concentrations of all radionuclides. The low radionuclide concentrations in this case made the determination of the activity distribution over particle size impossible.

In some cases (e.g. sample 7) the presense of fine disperse fraction of aerosols was observed also for Cs-134,Cs-137 (up to 20% of transported activity) and even for Ce-144,Ce-141, Zr-95, Nb-95(up to 5%). The compounds of Cs are more volatile as compared to those of Zr and Nb. This fact seems to explain the differences in their percentage content in fine and coarse

TABLE 1. The activity median aerodynamic diameters (in μm) for radioactive aerosols sampled during the first period (1986) in various points inside the 30-km zone

N	DATE	POINT	Ce-144	Ce-141	Ru-103	Ru-106	Cs-137	Cs-134	Zr-95	Nb-95	α
1	2	3	4	5	6	7	8	9	10	11	12
1.	03-04.07	pump station	2.8	2.8	2.8/80% 0.6/20%	-	1.7	3.0	3.0	3.0	
2.	05-06.07	purification structures	2.7	2.8	2.6/85% 0.6/15%	-	3.6	-	2.2	2.6	
3.	13.07	administration building of 1-2 blocks	2.5	2.4	2.5	2.6	2.5	2.1	2.5	2.5	
4.	16.07	-"-	3.3	4.0	3.3	4.0	3.7	4.0	4.0	3.4	
5.	20-21.08	administration building of 3-4 blocks	3.0	3.0	3.0/80% 0.2/20%	-	3.0/97% 0.2/3%	3.0/97% 0.2/3%	3.0	3.0	2.8
6.	24-27.08	-"-	-	-	0.15	0.2	-	-	-	-	-
7.	30.08	-"-	2.8/95% 0.2/5%	2.8/95% 0.2/5%	2.8/60% 0.2/40%	2.8/60% 0.2/40%	2.8/80% 0.2/20%	2.8/80% 0.2/20%	2.8/95% 0.2/5%	2.8/95% 0.2/5%	3.2
8.	13.07	machine room	1.9	1.9	2.0	1.7	2.2	1.9	1.9	1.9	1.6

TABLE 1 (continuation)

1	2	3	4	5	6	7	8	9	10	11	12
9.	13-14.07	machine room	3.8	2.2	2.1	2.4	2.2	2.4	2.2	2.2	1.7
10.	25.07	inside the building of block-4	2.7	-	-	2.3	2.7	2.6	2.5	2.5	
11.	25.07	machine room	2.2	-	1.8	1.9	2.2	2.7	2.0	2.0	
12.	23-24.09	ventillation station of block-4	3.6	3.4	3.2	3.0	3.2	3.2	3.6	3.6	
13.	10.09	over the damaged block-4	1.0	1.2	0.8	0.9	1.1	1.0	1.0	1.1	1.3
14.	10-11.09	-"-	1.8	1.8	1.8	1.7	1.6	1.6	1.8	1.8	1.7

NOTE: under the horisontal line the percentage of radionuclides on the coarse and fine disperse fractions is given.

TABLE 2. The activity median aerodynamic diameters (in μm) for radioactive aerosols sampled inside the 30-km zone during the second period (1987).

N	DATE	POINT	Ce-144	Cs-134	Cs-137	Ru-106	Ru-103	Zr-95	Nb-95	α
1.	17-19.09	ОВД -2	4.8	5.0	5.2	5.0	5.0	5.0	4.6	4.0
2.	17-20.09	Chernobyl	6.0	6.2	6.1	6.0	5.0	4.0	4.8	-
3.	18-21.09	ОРУ - 750	5.1	4.9	4.8	5.5	5.2	5.1	4.6	5.0
4.	19-21.09	ОВД - 2	6.2	6.3	6.3	4.2	4.2	6.2	-	-
5.	19-22.09	pump station	5.0	6.4	6.4	6.0	4.4	5.2	-	-
6.	20-22.09	ОРУ -750	5.2	6.2	3.9	5.5	5.5	5.5	5.1	5.5
7.	24-26.09	ОВД - 2	4.9	4.3	5.3	3.1	4.4	4.4	-	-
8.	25-26.09	industrial site	6.0	5.5	6.0	4.7	5.7	5.7	-	-
9.	18-21.11	ОРУ - 750	5.1	3.9	3.8	4.1	-	-	-	-
10.	18-21.11	pump station	5.0	3.5	5.0	4.0	-	-	-	-
11.	18-21.11	ОВД - 2	3.2	2.8	3.7	1.5	4.5	4.5	-	-

NOTE: The distances and directions of points of sampling relative to damaged block-4:
ОВД-2 (Pripyat) - 5 km, NW; Chernobyl - 18 km, SW; ОРУ-750 - 0.5 km, S; pump station-3 km, NE.

particles. The determination of the aerosol particle size distribution inside the station was performed while a lot of activities went on inside the building: shield partitions construction, welding, transport goods etc., which led to the intensive aerosol production(both inert and radioactive). The value of AMAD inside the building was usually less than that near the industrial site. The values of AMAD of these aerosols were essentially the same and varied from 1.6 to 2.7 μm for all radionuclides. But the particle diameters were equal to 3-3.6 μm inside the ventilation station of block-4 during the disactivation of filter cell of canyon(see sample 12).

The air sampling from the zone of damaged block-4 was accomplished before the completion of "Sarcofag". The filter holder with the filter pack was placed by a long rod over the disintegrated reactor. The same AMAD value(about 1 μm) was observed for the radioactive particles of Cs, Ce, Zr-95 and Nb-95 (see the sample 13) whereas it was a little smaller for Ru(0.8 μm). During the collection of sample 14 over the damaged block the concentration of radioactive aerosols increased by several times which led to variations in the particle size distribution. In this case all radionuclides were detected on the aerosols with AMAD more than 1.6 μm. The absence of radionuclide fractionation over this particle size interval can be accouned for by a low temperature of the nuclear fuel in the reactor disintegration when the sample was collected.

In some cases the distribution of total alpha-activity on the filter layers was measured. The aerosol particles carrying the alpha-activity were approximately the same size as the carriers of the gamma-active high-melting compounds.

The parameters of aerosol particle size distributions measured during the second period(inside the 30-km zone) are given in Table 2. All radionuclides were connected with the coarse aerosols. The AMAD values of these aerosols increased up to 3.0-6.4 μm. No essential differences were observed in the particle size spectra of alpha-betta-gamma active aerosols which was evidence in favour of similar mechanisms of their production. No essential correlations were observed of aerosol concentrations and other parameters of their size distributions.

CONCLUSION

The experimental technique described above has been found useful for measuring the particle size distributions in real conditions during the period after the Chernobyl accident. The filter pack technique was unique in most cases. The analysis of aerosol particle size by filter pack technique could be performed at the extremely low concentrations of radionuclides: down to 10^{-6} Bq/m^3.

ACKNOWLEDGEMENTS

We wish to thank the workers of PO "Kombinat" for their assistance in measuring some samples and making arrangements for the air sampling inside the 30-km zone around the Chernobyl nuclear power plant.

REFERENCES

1. Bogolapov, N.W., Kastchenko, N.N., Konstantinov, I.E., Ogorodnikov, B.I. and Skitovich, V.I., in Questions of dosimetry and radiation protection, ed. L.R.Kimel and V.K.Sacharov, no. 5, pp. 136-139, Atomizdat, Moscow, 1975.

2. Budyka, A.K., Ogorodnikov, B.I. and Skitovich, V.I., Hydrodynamics of the fan model of fibre filter and capture of aerosols with Stoks-numbers from 0,4 to 4, Doklady AN USSR, vol.284, no. 5, pp. 1160-1163, 1985.

3. Borisov, N.B., Ilin, L.A., Margulis, U.Ya., Parchomenko, G.M. and Hrustch, V.T., Radiation safety by polonium-210 working, pp. 76-79, Atomizdat, Moscow, 1980.

4. Karol, I.L., Radioactive isotops and transport in the atmosphere, pp. 49-54, Hydrometeoizdat, Leningrad, 1972.

Theoretical Analysis and Numerical Modeling of Heat Transfer and Fuel Migration in Underlying Soils and Constructive Elements of Nuclear Plants during an Accident Release from the Core

R. V. ARUTUNJAN, L. A. BOLSHOV, V. V. VITUKOV,
V. M. GOLOVIZNIN, A. M. DYKHNE, V. P. KISELEV,
S. V. KLEMENTOVA, I. E. KRAYUSHKIN, A. V. MOSKOVCHENKO,
V. D. PISMENNII, A. G. POPKOV, S. Y. CHERNOV, V. V. CHUDANOV,
O. V. KHORUZHII, and A. I. YUDIN
FIAE
Moscow, USSR

ABSTRACT

Migration of fuel fragments and core fission products during severe accidents on nuclear plants is studied analytically and numerically. The problems of heat transfer and migration of volume heat sources in construction materials and underlying soils are considered.

INTRODUCTION

A study of probable consequences of severe accdents on nuclear plants with core meltdown and release of fission products into the enviroment is carried out in the framework of goverment nuclear safety program . A review of works made in this direction in the USA was elaborated by the special group of APS[1].

The problem of migration heat sources in the enviroment is very important because there always exicts possibility spreading fuel fragments during the accident.Heating of significant amount of fuel resulted from residual radioactivity can be large enough to destroy surrounding materials by melting , evaporation , chemical reactions etc. Damage of containment constructions and underlying soils can lead to fuel migration upwards, downwards and in sides. The speed of migration and its appearance depend on initial fuel distribution, chemicalcomposition of the enviroment, constuction firmness, thermophysical properties of materials,correlation of fuel and enviroment densities, conditions forconvection and radiation heat transfer etc. The main part of these data is unknown during accident destruction of the reactor. Lacking full information simulation of heat and fuel transfer in the unit is of a probabilishe nature.Investigation of low probability events and elaboration of reliable methods to cease the fuel migration is necessary because of catastrophic consequences of accidents.

The problem of heat transfer and fuel migration in situations involved isessentialy nonnlinear and multidimensional, that's why an approximate solution can lead to wrong conclusions. So acomplete numerical experiment is needed, induding both analitical and computational efforts and based on mathematical models deduced from the basic physical principles and free from restrictive simplifications.

The present work includes a description of some mathematical models of radionuclide migration in soils, developed in the Kurchatov Institute of Atomic Energy in 1986, as well as some physical results, derived from these models.

INSTABILITY OF PLANE FUEL LAYER IN UNDERLYING GROUND.

1. The problem of migration of fuel fragments through underlying ground is an actual physical problem. Consider a model situation, where the fuel fragments forma plane layer. The layer pulling is a result of heating and melting of underlying ground with following filtration of lighter melted ground through the heavier fuel.

Assume that the fuel layer with thickness H and volume heat generation power q pulls down at a constant rate Vo . The fuel, the underlying ground and the melt occupy regions $0<Z<H$, $Z<0$ and $Z>H$, correspondingly, in the coordinate system moving with the layer. Equations of heat conduction are as follows:

$$
\begin{aligned}
&V_o T_1' = \chi_1 T_{11}', && Z>H, \\
&c_1 V_o T_2' = æ_2 T_2' + q, && 0<Z<H, \qquad (1) \\
&V_o T_3' = \chi_1 T_3', && Z<0
\end{aligned}
$$

where the prime denotes Z-derivative, χ_1, c_1 - arethermal diffusivity; $æ_2$ - the thermal conductivity of the fuel.

The temperature and heat-conduction fluxes are continuous through the layer boundaries Z=0 and Z=H. The bottom layer boundary coincides with melting isotherm $T=T_m$. The melting heat flux is neglected because of small parameter $\lambda_m / c_1 T_m$ (where λ_m- is the volume melting heat of the ground). System (1) have been easily integrated, and the rate Vo is determined by

$$1-\exp\left(-\frac{c_1 V_o H}{\text{æ}_2}\right)=\frac{c_1^2 V_o^2 T_m}{\text{æ}_2 q} \quad (2)$$

Determine a simpler expressions for the pull down rate in two limit cases.

The first one corresponds to a sufficiently thin fuel layer pulling down at a small rate. In this case one can neglected convection heat transfer due to melt filtration in comparison with layer heat conduction. It follow then

$$V_o=\frac{qH}{c_1 T_m}$$

The obtained expression is valid in approximation $H \ll (\text{æ}_2 T_m/q)^{1/2}$. Note, that characteristic time of stabilization of the rate of layer pull down is larger than the heat-diffusion time for a scale of the order of the layer thickness H. The maximum temperature of the layer is very close to the ground melting temperature Tm.

In a realistic situation the temperature of fuel melt is higher than the temperature of ground melt, therefore the fuel is a solid in a sufficient thin lay.

In another limit case, the fuel layer is thick and has pull dow rate.

If

$$H \gg \sqrt{\frac{\text{æ}_2 T_m}{q}},$$

then

$$V_o=\sqrt{\frac{\text{æ}_2 q}{c_1^2 T_m}}.$$

In this case the maximum layer temperature is higher, than the ground melt temperature. Thus the fuel can be believed considered as a solid, if

$$H \leqslant H_o=\sqrt{\frac{\text{æ}_2 T_m}{q}}.$$

2. Let's investigate the stability of lower boundary $z=\xi$ of thepulling down fuel layer with compactivity ε (in the undisturbed case $\xi=0$). Inhomogeneous disturbances are considered as large-scale in comparison with the layer thickness. The instability mechanism is connected with a modulation of the layer tickness caused by displacement of fuel. filling relief depression of the underlaing ground.

This result feeding back the inhomogeneous melting and futher modulation of ground relief. In this dynamic model we belive that at the initial stage of the instability a narrow layer of the fuel near the lower boundary, with a thickness of the order of the relief modulation amplitude takes part in the displacement. Such a case can be realized when the fuel layer adjascent to the underlying ground is less compact, so that fragments have higher mobility. Thus, the underlying ground relief is a slip surface. Fuel fragments in the depth of the layer are not displaced, because of sufficiently large internal friction. For this reason the upper boundary of the layer remains horizontaly. Thus, the temperature perturbation is caused inhomogeneous melt filtration and does not depend of fuel displacement, staying near the melt isoterm. Heat-conduction equation for ingomogeneous temperature are as follows:

$$\delta T_1+V_o\frac{\partial}{\partial z}(\delta T_1) = \chi_1\Delta(\delta T_1), \qquad z>H,$$

$$c_2\delta T_2+c_2V_o\frac{\partial}{\partial z}(\delta T_2)-(I-\varepsilon)c_1\xi T_1=\varkappa_2\Delta(\delta T_2), \qquad 0<z<H, \qquad (3)$$

$$\delta T_3+V_o\frac{\partial}{\partial z}(\delta T_3) = \chi_1\Delta(\delta T_3), \qquad z<0.$$

At the lower boundary $z=\xi$:

$$\delta T_2+\xi T_2=0,$$
$$\delta T_3+\xi T_3'=0,$$
$$\varkappa_2(\delta T_2'+\xi T_2'') - \varkappa_1(\delta T_3'+\xi T_3'') = -\lambda_{\text{пл}}\xi$$

Where the two first expressions reflect the fact that the lower boundary of the layer coinsides with the melt isoterm.The last expression describes continuity of heat-conduction fluxes with respect to the melt.

At the upper boundary $z=H$ the perturbation of temperature and heat- conduction fluxes are continuous.

Solution to equation (3) is searched in the form $\delta T \sim f(z)e^{\gamma t+ikx}$, where $f(z)$ vanishes at infinity. In thegeneral case this results a rather cumbersome dispersion expression. Consider two particular cases of practical intrest.

1.) $\varkappa_1=\varkappa_2=\varkappa, \qquad \chi_1=\chi_2=\chi$

If

$$\frac{qH^2}{\varkappa T_m} << 1,$$

then

$$\gamma(0) = A\frac{\chi}{H^2},$$

where A is the solution of the following equation

$$\frac{\lambda_m}{q}\cdot\frac{\chi}{H^2}\cdot A^2 + (3-\varepsilon)A + (I-\varepsilon)e^{-A} - 2 + \varepsilon = 0$$

Note, if $H<(\chi\lambda m/q)^{1/2}$ (very thin layer), then $\gamma \to q/\lambda m$.

If

$$\frac{qH^2}{\varkappa T_m} \gg I,$$

then

$$\gamma(0) = \frac{q}{(3-\varepsilon)cT_m}\cdot e^{-\frac{V_o H}{\chi}}$$

With an increase of parameter kH the increment γ decreases and if kH=0.5, then γ equals to zero.

2.) $c_1/c_2 \ll 1$. Volume heat capacity of the ground is essentially less than the volume heat-capacity of the fuel.

If $k = 0$, $H^2 \ll H_o^2$, then

$$\gamma(0)=\beta^2\frac{\chi_2}{H^2}$$

where β is the solution of equation

$$th(\beta) + \sqrt{\frac{\varkappa_2}{\varkappa_1}} - \frac{1}{\beta} = 0$$

If $\chi_2 \gg \chi_1$, then

$$\gamma=\frac{\chi_1}{H^2}(1-3k^2H^2).$$

3. Investigation of instability of the lower boundary of apulling down fuel layer demonstrates that the perturbation with k=0 has the maximum increment at fixed thickness of the layer. The obtained dependence of the increment on the layer thickness H canbe easily interpreted. At the region $H<(\chi\lambda m/q)^{1/2}$ modulationof the lower boundary with amplitude ξ causes additional surface heat source with intensity $\dot{\xi}q$. This heat source gives rise to inhomogeneous melting $\dot{\xi}\lambda m$ and heat conduction flux

$$\varkappa\sqrt{\frac{\gamma}{\chi}}\,\nabla T\xi$$

If $H \to 0$, then $\nabla T = \frac{V_o}{\chi} T_m \to 0$, because $\gamma \to q/\lambda m$.

If $\chi\lambda m < qH^2 \ll \varkappa Tm$, then the increment is determined from the balance equation

$$q\dot{\xi} = æ\sqrt{\gamma/\chi}\cdot\nabla T\xi.$$

This gives

$$\gamma \sim \frac{\chi}{H^2}$$

Note, that in this range of H the perturbation growth rate $\gamma\xi$ does not depend on parameters q and cTm like the pull down velocity. This is bacause the characteristic time of instability is less, than the time of pull down velocity stabilization H/Vo. Another case corresponds to large H>Ho. Here the increment is exponentially small and determinated by dependance of disturbance of homogeneous pull downing velocity with disturbance of lay thickness.

NUMERICAL SIMULATION OF THE PLANE LAYER INSTABILITY AND FUEL CONFINMENT IN THREE-DIMENSIONAL GEOMETRY.

The nonlinear phase of instability of ground melting by plane fuel layer and interaction of a single fuel fragment were investigated numerically. For this special thermalphysical codes were written in wich the following processes were taken into account:

1. Transient heat transfer in one,two and three dimensions caused by the dependences of the specific heat on temperature and spatial.
2. The motion of bodies pulling down through the melt, fuel intermixing in melt due to dissolving and convection, thermal convection,vaporiztion of active parts while heating the fuel, and subsequent condensation (absorbtion) while cooling the fuel.

In the framework of thermal model the motion of pouring bodies in a melt is simulated as fallig of a thermal sources on the melt bottom provided by a fuel volume and power dencity are constant values. Process of intermixing is simulated by distribution of volume thermal source in the melt.

Thermal convection caused by high temperatures(greater then the melt temperature) is described by effective heat transfer coefficient. The phenomena of vaporization and flux motion are described by introducing negative sources in the heat transfer equation. The influence of thenegative source is pronounced when the temperature is greater than boiling point.

We used two tecniques for numerical solution of two - dimensional heat transfer equation. These tecniques are described in [2,3]. The results otained with the help of these two tecnique are close each to ohter. For three-dimensional problem the method [3] was used.

The results of calculations show that the instability process very quickly (for the time of the order of $0.1t^*$, where t^* is the characteristic time of layer displacement to the depth compared with its thickness) comes to the nonlinear phase. Nonlinear phase of melting dynamics is essensialy depended on the initial temperature, amplitude, and disturbance scale and the directions of fuel fragment motion.

Based on numerical results consider some of the possible ways of fuel behavior in the nonlinear melting phase.

Cosine perturbation with amplitude $\delta=0.2$ cm and wave length $l=48$ cm is laid upon the lower boundary of the flat layer with thickness $h=1$cm and energy release power $q=0.3$ wt/cm^3 . It is assumed that heavy fuel parts fall on the melt bottom. According to the resuls of linear analysis such a layer should

be stable. But the linear theory considers perturbation of stady-state one-dimensional melting front. If the initial pertubation is not one-dimensional, the stabilization time can be equal or even greater then the evolution time of two-dimensional perturbations. In this case instability evolution is very sensitive to initial condition.

If initial conditions correspond to a stable phase of one dimensional melting the current perturbation decays with time. But if in the beginning the melting zone is absent then another picture is observed. In figure 1 three different moments of instability evolution are presented. Initial temperature of the fuel and the medium is $700°K$. One can see that melting begin in a zone of maximum thichness owing to the space heterogeniety of the layer. Fuel mirgration is possible only in the melting zone. Characteristic melting time is comparable with the time of heating the remaining fuel parts to the up melting point. Under these conditions a single drop is separated from the point of maximum layer thickness and its independent motion begin.

While heating the remaining "solid" fuel part, anohther drop can be generated. This drop moves in a hotter medium compared with the first one and at eventually overtake the first drop. Thus initially immovable perturbed layer has a tendency to splik into several drops which can later accumulate in more complex structures.

The size of the first separated drop practically does not depend on the perturbation wave length if the latter greater than *6* cm. If the wave lengty l equals *6* cm the first drop "consumes" the main part of the fuel layer. And as a result of thermal interaction the neighbor drops accumulate together and a stationary layer is formed again. The time of drop tearing off is equal to *4 days 1 hour 30min*. At the moment *4 days 10 hours 30min* the whole layer is restored again.

The three-dimensional simulation of confluence of two drops was carried too. Fig. 2 shows successive phases of confluence of two "three-dimensional drops". Initial form of the fuel drops is a cube with the side of *10.5* cm. The distance between them is *21* cm. Energy release power q equals *0.5* w/cm^3 , initial temperature is 321°K. The total confluence of two drops occures in about *25 hours*. Analogous calculations for two-dimensional case show that the confluence time is 16 hours.

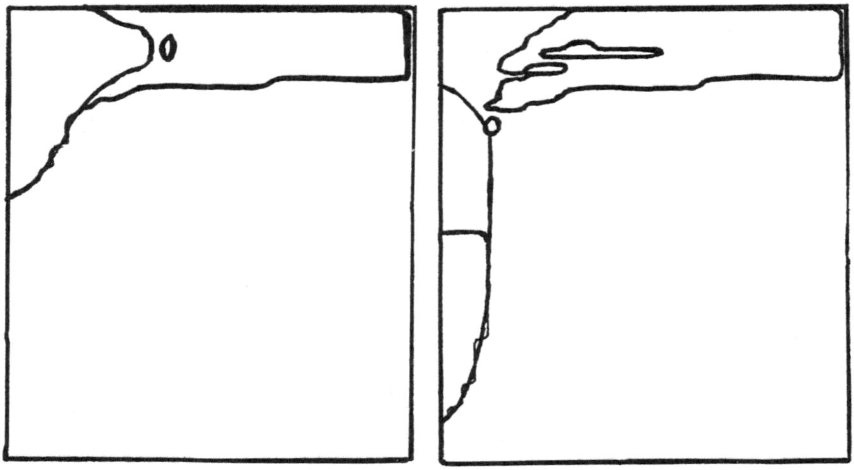

Figure 1

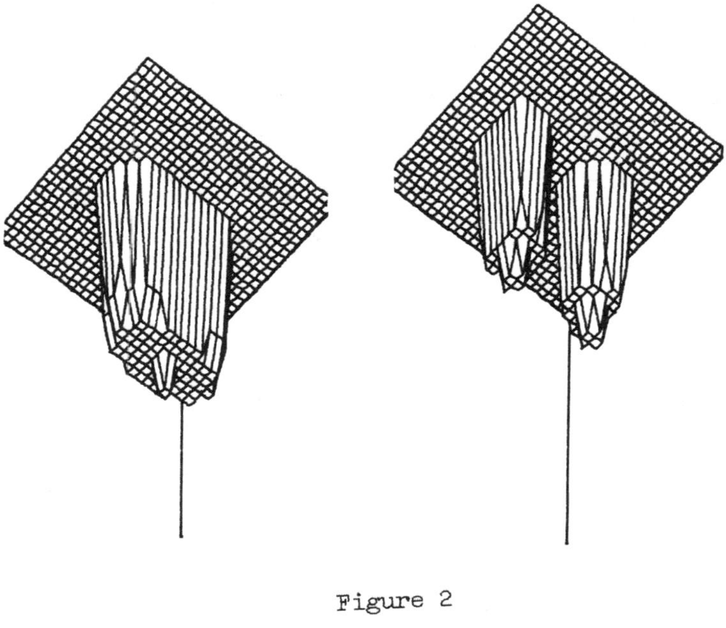

Figure 2

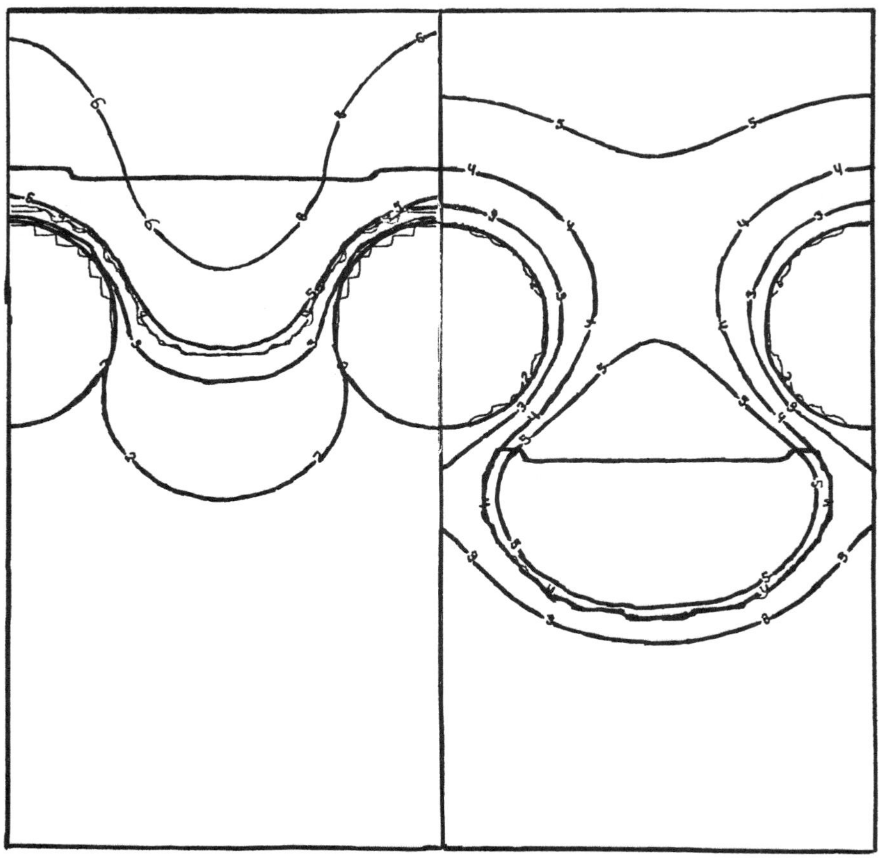

Figure 3

RELIABLE STOPPING SYSTEM STOPING AND SPREADING OF LARGE VOLUMES OF FUEL OVER COOLING PLATE.

The two-dimensional program for solving heat transfer equation with moving heat sources has been used to simulate fuel fall on a cooling trap plate. Figure 3 show the results of numerical simulation of fuel fall on the system of parallel pipes for spacing 30 cm.. In this case the break of safety system has taken place. If the piper spacing equals 26 cm the front of the melting fuel is stopped at the distance 1 cm from the surface of cooling pipes. It is assumed that the pipes are placed in sand or concrete with heat transfer conduction 0.02 w/cm·grad.

The absence of fuel spreading through the closely placed pipes is related with effective cooling the fuel "tongue" after its displacement by a distance of the pipe radius.

Thus, the fuel confinement by a system of parallel pipes essensially depends on the spacingbetween them. Even if this structure cantake awaythetotal heat flux from the falling uniform fuel layer, the fuel melt will spread down transforming into separate"tongues" provided that the spacing is greater then critical value, wich depends on the pipe diameter. For the diameter of 200 mm the spacing should be less than 26 cm.

Then dynamics of fuel falling on the pipe cooling safety system with requied spacing or on a flat cooling surface was investigated. After fuel melt front stops at adistance of the order ofseveral centimeters (it depends on the thermal characteristics of the medium: from 1 cm in sands to 4 - 5 cm in concretes) fuel spreads along the horizontal surface of thecooling plate due to concrete or sand melting at the side walls of the fuel.

REFERENCES

1. REVIEWS of MODERN PHYSICS, Vol. 57, No. 3, Part 2, July 1985.
2. V.M.Goloviznin, V.K.Korshunov et al. Preprint IPM AN USSR, Moscow, N 58 ,1985
3. R.V.Arutjunjan, L.A.Bolshov et al. Preprint IAE , Moscow, N 4121/16 , 1985

Radioactive Products over the Damaged Block-4 of Chernobyl Nuclear Power Plant before the Completion of "Sarcofag"

B. I. OGORODNIKOV
Karpov Institute of Physical Chemistry
Moscow, USSR

ABSTRACT

From the end of August - to the middle of September, 1986 radioactive samples were taken from the distance of 10-30 m over damaged block-4. Isotops Zr-95, Nb-95, Ru-103, Ru-106, Cs-134, Cs-137, Ce-141, Ce-144, Cm-242 and Pu-(238-242) were identified in all samples. In one sample from August 29 the concentration of gaseous form of I-131 was 0,6 Bq/m^3. The amount of ruthenium in gaseous form reached only 2-5% of amount of ruthenium in aerosols. Ba-14o, La-140, I-133, I-135 and other short-lived radionuclides were absent in the samples. It proved that in the nuclear fuel which remained in the damaged block-4 fission chain reactions did not proceed. On the basis of measurements and calculations of radioactive matter exhausts from the damaged block-4 we proposed making the roof of "Sarcofag" without breaking free convection but with the filtration station in store for protection.

INTRODUCTION

After the accident the temperature in the damaged reactor rose to 2500°C[1,2]. The basic heat return occured through the natural convection of air through the damaged reactor. Step by step the concentrations of the radioactive substances over the damaged block-4 diminished [1,3,4]. However before making the roof of "Sarcofag" the composition and the concentrations of the gases and the aerosols in the convectional flow was to be examined in detail so as to make the design of the ventilation system and the filtration station.

TECHNIQUE

At the end of August, 1986 a steel rope was throwen over the damaged block-4 from the south-east corner of that building to its west wall (Fig.1). The height of the rope over the damaged block-4 could be changed by stretching or loosening the rope. The distance was from the stretched rope to the damaged surface about 35 m. An electric car (funicular railway) was constructed to deliver the sampling apparatus to any point of the rope at the speed of 1 m/min.

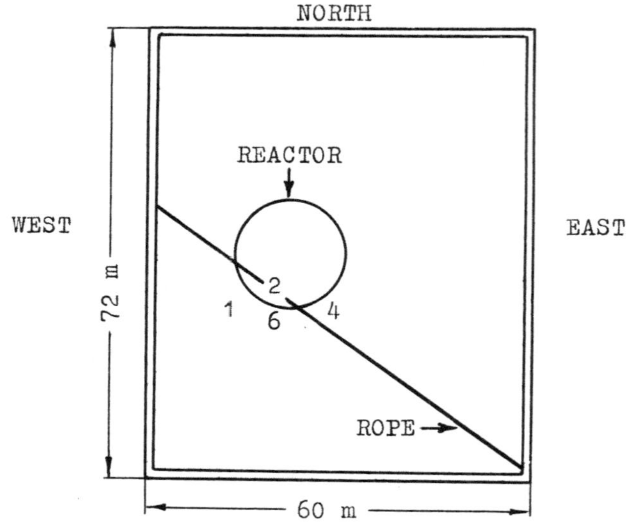

FIGURE 1. Positions of the steel rope and buoys no.1,2,4,6 over the damaged block-4.

In the first week of August several buoys were placed in the damaged block-4 by a helicopter [4]. Every buoy had the thermocouple, two thermoanemometers for measurements of the horizontal and vertical wind velocity and gamma-dosimeter.

Sampling of the radioactive products was made by the aspiration installations МР2-4Г, ШПА-2, ШПА-3, ПВ-20М and БАДБ-2 whose volume velocities were 0,01-0,1 m^3/min. A major part of the samples of aerosols was taken by the АФА fibre filters produced by ВО "Изотоп". In the БАДБ-2 sampler the СФЛ-И ribbon filter was used.

The filter pack technique [5] and the nuclear filters (the Flerov filters) [6] were used to define the particle sizes of aerosols. The contents of the gaseous forms of radioiodine and radioruthenium were measured by the АФАС-И sorbtion-filtration materials positioned in a filter pack behind АФА aerosol filters. The АФАС-И materials contained 2-5 mg/cm^2 of the Oy-A charcoal impregnated by $AgNO_3$ [7].

RESULTS

From the 28 of August to the 11 of September 11 radioactive samples were taken from the distance of 10-30 m over damaged block-4: 8 samples with the help of the devices which were moved along the rope and 3 samples with the help of the devices mounted on the end of a long stick. The conditions of sampling are given in Table 1. The samples as a rule were taken in the afternoon in dry weather, the air temperature being 20-22°C. Isotops Zr-95, Nb-95, Ru-103, Ru-106, Cs-134, Cs-137, Ce-141 and Ce-144 were identified in all samples.

TABLE 1. Conditions of sampling

N	Date 1986	Start	Duration, min	Place and height	Volume rate, m^3/min
1P	8/28	17^{25}	60	Over the reactor; 10 m	0.020
An	-"-	-"-	-"-	- " - - " -	0.013
2P	8/29	17^{30}	106	Movement to west wall and back; 20-30 m	0.020
3P	-"-	-"-	-"-	- " - - " -	0.013
An	9/03	15^{04}	90	Movement to reactor center, stop-over for 53 min and back movement; 20-30 m	0.013
5P	-"-	-"-	-"-	- " - - " -	0.020
6P	-"-	-"-	-"-	- " - - " -	0.100
7P	9/09	17^{18}	92	7 m from east wall; 30 m	0.049
8P	9/10	13^{30}	210	5 m from east wall; 20 m	0.020
9P	9/10-11	17^{45}	1035	- " - - " -	0.017
10P	9/11	13^{30}	120	- " - - " -	0.020

Their concentrations are tabulated in Table 2. In the 2P sample from August 29 I-131 was identified with the concentration of 0,6 Bq/m³. That isotop was contained in the air only in a gaseous form because it captured by АФАС-И layer with the charcoal placed in the filter pack behind 3 layers of aerosol filters. The amount of ruthenium in gaseous form reached only 2-5% the amount of ruthenium in aerosols.

Three measurements of alpha-activity of filters taken during 9 months (Fig.2) indicated that there was radionuclide with

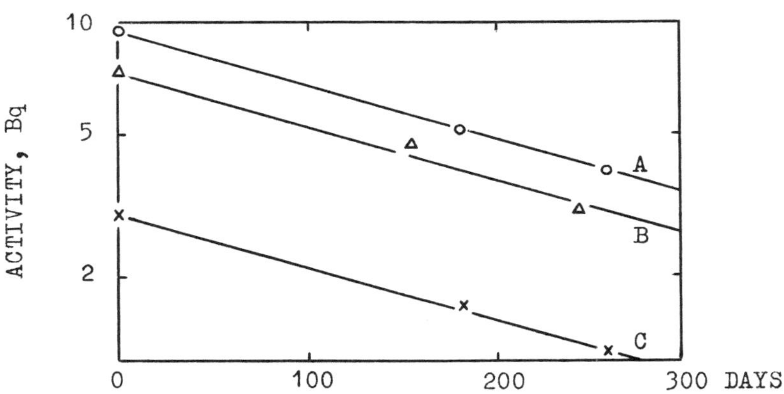

FIGURE 2. Alpha-activity of A,B,C filters depending on the time after sampling.

TABLE 2. Concentrations of radioisotops in aerosol samples, Bq/m^3

Nuclide	8/28		8/29		9/03			9/09	9/10	9/10–11	9/11
	1P	An	2P	3P	An	5P	6P	7P	8P	9P	10P
^{144}Ce	511	333	215	555	403	832	370	559	70	281	126
^{141}Ce	85	90	40	90	110	96	48	63	9	33	16
^{103}Ru	80	190	96	133	110	74	60	110	12	70	17
^{106}Ru	137	260	140	180	63	166	107	222	23	115	33
^{137}Cs	56	52	26	48	30	100	56	74	14	48	22
^{134}Cs	18	15	11	26	22	30	22	31	4	17	7
^{95}Zr	274	303	110	207	144	360	170	244	32	122	60
^{95}Nb	466	488	192	358	363	770	337	466	60	230	104
$\sum_{\gamma}$	1630	1630	830	1650	1150	2430	1170	1770	220	910	380

TABLE 3. Concentrations of alpha-activity in samples, Bq/m^3

N	Date 1986	$\sum_\gamma$	$\sum_\alpha$	$\sum_\gamma/\sum_\alpha$	^{242}Cm	$^{238-242}Pu$
1P	8/28	1 630	5.6	290	4.8	0.70
2P	8/29	830	1.8	440	1.6	0.23
3P	8/29	1 650	2.4	690	2.1	0.30
5P	9/03	2 430	4.1	600	3.6	0.51
6P	9/03	1 170	1.7	700	1.4	0.21
7P	9/09	1 770	7.4	240	6.5	0.92
8P	9/10	222	0.44	550	0.4	0.06
9P	9/10-11	914	1.1	830	1.0	0.14
10P	9/11	377	1.5	250	1.3	0.18

half-life about half a year. That was Cm-242 whose half-life is 162,8 days. The relation between alpha-activity of Cm-242 and long-lived isotops of Pu was to the end of August 7:1. This relation made it possible to calculate the concentrations of alpha-activity isotops in air over the damaged block-4 (Table 3).

Ba-140, La-140, I-133, I-135 and other short-lived radionuclides were absent in the samples. It proved that in nuclear fuel which remained in the damaged block-4 fission chain reactions did not proceed.

The concentrations of the radioactive aerosols were also investigated every day on the ground level near the АБК-2 building situated 300 m to the east of block-4. The sampling period ran between 12-36 hours. Fig.3 presents a plot of points resulting from these tests. During the period of the investigation the radioactive aerosol concentrations near the АБК-2 building did not exceed the limit level for the professionals given in USSR-s radiation safety standards НРБ-76 [8].

As can be seen from Fig.3 the sum of the gamma-activity of aerosols over the damaged block-4 was only ten times as big as that on the ground level near the АБК-2 building. Calculation of the aerosol transport based on the velocity and direction of the wind showed that the contribution of the exhaust emissions from the damaged block-4 consisted only of 1-10% of the radioactive aerosol concentrations observed near the АБК-2 building. The least concentrations of the radioactive aerosols were recorded near the АБК-2 building with the easterly or the calm. The biggest concentrations were with the westerly. So on August 21 the west wind velocity reached 10-15 m/s, on August 24 and 31 - 5-6 m/s. After the reactor accident at Chernobyl the largest land fallout took place on the west area of the 30-km zone therefore the west winds promoted the transport of large amounts of radioactive aerosols to the sampling point near the АБК-2 building.

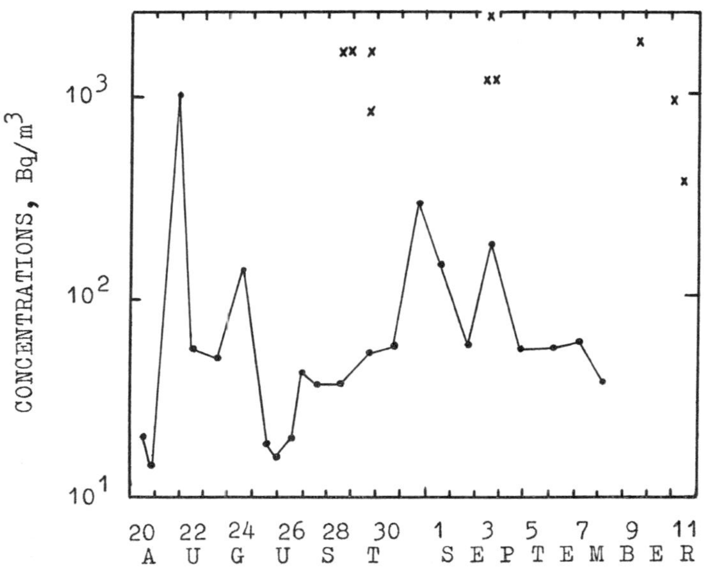

FIGURE 3. Concentrations of the sum of the aerosol gamma-activity over the damaged block-4 (points) and on the ground level near the AБK building (line).

The 8P and 9P samples from the zone of damaged block-4 were taken on September 10-11 before the completion of "Sarcofag". The filter holder with the filter pack was placed by a long stick over the desintegrated block-4. In the 8P sample the radioactive particles of Cs, Ce, Zr-95 and Nb-95 had the activity median aerodynamic diameter (AMAD) 0,95-1,15 μm. But the AMAD for Ru-103 and Ru-106 was a little smaller - only 0,7-0,9 μm.

The specific character of the particle size distribution for for ruthenium can be accounted for by the presence of volatile compound RuO_4 which is deposited from air on the most numerous particles with the diameters of 0,2 μm. This value usually corresponds to the maximum in the particle surface distribution.

During the collection of the 9P sample over the damaged block-4 the concentration of radioactive aerosols increased by several times. In this case all radionuclides were detected on the aerosols with AMAD more than 1,6 μm. The absence of radionuclide fractionations can be accounted for by a low temperature of the reactor desintegration by the middle of September.

The 10P probe was taken on September 11. The first layer in the sampling pack was a nuclear filter (the Flerov filter) with the pore diameter of 1 μm. The second and third layers were the AФA fibre filters. About 90-95% of radioactive materials were sampled by the Flerov filter. It meant that the greater part of radioactive particles had the diameter over

TABLE 4. Exhaust emission of aerosol gamma-activity over damaged block-4

Date 1986	Sampling period	Concentration, Bq/m^3	Vertical air velocity, (m/s). Data of buoys №				Exhaust, Bq/day
			1	2	4	6	
8/28	$17^{25}-18^{25}$	1 630	0.7	1.9	2.7	0.7	$5.2\ 10^{10}$
8/29	$17^{30}-19^{16}$	1 240	0.5	2.0	2.2	0.4	$3.3\ 10^{10}$
9/03	$15^{04}-16^{34}$	1 590	0.2	0.2	0.1	0.2	$0.7\ 10^{10}$
9/09	$17^{18}-18^{45}$	1 770	0.4	0.4	0.2	0.2	$1.1\ 10^{10}$

1 μm. That result confirmed the data obtained by the fibre filter pack technique.

The data on the aerosol concentrations (Table 2) and the velocity of updraft of heated air (Table 4) made it possible to define the radioactive exhaust emission from the damaged block-4. Since the reactor diameter is equal to 18 m it was the surface of the radioactive discharge was estimated 250 m^2. As can be seen in Table 4 at the beginning of September the summarized exhaust of aerosol gamma-activity did not exceed 10^{10} Bq per day.

CONCLUSION

On the basis of measurements and calculations of radioactive matter exhausts from the damaged block-4 we proposed making the roof of "Sarcofag" without breaking existing free convection but with the filtration station in store for protection.

REFERENCES

1. The accident at Chernobyl nuclear power plant and its cosequences, Information compiled for the IAEA experts' meeting, 25-29 August 1986, Vienna, Austria, USSR State committee on the utilization of atomic energy, August 1986.

2. Devell, L., Tovedal, H., Bergström, U., Appelgren, A., Chyssler, J. and Andersson, L., Initial observations of fallout from the reactor accident at Chernobyl, Nature, vol. 321, pp. 192-193, 15 May 1986.

3. Izrael, Yu.A., Petrov, V.N., Avdyushin, S.I., Gasilina, N.K., Rovinsky, F.Ya., Vetrov, V.A. and Vakulovsky, S.M., Radioactive contamination of the environment around the Chernobyl nuclear power station, Meteorology and hydrology, no.2, pp. 5-18, Moscow, 1987.

4. Asmolov, V.G., Borovoj, A.A., Demin, V.F., et al. The accident at Chernobyl nuclear power plant: after the first year, Atomnaja energuija, vol.64, no. 1, pp. 3-23, Moscow, 1988.

5. Bogolapov, N.W., Kastchenko, N.N., Konstantinov, I.E., Ogorodnikov, B.I. and Skitovich, V.I., in Questions of dosimetry and radiation protection, ed. L.R.Kimel and V.K.Sacharov, no. 5, pp. 136-139, Atomizdat, Moscow, 1975.

6. Flerov, G.N., Synthesis of superheavy elements and application of nuclear physics methods in adjacent spheres, Vestnik AN USSR, no. 4, pp. 35-48, 1984.

7. Borisov, N.B., New sorbtion-filtration materials for analyses of aerosols and vapor, Isotops in USSR, no. 52/53, pp. 66-67, Atomizdat, Moscow, 1978.

8. Radiation safety standard НРБ-76, Energoizdat, Moscow, 1981.

The TMI-2 Accident Evaluation Program

D. J. OSETEK, J. M BROUGHTON, and R. R. HOBBINS
Idaho National Engineering Laboratory
EG &G Idaho, Inc.
Idaho Falls, Idaho 83415, USA

INTRODUCTION

The accident at the Three Mile Island Unit 2 (TMI-2) reactor, now 10 years old, remains as the United States' worst commercial nuclear reactor accident. Although the consequences of the accident were restricted primarily to the plant itself, the potential consequences of the accident, should it have progressed further, are large enough to warrant close scrutiny of all aspects of the event.

The TMI-2 accident resulted in extensive oxidation and melting of the reactor core and in significant release of fission products from the fuel. At least 45% (62 metric tons) of the core melted and about 20 metric tons of molten core material relocated into the lower plenum of the reactor vessel. However, in spite of severe core damage, water in the reactor vessel prevented failure of the reactor vessel's lower head. The TMI-2 accident was similar to a station blackout or smallbreak LOCA accident sequence like those postulated in reactor safety studies. The accident progressed into the latter stages of core melt progression, that is, generally much further into the progression of a severe accident than core melt experiments have explored. As such, the accident greatly expands and enhances the current understanding of core melt progression and fission product behavior during severe accidents and provides a unique data base for assessment of severe accident computer models.

TMI-2 accident research is being conducted by the U.S. Department of Energy (DOE) and GPU Nuclear Corporation with support from many different organizations to provide a better understanding of severe reactor accidents [1]. Research objectives supporting this goal include developing a comprehensive understanding of the mechanisms that controlled the progression of core damage and subsequent fission product behavior during the TMI-2 accident, and applying that understanding to the resolution of important severe accident safety issues. Developing a best-estimate scenario of the core melt progression during the accident is the focal point of the research and involves analytical work to interpret and integrate: (1) data recorded during the accident from plant instrumentation, (2) the post-accident state of the core, (3) results of the examination of material from the damaged core, and (4) related severe accident research results.

This work was supported by the U.S. Department of Energy, Assistant Secretary for Nuclear Energy, Office of LWR Safety and Technology under DOE Contract Number DE-AC07-76ID01570.

This paper summarizes the TMI-2 Accident Evaluation Program that is being conducted for the U.S. DOE and briefly describes the important results that have been achieved. The Program is divided into four parts: Sample Acquisition and Plant Examination, Accident Scenario, Standard Problem Exercise, and Information and Industry Coordination. Each of these Program parts is described in the following sections. Details about the post accident condition of the core and the accident scenario are taken from Reference 2 supplemented by recent findings.

SAMPLE ACQUISITION AND PLANT EXAMINATION

Video inspections, debris sampling, and core boring data have provided the information that was used to reconstruct the post-accident condition of the TMI-2 core and vessel components. Figure 1 illustrates this end-state geometry. Physical examination and chemical analyses of material samples collected from various vessel regions have identified the types of interactions that occurred and the temperatures that were reached in the different parts of the core and in the relocated material. The DOE Program of core sampling is nearly complete; only a few additional samples of lower plenum debris will be collected before this part of the Program is finalized. Sample analysis by various laboratories in the U.S. and other countries is ongoing. Some results have been presented at the ANS Topical Meeting held in Washington, D.C. in November 1988 [3]. Further information will be published in 1989 and final results in 1990.

There is also a new program under way sponsored by the U.S. NRC and the Nuclear Energy Agency (NEA) of the Organization for Economic Cooperation and Development (OECD) to collect and analyze samples of steel from the inside surface of the TMI-2 vessel lower head and samples of the debris or prior molten material that is adjacent to the lower head. The results of this program will become available late in 1990. What has been learned up to this time about the post-accident condition of the TMI-2 reactor vessel and core is summarized in this section. Further details are provided in References 3-6.

Upper Plenum Region

The upper plenum assembly suffered little damage and was found to be essentially intact. Examination of the control rod drive lead screws taken from the upper plenum indicates that temperatures reached approximately 1250 K just above the central region of the core, approximately 1000 K near the core periphery, but only about 700 K at elevations above the outlet nozzles. The extent of melting and ablation of the upper grid was bi-nodal highly localized within two damage zones about 1.2 m in diameter, and did not progress significantly above the upper grid.

Core Region

Four distinct regions within the original core have been identified, as shown in Figure 1. These include: (1) a void region in the upper part of the core; (2) a region of loose, fragmented fuel debris immediately below the void region; (3) a region of consolidated, previously molten core material located in the mid-to-lower region of the core; and (4) partial to full length minimally damaged fuel rods in the lower and peripheral regions of the core.

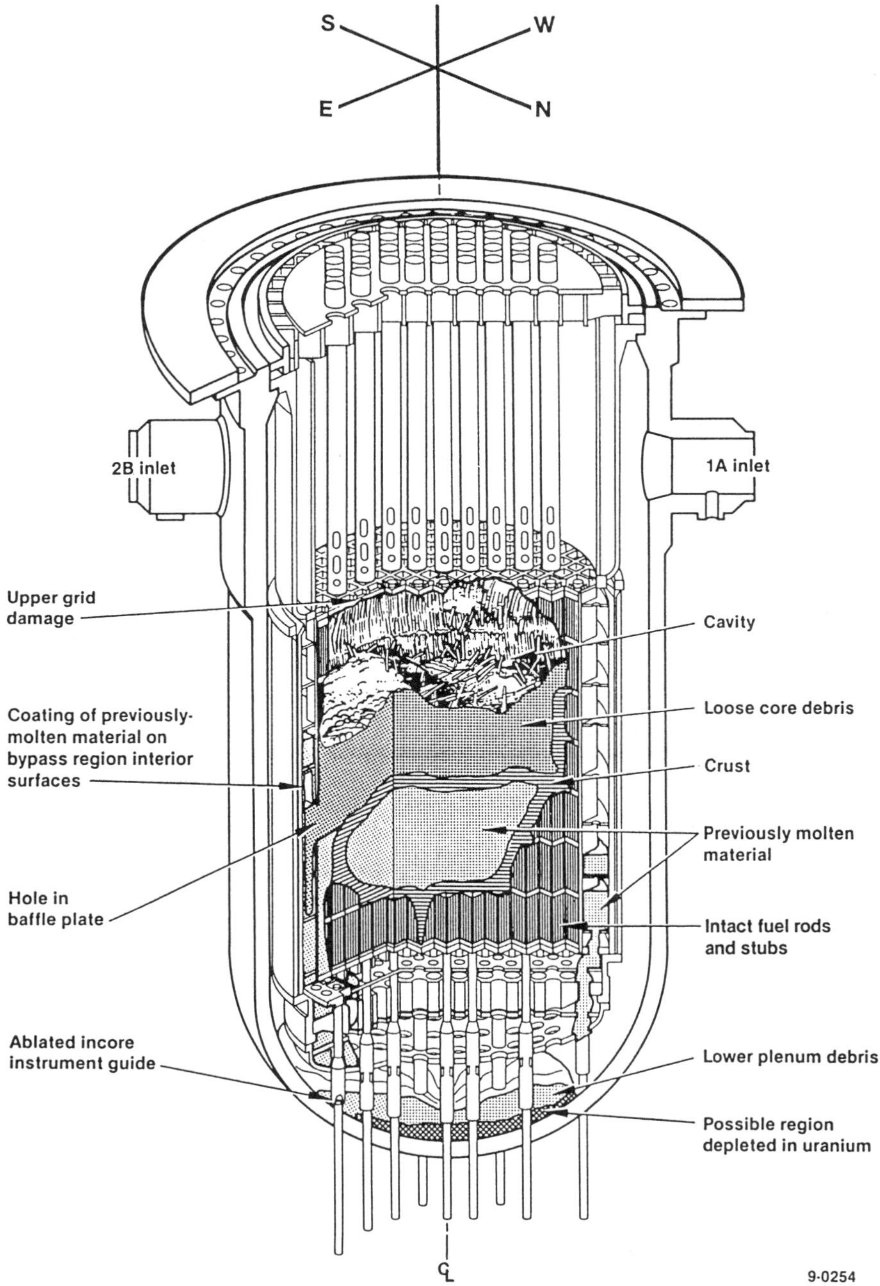

FIGURE 1. Post accident configuration of the TMI-2 core and vessel structures.

The volume of the cavity at the top of the core was approximately 9.3 m^3. The average depth of the cavity was approximately 1.5 m; in places, maximum depth approached 2 m. Standing fuel rods remained at the core periphery in 42 of the 177 fuel assemblies. In only two of these assemblies were 90% of the rods intact over the entire length; the other fuel assemblies suffered varying degrees of greater damage.

A debris bed, ranging from 0.6 to 1.0 m in depth, rested on top of the previously molten core material region located at about core mid-plane. The debris consisted of whole and fragmented fuel pellets, oxidized cladding shards, and previously molten metallic and ceramic particles. The debris was found to contain some particles of previously molten U-Zr-O, indicating peak temperatures greater than 2200 K. A few particles were previously molten $(U,Zr,)O_2$ and UO_2, indicating some regions of the upper core experienced temperatures between 2800 and 3100 K.

Beneath the upper core debris bed was a large consolidated mass, approximately 3 m in diameter, containing about 32.7 metric tons of previously molten core material. The central region of this consolidated mass was primarily previously molten ceramic $(U,Zr)O_2$ laced with previously molten metallic material. The metallic structures consisted primarily of silver and iron. A crust was identified around the consolidated region. The bottom, bowl-shaped portion of this crust, with an approximate thickness of 10 cm, was composed of previously molten metallics, primarily zirconium, silver, and iron, frozen in the flow channels between fuel rods. Standing columns of fuel pellets devoid of cladding were all that remained of the fuel rods in the lower crust. The top crust consisted of previously molten ceramic and metallic materials from 1 to 3 cm thick, with some embedded debris from the upper debris bed. The metallic material in the top crust was primarily iron and silver, with some nickel and indium.

Standing fuel rod stubs were found extending upward from the lower fuel assembly grid to the bottom of the consolidated region. These stubs varied in length from approximately 0.2 to 1.5 m. On the east side of the core, four adjacent fuel assemblies were almost completely replaced with previously molten core material, indicating that molten core material flowed downward through this region.

Core Support Assembly

In the upper core support assembly (CSA) fuel debris lay on the core former plates, between the baffle plates (vertical plates forming the peripheral boundary of the core) and the core barrel. The location of the debris varied in elevation from 1.5 m in the north quadrant of the vessel to almost zero near the southwest quadrant. The baffle plates in the east quadrant were partially melted, exposing the core barrel to potential damage. Molten core material entered the upper CSA through this hole, flowed circumferentially in the CSA, and drained into the lower CSA at nearly all locations around the core.

The lower CSA suffered little damage. Beneath several fuel assemblies in the east quadrant, resolidified material completely filled flow holes in the various plates, and columns of material between the plates suggested that a major pathway for flow of molten core material from the core into the lower plenum was in the southeast quadrant of the reactor vessel.

Lower Plenum Region

Extensive video examinations of the lower plenum were used to construct a contour map of the top surface of the lower plenum debris bed. Visual observations indicate that the size of debris particles varies widely, ranging from less than a few centimeters in diameter near the center of the vessel, to larger rock-like pieces tens of centimeters across near the vessel wall. The debris particles extend radially to the downcomer, except in the north quadrant, where an abrupt "wall" of previously molten material exists about 1.0 m from the vessel centerline.

Fission Product Distribution

The distribution of fission products in the TMI-2 reactor system has been determined by analysis of representative samples collected from the various locations in the reactor vessel and the balance of the reactor facility. The low volatility fission products that have been measured in TMI-2 samples include Ce-144, Eu-154, and Eu-155. The medium volatile species measurable are Sr-90, Sb-125, and Ru-106. The high volatiles are the noble gases, iodine and cesium; Kr-85, Cs-137, and I-129 were released from the fuel to a large extent at the temperatures found during the TMI-2 accident.

The data indicate that the low volatile fission products were not transported out of the reactor core and that only a small amount of fuel material was physically transported to the reactor coolant system. In the reactor vessel, the fuel material (and the associated low volatile fission product) distribution is similar to the overall mass distribution as might be expected.

The fission product distribution data for the medium volatiles indicate that only a small amount (3.5% of the core inventory) was released from the reactor vessel. Of this released quantity, almost all was retained in the reactor building with only 0.1% released to the auxiliary building where it was retained in the reactor coolant bleed tanks. In the reactor vessel, the distribution is again similar to the fuel material distribution which indicates that the bulk of the medium volatile fission product inventory was retained in the prior molten fuel material. The Sb-125 and Ru-106 data indicate high concentrations in the prior molten metallic portions of the lower reactor core. The metallic sample concentrations of Sb-125 are 6-20 times those found in the intact fuel.

The data for the high volatiles (Kr-85, I-129, and Cs-137) indicate significant relocation and transport of these radionuclides to the reactor coolant system and, consequently, to the reactor building atmosphere and basement. Approximately 85% of the core inventory of Kr-85 is accounted for by measurements of the containment building atmosphere.

Measurements of radioiodine concentrations in the reactor building basement, the major repository for iodine outside the reactor core, indicate that much of the core inventory of radioiodine could be contained in the reactor building basement. The data show the same retentions for iodine and cesium in the basement.

In the reactor vessel, the data indicate similar distributions in the debris for both the Cs-137 and I-129. The data for the upper core debris indicate that about 20% of the original inventories of both Cs-137 and I-129 were retained in the partially molten fuel debris. In the lower parts of the core there is evidence that suggests that some of the I-129 is found in association with metallic material.

Data from the consolidated melt region indicate an overall Cs-137 retention of 10% in prior molten material. This holdup of Cs-137 in high temperature material is surprising.

ACCIDENT SCENARIO

A postulated scenario of events that occurred during the accident was developed in the DOE Program from the post-accident core configuration, interpretation and analyses of data from on-line instrumentation, and analyses of various postulated core-damage-progression phenomena. The accident was divided into seven major periods marked in time after scram of the reactor (t=0). The reactor's coolant-system-pressure history, shown in Figure 2, provides a convenient reference with which to view these periods.

Loss of Coolant (0-100 min)

The accident was initiated by loss of feedwater to the steam generators, a "small break" loss-of-coolant accident, with limited ECCS. The resulting increase in the RCS pressure caused the pilot-operated relief valve (PORV) on the pressurizer to open and the reactor to scram. The PORV failed to close as the reactor cooling system (RCS) pressure decreased. Because of decreasing liquid inventory in the primary cooling system and continued heating by decay heat in the fuel, coolant void fraction increased sufficiently over the next hour to cause the reactor pumps to vibrate excessively from cavitation. At about 73 min, both pumps in the B-loop were shut off, resulting in reduced flow in the B-loop and loss of heat removal by the B-loop steam generator. At about 100 min, the A-loop pumps were shut off, terminating forced convection cooling of the core. The vessel's resulting liquid level is estimated to have been near the top of the core, shortly after the last coolant pump shutdown.

Initial Core Damage (100-174 min)

The liquid level in the reactor vessel decreased after 100 minutes because makeup to the primary cooling system was insufficient to compensate for coolant loss through the PORV. Core uncovery began shortly after 100 min, and temperatures at the top of the core started to increase. By approximately 140 min, the core liquid level had dropped to about mid-core. The upper regions of the core heated sufficiently (1100 to 1200 K) to result in cladding ballooning and rupture. Cladding failure and release of gaseous fission products were substantiated by significant increases in containment radiation levels commencing at 139 min. At approximately this same time, the reactor operators recognized coolant was being lost through the faulty PORV and manually closed the pressurizer block valve upstream of the PORV, thus temporarily terminating coolant loss and the release of gaseous fission products to the containment.

Core temperatures continued to increase between 150-165 min. Peak core temperatures probably reached at least 1500 K between 150 and 165 min. By 165 minutes, core cooling water probably dropped to within 1 m of the bottom of the core. This drop would have resulted in downward, as well as radial, progression of core damage.

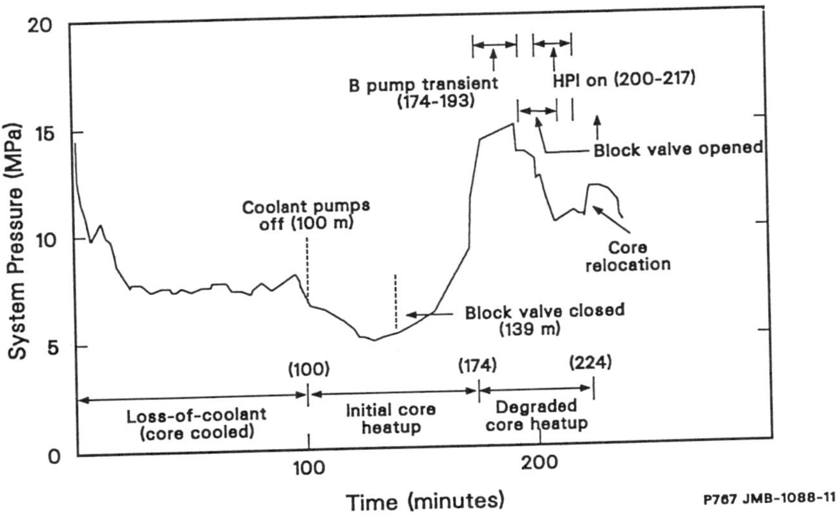

FIGURE 2. Pressure history of the TMI-2 primary coolant system during the accident.

As shown in Figure 2, the RCS pressure increased approximately 5.5 MPa between 150 and 174 minutes. This increasing RCS pressure was being driven primarily by the continued progression of core damage. Zircaloy cladding oxidation at temperatures above 1500 K was probably vigorous, producing significant quantities of hydrogen that subsequently degraded heat transfer in the steam generator to the secondary system, and also rapidly heated the core above temperatures required to melt control rods (1700 K) and fuel rod cladding (2150 K).

Eutectics of Ni-Zr and Fe-Zr would be the first liquid formed as a result of interaction between the Inconel grid spacers and zircaloy cladding, probably near the center of the core. The stainless steel control rod cladding would melt as soon as temperatures exceeded approximately 1700 K, releasing molten Ag-In-Cd control material (melting point 1100 K) and allowing it to flow to the liquid/steam interface and freeze with the Ni-Zr and Fe-Zr eutectics. Analysis of lower crust samples reveals that the crust is a Zr-Ag-In-Fe-Ni metallic mixture surrounding standing columns of fuel pellets.

The process of zircaloy oxidation, melting of core metallic components, UO_2 dissolution, and relocation of molten material downward to freeze and block coolant flow channels at or near the coolant/steam interface would continue throughout this phase of the accident. This process would be expected to gradually progress outward to the periphery of the core, forming essentially a complete blockage of the central region of the core by 174 min.

Above the lower crust a region of (at least partially) molten metallics and ceramics formed. Core heatup calculations indicate peak temperatures within this region of consolidated core materials may have reached fuel melting (i.e., 3100 K) and the average temperature of the material was probably between 2600 and 2800 K.

Undamaged rod stubs below the bottom crust indicate that coolant levels did not drop below 0.5 m for any significant period of time during the accident. Water covering the bottom of the core ensured that the lower supporting crust would be cooled and probably be structurally stable throughout the accident.

The dominant release mechanism for the volatile fission products during the initial core heatup between 100 and 174 min was diffusion of gaseous fission products through the fuel matrix until the fuel was liquefied by metallic zircaloy [7].

B-loop Pump Transient (174-180 min)

Activation of the 2B coolant pump at approximately 174 min resulted in the first significant addition of coolant to the reactor vessel following the A-loop pump shutdown at about 100 min. The 2B-pump was allowed to operate approximately 19 min; however, significant flow in the B-loop hot leg was measured only for the first 15 s. The reactor cooling system experienced a rapid repressurization as a result of the pump transient, as shown in Figure 2.

The thermal-mechanical forces resulting from partial quenching of the oxidized rod remnants in the top half of the core are believed to have fragmented the oxidized cladding and fuel pellets to form the upper debris bed. The measured fission product release from the upper core debris bed was somewhat higher on average than that calculated by diffusional release from fuel particles. The debris particles experienced extensive intergranular fracturing, most likely during formation of the debris bed during the 2B-pump transient. Fuel fragmentation during a rapid reflood has been correlated with enhanced release of fission products trapped at grain boundaries [9] and was observed during the PBF severe fuel damage scoping test [10].

Degraded Core Heatup (180-200 min)

During this period, core liquid decreased as decay heat from the degraded core boiled liquid from the reactor vessel. The low thermal diffusivity of the consolidated (primarily ceramic) core material, together with the region's relatively large size, would prevent the interior from cooling, even if the reactor vessel was filled with water. Degraded core heatup calculations indicate that a pool of molten material formed in the center of the consolidated region and increased in size during this period.

Degraded Core Heatup (200-224 min)

At 200 minutes, the high-pressure injection system was actuated and emergency cooling water was injected for the next approximately 17 min. As emergency cooling water filled the reactor vessel, water began to penetrate the upper debris bed, cooling and eventually quenching the debris bed. Calculations by Kuan [11], based on estimates of the debris bed mass, initial temperature, and particle size distribution from debris bed examination data, indicate the debris bed was coolable, i.e., heat generation within it was less than dryout heat flux. Subsequent cooldown and quenching of the debris bed is estimated to have taken approximately 25 min. Thus, final quenching of the debris bed is estimated to have occurred by approximately 230 min.

The consolidated region would have continued to heat up during this period even though the reactor vessel was probably filled with water. Most consolidated core material is estimated to have been molten and to have continued to heat up; water is estimated to have covered the core; and the upper debris bed is estimated to have been nearly, if not fully, quenched.

Fission product release during this phase of the accident was probably dominated by the thermal and hydrodynamic behavior of molten corium, by volatility of fission products, and by dynamics of bubble formation, coalescence, and mobility [7]. Calculations indicate that release of volatile fission products was probably rapid (approximately 5 min), due to formation of large buoyant bubbles from small fission gas bubbles as a result of bubble coalescence and rise and of convective mixing within the molten pool. These large gas bubbles would have been trapped inside the crust until crust failure. The medium and low volatile fission products were estimated to remain as oxides or metallic phases in the melt. Little release is expected because of the low vapor pressure and low mole fraction of these materials within the pool.

Crust Failure and Material Relocation into the Lower Plenum (224-226 min)

Relocation of about 20 metric tons of molten corium into the lower plenum of the reactor vessel occurred at approximately 224 min after reactor scram. The A-loop cold leg temperature increased to a maximum in less than 20 s, while the B-loop temperature reached a maximum in approximately 1 minute. The rapid increase in cold leg temperature was probably due to the flow of steam generated in the lower plenum. The RCS pressure also increased approximately 1.8 MPa (260 psi) between 223.5 and 225.5 min. This increase was probably also caused by steam rapidly produced during this limited time period. The rapid steam production occurs as a result of heat transfer from the molten core material to the water. Nothing in recorded data or post-accident core condition suggests an energetic steam explosion occurred as the tons of molten corium relocated into the lower plenum with the reactor vessel essentially full of water.

The crust failure appears to have been in the upper half of the consolidated region near the core periphery in the southeast quadrant of the reactor vessel, and material apparently flowed downward into the lower plenum through both upper CSA and peripheral fuel assemblies. Continued heating of the molten pool could lead to melting of the supporting crust. Thermal analysis that considered a range of initial and boundary conditions was performed and indicated that the steady state thickness of the supporting crust would be 6 to 10 cm at the bottom and only about 1 cm at the top. Heat transferred from the molten pool primarily to the top crust, and this accounts for a thinner top than bottom crust. The calculated steady state crust thicknesses are consistent with observed crust thicknesses obtained from core bore samples. A 1 cm crust would probably fail from the overlying weight of debris.

Another important aspect related to core relocation is interaction between molten corium and water in the reactor vessel. Both fine particulates and large structures of previously molten debris have been observed in the lower plenum, suggesting that some degree of interaction occurred between the molten material and the coolant during the relocation into the lower plenum.

Release of volatile fission products during molten corium relocation was probably governed primarily by breakup and fragmentation of the molten stream as it penetrated water in the lower CSA and lower plenum; temperature of the molten corium; fission product concentrations and chemical species; and the high oxidation potential of the steam environment. Results of examination of ceramic samples from the central consolidated region and lower plenum debris indicate that Sb and Ru concentrate in the molten metallic debris and that both I and Cs are retained to some extent in the previously molten ceramic debris, i.e., $(U,Zr)O_2$.

RCS Recovery Period (226 min-15.5 h)

The molten corium (ceramic and metallic phases) rapidly settled onto the reactor vessel lower head and was probably not cooled significantly by water during relocation. The RCS pressure pulse between 224 and 240 min, implies that heat transfer and steam generation within the debris bed was significant for at least 15 min. Re-formed thermocouple junctions probably embedded in the debris indicate temperatures greater than 1000 K existed within the debris bed up to three days. These high temperatures and the high RCS pressure may have seriously challenged the structural integrity of the reactor vessel lower head. However, the lower head was never breached. At 15.5 h after reactor scram, forced coolant flow through the core was reestablished upon the restart of one of the A-loop primary coolant pumps.

Calculation of the thermal/mechanical response of the reactor vessel lower head was performed using a bounding range of debris bed configurations and assumed a top down quench in 20 min. These calculations indicate that thermal stress in the lower head dominated and that some limited strain may have occurred in the structure's outer regions. This analysis indicates that cooling of the inside surface of the reactor vessel lower head must have been reestablished within about 30 minutes after the relocation. Otherwise, head temperatures would have exceeded 1100 K and creep failure of the lower head would have likely occurred. Despite the fact that this relatively simple analysis establishes an upper bound on vessel temperature, it provides little insight into thermal and mechanical processes that would have challenged the integrity of the reactor vessel lower head. The actual configuration and structure of lower plenum debris and detailed metallographic examinations of the lower head are needed to reduce the uncertainties in these analyses and define the margin-of-safety of the lower head.

STANDARD PROBLEM EXERCISE

As a means of assessing and improving the computer codes that are used to describe severe reactor accidents, the DOE Program also supports a Standard Problem Exercise. A Joint Task Group has been established in collaboration with the OECD Nuclear Energy Agency (NEA) to analyze various periods of the accident. At this time, the Joint Task Group includes analysis with eight different computer codes by representatives of eleven different organizations in eight countries. A detailed report of the analysis exercise comparing calculated results with measurements from the TMI-2 accident will be published on completion of the exercise in 1990; preliminary results are given in Reference 6.

Since there remains some uncertainty in the boundary conditions, it is not unreasonable that the codes produce results (e.g., in calculated reactor coolant system pressure) that deviate from the recorded data. While the calculated results for both reactor coolant system pressure and core temperatures are qualitatively similar to the accident data, the calculated results are quantitatively quite different from the data. Thus it has been concluded that although the codes have the capability to predict severe accidents, there is need for continued refinement and development.

INFORMATION AND INDUSTRY COORDINATION

The importance of the TMI-2 accident to the world nuclear industry was recognized early by the Accident Evaluation Program. To help achieve the maximum benefit of the lessons learned from the accident, the AEP includes a special task to collect, organize, and disseminate information about the accident and the results of the plant examinations and analyses to the world nuclear community, the news media, and other interested persons. This effort has provided many useful communications over the past few years and has culminated with the organization of the ANS Topical Meeting on the TMI-2 accident in Washington, D.C. in November 1988. Over 130 papers and two panel discussions were included in the meeting. Paper summaries appear in the ANS transactions [3], and the full papers from this meeting will be published as a series of journal articles in Nuclear Technology. Expected publication dates are August-November 1989.

CONCLUSION

The accident at Three Mile Island Unit 2 resulted in severe fuel damage, much beyond that expected to occur during design-basis accidents. Most aspects of core and RCS thermal/hydraulic responses during the accident have been evaluated, starting from accident initiation at reactor scram, to establishment of coolant flow about 15.5 h after scram. The TMI-2 accident provides valuable insight into phenomena and processes controlling core melt progression and fission product behavior through the "late phase" of a core melt progression accident.

Although not all aspects of the accident have been quantified with the degree of uncertainty desired from integral-effects experiments, the TMI-2 accident is the only source of data from a full size severe accident event for application to the resolution of technical issues. Progression of the accident to include a large scale thermal/mechanical challenge to the lower head is much beyond existing or planned experiments, both in duration and physical size. The first application of TMI-2 research results to technical issues has been the OECD/CSNI analysis exercise, which assesses severe accident analysis codes and methodologies.

Through the DOE TMI-2 research program, an inventory of core materials has been established to account for all major core components. The post-accident distribution of fission products within the plant has been determined and shows that, within the uncertainty of the determination, most fission products are accounted for. Application of these DOE research results is only beginning, and examination of reactor vessel lower head samples will be accomplished under sponsorship of the OECD/NRC cooperative research program. Thus, TMI-2 accident research will continue to provide new insight into reactor safety and severe accident phenomena for the next few years.

ACKNOWLEDGMENT

This paper has summarized the four principal parts of the U.S. DOE TMI-2 Accident Evaluation Program. However, it must be realized that the results presented here have been developed by many different organizations and that the principal reactor examination work has been accomplished by GPU Nuclear Corporation, the owners of the TMI-2 nuclear plant. The detailed examinations of core debris samples and much of the core melt progression analysis has been done by EG&G Idaho, and additional analyses have been provided by other DOE, NRC and EPRI contractors. For a better understanding of the contributions made by the various organizations the reader is referred to References 2 and 6 and the other references cited here.

REFERENCES

1. Tolman, E., TMI-2 Accident Evaluation Program, EGG-TMI-7048, EG&G Idaho, February 1986.

2. Eidam, G. R., Tolman, E. L., Broughton, J. M., McCardell, R. K., and Stratton, W. R., TMI-2 Defueling Conditions and Summary of Research Findings, Proceedings of the International Symposium on Severe Accidents in Nuclear Power Plants, Sorrento, Italy, March 21-25, 1988.

3. American Nuclear Society Transactions, 57, TANSAO-57, 1988, pp. 394-548.

4. Adams, J. P., TMI-2 Lower Plenum Video Data Summary, EGG-TMI-7429, EG&G Idaho, July 1987.

5. Akers, D. W., TMI Core Debris Grab Samples--Examination and Analysis, GEND-INF-0075, September 1986.

6. Broughton, J. M., Kuan, P., Petti, D. A., and Tolman, E. L., A Scenario of the TMI-2 Accident, Nuclear Technology, August 1989.

7. Golden, D. W., Summary of the TMI-2 Analysis Exercise, Nuclear Technology, August 1989.

8. Petti, D. A., Adams, J. P, Anderson, J. L., and Hobbins, R. R., Analysis of Fission Product Release Behavior During the TMI-2 Accident, Nuclear Technology, August 1989.

9. Fehrenbach, P., Hastings, I. J., Lipsett, J. J., Hunt, C. E. L. Rose, D. H. and Davidson, R. D., High Temperature Transient Fission Gas Release From UO_2 Fuel: Microstructural Observations, Proceedings of the International ANS/ENS Topical Meeting on Thermal Reactor Safety, San Diego, CA, February 1986.

10. Knipe, A., Ploger, S. A., and Osetek, D. J., PBF Severe Fuel Damage Scoping Test - Test Results Report, NUREG/CR-4683, August 1986.

11. Kuan, P., TMI-2 Core Debris Bed Coolability, EG&G Idaho Report EGG-TMI-7150, March 1986.

The TMI-2 Scenario

G. R. EIDAM
Three Mile Island
Bechtel National, Inc.
P.O. Box 72
Middletown, Pennsylvania 17057, USA

INTRODUCTION

The March 1979 accident of the Three Mile Island Unit 2 (TMI-2) pressurized water reactor is the most severe accident to occur in a commercial operating power reactor in the United States. The progression of the TMI-2 accident was mitigated by the presence or injection of cooling water. The reactor vessel did not fail. TMI-2 represents a source of information that can be used to expand upon the present knowledge of severe accident and core melt progression and the associated behavior of fission products, and to possibly mitigate severe accidents before the reactor vessel and containment are threatened. Although the consequences of the accident were restricted-- primarily to the reactor vessel itself--the potential consequences were significant enough to warrant close study of all aspects of the accident.

Although there was significant release of krypton and xenon during the accident, the extent of core damage, including fuel melting, was not confirmed until 1982, when a camera was first lowered into the core region, and a debris bed was observed at approximately 1.5 m below the top of the core region. Plant recovery was expected to require several years of effort and several hundred million dollars to complete. In response to this problem, the GEND organization was established to provide financial and technical support for plant recovery to GPU Nuclear Corporation (GPUN), the plant owner. The GEND organization consists of GPUN, the Electric Power Research Institute (EPRI), the US Nuclear Regulatory Commission (NRC), and the US Department of Energy (DOE). GPUN is responsible for ongoing cleanup operations. The DOE provides transportation and interim storage for the core until its permanent disposition is decided. The DOE also supports an extensive research program to learn what happened during the accident and to ensure that these results are available for resolution of severe accident source term technical issues. The NRC is coordinating a NRC/OECD cooperative research program for the examination of the reactor vessel lower head samples to determine actual conditions in the lower head during the accident. EPRI has provided research assistance in the areas of robotic development to assist in remote decontamination operations, chemical decontamination experiments, and mechanical equipment surviveability. To support this major research program, GPUN created a Data Management & Analysis organization in 1982. This organization has had the responsibility of performing data acquisition activities to support plant recovery activities and obtain samples to be analyzed by the research community. Examination programs performed by GPUN have been extensive and have produced a large amount of data used by the research community to understand the accident and enhance the current understanding of core melt progression and fission product behavior during severe accidents. This information provides a unique data base for the assessment of severe accident computer models.

By 1985, GPUN data acquisition results and results from DOE sample examinations indicated that damage to the core was more extensive and severe than first thought. Video inspections of the lower head showed that 20E3 kg of core materials, which appeared to have been once-molten, rested on the reactor vessel lower head. In addition, metallurgical examinations of particles taken from the debris bed at the top of the core indicated that some fuel had reached temperatures of at approximately 3100 K. Subsequent probing and video inspections of the interior of the lower core region provided evidence that at least 50% of the core had been molten at some time during the accident. Some of this molten core material had relocated from the core region into the lower core support assembly (CSA) with most reaching the reactor vessel lower head.

The task of defueling the reactor vessel and accounting for all fissile materials has been difficult because of the extensive damage to the core and its subsequent relocation to areas which are difficult to access, and because of the radiation fields in the containment. Fuel removal and handling have been challenging. Continuing development and application of the unique tools and methods are relied upon to defuel and characterize the reactor.

This paper summarizes the results of data acquisition activities and analytical evaluations performed at TMI and at several DOE laboratories. The successful completion of defueling depends on a clear and detailed understanding of the post-accident state of the plant. The current understanding of the relocation of molten core materials into the lower reactor vessel internals and possible damage to the reactor vessel lower head are discussed in some detail.

An accurate determination of the post-accident state of the facility is required to understand not only the accident progression and fission product transport and deposition, but to understand as well the properties and location of post-accident core materials in order to defuel the reactor. This information was developed from several sources: visual inspection of the reactor vessel internals, radionuclide examinations of samples acquired in the course of defueling, readings from on-line instrumentation, and DOE metallurgical/radiochemical examination of samples and experimental data developed from smaller scaled tests conducted at various facilities. The post-accident conditions of the plant are well characterized and many of the early post-accident conditions have been reported previously. (1-6)

REACTOR VESSEL

During the accident, peak temperatures ranged from approximately 3100 K at the center of the core (molten UO_2) to 1255 K immediately above the core to 723 K at hot leg nozzle elevations. The upper grid structure (stainless steel) was ablated in two major zones. Approximately 50% of the original core melted during the accident. Lower portions of three baffle plates on the east side of the core melted and some of the molten material flowed into the core bypass region. Approximately 30E3 kg of molten materials flowed from the core to the core bypass region and lower internals with approximately 20E3 kg, coming to rest on the reactor vessel lower head. Figure 1 illustrates the major reactor vessel components and the post-accident configuration of the core.

Post-accident conditions of the upper plenum assembly, the original core region, the core bypass region (upper CSA), the lower CSA, and the lower head region are described in the following sections.

Upper Plenum Assembly

The upper plenum assembly, which experienced no major structural damage, was removed intact. There were two damaged zones on the bottom of the plenum assembly. Localized variations of damage were evident in each zone. For example, in the limited area above one fuel assembly, ablation of the stainless steel structure was observed; however, grid structures adjacent to the ablated zone appeared to be undamaged. In some regions, the molten grid material had a foamy texture, which occurs when stainless steel oxidizes near its melting point. A once-molten mass close to this grid material appeared unoxidized, suggesting that some of the hot gases exiting the core were oxygen deficient. The damage to the upper assembly suggested that the composition and temperature of core exit gases varied significantly within the flow stream. Very little fuel debris was retained within the plenum assembly--this supports the conclusion that very little fuel debris was transported outside the reactor vessel because any fuel debris departing the reactor vessel would have had to pass through the plenum assembly.

Core Region

As illustrated in Figure 1, the end-state configuration of the original core region included a void or cavity at the top of the core region. Below that, a loose debris bed rested on a resolidified mass of material that was supported by standing fuel rod stubs that, in turn, were surrounded by portions of fuel assemblies. The once-molten, resolidified material was surrounded by a distinct crust of material in which other fragments, shards of cladding, etc., could be identified.

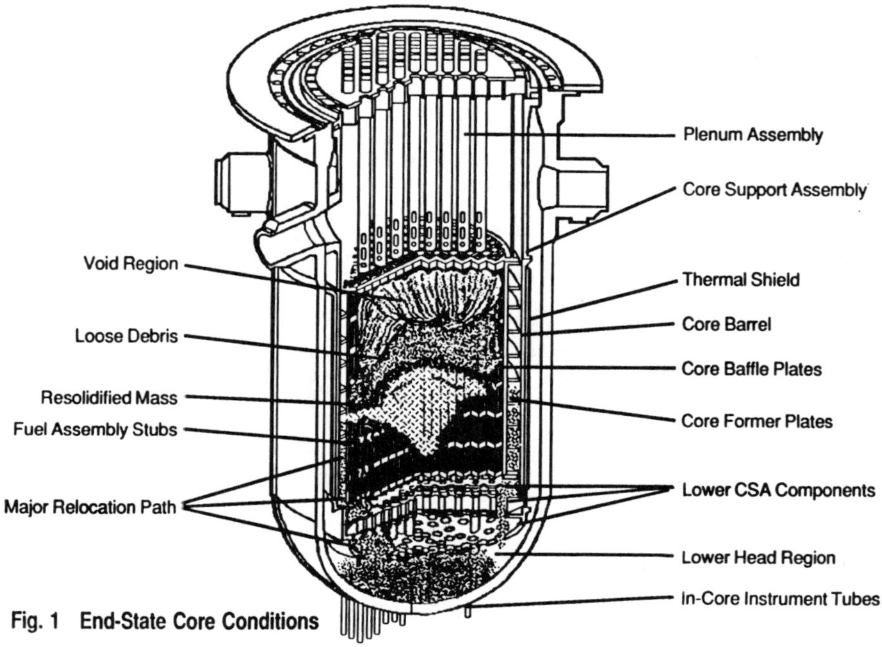

Fig. 1 End-State Core Conditions

Fig. 2 Post-Accident Estimated Core Material Distribution

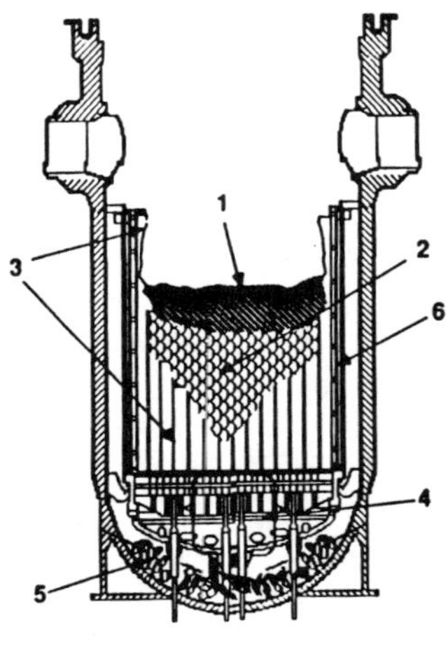

ZONE	DESCRIPTION	ESTIMATED QUANTITY (Kg)
1	Upper Debris Bed	26,400
2	Resolidified Mass	32,700
3	Intact Assemblies	44,500
4	Lower CSA (loose debris & resolidified mass)	5,800
5	Lower Head (loose debris) Resolidified Mass	12,500 6,800
6	Upper CSA (loose debris & resolidified mass)	4,200
7	Ex-Vessel (Table 1)	<300
	TOTAL	133,200

The void in the core was approximately 1.5 m deep with an overall volume of 9.3 m^3. Forty-two of the original 177 fuel assemblies were standing at the periphery of the void. Only two of these fuel assemblies contained more than 90% of their full-length cross sections with the majority of fuel rods intact. The other assemblies suffered varying degrees of greater damage ranging from ruptured fuel rods to partially dissolved fuel pellets surrounded by once-molten material.

The loose debris bed at the base of the cavity in the core ranged in depth from 0.6 to 1 m and consisted of whole and fractured fuel pellets, control rod spiders, end fittings, and resolidified debris totaling approximately 26 400 kg. Most of the loose debris bed consisted of shattered and resolidified material that contained regions of previously molten U-Zr-O, indicating peak temperatures greater than 2200 K. Many of the samples obtained from the debris bed were previously molten (U, Zr)O_2, indicating peak temperatures greater than 2800 K. There were also a few samples of once-molten material that were almost pure UO_2, indicating temperatures of 3100 K or greater. Metallographic examinations of individual particles indicated that most of the debris bed remained at temperatures below 2000 K or was exposed to higher temperatures for only a short time.

Beneath the loose debris bed was a large resolidified mass, approximately 3 m in diameter, 1.5 m thick at the center, and 0.25 m thick at the periphery, and containing approximately 32 700 kg of core debris. The center of this consolidated mass was primarily once-molten ceramic (U, Zr)O_2 laced with once-molten metallic material. The metallic structures consisted primarily of silver and iron. This material is believed to have been fully molten. The upper crust of this mass, which consisted of the same material and which also reached 2800 K, contained fuel pellets (unmelted) near the crust periphery. The center region consisted of once-molten ceramic and metallic materials varying from .001 to .003 m thick. The lower crust consisted of principally once-molten stainless steel, zircaloy cladding, and control rod materials frozen in flow channels surrounding intact and partially dissolved fuel pellets. The thickness of this lower crust based on video examinations is estimated to be approximately .01 m on the average. The peak temperature of the lower crust material was greater than 1400 but less than 2200 K. The resolidified mass was shaped like a funnel extending down to the lower end fitting at an assembly near the center of the core.

The standing, undamaged fuel assembly stubs extended upward from the lower grid plate to the bottom surface of the resolidified region of the once-molten materials. These stubs varied in length from approximately 0.2 to 1.5 m. Longer partial fuel assemblies stood at the periphery of the resolidified mass. On the east side of the core, four adjacent fuel assemblies were nearly completely replaced with once-molten core material, indicating a possible relocation path of molten material into the lower CSA and core bypass region. The standing fuel assembly stubs and peripheral assemblies constituted about 44 500 kg of core debris.

Core Bypass Region (Upper CSA)

This region consists of vertical core baffle plates that form the peripheral boundary of the core; horizontal core former plates, to which the baffle plates are bolted; the core barrel; and the thermal shield (Figure 1). A number of flow holes exist in the baffle and core former plates (0.035 and 0.033 m in diameter, respectively) through which coolant flowed during normal operations. Visual inspection (via closed-circuit television) of the baffle plates and core former plates indicated that they were generally in good condition, with one exception: on the east side of the core, there is a large hole approximately 0.6 m wide and 1.5 m high, extending across the lower portion of three baffle plates. The 0.019-m-thick baffle plates and sections of three horizontal core former plates (approximately 0.032 m thick) were melted in this region. Other baffle plates on the east and southeast were warped as a result of the high temperatures and the flow of molten material in the bypass region. Molten core material from the core region flowed through this hole into the upper CSA and flowed circumferentially in the upper CSA, and flowed downward through flow holes (.003 m) in the core former plates into the lower CSA at nearly all locations around the core. The majority of the molten material flowed into the lower CSA on the southeast side through assemblies, discussed previously, and through the southeast core former plate flow holes. The area behind the baffle plates contained loose debris all the way around the core region. The depth of debris within the upper CSA was approximately 1.5 m on the north side and a few millimeters thick on the southwest side. The entire volume between the lower core former plates was not totally filled with fuel debris. There appeared to be a resolidified crust on the upper horizontal surfaces of the three bottom core former plates; this crust varied in thickness from approximately 0.005 to 0.040 m. An estimated 4200 kg of core debris were retained in this region. The 0.025 m annulus between the core barrel and the thermal shield was visually inspected with only fine particulate was observed. No major damage to these components was seen.

Lower CSA

The lower CSA region consists of five stainless steel structures: four flow plates and a lower grid assembly. The structures vary in thickness from 0.025 to 0.34 m with 0.080 or 0.15 m diameter flow holes. Liquified material flowed through the periphery of these structures and came to rest on the lower head. There was resolidified material (5800 kg) at various locations on the circumference of these structures. In several places, resolidified material completely filled the flow holes, and columns of once-molten material were observed between the plates. The largest accumulation of resolidified material appeared to have flowed into the lower CSA from the east side of the core. Although most of this material was seen on the east to southeast sides, many columns of resolidified material were seen all the way around the core beneath the core bypass region.

Lower Head Region

The debris in the lower head region accumulated to a depth of 0.75 to 1 m above the lowest head elevation and to a diameter of 4 m. Inspections revealed that a large quantity of once-molten core material rested on the lower head. The spatial distribution of these materials was neither uniform nor symmetric. Particle sizes varied from large rocks (up to 0.20 m) to granular particles (less than .001 m). Visual inspections indicated that larger rocks, especially in the northeast and southwest areas, were located toward the periphery and the debris pile was lower at the vessel center than at the periphery. Granular or gravel-like material was observed to be in the central region of the vessel. A large agglomerated mass existed under the loose debris bed resting on the bottom head of the reactor vessel. This mass is approximately 0.5 m thick in the center and 1.7 m in diameter. A large cliff-like structure formed from once-molten core material existed in the northern region. The cliff face was approximately 0.38 m high and 1.25 m wide. It was estimated that 12,400 kg of loose core material and 6,800 kg of agglomerated material were in this region.

Significant structural damage consisting of several melted incore instrument guide tubes was observed in the lower head region. Nine of these incore instrument guide tubes were melted off within 0.5 m of the bottom plate of the lower CSA. Tests performed on the incore instrumentation after the accident showed that 12 of the incore thermocouples had new junctions in the lower head region; this has been confirmed by observations in the lower head region. (7)

REACTOR COOLANT SYSTEM

The major physical damage to the facility was limited to the reactor vessel internals. During the accident, small amounts of fuel debris and quantities of fission products (Table 1) were transported throughout the reactor coolant system (RCS). Reactor coolant pump RC-P-2B was started and operated for approximately 19 min. on March 28, 1979. The operations of this pump signified the completion of the initial core heat up and degradation period of the accident and is thought to be the major driving force for the relocation of fuel debris from the reactor vessel. The coolant injected into the reactor vessel by the operation of the pump was sufficient to have filled the reactor vessel. The importance of this event is as follows: as the coolant pump operated, the flow of the "B" loop was in a "forward" direction. This flow from the reactor vessel transported small amounts of fuel debris into the "B" hot leg, the majority of which settled out in the "B" hot leg and fell into the decay heat drop line, which comes off the bottom of the horizontal section of the "B" hot leg. The decay heat drop line had one of the largest collections of fuel debris as a result of this pump operation. The water then continued up over the "candy cane" and deposited fuel material on the "B" once-through steam generator (OTSG) upper tube sheet. This tube sheet acted as a "strainer" for the single largest collection of fuel debris transported outside the reactor vessel. Smaller amounts of fuel debris worked their way through the steam generator tubes and then were deposited on the lower head of the "B" steam generator and J-legs. As the coolant continued to flow, even

TABLE 1

STATUS OF FISSION PRODUCT INVENTORY a

Ex-Vessel Inventory (Percent)

	Noble Gases	I	Cs	Sr	Sb	Ru	Ce
Containment Atmosphere, Basement, and Tanks	54	47	47	2.1	0.7	0.5	.01
Reactor Coolant System	NM	1	3	1	0.2	ND	P
Auxiliary and Fuel Handling Building	1	7	5	0.1	0.7	ND	ND
Total Outside Vessel	55	55	55	3	2	1	0

In-Vessel Inventory (Percent)

	Fraction of Core Mass	Noble Gases	I	Cs	Sr	Sb	Ru	Ce
Intact Rods	35	30	30	30	30	30	30	30
Upper Debris Bed	20	6	5.3	4	19	24	16	20
Upper Crust		ND	.3	.4	.7	8.3	4.6	1.4
Consolidated Mass	25	ND	2.1	.8	8.3	10	11.2	24
Lower Crust		ND	3.5	1.4	4.5	43.4	29.7	5.9
Upper CSA	3	ND	.1	.5	4	.2	.2	3.4
Lower CSA	4	ND	.2	.6	5	.3	.3	4.7
Lower head	14	ND	0.5	2.1	18	1.0	1.1	16
Total In-Vessel Retention		36	42	40	90	117	93	105

Total Percent of Accounted-for Pre-accident Inventory

	Noble Gases	I	Cs	Sr	Sb	Ru	Ce
Pre-Accident Inventory	91	97	95	93	119	94	105

a. Some columns do not add properly due to rounding errors.
P = in progress
NM = not measured
ND = not detected

smaller amounts of fuel debris were then deposited in the "B" reactor coolant pump and cold legs. As a result of this pump action, reverse flow occurred in the "A" loop. The reactor vessel internals acted as a "strainer" for this material permitting only finely divided fuel debris to exit the reactor vessel into the "A" cold legs and to deposit fine fuel debris in the "A" reactor coolant loops and steam generator. This occurred because the loops are hydraulically arranged so that the higher relative pressure in the reactor vessel on the idle loop is in the cold leg downcomer versus the hot leg outlet, thus driving flow "backward." The 'A' loop of the RCS contained approximately 10 kg, the 'B' loop contained approximately 207 kg and 12 kg resided in the pressurizer. (8)

CONTAINMENT BUILDING

Reactor coolant was discharged from the RCS through the pilot-operated relief valve (PORV) located on top of the pressurizer. The PORV discharges to the reactor coolant drain tank in the basement of the containment building. The reactor coolant drain tank contains a relief valve that releases to the reactor building sump and a rupture disk, which ruptured early in the accident. Discharge from this rupture disk was directly to the containment building basement. At approximately 139 min. into the accident, the operators realized that the PORV had failed to close, although instrumentation indicated that it had closed. The operators manually closed the pressurizer block valve, thus limiting further loss of coolant and gaseous fission product release from the primary coolant system to the containment building. Fuel cladding rupture occurred at approximately the same time as the block valve was closed, thus limiting the amount of fuel debris and fission products to be discharged to the containment building. However, the block valve was cycled, opened and closed, to maintain system pressure. This opening and closing of the block valve permitted the transport of fission products, noble gases, and small quantities of fuel debris into the pressurizer, then through the PORV into the reactor coolant drain tank, and finally into the containment building. The largest concentration of fuel debris in the containment building was located at the discharge of the reactor coolant drain tank on the basement floor. (9, 10)

The reactor coolant makeup and purification (MUP) system inlet piping comes from the RCS on the discharge side of the reactor coolant pump 1A. The first components in this system are the letdown coolers, which are located in the basement of the containment building. The MUP system was operated during the accident, thus some fuel debris was transported into this system, which located fuel debris in the letdown coolers. A total of 11 kg of fuel debris was transported to the containment building.

AUXILIARY AND FUEL HANDLING BUILDING

A very small amount of fuel debris was transported to the auxiliary and fuel handling building during the accident with the majority of it being transported through the MUP system and finally coming to a stop in the reactor coolant bleed tanks. The inlet of this system comes off the RCS from the cold leg side of the reactor coolant pump 1A. The initial components in this system as described above are the letdown coolers, which are located in the containment building. Discharge of these coolers then exit the containment building and enter the auxiliary and fuel handling building. This system, with its adjoining systems, go through a large number of the cubicles in the auxiliary and fuel handling building. Only a small amount of fuel debris was transported into this system, as indicated by the fact that very little fuel debris was identified within the block orifice, MUP demineralizer filters, demineralizers, and the makeup filters. The block orifice, which is the normal pressure reduction device for flow rates up to 45 gpm through the system. The block orifice and its isolation valve became blocked during the accident, at which time they were bypassed. Very little fuel debris has been identified in the block orifice and its associated piping. The letdown flow was then directed to the letdown filters and purification demineralizers during the accident at very slow rates and was then routed to the reactor coolant bleed tank "A" and the makeup tank. Letdown flow was lost several times during the accident because of flow blockage. More than 24 hours after the initiation of the accident, the MUP demineralizers were bypassed and letdown was sent directly to the reactor coolant bleed tank "B". Because of the blockage of the letdown coolers and restriction of the block orifice, very little fuel was transported to the filters, demineralizers, and reactor coolant bleed tanks. As indicated in Table 1, very little fuel debris was transported to auxiliary and fuel handling building systems. A majority of the fuel transported to the building entered it as finely divided particles. Approximately 25 kg of fuel debris was transported to the building.

CORE MATERIAL INVENTORY

The original core inventory included approximately 94 000 kg of UO_2 and 35 500 kg of cladding, structural, and control materials. Accounting for oxidation of core materials during the accident and for portions of the upper plenum structure that melted, the total amount of post-accident core material is estimated to be 133 200 kg. The post-accident material balance is shown in Figure 2. The accuracy of these numbers varies from 5% to 40%.

FISSION PRODUCT INVENTORY

The objective of the fission product inventory is to determine the location and amounts of fission products in the plant, including those that remained in the RCS, the containment, and the auxiliary and fuel handling building. The best estimates of releases to the environment have been published. (11-13)

Fission product retention measurements for areas outside the reactor vessel were completed for the containment atmosphere, basement, and tanks; the RCS; and the auxiliary and fuel handling building. The results of these measurements are summarized in Table 1. Over half of the core inventory of noble gases, cesium, and iodine was measured outside of the reactor vessel. The large majority of the radionuclides released from the reactor vessel was retained in the containment building with little being transported to the auxiliary and fuel handling building or the environment.

To evaluate fission product content, the core was divided into eight distinct sections:

1. upper core debris
2. partially intact, undamaged fuel rods
3. lower head material
4. upper crust of the resolidified mass in the core region
5. center resolidified consolidated mass in the core region
6. lower crust of the resolidified mass in the core region
7. resolidified core material in the upper CSA
8. resolidified core material in the lower CSA.

Physical and radiochemical examinations were performed on samples of material from the first six regions to characterize fission product concentrations in these areas. Samples of the resolidified material in the lower core from the the eastern side near the melted baffle plates were obtained recently; results of examinations of these samples are not yet available. Samples from the two other regions (the upper and lower CSA) will be obtained when these regions are defueled. Also, additional samples have been obtained from the lower head region to better characterize the material resting there. Results of sample examinations of materials obtained and analyzed (Table 1) from various locations have been reported. (3, 6, 14-17)

It is assumed that the samples analyzed for fission products from all the regions except Nos. seven and eight of the reactor vessel are representative of the entire region. Samples from these regions still need to be obtained; consequently, the numbers given in Table 1 are only estimates. Additional samples of the lower reactor vessel head region also need to be obtained to better calculate the fission product inventory present there.

Examinations are under way to evaluate retention of fission products in selected intact and undamaged fuel pellets. These are being used to estimate the accuracy of the ORIGEN 2 code. It is assumed that the best ORIGEN 2 code results will define the fission product inventory of the whole core at the beginning of the accident. Eight of these fuel pellets from various core locations have been examined. Six were from partial fuel rods from the lower core region and two were from fuel pins from the upper periphery of the core. The upper core samples were bathed in a steam/hydrogen atmosphere while the samples from the bottom of the core were water cooled throughout the accident.

Measurements of the retention of noble gases (Kr and Xe) in fuel pellets from the partial fuel assemblies indicate that the intact fuel pellets in both the upper and lower core regions retained effectively all of the noble gases (at least 95%). These results, combined with measurements of the gases retained in the containment atmosphere, indicate that approximately 91% of the noble gases originally in the core were accounted for within the RCS and containment building. Because the noble gases seemed to have been retained so effectively, the assumption was made that these pellets allowed no escape of fission products. Other samples examined have determined that some noble gases were retained in the loose debris bed but no noble gases were detected in any of the once-molten material. Other high volatiles (I-129 and Cs-137) behaved in the same basic manner with the exception that some amounts were retained with the once-molten material. The data indicate similar distributions for both Cs-137 and I-129 in the different regions of the once-molten fuel debris. There is some evidence from the lower parts of the core suggesting that some of the I-129 is found in association with the metallic regions of the core. The fraction of core inventory found in association with once-molten regions could be significantly biased by the amount of metallic material present.

Examinations were also performed on a variety of samples throughout the facility for medium- and low-volatility fission products. Comparisons between these measurements and ORIGEN 2 calculations showed that for most fission products, the inventories calculated with the ORIGEN 2 code were in good agreement with experimental measurements. For medium-volatile fission products (Sr-90, Ru-106, and Sb-125), the data indicate that only a small amount was released from the reactor vessel: 3.2% for Sr-90, 1.6% for Sb-125, and .5% for Ru-106. The data also indicate that the bulk of the Sr-90 is located in the ceramic parts of the samples, with some transportation of the Sr-90 to the metallic region samples. The Sb-125 data indicate a very high concentration in the metallic portions of the once-molten material. Ceramic samples from the once-molten material indicate a relatively low concentration of Sb-125. The data for the lower crust in the central region are biased by the presence of metallic material in several mostly ceramic samples. The data indicate that the Sb-125 remained in a metallic state or was reduced during the accident because of a reaction mechanism. The Ru-106 has a similar distribution to the Sb-125 and the data again indicate that the majority of the Ru-106 within the once-molten material is an association with the metallic material.

The low-volatility fission products (Ce-144, Eu-154, and Eu-155) data indicate that these were not transported out of the reactor core and that only a small amount of fuel material was physically transported into the reactor coolant system. As indicated in Table 1, only 0.01% of the Ce-144 can be accounted for outside the reactor vessel. Some additional measurements are being made and calculations being performed to determine the amount of Ce-144 that was retained within fuel debris transported to the reactor coolant system. The data indicate that all low-volatile fission products were retained within the reactor vessel.

Fission product retention measurements from the degraded core materials inside the reactor vessel are summarized in Table 1. Note that the fission product percentage cannot be related to the debris mass because the actual ratio of fuel to structural material has not yet been measured. Thirty percent of the total inventory of fission products were retained in the intact fuel pellets. Based on the measurements completed to date, 91% of the noble gases, 97% of the iodine, and 95% of the cesium have been located. Thus, most of the fission products produced in the reactor have been located. For the more volatile fission products (fission gases, cesium, and iodine), most of the activity was retained in the containment or in the intact fuel rods within the reactor vessel.

CONCLUSIONS

The accident at TMI-2 has provided a benchmark for understanding severe accident scenarios. Data from the accident represent the only full-scale reactor system data relating to the understanding of core damage progression, fission product release, vessel margin-to-failure, and accident mitigation strategies. A comprehensive scenario of the accident is nearing completion. Major observations include the following:

- The integrity of the reactor vessel was maintained despite relocation of up to 20E3 kg of core material to the lower head. The TMI-2 accident demonstrates that once core melting has begun, the progression of a severe accident does not necessarily proceed to failure of the reactor vessel.

- The accident was effectively mitigated with the injection of relatively small quantities of cooling water. This demonstrates that once the core has begun to melt, its progression is not inevitable and, in fact, it can be controlled.

- Most of the released radionuclides did not escape from the plant and a large percentage did not escape the reactor vessel. The principle of defense in depth was shown to be effective in containing the effects of an accident and in minimizing the hazard to the public.

In addition, the accident provides the only data describing late-phase core melt progression, i.e., the relocation of molten core materials from the core and the interaction of molten core materials with the reactor vessel lower head. Thus, TMI-2 research can extend the understanding of severe accidents and provide data that is not available from other sources. These data are helping to resolve issues related to in-vessel core melt progression, fission product behavior, and source term.

REFERENCES

1. Tolman, E. L., et. al., TMI-2 Accident Evaluation Program, Rept. EG&G-TMI-7048, EG&G Idaho, Inc., 1986.

2. Eidam, G. R., Current TMI-2 Core Conditions and Defueling Update, Rept. NUREG/CP-0090, Washington, D.C., 1987.

3. Olsen, C. S., Akers, D. W., McCardell, R. K., TMI-2 Core Bore Examination Results, Rept. NUREG/CP-0090, Washington, D.C., 1987.

4. Neimark, L. A., Strain, R. V., Sanecki, J. E., The Microstructural and Microchemical Characterization of Samples from the TMI-2 Core, Rept. NUREG/CP-0090, Washington, D.C., 1987.

5. Toth, L. M., Malinauskas, A. P., Eidam, G. R., Burton, H. M., The Three Mile Island Accident, Diagnosis and Prognosis, (189th Meeting of America Chemical Society, Miami, 1985), ANS Symposium Series 293, ACS, Washington, D.C., 1986.

6. Eidam, G. R., Tolman, E. L., Broughton, J. M., McCardell, R. K., Stratton, W. R., Severe Accidents in Nuclear Power Plants, Vol. 1, 1988.

7. Yancey, M. E., Meininger, R. D., Hecker, L. A., TMI-2 In-Core Instrument Damage -- An Update, Rept. GEND-INF-031, Vol. II, U.S. Department of Energy, 1984.

8. Distenfeld, C. H., Summary of Fuel Quantities External to the Reactor Vessel, Rept. TB-86-38, GPU Nuclear, Middletown, PA, 1987.

9. Schmitt, C. R., Reactor Building Basement - History and Present Conditions, Rept. TPO/TMI-029, GPU Nuclear, Middletown, PA, 1982.

10. Eidam, G. R., Reactor Building Radiological Characterization, Rept. TPO/TMI-125, GPU Nuclear, Middletown, PA, 1985.

11. Rogovin, M., et al., Report of the Public Health and Safety Task Force, The President's Commission on the Accident at Three Mile Island, 1980.

12. Shuping, R. E., Use of Photographic Film to Estimate Exposure Near the Three Mile Island Nuclear Power Stations, Rept. FDA-81-8142, U.S. Department of Health and Human Services, Bureau of Radiological Health, 1981.

13. Rogovin, N., Three Mile Island--A Report to the Commissioners and to the Public, Rept. NUREG/CR-1250, Vol. 2, Pt. 2, U.S. Nuclear Regulatory Commission, 1980.

14. Akers, D. W., et al., TMI-2 Core Debris Grab Samples - Examination and Analysis, Rept. GEND-INF-075, U.S. Department of Energy, 1986.

15. Olsen, C. S., et al., Examination of Debris from the Lower Reactor Head of the TMI-2 Reactor (Draft), Rept. EGG-TMI-7573, EG&G Idaho, Inc., 1987.

16. G. R. Eidam, et al., U.S. Department of Energy and GPU Nuclear Corporation Research and Development Activities on Three Mile Island Unit Two Annual Report, 1988.

17. Schnitzler, B. G., Briggs, J. B., TMI-2 Isotopic Inventory Calculations, Rept. EGG-PBS-6798, EG&G Idaho, Inc. 1985.

Melt Progression Modeling Implications of the TMI-2 Accident

R. W. WRIGHT
Office of Research
U.S. Nuclear Regulatory Commission
Washington, D.C. 20555, USA

Abstract: The TMI-2 accident and core examination have provided unique and important information on LWR severe-accident behavior, particularly for the late stage of accidents that involve fuel melting and relocation. This accident was successfully terminated by reflooding the core even though the accident had progressed into substantial fuel melting. The phenomenology of the early stage of the TMI-2 accident for which few measurements are available is largely understood from the results of the PBF tests and other past research. The information on late-stage phenomenology that involves substantial fuel melting and relocation has nearly all come from the TMI-2 core examination. Phenomenological modeling has generally accounted for the TMI-2 accident behavior except for the absence of lower head failure under melt attack where relevant experimental data have not yet been obtained.

Introduction

The accident at TMI-2 has furnished a great deal of new and significant information on key phenomena in PWR core uncovery accidents (Ref. 1). For the late phase of in-vessel core melt progression (beyond the onset of substantial ceramic material relocation), the TMI-2 results provide the only integral data available on governing phenomena in PWR accident sequences. Phenomenological models developed from the TMI-2 results and other severe accident research have been incorporated into the mechanistic SCDAP/RELAP5 and MELPROG/TRAC in-vessel melt progression codes (Refs. 2, 3).

At the time of the TMI-2 accident, fission product release from the core and the pressure pulse from hydrogen combustion in the containment indicated that extensive core damage had occurred, although this was not then universally accepted. Analysis at that time by Marino and Picklesimer with their TMI-BOIL code (unpublished) indicated that there had been extensive steam oxidation of the zircaloy cladding, and that this had produced considerable hydrogen, core temperatures exceeding 2000K, considerable zircaloy melting and relocation, considerable fission-product release, and extensive core damage (Ref. 4). These results were presented to the Regovin Commission that reviewed the implications of the TMI-2 accident for the U.S. Nuclear Regulatory Commission (Ref. 5). Soon thereafter, a program of four Severe Fuel Damage tests was started in the PBF test reactor and development was begun on the mechanistic SCDAP fuel damage code (Refs. 2, 6). This program was subsequently expanded by the USNRC and its foreign program partners into a much more comprehensive program of severe-accident research (Ref. 7). In conjunction with the clean up of the damaged TMI-2 reactor, the U.S. Department of Energy has carried out an extensive examination of the damaged TMI-2 core (Ref. 8). An international standard problem exercise for the TMI-2 accident was also organized (Ref. 9).

The TMI-2 Accident Sequence

Although TMI-2 has furnished unique and important information, particularly on late-phase core-melt progression, it has some severe limitations when considered as an experiment on severe accident behavior. The accident sequence in TMI-2 was quite complex. The thermal-hydraulic boundary conditions for the core involve large uncertainties and also include several abrupt changes from operator intervention that strongly affected the course of the accident. In addition, no measurements exist for many of the most important core parameters including fuel temperatures in the severe damage regime, hydrogen generation rates, steam flow rates, material relocation etc. Nevertheless, considerable information on the boundary conditions and some on core parameters has been obtained by extensive, detailed, and often indirect analysis using the systems measurements that do exist (Ref. 11). The most important information on core behavior during the accident has come from the extensive post-accident core examination that also included detailed metallographic examination of the core debris (Ref. 8). The end-state core configuration, debris chemical composition, and some peak local temperatures were obtained in this way. The core examination gave much less information on the early phase of core melt progression, however. The early phase behavior is inferred from a combination of the end state core configuration, some reactor systems measurements, and extensive analysis. This early-phase behavior is consistent with that observed in the four Severe Fuel Damage tests performed in the PBF test reactor and with the results of other severe fuel damage and core-melt progression research. It also appears to be consistent with the early-phase modeling in the mechanistic SCDAP/RELAP5 and MELPROG/TRAC fuel-damage and melt-progression codes.

The overall reactor system behavior during the accident is indicated by the time history of the system pressure shown in Figure 1 (from Ref. 1). During the initial 100 minutes of the accident, coolant was lost through the open Pilot Operated Relief Valve (PORV) and the system depressurized to about 7 MPa. When the last of the primary system coolant pumps was turned off at 100 minutes, the water level had fallen nearly to the top of the core. The (two-phase) coolant level in the core as estimated by SCDAP/RELAP5 analysis is shown in Figure 2 (Ref. 1). Coolant boiloff and core heatup proceeded until 134 minutes when a block valve in series with the open PORV was closed. This stopped the coolant loss and the system started to repressurize. Core heat up continued until one of the B-loop primary cooling pumps was started at 174 minutes. At this time, the core water level is estimated to have fallen to about one meter in the 3.6 meter high core. Much of the steam oxidation of the core zircaloy and the hydrogen generation occurred during this time interval during which rapid heating into the 2500K range of local temperatures occured. A short flow transient occurred when the B-loop coolant pump was restarted at 174 minutes and the system pressure rapidly increased to about 15 MPa. (No water flow was measured after the first 15 seconds.) It is thought that this flow transient fragmented the damaged fuel as well as generating more hydrogen from unoxidized zircaloy. At 193 minutes the block valve for the PORV was reopened, decreasing the system pressure but renewing the coolant loss. At 200 minutes the high-pressure injection system was activated, injecting emergency cooling water into the core for 17 minutes, probably refilling the

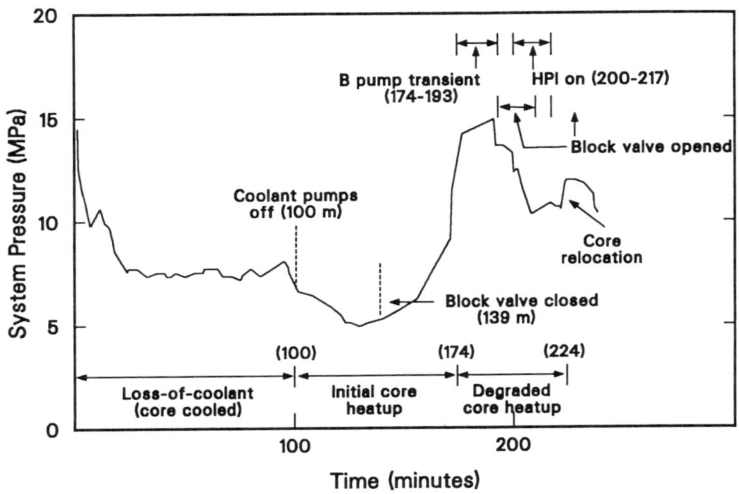

FIGURE 1. Measured TMI-2 System Pressure History

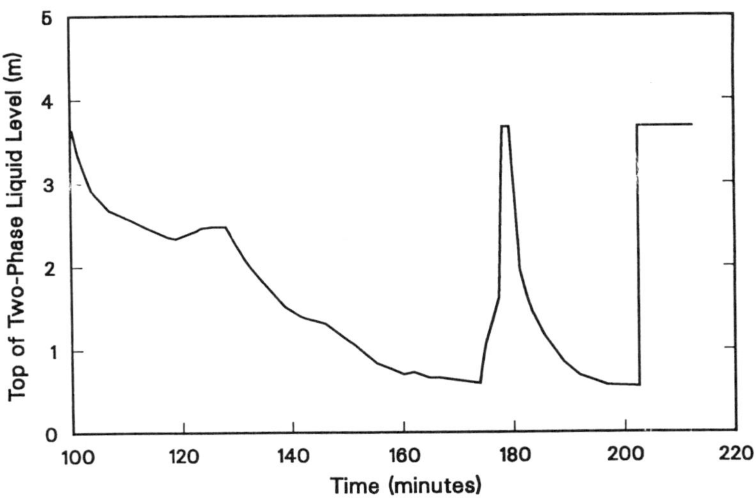

FIGURE 2. Estimated Two-Phase Coolant Level in the TMI-2 Core

core and also reducing the pressure to about 10 MPa. At 224 minutes about 20 metric tons of mostly ceramic melt relocated from the core region into the water-filled lower plenum and onto the vessel lower head over a period of about 100 seconds. No energetic thermal explosion (steam explosion) occurred, but a 2 MPa steam spike (above the 12 MPa ambient pressure) was generated and this lasted for about 15 minutes. The integrity of the lower head was maintained during and after the melt relocation, and final cooldown of the core by forced flow was established at 15.5 hours.

Detailed descriptions of the TMI-2 accident scenario including the events in the scenario are given in (Ref. 1) and in the ICHMT Seminar paper by Eidam (Ref. 10).

<u>Early Stage Accident Phenomenology</u> (prior to the B-pump transient)

The OECD Nuclear Energy Agency and the U.S. Department of Energy have undertaken a TMI-2 standard problem exercise to benchmark severe-accident computer codes against the TMI-2 accident (Ref. 9). Results have been obtained for the early phase of the accident prior to the B-pump transient (174 minutes), but results are not yet available for the late phase that includes ceramic core debris melting and relocation. The early-phase results do not appear to be very useful for assessing early-phase melt progression models for two reasons. First, the calculated core behavior appears to be very sensitive to uncertain thermal hydraulic boundary conditions, particularly the make up flow. Second, measurements for comparison with the analysis are not available for the most important early-phase melt-progression parameters, namely the time histories of the fuel temperatures and the steam flow and hydrogen generation rates. In contrast, much detailed late-phase melt progression information is available from the end-state examination of the damaged core. All the available information including that from the core examination indicates that the early-phase melt progression phenomenology of the TMI-2 accident is quite consistent with that observed in much more detail in the four PBF Severe Fuel Damage tests, the out-of-pile integral experiments and basic materials interactions experiments and analysis at KfK, and other early-phase melt progression experiments (Refs. 6, 12, 13, and 14). Therefore, the description here of early-phase melt progression in TMI-2 is based on this other information as well as on that obtained directly from the TMI-2 accident itself. Both the early-phase and the late-phase melt progression behavior are strongly dependent upon chemical interactions among the core materials and oxidation reactions with steam. Hofmann's chart from Reference 11 of the relevant threshold temperatures for liquid-phase formation from both melting and eutectic material interactions is shown in Figure 3.

After shutdown of the last coolant pump at 100 minutes and with the initiation of core uncovery, heatup began of the uncovered part of the core by the fission-product decay. The water level fell by boiloff through the open PORV from the top of the 3.6 meter core to about 1 meter from the bottom at the time of the B-coolant-pump transient, 174 minutes. Clad ballooning and rupture occurred as local clad temperatures reached about 1150K. Gaseous fission products were released from the cladding gap at this time as indicated by radiation monitors in the containment. The clad ballooning reduced the fuel-clad heat transfer coefficient, opened the cladding to two-sided oxidation, and reduced natural circulation steam flow through the core. Eutectic Fe-Zr and Ni-Zr interactions between the inconel grid spacers and the zircaloy cladding occurred at about 1200K. In addition, the Ag-In-Cd neutron absorber material within the intact stainless-steel control-rod cladding had melted at about 1100K.

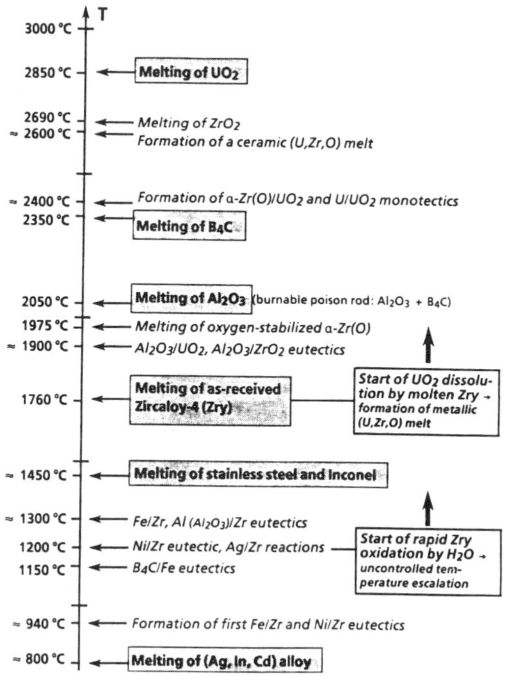

FIGURE 3. LWR Severe-Accident Relevant Liquid-Phase Formation Temperatures (from Ref. 12)

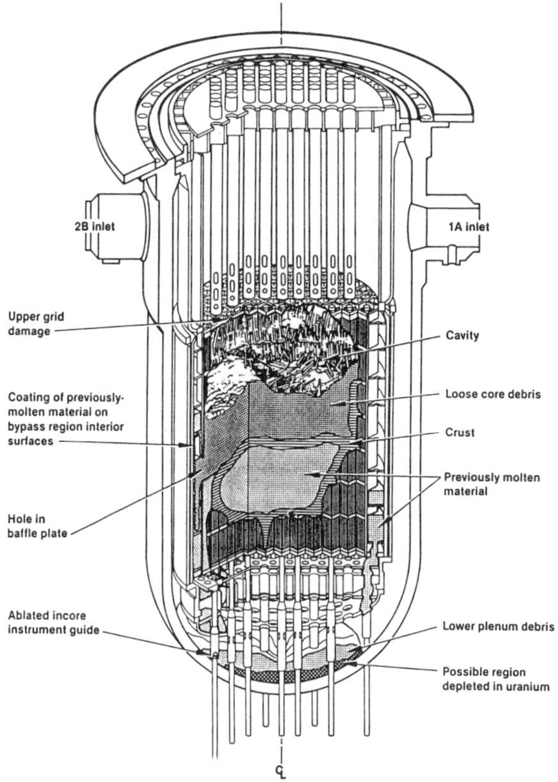

FIGURE 4. TMI-2 Core End State Configuration

As local clad temperatures increased to about 1500K, steam oxidation of the zircaloy cladding became significant, heating the cladding and generating hydrogen. Under these conditions the oxidation rate was limited by diffusion through the surface ZrO_2 layer according to a parabolic rate law (Ref. 11). The rate of oxidation and hydrogen generation increase rapidly with temperature, and, at about 160 minutes and 1700K, the local oxidation heating rate increased rapidly from about 0.5K per second from decay heating into the range of tens of K per second. All the boiloff steam flow was then reduced to hydrogen, a process known as "steam starvation", and a "burn front" propagated slowly downward in the core. During this rapid oxidation transient, local temperatures above 2500K were reached, and metallographic examination of debris fragments in the PBF tests (as well as in TMI-2) have indicated local (but not large volume) temperatures that exceeded the 3100K melting point of UO_2 (Refs. 6, and 8).

As local temperatures approached the 1720K melting point of the stainless-steel control-rod cladding, the cladding failed, releasing the molten silver, indium, and cadmium control-rod material. (Much of the cadmium vaporized and condensed outside the core as an aerosol.) The molten control rod material and later the molten stainless steel relocated downward and froze on the fuel rods (and grid spacers where present) to form a partial blockage. The molten stainless steel also liquified some of the zircaloy cladding because of the 1200K Fe-Zr and Ni-Zr eutectics, and this melt relocated downward (Ref. 11). This relocation of control-rod materials and alloyed zircaloy began formation of the lower metallic crust in the TMI-2 core that later supported a growing pool of molten ceramic core debris. A drawing of the end-state core configuration in Fig. 3 as determined from the post-accident core examination shows the metallic crust below the previously-molten ceramic pool material. This crust was cooled by water in the bottom of the core and by the fuel rods embedded in the crust. Results of full-length coolant-boildown tests in the NRU test reactor (Ref. 12) indicate that the forming metallic crust probably melted, relocated, and refroze several times in a repetitive process during the decrease in the water level prior to the B-pump transient.

During the rapid oxidation transient, melting of the cladding zircaloy occurred at temperatures that ranged from 2125K to 2245K, depending upon the increasing oxygen content of the zircaloy. Damaged fuel experiments in the ACRR test reactor have shown that a dense tin aerosol is released when the zircaloy melts (Ref. 13). Constraining sheaths of ZrO_2 on the outside of the zircaloy cladding prevented immediate relocation of the zircaloy upon melting. The molten metallic zircaloy dissolved a significant fraction of the UO_2 fuel before relocation occurred. As indicated by the pseudo-binary phase diagram for UO_2 and oxygen-stabilized alpha zirconium, (Ref. 11), molten zircaloy near its melting point of about 2200K can dissolve about 50% of its mass of solid UO_2 in a liquification process, a fraction that increases to about 85% as the temperature reaches the 2700K monotectic temperature. Considerably more UO_2 than this can also be dissolved in molten zircaloy as an inhomogeneous slurry (Ref. 11), and, at the monotectic temperature, the equilibrium mass ratio of soluble UO_2 to molten zircaloy can be as high as seven. When the molten zircaloy (with dissolved UO_2) failed the ZrO_2 sheaths holding it in place, the molten zircaloy relocated downward and froze to become the major part of the lower metallic blockage or crust. Information, particularly from the CORA tests, indicates that relocation of molten zircaloy is primarily by rivulet flow in a non-coherent, non-coplanar process, and not by coherent annular film flow (Ref. 14).

Experimental results, primarily from the PBF and NRU tests (Refs. 6 and 12), indicate that oxidation heating and hydrogen generation continued throughout the period up to the B-pump transient at 174 minutes. Oxidation of relocated and refrozen zircaloy had continued throughout these tests, and a complete blockage of steam flow to cut off the oxidation and hydrogen generation did not occur (Ref. 15). Analysis has indicated, however, that the concave-upward shape of the metallic lower crust in the TMI-2 accident as shown in Figure 4 was caused in part by the partial metallic blockage that originated near the core axis. This blockage diverted cooling steam flow away from the axis and increased the radial temperature gradient in the core (Ref. 16). The shape of the blockage proved to be significant later in the development of the accident.

Thermal analysis of the growth of the molten ceramic pool in the TMI-2 accident has indicated that there was substantial growth of a pool of molten ceramic material (UO_2, ZrO_2) prior to the B-pump transient at 174 minutes (Ref. 17). The melting and relocation of this ceramic material constitutes the start of late-phase core-melt progression.

The severe accident codes currently in use all model the intact-geometry oxidation transient and hydrogen generation with the reaction rate limited by diffusion through the ZrO_2 shell and by steam flow. The newer mechanistic codes such as SCDAP/RELAP5 and MELPROG/TRAC provide for separate treatment of first the metallic melt and later the ceramic melt. These two melts have differences in their melting points and eutectic liquifaction temperatures of the order of 1000K, as shown in Figure 3, and this produces a physical separation of these materials in the core. Such a separate treatment is necessary for a mechanistic analysis of severe LWR accidents as the TMI-2 core examination clearly illustrates. The older simplified parametric codes use a fictitious "corium" mixture of metallic and ceramic core material that melts and "slumps" as a single material at a prescribed temperature. The mechanisms determining the threshold for the failure of the ZrO_2 cladding sheaths and the relocation of the molten metallic cladding zircaloy (and dissolved UO_2) are currently uncertain. This threshold is usually taken as a parametric input temperature, although optional models for this threshold are available in the mechanistic SCDAP/RELAP5 and MELPROG codes.

Large uncertainties currently exist in the modeling of oxidation and hydrogen generation from relocating and relocated metallic debris, and also in the modeling of the relocation process and metallic blockage formation, particularly regarding the question of the completeness of the blockage. More generally, and of particular interest for BWR core geometries, it is important to know the severe accident conditions, if any, under which the metallic melt would drain from the core without forming a metallic blockage. This would produce a very different melt progression sequence from that in the TMI-2 accident.

Late Stage Accident Phenomenology

At 174 minutes, one of the B-loop coolant pumps was restarted. Measurements indicate, however, that significant flow from the pump lasted for only about 15 seconds (Ref. 1). Water quenching results from tests in the NSRR test reactor (Ref. 18) indicate that mechanical agitation and possibly thermal shock from the flow transient shattered the hot declad fuel pellets and ZrO_2 oxidized cladding shells in the TMI-2 core. The damaged fuel was largely in free-standing rod geometry at this time in a configuration similar to that in the post-test PBF fuel bundles, in particular the SFD 1-1 test bundle with trace-irradiated fuel (Ref. 6). The shattered pellets formed a largely ceramic particulate debris bed that was supported by the metallic crust across the lower fuel-rod stubs. This debris bed formation constituted the beginning of late-phase core-melt progression if substantial melting and relocation of ceramic fuel and ZrO_2 had not occurred previously. Most of the large central void in the upper region of the core in Figure 4, however, was probably generated later by densification as the molten ceramic pool grew and the debris bed above the pool subsided. It is noted that the TMI-2 fuel had very low burnup, (3,000 Mwd/TU) and that, in the PBF SFD 1-4 test with high burnup fuel (36,000 Mwd/TU) and no reflood quenching, the declad fuel largely fragmented to form an in-place debris bed with very little densification and slumping. The postulated absence of fuel fragmentation in TMI-2 prior to the reflooding in the B-pump transient may only be applicable to fresh fuel or low burnup fuel.

The copious supply of steam generated in the hot core by the B-pump flow transient rapidly oxidized much of the remaining zircaloy in the previously steam starved upper region of the core, generating considerable heat and hydrogen and increasing the core pressure (Ref. 1).

The hot gas flow from the core and in-vessel natural circulation produced very limited damage to the upper plenum structure and no observable damage to the hot-leg nozzles. It is thought that the heating of structure at TMI-2 was affected by flow channeling assymetries produced by the upper plenum structure and by water drainage both from the pressurizer and from the B-loop after the pump transient. The important question of why significant heating of upper-plenum structure did not occur at TMI-2 remains open, however.

During the 25 minutes after the B-pump flow transient, the core continued heating by fission-product decay. The pool of molten ceramic UO_2 and ZrO_2 at temperatures above the 2800K UO_2-ZrO_2 eutectic temperature continued to grow within the debris bed. The growing pool of ceramic melt was supported by the pre-existing mostly metallic crust across the fuel-rod stubs and was enclosed at the top by a weak, thin, mostly ceramic crust that underlay the mostly ceramic particulate debris bed. This is illustrated in the drawing of the end-state core configuration in Figure 3. The upper and side crusts formed as the pool of ceramic melt grew into the debris bed.

At 200 minutes the high-pressure injection system was activated for 17 minutes. The pressure vessel was refilled, including water that drained from the pressurizer as the primary-system pressure fell. Analysis indicates that this reflooding of the core probably quenched the particulate debris bed but that the surface cooling of the consolidated debris region that contained the molten ceramic pool and crusts was insufficient to stop the growth of the pool (Ref. 19). Thus reflooding the core did <u>not</u> terminate progression of the accident when it occured.

At 226 minutes the growing molten ceramic pool melted through the crust at one side of the pool, as shown in Figure 4. This side melt through was caused by natural circulation in the internally-heated melt pool. It is possible that the 0.5 MPa pressure decrease in the 4 minutes before the melt drainage event caused by opening the block valve contributed to the crust failure (Ref. 1). About 20% of the core mass (20 metric tons) drained from the side of the pool down through the edge of the core and into the lower plenum. At this time the water level in the core was above the top of the melt pool, but this did not prevent melt through of the supporting crust and drainage of the pool through water into the lower plenum. The melt mass at the time of the melt relocation event, as determined by the the post-accident core examination and supplemented by thermal analysis, was found to be about 50% of the total core mass. Readings from in-core self-powered neutron detectors indicate that drainage of the melt into the lower plenum took about 100 seconds. The weak thin ceramic crust above the pool and the debris bed above the crust subsided down to the level of the drainage opening as the ceramic melt drained from the pool. It is thought that this is the primary cause of the cavity in the upper central part of the core found in the post-accident core examination.

A model was developed for analysis of the thermal behavior of the consolidated debris region which contained the mostly ceramic (UO_2 and ZrO_2) molten pool and surrounding crusts and the bottom of the overlying particulate ceramic debris bed (Ref. 17). The geometry of this model in comparison with the end-state TMI-2 core configuration is shown in Figure 5. Initial temperatures for this analysis were taken from SCDAP/RELAP5 calculations with appropriate boundary conditions. Previous one-dimensional analyses had shown that surface cooling was not sufficient to prevent growth of the molten ceramic pool in the large-volume consolidated region by fission-product decay heating. In this model, heat loss at the pool walls was calculated from natural convection heat transfer coefficients for internally-healed pools which vary strongly around the pool periphery. This analysis showed that the most rapid pool growth occurred radially outward at the upper edge of the pool where the natural convection heat flux is the greatest. This is where the failure actually did occur at TMI-2. Because drainage occurred from near the top of the pool and the integrity of the bottom crust was maintained, about half of the molten debris mass remained in the pool and did not drain into the lower plenum, an effect that has general reactor safety significance.

Analysis with this model indicated that the calculated melt mass at 226 minutes (melt drainage) was very sensitive to the assumed pre-existing melt mass at 174 minutes (B-pump transient), was also sensitive to the volatile fission-product retention fraction, but was not sensitive to reasonable variations in the surface heat loss from the consolidated region (Ref. 17). The calculations showed that in order for the molten mostly ceramic melt pool in the consolidated region to contain 50% of the core mass at 226 minutes, the pool must have already existed and contained about 20% of the core mass at the time of the B-pump transient at 174 minutes. A corollary analysis of crust thickness was also performed (Ref. 19). Relocation of the 20 metric tons of fission-product-containing melt into the lower plenum water reduced the thermal heat source in the consolidated debris region of the core and effectively stabilized the gross core geometry. Nevertheless, thermal analysis with the model indicated that, following melt drainage from the pool and with the consolidated region of the core underwater, debris heating and molten pool growth continued for about two additional hours before the decrease in the fission-product decay heat source stopped growth of the pool (Ref. 17).

Thermal analysis has shown that relocation of 20 metric tons of melt in about 100 seconds through the observed passages along the side of the core can be reasonably accounted for. The TMI-2 core-examination showed that the melt relocation occured through parallel passages through the intact peripheral fuel assemblies and through the external core baffle plates, the latter by jet meltthrough (Ref.1).

No large energetic thermal explosion (steam explosion) occured when the 20 metric tons of mostly ceramic melt drained into the lower plenum water. This is thought to be because the vessel pressure at that time (12 MPa) was too

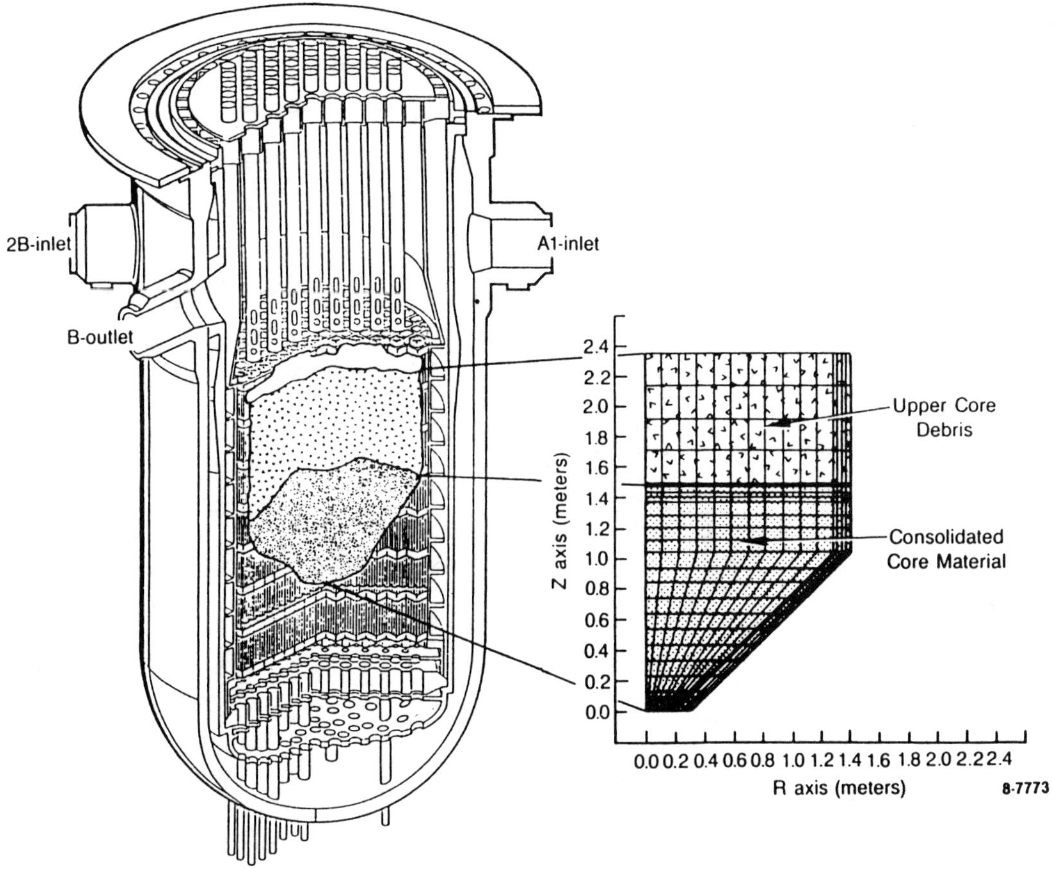

FIGURE 5. Late-Phase Consolidated Debris Thermal Model (from Ref. 18)

high for steam explosion triggering. As the melt stream broke up and cooled, however, there was non-explosive rapid steam generation (a steam spike) of about 2MPa above the 12 MPa ambient pressure that lasted for about 15 minutes. Thermal analysis showed that this pressurization is consistent with current knowledge of jet break up and heat transfer (Ref. 20). There is little direct information on the mean diameter (surface area) of the fragmented melt during relocation in part because of later agglomeration of much of the debris in the lower plenum.

Fast reactor safety experiments have demonstrated the effectiveness of frozen ceramic surface crusts in insulating metal structures during transient ceramic pours, and models of this process have been developed (Ref. 21). The relative lack of thermal attack by the ceramic melt on the core support plate and other lower plenum structures, (Ref. 1), is consistent with these results.

A most significant result of the TMI-2 accident is that the vessel lower head did not fail when the 20 metric tons of ceramic melt drained into the lower plenum water (Ref. 1). The pressure load on the lower head was then 12MPa (14 MPa during the 15-minute steam spike). Scoping analysis for gross creep rupture failure indicated that lower head temperatures exceeding 1100K through most of the wall thickness would have occurred if the ceramic melt had settled on the lower head as a cohesive non-porous structure (slab). Since the head would have suffered gross creep-rupture failure under these conditions, it was concluded that the actual thermal loading was less (Ref. 1). Penetration failure and jet impingement failure were also examined. A program has been started to provide samples of the lower head structure and penetrations for metallographic examination. Results of this examination should provide significant information on the margin to vessel failure in the TMI-2 accident and also on lower head failure mechanisms.

In general, the simplified parametric accident analysis codes do not model the late-phase core-melt progression part of severe accidents. The assumption is usually made that, when a parametric input temperature is reached, an assumed fraction of the core "slumps" into the lower plenum and soon fails the vessel. This assumed process clearly did not occur in the TMI-2 accident. TMI-2 was a very complex accident, and even the most mechanistic of the current codes do not include provisions for modeling all its significant complexities. The series of engineering models used to supplement to SCDAP/RELAP5 in analyzing the TMI-2 accident have done a reasonable job of analyzing what happened in that accident. The thermal analysis of the consolidated region (the ceramic pool and crust region) of the core in late-phase melt progression proved particularly fruitful (Ref. 17). With their narrow scope, however, there is no guarantee that these SCDAP/RELAP5 models are applicable to very different accident situations where the sequences might be quite different, a BWR accident with automatic depressurization for example. The MELPROG late-phase modeling in the DEBRIS and EJECT models and the mode of vessel failure analysis would appear to have considerably more ability to follow different accident sequences and conditions realistically. For example, selection of the key melt progression branch point of metallic blockage formation with later ceramic pool growth and drainage, as occurred at TMI-2, or metallic and ceramic melt drainage from the core as formed, has a built in choice of the TMI-2 path in SCDAP/RELAP5. (It has not yet been determined that the melt drainage path ever actually occurs under reactor accident conditions.)

Summary Conclusions

The TMI-2 accident and its corollary core examination have furnished unique and valuable reactor safety information, particularly on the late-phase of LWR in-vessel core-melt progression where very little other information exists. TMI-2 was a very complex accident, had uncertain thermal-hydraulic boundary conditions and was successfully terminated by core reflooding. The phenomenology of the accident as determined by the TMI-2 core examination and analysis of the available data has proved to be broadly applicable to unrecovered PWR accidents, particularly in their late phases, and to some aspects of BWR accidents. TMI-2 showed that: (1) severe accidents can be successfully terminated by core reflooding even when the accident has advanced into ceramic (fuel) melting; (2) core flooding does not immediately stop the progression of a severe accident, for most of the ceramic (fuel) melting in the severely damaged TMI-2 core occurred after the core had been flooded; and (3), a pour of a large mass (20 metric tons) of mostly ceramic melt at temperatures above 2800K into a water-filled lower plenum may not fail the vessel lower head. The phenomenology of the first two of these findings is currently understood, but that of the third is not. In addition, at TMI-2 drainage of about 20 metric tons of mostly ceramic melt into lower plenum water in about 100 seconds at 12 MPa pressure did not produce a large energetic steam explosion. The total hydrogen generated in the TMI-2 accident was equivalent to the oxidation of about 45% of the core zircaloy.

TMI-2 produced little new information for the phenomenological modeling of the early part of the accident up to the B-pump flow transient at 174 minutes. In part this is because of the absence of measurements of the core temperatures and the steam and the hydrogen flows as well as the complex and uncertain thermohydraulic boundary conditions. In this early phase, however, existing phenomenology based in the results of the four PBF Severe Fuel Damage tests and other research has provided an adequate and consistent description of the accident. In the late phase of the accident that involves substantial ceramic (fuel) melting and relocation, however, there was very little pre-existing phenomenological information. Here, the TMI-2 core examination and analysis have very largely provided the needed information, except on the question of lower-head failure where the TMI-2 investigation is just starting. Individual engineering models of significant accident phenomena constructed from the results of the core examination have been largely successful in accounting for what happened in the late phase of the TMI-2 accident. Application of accident analysis codes has shown the critical importance of adequately representing the thermal hydraulic boundary conditions, which in the TMI-2 accident were very complex and involved large transients and large uncertainties. This is particularly important for the early stage of accidents. At TMI-2, a principal remaining uncertainty is the relative lack of natural circulation heating of the upper plenum structure and the hot-leg nozzles, and this uncertainty may well be governed by boundary condition uncertainties. Finally, while the engineering models applied in the analysis of late stage accident behavior at TMI-2 were relatively successful, the degree to which these models or, on the other hand, more broadly-based first principles models can accurately predict late-phase severe accident behavior for the relevant range of severe accident conditions in PWRs and also in BWRs is currently uncertain. The modeling accuracy actually needed for severe-accident safety assessments and regulatory application is also currently uncertain.

REFERENCES:

1. Broughton, J. M. Kuan, P. Petti, D. A., and Tolman, E. L., "A Scenario of the TMI-2 Accident," Nuclear Technology, forthcoming.

2. Allison, C. M. Berna, G. A. Wagner, R. J., Chang, J. C., Siefken, L. J., and Hagerman, D. L., "Capabilities of the Integrated SCDAP/-RELAP5/TRAP-MELT Severe Accident Analysis Computer Code," Proc. IAEA International Symposium on Source Term Evaluation for Accident Conditions, Columbus, Ohio, October 1985.

3. Dosanj, S. S., Tomkins, J. L., Young, M. F., Dearing, J. F., Smith, R. C., Kelly, J. F., Grimley, A. T., Wheatley, C. J., Maudlin, P. J., Heames, T. J., and Camp, W. J., "MELPROG-PWR/MOD1: A Two-Dimensional, Mechanistic Code for Analysis of Reactor Core Melt Progression and Vessel Attack under Severe Accident Conditions," NUREG/CR-5193, March 1989.

4. Marino, G. P., and Picklesimer, M., unpublished.

5. Rogovin, M., and Frampton, G. T. Jr., "Three Mile Island, A Report to the Commissioners and to the Public," NUREG/CR-1250, January 1980.

6. Osetek, D. J., "Results of the Four PBF Severe Fuel Damage Tests," Proc. Fifteenth Water Reactor Safety Information Meeting, Gaithersburg, Maryland, NUREG/CP-0090, October 1987.

7. Wright, R. W., "Current Status of USNRC Research on In-Vessel Core-Melt Progression," Proc. International ENS/ANS Conference on Thermal Reactor Safety, 'NUCSAFE 88', Avignon, France, Vol. 4, p. 1324, October 1988.

8. Eidam, G. R., Tolman, E. L., Broughton, J. M., McCardell, R. K., and Stratton, W. R., "TMI-2 Defueling Conditions and Summary of Research Findings," Proc. IAEA/NEA Symposium on Severe Accidents in Nuclear Power Plants, Sorrento, Italy, Vol. 1, p. 207, March 1988.

9. Giessing, D., "The TMI Analysis Exercise," Nuclear Technology forthcoming.

10. Eidam, G. R., "The TMI-2 Scenario," Proc. ICHMT Seminar on "Fission Product Transport Processes in Reactor Accidents, Dubrovnik Yugoslavia, May 1989.

11. Hofmann, P., Hagen, S., Schanz, G. and Skokan, A., "Reactor Core Materials Interactions at Very High Temperatures," Nuclear Technology, forthcoming.

12. Lombardo, N. J., and Lanning, D. D., "Data Report FLHT Experiment 2," Pacific Northwest Laboratories, PNL-6551, April 1988.

13. Reil, K. O., Marshall, A. C., Gauntt, R. O., Ostensen, R. W. Pickard, P. S., Fryer, C. P., and Stalker, K. T., "Results of the ACRR DFR Experiments," Proc. ANS/ENS Topical Meeting on Thermal Reactor Safety, San Diego, California, February 1986.

14. Hagen, S. Sepold, L., Hofmann, P., and Schanz, G., "Out-of-Pile Experiments on Severe Fuel Damage Behavior of LWR Fuel Elements," Proc. IAEA/NEA Symposium on Severe Accidents in Nuclear Power Plants, Sorrento, Italy, Vol. 2, p. 275, March 1988.

15. Cronenberg, A., "Summary of Hydrogen Generation Data for Benchmarking Severe Accident Consequence Analysis," Proc. 24th National Heat Transfer Conf., Pittsburgh, Pennsylvania, p. 301, August 1987.

16. Kuan, P., "Damaged TMI-2 Core Consolidated Region Crust Behavior," Nuclear Technology, forthcoming.

17. Moore, R. L., Golden, D. W., and Tolman E. L., "TMI-2 Degraded Core Heatup and Cooldown Analysis," Nuclear Technology, forthcoming.

18. Saito, S., and Shiozawa, S., "Severe Fuel Damage in Steam and Helium Environments Observed in In-Reactor Experiments," Proc. Fifth International Conference on Thermal Nuclear Reactor Safety, Karlsruhe, FRG, KfK-3380, Vol. 2, p. 938, December 1984.

19. Kaun, P., and Tolman, E. L., "Damaged TMI-2 Consolidated Region Crust Behavior," Nuclear Technology, forthcoming.

20. Epstein, M., and Fauske, U. K., "The TMI-2 Core Relocation-Heat Transfer and Mechanisms," Fauske and Associates, FAI/87-49, July 1987.

21. Pilch, M. and Mast, P. K., "PLUGM: A Coupled Thermal-Hydraulic Computer Model for Freezing Melt Flow in a Channel," NUREG/CR-3190, April 1984.

7. CLOSING SESSION

Summaries of the Sessions and Panel Discussion

SESSION 1. INTRODUCTORY SESSION (D.F. TORGERSON)

The objective of this Seminar was to review the various phenomena associated with aerosol and fission product transport during severe reactor accidents and, in particular, to explore the coupling between thermalhydraulic phenomena and activity transport. The introductory session presented overviews on: 1) severe accident phenomena, 2) the nature of the interaction between thermalhydraulics and activity transport, 3) reactor cooling system (RCS) transport codes, and 4) containment transport codes.

In his introduction to the Seminar, Professor Teague reminded us of the evolution of safety principles that led to "defence-in-depth" and the design-basis accident concept (DBA). Although severe accidents did not form part of the safety environment, Professor Teague noted that TMI was a striking example of the ability of DBA safety systems to mitigate a severe accident. Professor Teague then reviewed the various severe accident phenomena, and stressed: 1) the need to develop the fundamentals of such accidents and 2) that scientific judgement is important - i.e., it is not possible to assemble everything into a code. Finally, since severe accidents are not well-defined, perhaps the most viable approach to enhance safety would be to extend the useful performance ranges of engineered safety features (ESFs).

Dr. McDonald discussed the importance of the interaction between fission product (FP) transport and thermalhydraulics. In particular, he raised the question of local thermalhydraulics effects that could have a large impact on FP transport transients. An example of this is convective heat transfer in the core - how accurately does this need to be known in order to characterize FP transport in the vessel and reactor cooling system? Dr. McDonald also stressed that we cannot assemble all the physics and chemistry into one single transport code. Thus, we need to focus on describing integrated events rather than coupling the separate physics. Finally, he noted the complexity of modelling severe accidents, where we must use complex equations to generate numbers that then have to be related to reality. I interpret this to mean that no matter how sophisticated our codes, the numbers generated are not a substitute for sound technical judgement.

Dr. Gieseke reviewed the major components of RCS FP/aerosol transport codes. Some of the most important items in the coding are fission product chemical forms and reactions, quantity, release, timing, RCS residence times, surface area/gas volume ratio, and revolatilization due to decay heating after vessel failure. The major uncertainties are: 1) accident progression, 2) re-suspension/re-entrainment, 3) recirculation, 4) aerosol nucleation and re-vaporization, and 5) geometry effects. In closing, Dr. Gieseke repeated a question posed by the previous speakers: "How much detail do we really need in the codes?"

Dr. Schikarski reviewed containment aerosol codes. He noted that in simulating a Demona experiment in which aerosol concentrations were processed as a function of time, the codes agreed with each other within a factor of 2, and with the experimental data within a factor of 3. He attributed this to the fact that the physics used in the various codes is essentially the same. However, a major difference is how steam condensation is treated, and Dr. Schikarski stressed that steam condensation has a large effect on aerosol suspension. For example, in Demona, dry aerosols were reduced to the detection limits in 2 days, while in a condensing steam atmosphere, the detection limit was reached in 7-8 hours. Dr. Schikarski felt that new thermal hydraulic codes will lead to significant improvements. Finally, he pointed out the need to describe correctly iodine chemistry and to characterize long-term aerosol releases.

There were also some comments from the audience that are interesting to note. Dr. Bowsher pointed out that particle sizes in containment may be affected by chemistry in the RCS. An example of this is the LACE experiments where the MnO particles were different from the CsOH aerosols.

Mr. Warman noted that the composition of the aerosols could have a large impact on suspension times. If the aerosols are hygroscopic, the removal rates could be enhanced by a factor of 3-4. This is important, since experiments have shown that the surface of aerosols may be enriched in hygroscopic chemicals. Mr. Warman also cautioned us not to neglect effects that are perceived to be small, since, in combination, the small effects can become significant.

SESSION 2. FUEL BEHAVIOR AND FISSION PRODUCT RELEASE (F.C. IGLESIAS)

For years, the consequences of reactor accidents have been assessed using bounding calculations. Since TMI-2 and Chernobyl, these bounding calculations have been perceived to be insufficient. In order to get a better understanding of accident consequences, more detailed analyses are being sought. This involves a greater knowledge of fuel behaviour and of fission product release.

Three issues were identified in the opening lecture by Dana Powers:

1) release from the fuel,
2) transport and deposition in the reactor coolant system, and
3) re-vapourization of the deposited material.

The first of these was the one primarily treated in this session.

The importance of thermodynamics and kinetic factors, for the determination of the rates of the release processes, was emphasized in several papers. Potter and Alexander addressed some of the thermodynamic features of reactor fuel. An interesting technique was presented for the determination of the temperature dependence of the energy of formation of chemical species. Coupled with the determination of the energy formation at 298 K, this technique will be of great help in enlarging our data base.

An important evolution has occurred in our understanding of core degradation processes in recent years. One important finding is that the core does not behave uniformly. According to certain core degradation models a substantial fraction of the core will remain as intact rods after other parts of the core have melted and penetrated the reactor vessel. The remaining fuel will be subjected to continued heating, but in an atmosphere of air drawn from the reactor containment into the reactor vessel. This will be in contrast to the steam/H_2 mixture usually assumed.

Two interesting features appear from the study of releases under oxidizing conditions:

1) Fast releases of noble gases and volatiles during the UO_2 oxidation process.

2) Under the reducing conditions prevalent prior to vessel rupture, ruthenium is not released. However, under the oxidizing conditions after vessel rupture, extensive Ru release is to be expected. The radiological consequences of Ru release then become important. The Ru radiological hazard, based on lung dose, will then become 3 times more important than iodine, and the incidence of latent cancers from Ru are similar to those induced by Cs as pointed out in the introductory lecture.

The results of the HEVA-6 test, for noble gases, I and Cs releases in reducing conditions with 2/3 of the clad oxidized seems to be in agreement with the predictions of diffusion models. The same is not evident for releases measured under oxidizing conditions. This seems to indicate that the rate controlling mechanisms, at least in air, are different from the rate controlling mechanisms in steam/hydrogen mixtures.

The steam/hydrogen ratio in the gas surrounding the degrading fuel together with the extent of Zircaloy and UO_2 oxidation are the important factors in determining the release of fission products from fuel, their volatility, and their chemical forms in the gas and condensed phases.

Experimental studies and the analyses of in-reactor and out-of-reactor tests of fission product release during core degradation are continuing. Modelling and new ways of correlating data from past tests of radionuclide release were discussed. The combination of in-reactor tests, out-reactor tests, TMI-2 analyses, thermodynamic calculations and reaction kinetics are greatly improving our understanding of the complex nature of fission product release from UO_2 fuel. Models of these processes are also reflecting this increase in understanding. Models of varying degrees of complexity are becoming available to treat fission product release. The user has to decide on the minimal level of complexity required for a given task.

SESSION 3. TRANSPORT PHENOMENA IN THE REACTOR COOLANT SYSTEM (A.P. MALINAUSKAS)

Sophisticated mathematical descriptions of the thermo-physical and physico-chemical processes that would occur in the reactor coolant system (RCS) during a severe core damage accident have been developed. Nonetheless, the need for considerably more research has been indicated during this Seminar to reduce the uncertainties in the calculated results.

Foremost among the improvements that have been identified is the need to treat thermalhydralics and materials transport in an integrated way. Formerly, thermalhydraulics models and mass transport models were developed independently, and the two sets of phenomena were artificially coupled through the boundary and initial conditions in the two independent sets of computations. This approach can lead to significant error.

In a related vein, it has become obvious that the current suite of thermalhydraulics and mass transport computer codes, which had been developed almost exclusively for purposes of analyzing accidents in pressurized water reactors (PWRs), is inadequate for use in the analysis of boiling water reactor (BWR) accidents. Not only are BWRs very different from PWRs in terms of geometry, but also in modes of pressure control and in the employment of accident management procedures. Thus, considerable error can result if BWR accident sequences are described using crude modifications of the PWR computational programs; the entire code framework must be modified.

Several phenomena have been identified during this Seminar that require additional research, of a theoretical nature, in order to provide adequate mathematical descriptions. One of the most significant concerns natural circulation phenomena in the primary circuit, particularly for PWRs, since natural circulation processes in these reactors can theoretically preclude threats to the reactor containment resulting from high pressure melt ejection from the reactor pressure vessel. Additional theoretical work is also required to elucidate further the nature of a number of mass transport phenomena; these include the deposition behavior of aerosols in pipe bends, the interaction of gas-borne fission product species with aerosols, physical re-suspension and re-vaporization phenomena.

In the more usual situation, however, the mathematical treatments appear to have overtaken the supporting experimental programs; in a preponderant number of cases, the accuracy of the results is not limited by the theoretical treatment, but rather by the paucity of input data that are required to perform the calculations. As examples, thermodynamics and kinetics data which are required to calculate effects of fission product-aerosol and fission product-surface interactions are largely unavailable, and particle adhesion characteristics remain as a major limitation to the accurate assessment of particle re-suspension behavior.

The description of chemical phenomena which can occur in the reactor coolant system in the course of a severe core damage accident is in a relative state of infancy. Although approaches based solely on chemical thermodynamics considerations are deficient only in the lack of requisite thermodynamics data, and are currently constrained by considerations of what is necessary to define the system completely and what is important, the entire approach could be severely compromised if chemical kinetics, as well as thermodynamics, is shown to determine the chemical behavior. Thus far, the role of chemical kinetics has been largely ignored. Another complicating factor involves the effects of radiation on the RCS chemistry; fortunately, radiation effects appear to be relatively unimportant.

Lastly, the ability to translate results developed from bench-scale experiments or with larger, integral-type tests to actual reactor conditions requires additional examination. Relatively little attention has been given to the development of appropriate scaling laws, although this clearly has been a major concern in the development of the PHEBUS fission product experiments.

The understanding that has been developed of the processes which can occur within the RCS during a severe core damage accident has had little impact on reactor design or on the development of accident management procedures. The state of the technology has become sufficiently mature, however, that new management strategies, and new reactor design features, can now be developed on the basis of this technology.

SESSION 4. TRANSPORT INTO CONTAINMENT (E.F. HICKEN)

The transport of fission products into the containment is sequence and design dependent. In session 4 the areas of (1) failure of the head of PWR and BWR Reactor Pressure Vessels (RPVs), (2) direct containment heating, (3) ex-vessel fuel/coolant interaction and (4) long-term resuspension of fission product aerosols from a boiling sump were presented.

The bottom head of a BWR reactor vessel has more that 200 penetrations, for some designs. Of further influence on the RPV failure are the lower plenum internals and the support of the fuel elements. Due to these internals, the melt-down process is difficult to predict and results in some uncertainties. The earliest relocation of material from the core region is expected to be from control rod blades, channel boxes and cladding. The subsequent debris or melt will evaporate the water in the lower plenum and weaken or even melt the internals. It is highly probable that the bottom head will fail first at a penetration, allowing most of the debris or melt to leave the reactor pressure vessel. However, it cannot be excluded that a gross failure of the bottom head occurs at a later stage and after most of the debris or melt have left the vessel.

The bottom head of a PWR reactor pressure vessel can be divided into two categories, with and without penetration. If core melt or debris slumps into the lower plenum, the residual water will evaporate fast and the inner surface of the lower head will be exposed to hot debris. Thermal stresses and weakening of the wall may result in a failure of the bottom head before any melting, if the pressure in the primary circuit is high. Beacuse a gross failure cannot be excluded, reaction forces on the RPV supports may be high. At high system pressures, the surge line or a hot or cold leg may fail, if heated to a temperature of about 700°C. While the strength of the pipes can be calculated quite accurately, the calculations of the temperature show a large uncertainty due to in-vessel natural circulation and core degradation processes.

In his lecture, Professor Corradini stated that in direct containment heating, the scenario can be grouped into four stages: (1) forceful expulsion of the melt from the RPV, (2) dispersal of a large fraction of this melt into the containment atmosphere, (3) rapid exothermic oxidation of the metallic components of the melt, (4) transfer of energy into the containment atmosphere resulting in a rapid temperature and pressure increase. The processes during the first two stages are highly uncertain, namely the RPV failure mode, the character of the expelled melt, and the influence of the shape of the cavity and the effect of water in the cavity. It seems that decreasing the primary system pressure is the most effective mitigation measure.

A rapid fuel/coolant interaction may occur if (fluid) melt contacts water. This process is often called "steam explosion". The cavity below a reactor pressure vessel can be dry or filled with water. The steam production resulting from the energy transfer from melt to water may threaten containment internals or structures. In addition, shock waves will give a load on surrounding walls. These loads are evidently influenced by the injected mass flow rate and available water pool. One of the major uncertainties is the pre-fragmentation and the pre-mixing processes in a large water pool. More research in the area of ex-vessel melt/water interaction is necessary.

Re-suspended aerosols from a boiling sump constitute a long-term but weak source term. Experimental investigations as well as code calculations have resulted in a rather clear picture of this process. This source term is only important in case of failed containments.

The lecture and the four papers in this session have shown the dependance of containment loads on the RPV failure mode and the expelled mass. Because these loads may result in an early containment failure, appropriate research is necessary to understand the processes. It seems that a depressurization of the RPV may be the most effective mitigation measure.

SESSION 5. CONTAINMENT TRANSPORT PHENOMENA (F. ABBEY)

Session 5 of the seminar included two lectures and ten papers, of which both the lectures and eight of the papers (5.3 - 5.8, 5.11 and 5.12) were presented orally.

The session began with a keynote lecture (5.1) by Dr. Kress. This set the scene for the remainder of the session by reviewing the history of LWR containment designs. In the speaker's view it had been a mistake to design containments to withstand only what is now called the design basis LOCA. If it had been possible to estimate severe accident loads, and if containments had then been designed against these loads, severe accident source terms would not have been an issue. Dr. Kress did not accept the view sometimes expressed that enough is known about containment transport phenomena; most of the topics which he highlighted as in need of further investigation are topics on which there were papers in the session, including:

- thermal-hydraulics
- hydrogen behaviour
- aerosol behaviour
- core-concrete interaction
- iodine chemistry and behaviour

Dr. Kress concluded with the interesting proposition that many of the problems in the containment area are exacerbated by the use of Zircaloy in the reactor core and that it is the Zircaloy and not the fission products which drives the risk. He recommended that there should be research into less reactive replacement materials.

There was a total of four papers on thermal-hydraulics, if hydrogen distribution is included under that heading, and on the interaction between thermal-hydraulics and aerosol behaviour.

Between them, Dr. Wolf (5.6) and Dr. Kanzleiter (5.11) reviewed existing or planned experiments and model developments in the thermal-hydraulics area. They noted how the experiments have moved to larger and larger facilities and from the period of the initial blowdown, of interest to design basis accident studies, to the longer time scales relevant to severe accidents. They concluded that a substantial data base is, or will soon become, available. Other conclusions were that for multi-compartment containment geometries, models with 30-50 nodes and codes such as FIPLOC, CONTAIN and COBRA-NC are needed if phenomena such as natural convection patterns, thermal stratification and the distribution of non-condensible gases are to be properly evaluated. Very stable numerical procedures are required to handle the small pressure differentials in the buoyancy effects involved.

In his review (5.2) of the interface between containment thermal-hydraulics and fission product transport Dr. Dunbar identified three phenomena in which there could be feedback from fission product transport to thermal-hydraulics: decay heat, aerosol interaction with thermal radiation, and steam condensation onto aerosols. The rest of the talk then concentrated on the latter topic. One of the items of thermal-hydraulic information Dr. Wolf identified as still subject to major uncertainty was the extent of steam condensation in the bulk of the containment atmosphere. Dr. Dunbar emphasized both the importance of the quantity as a determinant of aerosol behaviour, and its close coupling to the nature of the aerosol. He raised the prospect that it may never be possible to calculate the thermal-hydraulics accurately enough to predict the details of condensation onto aerosols with confidence, but also suggested that the overall rate of condensation may simply be governed by heat transfer to the containment walls.

Dr. van Rij implicitly supported this view, specifically identifying the sensible heat removal at the walls as the important factor, and pointing out that the sensible heat is only a small proportion of the total heat removed so that it may be difficult to calculate accurately even when the total heat removal is known. Dr. van Rij claimed that the heat/mass transfer analogy used in codes has a number of potentially serious limitations.

The three papers on core debris-concrete interactions all dealt with the thermal-hydraulics aspects rather than fission product release. Recent developments were reviewed. Dr. Powers noted that, whereas models for axial heat transfer from the debris into the concrete yield good agreement, those for radial heat transfer and for heat transfer into overlying water still need some attention: He also highlighted the role of condensed-phase reactions between metallic phases of the melt, especially zirconium, and oxidic phases as a source of heat. When models for these reactions are included in the CORCON code, agreement with the temperature history in the SURC-4 test greatly improved.

Dr. Alsmeyer (5.4) described a comparison between experiments in the BETA series and the predictions of the WECHSL code. Agreement on the rates of concrete erosion were said to be good. The code was then used to predict base-mat melt-through in a severe accident sequence in a German standard PWR. Final penetration of the six metre thick base-mat and the need for mitigating actions such as containment venting, was predicted after five days. Flooding the debris with sump water was predicted to have very little impact on the rate of penetration. Cooling of the debris by the water was limited by the formation of an oxidic crust on top of the debris. This effect was said to have been seen in the SWISS experiments and in model experiments at KfK.

Dr. Adroguer (5.5) described a new code, CALTHER, which has been developed by the CEA to enable a more mechanistic calculation of the upward heat flux to be linked to the WECHSL-MOD2 code. The combined code models feedback effects from conditions in the cavity to the core debris-concrete interaction and the fraction of the upward heat flux added directly to the cavity atmosphere. Preliminary calculations showed that for limestone concrete the addition of CALTHER leads to a 30% reduction in the base-mat penetration time. The gas release was also increased with Calther due to increased radial erosion of the concrete and the thermal attack on upper cavity concrete. The formation of a crust on top of the melt was suppressed.

Points raised in discussion of the core debris-concrete interaction papers included differing views as to the role and effects of water flooding in a real reactor accident, and the inability of the codes accurately to predict gas and aerosol evolution in International Standard Problem exercises.

There were also three papers on iodine behaviour in the containment. Dr. Beahm (5.8) described the Oak Ridge code TRENDS, one of the codes being developed worldwide which attempt to draw all the complexities of iodine behaviour into a single calculational vehicle suitable for incorporation into containment codes such as CONTAIN. TRENDS is a best-estimate, mechanistic code. Models to be added to the code to calculate the pH of solutions, the hydrolysis and radiolysis reactions and effects of radiation were also described. In the discussion of Dr. Beahm's paper it was suggested that a valuable use of TRENDS would be to examine the sensitivity of the iodine release to the environment in severe accidents to the form in which the iodine enters the containment. Another topic raised in the discussion was the need to include adventitious materials in calculations with codes such as TRENDS, and this was the subject of Paper 5.10 by J.O. Liljenzin and O. Lindquist. The paper notes that in a typical Swedish reactor there are large amounts of aluminum, copper, zinc and various synthetic rubbers around and below the vessel and identifies a number of related effects which may be possible. Pyrolysis of organic polymers will give rise to acids which will lower the pH of water pools and to unsaturated organic compounds from which organic iodides could be formed. Molecular iodine is likely to be captured by finely divided copper from vaporized electric wiring. Aluminum and zinc could be dominant sources of hydrogen to the containment atmosphere and their compounds could become significant pH buffers in pools. Paper 5.9 by K Ishigure and H Shiraishi described the results of experiments in a loop which allows gamma irradiation of a sample, followed 30 seconds later by chemical analysis of the iodine species produced. Both iodide and molecular iodine solutions were studied, and concentration, pH, dissolved oxygen levels, and temperature were varied.

In the final paper (5.12), Mr. Abbey described the BUSGAS and BUSCA codes developed in SRD to model the removal of gases and aerosols respectively from spherical cap gas and vapour bubbles rising through water pools. He also reported a favourable comparison between BUSCA and SPARC predictions for the test matrix of the proposed SPARTA experiments.

SESSION 6. THE ACCIDENTS AT THREE MILE ISLAND AND CHERNOBYL (D.F. ROSS)

Several speakers noted the difficulties in modeling core disruptive events such as Chernobyl and TMI-2. Analysis techniques are still being developed which are quite complex, and are probably well ahead of the data available from the disruptive events.

Several speakers noted that post-mortems of core disruptive events are quite expensive, as well as difficult. In both cases, special tools had to be developed for the purpose. Some support for pre-planning of remote disassembly was evidenced. In both reactor events, the interests of securing the residue of the disrupted core competed with the research interests of studying the damage.

Some concern was expressed about the preoccupation of studying the past event, without an equivalent effort towards prevention of the next event. The studies of damaged cores is quite expensive, and tends to divert resources from other work. Some questioners asked how long the research community should study these two events.

The talks presented varied widely in style; some were graphic descriptions of the event, while others were analytic in terms of needed generic research on damaged cores. There was broad interest in theories of modeling, and some regret that there was not more time to pursue this topic.

Index